Second Edition

Handbook of
CELL-PENETRATING
PEPTIDES

Second Edition

Handbook of
CELL-PENETRATING PEPTIDES

Edited by
Ülo Langel

CRC Press
Taylor & Francis Group
Boca Raton London New York

CRC Press is an imprint of the
Taylor & Francis Group, an **informa** business

CRC Press
Taylor & Francis Group
6000 Broken Sound Parkway NW, Suite 300
Boca Raton, FL 33487-2742

First issued in paperback 2019

© 2007 by Taylor & Francis Group, LLC
CRC Press is an imprint of Taylor & Francis Group, an Informa business

No claim to original U.S. Government works

ISBN-13: 978-0-8493-5090-0 (hbk)
ISBN-13: 978-0-367-39048-8 (pbk)

Library of Congress Cataloging-in-Publication Data

Handbook of cell-penetrating peptides / edited by Ülo Langel. -- 2nd ed.
   p. ; cm.
Rev. ed. of: Cell-penetrating peptides / edited by Ülo Langel. c2002.
Includes bibliographical references and index.
ISBN-13: 978-0-8493-5090-0 (hardcover : alk. paper)
ISBN-10: 0-8493-5090-5 (hardcover : alk. paper)
   1. Peptides--Physiological transport. 2. Drug carriers (Pharmacy) I. Langel, Ülo. II. Cell-penetrating peptides.
   [DNLM: 1. Peptides--pharmacokinetics. 2. Biological Transport--physiology. 3. Drug Carriers. QU 68 H236 2007]

QP552.P4L364 2007
572'.65--dc22
          2006018100

Visit the Taylor & Francis Web site at
http://www.taylorandfrancis.com

and the CRC Press Web site at
http://www.crcpress.com

Preface

Cell-penetrating peptides (CPPs), alternatively designated as protein transduction domains (PTDs), Trojan peptides, model amphipathic peptides (MAPs), or membrane translocating sequences (MTSs), have the ability to translocate the cell membrane and gain access to the cellular interior. Unfortunately, we currently lack both a common terminology and a uniform explanation for the penetrative mechanism of CPPs. These deficiences confound the current situation in the field and reflect clearly the uncertainty in defining this novel class of peptides in accordance with modern research achievements. Hence, this book summarizes the diffuse area of CPPs, as we understand the field in late 2005, with the anticipation that a universal terminology together with a defined mechanism of cellular penetration will be forthcoming.

When the first book on CPPs was prepared in 2001, the prevailing situation was less complex and the field apparently better defined.[1] CPPs were then characterized as short peptides, consisting of fewer than 30 amino acids, that demonstrated seemingly receptor- and/or protein-independent cellular internalization. Internalization occurred even at 4°C, and was applicable for the cellular delivery of cargoes.[2] Such a definition was based on early studies of CPP translocation mechanisms suggesting that the internalization of these peptides was not inhibited by depletion of the cellular adenosine triphosphate (ATP) pool, low temperature ($+4$°C), or by inhibitors of endocytosis.[2] Many modified peptide sequences, including all-D-analogs, also demonstrated effective internalization properties, suggesting that selective cell surface receptors were not involved in the uptake mechanisms. Based on these observations, translocation was believed to be the consequence of a direct, physical transfer through the lipid bilayer of the cell membrane. This view is still supported by the proven inverted micelle model[3] of CPP uptake and, today, seems to hold for at least several individual CPPs, e.g., the penetratins.

As summarized in this second edition, recent research reports have started to re-evaluate earlier studies of the mechanisms utilized by CPPs. However, the definition for CPPs is still diffuse and newcomers might feel confused. The most meaningful definition today seems to be:

> A cell-penetrating peptide is a relatively short peptide, 5–40 amino acids, with the ability to gain access to the cell interior by means of different mechanisms, including endocytosis, and with the capacity to promote the intracellular delivery of covalently or noncovalently conjugated bioactive cargoes.

The CPP field has been subject to constant internal change during more recent years. The establishment of a concrete definition of CPPs is still in progress, hampered in part by apparent ambiguities in cellular uptake mechanism(s). Hence, I prefer not to specify these mechanisms, but rather refer to "different mechanisms, including endocytosis." I have chosen to adopt this stance since different peptides might utilize different uptake pathways simultaneously, and recent studies show that CPP translocation is mostly an energy-dependent process involving extracellular heparan sulfate binding and different types of endocytosis, such as macropinocytosis, clathrin-dependent, caveoale-dependent, or clathrin- and caveoale-independent endocytosis. Moreover, it seems that endocytotic processes of CPP uptake are even more obvious for CPP–cargo complexes than for naked CPPs, where several reports show a preferred endocytosis-independent uptake.[4–6]

Since the pioneering work in 1994 by Alain Prochiantz's group, numerous peptides have been reported to have cell-penetrating properties. As examples, older and newer CPPs are introduced in Part I of this book. Penetratins, TAT, and MAP peptides, as well as the new class of *cis*-γ-amino-L-proline-derived peptides, are each described in separate chapters. Additionally, chapter 5 describes a novel technology for the prediction of CPPs by the application of descriptors for individual amino acids in the CPP sequence. This prediction reveals a myriad of putative novel peptides, derived from natural proteins or random amino acid sequences, with CPP properties.

Part II is dedicated to the CPP uptake mechanisms from early inverted lipid models to more recently proposed endocytotic pathways. Chapters describing CPP interactions with model membranes are included due to the high impact of these in the initial stage of CPP binding to the plasma membrane as well as in the stage of endosomal escape of the CPP uptake process. I also find it necessary that CPP uptake is compared to CPP–cargo uptake, and that an in vivo uptake report is included in this section.

The functionality of shuttling proteins and of protein-derived peptides is the major topic of Part III. A small collection of the papers here clearly demonstrate that, when delivered into cells, both small peptides and large proteins with biological activity can be applied as valuable research tools and will, hopefully, serve as future drug leads.

The peptide-mediated delivery of bioactive molecules is characterized by rapid and high-yield transfer to the cytoplasm and/or nucleus and comparatively low toxicity. The flexibility to modify CPP vectors using peptide chemistry is a further advantage over other commonly used delivery technologies and has, in recent years, shown great potential in the field of drug delivery. A wide range of bioactive molecules, including proteins, peptides, oligonucleotides, and nanoparticles, have been delivered to a variety of cell types and to different cellular compartments, both in vivo and in vitro. Part IV mainly reports applications of CPPs in gene modulation; the collected articles describe the improvement of cellular uptake and biological effects of short oligonucleotides. Applications of antisense, siRNA, and decoy oligonucleotides are the most popular gene silencing/regulating technologies today and, hence, the CPP-aided improvement of these is described herein.

The selective targeting of tumors and other tissues with CPPs and their in vivo applications is a rapidly growing research area summarized in Part V. When the first edition was published, these questions were only starting to be addressed. Today, the progress is obvious and several tissue-specific targeting vectors are now available for use in vivo. I believe that the future of CPPs as drug delivery vectors largely depends on the development of these areas.

Part VI is included to yield insight into the most commonly used CPP study methods that, in our opinion, should be standardized by every research group to provide the basis for modern research methods free from misinterpretations.

In summary, there is an increasing need for efficient, nontoxic and non-hazardous transport vectors. CPPs fulfill all of these criteria and may in the future be an important tool in pharmaceutical research and industry. In recent years, several biotech companies have shown interest in CPP technologies, yielding several commercially available CPPs like the transport system Chariot (Active Motif, www.activemotif.com), formally known as Pep-1; the SynB vectors[7] (based on protegrin 1, PG-1) from Synt:em (www.syntem.com); and Express-si Delivery (based on the MPG peptide) from Genospectra (www.genospectra.com). A small start-up company, CePeP III (www.cepep.se), has very recently started to commercialize the peptide vectors pVEC and YTA2 for delivery of several cargoes. These examples of commercially available CPPs, together with published reports from the pharma-industry on CPP applications, clearly demonstrate the future perspective of CPPs in drug development.

REFERENCES

1. Langel, Ü., Ed., *Cell Penetrating Peptides. Processes and Applications*, CRC Press, Boca Raton, 2002.
2. Langel, Ü., Ed., Preface, *Cell Penetrating Peptides. Processes and Applications*, CRC Press, Boca Raton, 2002.
3. Berlose, J.-P., et al., Conformational and associative behaviours of the third helix of antennapedia homeodomain in membrane-mimetic environments, *Eur. J. Biochem.*, 242, 372, 1996.
4. Thorén, P. E., et al., Uptake of analogs of penetratin, Tat(48–60) and oligoarginine in live cells, *Biochem. Biophys. Res. Commun.*, 307, 100, 2003.
5. Henriques, S. T., Costa, J., and Castanho, M. A., Translocation of beta-galactosidase mediated by the cell-penetrating peptide Pep-1 into lipid vesicles and human HeLa cells is driven by membrane electrostatic potential, *Biochemistry*, 44, 10189, 2005.
6. Deshayes, S., et al., Insight into the mechanism of internalization of the cell-penetrating carrier peptide Pep-1 through conformational analysis, *Biochemistry*, 43, 1449, 2004.
7. Drin, G., et al., Peptide delivery to the brain via adsorptive-mediated endocytosis: Advances with SynB vectors, *AAPS Pharm. Sci*, 4, E26, 2002.

Editor

Ülo Langel is chairman of the Department of Neurochemistry, Stockholm University, Stockholm, Sweden. Prof. Langel earned a Ph.D. in bio-organic chemistry from Tartu University, Tartu, Estonia (1980) and a Ph.D. in biochemistry/neurochemistry from Tartu University/Stockholm University (1993). His professional experience includes junior research fellow, associate professor, and visiting professor at Tartu University (1974–present); associate professor at The Scripps Research Institute, La Jolla, California (2000–2001); and research fellow, associate professor, professor, and chairman at Stockholm University (1987–present). He is also an honorary professor at Ljubljana University, Slovenia.

Prof. Langel has been selected as a Fellow Member of the International Neuropeptide Society (1995), and is a member of the International Society for Neurochemistry, the European Peptide Society, the Swedish Biochemical Society, and the Estonian Biochemical Society. He has been awarded a White Star Order, 4th class, by the Estonian Republic. He has been an invited lecturer at numerous international conferences, and is a co-author of more than two hundred scientific articles and eight patents.

Contributors

Saïd Abes
UMR 5124 CNRS
Université Montpellier 2
Montpellier, France

Stephen R. Adams
Department of Pharmacology
University of California–San Diego
La Jolla, California

Fernando Albericio
Institute of Research in Biomedicals
Barcelona Science Park
Barcelona, Spain

Gudrun Aldrian-Herrada
CRBM-CNRS
Department of Molecular Biophysics
and Therapeutics
Montpellier, France

Sophie Amsellem
INSERM U567, CNRS UMR 8104
Université Paris V
Paris, France

Andrey A. Arzumanov
Medical Research Council
Laboratory of Molecular Biology
Cambridge, United Kingdom

Jesus Ayala-Sanmartin
INSERM U538
Faculté de Médecine Saint-Antoine
Paris, France

Mattias Belting
Department of Clinical Sciences
Lund University
Lund, Sweden

Michael Bienert
Institute of Molecular Pharmacology
Berlin, Germany

Gérard Bolbach
UMR CNRS 7613
Université Paris VI
Paris, France

Roland Brock
Interfaculty Institute for Cell Biology
University of Tübingen
Tübingen, Germany

Fabienne Burlina
UMR CNRS 7613
Université Paris VI
Paris, France

Gérard Chassaing
UMR CNRS 7613
Université Paris VI
Paris, France

Philippe Clair
UMR 5124 CNRS
Université Montpellier 2
Montpellier, France

Christopher H. Contag
Department of Pediatrics
Stanford University Medical School
Stanford, California

Lidija Covic
New England Medical Center
Tufts University School of Medicine
Boston, Massachusetts

Margitta Dathe
Institute of Molecular Pharmacology
Berlin, Germany

Sébastien Deshayes
CRBM-CNRS
Department of Molecular Biophysics
and Therapeutics
Montpellier, France

Gilles Divita
CRBM-CNRS
Department of Molecular Biophysics
and Therapeutics
Montpellier, France

Steven F. Dowdy
Howard Hughes Medical Institute
University of California–San Diego
La Jolla, California

Edmond Dupont
CNRS UMR 8542
Ecole Normale Supérieure
Paris, France

Samir El-Andaloussi
Department of Neurochemistry
Stockholm University
Stockholm, Sweden

Elin K. Esbjörner
Department of Chemistry and
Biotechnology
Chalmers University of Technology
Gothenburg, Sweden

Martin Fabani
Medical Research Council
Laboratory of Molecular Biology
Cambridge, United Kingdom

Josep Farrera-Sinfreu
Institute of Research in Biomedicals
Barcelona Science Park
Barcelona, Spain

Serge Fichelson
INSERM U567, CNRS UMR 8104
Université Paris V
Paris, France

Mariola Fotin-Mleczek
Interfaculty Institute for Cell Biology
University of Tübingen
Tübingen, Germany

Michael J. Gait
Medical Research Council
Laboratory of Molecular Biology
Cambridge, United Kingdom

Vivian Gama
Division of Hematology and Oncology
Case Western Reserve University
Cleveland, Ohio

Anita Ganyu
Center for Genomics and
Bioinformatics
Karolinska Institutet
Stockholm, Sweden

Jean Gariépy
Department of Medical Biophysics
University of Toronto and Ontario
Cancer Institute
Toronto, Canada

Ernest Giralt
Institute of Research in Biomedicals
Barcelona Science Park
Barcelona, Spain

Jose A. Gomez
Division of Hematology and Oncology
Case Western Reserve University
Cleveland, Ohio

Liam Good
Center for Genomics and
Bioinformatics
Karolinska Institutet
Stockholm, Sweden

Elena Goun
Department of Chemistry
Stanford University
Stanford, California

Astrid Gräslund
Department of Biochemistry and
 Biophysics
Stockholm University
Stockholm, Sweden

Bhawna Gupta
Department of Pharmaceutical
 Sciences
Northeastern University
Boston, Massachusetts

Mattias Hällbrink
Department of Neurochemistry
Stockholm University
Stockholm, Sweden

Frédéric Heitz
CRBM-CNRS
Department of Molecular Biophysics
 and Therapeutics
Montpellier, France

John Howl
Research Institute in Healthcare
 Sciences
University of Wolverhampton
Wolverhampton, United Kingdom

Gabriela Ivanova
Medical Research Council
Laboratory of Molecular Biology
Cambridge, United Kingdom

Suzanne Jacques
New England Medical Center
Tufts University School of Medicine
Boston, Massachusetts

Theodore C. Jessop
Department of Chemistry
Stanford University
Stanford, California

Yang Jiang
Department of Neurochemistry
Stockholm University
Stockholm, Sweden

Henrik Johansson
Department of Neurochemistry
Stockholm University
Stockholm, Sweden

Alain Joliot
CNRS UMR 8542
Ecole Normale Supérieure
Paris, France

Lisa Jones
Department of Chemistry
Stanford University
Stanford, California

Sarah Jones
Research Institute in
 Healthcare Sciences
University of Wolverhampton
Wolverhampton, United Kingdom

Sandro Keller
Institute of Molecular Pharmacology
Berlin, Germany

Athan Kuliopulos
New England Medical Center
Tufts University School of Medicine
Boston, Massachusetts

Pirjo Laakkonen
Biomedicum Helsinki
University of Helsinki
Helsinki, Finland

Antonin Lamaziere
INSERM U538
Faculté de Médecine Saint-Antoine
Paris, France

Ülo Langel
Department of Neurochemistry
Stockholm University
Stockholm, Sweden

Bernard Lebleu
UMR 5124 CNRS
Université Montpellier 2
Montpellier, France

Olivier Lequin
UMR CNRS 7613
Université Paris VI
Paris, France

Maria Lindgren
Department of Neurochemistry
Stockholm University
Stockholm, Sweden

Mark A. Lindsay
National Heart and Lung Institute
Imperial College
London, United Kingdom

Pontus Lundberg
Department of Neurochemistry
Stockholm University
Stockholm, Sweden

Shigemi Matsuyama
Division of Hematology and Oncology
Case Western Reserve University
Cleveland, Ohio

May C. Morris
CRBM-CNRS
Department of Molecular Biophysics
   and Therapeutics
Montpellier, France

Sterghios Moschos
National Heart and Lung Institute
Imperial College
London, United Kingdom

Bengt Nordén
Department of Chemistry and
   Biotechnology
Chalmers University of Technology
Gothenburg, Sweden

Johannes Oehlke
Institute of Molecular Pharmacology
Berlin, Germany

Manfred Ogris
Department of Pharmacy
Ludwig Maximilians Universitaet
Munich, Germany

Kärt Padari
Institute of Zoology and Hydrobiology
University of Tartu
Tartu, Estonia

Margus Pooga
Estonian Biocentre
Tartu, Estonia

Gregory M.K. Poon
Department of Medical Biophysics
University of Toronto and Ontario
   Cancer Institute
Toronto, Canada

Alain Prochiantz
CNRS UMR 8542
Ecole Normale Supérieure
Paris, France

Jean-Philippe Richard
National Institutes of Health
Bethesda, Maryland

Jonathan B. Rothbard
Department of Chemistry
Stanford University
Stanford, California

Miriam Royo
Combinatorial Chemistry Unit
Barcelona Science Park
Barcelona, Spain

Erkki Ruoslahti
Burnham Institute for Medical Research
La Jolla, California

Külliki Saar
Department of Neurochemistry
Stockholm University
Stockholm, Sweden

Sandrine Sagan
UMR CNRS 7613
Université Paris VI
Paris, France

Staffan Sandgren
Department of Clinical Sciences
Lund University
Lund, Sweden

Ines Sauer
Institute of Molecular Pharmacology
Berlin, Germany

Wenge Shi
Howard Hughes Medical Institute
University of California–San Diego
La Jolla, California

Rajesh Shinde
Department of Pediatrics
Stanford University Medical School
Stanford, California

Federica Simeoni
CRBM-CNRS
Department of Molecular Biophysics
    and Therapeutics
Montpellier, France

Ursel Soomets
Department of Biochemistry
University of Tartu
Tartu, Estonia

Boris Tchernychev
New England Medical Center
Tufts University School of Medicine
Boston, Massachusetts

Jamal Temsamani
CLL Pharma
Nice, France

Alain Thierry
UMR 5124 CNRS
Université Montpellier 2
Montpellier, France

Vladimir P. Torchilin
Department of Pharmaceutical
    Sciences
Northeastern University
Boston, Massachusetts

Germain Trugnan
INSERM U538
Faculté de Médecine Saint-Antoine
Paris, France

Roger Y. Tsien
Howard Hughes Medical Institute
University of California–San Diego
La Jolla, California

John J. Turner
Medical Research Council
Laboratory of Molecular Biology
Cambridge, United Kingdom

Söhnke Voss
Interfaculty Institute for Cell Biology
University of Tübingen
Tübingen, Germany

Ernst Wagner
Department of Pharmacy
Ludwig Maximilians Universitaet
Munich, Germany

Stephen A. Waschuk
Department of Medical Biophysics
University of Toronto and Ontario
    Cancer Institute
Toronto, Canada

Paul A. Wender
Department of Chemistry
Stanford University
Stanford, California

Burkhard Wiesner
Institute of Molecular Pharmacology
Berlin, Germany

Andrew Williams
National Heart and Lung Institute
Imperial College
London, United Kingdom

Anders Wittrup
Department of Clinical Sciences
Lund University
Lund, Sweden

Contents

Part VI
CPP Methods 535

Part I

Classes and Prediction of Cell-Penetrating Peptides

Ülo Langel

Since the discovery of pAntp/penetratins by Alain Prochiantz' group,[1] numerous peptides have been reported to have cell-penetrating properties (CPPs). In Part I of this book, the reader will find five chapters on different representatives of CPPs. Penetratins, Tat peptides, MAPs, and transportans are certainly among the most studied CPPs today, and their inclusion in this book was not a difficult decision. However, the selection among more recently discovered/designed CPPs was considerably more difficult due to the enormous number of interesting newcomers. The choice was made to favor *cis*-γ-amino-L-Pro-derived peptides and the CPP prediction approach. Both approaches are especially innovative in that they enable the application, besides traditional/coded L-amino acids, of noncoded amino acids in the efficient CPP sequences. This is likely to be an important step towards the commercialization of CPPs, as well as towards their applications in drug delivery. Additionally, the CPP prediction algorithm presented in chapter 5 may enable the exciting possibility of using prodrug concepts for the targeted delivery of drugs in future. Certainly, the first steps in this direction are being made, though many unsolved questions will probably arise along the way. Nonetheless, we believe that it is a good start in rationalizing the CPP design and for understanding of the CPP transfer mechanisms.

Due to our lack of understanding of the true mechanisms of CPP cellular uptake (see the Foreword in this book), the classification of CPPs still remains to be clarified. Hence, the classification presented in Table 1 is somewhat arbitrary and highly debatable, awaiting to be taken forward by new information. Perhaps the most important issue at hand has to do with the clarification of the nomenclature of efficient peptide delivery vectors, variously named CPPs, PTDs (see the Foreword), or otherwise. My strong personal feeling is that the term PTD is the most appropriate for describing peptide delivery vectors derived from proteins or responsible for the protein transduction in general, and that the term CPP (with its inherent drawbacks) should only be applied until studies of the uptake mechanisms have made progress (see above).

Other classifications are available based on different criteria. In most, the requirement of positively charged side chains of amino acids in the CPP sequence seems to apply. Penetratins are classified in chapter 1 according to their physicochemical characteristics and peptide–lipid interactions. Chirality, helicity, amphiphilicity, sequence length, and individual residue in the penetratin sequences all play important roles for the CPP properties. Translocation of the penetratin peptides requires more than one mechanism, including endocytosis, strongly dependent on cell lines. However, it seems that this classification is valid within the penetratins and broader generalization requires the long-desired clarification of the CPP uptake mechanisms. The latter very likely holds for the whole CPP family.

Another possible way to classify CPPs is according to their arbitrary structural properties. This would create three classes which behave differently, especially in their endocytotic uptake:

1. Peptides with low amphipaticity where the charge contribution originates mostly from arginine residues (penetratin and tat)
2. Peptides with a high degree of amphipaticity, where the charge contribution originates mainly from lysine residues (MAP and transportan)
3. Peptides where the charged and hydrophobic residues are separated lengthwise in the chain (pVEC and pep-1)

Even this classification of CPPs lacks the power of generalization among the plethora of available cell-penetrating peptides.

TABLE 1
Possible Classification of Cell-Penetrating Peptides

1. Protein/peptide derived

   a. Protein transduction domains (PTD), responsible for the parent protein translocation penetratin, Tat, pVP22

   b. Protein derived: pVEC, pIsl-1, LL-37, mPrP(1-28), hCT(9-32), NLS, mimicking peptides

2. Designed

   Amphiphilic: MAP

   Chimeric: transportan, MPG

   Homing peptides: Lyp-1, F3, CGKRK, RGR

   Synthetic: Pep-1, KALA, KFFK, Pro-rich, Arg9

A family tree of CPPs is proposed by Roland Brock's group,[2] where most of the above classification principles are united. It is noteworthy that antimicrobial peptides have recently been shown to translocate the plasma membrane moderately efficiently and, hence, they could become a special category of cell interacting/permeating peptides.

In chapter 5, we present the summary of our recent data on a possible way of predicting novel CPPs. Although not extensively studied and verified, this new CPP prediction strategy possibility may enable the design of novel protein mimics with a variety of intracellular bioeffects. It remains to be seen how comprehensive and useful this prediction method will be in the long term. However, it is our hope that it might open up improved ways of designing novel approaches to drug delivery and advance our understanding of CPP classification and mechanisms.

REFERENCES

1. Derossi, D. et al., The third helix of the Antennapedia homeodomain translocates through biological membranes, *J. Biol. Chem.*, 269, 10444, 1994.
2. Fischer, R. et al., Break on through to the other side — biophysics and cell biology shed light on cell-penetrating peptides, *Chembiochem*, 6(12), 2126, 2005.

1 Penetratins

Edmond Dupont, Alain Prochiantz, and Alain Joliot

CONTENTS

1.1   INTRODUCTION

Homeoproteins are transcription factors involved in several biologic processes, primarily during development. They were first discovered in Drosophila, and later in all metazoans and plants. They bind DNA through a 60 amino acid-long sequence, the homeodomain. This homeodomain is highly conserved across homeoproteins and species, and is composed of three α-helices, the third helix being more particularly dedicated to the recognition of the DNA target site.[1] In addition to their role in pattern formation during development, homeoproteins can act later to refine or maintain neuronal connections, both in invertebrates and vertebrates. Indeed, mutations in homeogenes lead to modifications in axon pathfinding[2–4] and synapse formation.[5–7] This suggests that homeoproteins probably help to define the important plasticity that characterizes the vertebrate nervous system.

1.2   HOMEOPROTEIN-DERIVED PEPTIDIC VECTORS

1.2.1   The Antennapedia Homeodomain

To analyze the role of homeoproteins in neuronal morphogenesis we developed a protocol aimed at antagonizing transcriptional activity of endogenous homeoproteins. This was achieved through the mechanical internalization of FITC-labeled homeodomains into live postmitotic neurons.[8–10] The addition of the exogenous Drosophila Antennapedia homeodomain induced strong neurite outgrowth that was attributed to competition between the homeodomain and endogenous homeoproteins for their binding sites.[10] It was during these experiments that the translocation of the Antennapedia homeodomain (AntpHD) across biological membranes, followed by its nuclear addressing, was discovered. This observation was later extended to several homeodomains and full-length homeoproteins, leading to the concept of messenger proteins.[11,12]

In an attempt to analyze the neurite-promoting function of the homeodomain and its mechanism of action, three different point mutations have been introduced (Table 1.1).[13–15] AntpHD 50A has a glutamine/alanine substitution in position 50 of the homeodomain (position 9 in the third helix); this position is important for the specificity of protein–DNA interactions. In AntpHD 48S, a single serine residue replaces three amino acids (tryptophan 48, phenylalanine 49, and glutamine 50). Tryptophan 48 (Trp 48) and phenylalanine 49 (Phe 49) are conserved in all homeodomains and are important for the homeodomain structure. PAntp40P2 has a substitution of two amino acids located in the turn between

TABLE 1.1
Behavior of AntpHD Mutants

Homeodomain	Sequence[a]	Interna-lization	DNA Binding	Biological Effect
	35 60			
AntpHD	-AHALCL*TERQIKIWFQNRRMKWKKEN*	+ + +	+ + +	+ + +
AntpHD 50A	-A**Y**ALCL*TERQIKIWF**A**NRRMKWKKEN*	+ + +	+	−
AntpHD 40P2	-AHALC**PP***ERQIKIWFQNRRMKWKKEN*	+ +	−	−
AntpHD 48S	-AHALCL*TERQIK*------*SNRRMKWKKEN*	−	−	−

[a] Only residues 35–60 are shown. The third helix is in italics; mutations and deletions are shown in bold.

helices 2 and 3 (leucine 40 and threonine 41) by two prolines. The DNA-binding capacity of the three mutants is either decreased (AntpHD 50A) or completely abolished (AntpHD 48S and AntpHD 40P2) and translocation into live cells is lost only in the AntpHD 48S mutant.[14] Biological activity (neurite outgrowth stimulation) is lost in all cases.[13–15]

1.2.2 THE PENETRATIN PEPTIDE

The results with AntpHD48S suggested the presence of a cell translocation sequence in the third helix. The 16 amino acids of the helix (amino acids 43–58, Table 1.2) were synthesized, and internalization into live cells was followed and observed thanks to an N-terminal biotin.[16] Shorter versions of the same peptide, with N-ter or C-ter deletions, are not internalized, suggesting that this peptide, hereafter penetratin, is necessary and sufficient for internalization. By deleting more amino acids starting from the N-terminus, Fischer et al. have described a shorter (seven amino acids) penetratin-derived peptide with internalization properties[17] (see Section 1.3.1.4).

Similarly to AntpHD, penetratin is internalized by an energy-independent mechanism at both 4 and 37°C, and has access to the cytoplasm and nucleus from which it can be retrieved without apparent degradation.[16] Translocation has been observed in all cell types and in 100% of the cells. It is not concentration dependent between 10 pM and 100 μM and below 10 μM toxicity is rare and cell-type dependent. The presence of three arginine and three lysine residues confers to the peptide an isoelectric point above 12. Finally, although penetratin can be represented under the form of an amphipathic helix, circular dichroism experiments have shown that it is poorly structured in water, and adopts an α-helical structure only in a hydrophobic environment.[16] Increasing the amphipaticity of penetratin by mutations increases the toxicity of the peptide rather than its translocation efficiency.[18]

TABLE 1.2
Internalization of Penetratin and of Its Derivatives

Peptide	Sequence	Internalization	Subcellular Distribution
43–58	RQIKIWFQNRRMKWKK	+++	N > C
58–43	KKWKMRRNQFWIKIQR	+++	N > C
D43–58	RQIKIWFQNRRMKWKK	+++	N > C
Pro50	RQIKIWFPNRRMKWKK	+++	N > C
3PRO	RQPKIWFPNRRMPWKK	+++	C > N
Met-Arg	RQIKIWFQNMRRKWKK	+++	N > C
7Arg	RQIRIWFQNRRM**RWRR**	+++	N > C
W/R	RRWRRWWRRWWRRWRR	+++	N > C
45–58	IKIWFQNRRMKWKK	++	N > C
52–58	RRMKWKK	++	N > C
48–58	WFQNRRMKWKK	+	N > C
47–58	IWFQNRRMKWKK	+	N > C
46–58	KIWFQNRRMKWKK	+	N > C
43–57	RQIKIWFQNRRMKWK	−	−
43–54	RQIKIWFQNRRM	−	−
2Phe	RQIKIFFQNRRMKFKK	−	−
41–55	TERQIKIWFQNRRMK	−	−
46–60	KIWFQNRRMK	−	−

1.3   HOW DO PENETRATIN PEPTIDES CROSS CELLULAR MEMBRANES?

Penetratin internalization mechanisms were investigated through a chemical approach. Two groups of data have emerged, gathered from structure/function studies in biological systems and from biophysical/biochemical approaches in purely artificial systems.

1.3.1   INFLUENCE OF STRUCTURAL PARAMETERS ON PEPTIDE TRANSLOCATION

Several derivatives from the original penetratin sequence have been chemically synthesized (Table 1.2). They were generated to address the following points. Is there a chiral recognition mechanism between penetratins and a membrane receptor? Do helicity and amphiphilicity influence translocation? What is the minimal sequence required for internalization? What is the relative importance of each residue of penetratin in the internalization process?

1.3.1.1   Chirality

Two peptides were synthesized, a 43–58 peptide composed of D-amino acids (D-penetratin) and an *inverso* form of the 43–58 peptide: the 58–43 peptide. These peptides are internalized as efficiently as penetratin at 4 and 37°C, demonstrating that a chiral membrane receptor is not required for cellular translocation.[16,19] Additionally, accumulation of D-penetrin was increased because of the resistance of D-peptides to proteolysis. Fluid-phase endocytosis does not require a membrane receptor, cannot be saturated, and is not inhibited at 4°C, three characteristics resembling those of penetratin uptake. However, peptide localization by electronic microscopy showed no endocytotic figures and demonstrated an accumulation in the cytoplasm and nucleus, precluding fluid-phase endocytosis.[19]

1.3.1.2   Helicity

Peptide helicity was broken by introducing one (Pro 50) or three prolines (3Pro) within the sequence (Table 1.2).[19] In Pro50, glutamine in position 50 is replaced by a proline. In 3Pro, glutamine 50, isoleucine 45, and lysine 55 are replaced by prolines. Neither modification hampered peptide internalization at 4 or 37°C, suggesting that a helical structure is not required. However, the subcellular localization of 3Pro differed from that of penetratin as it was not conveyed to the cell nucleus. This suggested that nuclear addressing and accumulation is sequence dependent and does not simply reflect the small size of the peptides.

Recently, Fischer et al. have analyzed the importance of the penetratin secondary structure by constraining its conformation.[17] Cyclic peptides were obtained through the addition of N- and C-terminal cysteines followed by air oxidation at an elevated pH. The linear form of the mutant peptide is internalized, whereas the cyclic one is not. This suggests that the secondary structure is important for translocation.

1.3.1.3   Amphiphilicity

To address the role of amphiphilicity, a penetratin mutant with two phenylalanines in place of Trp 48 and trytophan 56 (Trp 56) (Table 1.2) was tested.[16] This double mutant was not internalized, demonstrating that amphiphilicity is not sufficient to mediate internalization. This experiment also suggested that one or both tryptophans are crucial. More recent data[17,20] and the internalization of the homeodomain of Engrailed, which lacks Trp 56,[21] demonstrate that Trp 48 is a key residue. It is noteworthy that this residue has been conserved in all homeodomains.

1.3.1.4   Sequence Length

Two shorter peptides encompassing amino acid 41–55 or 46–60 (Table 1.2) do not translocate into live cells.[16,19] This suggests that at least the N- and C-terminal residues are crucial for translocation into neuronal cells. Fischer et al. studied the effect of successive truncations of the 43–58 peptide.[17] C-terminal truncations had a dramatic negative effect on cellular translocation (43–57 and 43–54 mutants; Table 1.2), the deletion of only the last lysine residue strongly impairing peptide penetration. In contrast, N-terminal truncations were less detrimental, allowing the definition of a heptamere (RRMKWKK; Table 1.2) internalized with 60% efficiency, compared to full-length penetratin. The latter result, at odds with earlier reports,[16,19] could be explained by the different types of cells used (immortalized

HaCat fibroblasts and lung cancer A549 cell lines instead of neurons). Another unexplored possibility is the different spacers (amino-pentanoic acid or β-alanine) used to separate the peptide from the biotin moiety. More recently, it has been shown that a shorter version of penetratin, generated by the internal deletion of four amino acids, has unaltered translocation acitvity.[22]

1.3.1.5 Relative Importance of Each Individual Residue

In two separate studies, each amino acid of the penetratin sequence was substituted by alanine scanning.[17,20] These two studies differ both in cell type (human HaCat and A549 vs. human transformed leukemia cells K562) and the mode of detection (biotin and NBD fluorochrome (7-nitrobenz-2-oxa-1, 3-diazol-4-yl)). The two reports confirm the key role of several basic residues: Lys 58, Lys 57, Lys 55, Arg 53, Arg 52, Lys 46. The role of hydrophobic residues is less clear, with a distinct controversy about the function of Trp 48 and 56.[16,17,20] Double substitutions of specific amino acids to alanine demonstrate the importance of the basic residues, but not the Trp 48 and 56, in the initial electrostatic interaction of the peptide with the lipid bilayer.[23]

Taken together, the results from all groups suggest the involvement of basic residues and of at least one tryptophan in the translocation process. This proposition has been confirmed by biophysical experiments demonstrating a role of charged groups in lipid binding and of one tryptophan in lipid destabilization and peptide translocation.

1.3.2 PEPTIDE–LIPID INTERACTIONS

Translocation of the penetratin peptides does not require chiral receptors and does not involve classical endocytosis. Fluid-phase pinocytosis can be excluded on the basis of ultrastructural studies that failed to reveal any association of penetratins with vesicular structures. Internalization through inverted micelles was proposed and is further detailed in Chapter 6. This model is based on the induction of inverted micelles by tryptophan residues[24] and by penetratin,[25] and on the capacity of penetratins to form multimer in the presence of SDS and to lower the critical micellar concentration in the presence of lipids.[16,25]

Structural studies achieved with synthetic lipids showed that although penetratin adopts a random coil structure in an aqueous environment, it becomes structured in the presence of negatively charged phospholipids. At a low peptide/lipid ratio (1/325), the peptide adopts an α-helical conformation.[16,20,25–27] At a high peptide/lipid ratio (1/10), the peptide forms an antiparallel β-sheet and induces vesicle aggregation.[27–29] Although penetratin directly interacts with the lipid bilayer, it is mainly located in the lipid headgroups in a parallel orientation, and not in the acyl chains.[30,31] Its hydrophylicity thus precludes its translocation across the hydrophobic core of the bilayer. Interestingly, subtle but significant differences are observed between penetratin–lipid and penetratin2Phe–lipid interactions, the mutated peptide being inserted more deeply in the hydrophobic core.[32,33]

1.3.3 PEPTIDE TRANSLOCATION

Direct translocation of penetratin across artificial lipid vesicles has been studied by several groups, leading to contradictory results. The first observation of penetratin translocation across giant unilamelar vesicles, reported by Thorén et al., was subsequently attributed by the same group to the high permeability of this kind of vesicle.[34,35] Penetratin translocation in large unilamelar vesicles has since been reported in the presence of a transbilayer potential (negative inside), either applied extrinsically, or directly created by the asymetrical distribution of the peptide itself.[36,37] The critical role of the lipid composition in the translocation ability of penetratin could account for the absence of penetratin translocation reported by other groups.[34,37,38]

1.3.4 NEW INSIGHTS INTO THE MECHANISM OF PENETRATIN
INTERNALIZATION

Recent studies on the mechanism of internalization of penetratin in live cells have revealed a more complex picture than previously thought, due in part to the multiplication of the experimental protocols used for these studies. A critical point is the distinction between the intracellular and the bound extracellular peptide. For example, the use of organic fixation can lead to the artifactual intracellular accumulation of the peptide. However, distinct translocation abilities of penetratin and penetratin2Phe, observed after organic fixation, have been confirmed by other techniques.[16] Similarly, the effects of drug treatments on cell-penetrating peptide (CPP) internalization should be analyzed cautiously as they can interfere with secondary events, such as peptide stability.[39]

It is now clear that penetratin is internalized by more than one mechanism, including endocytosis.[22,23,40] It is thus reasonable to assume that, depending on the experimental protocol, the relative contribution of each mechanism could differ. For example, the intracellular distribution of internalized penetratin differs greatly between Hela and MC57 cell lines, and the macropynocitosis inhibitor ethylisopropylamiloride (EIPA) decreases penetratin uptake added at high (50 μM) but not low (10 μM) concentrations.[41,42] Another important parameter is the nature of the reporter used for internalization quantification, from a biotin residue (the smallest) to large cargo proteins. In the latter case, the contribution of the endocytotic component in the internalization process seems to be prevalent.[43]

1.3.5 PROPOSED MODEL OF PENETRATIN-SPECIFIC INTERNALIZATION

Due to the diversity of penetratin internalization mechanisms, we focus on those specific for the penetratin family. In particular, endocytosis appears to be a general internalization mechanism common to highly basic peptides, including Tat, with little specificity for penetratin. Indeed, W→F substitution within penetratin has minor effects on endocytosis-based internalization but affects nonendocytotic internalization significantly.[23,44] A model can be proposed that takes into account all structure/function and biophysical/biochemical studies on penetratin and its derivatives. Our working hypothesis is that penetratin peptides interact directly with negatively charged lipids or sugar components, presumably through an electrostatic interaction, and that this interaction is followed by a destabilization of the lipid bilayer and the formation of inverted micelles in which the peptides are trapped. Eventually, a fraction of the micelles will open on the cytoplasmic side, thus allowing intracellular delivery. In this model, penetratin is always kept in a hydrophilic environment, included in the cavity of the micelle, and delivered from the extracellular medium to the cytoplasm of the cell. This translocation mechanism led us to speculate that hydrophilic molecules linked to penetratin peptides would also be internalized. As stated previously, the latter point has to take into account a possible and hardly predictable influence of the cargo on the internalization mechanism.

1.4 DELIVERY OF HYDROPHILIC CARGOES IN LIVE CELLS WITH PENETRATIN VECTORS

Since the first reports demonstrating the efficient cell delivery of hydrophilic molecules linked to penetratin or AntpHD,[45–47] several applications have been developed thanks to the Antennapedia homeodomain, its third helix, and several variants (for reviews see Refs. [48–52]). Penetratin and penetratin-derived peptides have been used successfully to deliver chemical molecules, proteins, oligopeptides, oligonucleotides, peptide nucleic acids, siRNA into live cells, both in vitro and insections describe the mode of linkage and highlights some typical applications. A more comprehensive list of cargoes delivered by this vector system is presented in Table 1.3.

1.4.1 PRINCIPLES OF CARGO–VECTOR LINKAGE

One of the first applications was to link and internalize oligonucleotides bearing a free thiol group to the cysteine naturally present in the β-turn between the second and the third helix of AntpHD.[53] AntpHD has also been used to deliver recombinant polypeptides of different lengths designed by in vitro recombination (Table 1.3). However, most experiments were developed with the penetratin peptide. Two main types of coupling protocols were used.

In the first protocol, penetratin is synthesized with an additional N-terminal-activated cysteine, protected by a nitropyridinium group to prevent peptide homodimerization. The cargo is probably released freely in the cytoplasm as a result of disulfide bond breakage in the reductive cytoplasmic milieu.

In the second method, the cargo and the vector are chemically synthesized in continuity. In this case, no coupling reaction is necessary. These first two methods also allow the internalization of modified peptides, such as phosphopepetides or peptide nucleic acids (PNA).

1.4.2   Vectorization with AntpHD

1.4.2.1   AntpHD-Mediated Internalization of Oligonucleotides

The first applications were developed with the entire homeodomain. AntpHD has been used to internalize antisense oligonucleotides against the β-amyloid precursor protein (β-APP) into neurons in culture.[46] Internalized antisense oligonucleotides cause a transient inhibition of β-APP synthesis and of neurite elongation at concentrations in the nanometer range. A more recent example is the overcoming of radioresistance in human gliomas by AntpHD-coupled p21$^{WAF1/CIP1}$ antisense oligonucleotides.[54] Malignant gliomas are highly resistant tumors against γ-irradiation and overexpress the cyclin-CDK inhibitor protein p21. This over-expression enhances survival and suppresses apoptosis after γ-irradiation. The pretreatment of cells with antisense oligonucleotides coupled with AntpHD enhanced γ-irradiation-induced apoptosis and cytotoxicity in radioresistant glioma cells.

However, the success of strategies based on the internalization of oligonucleotides is highly dependent on the stability of the oligonucleotides, and on the half-life of the target protein. Therefore, we believe that, when possible, internalizing a peptide with direct target antagonizing activity represents a better approach.

1.4.2.2   AntpHD-Mediated Internalization of Polypeptides

1.4.2.2.1   In Vitro

In a study aimed at understanding the role of small GTP-binding proteins in prolactin exocytosis, several fusion peptides were constructed linking AntpHD to 30–40 amino acids derived from the C-terminal domains of rab1, rab2, and rab3.[47] After internalization by anterior rat pituitary cells, only rab3 C-terminus blocked prolactin exocytosis. AntpHD-coupled peptides have also been used to inactivate tyrosine-kinase membrane receptors such as PDGF-receptor, IGF-I receptor, and insulin receptor. To that end, Grb10 binding to the activated cytoplasmic domain of the receptors has been inhibited by internalizing phosphopeptides that bind Grb10 SH2 and SH3 adaptor domains.[55] A similar protocol was used to study the function of another cellular partner of PDGF, IGF-I, and insulin receptors: PSM (proline-rich, PH SH2 domain-containing signaling mediator).[56] Similarly to Grb10, PSM possesses SH2 and SH3 adaptor domains, and an AntpHD-coupled peptide mimetic of the proline-rich putative SH3 domain-binding region interfered with PDGF, IGF-I, and insulin, but not with EGF-induced DNA synthesis. As in the case of Grb10, PSM is a positive effector for mitogenic signals triggered by PDGF, IGF-I, or insulin, and SH2 and SH3 domains determine the specificity of PSM action in each mitogenic pathway.

1.4.2.2.2   In Vivo

The first in vivo application of AntpHD-mediated vectorization was the induction of T-cell responses by a peptide derived from the HLA-cw3 cytotoxic T-cell epitope. AntpHD-based fusion peptides expressing the 170–179 amino acids from HLA-Cw3 or 147–156 amino acids from influenza nucleoprotein with flanking proteasome

TABLE 1.3
Cargoes Internalized In Vitro and In Vivo with AntpHD-Derived Peptidic Vectors

Type of Cargo	Length	Vector	Cargo Effect	Refs.
Particle				
Adenovirus		Penetratin	Increases viral cell entry	107
Entire protein				
p16^{INK4A}	156 aa	Penetratin	Induces replicative senescence	94
HoxB4	251 aa	HoxB4HD	Stem cell proliferation	95
Engrailed2	290 aa	En2HD	Retinal axon guidance	96
ScFv	560 aa	Penetratin	Increase tumor retention of ScFv	106
Oligopeptide				
PKC	14 aa	AntpHD	Inhibits PKC activity in neurons	45
PKC-ε	9 aa	Penetratin	Inhibits PKC-ε activity in rabbit cardiomyocytes	78
cGPK	8 aa	Penetratin	Inhibits cGMP kinase in cerebral arteries	79
55,56SSeKS	16 aa	Penetratin	Inhibits SSeKS binding to cyclinD1	71
ICE	6 aa	Penetratin	Blocks ICE proteases	60
EGF-R	9 aa	Penetratin	Inhibits EGF and PDGF-stimulated mitosis	80
FGF-R1	9 aa	Penetratin	Inhibits FGF-R signaling in neurons	80
p75^{NTR}	35 aa	Penetratin	Suppresses neuronal death promoted by the chopper domain of p75^{NTR}	110
Pcw3	41 aa	AntpHD	T-cell activation	57,58
OVA-K	8 aa	Penetratin	T-cell activation and protection against ovalbumin expressing tumor cell line	99
Rab 1	41 aa	AntpHD	Inhibits prolactin exocytosis	47
Grb10	16 aa	Penetratin	Blocks PDGF and insulin-stimulated mitogenesis	55
PSM	16 aa	Penetratin	Blocks PDGF and insulin-stimulated mitogenesis	56
Cav1	19 aa	Penetratin	Inhibits NO synthesis and inflammation	100
Met-R	19 aa	Penetratin	Inhibits Met kinase activity	83
CD44	13 aa	Penetratin	Inhibits CD44 activation	81
p16^{INK4A}	20 aa	Penetratin	Inhibits cyclin D1-CDK4 activity	65,66

p16[INK4A]	20 aa	Penetratin	Inhibits cyclin D1-CDK4 mediated cell-spreading on vitronectin	68
p16[INK4A]	20 aa	Penetratin	Inhibits transformation of pancreatic tumor cells	66
p21[Waf1/Cip1]	20 aa	Penetratin	Inhibits PCNA activity and promotes growth arrest of p53$^{-/-}$ cells	108
p21[Waf1/Cip1]	26 aa	Penetratin	Inhibits cyclin D1/CDK4 and cyclin E/CDK2 activity and kills human lymphoma cells	111
eIF4e	12 aa	Penetratin	Blocks eIF4e-induced transformation	72
p53/E1B	17 aa	Penetratin	Inhibits E1B binding to p53	73
p53	95 aa	Penetratin	Inhibits p53-mediated apoptosis	74
cMyc	14 aa	Penetratin	Inhibits cancerization of human breast cancer cells	75,76
E2F-1	8 aa	Penetratin	Inhibits cyclin A/E-CDK2 activity	69
APP	15 aa	Penetratin	Induces apoptosis of neurons	84
Oligonucleotide				
SOD1	21 mer	Penetratin	Inhibits SOD1 in PC12 cells	53
β-APP	25 mer	AntpHD	Inhibits β-APP in neurons	46
p21[Waf1/Cip1]	21 mer	AntpHD	Inhibits p21[Waf1/Cip1] in human gliomas	54
HAND	18 mer	Penetratin	Inhibits the bHLH factor HAND in neural crest cells	109
APP	20 mer	Penetratin	Inhibits proliferation of adult neuronal progenitors	102
PNA				
GalR1	21 mer	Penetratin	Downregulates galanin receptor R1 in vivo	89
Telomerase	13 mer	Penetratin	Inhibits telomerase in melanoma cells	87,88
HIV TAR	16 mer	Penetratin	Inhibits HIV replication	90
siRNA				
SOD	2×19 mer	Penetratin	Inhibits SOD in neurons	61
Caspase 3	2×19 mer	Penetratin	Inhibits caspase 3 in neurons	61
Chemical drug				
Doxorubicin		Penetratin	Doxorubicin crosses BBB	103

recognition sites were directed into the cell cytoplasm to allow their presentation in the MHC-I context.[57] Peptide presentation led to the priming of cytotoxic T-cells in vitro and in vivo (after intraperitoneal injection). In vivo efficiency in antigen presentation required that the peptide be associated with negative charges under the form of SDS or polysialic acid. Although this was not investigated, it is believed that negative charges protect the peptide from degradation and favor its diffusion within the organism.

In a more recent report, Chikh et al. have shown that the addition of liposomes stabilizes AntpHD-Cw3 and protects it against degradation.[58] Association with liposomes is spontaneous regardless of lipid composition, and approximately 50% of the tethered peptide can also be exchanged from these liposomes. Moreover, this association with liposomes confers a partial resistance against protease degradation, and does not alter the capacity of the recombinant peptide to be internalized in macrophage and dendritic cells and to prime T-cell response in vivo through the MHC-I pathway.

1.4.3   VECTORIZATION WITH PENETRATIN PEPTIDE

1.4.3.1   Oligonucleotide and Oligopeptide Delivery

In this section we primarily consider the in vitro delivery of oligopeptides and oligonucleotides, with a specific section being devoted to full-length proteins and to in vivo approaches. Two main domains of application (cell death/cell cycle and signal transduction) are developed.

1.4.3.1.1   Cell Death and Cell-Cycle Regulation

Several pathways lead to apoptosis in PC12 cells (nitric oxide (NO), withdrawal of trophic factors). Incubation of PC12 cells with an antisense oligonucleotide against superoxyde dismutase 1 (SOD-1) mRNA efficiently decreases SOD-1 activity (50 to 60%) and promotes apoptotic cell death specifically induced by an accumulation of NO.[59] A 100-fold increase in antisense efficiency on both effects was observed after linkage to penetratin by a disulfide bridge.[53] When linked to penetratin, oligonucleotide internalization became insensitive to the presence of serum during the incubation.

In the same experimental model, incubation of PC12 cells with a penetratin-linked oligopeptide, which mimics the catalytic site of interleukin 1β-converting enzyme (ICE)-like proteases, protected cells from apoptosis.[60] Equivalent effects were obtained with the permeant competitive ICE inhibitor ZVAD-FMK, but at much higher concentrations (200-fold). It should be mentioned that two different penetratin-linked cargoes (SOD antisense, ICE oligopeptide) have been used simultaneously without interference.

More recently, the same group has used penetratin to internalize SOD1 and caspase-3 siRNA into neurons in primary culture. Internalization occurs in almost all cells with a strong effect on neuronal mRNA translation and neuronal survival and without nonspecific toxicity.[61] This use of penetratin-1 to internalize si-RNA into live cells confirmed previous studies demonstrating the interest of associating si-RNA approaches to CPPs.[62,63]

Penetratin peptides have been used to interfere with the cell cycle, in particular to antagonize the interaction between cyclins and cyclin-dependent kinases (CDK). Cell-cycle inhibitors, also tumor suppressors, are grouped into two families: the $p21^{Cip1/Waf1}$ family (including p21, p27, and p57) and the p16 family (including $p15^{INK4b}$, $p16^{INK4b}$, $p18^{INK4c}$, and $p19^{INK4d}$).[64] p16 CDK2/INK4A is deleted or mutated in a large number of human cancers and its overexpression blocks the transition through the G1/S phase of the cell cycle. p16 binding blocks CDK4, CDK6 activity, and pRb phosphorylation. Peptides derived from the p16-binding domain to CDK4, CDK6, and internalized after penetratin linkage, mimicked the p16 inhibitory action on CDK4 and CDK6 and prevented pRb phosphorylation and cell-cycle progression.[65,66]

Inhibition of pRb phosphorylation in pancreatic cancer cells devoid of endogenous p16 activity was also obtained by Fujimoto et al.[67] Indeed, they restored p16 function using the same peptide as Fahraeus et al.[65,66] Consequently, the growth of cancer cells was inhibited through arrest in G_1. Another study with p16-derived peptides also demonstrated that p16 inhibits $\alpha_v\beta_3$ integrin-mediated cell spreading on vitronectin.[68]

Penetratin vectors have been also directed against downstream targets of pRb. Members of the E2F transcription factor family are positive effectors of the cell cycle, and are downstream targets of pRb. Unphosphorylated pRb is able to sequester E2F and converts it from a transcriptional activator to a transcriptional repressor. Consequently, disruption of the pRb/E2F can lead to oncogenesis. Among proteins regulating E2F, cyclin A/CDK2 complexes neutralize E2F DNA binding. Chen et al. have designed peptides corresponding to the cyclin A/CDK2 binding site on E2F.[69] Coupled to penetratin, they were able to kill transformed cells in which E2F was already deregulated by pRb inactivation. It is noteworthy that the same team has recently identified about 12 peptides able to antagonize in vitro E2F-1 and cyclin A activity.[70] The potential anticancer effect of cell-permeant versions of the peptides (coupled to penetratin) is currently being tested on tumor cells.

The tumor suppressor (sarc-suppressed C kinase substrate) SSeCKS also blocks the G1/S transition through an interaction with cyclin D1. Internalization with penetratin of the SSeCKS sequence responsible for this interaction demonstrated that SSeCKS anchors cyclin D1 in the cytoplasm, thus giving a molecular basis to G1/S arrest induced by SSeCKS.[71]

Another study with penetratin-coupled antagonizing peptides was conducted to evaluate the role of the translation initiation factor eIF4E in cellular transformation.[72] Overexpression of eIF4E is oncogenic and is found in a number of human cancers, and eIF4E activity can be blocked by eIF4E-binding proteins (4E-BPs). Peptides designed after eIF4E-binding motifs found in 4E-BPs and introduced into cells bound eIF4E and induced massive apoptosis, thus revealing an involvement of this protein in the control of cell death, unrelated to its known role in mRNA translation.

The tumor suppressor protein p53 is frequently targeted by oncoproteins in transformed cells. This is typically the case for adenovirus early region 1B protein (E1B), which binds and inactivates p53 through its sequestration in the cytoplasm. Penetratin-coupled peptides corresponding to the p53-binding site of E1B disrupted p53–E1B interaction allowing p53 nuclear addressing and restoring cell-cycle arrest.[73]

Chemotherapy or radiation therapy activates p53-mediated processes in sensitive tissues. Thus, p53 may be an appropriate target for reducing damage done to normal tissues during cancer therapies. Using a selection of genetic suppressor elements (GSE) from a library of randomly fragmented p53 cDNA, Mittelman and Gudkov[74] have isolated p53-derived antagonizing peptides. One of them, GSE56, was fused to penetratin, and the fusion protein was shown to attenuate p53-mediated transactivation and to reduce anticancer treatment side effects.[75]

Transcription factor c-Myc is a positive effector of cell proliferation. The misregulation of its activity can lead to cell transformation. With the aim of counteracting its activity, Giorello et al. have developed penetratin-coupled inhibitory peptides.[76] These peptides were efficiently internalized in Myc-transformed human breast cancer cells and suppressed their transformation characteristics. Efficiency was improved by synthesizing their retro-inverso forms; these new peptides were both more potent (5–10 times) and more stable (30–35 times) than their natural counterparts.[77]

1.4.3.1.2 Regulation of Signal Transduction

Intracellular protein kinase C (PKC) activity was downregulated in neurons by penetratin-coupled PKC pseudosubstrate provoking growth cone collapse.[45] This confirmed PKC involvement in the transduction of extracellular signals regulating axon elongation. Penetratin-coupled inhibitory peptides targeting protein kinase C epsilon (PKC-ε) have highlighted its role in the protection of ischemic rabbit cardiomyocytes against osmotic shock.[78]

Another intracellular kinase, the cGMP-dependent protein kinase I (cGPKI) was impaired by peptide blockers fused to penetratin. This inhibition into cerebral arteries cultured in vitro revealed a central role of the kinase in promoting NO-induced vasodilatation.[79]

The intracytoplasmic interactions between fibroblast growth factor receptor 1 (FGFR1) and the (sarc-homology 2) SH2 domain of phospholipase C have been antagonized with a penetratin-coupled phosphopeptide corresponding to the phosphorylated site of the cytoplasmic domain of FGFR1.[80] This inhibited phospholipid hydrolysis stimulated by basic FGF, and in consequence, neurite outgrowth stimulated by basic FGF.

Peck and Isacke have analyzed the signaling pathway induced by an extracellular matrix glycosaminoglycan, the hyaluronan.[81] This pathway leads to cell migration. Protein CD44, the hyaluronan transmembrane receptor, plays an essential role in many physiological events, including tumor progression, lymphocyte homing, and tissue morphogenesis. Hyaluronan binding to CD44 induces its phosphorylation on a serine crucial for hyaluronan-dependent cell migration. Penetratin-linked peptides mimicking endogenous phosphorylation sites blocked CD44-induced cell migration without reducing its expression or its ability to bind hyaluronan. Penetratin-coupled peptides have also been used to analyze the signaling pathways of other glycosaminoglycans and their role in neuronal polarity.[82]

Still in the cell migration area, the binding of hepatocyte growth factor (HGF) to its receptor (c-Met) leads to invasive growth, an essential developmental event also implicated in tumor metastasis. The first event in this cascade is receptor autophosphorylation. Peptides interfering with HGF-induced c-Met autophosphorylation coupled to penetratin, once internalized in normal and transformed epithelial cells, blocked c-Met kinase and invasive growth.[83]

In the field of Alzheimer's disease, Allinquant and colleagues showed that a small sequence present in the cytoplasmic domain of the β-amyloid precursor (βAPP), once internalized in vitro and in vivo through linkage to penetratin, induces apoptosis.[84] More recently, using a biotinylated version of the same peptide, they have retrieved several proteins attached to this cytoplasmic bait after its internalization and identified SET-1 as a new regulator of cell death, possibly connected to Alzheimer's disease.[85]

1.4.3.2 Delivery of Peptide Nucleic Acids (PNA)

PNAs are oligonucleotides in which the sugar–phosphate backbone has been replaced by a neutral peptidic backbone.[86] PNAs bind complementary RNA and DNA in a parallel or antiparallel orientations. Compared to deoxyoligonucleotide they have the advantage of forming extremely stable PNA–DNA or PNA–RNA duplexes. Moreover, PNAs are highly resistant to proteases and nucleases and inhibit gene expression at transcriptional and translational levels. Unfortunately, PNAs are only poorly internalized by live cells. Three reports have demonstrated that eukaryotic cells efficiently take up penetratin-coupled PNAs in vitro[87,88] and in vivo.[48,89]

In vitro delivery of PNA using penetratin has been demonstrated in human prostate tumor cells[88] and human melanoma cell lines.[87] The aim was to investigate the effect of two inhibitory PNAs on the catalytic activity of human telomerase. The function of telomerase in normal cells is to maintain chromosome integrity, but several reports have raised the question of telomerase implication in cell immortality and cancer. From a therapeutic point of view, it is of interest to block telomerase activity in transformed cells. To add telomeric repeats to the end of chromosomes, telomerase copies a small RNA sequence complementary to DNA. The strategy used here was to internalize two PNAs, an 11mer and a 13mer designed to cover the 5′-proximal region of the RNA and thus to inhibit human telomerase activity. The 13mer PNA coupled to penetratin through a disulfide-bound downregulated telomerase activity in intact live cells but the doubling time of the treated cells was only slightly decreased, and telomeres were not shortened.[87]

A more successful use of PNAs linked to penetratin was the antagonizing of the biological activity of galanin, a peptide neuromediator involved in pain transduction (see Section 1.4.4, devoted to the in vivo use of penetratin-derived peptides).

More recently, penetratin has been used to internalize an anti-TAR PNA. This treatment was found extremely efficient (IC50 in the nanometer range), showing strong anti-HIV activity.[90]

1.4.3.3   Delivery of Entire Proteins Fused to Penetratin

AntpHD and penetratin are able to deliver large molecular weight protein into live cells; among them are homeoproteins such as Engrailed, HoxA-5, HoxB-4, HoxC-8, Pax 6, HoxC-4, Emx2, and Otx2 (Refs. [91–93] and unpublished results).

A good example of the successful delivery of entire proteins fused to penetratin is the study of the replicative senescence phenomenon. Genetically programmed senescence is induced when cells approach their limit of population doubling. Interestingly, p16^{INK4b} (see above) RNA and protein levels rise as cells enter senescence, but there was no evidence that p16 accumulation drives senescence. Using penetratin to internalize p16^{INK4b} into human diploid fibroblasts, Kato et al. have shown that p16^{INK4b}, but not a functionally compromised variant, provokes G1 arrest.[94] This arrest is certainly due to an inhibition of pRb phosphorylation, and the arrested cells display a senescent phenotype, strongly suggesting that p16^{INK4b} is implicated in replicative senescence.

1.4.3.4   Full-Length Homeoproteins

A special case is the internalization of full-length homeoproteins in which the protein carries its own penetratin vector in the form of the third helix of its homeodomain. In fact, homeoprotein internalization has been demonstrated several times (reviewed in Ref. [12]), but it is only recently that internalization could be associated with a function. A first case was provided by the work of Amsellem and colleagues on the stimulation of CD34-positive hematopoietic stem cell division by HoxB4.[95] The capture of HoxB4 by these cells mimics the physiological and cell-autonomous function of the homeoprotein, suggesting that homeoprotein transcription factors could be used as therapeutic proteins.

The example given by the internalization of Engrailed by the axons of retinal ganglion cells (RGC) is particularly interesting as it suggests that the intercellular transport is physiologically relevant. Indeed, extracellular Engrailed, once internalized, attracts and repels the nasal and temporal cones, respectively. This corresponds to the expression gradient of Engrailed in the tectum and to the retinotectal topology. An important point in the latter study is that the internalized protein guides the axons by regulating the translation of local mRNAs.[96]

1.4.3.5   Chemical Drug Delivery

Doxorubicin, an anthracyclin, is a commonly used anticancer drug. Anthracyclins are among the most active antitumor compounds. Not only do they, as intercalating agents, induce DNA damage but they also bind cell surface acidic phospholipids, such as cardiolipids, and induce necrosis and apoptosis via the sphingomyelin–ceramide pathway. However, anthracyclin treatments often lead to multidrug-resistant tumors through enhanced expression of a multidrug-resistance gene (*mdr*). This gene encodes a 170-kD protein called the P-glycoprotein (Pgp) that actively transports several cytotoxic agents out of the cells.[97]

To overcome this problem, Mazel et al. have covalently linked doxorubicin to the penetratin N terminus and tested this conjugate in human erythroleukemia K562 cells.[98] Doxo-penetrin was approximately 20-fold more effective than doxorubicin alone in killing K562 doxorubicin-resistant cells. This suggests that the mode of internalization of penetratin, very probably through inverted micelles, allows the conjugate to escape Pgp activity.

1.4.4 IN VIVO INTERNALIZATION WITH PENETRATIN-DERIVED VECTORS

1.4.4.1 Immune System

Similarly to Schutze-Redelmeier et al. with AntpHD,[57] Pietersz et al. have demonstrated that penetratin efficiently targets antigenic peptides to the MHC-I complex.[99] Penetratin-coupled 9mer peptides corresponding to CTL epitopes have been successfully and rapidly imported into the cytosol of peritoneal exudate cells (PEC). Moreover, an 8mer peptide derived from ovalbumin (SIINFKEL) and coupled to penetratin induced a T-cell response after injection in mice, and protected them against the development of tumor cell line E.G7-OVA expressing ovalbumin.

1.4.4.2 Delivery of Peptides in Blood Vessels

To appreciate further the role of caveolin-1 (the primary coat protein of caveolae) in signal transduction, Bucci et al. synthesized a hybrid peptide containing penetratin and the caveolin-1 scaffolding domain (19 aa).[100] Mice blood vessels and endothelial cells efficiently took up the peptide ex vivo and in vivo. Consequently, acetylcholine-induced vasodilatation and NO production in isolated mouse aortic rings were selectively inhibited. Moreover, systemic administration of the peptide in mice suppressed acute inflammation and vascular leak as efficiently as a glucocorticoid or endothelial nitric oxide synthase (eNOS) inhibitors.[100]

1.4.4.3 Direct Perfusion Inside the Central Nervous System

Penetratin has been used in vivo to deliver PNAs into the central nervous system.[48,89] Galanin is a widely distributed neuropeptide involved in several biological effects. Galanin receptor type 1 (Gal R1) is highly conserved between species and is abundant in the hypothalamus, hippocampus, and spinal cord.[101] PNAs designed to downregulate the expression of the *Gal R1* gene were covalently coupled to the N terminus of penetratin by a disulfide bond. These PNAs downregulated Gal R1 receptor expression by human Bowles melanoma cells and in vivo after intrathecal delivery at the level of the dorsal spinal cord. Downregulation in the spinal cord decreases galanin binding and inhibits the C-fibers stimulation-induced facilitation of the rat flexor reflex, a response monitoring Gal R1 activity.

Caillé and colleagues have achieved the same downregulation of specific targets by infusing penetratin-linked antisense constructs directly into the lateral ventricles.[102] By doing this, they could establish that this molecule is a true regulator of neural stem cell division in vivo. Unpublished results from our group have also demonstrated that siRNA linked to penetratin can reduce by more than

80% the amount of mRNA encoding homeoprotein Otx2 in the retina and in the basolateral septum.

1.4.4.4   The Blood–Brain Barrier (BBB)

Following the intravenous injection of a doxorubicin–penetratin peptide, Rousselle et al. have observed its passage into rat brain parenchyma.[103] These data are at odds with the absence of the BBB crossing reported by Bolton et al.[104] However, the conditions between the two experiment are very different. In particular, Rousselle et al. used radioactive doxorubicin coupled to a penetratin peptide composed of D-amino acids (2.5 mg/kg), and quantified brain accumulation 30 min after injection. In contrast, Bolton et al. used a fluorescent compound (2.0 mg/kg) linked to normal penetratin, and quantified its presence in the brain 24 h after injection. The two approaches thus differ not only in term of protocols but also of sensitivity, not to mention the modification of the vector (D- vs. L-amino acids).

Drug delivery to the brain is often restricted by the BBB, which regulates the exchange of substances between brain and blood. The protein Pgp, which participates in the multidrug-resistance phenomenon (see above), has been detected at the luminal site of the endothelial cells of the BBB.[105] As a consequence, the brain availability of several drugs and, in particular, of anticancer drugs is extremely low, a likely explanation for the failure of brain tumor chemotherapy. The crossing of doxorubicin was analyzed either by the in situ brain perfusion technique or by intravenous injection.[103] The amount of penetratin-coupled doxorubicin transported into the brain was 20-fold higher than that of free doxorubicin, without BBB disruption. These results demonstrate that penetratin peptides might be used as a very efficient and safe means to deliver drugs across the BBB.

1.4.4.5   Single-Chain Antibodies

Antibodies appear more and more like interesting therapeutic molecules not only because they can neutralize toxic epitopes, as proposed in Alzheimer's disease, but also because they provide a way to target toxic compounds to specific cells, for example tumor cells expressing disease-associated epitopes at their surface. More recently, phage display technologies have permitted the development of single-chain antibodies that can recognize not only extracellular but also intracellular targets. These intrabodies can be very useful if they keep their structure within the reductive intracellular milieu. However, they must be internalized by the cells. Fusion with penetratin or Tat peptide of such an intrabody and the resulting accumulation in solid tumors have recently been demonstrated by Jain and colleagues.[106]

1.4.4.6   Huge Cargoes

Finally, it seems that through a mechanism not fully understood, adenoviruses soaked in penetratin can gain access to the cell interior allowing the use of much lower concentrations of the viruses in vitro and in vivo.[107] Similarly, the same technology was used in our group with bacterial phages. We could induce the expression, in vitro and in vivo, of high molecular weight proteins. The fact that such phages can carry up to 20 kb of exogenous genetic information is possibly a new and friendly usage of CPP.

REFERENCES

1. Gehring, W.J. et al., Homeodomain-DNA recognition, *Cell*, 78(2), 211, 1994.
2. Doe, C.Q. and Scott, M.P., Segmentation and homeotic gene function in the developing nervous system of Drosophila, *Trends Neurosci.*, 11(3), 101, 1988.
3. Doe, C.Q., Smouse, D., and Goodman, C.S., Control of neuronal fate by the Drosophila segmentation gene even-skipped, *Nature*, 333(6171), 376, 1988.
4. Le Mouellic, H., Lallemand, Y., and Brulet, P., Homeosis in the mouse induced by a null mutation in the Hox-3.1 gene, *Cell*, 69(2), 251, 1992.
5. Miller, D.M. et al., C. elegans unc-4 gene encodes a homeodomain protein that determines the pattern of synaptic input to specific motor neurons, *Nature*, 355(6363), 841, 1992.
6. Tiret, L. et al., Increased apoptosis of motoneurons and altered somatotopic maps in the brachial spinal cord of Hoxc-8-deficient mice, *Development*, 125(2), 279, 1998.
7. White, J.G., Southgate, E., and Thomson, J.N., Mutations in the *Caenorhabditis elegans* unc-4 gene alter the synaptic input to ventral cord motor neurons, *Nature*, 355(6363), 838, 1992.
8. Ayala, J. et al., The product of rab2, a small GTP binding protein, increases neuronal adhesion, and neurite growth in vitro, *Neuron*, 4(5), 797, 1990.
9. Borasio, G.D. et al., ras p21 protein promotes survival and fiber outgrowth of cultured embryonic neurons, *Neuron*, 2(1), 1087, 1989.
10. Joliot, A. et al., Antennapedia homeobox peptide regulates neural morphogenesis, *Proc. Natl Acad. Sci. U.S.A.*, 88(5), 1864, 1991.
11. Prochiantz, A., Messenger proteins, *J. Soc. Biol.*, 194(3–4), 119, 2000.
12. Prochiantz, A. and Joliot, A., Can transcription factors function as cell–cell signalling molecules?, *Nat. Rev. Mol. Cell Biol.*, 4(10), 814, 2003.
13. Bloch-Gallego, E. et al., Antennapedia homeobox peptide enhances growth and branching of embryonic chicken motoneurons in vitro, *J. Cell Biol.*, 120(2), 485, 1993.
14. Le Roux, I. et al., Neurotrophic activity of the Antennapedia homeodomain depends on its specific DNA-binding properties, *Proc. Natl Acad. Sci. U.S.A.*, 90(19), 9120, 1993.
15. Le Roux, I. et al., Promoter-specific regulation of gene expression by an exogenously added homedomain that promotes neurite growth, *FEBS Lett.*, 368(2), 311, 1995.
16. Derossi, D. et al., The third helix of the Antennapedia homeodomain translocates through biological membranes, *J. Biol. Chem.*, 269(14), 10444, 1994.
17. Fischer, P.M. et al., Structure-activity relationship of truncated and substituted analogues of the intracellular delivery vector penetratin, *J. Pept. Res.*, 55(2), 163, 2000.
18. Drin, G. et al., Translocation of the pAntp peptide and its amphipathic analogue AP-2AL, *Biochemistry*, 40(6), 1824, 2001.
19. Derossi, D. et al., Cell internalization of the third helix of the Antennapedia homeodomain is receptor-independent, *J. Biol. Chem.*, 271(30), 18188, 1996.
20. Drin, G. et al., Physico-chemical requirements for cellular uptake of pAntp peptide. Role of lipid-binding affinity, *Eur. J. Biochem.*, 268(5), 1304, 2001.

21. Mainguy, G. et al., An induction gene trap for identifying a homeoprotein-regulated locus, *Nat. Biotechnol.*, 18(7), 746, 2000.

22. Letoha, T. et al., Investigation of penetratin peptides. Part 2. In vitro uptake of penetratin and two of its derivatives, *J. Pept. Sci.*, 11(12), 805, 2005.

23. Christiaens, B. et al., Membrane interaction and cellular internalization of penetratin peptides, *Eur. J. Biochem.*, 271(6), 1187, 2004.

24. de Kruijff, B. et al., Molecular aspects of the bilayer stabilization induced by poly(L-lysines) of varying size in cardiolipin liposomes, *Biochim. Biophys. Acta*, 820(2), 295, 1985.

25. Berlose, J.P. et al., Conformational and associative behaviours of the third helix of antennapedia homeodomain in membrane-mimetic environments, *Eur. J. Biochem.*, 242(2), 372, 1996.

26. Lindberg, M. and Graslund, A., The position of the cell penetrating peptide penetratin in SDS micelles determined by NMR, *FEBS Lett.*, 497(1), 39, 2001.

27. Magzoub, M. et al., Interaction and structure induction of cell-penetrating peptides in the presence of phospholipid vesicles, *Biochim. Biophys. Acta*, 1512(1), 77, 2001.

28. Persson, D., Thorén, P.E.G., and Norden, B., Penetratin-induced aggregation and subsequent dissociation of negatively charged phospholipid vesicles, *FEBS Lett.*, 25245, 1, 2001.

29. Bellet-Amalric, E. et al., Interaction of the third helix of Antennapedia homeodomain and a phospholipid monolayer, studied by ellipsometry and PM-IRRAS at the air–water interface, *Biochim. Biophys. Acta*, 1467(1), 131, 2000.

30. Fragneto, G. et al., Neutron and X-ray reflectivity studies at solid–liquid interfaces: The interactions of a peptide with model membranes, *Physica B*, 276, 2000.

31. Fragneto, G. et al., Interaction of the third helix of Antennapedia homeodomain with a deposited phospholipid bilayer: A neutron reflectivity structural study, *Langmuir*, 16, 4581, 2000.

32. Lindberg, M. et al., Structure and positioning comparison of two variants of penetratin in two different membrane mimicking systems by NMR, *Eur. J. Biochem.*, 270(14), 3055, 2003.

33. Magzoub, M., Eriksson, L.E., and Graslund, A., Comparison of the interaction, positioning, structure induction and membrane perturbation of cell-penetrating peptides and nontranslocating variants with phospholipid vesicles, *Biophys. Chem.*, 103(3), 271, 2003.

34. Persson, D. et al., Vesicle size-dependent translocation of penetratin analogs across lipid membranes, *Biochim. Biophys. Acta*, 1665(1–2), 142, 2004.

35. Thorén, P.E. et al., The antennapedia peptide penetratin translocates across lipid bilayers — the first direct observation, *FEBS Lett.*, 482(3), 265, 2000.

36. Binder, H. and Lindblom, G., Charge-dependent translocation of the Trojan peptide penetratin across lipid membranes, *Biophys. J.*, 85(2), 982, 2003.

37. Terrone, D. et al., Penetratin and related cell-penetrating cationic peptides can translocate across lipid bilayers in the presence of a transbilayer potential, *Biochemistry*, 42(47), 13787, 2003.

38. Barany-Wallje, E. et al., A critical reassessment of penetratin translocation across lipid membranes, *Biophys. J.*, 89(4), 2513, 2005.

39. Hällbrink, M. et al., Uptake of cell-penetrating peptides is dependent on peptide-to-cell ratio rather than on peptide concentration, *Biochim. Biophys. Acta*, 1667(2), 222, 2004.

40. Letoha, T. et al., Membrane translocation of penetratin and its derivatives in different cell lines, *J. Mol. Recognit.*, 16(5), 272, 2003.

41. Fischer, R. et al., A quantitative validation of fluorophore-labelled cell-permeable peptide conjugates: Fluorophore and cargo dependence of import, *Biochim. Biophys. Acta*, 1564(2), 365, 2002.

42. Nakase, I. et al., Cellular uptake of arginine-rich peptides: Roles for macropinocytosis and actin rearrangement, *Mol. Ther.*, 10(6), 1011, 2004.

43. Saalik, P. et al., Protein cargo delivery properties of cell-penetrating peptides. A comparative study, *Bioconjug. Chem.*, 15(6), 1246, 2004.

44. Thorén, P.E. et al., Uptake of analogs of penetratin, Tat(48–60) and oligoarginine in live cells, *Biochem. Biophys. Res. Commun.*, 307(1), 100, 2003.

45. Theodore, L. et al., Intraneuronal delivery of protein kinase C pseudosubstrate leads to growth cone collapse, *J. Neurosci.*, 15(11), 7158, 1995.

46. Allinquant, B. et al., Downregulation of amyloid precursor protein inhibits neurite outgrowth in vitro, *J. Cell Biol.*, 128(5), 919, 1995.

47. Perez, F. et al., Rab3A and Rab3B carboxy-terminal peptides are both potent and specific inhibitors of prolactin release by rat cultured anterior pituitary cells, *Mol. Endocrinol.*, 8(9), 1278, 1994.

48. Prochiantz, A., Peptide nucleic acid smugglers, *Nat. Biotechnol.*, 16(9), 819, 1998.

49. Prochiantz, A., Messenger proteins: Homeoproteins, TAT and others, *Curr. Opin. Cell Biol.*, 12(4), 400, 2000.

50. Lindgren, M. et al., Cell-penetrating peptides, *Trends Pharmacol. Sci.*, 21(3), 99, 2000.

51. Derossi, D., Chassaing, G., and Prochiantz, A., Trojan peptides: The penetratin system for intracellular delivery, *Trends Cell Biol.*, 8(2), 84, 1998.

52. Joliot, A. and Prochiantz, A., Transduction peptides: From technology to physiology, *Nat. Cell Biol.*, 6(3), 189, 2004.

53. Troy, C.M. et al., Downregulation of Cu/Zn superoxide dismutase leads to cell death via the nitric oxide-peroxynitrite pathway, *J. Neurosci.*, 16(1), 253, 1996.

54. Kokunai, T. et al., Overcoming of radioresistance in human gliomas by p21WAF1/CIP1 antisense oligonucleotide, *J. Neurooncol.*, 51(2), 111, 2001.

55. Wang, J. et al., Grb10, a positive, stimulatory signaling adapter in platelet-derived growth factor BB-, insulin-like growth factor I-, and insulin-mediated mitogenesis, *Mol. Cell Biol.*, 19(9), 6217, 1999.

56. Riedel, H. et al., PSM, a mediator of PDGF-BB-, IGF-I-, and insulin-stimulated mitogenesis, *Oncogene*, 19(1), 39, 2000.

57. Schutze-Redelmeier, M.P. et al., Introduction of exogenous antigens into the MHC class I processing and presentation pathway by Drosophila antennapedia homeodomain primes cytotoxic T cells in vivo, *J. Immunol.*, 157(2), 650, 1996.

58. Chikh, G., Bally, M., and Schutze-Redelmeier, M.P., Characterization of hybrid CTL epitope delivery systems consisting of the Antennapedia homeodomain peptide vector formulated in liposomes, *J. Immunol. Methods*, 254(1–2), 119, 2001.

59. Troy, C.M. and Shelanski, M.L., Down-regulation of copper/zinc superoxide dismutase causes apoptotic death in PC12 neuronal cells, *Proc. Natl Acad. Sci. U.S.A.*, 91(14), 6384, 1994.

60. Troy, C.M. et al., The contrasting roles of ICE family proteases and interleukin-1beta in apoptosis induced by trophic factor withdrawal and by copper/zinc superoxide dismutase down-regulation, *Proc. Natl Acad. Sci. U.S.A.*, 93(11), 5635, 1996.

61. Davidson, T.J. et al., Highly efficient small interfering RNA delivery to primary mammalian neurons induces microRNA-like effects before mRNA degradation, *J. Neurosci.*, 24(45), 10040, 2004.

62. Muratovska, A. and Eccles, M.R., Conjugate for efficient delivery of short interfering RNA (siRNA) into mammalian cells, *FEBS Lett.*, 558(1–3), 63, 2004.
63. Xu, Z., Kukekov, N.V., and Greene, L.A., POSH acts as a scaffold for a multiprotein complex that mediates JNK activation in apoptosis, *EMBO J.*, 22(2), 252, 2003.
64. Sherr, C.J. and Roberts, J.M., CDK inhibitors: Positive and negative regulators of G1-phase progression, *Genes Dev.*, 13(12), 1501, 1999.
65. Fahraeus, R. et al., Inhibition of pRb phosphorylation and cell-cycle progression by a 20-residue peptide derived from p16CDKN2/INK4A, *Curr. Biol.*, 6(1), 84, 1996.
66. Fahraeus, R. et al., Characterization of the cyclin-dependent kinase inhibitory domain of the INK4 family as a model for a synthetic tumour suppressor molecule, *Oncogene*, 16(5), 587, 1998.
67. Fujimoto, K. et al., Inhibition of pRb phosphorylation and cell cycle progression by an antennapedia-p16(INK4A) fusion peptide in pancreatic cancer cells, *Cancer Lett.*, 159(2), 151, 2000.
68. Fahraeus, R. and Lane, D.P., The p16(INK4a) tumour suppressor protein inhibits alphavbeta3 integrin-mediated cell spreading on vitronectin by blocking PKC-dependent localization of alphavbeta3 to focal contacts, *EMBO J.*, 18(8), 2106, 1999.
69. Chen, Y.N. et al., Selective killing of transformed cells by cyclin/cyclin-dependent kinase 2 antagonists, *Proc. Natl Acad. Sci. U.S.A.*, 96(8), 4325, 1999.
70. Sharma, S.K. et al., Identification of E2F-1/Cyclin A antagonists, *Bioorg. Med. Chem. Lett.*, 11, 2449, 2001.
71. Lin, X., Nelson, P., and Gelman, I.H., SSeCKS, a major protein kinase C substrate with tumor suppressor activity, regulates $G(1) \rightarrow S$ progression by controlling the expression and cellular compartmentalization of cyclin D, *Mol. Cell Biol.*, 20(19), 7259, 2000.
72. Herbert, T.P. et al., Rapid induction of apoptosis mediated by peptides that bind initiation factor eIF4E, *Curr. Biol.*, 10(13), 793, 2000.
73. Hutton, F.G. et al., Consequences of disruption of the interaction between p53 and the larger adenovirus early region 1B protein in adenovirus E1 transformed human cells, *Oncogene*, 19(3), 452, 2000.
74. Mittelman, J.M. and Gudkov, A.V., Generation of p53 suppressor peptide from the fragment of p53 protein, *Somat. Cell Mol. Genet.*, 25(3), 115, 1999.
75. Komarov, P.G. et al., A chemical inhibitor of p53 that protects mice from the side effects of cancer therapy, *Science*, 285(5434), 1733, 1999.
76. Giorello, L. et al., Inhibition of cancer cell growth and c-Myc transcriptional activity by a c-Myc helix 1-type peptide fused to an internalization sequence, *Cancer Res.*, 58(16), 3654, 1998.
77. Pescarolo, M.P. et al., A retro-inverso peptide homologous to helix 1 of c-Myc is a potent and specific inhibitor of proliferation in different cellular systems, *FASEB J.*, 15(1), 31, 2001.
78. Liu, G.S. et al., Protein kinase C-epsilon is responsible for the protection of preconditioning in rabbit cardiomyocytes, *J. Mol. Cell Cardiol.*, 31(10), 1937, 1999.
79. Dostmann, W.R. et al., Highly specific, membrane-permeant peptide blockers of cGMP-dependent protein kinase $I\alpha$ inhibit NO-induced cerebral dilation, *Proc. Natl Acad. Sci. U.S.A.*, 97(26), 14772, 2000.
80. Hall, H. et al., Inhibition of FGF-stimulated hosphatidylinositol hydrolysis and neurite outgrowth by a cell-membrane permeable phosphopeptide, *Curr. Biol.*, 6(5), 580, 1996.
81. Peck, D. and Isacke, C.M., Hyaluronan-dependent cell migration can be blocked by a CD44 cytoplasmic domain peptide containing a phosphoserine at position 325, *J. Cell Sci.*, 111(Pt 11), 1595, 1998.

82. Calvet, S., Doherty, P., and Prochiantz, A., Identification of a signaling pathway activated specifically in the somatodendritic compartment by a heparan sulfate that regulates dendrite growth, *J. Neurosci.*, 18(23), 9751, 1998.

83. Bardelli, A. et al., A peptide representing the carboxyl-terminal tail of the met receptor inhibits kinase activity and invasive growth, *J. Biol. Chem.*, 274(41), 29274, 1999.

84. Bertrand, E. et al., A short cytoplasmic domain of the amyloid precursor protein induces apoptosis in vitro and in vivo, *Mol. Cell Neurosci.*, 18(5), 503, 2001.

85. Madeira, A. et al., SET protein (TAF1beta, I2PP2A) is involved in neuronal apoptosis induced by an amyloid precursor protein cytoplasmic subdomain, *FASEB J.*, 19(13), 190, 2005.

86. Nielsen, P.E., Peptide nucleic acids as therapeutic agents, *Curr. Opin. Struct. Biol.*, 9(3), 353, 1999.

87. Villa, R. et al., Inhibition of telomerase activity by a cell-penetrating peptide nucleic acid construct in human melanoma cells, *FEBS Lett.*, 473(2), 241, 2000.

88. Simmons, C.G. et al., Synthesis and membrane permeability of PNA-peptide conjugate, *Bioorg. Med. Chem. Lett.*, 7(23), 3001, 1997.

89. Pooga, M. et al., Cell penetrating PNA constructs regulate galanin receptor levels and modify pain transmission in vivo, *Nat. Biotechnol.*, 16(9), 857, 1998.

90. Tripathi, S. et al., Anti-HIV-1 activity of anti-TAR polyamide nucleic acid conjugated with various membrane transducing peptides, *Nucleic Acids Res.*, 33(13), 4345, 2005.

91. Maizel, A. et al., A short region of its homeodomain is necessary for engrailed nuclear export and secretion, *Development*, 126(14), 3183, 1999.

92. Joliot, A. et al., Identification of a signal sequence necessary for the unconventional secretion of Engrailed homeoprotein, *Curr. Biol.*, 8(15), 856, 1998.

93. Chatelin, L. et al., Transcription factor hoxa-5 is taken up by cells in culture and conveyed to their nuclei, *Mech. Dev.*, 55(2), 111, 1996.

94. Kato, D. et al., Features of replicative senescence induced by direct addition of antennapedia-p16INK4A fusion protein to human diploid fibroblasts, *FEBS Lett.*, 427(2), 203, 1998.

95. Amsellem, S. et al., Ex vivo expansion of human hematopoietic stem cells by direct delivery of the HOXB4 homeoprotein, *Nat. Med.*, 9(11), 1423, 2003.

96. Brunet, I. et al., Engrailed-2 guides retinal axons, *Nature*, 438(7064), 94, 2005.

97. Broxterman, H.J., Giaccone, G., and Lankelma, J., Multidrug resistance proteins and other drug transport-related resistance to natural product agents, *Curr. Opin. Oncol.*, 7(6), 532, 1995.

98. Mazel, M. et al., Doxorubicin–peptide conjugates overcome multidrug resistance, *Anticancer Drugs*, 12(2), 107, 2001.

99. Pietersz, G.A., Li, W., and Apostolopoulos, V., A 16-mer peptide (RQIKIWFQNR-RMKWKK) from antennapedia preferentially targets the Class I pathway, *Vaccine*, 19(11–12), 1397, 2001.

100. Bucci, M. et al., In vivo delivery of the caveolin-1 scaffolding domain inhibits nitric oxide synthesis and reduces inflammation, *Nat. Med.*, 6(12), 1362, 2000.

101. Burgevin, M.C. et al., Cloning, pharmacological characterization, and anatomical distribution of a rat cDNA encoding for a galanin receptor, *J. Mol. Neurosci.*, 6(1), 33, 1995.

102. Caille, I. et al., Soluble form of amyloid precursor protein regulates proliferation of progenitors in the adult subventricular zone, *Development*, 131(9), 2173, 2004.

103. Rousselle, C. et al., New advances in the transport of doxorubicin through the blood–brain barrier by a peptide vector-mediated strategy, *Mol. Pharmacol.*, 57(4), 679, 2000.

104. Bolton, S.J. et al., Cellular uptake and spread of the cell-permeable peptide penetratin in adult rat brain, *Eur. J. Neurosci.*, 12(8), 2847, 2000.

105. Cordon-Cardo, C. et al., Multidrug-resistance gene (P-glycoprotein) is expressed by endothelial cells at blood–brain barrier sites, *Proc. Natl Acad. Sci. U.S.A.*, 86(2), 695, 1989.

106. Jain, M. et al., Penetratin improves tumor retention of single-chain antibodies: A novel step toward optimization of radioimmunotherapy of solid tumors, *Cancer Res.*, 65(17), 7840, 2005.

107. Gratton, J.P. et al., Cell-permeable peptides improve cellular uptake and therapeutic gene delivery of replication-deficient viruses in cells and in vivo, *Nat. Med.*, 9(3), 357, 2003.

108. Cayrol, C., Knibiehler, M., and Ducommun, B., p21 binding to PCNA causes G1 and G2 cell cycle arrest in p53-deficient cells, *Oncogene*, 16(3), 311, 1998.

109. Howard, M., Foster, D.N., and Cserjesi, P., Expression of HAND gene products may be sufficient for the differentiation of avian neural crest-derived cells into catecholaminergic neurons in culture, *Dev. Biol.*, 215(1), 62, 1999.

110. Coulson, E.J. et al., Chopper, a new death domain of the p75 neurotrophin receptor that mediates rapid neuronal cell death, *J. Biol. Chem.*, 275(39), 30537, 2000.

111. Mutoh, M. et al., A p21(Waf1/Cip1)carboxyl-terminal peptide exhibited cyclin-dependent kinase-inhibitory activity and cytotoxicity when introduced into human cells, *Cancer Res.*, 59(14), 3480, 1999.

2 Tat-Derived Cell-Penetrating Peptides: Discovery, Mechanism of Cell Uptake, and Applications to the Delivery of Oligonucleotides

Saïd Abes, Jean-Philippe Richard, Alain Thierry,
Philippe Clair, and Bernard Lebleu

CONTENTS

2.1   INTRODUCTION

Peptide and nucleic acid-based drugs offer a large array of strategies to regulate very specifically gene expression or to rescue deficient gene expression. A major limitation is, however, the poor efficiency with which these large hydrophilic biomolecules cross biological barriers.

For a long time cationic vectors have been proposed to improve the bioavailability of nucleic acid-based drugs (antisense oligonucleotides (ON), ribozymes, siRNA, plasmid DNA) since (1) they associate spontaneously with negatively charged nucleic acids through electrostatic interactions to form polyplexes and (2) these positively charged complexes allow binding to the negatively charged cell surface and further internalization.

These cationic carriers include cationic lipids, synthetic polymers as polyethyleneimine, dendrimers, and basic peptides.

Homopolymers and, in particular, poly (L-lysine) have been extensively investigated as nucleic acid delivery systems following pioneering work by Ryser et al. (see Lochmann et al.[1] for a recent review). Our own group has documented the enhanced cellular uptake via adsorptive endocytosis of antisense ON covalently bound to poly (L-lysine) in several in vitro models.[2] As an example, a sequence-specific antiviral activity has been achieved with an IC50 in the low micromolar concentration range in an HIV-1 acute infection cell assay. Further work by several groups has indicated that cell addressing can be achieved and that targeted in vivo delivery of the transported antisense ON or plasmid DNA was possible. As an example, mannosylated poly (L-lysine) has been successfully used to deliver antisense ON to macrophages in cell culture and in mice following binding to mannose-specific membrane receptors in these cells.[3]

Despite encouraging preliminary data these cationic homopolymers are rarely used nowadays due to their cytotoxicity in some cell types, their propensity to activate complement, and their heterodispersity.

A new era began when it was realized that purified proteins, such as the Drosophila Antennapedia transcription factor[4] or the HIV-1 Tat transactivating protein,[5,6] were able to cross cellular membranes and find their way to the nucleus. These experiments have paved the way to the first cell-penetrating peptides (CPPs), also named protein transduction domain (PTD). Before describing the history of Tat (48–60) PTD discovery and investigations on its still controversial mechanism of internalization we felt it useful to summarize a few relevant properties of the HIV-1 Tat protein and its mechanism of cellular uptake.

2.2   THE HIV-1 TAT PROTEIN: GENERAL PROPERTIES AND MECHANISM OF CELL UPTAKE

Tat protein transactivation requires binding to a conserved TAR element at the $5'$ end of HIV-1 coded mRNAs. RNA binding is due to a stretch of basic amino acids (RKKRRYRRR) within Tat domain 4, while nuclear tropism depends on its GRKKR nuclear localization signal (NLS).[7] Early studies by Mann and Frankel[8] had already proposed an endocytotic mechanism of Tat uptake involving heparan sulfate binding followed by adsorptive endocytosis. The role of cell surface proteoglycans in the uptake of the Tat protein has been largely confirmed in recent studies.[9] Alternative binding sites have, however, been described, such as the low-density lipoprotein receptor-related protein in neurons[10] or the Flk-1/KDR receptor on vascular endothelial cells.[11] Such observations might explain Tat-associated toxicity for some tissues. Tat is thereafter internalized through an endocytotic mechanism, whose details are still controversial. In a recent comprehensive study using both genetic and pharmacological tools available for the characterization of endocytosis, Vendeville et al.[12] proposed a clathrin-coated pits-dependent uptake followed by early escape from endocytic vesicles.

2.3   FROM FULL-SIZE TAT PROTEIN TO TAT PTD

A first step towards the molecular dissection of the Tat protein was provided by Fawell et al.[13] when they discovered that a 36 amino acid-long Tat fragment (Tat 37–72) could be conjugated to β-galactosidase and was able to promote the uptake of this 177 kDa protein in mammalian cells. This Tat peptide included two potentially interesting domains with respect to delivery, namely the basic domain and the adjacent α-helix.[14] Interestingly, the Antennapedia-derived PTD also included a cluster of basic amino acids and its α-helix conformation was thought at that time to be important for cellular uptake.[4]

A series of fluorochrome-tagged Tat peptides with deletions in the basic domain or in the α-helix domain were synthesized, and their cellular uptake was monitored by fluorescence microscopy. Cellular internalization was clearly associated with the cationic domain and not with the α-helix, thus leading to the prototype GRKKRRQRRR Tat PTD,[15] which includes the Tat RNA binding and NLS motifs mentioned in Section 2.2. Further trimming of this Tat-derived PTD or replacements of any one of its basic amino acids by alanine rapidly decreases translocation efficiency.[16,17] Interestingly, removal of the α-helix domain abolished the cytotoxicity of Tat, as monitored by MTT assays,[15] a key feature for any potential development of Tat PTD as a delivery vector. This SAR study was performed on formaldehyde-fixed cells but its key elements have now been confirmed in experimental conditions that did not lead to artifactual peptide redistribution (J.P. Richard et al., unpublished data).

The core GRKKRRQRRR Tat PTD has often been used with extensions allowing chemical conjugation to various biomolecules. As an example, several groups, including our own, have used a GRKKRRQRRRPPQC whose C-terminal extension includes the PPQ sequence from the Tat protein and a terminal cysteinyl residue to allow coupling to peptides or to ON through a disulfide bridge.[18] Since disulfide bridges are unstable within the reductive intracellular environment, they should open after cell entry and release their cargo, as demonstrated elegantly by Hällbrink et al.[19] Whether dissociation of the cargo from Tat PTD takes place in endocytic vesicles or in the cytoplasm has not, to our knowledge, been established.

Other versions of Tat PTD carrying various N- or C-terminal extensions have been used for the delivery of various cargoes, as reviewed in Brooks et al.[20]

Whether a stable or a labile (as the disulfide bridge) bond between Tat PTD and the transported biomolecules is preferable has seldom been investigated and will anyway depend on the nature of the cargo, as discussed extensively by Brooks et al.[20] In some instances, Tat PTD may conceivably impair target recognition by the attached cargo. In other instances, a positively charged entity might reinforce target recognition. This could become an advantage when delivering steric block ON analogs, such as peptide nucleic acids (PNAs) or morpholino ON derivatives (PMOs).[21]

2.4   TAT-MEDIATED DELIVERY OF OLIGONUCLEOTIDES

Chemical conjugation or fusion to Tat PTD has been exploited by an increasing number of groups in recent years in order to improve the cellular uptake or the bioavailability of low molecular weight drugs, biomolecules, and even large molecular weight material (such as liposomes or nanoparticles for imaging), as reviewed in Snyder and Dowdy.[22] In their comprehensive review, Dietz and Bähr[23] listed 124 examples of Tat PTD-mediated transport, and the number of published applications has increased exponentially over the last 2 years. Several chapters in this book are concerned with various applications of the CPP concept to macromolecular drugs delivery.

We will therefore not attempt to review comprehensively this aspect of the field and will thus focus on ON delivery. As stated in Section 2.1, antisense ON and related strategies are valuable tools to regulate gene expression in a very specific way and have become routine tools in functional genomics. Cellular delivery in cell culture experiments has been achieved by electroporation or by complexation with commercially available cationic lipids or PEI.[24,25] While easy to implement in vitro with most established cell lines, these strategies proved to be more cumbersome in vitro with some primary cells and in vivo for problems of toxicity or poor efficiency in the presence of serum proteins, as reviewed in Refs. [26,27]. The low toxicity of CPPs, such as Tat PTD, and the possibility of delivering a protein cargo to its intracellular target in vivo in various organs, has fostered searches for applications in ON delivery.

A comprehensive survey of the literature reveals less than a dozen publications describing the use of Tat PTD for ON delivery, which is rather low when compared to the large number of publications dealing with peptides and proteins delivery (see reviews by Lindsay[28] for peptide delivery and by Wadia and Dowdy[29] for protein transfection).

Key initial data have been provided by Pooga and colleagues[30] for transportan-conjugated PNAs. The PNA antisense–CPP conjugate was delivered into cultured neuronal cells and was able to downregulate a galanin receptor in a sequence-specific manner. Most impressively, these same constructions were effective after injection in mice, thus indicating that the transportan CPP was able to cross the blood–brain barrier together with its PNA cargo.

Astriab-Fisher and colleagues[18] were the first to demonstrate a sequence-specific and energy-dependent antisense response with Tat PTD-conjugated 2'Omet phosphorothioate antisense ON.

PMO are steric block ON analogs that have been widely used in gene development analysis (reviewed in Heasman[31]). Despite their interest, however, cellular delivery remains problematic for these uncharged ON analogs. The ability to deliver PMO after conjugation to the Tat PTD has been analyzed in a splicing-correction assay, described by Kang et al.[32] (*vide infra*), and in an assay monitoring the downregulation of a c-myc reporter gene expression.[33] Sequence-specific upregulation of luciferase and downregulation of c-myc expression were achieved with the appropriate peptide-conjugated PMO. Tat conjugates were 10 to 20 times more efficient than Pep-1 or NLS conjugates while free PMOs were almost not active in these assays.[21] Requested Tat conjugate concentrations remained, however, relatively high in keeping with entrapment of internalized material in endocytic vesicles.

The potential of CPP conjugation for steric block ON delivery has been extensively evaluated by Gait and colleagues.[34] They have capitalized on a well-controlled assay monitoring the inhibition of Tat-dependent transactivation by 12-mer 2'OMet/LNA mixmer ON analogs complementary to the TAR region of a HIV-1 LTR promoter. Fluorescein-labeled ON mixmers were conjugated to various CPPs (including Tat PTD) through a disulfide bridge. Cellular uptake of the conjugates was largely increased as compared to free ON but was confined to cytoplasmic vesicles, at variance with previous data.[18] No nuclear delivery was detected, and accordingly, no specific inhibition of transactivation could be monitored.[35] As a control, these ON mixmers could be delivered to the nuclei and could promote a sequence-specific transactivation inhibition when delivered with cationic lipids.

The paucity of data obtained with Tat PTD conjugation of antisense ON or PNA could be due to numerous reasons. Among these, poor escape from endocytic vesicles and degradation by nucleases appear the most plausible explanations, as noted above. Likewise, we have shown[36,37] that Tat PTD conjugated to a fluorescent PNA derivative rapidly accumulated within endocytotic vesicles in unfixed HeLa or HUVEC cells, and could barely be detected in the cytosol or in the nuclei. It should be pointed out, however, that fluorescence microscopy may not be able to detect a small proportion of antisense ON (or PNA) escaping from the endocytic vesicles or entering the cytoplasm through a nonendocytotic mechanism. A rather different distribution has, on the other hand, been observed when monitoring in parallel the intracellular distribution of antisense ON delivered with cationic DLS lipoplexes.[38] Antisense ON initially localized in endocytic vesicles and redistributed thereafter to the cytoplasm and the nuclei.[37]

Monitoring an unequivocal and easy-to-quantify biological response seems critical for the assessment of nuclear or cytoplasmic delivery of antisense ON. Most antisense ON assays suffer from the following drawbacks. First, it has proved difficult to delineate whether an antisense ON has been interacting with its target in the nuclei (thereby interfering with pre-mRNA processing or with mRNA nucleocytoplasmic transport) or in the cytoplasm (interfering with mRNA translation and inducing mRNA degradation). Second, antisense ON (or siRNA) action leads, in general, to the downregulation of the target RNA, and it has often been difficult to discriminate between a bona fide antisense effect and side effects.

Recent work by Kole et al.[39] has provided an elegant assay with a positive readout that is now considered the most reliable to assess the nuclear delivery of an antisense ON analog and also to assay for new ON delivery vectors. It capitalizes on studies dealing with abnormalities in the splicing of a human thalassemic β-globin gene. Intron 2 mutations lead to the activation of an intronic cryptic splice site, and as a consequence, to the incomplete removal of the mutated intron. Masking of these cryptic splice sites by RNase H-incompetent ON analogs restores, at least in part, normal splicing and allows the production of a functional globin mRNA.[40] To convert these observations in antisense in vitro and in vivo assays, this mutated intron has been introduced in the coding region of luciferase or EGFP reporter genes, respectively.[32,41] The nuclear delivery of RNase H-incompetent ON (as 2′OMet ON analogs, PNA or PMO) leads to the production of functional luciferase or EGFP, which can be quantitated by biochemical assays or by FACS analysis, respectively.

In a series of recent publications, splicing correction has been documented using this assay both in cell culture experiments[41] and in vivo in a transgenic mouse model expressing the EGFP construction described above.[42] Impressively, appending as few as four lysine residues to the splice correcting PNA allowed functional delivery. A systematic further survey in a slightly different biological model for splicing correction pointed to an optimal length of eight lysines for PNA delivery.[43]

In our hands, however, similar (Lys)$_4$–PNA–Lys (unpublished observations) or (Lys)$_8$–PNA–Lys conjugates[44] were only slightly efficient in splicing correction although they were efficiently taken up by cells. Likewise, a (Lys)$_8$–PNA–Lys construct was ineffective in a Tat/TAR transactivation assay.[45] These disappointing data strongly suggest that the conjugates were taken up by endocytosis and remained entrapped in endocytotic vesicles, as directly evidenced by fluorescence microscopy. In keeping with this hypothesis, a lysosomotropic agent, such as chloroquine, significantly increased biological responses in the splicing-correction assay.[44] That endosome entrapment was limiting ON availability in the nuclei has also been substantiated in recent work by Turner et al.[45] Treatment with chloroquine, according to the protocol defined by Abes et al.[44] did promote transactivation inhibition by ON mixmers and led to significant redistribution of endosome-entrapped material.[45]

Chloroquine has been frequently used to improve the delivery of various drugs entrapped in endocytotic vesicles. For example, it has been shown to improve gene transfer by various nonviral vectors including CPPs.[46]

Likewise, Koppelhus and colleagues[47] have carefully investigated the uptake of PNAs conjugated to CPPs (pTat or pAnt) via a stable maleimide or via a reducible disulfide bridge in a panel of cell lines using confocal scanning microscopy. An energy- and concentration-dependent uptake was clearly documented but little if any material was found outside of endocytic vesicles, at variance with the data reported in Astriab-Fisher et al.[18] In agreement with these observations, free PNA or Tat-conjugated PNAs did not promote any significant splicing-correction in HeLa pLuc 705 cells. Interestingly, coincubation of the Tat–PNA conjugate and 6 mM Ca^{2+} led to a large increase (44-fold) in luciferase expression and to a significant redistribution of fluorochrome-labeled dextran from endocytic vesicles towards the cytosol, in keeping with the Ca^{2+}-associated increase of plasmid DNA transfection efficiency by nonviral vectors.[48] Likewise, an effect of chloroquine treatment on luciferase expression and a redistribution of dextran has been documented by Shiraishi et al.[49]

All together, these data point to a bottleneck in the delivery of nucleic acid-based drugs by most basic amino acid-rich CPP, namely, escape from endocytic vesicles, a problem also encountered in gene therapy with nonviral vectors.

None of the endosome-destabilizing tools described above will be easily implemented in an in vivo situation. Alternative strategies might include co-treatment with fusogenic or membrane-destabilizing peptides and second-generation CPPs with intrinsic membrane-destabilizing properties. A large number of natural or synthetic fusogenic peptides have been described and some of them have been used to improve the expression of plasmid DNA by nonviral delivery vectors. The most interesting ones are peptides, whose fusogenic (or membrane-destabilizing) potential is pH dependent.

One of the most studied families of pH-sensitive fusogenic peptides is derived from the N-terminal region of the influenza virus hemagglutinin. This region of the viral protein is buried at neutral pH and reorganizes in an amphipathic helix at the slightly acidic pH of the endosomes. Although details of the mechanism are still not understood, these conformational changes ultimately lead to the cytoplasmic release of the viral nucleocapsid.[50] Along these lines, a significant increase of Tat-Cre recombinase activity has been obtained when cotreating cells in culture with the fusion protein and with the influenza hemagglutinin fusogenic peptide.[51] A series of synthetic peptides modeled on that of the influenza hemagglutinin one has been proposed in order to increase their fusogenic potential for this type of application, as comprehensively reviewed in Lochmann et al.[1]

A series of synthetic peptides undergoing pH-dependent conformational rearrangements from random coil to amphipathic α-helix has also been proposed by Wyman and colleagues.[52]

Several natural peptides lead to membrane fusion or destabilization at neutral pH but they are generally rather cytotoxic and might be difficult to use for delivery purposes. An interesting derivative of the highly potent melittin peptide with pH-dependent membrane-destabilizing properties has recently been described and has been successfully incorporated in a plasmid DNA delivery vector.[53]

Many options to optimize nucleic acids delivery vectors thus remain open. The main problem will be to introduce, within a single entity, determinants required for cell binding (and eventually for cell targeting), for cell endocytosis, and for endosomal escape without compromising target recognition in the cytoplasm or in the nucleus and without becoming too complicated.

2.5   MECHANISM OF INTERNALIZATION OF TAT PTD AND TAT–CARGO CONJUGATES

2.5.1   FROM AN ENERGY-INDEPENDENT TO AN ENDOCYTOTIC MECHANISM OF UPTAKE

A receptor- and energy-independent mechanism of cell uptake was initially proposed for the cellular uptake of Tat PTD and Tat–cargo conjugates, as for all cell penetrating peptides.

This model was based on several experimental arguments (as reviewed in Vives et al.[54]), namely:

1. CPP analogs including amino acids in D-configuration or retro-inverso forms are internalized as efficiently as the parent peptide, thus indicating that no chiral receptor is involved.
2. Inhibitors of endocytosis do not significantly alter cell uptake.
3. CPPs interact with the charged heads of phospholipids in model lipid bilayers.

These early conclusions have been questioned for methodological reasons. Indeed, fluorescence microscopy and fluorescence-activated cell sorter (FACS) analysis proved to lead to unforeseen artifacts when dealing with these highly cationic CPPs. Basic CPPs strongly bind to negatively charged cell surface determinants (proteoglycans essentially) as well as to plastic and glass surfaces. Extensive washing before FACS analysis is therefore not sufficient to eliminate membrane-bound material.[36] Early experimental data have thus addressed cell-bound as well as cell-internalized CPPs. Likewise, supposedly mild fixation protocols (with paraformadehyde) lead to artifactual redistribution of cell surface-bound peptides and peptide conjugates.[55] Finally, model lipid bilayers used to assess membrane interaction and reorganization are far from representative of the complexity of a biological membrane.

Two elements led us to challenge the validity of the then-prevailing nonendocytotic mechanism of Tat uptake. First, cell fractionation experiments with an iodinated radioactive derivative of Tat PTD indicated that most of the material was membrane bound and that little (if any) was associated with the cell nuclei in Hep G2 hepatoma cells (Courtoy et al., unpublished observations). Second, it became rapidly evident, in keeping with the proposal made by Lundberg and Johanson[55] for the VP22 protein PTD, that paraformaldehyde treatment leads to an artifactual redistribution of membrane-bound material and to its nuclear concentration, probably through nonspecific electrostatic interactions with nucleic acids.[36] Various solutions were thereafter proposed to overcome these methodological problems. They include enzymatic removal (through trypsin or pronase treatment)[36,56] or fluorescence quenching[57] of cell surface-bound peptides.

In order to avoid artifacts linked with fixation protocols, most recent studies rely only on live-cells imaging. Unfortunately, this complicates protocols and precludes interesting strategies such as the use of Tat-specific antibodies to reveal the intracellular distribution of Tat PTD while avoiding the conjugation of a fluorochrome to the Tat peptide.[15] We still have to keep in mind possible artifacts linked to the conjugation of a bulky fluorochrome to the CPPs.

Recent developments have capitalized on the conjugation of cargoes whose biological activity could easily be monitored. Examples include fusion of Tat PTD to the Cre recombinase[51] or Tat conjugation to a splicing-correcting PNA as described in the previous section.[44] In both cases, nuclear delivery of the Tat-associated cargo can be quantified with sensitive biochemical assays providing a positive read-out over a low background.

2.5.2 CELL SURFACE BINDING

As mentioned in Section 2.4 and Section 2.5.1, most early studies on the involvement of proteoglycans were biased by methodological problems. For example, we erroneously concluded that a GST–Tat–GFP construct behaved differently from Tat PTD[58] using FACS analysis, most probably because high ionic strength washings were able to remove cell-bound Tat–GST efficiently while not allowing complete removal of Tat PTD. Indeed, inclusion of a brief proteolytic treatment before FACS analysis led us to conclude that cell surface proteoglycans were involved in Tat–PTD cellular internalization using two commonly used tools, for example, CHO-mutant cells altered in proteoglycan biosynthesis and treatment with heparan sulfate lyases.[56] Several independent studies, as extensively reviewed,[20,59] have led to similar conclusions over the past few years, whether dealing with fluorochrome-linked Tat PTD or with various Tat constructions.

This is unsurprising, because cell surface (and, in particular, heparan sulfate) proteoglycans play a key role in the internalization of molecular entities as diverse as viruses, growth factors, cationic lipoplexes, and basic CPPs (for a recent review see Ref. [60]). While cell surface heparan sulfate proteoglycans (HSPG) serve as a major (but not necessarily exclusive) binding site for Tat PTD, for Tat–cargo conjugates, and importantly for the Tat protein itself, biological implications are still far from being understood. Does initial docking to HSPG precede transfer to higher-affinity receptors, as established for bFGF,[61] or to plasma membrane-charged lipids? Does HSPG binding trigger endocytosis and subsequent nuclear translocation, as suggested by Sandgren et al.,[62] and in this case, how does HSPG-bound Tat escape from endocytotic vesicles? These are important questions whose solutions might help to define more efficient second-generation Tat-derived peptides.

HSPGs are rather ubiquitous, which might explain why Tat-based delivery vectors are able to deliver their associated cargoes in a large number of cell types. On the other hand, the large variability of glycosaminoglycan motifs might ultimately allow tissue-targeting with CPP–cargo conjugates.

2.5.3 WHICH PATHWAY(S) OF ENDOCYTOSIS ARE INVOLVED IN THE UPTAKE OF TAT PTD TAT–CARGO CONJUGATES?

Endocytosis can occur through various mechanisms, including phagocytosis, clathrin-dependent and -independent pathways as reviewed in Ref. [63]. Recently, several studies have addressed this point for the Tat protein, Tat PTD, or for various Tat–cargo constructions. Little consensus has emerged and the underlying reasons could be several. First, discrimination between various endocytic pathways mostly capitalizes on the use of pharmacological agents whose specificity is rarely complete. Second, the nature of the cargo and the cell type might influence the fate of the conjugates.

Our own group has investigated the uptake of Tat PTD conjugated to fluoro-chromes[56] and of Tat–PNA conjugates (Richard et al., unpublished observations). Pharmacological agents inhibiting clathrin-dependent endocytosis (such as chlorpromazine and potassium depletion) decrease the uptake of Tat and transferrin, a specific marker of this pathway. In contrast, inhibitors of lipid raft-dependent endocytosis, such as nystatin or filippin, did not significantly affect Tat uptake while reducing the internalization of a fluorochrome-labeled lactosylceramide-specific marker of this pathway. Similar conclusions have been reached independently by Potocky et al.[64] when investigating the fate of the Tat PTD and by Vendeville et al.[12] for the full-size Tat protein. Different conclusions have been reached upon studying the uptake of Tat PTD fused to Cre recombinase[65] or of a GST–Tat–GFP construct.[66] Macropinocytosis and caveolin-dependent endocytosis have been proposed in these two cases.

Altogether, these seemingly conflicting data probably reflect the possibility for basic CPPs to interact with various cell membrane microdomains and thereafter to be internalized by any type of endocytic vesicle.[20,56,59]

Conflicting data can also be found concerning the major issue of escape from endocytic vesicles. A few publications have reported cytoplasmic or nuclear accumulation of Tat-conjugates,[64,67] while in other studies,[35,51,56] no material could be detected outside of endocytic vesicles. These latter data do not necessarily eliminate the possibility of a small proportion of the endocytosed cargoes escaping endocytic vesicles before being destroyed in liposomes or recycled to the cell surface. Whatever the case, it seems that endosome entrapment limits the efficiency of Tat-mediated delivery. In keeping with this hypothesis, lysomotropic agents such as chloroquine or endosome-disrupting agents such as the influenza hemagglutinin fusogenic peptide significantly increase the functional delivery of Tat-conjugated oligonucleotides[35,44,49] or of Tat PTD-fused proteins,[51,68] as already noted in Section 2.4. A notable exception to endosome segregation of Tat-conjugated material is dendritic cells, in keeping with efficient antigen presentation by DCs after Tat PTD-mediated peptide transport.[69]

ACKNOWLEDGMENTS

Work in the authors' laboratory has been financed by the CNRS and by EEC grant QLK3-CT-2002-01989. S.Abes holds a predoctoral fellowship from the Ligue contre le Cancer.

REFERENCES

1. Lochmann, D., Jauk, E., and Zimmer, A., Drug delivery of oligonucleotides by peptides, *Eur. J. Pharm. Biopharm.*, 58, 237, 2004.
2. Leonetti, J.P. et al., Biological activity of oligonucleotide-poly(L-lysine) conjugates: Mechanism of cell uptake, *Bioconjug. Chem.*, 1, 149, 1990.
3. Mahato, R.I. et al., Physicochemical and disposition characteristics of antisense oligonucleotides complexed with glycosylated poly(L-lysine), *Biochem. Pharmacol.*, 53, 887, 1997.
4. Derossi, D. et al., The third helix of the Antennapedia homeodomain translocates through biological membranes, *J. Biol. Chem.*, 269, 10444, 1994.
5. Frankel, A.D. and Pabo, C.O., Cellular uptake of the tat protein from human immunodeficiency virus, *Cell*, 55, 1189, 1988.
6. Green, M. and Loewenstein, P.M., Autonomous functional domains of chemically synthesized human immunodeficiency virus tat trans-activator protein, *Cell*, 55, 1179, 1988.
7. Jeang, K.T., Xiao, H., and Rich, E.A., Multifaceted activities of the HIV-1 transactivator of transcription, Tat, *J. Biol. Chem.*, 274, 28837, 1999.
8. Mann, D.A. and Frankel, A.D., Endocytosis and targeting of exogenous HIV-1 Tat protein, *EMBO J.*, 10, 1733, 1991.
9. Tyagi, M. et al., Internalization of HIV-1 tat requires cell surface heparan sulfate proteoglycans, *J. Biol. Chem.*, 276, 3254, 2001.
10. Liu, J. et al., Expression of low and high density lipoprotein receptor genes in human adrenals, *Eur. J. Endocrinol.*, 142, 677, 2000.
11. Albini, A. et al., The angiogenesis induced by HIV-1 tat protein is mediated by the Flk-1/KDR receptor on vascular endothelial cells, *Nat. Med.*, 2, 1371, 1996.
12. Vendeville, A. et al., HIV-1 Tat enters T cells using coated pits before translocating from acidified endosomes and eliciting biological responses, *Mol. Biol. Cell.*, 15, 2347, 2004.
13. Fawell, S. et al., Tat-mediated delivery of heterologous proteins into cells, *Proc. Natl Acad. Sci. U.S.A.*, 91, 664, 1994.
14. Loret, E.P. et al., Activating region of HIV-1 Tat protein: Vacuum UV circular dichroism and energy minimization, *Biochemistry*, 30, 6013, 1991.
15. Vives, E., Brodin, P., and Lebleu, B., A truncated HIV-1 Tat protein basic domain rapidly translocates through the plasma membrane and accumulates in the cell nucleus, *J. Biol. Chem.*, 272, 16010, 1997.
16. Vives, E. et al., Structure activity relationship study of the plasma membrane translocating potential of short peptide from HIV-1 Tat protein, *Lett. Pept. Sci.*, 4, 429, 1997.
17. Suzuki, T. et al., Possible existence of common internalization mechanisms among arginine-rich peptides, *J. Biol. Chem.*, 277, 2437, 2002.
18. Astriab-Fisher, A. et al., Conjugates of antisense oligonucleotides with the Tat and antennapedia cell-penetrating peptides: Effects on cellular uptake, binding to target sequences, and biologic actions, *Pharm. Res.*, 19, 744, 2002.
19. Hällbrink, M. et al., Cargo delivery kinetics of cell-penetrating peptides, *Biochim. Biophys. Acta*, 1515, 101, 2001.
20. Brooks, H., Lebleu, B., and Vives, E., Tat peptide-mediated cellular delivery: Back to basics, *Adv. Drug Deliv. Rev.*, 57, 559, 2005.
21. Moulton, H.M. and Moulton, J.D., Peptide-assisted delivery of steric-blocking antisense oligomers, *Curr. Opin. Mol. Ther.*, 5, 123, 2003.

22. Snyder, E.L. and Dowdy, S.F., Cell penetrating peptides in drug delivery, *Pharm. Res.*, 21, 389, 2004.

23. Dietz, G.P. and Bahr, M., Delivery of bioactive molecules into the cell: The Trojan horse approach, *Mol. Cell Neurosci.*, 27, 85, 2004.

24. Mir, L.M. et al., Electric pulse-mediated gene delivery to various animal tissues, *Adv. Genet.*, 54, 83, 2005.

25. Lungwitz, U. et al., Polyethylenimine-based non-viral gene delivery systems, *Eur. J. Pharm. Biopharm.*, 60, 247, 2005.

26. Martin, B. et al., The design of cationic lipids for gene delivery, *Curr. Pharm. Des.*, 11, 375, 2005.

27. Thierry, A.R. et al., Cellular uptake and intracellular fate of antisense oligonucleotides, *Curr. Opin. Mol. Ther.*, 5, 133, 2003.

28. Lindsay, M.A., Peptide-mediated cell delivery: Application in protein target validation, *Curr. Opin. Pharmacol.*, 2, 587, 2002.

29. Wadia, J.S. and Dowdy, S.F., Modulation of cellular function by TAT mediated transduction of full length proteins, *Curr. Protein Pept. Sci.*, 4, 97, 2003.

30. Pooga, M. et al., Cell penetrating PNA constructs regulate galanin receptor levels and modify pain transmission in vivo, *Nat. Biotechnol.*, 16, 857, 1998.

31. Heasman, J., Morpholino oligos: Making sense of antisense?, *Dev. Biol.*, 243, 209, 2002.

32. Kang, S.H., Cho, M.J., and Kole, R., Up-regulation of luciferase gene expression with antisense oligonucleotides: Implications and applications in functional assay development, *Biochemistry*, 37, 6235, 1998.

33. Hudziak, R.M. et al., Antiproliferative effects of steric blocking phosphorodiamidate morpholino antisense agents directed against c-myc, *Antisense Nucleic Acid Drug Dev.*, 10, 163, 2000.

34. Arzumanov, A. et al., Inhibition of HIV-1 Tat-dependent trans activation by steric block chimeric 2'-O-methyl/LNA oligoribonucleotides, *Biochemistry*, 40, 14645, 2001.

35. Turner, J.J., Arzumanov, A.A., and Gait, M.J., Synthesis, cellular uptake and HIV-1 Tat-dependent trans-activation inhibition activity of oligonucleotide analogues disulphide-conjugated to cell-penetrating peptides, *Nucleic Acids Res.*, 33, 27, 2005.

36. Richard, J.P. et al., Cell-penetrating peptides. A reevaluation of the mechanism of cellular uptake, *J. Biol. Chem.*, 278, 585, 2003.

37. Thierry, A.R. et al., Comparison of basic peptides- and lipid-based strategies for the delivery of splice correcting oligonucleotides. *BBA — Biomembranes*, in press.

38. Lavigne, C. et al., Cationic liposomes/lipids for oligonucleotide delivery: Application to the inhibition of tumorigenicity of Kaposi's sarcoma by vascular endothelial growth factor antisense oligodeoxynucleotides, *Methods Enzymol.*, 387, 189, 2004.

39. Kole, R., Vacek, M., and Williams, T., Modification of alternative splicing by antisense therapeutics, *Oligonucleotides*, 14, 65, 2004.

40. Lacerra, G. et al., Restoration of hemoglobin A synthesis in erythroid cells from peripheral blood of thalassemic patients, *Proc. Natl. Acad. Sci. U.S.A.*, 97, 9591, 2000.

41. Sazani, P. et al., Nuclear antisense effects of neutral, anionic and cationic oligonucleotide analogs, *Nucleic Acids Res.*, 29, 3965, 2001.

42. Sazani, P. et al., Systemically delivered antisense oligomers upregulate gene expression in mouse tissues, *Nat. Biotechnol.*, 20, 1228, 2002.

43. Siwkowski, A.M. et al., Identification and functional validation of PNAs that inhibit murine CD40 expression by redirection of splicing, *Nucleic Acids Res.*, 32, 2695, 2004.

44. Abes, S. et al., Endosome trapping limits the efficiency of splicing correction by PNA–oligolysine conjugates, *J. Controlled Release*, 110, 595, 2006.

45. Turner, J.J. et al., Cell-penetrating peptide conjugates of peptide nucleic acids (PNA) as inhibitors of HIV-1 Tat-dependent trans-activation in cells, *Nucleic Acids Res.*, 33, 6837, 2005.

46. Manickam, D. et al., Influence of TAT-peptide polymerization on properties and transfection activity of TAT/DNA polyplexes, *J. Controlled Release*, 102, 293, 2005.

47. Koppelhus, U. et al., Cell-dependent differential cellular uptake of PNA, peptides, and PNA-peptide conjugates, *Antisens Nucleic Acid Drug Dev.*, 12, 51, 2002.

48. Zaitsev, S. et al., Histone H1-mediated transfection: Role of calcium in the cellular uptake and intracellular fate of H1–DNA complexes, *Acta Histochem.*, 104, 85, 2002.

49. Shiraishi, T., Pankratova, S., and Nielsen, P.E., Calcium ions effectively enhance the effect of antisense peptide nucleic acids conjugated to cationic tat and oligoarginine peptides, *Chem. Biol.*, 12, 923, 2005.

50. Skehel, J.J. and Wiley, D.C., Influenza haemagglutinin, *Vaccine*, 20, S51, 2002.

51. Wadia, J.S., Stan, R.V., and Dowdy, S.F., Transducible TAT-HA fusogenic peptide enhances escape of TAT-fusion proteins after lipid raft macropinocytosis, *Nat. Med.*, 10, 310, 2004.

52. Wyman, T.B. et al., Design, synthesis, and characterization of a cationic peptide that binds to nucleic acids and permeabilizes bilayers, *Biochemistry*, 36, 3008, 1997.

53. Boeckle, S., Wagner, E., and Ogris, M., C- versus N-terminally linked melittin–polyethylenimine conjugates: The site of linkage strongly influences activity of DNA polyplexes, *J. Gene. Med.*, 7, 1335, 2005.

54. Vives, E. et al., TAT peptide internalization: Seeking the mechanism of entry, *Curr. Protein. Pept. Sci.*, 2, 125, 2003.

55. Lundberg, M. and Johansson, M., Positively charged DNA-binding proteins cause apparent cell membrane translocation, *Biochem. Biophys. Res. Commun.*, 291, 367, 2002.

56. Richard, J.P. et al., Cellular uptake of unconjugated TAT peptide involves clathrin-dependent endocytosis and heparan sulfate receptors, *J. Biol. Chem.*, 280, 15300, 2005.

57. Drin, G. et al., Studies on the internalization mechanism of cationic cell-penetrating peptides, *J. Biol. Chem.*, 278, 31192, 2003.

58. Silhol, M. et al., Different mechanisms for cellular internalization of the HIV-1 Tat-derived cell penetrating peptide and recombinant proteins fused to Tat, *Eur. J. Biochem.*, 269, 494, 2002.

59. Melikov, K. and Chernomordik, L.V., Arginine-rich cell penetrating peptides: From endosomal uptake to nuclear delivery, *Cell. Mol. Life. Sci.*, 62, 2739, 2005.

60. Belting, M., Heparan sulfate proteoglycan as a plasma membrane carrier, *Trends. Biochem. Sci.*, 28, 145, 2003.

61. Colin, S. et al., In vivo involvement of heparan sulfate proteoglycan in the bioavailability, internalization, and catabolism of exogenous basic fibroblast growth factor, *Mol. Pharmacol.*, 55, 74, 1999.

62. Sandgren, S., Cheng, F., and Belting, M., Nuclear targeting of macromolecular polyanions by an HIV-Tat derived peptide. Role for cell-surface proteoglycans, *J. Biol. Chem.*, 277, 38877, 2002.

63. Pelkmans, L. and Helenius, A., Insider information: What viruses tell us about endocytosis, *Curr. Opin. Cell. Biol.*, 15, 414, 2003.

64. Potocky, T.B., Menon, A.K., and Gellman, S.H., Cytoplasmic and nuclear delivery of a TAT-derived peptide and a beta-peptide after endocytic uptake into HeLa cells, *J. Biol. Chem.*, 278, 50188, 2003.

65. Kaplan, I.M., Wadia, J.S., and Dowdy, S.F., Cationic TAT peptide transduction domain enters cells by macropinocytosis, *J. Controlled Release*, 102, 247, 2005.

66. Fittipaldi, A. et al., Cell membrane lipid rafts mediate caveolar endocytosis of HIV-1 Tat fusion proteins, *J. Biol. Chem.*, 278, 34141, 2003.

67. Fischer, R. et al., A stepwise dissection of the intracellular fate of cationic cell-penetrating peptides, *J. Biol. Chem.*, 279, 12625, 2004.

68. Caron, N.J., Quenneville, S.P., and Tremblay, J.P., Endosome disruption enhances the functional nuclear delivery of Tat-fusion proteins, *Biochem. Biophys. Res. Commun.*, 319, 12, 2004.

69. Loison, F., A ubiquitin-based assay for the cytosolic uptake of protein transduction domains, *Mol. Ther.*, 11, 205, 2005.

3 Model Amphipathic Peptides

Johannes Oehlke, Burkhard Wiesner, and Michael Bienert

CONTENTS

3.1 INTRODUCTION

Peptides, due to their relatively high polarity and hydrogen bridge forming propensity, are normally unable to cross lipidic membranes.[1] Until the 1990s, therefore, it was generally believed that peptides would also be unable to enter cells, and thus the complex nature of the plasma membrane was neglected. This belief was supported by the experience that no compelling evidence for a cellular import of peptides had been provided until that time. One of the main reasons for this failure was the inability of the available protocols to unambiguously discriminate between the minor internalized and the overwhelming surface adsorbed peptide portions. This problem was first solved at the end of the 1980s when a new technique, confocal laser scanning microscopy (CLSM), became available. The discovery of the cell-penetrating ability of peptides by several groups in the following years,[2–4] therefore, appears closely related to the introduction of this new technique.

In our group, in the early 1990s, the phenomenon of the cell-penetrating ability of peptides was approached in two independent ways.[5,6] These studies provided the fundamentals for the rather accidental discovery of the cell-penetrating ability of a further peptide group, later termed model amphipathic peptides (MAPs), by Lindgren et al.[7] The MAP group comprised simple synthetic amphipathic peptides, able to enter cells in a similar manner, as was now known, as special sequences selected from the HIV-TAT protein,[4] the third helix of the Antennapedia homoeo domain,[2] and the fibroblast growth factor.[3] This finding implied that the cell permeability was not confined to special proteinaceous sequences, and promised a broader basis for searching for further cell-penetrating peptides (CPPs).

Moreover, as outlined in the following history of discovery, our ongoing studies suggested that the particularity of the group called CPPs did not consist of their ability to cross cellular membranes, which appeared to be shared by other peptides, but in an extensive sequestering within cells.

3.2 HISTORY OF THE DISCOVERY OF MAP PEPTIDES

3.2.1 Preconditions Created During Studies on the Internalization of Substance P into Mast Cells

Our group was first confronted with the phenomenon of the cell-penetrating ability of peptides during mechanistic investigations concerning the mast cell-degranulating activity of the neuropeptide substance P (SP; RPKPQQFFGLM-NH$_2$).[8] Studies on structure-activity relationships and receptor autoradiography had both failed to detect specific peptide receptors on mast cells.[8–10] Thus an alternative mechanism, the direct interaction of SP with heterotrimeric G proteins, as proposed by Mousli et al.,[9] was considered (Figure 3.1). This process, however, required insertion into and translocation across the plasma membrane of SP. In order to clarify whether a translocation of such cationic peptides into mast cells is possible, a CLSM device, which became accessible to us at that time, was exploited for uptake studies. These investigations indeed revealed a translocation directly into the cytosol of a fluorescein-labeled SP derivative.[5] The uptake proceeded within a few seconds, and also was observed with an analogous all-D-SP derivative, ruling out that intracellular fluorescence was caused by metabolic degradation products.[5]

These results, due to their disagreement with general belief, were received with serious doubts. Mainly to rule out these doubts, a CLSM protocol to measure the

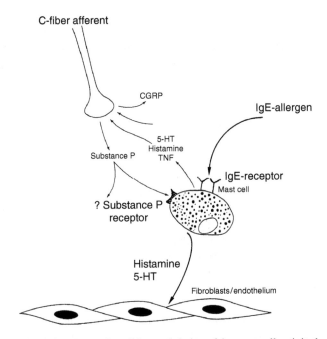

FIGURE 3.1 Schematic presentation of the modulation of the mast cell activity by substance P. The allergen-IgE-induced release of histamine/serotonin from mast cells is a key event in the pathophysiology of allergies, asthmatic attacks, and autoimmune diseases. Mast cell degranulation may be modulated by neuropeptides, such as substance P, RPKPQQFFGLM-NH$_2$, thus linking the central nervous system and the immune system.

intensity of the cytosolic and nuclear fluorescence without interference with the extracellular and the intracellular vesicular fluorescence was developed,[5,11] which likewise also provided a decisive basis for scrutinizing the cellular uptake of other peptides, including those of the MAP series. The intracellular fluorescence intensities measured in this way confirmed the results previously obtained, indicating a plasma membrane passage of SP directly into the cytosol of mast cells.

3.2.2 PRECONDITIONS CREATED BY THE HIGH-PERFORMANCE LIQUID CHROMATOGRAPHY-BASED MONITORING OF THE CELLULAR UPTAKE OF SMALL UNSTRUCTURED 3–11MER PEPTIDES

Parallel to the study of SP internalization, the cellular uptake of a series of peptides derived from leucine-enkephalin was investigated. The starting point in this case was a report by Smith et al.[12] which suggested the possibility that amino acid transporters might tolerate small peptides bearing free α-amino as well as carboxy groups. To test this possibility, the cellular uptake of enkephalin-derived peptides linked to the ε-amino group of lysine was studied. Internalization and metabolism of the peptides were assessed by means of high-performance liquid chromatography (HPLC)-based approaches relying on Fluos-labeling or the tryptophane fluorescence of the peptides.[6]

These experiments provided evidence for an extensive cellular uptake, accompanied by only a moderate metabolism (Figure 3.2). Surprisingly, a similar uptake was also found for the corresponding unmodified peptides intended to serve as negative controls (Figure 3.2). The latter results, as in the case of the aforementioned SP studies, were not reconcilable with the belief that peptides, for thermodynamic reasons, would be unable to cross cell membranes and, therefore, were received with serious doubts.

However, additional experiments conducted with various tritium-labeled peptides at 100-fold lower concentrations than required for the HPLC-based protocols provided similar results.[6] Further support came from CLSM investigations, which revealed extensive cytosolic and nuclear fluorescence of cells treated with fluorescently labeled analogs of these peptides.[6]

Thus, it became apparent that in contrast to the common belief, the peptides indeed must have been taken up by the cells.

Characteristics of this type of cellular uptake were, in contrast to those found later on for MAP peptides, an intracellular peptide quantity approaching the external concentration within about 30 min and a very rapid reexport of more than 70% of the internalized peptide within 5 min after replacing the incubation solution by fresh buffer.[6]

Systematic structural alterations performed with Leu-enkephalin did not reveal structural preferences for entry into the cells (Table 3.1), ruling out specific peptide transporters to be involved. The translocation mode, however, in analogy to that of CPPs, remained unclear.

The most simple interpretation, assuming an exclusively endocytic mechanism, was questioned by the relatively high extent of the uptake and a lack of effects of

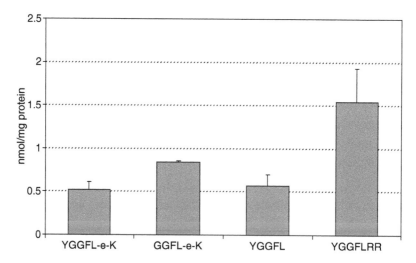

FIGURE 3.2 Quantity of cell-associated peptide after exposure of bovine aortic endothelial cells for 30 min at 37°C to 2 mM solutions of enkephalin-derived peptides. Each bar represents the mean of three samples $\pm$ SEM. (Exposure at 0°C yielded values <0.1 nmol/mg protein, ruling out surface adsorption to contribute decisively to the 37°C values.)

TABLE 3.1

Cell-Associated Peptide After 2 mM Exposure to Bovine Aortic Endothelial Cells for 30 min at 37°C

Peptide	nmol/mg Protein (SEM)[a]	Peptide	nmol/mg Protein (SEM)
YGGFL	0.57 (0.13)	AcKYGGWL-NH$_2$	1.85 (0.27)
YGGWL	0.60 (0.17)	RYGGWL-NH$_2$	1.65 (0.56)
YGGWL-NH$_2$	1.12 (0.11)	EYGGWL-NH$_2$	1.30 (0.13)
All-D-YGGWL-NH$_2$	1.48 (0.27)	Succinyl YGGWL-NH$_2$	0.80 (0.04)
All-D-LWGGY-NH$_2$	2.56 (0.36)	YGGWLG-NH$_2$	2.40 (0.26)
AcYGGWL-NH$_2$	0.97 (0.22)	YGGWLR-NH$_2$	1.98 (0.23)
GYGGWL-NH$_2$	1.29 (0.26)	YGGWLE-NH$_2$	0.67 (0.13)
N-Methyl-GYGGWL-NH$_2$	0.71 (0.05)	YGGGWL-NH$_2$	2.99 (0.24)
N-Dimethyl-YGGWL-NH$_2$	0.63 (0.19)	YGGRWL-NH$_2$	2.12 (0.65)
KYGGWL-NH$_2$	1.90 (0.08)	YGGEWL-NH$_2$	1.61 (0.19)

[a] Exposure at 0°C yielded values <0.1 nmol/mg protein, ruling out surface adsorption to contribute decisively to the 37°C values.

agents known to affect endocytosis, such as chloroquine or brefeldin A.[6] Moreover, reduced temperature and energy depletion, which both suppress endocytosis, influenced the internalization of various peptides differently (Figure 3.3).

Only insignificant effects were observed for agents known to affect active transport processes, such as vincristine, DNP-S-glutathione, or bromosulphophthalein,[6] suggesting yet known energy-consuming transporters not to be decisively involved.

3.2.3 Cell-Penetrating Characteristics of MAP Peptides

As stated previously, rather accidentally, a further peptide group, termed MAP by Lindgren et al.[7] was included in the aforementioned cellular uptake studies. A series of double-D-amino acid anaologs of the α-helical amphipathic model peptide KLALKLALKALKAALKLA-NH$_2$ (I; Figure 3.4),[13-16] was made available for cell experiments by the courtesy of colleagues. This model peptide, originally designed by Steiner et al.[17] was initially used in our group for biophysical studies on the interactions of bioactive helical amphipathic peptides with lipidic interfaces.[13-16,18]

Because of the molecule size and membrane affinity, merely different degrees of surface adsorption and endocytotic uptake were expected for this series, depending on the impairment of the amphipathicity. HPLC analysis of the cell lysates indeed revealed correlation of the cell-associated quantity with the degree of amphipathicity.[19] Cells treated with the double-D-amino acid derivatives of I and its nonamphipathic analog II displayed a negligible intracellular fluorescence during CLSM, consistent with only an endocytotic uptake. Surprisingly, however, cells exposed to the amphipathic all-L-parent peptide I exhibited a bright diffuse cytosolic as well as nuclear fluorescence, suggesting an extensive nonendocytotic uptake.

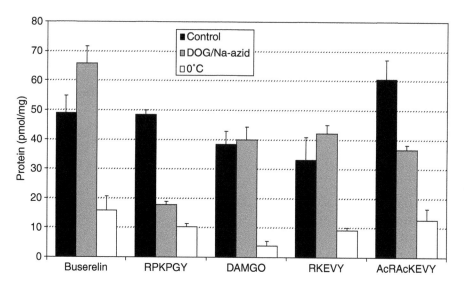

FIGURE 3.3 Quantity of cell-associated peptide after exposure of bovine aortic endothelial cells for 60 min at 37 and 0°C without and with energy depletion (25 mM 2-deoxyglucose/10 mM Na-azide) to 10 μM solutions of various tritiated peptides and subsequent acid wash. Each bar represents the mean of three samples $\pm$ SEM.

Quantification of the internalized KLA revealed a more than tenfold intracellular enrichment for I, whereas the internalized amount of its double-D analogs had remained below the quantitation limit of the approach used.[20] A precondition for performing the quantitation was the development of a protocol for discriminating the actually internalized from the at least tenfold larger surface-bound peptide portions by means of diazotized *o*-nitraniline.[20] Figure 3.5 illustrates the problems posed by the overwhelming extent of surface adsorption and likewise displays further characteristics of the MAP-type of cellular uptake, a virtually negligible reexport of internalized peptide. Both characteristics differed greatly from

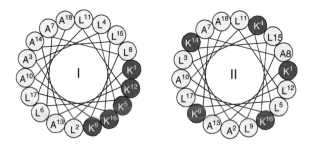

FIGURE 3.4 Helical wheel projections of the α-helical amphipathic model peptide KLALKLALKALKAALKLA-NH$_2$ (I) and its nonamphipathic analog (II) possessing the same amino acid composition but evenly around the helix arranged polar and unpolar side chains.

the very rapid reexport and the low intracellular levels observed for the small unstructured peptides.

In analogy to the uptake of small unstructured peptides, however, conflicting effects of factors known to affect endocytosis or active transport processes suggested a complex mode of uptake comprising various energy-dependent as well as independent mechanisms[6,20] (Figure 3.6). The extent of these effects proved cell-type dependent, as Figure 3.7 illustrates for the influence of reduced temperature.

Virtually identical uptake results were obtained with the all-D-analog of I, suggesting structure-specific interactions did not play a decisive role.[20]

Variation of the structure of I and inclusion of additional pairs of amphipathic–nonamphipathic analogs (Table 3.2) suggested that amphipathicity, the presence of basic side chains, and a minimum chain length of 16 amino acids are seemingly essential structural requirements for the uptake of MAP peptides. The number of positive charges and even their compensation by negative side chains seemed to be without significant importance (Table 3.2). For peptides devoid of these structural elements, no significant uptake could be detected either by the HPLC-based approach or by CLSM.

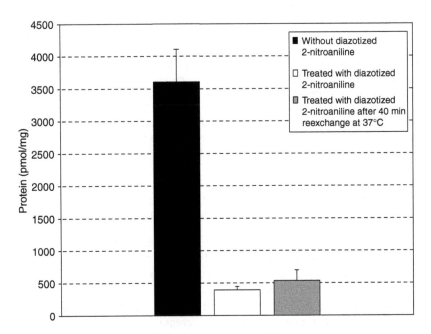

FIGURE 3.5 Nonmetabolized fractions of cell-associated peptide after exposing bovine aortic endothelial cells to 4 μM I for 60 min at 37°C in the presence of 0.2% bovine serum albumin without and with reexchange into fresh buffer for 30 min at 37°C and treatment with diazotized 2-nitroaniline. Each bar represents the mean of three samples $\pm$ SD.

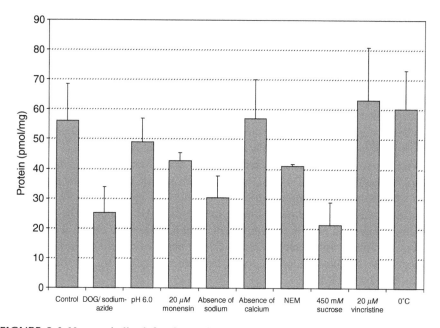

FIGURE 3.6 Nonmetabolized fractions of cell-associated peptide after exposing bovine aortic endothelial cells for 30 min at 37°C to 1.8 μM I dissolved in DPBSG (control and NEM, respectively), in DPBSG adjusted with HCl to pH 6.0, in sodium-free buffer (DPBSG with NaCl and phosphates replaced by 137 mM choline chloride and 10 mM HEPES, respectively), in DPBSG without calcium, in DPBS containing 25 mM 2-deoxyglucose/10 mM sodium azide (DOG/sodium azide), or 20 μM monensin or 20 μM vincristine, respectively, and in DPBSG at 0°C, followed by treatment with diazotized 2-nitroaniline. Before exposure to the peptide, the cells used for the NEM, DOG/sodium azide, and 0°C experiments were incubated at 37°C for 5 min in DPBSG containing 1 mM N-ethylmaleimide and subsequently rinsed three times with DPBSG (NEM) or incubated for 60 min at 37°C in DPBS containing 25 mM 2-deoxyglucose/10 mM sodium azide (DOG/sodium azide) or in DPBSG for 60 min at 0°C, respectively. Each bar represents the mean of three samples ± SD.

3.2.4 MECHANISTIC CONSIDERATIONS CONCERNING THE CELLULAR UPTAKE OF AMPHIPATHIC AND NONAMPHIPATHIC PEPTIDES

Seen in context, the differences found concerning structural requirements, achievable intracellular concentration, and reexport suggested distinct modes of uptake for the amphipathic and nonamphipathic members of the MAP series as well as for the enkephalin-derived peptides.

On the other hand, a common translocation mode for these peptide groups also appeared conceivable, if assuming that the different uptake patterns relied primarily on different rates of reexport. According to such an interpretation, the amphipathic members of the MAP series bearing basic side chains would, in part, escape the reexport by binding to intracellular structures and become enriched within the cell. The internalized nonamphipathic members of the MAP series as well as the

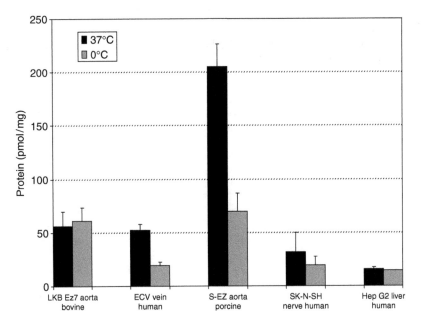

FIGURE 3.7 Nonmetabolized fractions of cell-associated peptide after exposing various cell types from different species and organs for 30 min at 37 and 0°C to 1.8 μM I and subsequent treatment with diazotized 2-nitroaniline. Before exposure to the peptide at 0°C the cells were incubated in DPBSG for 60 min at 0°C. Each bar represents the mean of three samples $\pm$ SD.

enkephalin-derived peptides, on the other hand, due to unhindered reexport, would remain in equilibrium with the external medium and, therefore, would be washed out after replacing the incubation solution by fresh buffer.

Discrimination between these conflicting interpretations became possible by means of a CLSM protocol which, as mentioned above, was developed for confirming the internalization of SP into mast cells.[5,11] This protocol enabled assessment of the intracellular fluorescence intensity without noticeable interference with even an extensive external one. Therefore, measurements could be performed in the presence of the incubation solution without washing and consequent loss of intracellular peptide by wash out.

By assessing the cellular uptake of the MAP series in this way, extensive intracellular fluorescence became apparent in all cases irrespective of the structural properties of the peptides[21] (Figure 3.8A through Figure 3.10A).

These findings favored the second interpretation proposing a common mode of uptake and different rates of reexport. In further support of the second interpretation, the intracellular fluorescence of cells exposed to peptides lacking the structural elements regarded to be characteristic of CPPs, disappeared after washing[21] (Figure 3.8B through Figure 3.10B). In contrast, the intracellular fluorescence of cells incubated with amphipathic peptides bearing basic side chains remained widely uninfluenced[21] (Figure 3.8B through Figure 3.10B).

TABLE 3.2
Internalization of Peptide into Aortic Endothelial Cells after 4.5 μM Exposure for 30 min at 37°C

Peptide	Peptide Composition	pmol Internalized Peptide/mg protein	H	μ	Φ (°)	α-Helix (%) TFE	α-Helix (%) SDS
I	KLALKLALKALK-AALKLA-NH$_2$	228 ± 54	-0.0161	0.3339	80	60	68
III	KLALKLALKAL-KAALK-NH$_2$	218 ± 23	-0.0161	0.3339	80	52	61
IV	KLALKALKAAL-KLA-NH$_2$	<30	-0.0317	0.3358	60	45	51
V	KLALKLALKAL-KAA-NH$_2$	<30	-0.0317	0.330	80	50	59
VI	KLGLKLGLKGLK-GGLKLG-NH$_2$	<30	-0.0461	0.313	80	13	8
VII	KLALKLALKALQ-AALQLA-NH$_2$	42 ± 12.4	0.0294	0.2912	80	53	66
VIII	KLALQLALQALQ-AALQLA-NH$_2$	461 ± 44	0.075	0.2505	80	78	78
IX	QLALQLALQALQ-AALQLA-NH$_2$	5670 ± 3971^a	0.0978	0.234	80	72	72
X	ELALELALEALE-AALELA-NH$_2$	135 ± 14.6^a	0.117	0.216	80	84	21
XI	LKTLATALTKLA-KTLTTL-NH$_2$	539 ± 80	-0.025	0.3414	160	95	79
XII	LLKTTALLKTTA-LLKTTA-NH$_2$	n.d.^{b}	-0.025	0	360	74	39
XIII	LKTLTETLKELT-KTLTEL-NH$_2$	141 ± 54	-0.170	0.387	220	78	56
XIV	LLKTTELLKTTE-LLKTTE-NH$_2$	n.d.^{b}	-0.170	0	360	52	17
XV	RQIKIWFQNRR-MKWKK-NH$_2$	216 ± 58	-0.4822	0.1652	280	29	23

^{a} No discrimination between internalized and surface bound peptide possible with diazotized 2-nitroaniline because of the lack of modifiable side chain amino groups.
^{b} Nondetectable.

3.3 CONCLUSIONS AND OUTLOOK

The discovery that synthetic amphipathic model peptides are able to enter cells in a similar manner to protein-derived CPPs has broadened the basis for searching for further membrane permeable peptide sequences. Numerous such sequences of proteinaceous as well as of synthetic origin became known in the decade following

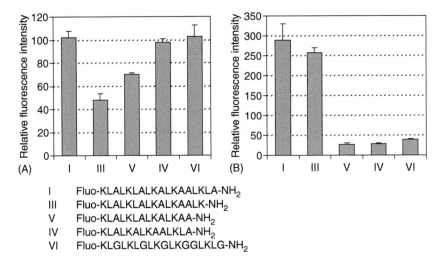

I   Fluo-KLALKLALKALKAALKLA-NH₂
III   Fluo-KLALKLALKALKAALK-NH₂
V   Fluo-KLALKLALKALKAA-NH₂
IV   Fluo-KLALKALKAALKLA-NH₂
VI   Fluo-KLGLKLGLKGLKGGLKLG-NH₂

FIGURE 3.8 Cytosolic fluorescence intensity measured by CLSM after exposing bovine aortic endothelial cells cells for 30 min at 37°C to 1.8 μM I and I-derived peptides showing altered molecule size and structure forming propensity without (A) and with (B) subsequent washing. The fluorescence intensities were normalized to those of the external peptide solutions. Each bar represents the mean of three samples $\pm$ SEM.

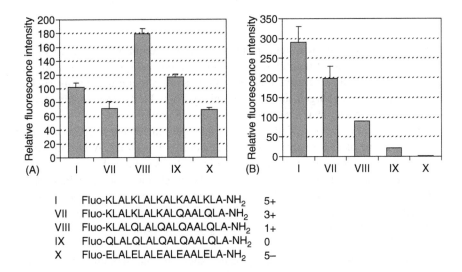

I   Fluo-KLALKLALKALKAALKLA-NH₂   5+
VII   Fluo-KLALKLALKALQAALQLA-NH₂   3+
VIII   Fluo-KLALQLALQALQAALQLA-NH₂   1+
IX   Fluo-QLALQLALQALQAALQLA-NH₂   0
X   Fluo-ELALELALEALEAALELA-NH₂   5−

FIGURE 3.9 Cytosolic fluorescence intensity measured by CLSM after exposing bovine aortic endothelial cells cells for 30 min at 37°C to 1.8 μM I and I-derived peptides showing altered charge without (A) and with (B) subsequent washing. The fluorescence intensities were normalized to those of the external peptide solutions. Each bar represents the mean of three samples $\pm$ SEM.

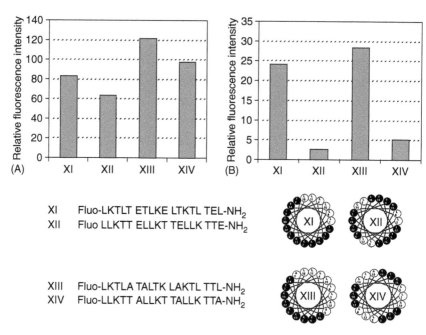

XI   Fluo-LKTLT ETLKE LTKTL TEL-NH$_2$
XII   Fluo LLKTT ELLKT TELLK TTE-NH$_2$

XIII Fluo-LKTLA TALTK LAKTL TTL-NH$_2$
XIV   Fluo-LLKTT ALLKT TALLK TTA-NH$_2$

FIGURE 3.10 Cytosolic fluorescence intensity measured by CLSM after exposing bovine aortic endothelial cells cells for 30 min at 37°C to 1.8 μM amphipathic (XI, XIII)/nonamphipathic (XII, XIV) pairs of I-nonrelated peptides showing altered net charge without (A) and with (B) subsequent washing. The fluorescence intensities were normalized to those of the external peptide solutions. Each bar represents the mean of three samples $\pm$ SEM.

this discovery and, along with their benefits, are presented in other chapters of this book. Mechanism and structural requirements for the cell-penetrating ability of CPPs, however, remained elusive until now. Only a few structural elements, such as a cluster of positively charged side chains, high hydrophobicity, and amphipathicity, have emerged as being characteristic of this peptide group.

The history of discovery of the MAP peptides presented here suggests a further broadening of the basis for searching optimized delivery peptides. According to the resulting considerations, the penetration of cell membranes should be a more general ability of peptides, and the structural elements being characteristic of CPPs appear not to be mandatory for entry into the cell interior but should merely mediate an enhanced sequestering within the cell. Thus, the search for further delivery peptides should not be restricted to sequences bearing the structural elements regarded as typical of CPPs.

With respect to CPP-cargo conjugates, such an interpretation means that their benefits should be related to strong binding to intracellular structures rather than to facilitation of the cargo import, which should be achievable by conjugation with peptides lacking the structural elements being characteristic of CPPs as well. Delivery studies performed with conjugates of oligonucleotides and peptide nucleic acids with MAP peptides and other CPPs as well as with peptides lacking the typical structural properties of CPPs revealed, in support of this interpretation, comparable cellular uptake in all cases.[22,23] On the other hand, irrespective of the seemingly structure-independent net uptake, significant influence of the peptide structure on the time course and energy dependence of the uptake as well as on the compartmentalization of the conjugates was found.[22,24] Likewise, the benefits reported for CPP conjugates, such as enhanced biological activity or resistance against the presence of serum, could be corraborated in these studies.

Hence, the presented considerations do not question the value of peptides bearing the typical structural elements of CPPs for eliciting enhanced effects of linked bioactive agents but promise more degrees of freedom for designing optimized peptide–cargo conjugates.

3.4   EXPERIMENTAL

3.4.1   Cell Culture

Cells were seeded at an initial density 5×10^4 cells/cm^2 in 24-well culture plates and cultured at 37°C in a humidified 5% CO_2 containing an air environment for 4 days without replacing the medium. For CLSM, 10^4 cells on 22×22 mm coverslips were cultured analogously.

3.4.2   Uptake Experiments

After removal of the medium, the cell layers were rinsed twice at 37°C with Dulbecco's phosphate-buffered saline (DPBS; Biochrom KG, Berlin, Germany) supplemented with 1 g/l D-glucose (DPBSG) and subsequently exposed, unless indicated otherwise, at 37°C for 30 min to 200 μL of the peptide solutions in DPBSG. Thereafter the incubation solutions were removed and the cells were washed four times with ice-cold PBS. Alternatively, in the case of the MAP series, the cells were washed twice with ice-cold PBS, then incubated with 200 μL of ice-cold PBS and treated with diazotized 2-nitroaniline, as described previously,[20] in order to modify any surface-bound peptide. In brief: 50 μL 0.6 M NaNO$_2$ were added to 400 μL ethanol/water 1/1 v/v containing 2-nitroaniline (0.06 M) and HCl (0.125 M). After standing for 5 min at ambient temperature, 4 μL of this reagent were added to the ice-cold PBS covering the cell layer and allowed to react for 10 min at 0°C. After aspiration of the diazo reagent the cells were washed twice with ice-cold PBS. Finally, lysis was performed with 0.2 mL 0.1% Triton X-100 containing 10 mmol/L trifluoroacetic acid for 2 h at 0°C. The resulting lysate was used for HPLC analysis and for protein determination according to Bradford.[25]

Integrity of the cells was assessed by means of lactate dehydrogenase release, the MTT test, or trypan blue exclusion, as described,[20] and was generally higher than 90%.

3.4.3 HPLC Analysis

HPLC was performed using a Bischoff HPLC-gradient system (Leonberg, Germany) with a Polyencap A 300, 5 μm column (250$\times$4 mm I.D.), with precolumns containing the same adsorbent and a Fluorescence HPLC-Monitor RF-551 (Shimadzu Duisburg, Germany).

Up to 200 μL of the cell lysates were passed through a precolumn containing 60 mg of Polyencap (A 300, 5 μm), which was subsequently connected to the HPLC system. The elution was carried out with 0.01 M TFA (A) and acetonitrile/water 9/1 (B) at a flow rate of 1.0 mL/min with gradients from 30 to 60% B within 20 min for enkephalin-derived peptides and from 30 to 45% B (0–10 min) and from 45 to 80% B (15–20 min) for MAP peptides. Quantitation was performed by fluorescence measurement at 520 nm after excitation at 445 nm for FLUOS-labeled peptides and at 350 nm after excitation at 285 nm for tryptophan-containing peptides by means of calibration lines obtained with the parent peptides under identical conditions. For unknown metabolites of I, a molar fluorescence intensity twice that found during HPLC for I was used for quantitation, based on the generally two- to threefold higher molar fluorescence intensities observed for HPLC peaks of characterized derivatives of I, which had a lower structure-forming propensity.

Cells' own fluorescently active components did not interfere with the detection of peptides, even in the case of the tryptophane-containing derivatives, so that pretreatment of the cell lysates could be avoided at the cost of a somewhat more frequent replacement of the precolumn.

3.4.3.1 FLUOS Derivatization of Enkephalin-Derived Peptides Directly in the Cell Lysates

To 70 μL cell lysate, 10 μL of 2 M K$_2$HPO$_4$, 10 μL of 23 μM cyclohexyl-alanine solution (internal standard), and 5 μL of a solution of 4.9 mg of 5(6)-carboxyfluorescein-N-hydroxysuccinimide ester (Boehringer, Mannheim, Germany) in 100 μL dimethylsulfoxide were added and allowed to react for approximately 20 h at room temperature. Subsequently, the solution was mixed with 4 μL of 8 M NaOH and, after standing for at least 30 min, with 11 μL of 1.4 M H$_3$PO$_4$. HPLC was carried out at first by eluting with 0.04 M phosphate buffer pH 7.4 (A) and acetonitrile–water 9/1 (B) with gradients from 9 to 28% B (0–20 min) and from 28 to 85% B (20–30 min) at a flow-rate of 1.0 mL/min. The detection was performed by UV at 242 nm. The eluates containing the derivatized peptides and their aminopeptidase-degradation products (16–21 min) were collected and concentrated to dryness by means of a vacuum centrifugal evaporator. Residuals corresponding to 1 mL of the eluate were dissolved in 100 μL of 0.5 M trifluoracetic acid containing 25% acetonitrile and subjected to the HPLC protocol described above for quantitation.

3.4.4   Confocal Laser Scanning Microscopy

3.4.4.1   Fluorescence Imaging at 37°C, Online

The coverslips carrying coherent cell formations as well as single cells were overlayed with 1 mL PBS and the initial microscopy was performed without peptide solution (control image). Thereafter the buffer was aspirated and the cells were exposed to 1 mL of a 1 μM solution of the peptide in DPBS for 30 min; fluorescence images were monitored at 5-min intervals.

Fluorescence imaging after incubation for 30 min at 37°C and subsequent washing:

The cells were exposed for 30 min at 37°C to 1 μM peptide in DPBS and subsequently washed four times with ice-cold PBS. The intracellular fluorescence signal was measured by CLSM.

Fluorescence imaging after incubation for 30 min at 4°C without subsequent washing:

The cells were precooled for 30 min at 0–4°C in DPBS and then exposed to 1 mL of 1 μM peptide in DPBSG for 30 min at 0–4°C. Fluorescence images were monitored at 5-min intervals.

At the end of each protocol the integrity of the plasma membrane was verified by addition of trypan blue and measurement of the Trypan Blue fluorescence ($\lambda_{exc} = 543$ nm, $\lambda_{em} > 590$ nm).

To monitor the fluorescence images and measure the fluorescence intensities an LSM 410 invert confocal laser scanning microscope (Carl Zeiss, Jena GmbH, Jena, Germany) was used. Excitation was performed at 488 nm (Fluos) or 543 nm (Trypan Blue) using an argon–krypton or helium–neon laser and a dichroitic mirror FT 510 or NT 543/80/20 for respective wavelength selection. Emission was measured at wavelengths > 515 or > 570 nm with cutoff filters LP 515 and LP 570, respectively, in front of the detector. The fluorescence intensities inside and outside the cell were recorded in predetermined regions of interest (ROIs, 16×16 pixel; 30 scans with a scan time of 2 sec with double averaging) as described by Lorenz et al.[5,11] The background fluorescence intensity was determined before addition of the peptide solution for 300 sec. For measurement of the intracellular fluorescence, three ROIs were selected in the cytosol and one in the nucleus of three cells in such a manner that the diffuse fluorescence could be recorded without interference with vesicular fluorescence.

The intracellular fluorescence signal was corrected for contributions from the extracellular fluorescence arising from nonideal confocal properties of the CLSM by estimating the distribution function of sensitivity in the Z direction of the microscope according to Wiesner et al.[11]

3.4.4.2   Measurement of the Axial Response of the Confocal System

A dye solution of 400 μL (carboxyfluorescein or fluorescein-labeled peptide; 1.25 μM) was transferred to the cover slip. The fluorescence of the dye and the reflection signal of the excitation source were detected in a two-channel measurement. The scan plane (XY plane) was moved in 1 μm steps in the Z

direction up to 200 μm, beginning at the middle of the cover slip. Adjacent to the optical boundary (cover slip — dye) the step width was decreased to 0.2 μm. At each step a two-channel image (512 × 512 pixel) with a scan time of 2 sec and a double averaging was captured. The intensities of the fluorescence and the reflection signals were obtained directly from the single-channel image after each scan. The values of the intensities were stored according to the Z position. The registration of the reflection signal was used for determination of the Z position of the optical boundary.

3.4.4.3 Calculation of the Contributions from Out-of-Focus Planes to the Intracellular Fluorescence Intensity

After preincubating the cells for 10 min with 1.25 μM solution of the dye or the fluorescently labeled peptide, respectively, and subsequent transfer of 400 μL of the cell suspension to the cover slip, the optical measurements were performed using the following experimental setup.

The XY plane was positioned in the middle of a cell in the Z axis using both the reflection and transmission modes of the microscope. Regions of interest (ROIs, 16 × 16 pixel) were fixed between the center of the cell and the extracellular medium. The fluorescence intensity in each ROI was recorded as the mean of 30 scans with a scan time of 2 sec with double averaging. The reflection mode was used to determine the cell geometry and each different ΔZ_n in the Z direction from the center of ROI to the extracellular medium. Thereby the distances ΔZ_n were obtained, with the dimensions of the cell, the Z position at the cell, and the positions of ROIs at the focal plane found by simple mathematical calculation. The values of the fluorescence intensities of the ROIs and the distances of the ΔZ_n values were stored for later computation.

REFERENCES

1. Burton, P.S., Conradi, R.A., and Hilgers, A.R., Mechanisms of peptide and protein-absorption 2. Transcellular mechanism of peptide and protein-absorption — passive aspects, *Adv. Drug Deliv. Rev.*, 7, 365, 1991.
2. Derossi, D., Joliot, A.H., Chassaing, G., and Prochiantz, A., The third helix of the Antennapedia homeodomain translocates through biological membranes, *J. Biol. Chem.*, 269, 10444, 1994.
3. Lin, Y.Z., Yao, S.Y., Veach, R.A., Torgerson, T.R., and Hawiger, J., Inhibition of nuclear translocation of transcription factor NF-kappa B by a synthetic peptide containing a cell membrane-permeable motif and nuclear localization sequence, *J. Biol. Chem.*, 270, 14255, 1995.
4. Vives, E., Brodin, P., and Lebleu, B., A truncated HIV-1 Tat protein basic domain rapidly translocates through the plasma membrane and accumulates in the cell nucleus, *J. Biol. Chem.*, 272, 16010, 1997.
5. Lorenz, D., Wiesner, B., Zipper, J., Winkler, A., Krause, E., Beyermann, M., Lindau, M., and Bienert, M., Mechanism of peptide-induced mast cell degranulation — translocation and patch-clamp studies, *J. Gen. Physiol.*, 112, 577, 1998.

6. Oehlke, J., Beyermann, M., Wiesner, B., Melzig, M., Berger, H., Krause, E., and Bienert, M., Evidence for extensive and nonspecific translocation of oligopeptides across plasma membranes of mammalian cells, *Biochim. Biophys. Acta*, 1330, 50, 1997.

7. Lindgren, M., Hällbrink, M., Prochiantz, A., and Langel, Ü., Cell-penetrating peptides, *Trends Pharmacol. Sci.*, 21, 99, 2000.

8. Repke, H. and Bienert, M., Mast-cell activation — a receptor-independent mode of substance-P action, *FEBS Lett.*, 221, 236, 1987.

9. Mousli, M., Hugli, T.E., Landry, Y., and Bronner, C., Peptidergic pathway in human skin and rat peritoneal mast cell activation, *Immunopharmacology*, 27, 1, 1994.

10. Oflynn, N.M., Helme, R.D., Watkins, D.J., and Burcher, E., Autoradiographic localization of substance-P binding-sites in rat footpad skin, *Neurosci. Lett.*, 106, 43, 1989.

11. Wiesner, B., Lorenz, D., Krause, E., Beyermann, M., and Bienert, M., Measurement of intracellular fluorescence in the presence of a strong extracellular fluorescence using confocal laser scanning microscopy, *LAMSO*, 3, 1, 2002.

12. Smith, Q.R., Drug delivery to brain and the role of carrier-mediated transport, *Adv. Exp. Med. Biol.*, 331, 83, 1993.

13. Rothemund, S., Krause, E., Beyermann, M., Dathe, M., Engelhardt, H., and Bienert, M., Recognition of alpha-helical peptide structures using high-performance liquid chromatographic retention data for D-amino acid analogues: Influence of peptide amphipathicity and of stationary phase hydrophobicity, *J. Chromatogr.*, 689, 219, 1995.

14. Krause, E., Beyermann, M., Dathe, M., Rothemund, S., and Bienert, M., Location of an amphipathic alpha-helix in peptides using reversed-phase HPLC retention behavior of D-amino acid analogs, *Anal. Chem.*, 67, 252, 1995.

15. Dathe, M., Wieprecht, T., Nikolenko, H., Handel, L., Maloy, W.L., MacDonald, D.L., Beyermann, M., and Bienert, M., Hydrophobicity, hydrophobic moment and angle subtended by charged residues modulate antibacterial and haemolytic activity of amphipathic helical peptides, *FEBS Lett.*, 403, 208, 1997.

16. Cross, L.J., Ennis, M., Krause, E., Dathe, M., Lorenz, D., Krause, G., Beyermann, M., and Bienert, M., Influence of alpha-helicity, amphipathicity and D-amino acid incorporation on the peptide-induced mast cell activation, *Eur. J. Pharmacol.*, 291, 291, 1995.

17. Steiner, V., Schar, M., Bornsen, K.O., and Mutter, M., Retention behaviour of a template-assembled synthetic protein and its amphiphilic building blocks on reversed-phase columns, *J. Chromatogr.*, 586, 43, 1991.

18. Rothemund, S., Krause, E., Beyermann, M., Dathe, M., Bienert, M., Hodges, R.S., Sykes, B.D., and Sonnichsen, F.D., Peptide destabilization by two adjacent D-amino acids in single-stranded amphipathic alpha-helices, *Pept. Res.*, 9, 79, 1996.

19. Oehlke, J., Krause, E., Wiesner, B., Beyermann, M., and Bienert, M., Nonendocytic, amphipathicity dependent cellular uptake of helical model peptides, *Protein Pept. Lett.*, 3, 393, 1996.

20. Oehlke, J., Scheller, A., Wiesner, B., Krause, E., Beyermann, M., Klauschenz, E., Melzig, M., and Bienert, M., Cellular uptake of an alpha-helical amphipathic model peptide with the potential to deliver polar compounds into the cell interior nonendocytically, *Biochim. Biophys. Acta*, 1414, 127, 1998.

21. Scheller, A., Wiesner, B., Melzig, M., Bienert, M., and Oehlke, J., Evidence for an amphipathicity independent cellular uptake of amphipathic cell-penetrating peptides, *Eur. J. Biochem.*, 267, 6043, 2000.

22. Wolf, Y., Wallukat, G., Wiesner, B., Pritz, S., Bienert, M., and Oehlke, J. Submitted.

23. Oehlke, J., Birth, P., Klauschenz, E., Wiesner, B., Beyermann, M., Oksche, A., and Bienert, M., Cellular uptake of antisense oligonucleotides after complexing or conjugation with cell-penetrating model peptides, *Eur. J. Biochem.*, 269, 4025, 2002.

24. Oehlke, J., Wallukat, G., Wolf, Y., Ehrlich, A., Wiesner, B., Berger, H., and Bienert, M., Enhancement of intracellular concentration and biological activity of PNA after conjugation with a cell-penetrating synthetic model peptide, *Eur. J. Biochem.*, 271, 3043, 2004.

25. Bradford, M.M., A rapid and sensitive method for the quantitation of microgram quantities of protein utilizing the principle of protein-dye binding, *Anal. Biochem.*, 72, 248, 1976.

4 Cell-Penetrating *cis*-γ-Amino-L-Proline-Derived Peptides

Josep Farrera-Sinfreu, Ernest Giralt, Miriam Royo, and Fernando Albericio

CONTENTS

4.1 INTRODUCTION

Cell membranes protect the cell against invasions from the extracellular media. Large hydrophilic compounds, such as polypeptides and oligonucleotides, are unable to pass through this barrier and thus have limited therapeutic value. In recent years, several membrane-permeable peptides, namely cell-penetrating peptides (CPPs), have been described. The capacity of CPPs to penetrate cells indicates their potential application as new agents for cellular delivery of drugs that are unable to pass through the cell membrane. CPPs may provide an alternative to other agents, such as viral delivery systems,[1] liposomes,[2,3] encapsulation in polymers,[4] electroporation,[5] or receptor-mediated endocytosis,[6] none of which is sufficiently efficient and they very often cause cellular toxicity. Although CPPs offer exciting possibilities for drug delivery, several known drawbacks must be considered for their use in life sciences, such as their metabolic stability or toxicity. These aspects should not be interpreted as limitations for their use in vivo. In fact, studies to improve these pitfalls have helped us to develop new and better CPPs. In this regard, several authors have described nonnatural CPPs, such as those formed with D-amino acids,[7,8] β-peptides,[9–12] and peptoids,[13] that skip enzymatic degradation, although in some cases the toxicity still remains.

The brief history of CPPs began in 1988 when Green and Loewenstein[14] realized that proteins might contain sequences responsible for internalization into cells, discovering that a fragment from the HIV-1 Tat protein (Tat 86) with a length of 86 amino acids could be internalized by cells. Later, in 1991, Joliot et al.[15] found that the homeodomain of the Drosophila antennapedia protein could also enter the cell, a fact leading to the discovery that only its third helix was required to pass across cell membranes, yielding the CPP penetratin or pAntp as a result.[16] In 1997, Vivès et al.[17] reported the basic sequence of Tat peptide, and during 1996 and 1997 several signal-sequence-based peptides were described as CPPs.[18–20] In 1998, Langel and coworkers[21] described a chimeric biotinyltransportan peptide and Oehlke et al.[22] a synthetic 18-mer amphipathic model peptide.

In addition to being derived from natural sources, all these CPPs share two structural features: they are either cationic peptides with at least six charged amino acids, such as HIV-1 TAT peptide, penetratin, the chimeric transportan, and Arg$_9$,[23] or they are hydrophobic peptides, such as those based on the H-region of signal-sequence proteins. Strikingly, bactericidal peptides are charged and contain hydrophobic regions in their primary or secondary structure,[24–27] and therefore their possible antibacterial peptide characteristics should be considered for the design of novel CPPs. In this regard, selective modulation of the bactericidal activity of peptides has yielded new proline-rich CPPs.[28]

Despite certain hydrophobicity provided by proline, the most important advantage of proline-rich peptides in biological systems is their high degree of water solubility, an essential property for their use in life sciences. In this regard, our group has reported proline-rich peptides,[29,30] polyproline dendrimers,[31–35] and nonnatural proline-derived γ-peptides (from the *cis*-γ-amino-L-proline)[36,37] with the capacity to be taken up into eukaryotic cells. Other authors have recently described modified polyproline helix-based scaffolds with this capacity.[38] Here, we focus on the design and synthesis of these γ-peptides, with the aim of overcoming several drawbacks presented by natural CPPs.

In general, it is accepted that CPPs offer several advantages over other known cellular delivery systems, including low toxicity, high efficiency towards distinct cell lines, and even inherent therapeutic potential. However, it has been demonstrated that CPPs have some elements and pitfalls that can limit their use as drug carriers,[39] including low protease resistance,[8,40–44] low membrane permeability,[43,45] poor cellular uptake, cell-to-cell variability, toxicity, and immunogenicity. Nevertheless, some of these features, such as low membrane permeability, are not completely negative as they can be used to advantage to test for selectivity. In this context, this chapter focuses on three of these points: (1) metabolic degradation, (2) toxicity, and (3) biological barrier permeability.

One of the greatest limitations of CPPs as a drug delivery agent is their low protease resistance. Cargo-carrying CPPs should enter the cell and then be cleaved from it, after which the cargo must perform its work. However, the metabolic degradation of several CPPs in the extracellular media is considerable and therefore these do not fulfill their role as carriers.[8,40–44] Nevertheless, the metabolic degradation of the CPP and the CPP cargo is necessary for the delivery of the cargo after internalization and the final elimination of the CPP to avoid chronic toxicity. An equilibrium between these two aspects is required in order to avoid the premature release of the cargo and its cleavage once internalized.

In terms of toxicity, Tat-derived peptides are neurotoxic and the level of toxicity depends on the length of exposure and on peptide length. It has been demonstrated that Tat(31–61) produces more neurotoxicity than the full-length protein.[46] Other studies report that several Tat-derived peptides decrease HeLa cell viability.[17] Penetratin (43–58) also shows a range of negative effects, such as neurotoxic cell death,[47] cytotoxic effects in some cell lines,[48] or the induction of cell lysis.[49] Metabolic degradation and toxicity are directly related. If enzymatic degradation implies increasing the therapeutic dose, then the toxicity levels of CPPs may be high, which clearly prevents their use in life sciences.

The third aspect that should be considered for CPPs is their low membrane permeability. The literature indicates that this feature depends on the cell line. Thus, while Tat-derived peptides pass through cell membranes irrespective of cell type,[50,51] even across the blood–brain barrier (BBB),[52] later studies have shown that some cell lines are nonpermeable to these peptides under physiologic conditions.[45] Reinforcing this idea, other authors have demonstrated that the cellular uptake and the cargo delivery by CPP depend on the nature of the peptides themselves and on the lipid composition of the membrane, which varies with each cell line.[53] Given that not all CPPs show the same efficiency for cell penetration, the development of new CPPs, which show selectivity for certain cell membranes, is required. The development of this selectivity would result in a wide range of CPPs for application in drug delivery. These would be chosen for the purpose of the problem to be solved.

To overcome the drawbacks presented by natural CPPs, a number of peptidomimetics with greater proteolytic resistance such as loligomers,[54,55] peptides with ʟ-amino-acid residues replaced by their nonnatural ᴅ-analogs,[7,8] as well as β-peptides[9–12] and peptoids[13] have been evaluated as nonnatural CPPs for drug delivery purposes. These molecules are not degraded enzymatically as they are not recognized by enzymes. The contribution of this laboratory to the field has been the development of a new *cis*-γ-amino-ʟ-proline-derived peptidomimetic (Figure 4.1) with the peptide backbone formed through the γ-amino function.[37] These cell-penetrating *cis*-γ-amino-ʟ-proline-derived γ-peptides are nonsensitive to both trypsin and serum, thereby overcoming one of the main drawbacks of natural CPPs, the enzymatic degradation. Another interesting feature is their low or nonexistent toxicity in COS-1 and HeLa cells. Furthermore, permeability can be modulated by taking advantage of the amino side chain, which, in this case, is the

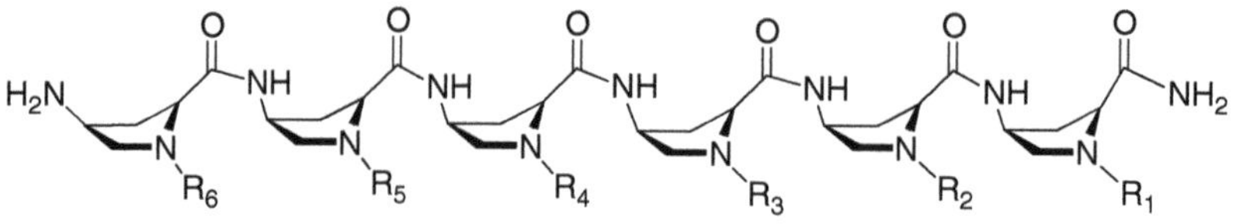

FIGURE 4.1 Chemical structure of *cis*-γ-amino-L-proline oligomers.

α-amino function of the proline. Thus, through this side chain, several parameters of the CPP can be altered, such as hydrophilicity/hydrophobicity characteristics.

4.2 RESULTS

4.2.1 SYNTHESIS

The γ-peptide syntheses were performed following state-of-the-art solid-phase protocols. All these peptides were constructed using a common polymeric backbone (hexamer) of *cis*-γ-amino-L-proline. The skeleton was elongated through the γ-amino group of the proline by repetitive monomer couplings. Removal of the Boc group released the α-amino function, which was further manipulated to confer diversity to the CPP, mimicking the side chains of natural amino acids. When the side chain introduction was performed immediately after the stepwise elongation, homo-oligomers (identical side chains) were formed (Figure 4.2). In contrast, heterodimers (different side chains) were obtained when each side chain was introduced step by step, after the incorporation of each monomer (Figure 4.3). Three γ-peptide families were synthesized: (i) N^{α}-alkyl-γ-peptides (side chains introduced through an amino function); (ii) N^{α}-acyl-γ-peptides (side chains introduced through an amide function); and (iii) N^{α}-guanidylated-γ-peptides (guanidinium function on the α position) (Figure 4.4).

The preparation of these CPPs, which are readily accessible, requires the concourse of two orthogonal protecting groups for the two amino functions. As Fmoc and Boc are the most common protecting groups for this function, the former was used as the temporary protecting group for the γ-amino group of each monomer

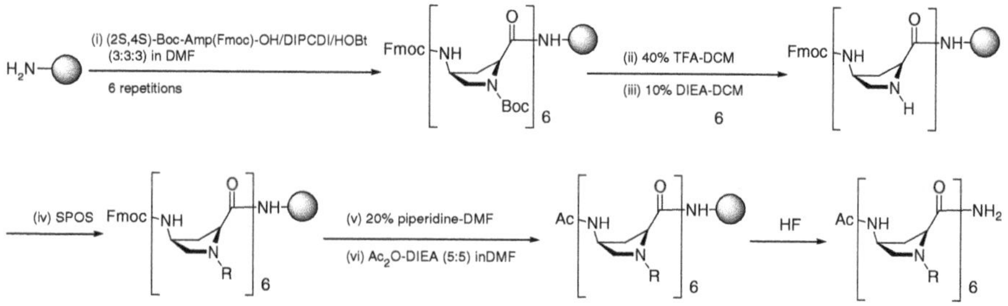

FIGURE 4.2 Synthesis of the homo-oligomeric systems (synthesis of N^{α}-alkyl-γ-hexapeptides and the N^{α}-acyl-γ-hexapeptides).

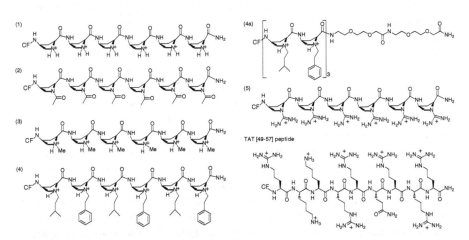

FIGURE 4.3 Synthetic scheme of the hetero-oligomers (synthesis of N^{α}-alkyl-γ-hexapeptides and the N^{α}-acyl-γ-hexapeptides).

and the latter as the semipermanent protecting group for the α-amino group through which the side chain was introduced. The syntheses were carried out using N^{a}-Fmoc-N^{a}-Boc-*cis*-g-amino-L-proline as the building block on a *p*-methylbenzy-drylamine (MBHA) resin, which is stable at conditions used to remove the Boc and the Fmoc groups, and requires anhydrous HF or trifluoromethanesulfonic acid (TFMSA) to cleave the peptide from the resin.[55] The stepwise elongation of the peptide chain was performed smoothly using this combined Fmoc/Boc strategy and *N,N*-diisopropylcarbodiimide (DIPCDI) in the presence of *N*-hydroxybenzotriazole (HOBt) as the coupling method.

α-Amino groups of the amino-proline were acylated using the corresponding carboxylic acid and DIPCDI/HOBt, or alkylated using a reductive amination with the corresponding aldehyde and NaBH₃CN as a reductive agent. The N^{α}-amino guanidylation was performed using *N,N'*-di-Boc-*N''*-trifluoromethanesulfonyl

FIGURE 4.4 γ-Aminoproline monomer-based γ-peptides labeled with 5(6)-carboxyfluorescein (CF).

guanidine[57] and *N,N,N*-triethylamine (TEA). The last step in the synthesis before cleavage was the removal of Fmoc and acetylation of this terminal amino group. When the peptide was required with a fluorescent label, this step was substituted by introducing 5(6)-carboxyfluorescein (CF) using DIPCDI/HOBt, followed by piperidine washes just before cleavage of the peptide from the resin. These washes were required to remove overincorporated CF.[58] When these washes were not carried out, compounds containing two and three extra units of CF were observed in the crude product.

Peptides were finally cleaved from the resin by acidolytic treatment with anhydrous HF. The purity of the CF-γ-peptide crude products, as determined by high-performance liquid chromatography (HPLC), ranged from 65 to 90%. Compounds (Figure 4.4) were purified to more than 95% homogeneity by preparative reverse-phase HPLC and characterized by electrospray and/or MALDI-TOF mass spectrometry.

4.2.2 CELLULAR UPTAKE

A preliminary evaluation of the cellular uptake of the γ-peptide family was performed in COS-1 and HeLa cells using plate fluorimetry. Although this technique is commonly used to quantify the cellular uptake of CPPs, several problems related to γ-peptide binding to the plastic surface of the plates were unable to ensure that the results of the quantifications were from the internalized peptides. Having completed the aforementioned experiments, flow cytometry was then used to differentiate peptide uptake from the cell surface or plastic surface binding.

Flow cytometry appears to provide more reliable quantitative data than plate fluorimetry. The main source of imprecision in plate fluorimetry when measuring the cell uptake of highly hydrophobic compounds is an artifactual increase in the apparent uptake caused by nonspecific binding of the ligand to the plastic surface of the plate.[59] Flow cytometry experiments focused on the two most promising compounds, **4a** and **5**, using the well-known cell-penetrating peptide TAT 49–57 (CF-RKKRRQRRR-NH$_2$) as a positive control. Flow cytometry analysis quantifies cellular association of CF-tagged peptides, excluding any unspecific binding of the peptides to the plastic surface of the plate. The uptake levels of **4a** and **5** were lower than those found using plate fluorimetry, indicating that these compounds had been partially bound to the plastic plate. Peptide **4a** was taken up very efficiently by COS-1 cells, even at 4°C, while the internalization of peptide **5** was lower, especially at 4°C. In comparison with **TAT** peptide, cells incubated with **4a** fluoresced 3.5 times less than those incubated with **TAT**.

To determine the contribution of cell surface-bound CF peptides in the peptide uptake measurements, additional flow cytometry experiments were performed. For this purpose, cells could not be treated with trypsin to remove cell surface-bound peptide before flow cytometry analysis because these γ-peptides are not trypsin sensitive. We applied an alternative method based on the pH dependence of fluorescence emission in receptor-endocytosis studies.[60] Flow cytometry analysis revealed that the amount of fluorescence was very similar at pH 7.4 and pH 6 for peptide **4a** and **TAT**, and only a small decrease in fluorescence was observed, indicating that the fluorescence detected by the cytometer was mainly caused by internalized peptides.

4.2.3   MTT Cytotoxicity Assay and Enzymatic Stability

Cell viability assays were performed to determine the toxicity of the proline-derived γ-peptides. The toxicity of these peptides to COS-1 and HeLa cells was determined by MTT assays and estimated to be lower than that of **TAT** peptide at the same concentration (**TAT** peptide, consistent with literature reports,[61] exhibited cytotoxicity even at low concentrations). The viability of COS-1 cells after treatment with γ-peptides for 24 h was always higher than 95%, even at the highest concentration used (25 μM). Only peptide **5** exhibited slight cytotoxicity at 25 μM (89–92%) in HeLa cells.

Enzymatic stability was tested with peptide **4a** in trypsin and human serum and checked by HPLC. This peptide showed very high stability to enzymes, even during long incubation times (up to 120 h).

4.2.4   Confocal Laser Scanning Microscopy

The intracellular distribution of CF peptides was examined by confocal microscopy in fixed and unfixed COS-1 cells, recording optical sections that allowed three-dimensional reconstruction of the cell. All γ-peptides were located inside the cells and were not attached to the membrane. In cells treated with peptides **1, 2,** and **3** at 37°C, a slight vesicular distribution of the peptides was observed. This distribution, similar to the localization of the endocytotic marker used, indicated that an endocytotic mechanism was involved. Peptides **5** and **TAT** were found throughout the cytoplasm and in the nuclei, and no punctuated pattern was observed. Similar distribution of the two peptides was found when cells were incubated at 4°C. Cells treated with peptide **4a** showed a similar but more intense punctuated distribution, as well as a diffused distribution throughout the cytoplasm.

The cellular distribution of **TAT** can vary for live or paraformaldehyde-fixed cells.[11] Confocal microscopy pictures of **TAT** internalized in paraformaldehyde-fixed cells revealed distribution in the nucleus and endocytotic vesicles of the cytoplasm, whereas live-cell images indicated that endocytosis is the predominant uptake mechanism of this peptide. The difference in cellular distribution can be attributed to fixation of the cells with paraformaldehyde. Arginine-rich peptides, such as **TAT** and (Arg)$_9$, arbitrarily redistribute in the nucleus in a number of common cell fixation protocols.[11]

To determine whether the CF peptides were endocytocized or whether they diffused across the cell membrane, time course uptake experiments using the fluid-phase endocytotic marker Texas Red-Dextran (TR-DX) were performed. HeLa cells were incubated with CF peptides **2** and **4a** (10 μM, 37°C) in the presence of TR-DX (Figure 4.5) for 15 min or 2 h. To avoid formaldehyde-induced changes in the distribution of these peptides, confocal images of live cells were obtained. After 15 min of incubation, peptide **2** almost totally

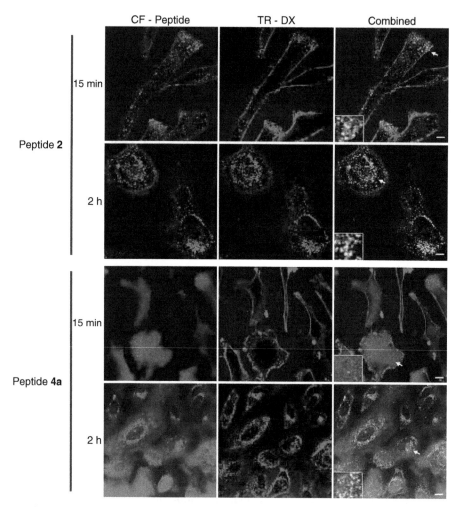

FIGURE 4.5 Live-cell imaging of carboxyfluorescein (CF) peptide uptake. Internalization experiments of CF peptide (10 μM, 37°C) in HeLa cells were performed with the fluid-phase (endocytotic) marker TR-DX; xy confocal sections obtained from the middle part of cells. Scale bars, 10 μm. Inserts: 3.2×.

colocalized with TR-DX, indicating a fluid-phase endocytotic mechanism. Peptide **4a** also colocalized in TR-DX-containing vesicles after 15 min of incubation, but it was also distributed in the nucleus and cytoplasm. After 2 h of incubation, there were more vesicles containing peptide **4a** and TR-DX but they were more concentrated in the perinuclear area of the cells. Cytoplasmic and punctuated or vesicular distribution of peptide **4a** was also observed. Although **4a** did not concentrate exclusively in vesicular structures, it cannot be affirmed that it translocates across the cell membrane. An alternative explanation could be that **4a** was taken up by endocytosis and later released from the vesicles into the cytoplasm in a manner similar to that demonstrated for the human antimicrobial peptide LL-37[62] and other cell-penetrating peptides.[63]

4.3 DISCUSSION AND PERSPECTIVES

γ-Amino-ʟ-proline is a building block of interest for the preparation of proline-based cell-penetrating γ-peptides. The use of two orthogonal protecting groups for the two amino functions of the building block leads to a flexible synthetic strategy for diverse substituted γ-peptides, thereby providing the possibility of incorporating a covalently linked drug in the last step of the synthesis. The diversity introduced through the γ-amino position of the proline alters the uptake properties of each peptide. With this accessible synthesis, libraries of γ-peptides could be easily synthesized, modulating permeability to certain cell line membranes and, as a consequence, improving targeting to certain cells.

Thus, γ-peptides with the same proline-mimetic skeleton but distinct side chains are taken up into cells via diverse mechanisms. When these γ-peptides were incubated in both COS-1 and HeLa cells at 37°C and the uptake rates were quantified by plate fluorimetry and flow cytometry, the results indicated that the γ-peptide family entered at different rates. When they were incubated at 4°C, only **4a** translocated across the cell membrane, demonstrating that the internalization of **4a** was not completely dependent on the endosomal pathway. Our hypothesis is that this peptide translocates across the cell membrane (passive diffusion), suggesting that at least two mechanisms are involved in the entry of **4a**.

Confocal microscopy revealed that all of the peptides either congregated in the cytoplasm, distributed throughout vesicles when incubated at 37°C (peptides **1**, **2**, **3**, **4a**, and **5**), or were widespread in the cytoplasm when incubated at 37 and 4°C (peptides **4a** and **5**). This punctuated distribution implies that the γ-peptide family undergoes endocytosis at 37°C. The translocation observed in the flow cytometry quantifications at 4°C reinforces the idea that peptide **4a** translocates across the cell membrane. Thus, two distinct routes appear to be involved in the transduction mechanism of peptide **4a**, one including endocytosis and the other an energy-independent mechanism. Nevertheless, the possibility that γ-peptides enter the cell via endocytosis and are thereafter released from the vesicles into the cytoplasm cannot be completely excluded. Furthermore, peptides **4a** and **5** reached the nucleus, indicating that they could be used for the delivery of molecules that act in this compartment.

In conclusion, given the capacity of these γ-peptides to enter into eukaryotic cells, they offer several interesting features for use in life sciences. These peptides display low toxicity, proteolytic stability and, in general, good aqueous solubility, all properties that indicate their potential utility as agents for drug delivery. These features are of great interest in order to improve the behavior of current CPPs when used in life sciences. Furthermore, the flexible synthetic strategy described for these peptides may allow modulation of the permeability by changing the side chains. However, the best candidate to act as a drug delivery agent was not highly soluble in physiological conditions, although it was functionalized with a short PEG to improve its solubility. As compound solubility is a fundamental issue for potential therapeutic applications, a candidate with increased solubility and improved translocation properties is required. Thus, to gain a better understanding of peptide translocation across cell membranes, future research efforts must focus on the synthesis of new analogs by means of a wide range of functional groups.

4.4 EXPERIMENTAL

4.4.1 GENERAL

Protected amino acids were obtained from Neosystem (Strasbourg, France) and MBHA resin (0.7 mmol/g) was supplied by Calbiochem-Novabiochem AG (Länfelfingen, Switzerland). Solvents for peptide synthesis and other reagents (DIPCDI, HOBt, $NaBH_3CN$, carboxylic acids, and aldehydes) were purchased from several commercial sources. HF was obtained from Air Products and Chemicals, Inc. (Allentown, Canada), and related equipment was obtained from Peptide Institute Inc. (Minoh, Osaka, Japan).

4.4.2 PEPTIDE SYNTHESIS

Peptide syntheses were performed manually in a polypropylene syringe fitted with a polyethylene porous disk. Solvents and soluble reagents were removed by suction. Amino acid couplings were carried out using DIPCDI in the presence of HOBt. Acylations of the α-amino positions were carried out using the same protocol with the corresponding carboxylic acid. Alkylations of the α-amino group of the proline were carried out with a reductive amination with the corresponding aldehyde and sodium cyanoborohydride. Solid-phase guanidylations were carried out using N,N'-di-Boc-N''-trifluoromethanesulfonyl guanidine and TEA in DCM. CF was introduced in the γ-amino terminal position using the same coupling reagents mentioned above. Finally, peptides were cleaved from the resin with HF. Products were analyzed by analytical RP-HPLC using Waters (Milford, MA) chromatography systems using reverse-phase Symmetry C_{18} (150×4.6 mm, 5 μm columns) with UV detection at 220 nm. Semipreparative RP-HPLC was performed on a Waters chromatography system using Symmetry C_8 (3×10 cm, 5 μm) columns. Compounds were detected by UV absorption at 220 nm. Mass spectra were recorded on a MALDI Voyager DE RP time-of-flight (TOF) spectrometer (Applied Biosystems, Framingham, MA).

4.4.3 CELL CULTURE

Cell lines were routinely grown at 37°C in a humidified atmosphere with 5% CO_2. HeLa and COS-1 cells were maintained in DMEM (1000 mg/mL glucose for HeLa and 4500 mg/mL for COS-1) culture medium containing 10% fetal calf serum (FCS), 2 mM glutamine, 50 U/mL penicillin, and 0.05 g/mL streptomycin. For the different techniques, cells were cultured analogously in the appropriate container. The γ-peptides were dissolved in PBS buffer and their concentration determined by measuring the fluorescence of 5(6)-CF at $\lambda_{excitation} = 485/20$ nm and at $\lambda_{emission} = 530/25$ nm.

4.4.4 Uptake Experiments with FACS

COS-1 and HeLa cells were seeded onto 35-mm plates at a concentration of 21.4×10^3 cells/cm^2. After 24 h, cells were incubated with CF peptides at a range of concentrations (from 0.01 to 25 μM). After various incubation times, cells were washed three times with PBS, detached with tripsine-EDTA 0.25%, centrifuged, and washed again. Finally, they were resuspended in PBS containing 0.1 mM of propidium iodide (PI) and the florescence was analyzed with an Epics XL flow cytometer (Coulter). To remove fluorescence of CF or CF peptides bound to the plasma membrane, the pH of the PBS/PI solution was brought down to 6 by the addition of 1 N HCl just before measuring fluorescence (at pH=6, extracellular fluorescence of CF is quenched without altering cell mechanisms[60]), and was analyzed immediately. Cells stained with PI were excluded from further analysis. Triplicates of each sample were performed for each condition and results from independent experiments were normalized by subtraction of the autofluorescence control value from each value and dividing by the fluorescence value obtained from the CF control under the same experimental conditions.

4.4.5 Enzymatic Stability

Enzymatic degradation with trypsin was performed by incubating the γ-peptides at 37°C with the enzyme in 100 mM Tris–HCl at pH 8. The trypsin/peptide ratio was 1:100, using 3.43 μL of a solution of trypsin from bovine pancreas E.C. 3.4.21.4 (Roche, Basel, Switzerland) in glycerol-H$_2$O (1:1) (6.25 mg/mL). Aliquots (50 μL) were periodically taken at 2–120 h, 150 μL of 1 N HCl were added and the resulting solution was cooled with ice. Degradation was monitored by HPLC.

Enzymatic degradation using human serum (Aldrich, Milwaukee, WI) was carried out by incubating the γ-peptides at 37°C with the serum (diluted 9:1 in HBSS buffer). Peptides were used at a final concentration of 125 μM (added to the serum dissolved in the buffer). Aliquots (50 μL) were periodically taken at 2–120 h, poured into 200 μL of MeOH to precipitate the proteins, and cooled on ice. After 30 min, the sample was centrifuged and the supernatant was analyzed by HPLC.

4.4.6 Cytotoxicity Assay

The viability of COS-1 and HeLa cells in the presence of the peptides was tested using the 3-(4,5-dimethylthiazol-2-yl)-2,5-diphenyltetrazolium bromide (MTT) assay. To avoid saturation in cell growth after 24 h of peptide incubation, 7×10^3 cells/well were seeded on a 96-well plate (Nalge Nunc, Rochester, USA) for each assay. After 24 h, the culture medium was discarded and replaced by new medium containing a range of γ-peptide concentrations. Cells were incubated for 2, 8, and 24 h at 37°C under a 5% CO$_2$ atmosphere, and MTT (0.5 mg/mL) was added 2 h before the end of incubation. After 2 h of incubation with MTT, the medium was discarded by aspiration and isopropanol was added to dissolve formazan. Absorbance was measured at 570 nm in a spectrophotometric El×800 Universal Microplate Reader (Bio-Tek Instruments, Vermont, USA), 30 min after the addition of isopropanol. Cell viability was expressed as a percent ratio of cells treated with peptide to untreated cells, which were used as a control.

4.4.7 CONFOCAL LASER SCANNING MICROSCOPY

COS-1 and HeLa cells were seeded onto glass coverslips at 21.4×10^3 cells/cm^2 and after 24 h were incubated with CF peptides, as described above. For endocytosis experiments, TR-DX (3 mg/mL, MW = 10,000, Molecular Probes, Carlsbad, USA) was incubated together with the CF peptide. After CF peptide incubation, cells were washed three times with PBS and fixed in 3% paraformaldehyde–2% sucrose in 0.1 M phosphate buffer (PB) for 15 min, washed three times in PBS and mounted in Mowiol with 2.5% DABCO. PI (1 μg/mL) staining was performed at room temperature for 15 min in the presence of RNAsa in PBS (1 mg/mL).

Similar experiments were performed in cells plated onto glass-bottom Lab-Tek™ (Campbell, USA) chambers for live-cell imaging. After 2 h of incubation, cells were washed three times with PBS containing 1.1 mM CaCl$_2$, 1.3 mM MgCl$_2$, and 25 mM Hepes. Images were subsequently acquired within the next 30 min.

Confocal laser scanning microscopy was performed with an Olympus Fluoview 500 confocal microscope using a 60$\times$/1.4 NA plan-apochromatic objective. CF fluorescence was excited with the 488-nm line of an argon laser and its emission was detected in a range of 515–530 nm. The microscope settings were identical for each peptide and dose. PI and TR-DX were excited at 543 nm and detected with a 560-nm long pass filter. To avoid crosstalk, the two-fluorescence scanning was performed in a sequential mode.

ACKNOWLEDGMENTS

The work performed by the authors was partially supported by funds from Ministerio de Ciencia y Tecnología (BQU2002-02047 and BQU2003-00089), the Generalitat de Catalunya (Grup Consolidat and Centre de Referència en Biotecnologia), and the Barcelona Science Park. Josep Farrera-Sinfreu received a pre-doctoral fellowship from the University of Barcelona.

REFERENCES

1. Davidson, B.L. and Breakefield, X.O., Neurological diseases: Viral vectors for gene delivery to the nervous system, *Nat. Rev. Neurosci.*, 4, 353, 2004.
2. Connor, J. and Huang, L., Efficient cytoplasmatic delivery of a fluorescent dye by pH-sensitive immunoliposomes, *J. Cell. Biol.*, 101, 582, 1985.
3. Foldvari, M., Mezei, C., and Mezei, M., Intracellular delivery of drugs by liposomes containing P0 glycoprotein from peripheral nerve myelin into human M21 melanoma cells, *J. Pharm. Sci.*, 80, 1020, 1991.
4. Gentile, F.T. et al., Polymer science for macroencapsulation of cells for central nervous system transplantation, *React. Polym.*, 25, 207, 1995.
5. Chakrabarti, R., Wylie, D.E., and Schuster, S.M., Transfer of monoclonal antibodies into mammalian cells by electroporation, *J. Biol. Chem.*, 264, 15494, 1989.

6. Leamon, C.P. and Low, P.S., Delivery of macromolecules into living cells: A method that exploits folate receptor endocytosis, *Proc. Natl Acad. Sci. U.S.A.*, 88, 5572, 1991.

7. Rothbard, J.B. et al., Conjugation of arginine oligomers to cyclosporin A facilitates topical delivery and inhibition of inflammation, *Nat. Med. (New York)*, 6, 1253, 2000.

8. Elmquist, E. et al., VE-cadherin-derived cell penetrating peptide, pVEC, with carrier functions, *Exp. Cell Res.*, 269, 237, 2001.

9. Umezawa, N. et al., Translocation of a β-peptide across cell membranes, *J. Am. Chem. Soc.*, 124, 368, 2002.

10. Rueping, M. et al., Cellular uptake studies with β-peptides, *ChemBioChem*, 3, 257, 2002.

11. Potocky, T.B., Menon, A.K., and Gellman, S.H., Cytoplasmic and nuclear delivery of a TAT-derived peptide and a β-peptide after endocytic uptake into HeLa cells, *J. Biol. Chem.*, 278, 50188, 2003.

12. Garcia-Echeverria, C. and Ruetz, S., β-Homolysine oligomers: A new class of Trojan carriers, *Bioorg. Med. Chem. Lett.*, 13, 247, 2003.

13. Wender, P.A. et al., The design, synthesis, and evaluation of molecules that enable or enhance cellular uptake: Peptoid molecular transporters, *Proc. Natl Acad. Sci. U.S.A.*, 97, 13003, 2000.

14. Green, M. and Loewenstein, P.M., Autonomous functional domains of chemically synthesized human immunodeficiency virus tat trans-activator protein, *Cell*, 55, 1179, 1988.

15. Joliot, A. et al., Antennapedia homeobox peptide regulates neural morphogenesis, *Proc. Natl Acad. Sci. U.S.A.*, 88, 1864, 1991.

16. Derossi, D. et al., The third helix of the Antennapedia homeodomain translocates through biological membranes, *J. Biol. Chem.*, 269, 10444, 1994.

17. Vivès, E., Brodin, P., and Lebleu, B., A truncated HIV-1 Tat protein basic domain rapidly translocates through the plasma membrane and accumulates in the cell nucleus, *J. Biol. Chem.*, 272, 16010, 1997.

18. Lin, Y., Yao, S., and Hawiger, J., Role of the nuclear localization sequence in fibroblast growth factor-1-stimulated mitogenic pathways, *J. Biol. Chem.*, 271, 5305, 1996.

19. Liu, X. et al., Identification of a functionally important sequence in the cytoplasmatic tail of integrin β3 by using cell-permeable peptide analogs, *Proc. Natl Acad. Sci. U.S.A.*, 93, 11819, 1996.

20. Morris, M.C. et al., A new peptide vector for efficient delivery of oligonucleotides into mammalian cells, *Nucleic Acids Res.*, 25, 2730, 1997.

21. Pooga, M. et al., Cell penetration by transportan, *FASEB J.*, 12, 67, 1998.

22. Oehlke, J. et al., Cellular uptake of an alpha-helical amphipathic model peptide with the potential to deliver polar compounds into the cell interior non-endocytically, *Biochim. Biophys. Acta*, 1414, 127, 1998.

23. Mitchell, D.J. et al., Polyarginine enters cells more efficiently than other polycationic homopolymers, *J. Pept. Res.*, 56, 318, 2000.

24. Boman, H.G., Peptide antibiotics and their role in innate immunity, *Annu. Rev. Immunol.*, 13, 61, 1995.

25. Gallo, R.L. and Huttner, K.M., Antimicrobial peptides: An emerging concept in cutaneous biology, *J. Invest. Dermatol.*, 111, 739, 1998.

26. Bulet, P. et al., Antimicrobial peptides in insects; structure and function, *Dev. Comput. Immunol.*, 23, 329, 1999.

27. Shai, Y., Mechanism of the binding, insertion and destabilization of phospholipid bilayer membranes by α-helical antimicrobial and cell non-selective membrane-lytic peptides, *Biochim. Biophys. Acta (Biomembr.)*, 1462, 55, 1999.

28. Sadler, K. et al., Translocating proline-rich peptides from the antimicrobial peptide bactenecin 7, *Biochemistry*, 41, 14150, 2002.

29. Fernández-Carneado, J. et al., Potential peptide carriers: Amphipathic proline-rich peptides derived from the N-terminal domain of γ-zein, *Angew. Chemie Int. Ed.*, 43, 1811, 2004.

30. Fernández-Carneado, J. et al., Amphipathic peptides and drug delivery, *Biopolymers*, 76(2), 196, 2004.

31. Crespo, L. et al., Branched poly(proline) peptides: An efficient new approach to the synthesis of repetitive branched peptides, *Eur. J. Org. Chem.*, 11, 1756, 2002.

32. Crespo, L. et al., Peptide dendrimers based on polyproline helices, *J. Am. Chem. Soc.*, 124, 8876, 2002.

33. Sanclimens, G. et al., Solid-phase synthesis of second-generation polyproline dendrimers, *Biopolymers (Pept. Sci.)*, 76, 283, 2004.

34. Sanclimens, G. et al., Preparation of de novo globular proteins based on proline dendrimers, *J. Org. Chem.*, 70, 6274, 2005.

35. Sanclimens, G. et al. Synthesis and screening of a small library of proline based biodendrimers for use as delivery agents, *Biopolymers (Pept. Sci.)*, 80, 800, 2005.

36. Farrera-Sinfreu, J. et al., A new class of foldamers based on cis-γ-amino-L-proline, *J. Am. Chem. Soc.*, 126, 6048, 2004.

37. Farrera-Sinfreu, J. et al., Cell penetrating cis-γ-amino-L-proline-derived peptides, *J. Am. Chem. Soc.*, 127, 9459, 2005.

38. Fillon, Y.A., Anderson, J.P., and Chmielewski, J., Cell penetrating agents based on a polyproline helix scaffold, *J. Am. Chem. Soc.*, 127, 11798, 2005.

39. Tréhin, R. and Merkle, H.P., Chances and pitfalls of cell penetrating peptides for cellular drug delivery, *Eur. J. Pharm. Biopharm.*, 58, 209, 2004.

40. Elmquist, A. and Langel, U., In vitro uptake and stability study of pVEC and its all-D analog, *Biol. Chem.*, 384, 387, 2003.

41. Soomets, U. et al., Deletion analogues of transportan, *Biochim. Biophys. Acta*, 1467, 165, 2000.

42. Lindgren, M.E. et al., Passage of cell-penetrating peptides across a human epithelial cell layer in vitro, *Biochem. J.*, 377, 69, 2004.

43. Tréhin, R. et al., Cellular uptake but low permeation of human calcitonin-derived cell penetrating peptides and Tat(47–57) through well-differentiated epithelial models, *Pharm. Res.*, 21, 1248, 2004.

44. Tréhin, R. et al., Metabolic cleavage of cell-penetrating peptides in contact with epithelial models: Human calcitonin (hCT)-derived peptides, Tat(47–57) and penetratin(43–58), *Biochem. J.*, 382, 945, 2004.

45. Violini, S. et al., Evidence for a plasma membrane-mediated permeability barrier to Tat basic domain in well-differentiated epithelial cells: lack of correlation with heparan sulfate, *Biochemistry*, 41, 12652, 2002.

46. Nath, A. et al., Identification of a human immunodeficiency virus type 1 Tat epitope that is neuroexcitatory and neurotoxic, *J. Virol.*, 70, 1475, 1996.

47. Bolton, S.J. et al., Cellular uptake and spread of the cell-permeable peptide penetratine in adult rat brain, *Eur. J. Neurosci.*, 12, 2847, 2000.

48. Garcia-Echevarria, C. et al., A new antennapedia-derived vector for intracellular delivery of exogenous compounds, *Bioorg. Med. Chem. Lett.*, 11, 1363, 2001.

49. Drin, G. et al., Studies on the internalisation mechanism of cationic cell-penetrating peptides, *J. Biol. Chem.*, 278, 31192, 2003.
50. Schwarze, S.R., Hruska, K.A., and Dowdy, S.F., Protein transduction: Unrestricted delivery into all cells?, *Trends Cell Biol.*, 10, 290, 2000.
51. Bogoyevitch, M.A. et al., Taking the cell by stealth or storm? Protein transduction domains (PTDs) as versatile vectors for delivery, *DNA Cell Biol.*, 21, 879, 2002.
52. Schwarze, S.R. et al., In vivo protein transduction: Delivery of a biologically active protein into the mouse, *Science*, 285, 1569, 1999.
53. Hällbrink, M. et al., Cargo delivery kinetics of cell-penetrating peptides, *Biochim. Biophys. Acta*, 1515, 101, 2001.
54. Singh, D. et al., Penetration and intracellular routing of nucleus-directed peptide-based shuttles (loligomers) in eukaryotic cells, *Biochemistry*, 37, 5798, 1998.
55. Brokx, R.D., Bisland, S.K., and Gariépy, J., Designing peptide-based scaffolds as drug delivery vehicles, *J. Controlled Release*, 78, 115, 2002.
56. Lloyd-Williams, P., Albericio, F., and Giralt, E., Chemical Approaches to the Synthesis of Peptides and Proteins, CRC, Boca Raton, FL, 1997.
57. Feichtinger, K. et al., Diprotected triflylguanidines: A new class of guanidinylation reagents, *J. Org. Chem.*, 63, 3804, 1998.
58. Fischer, R. et al., Extending the applicability of carboxyfluorescein in solid-phase synthesis, *Bioconjugate Chem.*, 14, 653, 2003.
59. Chico, D.E., Given, R.L., and Miller, B.T., Binding of cationic cell-permeable peptides to plastic and glass, *Peptides*, 24, 3, 2003.
60. Chambers, J.D. et al., Endocytosis of beta 2 integrins by stimulated human neutrophils analyzed by flow cytometry, *J. Leukoc. Biol.*, 53, 462, 1993.
61. Pooga, M., Elmquist, A., and Langel, Ü., *Cell-Penetrating Peptides, Processes and Applications*, CRC Press, Boca Raton, FL, 2002.
62. Sandgren, S. et al., The human antimicrobial peptide LL-37 transfers extracellular DNA plasmid to the nuclear compartment of mammalian cells via lipid rafts and proteoglycan-dependent endocytosis, *J. Biol. Chem.*, 279, 17951, 2004.
63. Foerg, C. et al., Decoding the entry of two novel cell-penetrating peptides in HeLa cells: Lipid raft-mediated endocytosis and endosomal escape, *Biochemistry*, 44, 72, 2005.

5 Prediction of Cell-Penetrating Peptides

Mattias Hällbrink and Ülo Langel

CONTENTS

5.1 INTRODUCTION

Cell-penetrating peptides (CPPs) are accepted as a valuable tool in biological research. However, several basic questions regarding the behavior of the CPPs remain to be answered. What is clear is that the CPP phenomenon is universal: data from our laboratory show that uptake occurs in several species of yeast and bacteria, as well as insect cells and plant cells.[1–3] Thus, the outer membrane of cells can vary widely and still be permeable to CPPs.

Does the translocation mechanism of CPPs represent a new biological principle or is it a special case of the already known? Whatever the mechanism, seemingly cell penetration by macromolecules can be a distinct evolutionary disadvantage as it would enable an efficient means of toxin delivery. So why do cells allow cell penetration to proceed? Since several CPPs have been isolated from membrane-spanning proteins, a possibility is that the cell-penetration phenomenon is necessary for correct intermembrane folding.

Predicting function from any object without knowing the acting mechanism poses problems. Nevertheless, the prediction of new receptor ligands, without knowing the mode of binding, has met with considerable success. However, such structure-activity relationship (SAR) study requires a starting library of compounds with a corresponding measurement of a clean constant, such as receptor affinity.

Any quantitative parameter of CPP uptake characterized so far, either equilibrium or kinetic constants, is unfortunately a function of several independent interactions. Even simplified kinetic schemes, such as the one presented in Figure 5.1, are difficult to quantify experimentally. Thus, any prediction of CPP functionality would necessitate the independent measurement of several variables,

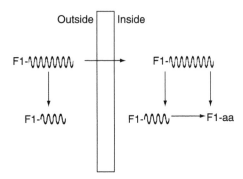

FIGURE 5.1 Simplified scheme of CPP uptake. The peptide degradation includes both an energy-dependent and -independent component.[4] The energy-dependent degradation process gives rise to a membrane impermeable metabolite, e.g., fluoresceinyl-Lys-OH, and thus contributes to the accumulation of intracellular label, while energy-independent processes produce various peptide fragments. Moreover, the peptide can be degraded to inactive fragments in the extracellular space. In addition, it is possible that any peptide fragment can pass the plasma membrane.

dependent on different structural features of the peptide, that together lead to the CPP phenomenon. To make the matter more complex, the different variables will differ for different cells, the method of measurement, etc. It is also not clear if all the peptides that are recognized as CPPs (or PTDs) use the same mechanism for cell entry (see Foreword).

Moreover, reviewing the literature on CPPs, it is immediately evident that results on uptake efficiency vary widely between different laboratories (e.g., from 100-fold enrichment of the peptide in the cytoplasm to equilibrium concentrations, comparable to the extracellular concentration). It is possible that the importance of the ratio of intracellular and extracellular volume for uptake, indicated in Ref. [4], can explain some of the large differences seen in accumulation of label.

In addition, a consensus on the contribution of classical endocytosis of extracellular material to the uptake efficiency is far from achieved. We have proposed that CPPs are in equilibrium over the plasma membrane, but that an irreversible step, i.e., degradation, significantly contributes to intracellular labeling.[4] However, information concerning the extracellular degradation of CPPs is scarce. Trehin and colleagues have characterized the degradation of some CPPs in the presence of differentiated epithelial cells, and found that the half-life varied from approximately 15 to 1500 min depending on peptide and cell type.[5] Thus, the intra- and extracellular degradation are probably critical determinants of CPP function.

One way to simplify the prediction process is to reduce the predicted variable to a binary, i.e., functional/nonfunctional representation. Thus, the model would be easier to construct but would, in turn, not yield a particularly detailed prediction.

Both for reasons of simplicity as well as the lack of any easy identifiable sequence similarities in the CPPs, it is fruitful to assume that the CPP property of the peptide is a bulk property, i.e., that it depends not on specific motifs, but rather on the net property of the amino acid residues in the peptide.

In general terms, the properties that are important for the membrane interaction(s) of the peptides are established.[6,7] The membrane interface region consists of the charged and neutral head groups of lipids, while deeper in the membrane interior, the properties are mostly dependent on the hydrocarbon chains of the lipids. Thus, it is reasonable to speculate that electrostatic and hydrophobic interactions should contribute to the initial process of cellular penetration.[8,9] This has been used practically to produce carriers through conjugation of lipophilic moieties with charged peptides.[7] No such general agreement exists on structural features that determine the intra- and extracellular proteolytic stability.

A general question in predicting CPPs function is what is to be considered the 'core' CPP and what is to be considered additional cargo. For instance, successive truncations performed on transportan[10] show that the peptide contains at least six active sequences, the shortest being transportan (7–27). A similar phenomenon has been seen for penetratin.[11] Thus, searching a protein, sequences corresponding to CPPs will appear clustered in blocks due to the existence of 'transport motors', i.e., shorter sequences with CPP characteristics, in the search window (a practical example is shown in Table 5.1).

Structurally, CPPs can be differentiated into three classes: the first are peptides with low amphipaticity where the charge contribution originates mostly from arginine residues. Examples of this class are penetratin[12] and Tat[13] peptides. The second class is peptides with a high degree of amphipaticity, where the charge contribution originates mainly from lysine residues. Examples of this class are MAP[14] and transportan.[15] In the third class, the charged and hydrophobic residues are separated lengthwise in the chain, exemplified by pVEC[16] and Pep-1.[17] There is

TABLE 5.1
Example of Transport Motors Found in the Fourth Intracellular Part of the Angiotensin 1A Receptor

Position	Sequence	Block Start
FKKYFLQ	Transport motor	
298	NPLFYGFLGKKFKKYFLQ	
299	PLFYGFLGKKFKKYFLQL	
300	LFYGFLGKKFKKYFLQLL	
301	FYGFLGKKFKKYFLQLLK	
302	YGFLGKKFKKYFLQLLKY	
303	GFLGKKFKKYFLQLLKYI	
304	**FLGKKFKKYFLQLLKY**IP	M511
305	LGKKFKKYFLQLLKYIPP	
306	GKKFKKYFLQLLKYIPPK	
307	KKFKKYFLQLLKYIPPKA	
308	KFKKYFLQLLKYIPPKAK	
309	FKKYFLQLLKYIPPKAKS	

some evidence in the literature that these peptide classes behave differently, especially concerning endocytotic uptake.

Reviewing the literature, sequences of CPPs have some similarities in terms of chemical properties. The first is related to the unique and surprising ability of these hydrophilic molecules to readily pass plasma membranes. If the CPPs were in fact highly hydrophobic, the function would be somewhat easier to explain as similar to the transport increase seen for lipid-conjugated biomolecules.[18] Hence, the first restriction that can be imposed is that the peptide should overall be fairly water-soluble, e.g., a $\Delta G_{W \to OCT} > 0$. This can quite accurately be determined using the whole residue hydrophobicity scales outlined in Ref. [19]. However, it has been proposed that the uptake efficiency of analogs of penetratin and tat, both highly charged, is mainly dependent on lipophilicity.[20] Besides the hydrophilicity of the peptide, a positive net-charge seems to be a requisite for CPPs function. Examination of the published peptides shows that the net charge varies between $+2$ and $+9$ units, not counting the contribution of the N-terminal ammonium ion. Indeed, CPPs have been isolated from proteins based on net-charge alone.[21] Specifically, Arg-rich domains from RNA binding proteins were isolated. A correlation of the Arg content of the peptide with the uptake efficiency was noted. Interestingly, peptides containing fewer than five Arg residues were much less active than CPPs. Although clearly successful, this method does not predict such well-defined CPPs as MAP or transportan.

In a quantitative structure-activity relationship (QSAR) study, the physico-chemical characteristics of compounds are translated into sets of descriptor variables. For peptides, it would be preferable if this could be done on an amino acid residue basis.

In 1991, Hellberg and colleagues presented three descriptive scales for the 20 coded amino acids, called z1, z2, and z3.[22] These scales were produced using a multitude of physicochemical variables, such as molecular weight, pKa, partition coefficients, etc.

With the three z-scales, it is possible to numerically quantify the structural variation within a series of related peptides by arranging the z-scales according to the amino acid sequence. Thus, QSARs relate positional information and physicochemical properties with a particular response. However, the large differences in the primary sequences of the CPP family make this less fruitful and perhaps there is no positional requirement for CPPs.

Reasoning that the CPP characteristic of the peptide is a bulk property, i.e., that it depends not on specific motives, but rather on the net property of the amino acid residues in the peptide, we assembled the averaged sums of various amino acid descriptors for published CPPs (Table 5.2). In order to restrict the descriptor intervals further, we used the averaged descriptors of a range of CPP analogs reported not to internalize (Table 5.2).

More recently, Sandberg and colleagues, building on Hellberg's classic work, published five such descriptor sets, Z_1–Z_5, covering 87 coded and noncoded amino acids.[23] In addition to these descriptors, the bulk of the amino acid, calculated as the number of heavy atoms (C, N, S, and O) in the side chains of the amino acids, was used. Moreover, the net donated hydrogen bonds of the amino acid, calculated as the

TABLE 5.2
Peptide Sequences Used for the Assembly of Descriptor Intervals

Sequences of Previously Reported CPPs	Sequences of Previously Reported Inactive CPP Analogs
GWTLNSAGYLLGKINLKALAALAKKIL	KKLSECLKRIGDELDS
RQIKIWFQNRRMKWKK	PVVHLTLRQAGDDFSR
KLALKALKALKAALKLA	EILLPNNYNAYESYKYPGMFIALSK
LLIILRRRIRKQAHAHSK	KKKQYTSIHHGVVEVD
AGYLLGKINLKALAALAKKIL	GWTNLSAGYLLGPPPGFSPFR
KLALKLALKAWKAALKLA	QNLGNQWAVGHLM
NAKTRRHERRRKLAIER	RPPGFSPFR
TRRNKRNRIQEQLNRK	LLKTTELLKTTELLKTTE
GGRQIKIWFQNRRMKWKK	LNSAGYLLGKALAALAKKIL
MGLGLHLLVLAAALQGAKKKRKV	LNSAGYLLGKLKALAALAK
KMTRAQRRAAARRNRWTAR	GWTLNSAGYLLGKINLKAPAALAKKIL
RVIRVWFQNKRCKDKK	LLKTTALLKTTALLKTTA
RKSSKPIMEKRRRAR	LRKKKKKH
VQAILRRNWNQYKIQ	RQIKIFFQNRRMKFKK
KRPAATKKAGQAKKKKL	GWTLNSAGYLLGKFLPLILRKIVTAL
MDAQTRRRERRAEKQAQWKAAN	KLALKALKAALKLA
TAKTRYKARRAELIAERR	
RQGAARVTSWLGRQLRIAGKRLEGRSK	
GRQLRIAGKRLEGRSK	
KCRKKKRRQRRRKKLSECLKRIGDELDS	
AAVALLPAVLLALLAPVQRKRQKLMP	
ALWKTLLKKVLKA	
PKKKRKVALWKTLLKKVLKA	
RQARRNRRRALWKTLLKKVLKA	
RGGRLSYSRRRFSTSTGR	

accepted hydrogen bonds of the side chains subtracted from the donated hydrogen bonds, were necessary in order to predict CPPs.[24] The value intervals for the descriptors, with the most predictive values for CPPs, are shown in Table 5.4.

TABLE 5.4
Descriptor Intervals Used for Searching CPPs

Value	Low	High
Bulk	3.2	5.9
$Z_1\Sigma/n$	−1.25	1.92
$Z_2\Sigma/n$	−1.22	1.29
$Z_3\Sigma/n$	−0.5	−1.94
$Z_{\Sigma hdbn}$	0.28	2

Despite this simplistic approach, the search criteria proved surprisingly useful for finding CPP properties in either proteins or random generated sequences. A number of protein and random sequences were searched for the sequences falling within the bulk property value ($Z\Sigma/n$) interval obtained from the training set. Using these search criteria we selected and tested several novel cell-penetrating sequences derived from natural proteins.

Synthetic peptides, derived from the intracellular loops of GPCRs, can influence receptor–G protein interactions with the original receptor in membrane preparations.[25–29] For the AT1AR, peptides corresponding to fragments of the C-terminal tail,[29,30] the third,[31] and even the second[32] intracellular loops are found by this method to affect signal transduction. For the GLP-1R, the third intracellular loop seems to be involved in transmitting the signal, whereas the first and second intracellular loops exhibit modulatory roles.[25,26] To allow entry of the peptides derived from GPCRs into intact cells, they have previously been coupled to CPPs.[33] In contrast, the method outlined here enables an all-in-one approach, where the receptor-derived peptide is in itself a CPP.

As proofs of principle, we synthesized and tested CPPs derived from 7 TM receptors.[34] Our results indicate that short CPP can be used to mimic activated receptors. Thus, the peptides require no additional carrier in order to shuttle into cell cytoplasm and exert their effect. Moreover, while commonly synthetic peptides act as decoys to inhibit functional protein interactions and inhibit signal transduction, the peptides we presented activate the target G proteins. This opens a new possibility for studying receptor functions, protein–protein interactions, and intracellular signal transduction.

5.2 METHODS

The published CPP prediction methods assume that the important properties of the peptides are bulk properties and not dependent on a specific motif. Choosing one method over the other is dependent on which type of CPP is sought. If a Tat-like peptide is the desired goal, a search for peptides encompassing a large positive net charge (e.g., >4) and as many Arg residues as possible should be performed. If transportan-like peptides are sought, the method outlined below should be used.

5.2.1 Peptide Search Criteria

First, we used the descriptors published by Sandberg and colleagues.[23] In addition, we found that the bulk of the amino acid side chain, calculated as the number of heavy atoms (C, N, S, and O) in the side chains of the amino acids, and the net donated hydrogen bonds of the side chain, calculated as the accepted hydrogen bonds of the side chains subtracted from the donated hydrogen bonds, were necessary in order to predict CPPs. The averaged sums of various amino acid descriptors, $z1$ through $z5$, for previously reported CPPs (Table 5.2) were assembled. In order to further restrict the descriptor intervals, the averaged descriptors of a range of CPP analogs reported not to internalize (Table 5.2) were used as negative controls. We found the optimal value intervals that included most CPPs and excluded a majority of the non-CPP analogs, and thus the most predictive values, to be those shown in Table 5.4.

Analysis of the CPP search among the previously reported CPP (for references see Ref. [35]) yields controversial results. Most of the reported CPPs fall into the CPP class according to our search criteria here, for example, but, the Tat peptide does not. Including the tat peptides in the training set yields descriptor intervals with little prediction power. Overall prediction success (calculated as correctly predicted positive CPPs subtracted by the percentage of wrongly predicted negative CPPs) is 90%.[24]

5.3   RESULTS

We decided to search for CPP sequences in two protein families: the G protein-coupled receptors and human transcription factor sequences. For the receptors, the predicted CPPs occur almost exclusively in the four intracellular parts of the 7TM receptors, with the largest part in the third intracellular loop (IC3).[24]

We experimentally verified the cell-penetrating property of sequences derived from seven different membrane spanning proteins: the GLP-1, AT1, D2(long), mGlu, and galanin receptors; the amyloid precursor protein and presenilin-1. We also searched random peptide sequences to produce novel CPPs.[24]

In an attempt to compare the uptake efficiencies of the CPPs, quantitative uptake measurements were performed. The labeling of the cells varied; however, all peptides reached intracellular concentrations that are above the extracellular concentration, thus indicating the diagnostic accumulation seen for CPPs. Possibly a partial explanation for the varied uptake of the peptides could be the differences in degradation, as been pointed out in Ref. [4].

The membrane disturbance of the peptides was investigated by examining the concentration and time dependence of LDH leakage. Here, only peptides with comparatively high amphipatic moment caused leakage above baseline.[24]

5.4   CONCLUSIONS

Here, using the expanded descriptor scales, bulk property values, Z_Σ, were assembled for 24 published CPP and 17 nonpenetrating analogs (the training set), and averaged over the total number of amino acids in the sequence. Using the criteria outlined by our approach, most of the known CPP sequences can be predicted. A number of protein and random sequences were searched for sequences falling within the bulk property value ($Z\Sigma/n$) interval obtained from the training set. Using these search criteria, we selected and tested several novel cell-penetrating sequences derived from natural proteins, indicating that the method can be used for de novo design of CPPs.

REFERENCES

1. Mäe, M. et al., Internalisation of cell-penetrating peptides into tobacco protoplasts, *Biochim. Biophys. Acta*, 1669(2), 101, 2005.
2. Holm, T. et al., Uptake of cell-penetrating peptides in yeasts, *FEBS Lett.*, 579(23), 5217, 2005.
3. Nekhotiaeva, N. et al., Cell entry and antimicrobial properties of eukaryotic cell-penetrating peptides, *FASEB J.*, 18(2), 394, 2004.
4. Hällbrink, M. et al., Uptake of cell-penetrating peptides is dependent on peptide-to-cell ratio rather than on peptide concentration, *Biochim. Biophys. Acta*, 1667(2), 222, 2004.
5. Trehin, R. et al., Metabolic cleavage of cell-penetrating peptides in contact with epithelial models: human calcitonin (hCT)-derived peptides, Tat (47–57) and penetratin (43–58), *Biochem. J.*, 382(Pt 3), 945, 2004.
6. Ladokhin, A.S. and White, S.H., Protein chemistry at membrane interfaces: nonadditivity of electrostatic and hydrophobic interactions, *J. Mol. Biol.*, 309(3), 543, 2001.
7. Carrigan, C.N. and Imperiali, B., The engineering of membrane-permeable peptides, *Anal. Biochem.*, 341(2), 290, 2005.
8. Magzoub, M. et al., Interaction and structure induction of cell-penetrating peptides in the presence of phospholipid vesicles, *Biochim. Biophys. Acta*, 1512(1), 77, 2001.
9. Magzoub, M. et al., Comparison of the interaction, positioning, structure induction and membrane perturbation of cell-penetrating peptides and nontranslocating variants with phospholipid vesicles, *Biophys. Chem.*, 103(3), 271, 2003.
10. Soomets, U. et al., Deletion analogues of transportan, *Biochim. Biophys. Acta*, 1467(1), 165, 2000.
11. Fischer, P.M. et al., Structure–activity relationship of truncated and substituted analogues of the intracellular delivery vector Penetratin, *J. Pept. Res.*, 55(2), 163, 2000.
12. Derossi, D. et al., The third helix of the Antennapedia homeodomain translocates through biological membranes, *J. Biol. Chem.*, 269(14), 10444, 1994.
13. Vivès, E. et al., A truncated HIV-1 Tat protein basic domain rapidly translocates through the plasma membrane and accumulates in the cell nucleus, *J. Biol. Chem.*, 272(25), 16010, 1997.
14. Oehlke, J. et al., Cellular uptake of an alpha-helical amphipathic model peptide with the potential to deliver polar compounds into the cell interior nonendocytically, *Biochim. Biophys. Acta*, 1414(1–2), 127, 1998.
15. Pooga, M. et al., Cell penetration by transportan, *FASEB J.*, 12(1), 67, 1998.
16. Elmquist, A. et al., VE-cadherin-derived cell-penetrating peptide, pVEC, with carrier functions, *Exp. Cell Res.*, 269(2), 237, 2001.
17. Morris, M.C. et al., A peptide carrier for the delivery of biologically active proteins into mammalian cells, *Nat. Biotechnol.*, 19(12), 1173, 2001.
18. Ekrami, H.M. et al., Water-soluble fatty acid derivatives as acylating agents for reversible lipidization of polypeptides, *FEBS Lett.*, 371(3), 283, 1995.
19. White, S.H. and Wimley, W.C., Membrane protein folding and stability: physical principles, *Annu. Rev. Biophys. Biomol. Struct.*, 28, 319, 1999.
20. Li, Y. et al., Correlation between hydrophobic properties and efficiency of carrier-mediated membrane transduction and apoptosis of a p53 C-terminal peptide, *Biochem. Biophys. Res. Commun.*, 298(3), 439, 2002.

21. Futaki, S. et al., Arginine-rich peptides. An abundant source of membrane-permeable peptides having potential as carriers for intracellular protein delivery, *J. Biol. Chem.*, 276(8), 5836, 2001.

22. Hellberg, S. et al., Minimum analogue peptide sets (MAPS) for quantitative structure–activity relationships, *Int. J. Pept. Protein Res.*, 37(5), 414, 1991.

23. Sandberg, M. et al., New chemical descriptors relevant for the design of biologically active peptides. A multivariate characterization of 87 amino acids, *J. Med. Chem.*, 41(14), 2481, 1998.

24. Hällbrink, M. et al., Prediction of cell-penetrating peptides, *Int. J. Pept. Res. Ther.*, 11(4), 249, 2005.

25. Bavec, A. et al., Different role of intracellular loops of glucagon-like peptide-1 receptor in G-protein coupling, *Regul. Pept.*, 111(1–3), 137, 2003.

26. Hällbrink, M. et al., Different domains in the third intracellular loop of the GLP-1 receptor are responsible for Galpha(s) and Galpha(i)/Galpha(o) activation, *Biochim. Biophys. Acta*, 1546(1), 79, 2001.

27. Rezaei, K. et al., Role of third intracellular loop of galanin receptor type 1 in signal transduction, *Neuropeptides*, 34(1), 25, 2000.

28. Bommakanti, R.K. et al., Extensive contact between Gi2 and N-formyl peptide receptor of human neutrophils: Mapping of binding sites using receptor-mimetic peptides, *Biochemistry*, 34(20), 6720, 1995.

29. Shirai, H. et al., Mapping of G protein coupling sites of the angiotensin II type 1 receptor, *Hypertension*, 25(4 Pt 2), 726, 1995.

30. Franzoni, L. et al., Structure of the C-terminal fragment 300–320 of the rat angiotensin II AT1A receptor and its relevance with respect to G-protein coupling, *J. Biol. Chem.*, 272(15), 9734, 1997.

31. Conchon, S. et al., The C-terminal third intracellular loop of the rat AT1A angiotensin receptor plays a key role in G protein coupling specificity and transduction of the mitogenic signal, *J. Biol. Chem.*, 272(41), 25566, 1997.

32. Thompson, J.B. et al., Cotransfection of second and third intracellular loop fragments inhibit angiotensin AT1a receptor activation of phospholipase C in HEK-293 cells, *J. Pharmacol. Exp. Ther.*, 285(1), 216, 1998.

33. Howl, J. et al., Intracellular delivery of bioactive peptides to RBL-2H3 cells induces beta-hexosaminidase secretion and phospholipase D activation, *Chembiochem*, 4(12), 1312, 2003.

34. Östlund, P. et al., Cell-penetrating mimics of agonist activated G-protein coupled receptors, *Int. J. Pept. Res. Ther.*, 11(4), 237, 2005.

35. Langel, Ü., ed., *Cell Penetrating Peptides. Processes and Applications*, CRC Press, Boca Raton, FL, 2002.

Part II

Mechanisms of Cell-Penetrating Peptides

Astrid Gräslund

The mechanism(s) by which cell-penetrating peptides (CPPs) enter living cells and mediate the entry of large cargo molecules which perform their prescribed biological functions inside the cell have been the subject of much speculation and many studies since the first reports on CPPs around 1990. The mechanism(s) are still not understood in any detail. A brief history of the CPP mechanism field has the following stages: In the early 1990s, the discoveries of the CPP phenomena led to proposals of direct membrane passage by more or less mysterious mechanisms. Endocytotic mechanisms were almost ruled out because the translocation could be observed under conditions of low temperature and ATP depletion. Membrane model systems came into use towards the 2000s, but observations there at first did not shed much light on the mechanisms.

Around 2002 came the sobering news[1,2] that many reports on the visualization of cellular uptake mediated by CPPs could have been affected by artifacts caused by cell fixation that was previously routinely employed to facilitate the studies. The field moved towards observations using live cells, and endocytotic mechanisms like lipid raft-mediated macropinocytosis[3] came into focus for the first time. As will be seen in the chapters in this section of the book, the results with different peptides and peptide/cargo systems suggest that a variety of mechanisms may be in operation, sometimes even in parallel. They are often driven by cell biology, such as the endocytotic mechanisms, but some obviously involve direct membrane interactions similar to, but more benign than, those described for antimicrobial peptides.

The former suggest a link to virus-mediated mechanisms, where a number of endocytotic pathways have been implicated.[4] The latter involving antimicrobial peptides have been subject to a large number of mechanistic studies leading to a growing understanding of the processes involved, including the formation of transient pores in the membrane.[5]

Recently, studies of CPPs in model membrane systems have begun to be better correlated with cell biology studies. Endosomal escape has become one key issue. Questions about the importance of an electrochemical or pH gradient have been asked and answers have started to come in from different systems. The roles of cell surface-bound glucosaminoglycans have also come into focus. The following chapters address the mechanistic aspects of CPPs from a variety of examples and techniques in cells and in model systems. They represent the current views of the authors and show that although many important questions remain unanswered, there is a growing consensus about which questions to ask and how to go about trying to answer them. The complexity of the issues also brings hope for achieving more specific responses and targeting of the CPP–cargo complexes in future applications.

An additional interesting issue deals with the dual concept of the CPP as a biotechnological tool versus mediator of a biologically important function that may be carried by a CPP sequence in its natural setting. In both cases, the questions about mechanisms are crucial, either for the further development of the tool for specific purposes or for the further understanding of the biological role of this peptide sequence with CPP properties. In the latter case, one could possibly distinguish between a case where the peptide is on its own, such as the dynorphin neuropeptides,[6] or where it is part of a protein (a messenger protein), like the N-terminus of the unprocessed prion protein[7] or the third helix of the homeodomain transcription factors (penetratin).[8]

REFERENCES

1. Lundberg, M. and Johansson, M., Positively-charged DNA-binding proteins cause apparent cell membrane translocation, *Biochem. Biophys. Res. Comm.*, 291, 367, 2002.
2. Richard, J.P., Melikov, K., Vives, E. et al., Cell penetrating peptides. A reevaluation of the mechanism of cellular uptake, *J. Biol. Chem.*, 278, 585, 2003.
3. Wadia, J., Stan, R.V., and Dowdy, S.F., Transducible TAT-HA fusogenic peptide enhances escape of TAT-fusion proteins after lipid raft macropinocytosis, *Nat. Med*, 10, 310, 2004.
4. Pelkmans, L. and Helenius, A., Insider information: What viruses tell us about endocytosis, *Curr. Op. Cell Biol.*, 15, 414, 2003.
5. Lee, M.-T., Hung, W.-C., Chen, F.-Y., and Huang, H.W., Many-body effect of antimicrobial peptides: On the correlation between lipid's spontaneous curvature and pore formation, *Biophys. J.*, 89, 4006, 2005.
6. Marinova, Z., Vukojevic, V., Surcheva, S. et al., Translocation of dynorphin neuropeptides across the plasma membrane, *J. Biol. Chem.*, 280, 26360, 2005.
7. Lundberg, P., Magzoub, M., Lindberg, M. et al., Cell membrane translocation of the N-terminal (1–28) part of the prion protein, *Biochem. Biophys. Res. Comm.*, 299, 85, 2002.
8. Joliot, A. and Prochiantz, A., Transduction peptides: From technology to physiology, *Nat. Med.*, 10, 310, 2004.

6 Inverted Lipid Models: A Pathway for Peptide Internalization?

Gérard Chassaing, Sandrine Sagan,
Olivier Lequin, Antonin Lamaziere,
Jesus Ayala-Sanmartin, Germain Trugnan,
Gérard Bolbach, and Fabienne Burlina

CONTENTS

6.1   INTRODUCTION

A reductionist and systematic study of endogenous and exogenous proteins has demonstrated the presence of protein transduction domains (PTDs) implicated in protein uptake. The synthesis and study of the peptides corresponding to these domains further allowed their identification as members of the cell-penetrating peptide (CPP) family. This is exemplified by the first studies of molecular dissections of TAT and Antennapedia proteins that led to the definition of small proteins able to enter cells: 60 residues for Antennapedia homeodomain (AntpHD)[1] and 72 residues for TAT.[2] Further dissections of these small proteins revealed that shorter and contiguous domains were sufficient to promote internalization: 16 residues for the helix III of AntpHD (pAntp(43–58), named penetratin)[3] and 36 residues for TAT (TAT(37–72)).[4] From these original data, many other shorter peptides were synthesized, containing either nine or 10 residues for TAT[5] and seven residues for pAntp(52–58).[6] Since these pioneer works, numerous short peptides derived from various proteins have been identified as CPPs with the use of fluorescent or biotinyl tags and detection by confocal microscopy or flow cytometry.[6] Some of the CPPs still translocate when coupled to hydrophobic or acidic cargos (e.g., phosphopeptides, oligonucleotides, peptide nucleic acids, drugs, etc.). Therefore, these peptides can be considered as Trojan peptides for their ability to deliver bioactive molecules inside cells.[7,8]

6.2   CONFLICTING DATA ON THE CELLULAR UPTAKE OF CPPS

6.2.1   CPP QUANTIFICATION

Although the concept of internalization is almost consensual and accepted,[9] numerous contradictory data have been reported for the uptake efficiencies and the intracellular locations of CPPs. In order to simplify the comparison of the literature data, we have defined a translocation coefficient, K_T, that corresponds to the CPP concentration ratio between the intra- and the extracellular media after 1 h of incubation: $K_T = [CPP]_{intra}/[CPP]_{extra}$. An intracellular volume of 1.5 pL (measured for CHO-K1 cells) was taken whatever the cell type. A K_T value greater than 1 reflects an intracellular accumulation of the CPP. The analysis of these K_T values calculated from the literature[6] revealed a wide variation from one study to another. For example, K_T values vary by two orders of magnitude for penetratin and three orders of magnitude for TAT. These contradictory data mainly result from an overestimation of the internalized peptide amount in some of the studies. This is exemplified by penetratin, for which it is rather surprising that pAntp(43–58) is internalized, but not pAntp(43–57).[6] Indeed, two independent alanine-scanning experiments led to different results on the role of basic and hydrophobic residues mainly in the N-terminal domain.[6]

6.2.2   INTRACELLULAR LOCATION OF THE CPPS

Data concerning the intracellular location of the CPPs also vary between studies. It is now clear that the fixation methods used for in vitro confocal microscopy assays may cause some peptide relocalization leading to an artefactual cytoplasmic accumulation. In living cells, most of the fluorophore-labeled CPPs are visualized in punctuated vesicular structures. However, comparison of the cellular distribution of fluorescein- and TAMRA-labeled non-CPP sequences has shown that the TAMRA probe by itself may promote accumulation in vesicular structures.[10]

In summary, at least six parameters could explain the major differences between data reported in the literature: (1) the detection methods that require chemical modifications (fluorescent probes, biotin, ...), which more or less modify the physical properties of the peptides, such as their membrane affinity; (2) the inaccurate distinction between the internalized and membrane-associated peptides, discrimination being only possible if the exchange between free and membrane-associated peptide is fast enough compared to the kinetic rates of the enzymatic or chemical reactions used to eliminate the membrane fraction; (3) the use of fixed or nonfixed cells, since during the fixation step, membrane impermeability can be lost and the peptide may diffuse, leading to an overestimation of the internalized peptide; (4) the potential peptide degradation that may occur either in lysosomes or in the cytoplasm by the proteasome; (5) the physicochemical properties of the CPP that may affect the detection of the fragments, as cellular efflux of CPP or CPP fragments may occur during the washing cycles; (6) competition between different uptake pathways that can vary with the cellular types and the CPP functionalization.

6.3   NEW METHODS FOR CPP UPTAKE QUANTIFICATION

6.3.1   CPP QUANTIFICATION

The methods of quantification based on radioactivity counting, biotinylation/cell-ELISA, fluorescence-labeling/spectrophotometer/FACS, resonance energy transfer, and HPLC detection and immunodetection have already been reviewed in chapter 12 of the previous book on cell-penetrating peptides.[6] Three new methods of quantification have been reported since. The first one is fluorescence correlation microscopy (FCM) carried out for quantification of fluorescent peptides. In contrast to other techniques, FCM allows the direct determination of the number of intracellular molecules of CPP when the extracellular concentration is around the nanomolar.[10] Using this technique, a K_T value of 0.6 was obtained for penetratin. The second method is the laser micropipette system (LMS), also called cell activity by capillary electrophoresis (CACE), that has been used to quantify the amount of PKB_S–TAT(49–57) and $CamK_{IIS}$–TAT(49–57) conjugates (PKB_S = GRPRAAT-FAEGC, a substrate of the protein kinase B; $CamK_{IIS}$ = KKALHRQETVDALC, a substrate of the Ca^{2+}/calmodulin-activated kinase).[11] A K_T value of 0.1 was obtained for PKB_S–TAT(49–57). In our initial internalization study, the penetratin was labeled by biotin and detected by FITC streptavidin.[12] We have also reported a method to directly analyze by MALDI-TOF MS biotinylated peptides that are extracted from a complex mixture using streptavidin-coated magnetic beads.[13] We have recently combined this technique with isotope labeling to quantify CPP cellular uptake.[14] The use of MALDI-TOF MS offers the advantage of direct peptide detection. Therefore, CPP intracellular degradation can be studied. In addition, the direct comparison of several CPPs cellular uptake efficiencies can be performed by using CPPs mixtures. The discrimination between internalized and membrane-bound peptide is the most delicate problem in CPP quantification. In the methods presented in the literature,[5,6] distinction has been performed by: (1) acid and high ionic strength washes to eliminate the membrane-adsorbed peptide, chemical modification of CPP amines by diazotized 2-nitroaniline followed by HPLC analysis, or trypsin digestion; (2) measurement of fluorescence differences between sodium azide-treated and control cells; (3) modifications of fluorescent properties of adsorbed peptides by dithionite quenching of fluorescence and fluorescence resonance energy transfer (FRET) between two fluorophores linked by disulfide bridge to the CPP (2-amino benzoic acid and 3-nitro-tyrosine). In the method based on MALDI-TOF MS, a trypsin treatment is used for the discrimination between internalized and membrane-bound peptide. Peptide quantification by MALDI-TOF MS can only be achieved by using as an internal standard (calibrant) the same peptide labeled with a stable isotope.[15] Indeed, both peptides having identical chemical structure exhibit the same efficiencies of desorption/ionization and detection. The relative intensity of the peptides $[M + H]^+$ signals therefore corresponds to the peptides relative proportion in the sample.[15] CPPs are functionalized on their N-terminus with biotin and an isotope tag composed of four nondeuterated glycine residues for the internalized species (H-CPP) and four deuterated glycine residues for the calibrant (D-CPP). Both peptide signals are separated by $\Delta m = 8$ on the mass spectra (Figure 6.1A). The amount of intact internalized peptide is calculated from the ratio between the areas of H-CPP and D-CPP signals ($[M + H]^+$ peak of the intact peptides) (Figure 6.1A). This method was first applied to the study of the three widely used CPPs: penetratin, TAT(48–59), and (Arg)$_9$.[14] The K_T values are 0.1, 0.6, and 0.9 for TAT(48–59), penetratin, and (Arg)$_9$, respectively (Figure 6.1B). These K_T values are in good agreement with the values reported by FCM and CACE techniques. When cells were incubated with a mixture of the three CPPs, lower K_T values were obtained but the relative CPPs internalization efficiencies remained unchanged compared to those calculated from the single CPP incubation experiments (Figure 6.1B). No cell lysis during incubation was observed. These data show that no competition occurs between the three CPPs for internalization and suggest that a limited number of internalization domains are available. Affinity of the CPPs for the membrane was evaluated by omitting in the protocol the trypsin treatment. Depending on the CPP, the amount measured corresponding to the sum of internalized and membrane-bound peptide was 5 to 60 times higher than the amount of internalized peptide (Figure 6.1C). These results show that there is a strong accumulation of peptide at the membrane surface. They stress the importance of the trypsin treatment, although there is no definitive proof that trypsin is able to digest all the membrane-associated peptide. These data also show that CPP internalization efficiency is not directly related to membrane affinity. Interestingly, the binding of TAT(48–59) to the cell surface was found to be very strongly inhibited when the two other CPPs were also present in the incubation

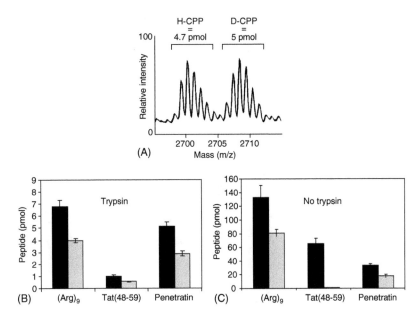

FIGURE 6.1 (A) Partial mass spectrum showing the $[M+H]^+$ signals of intact peptides and corresponding to the uptake of penetratin. (B) Total amount of intact internalized CPPs in 10^6 CHO cells. (C) Total amount of membrane-bound and internalized CPP for 10^6 CHO cells. In both cases the black bars correspond to cell incubation with each CPP independently (7.5 μM) and gray bars to cell incubation with a mixture of the three CPPs (7.5 μM each) for 75 min at 37°C.

mixture (Figure 6.1C). Using the MALDI-TOF MS method, we then compared the internalization efficiencies of four different homeoprotein-derived peptides: the third helix of Antennapedia or penetratin (H3Antp), Engrailed 2 (H3Eng), Hox A13 (H3Hox), and Knotted 1 (H3Kno) (Figure 6.2).[16] The helix III of Knotted was internalized seven times more efficiently than penetratin, corresponding to a K_T value of 3.5.

At this point, new problems of quantification related to the chemical nature of the CPP emerge and have to be highlighted. Indeed, since this quantification method requires trypsin degradation, membrane-bound D-amino acids-containing peptides and pseudo-peptides are hardly removed. Furthermore, since there is no way to prove that all the membrane-bound peptide has been cleaved by trypsin, an increase of the detected amount of one CPP compared to another may only reflect a greater affinity of the peptide for the membrane than its real cellular uptake. These problems can be solved by quantifying the cargo delivery instead of the free CPP uptake.

6.3.2 CARGO QUANTIFICATION, TROJAN PEPTIDES EFFICIENCIES

To demonstrate the role of CPPs as Trojan peptides, we proposed to quantify the cellular delivery of the cargo corresponding to the protein kinase C pseudo-substrate PKC peptide phosphorylation inhibitor (PKCi).[17] The cargo is functionalized by the biotin and the isotope tag composed of four nondeuterated glycine residues and is

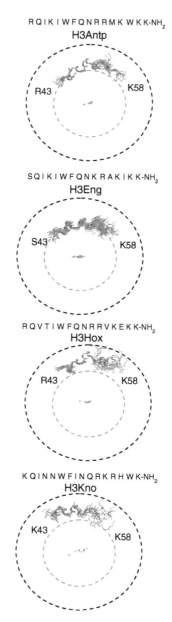

FIGURE 6.2 Sequences and schematic models of the four H3 helices of Antennapedia, Engrailed-2, Hox A13, and Knotted 1 corresponding to the superimposition of the best structures generated by X-PLOR and to their positions with respect to SDS micelle. The inner and outer circles correspond to 16 and 25 Å distances from micelle mass center (MMC), respectively. The + sign corresponds to the fluctuation of MMC during the fitting of backbone. Tangential average orientations were obtained for the four H3 helices. These locations agree with the parallel orientation observed for Antp in bicelles.

linked via a disulfide bridge to the CPP.[18] The cargo is expected to be released from the CPP after disulfide bond reduction inside cells. The species quantified by MALDI-TOF MS in this experiment thus corresponds to the free cargo (PKCi = biotin-GGGG-CRFARKGALRQKNV-NH$_2$). The PKCi peptide contains basic residues in its sequence so that distinction between the internalized and the membrane-bound peptide can still be performed by trypsin digestion. This method is used to compare the efficiency of different carriers to transport the same cargo. Since in this case it is always the same species that is quantified, problems related to differences of accessibility of the membrane-bound peptide to trypsin that can be encountered when quantifying the free CPP uptake are avoided here.

Surprisingly, the 0.06 K_T value obtained for the free PKCi peptide uptake is greater than the mean value expected for pinocytosis (0.01), but similar to the K_T value obtained for TAT(48–59) (0.1). Nonetheless, inhibition of phosphorylation in cultured cells (neurons) was only observed for PKCi–penetratin conjugates,[17] showing that the uncoupled PKCi peptide does not reach its cytosolic target. The attachment via a disulfide bridge of the PKCi peptide to (Arg)$_9$, penetratin, or H3Kno specifically increased its cellular internalization. The best K_T value (2.5) was obtained for the PKCi–H3Kno conjugate.[18]

One could emphasize that FCM,[10] CACE,[11] and MALDI-TOF MS[14,16,18] lead to lower intracellular concentrations of CPP compared to other methods. However, there is good agreement between the values obtained by these three techniques. The amount of peptide determined by FCM and MALDI-TOF analyses corresponds to the totality of peptide internalized in the cell, whereas the CACE method gives the cytosolic content. By subcellular fractionation, Zaro and Shen measured the free and vesicular compartment-associated amounts of radioactive CPP.[19]

6.4 MULTIPLE PATHWAYS OF INTERNALIZATION

6.4.1 FROM THE PLASMA MEMBRANE

The local disturbance of lipid assemblies of the plasma membrane (translocation) directly addresses peptides to the cytosol, such as antimicrobial peptides (APs). Four models have been proposed to rationalize the uptake and the lytic properties (APs): the helical bundle or barrel stave models for alamethicin; the carpet model for magainins, cecropins, etc., the detergent-like model for melittin, signal peptides, and amphipathic basic peptides and the lipid channel formation model for mastoparan and cecropin–melittin hybrids.[20] Except for the lipid channel formation, all these models describe a cytosolic internalization concomitant with a loss of membrane impermeability. Some APs, such as dermaseptin,[21] are internalized by mammalian cells without lytic activity. CPPs are also able to interact strongly with membranes without inducing membrane permeability. By analogy with the translocation of apocytochrome c[22] in mitochondria, we have proposed that the deformation of the bilayer plane leads to unstable reverse micelles.[12] During return to the initial state, part of the CPP associated with reverse micelles can be internalized. A spatial rearrangement of the reverse micelle could also lead to formation of peptide–lipid aggregates in the cytosol.[23] Similar transient complexes have already been observed in the presence of melittin resulting in equilibrium between discoidal structures and vesicles.[24] The strong peptide–membrane interaction can also induce membrane fusion by forming reverse micelles or H_{II} phase between two bilayers.[25] During the fusion process, the contents of the intermediate nonbilayer structures should be poured into the plasma of both cells. The absence of chiral recognition by a receptor is a common behavior shared by the CPP and AP families. The different models postulated for APs could also be valid for some CPPs; the major difference should concern the concentrations required to promote membrane permeability.

6.4.2 FROM THE ENDOSOMES

The intracellular location of CPPs after internalization by pinocytosis depends on the turnover and the traffic of the different types of endosomes and on the ability of the CPP to escape from these compartments. After internalization, the endocytosed molecules are delivered rapidly to sorting endosomes that return to the cell surface with a half-life of 2 min. A major recycling pathway involves a passage through the endocytic recycling compartment (ERC) that goes back to the plasma membrane with a half-life of 12 min.[26] Therefore, the rates of membrane turnover are of the same order as the kinetic of CPP internalization.[6] The different types of pinosomes are characterized by the initial recruitment of various intracellular proteins (AP2, AP180, EPS15, caveolin, clathrin, actin, syndapin, dynamin) and by their traffics.[27] Using a panel of specific inhibitors, the contribution of intracellular proteins to the endocytic pathways has been analyzed (see the corresponding chapters of this book). The effects of metabolic inhibitors on chlathrin-, caveolin-, actin-, or dynamin-dependent endosomes vary strongly with peptides or protein constructions, suggesting that different types of endosomes may simultaneously be involved in CPP uptake.

6.5 EFFECTS OF PEPTIDE–MEMBRANE INTERACTIONS IN MODEL SYSTEMS

The association of APs and CPPs with membranes induces the modification of several physical properties, such as the surface pressure of monolayer (Langmuir Blodgett), the secondary structure of the peptide (FT-IR, PM-IRRAS, CD, NMR), the deformation of the lipid assemblage due to an optimization of interactions between the hydrophobic and hydrophilic components (^{31}P and ^{2}H NMR, neutron reflectivity, X-ray diffraction, coupled plasmon waveguide resonance),[6,28] and the peptide location resulting in reciprocal adaptation (EPR, NMR, FRET).[6]

6.5.1 FOLDING OF PEPTIDES AT THE WATER–LIPID INTERFACE

Most of the APs and CPPs described in the literature adopt α-helical structures in contact with membrane mimics as shown by CD and NMR spectroscopy.

NMR studies have been performed with various membrane models extending from organic solvents to micelles and bicelles. The methanol and trifluoro-ethanol are sometimes considered to reproduce the dielectric constants at a water–membrane interface. Micelle systems are better models of water–lipid interface, but are characterized by a high curvature. Membrane bilayers are best mimicked by bicelle systems. The mixture of short- and long-chain phosphatidyl–choline lipids leads to a discoidal shape for the bicelles instead of the spherical shape for micelles. Only small, nonoriented bicelles are amenable to high-resolution structural studies by NMR. In contrast, CD studies are compatible with the use of these membrane models and also small unilamellar vesicles (SUVs). Polarized linear CD allows the orientation of amide groups in the plane or perpendicular to the bilayer surface to be distinguished. All NMR studies have shown that penetratin adopts a helix at the hydrocarbon–water interface (methanol, SDS micelles, and bicelle models).[33] CD studies of penetratin in bilayers (SUV) unveiled an equilibrium between α-helix and a β-structure that is shifted to β-structure with the increase in peptide concentration. We have recently studied by NMR different homeodomain-derived peptides to analyze the effects of sequence divergence on the peptide conformation and orientation in a membrane environment.[16] They correspond to the third helices of Antennapedia, Engrailed 2, Hox A13, and Knotted-1, a maize homeoprotein,[33] which possesses the most divergent sequence compared to the others (Figure 6.2). These peptides were found to exhibit different efficiencies of cellular uptake. NMR analysis indicated that they tend to adopt helical conformations in SDS micelles. The helical propensity decreases in the order H3Eng > H3Antp > H3Hox > H3Kno.[18§] Depending on the membrane models, the third helix of Antennapedia was found to adopt a β-hairpin (lipid film) or an α-helical structure in isotropic solvents (TFE, methanol) or in membrane–mimicking systems (micelle, bicelle, and SUV). These structural behaviors are also observed for many APs. For example, dermaseptin B2 adopts an α-helical structure in TFE and SDS micelle[34] and a β-structure on synthetic supports at the aqueous–solid interface.[35] Structural analysis by NMR spectroscopy and other biophysical techniques indicates that many antibiotic peptides (cecropins or magainins) strongly interact with lipid membranes. In bilayer environments, these peptides exhibit amphipathic α-helical conformations.[36]

The stabilization or destabilization of the peptide conformation by means of chemical modifications is the unique way to probe the structure requirements for peptide uptake. For instance, the introduction of a proline in position 50 of penetratin, [Pro^{50}]-Antp(43–58), induces a decrease in helix propensity, but the CD spectrum remains compatible with the presence of a β-turn stabilized by the Pro 50. The introduction of three prolines every five residues prevented both helix and β-hairpin formations and yet the [Pro45,50,55]-Antp(43–58) analog was still efficiently internalized.[12] A large panel of chemical modifications described in the literature is compatible with peptide uptake.[6,37–39] These data underline the fact that a particular three-dimensional structure should not be required for the peptide translocation.

6.5.2 Location of Peptide in Model Systems

6.5.2.1 In SDS Micelles and Bicelles Using Paramagnetic Probes

Many reports suggest that peptide location in the membrane may be a critical factor for internalization or membrane disruption. SDS micelles were taken as a water–membrane interface and the peptide location with respect to micelle surface was investigated using charged aqueous paramagnetic probes. Three paramagnetic probes, Mn^{2+}, 5-doxyl, and 12-doxyl stearic acids, have been used in several studies to locate penetratin in micelles and bicelles. Penetratin was previously found to adopt two different orientations, perpendicular and parallel, with respect to the surface of SDS micelles and bicelles, respectively.[6,29–32] However, the dynamics of the fatty acid chains leads to difficult interpretation of relaxation effects induced by lipid-containing doxyl groups. We have therefore chosen to analyze the proton relaxation enhancements using a progressive increase in Mn^{2+} concentration.[18] The Mn^{2+} ion is located in the aqueous phase in the vicinity of anionic head groups of SDS. The addition of this paramagnetic ion is known to cause selective broadening of resonances for residues exposed to solvent or close to the water–micelle interface. From the relaxation effects of Mn^{2+} in TOCSY experiments, it can be inferred that the depth of immersion within the SDS micelle increases in the order H3Eng > H3Antp > H3Hox > H3Kno.[16] Comparison of the decay of HN–H$^{\alpha}$ cross-peaks of individual residues also yields information on the peptide orientation with respect to the SDS micelle surface. For each peptide, the HN–H$^{\alpha}$ correlations for the N-terminal and C-terminal residues are not observed at a Mn^{2+} concentration of 4 mM, indicating that the peptide extremities are not deeply buried within the SDS micelle. Therefore, perpendicular orientations of all these peptides with respect to the SDS micelles can be dismissed. Schematic models of peptide position micelles are shown in Figure 6.2. Surprisingly, the least buried peptide H3Kno is internalized in cells most efficiently.[16] Similarly, the less efficient CPP [Phe48,56] Antp(43–58) was previously found to be more deeply inserted in the bicelle than Antp(43–58).[31]

6.5.2.2 In Bilayers by Photocrosslinking

We have applied the method of photocrosslinking to study peptide location. Under irradiation with UV light, the benzophenone and diazirine probes give highly reactive species, triplet radical and carbene, respectively. These probes should crosslink with any molecular species in their vicinity. Many studies have used photocrosslinking to determine the topography of the binding site of a peptide ligand in a protein receptor.[40,41] More recently, we have demonstrated the photocrosslinking between peptides and carbohydrates (β-cyclodextrin).[42] Photocrosslinking between peptides and lipids has been rarely used although the potentiality of this approach was demonstrated a long time ago.[43] Initially, peptide and cholesterol locations in bilayers have been analyzed using lipids bearing a photoprobe.[44,45] The location of transmembrane peptides into bilayers has been investigated by incorporating a photoprobe in the hydrophobic segment of a model transmembrane peptide.[46] We have recently studied two photo-activable derivatives of penetratin, PenBz1 (biotin-GGGG-(Bbz)KRQIKIWFQNRRMKWKK-NH$_2$) and PenBz2 (biotin-GGGG–RQI(Bbz)KIWFQN-RRMKWKK-NH$_2$) (Bbz = 4-Benzoylbenzoic acid). So far, in bilayers made of either PC and PE, the only crosslinked lipids identified by MALDI-TOF for PenBz1 were choline or ethanolamine heads, whereas PenBz2 crosslinked to glycerol (unpublished data). The observed photo-adducts agree with a penetratin location in the domain of the polar head of phospholipids and with an orientation of the peptide parallel to the bilayer surface.

Whatever the techniques and the membrane models used, all the data converge towards a location of penetratin at the water–membrane interface with a weak penetration in the hydrocarbon phase, its peptide backbone lying nearly parallel to the surface. A similar alignment of the helix axis parallel to the membrane surface has been reported for cecropins, magainins, and dermaseptins. This behavior is associated with carpets and detergent-like effects, and contrasts with the trans-membrane orientations observed for alamethicin or gramicidin A associated with the transmembrane helical bundle.[47]

6.5.3 LIPID POLYMORPHISM

6.5.3.1 Supramolecular Association of Lipids

The concept of lipid shape-structure allows prediction of the type of supramolecular assemblage of pure lipids.[47] When the cross-sectional area of the lipid headgroup is similar to the cross-sectional area of the acyl chains, these lipids self-assemble into a liquid lamellar phase (Lα), corresponding to group I (zwitterionic phosphatidyl-cholines [PC]; phosphatidylserines [PS]; cardiolipin; phosphatidyl glycerol [PG]; phosphatidic acid [PA]; and sphingomyelins [SM]). When the cross-sectional area of the head group is smaller, resulting in an conical shape (type II lipids: anionic phosphatidyl- and plasmogens-ethanol amines [PE]), the assemblage leads to a negative curvature such as in the inverted hexagonal phase (H$_{II}$). If the cross-sectional area of the head group is larger than that of the acyl chains (type III lipids: anionic glycosphingolipids, lysophospholipids), a positive curvature is favored, such as in micelles. The equilibria between lamellar and nonlamellar phases depend on the intrinsic properties of lipids, the nature of counterions, the temperature, the pH, and the hydratation of polar heads. A transition temperature T_H characterizes them. Even if biological membranes contain substantial amounts of group II and III lipids, it is postulated that they are included into a continuous bilayer. The NMR spectra of intact tumor cells exhibit proton signals corresponding to isotropically mobile lipids of still uncertain origin. Recent experiments suggest the contribution of plasma membrane microdomains.[31] A high local concentration of lipids of group II and III should tend to modify the surface curvature leading to a frustrated bilayer domain which could be in equilibrium with a nonlamellar phase. Exogenous molecules could shift this equilibrium towards inverted structures providing a pathway for lipids and polar molecules uptake or membrane fusion. For several peripheral membrane proteins it has been shown that their activities are increased in the presence of nonbilayer lipids.[47]

6.5.3.2 Peptide Modulation of Polymorphism

X-ray diffraction, ^{31}P-, ^{2}H-NMR, and freeze–fracture procedures are the classical techniques allowing a detailed structure analysis. Differential scanning calorimetry (DSC) measurements allow access to endothermic (or exothermic) phase transitions due to the first gel–liquid crystalline (T_C) and liquid crystalline to nonbilayer transition (T_H). ^{31}P-NMR is highly sensitive to lipid polymorphism, leading to quite different spectra depending on whether the lipid bilayers (Lα) exhibit a low-field shoulder and high-field peak separated by about 40 ppm. The hexagonal phases signal (H_{II}) is narrower by a factor of two with reversed asymmetry. Small particles (small unilamellar vesicles [SUV], micelles, inverted micelles, cubic, and rhomboic phases) in which isotropic motion occurs exhibit rather sharp signals (Ip). The hydrophobicity of the peptide is one of the major factors affecting the lipid polymorphism.[47] From the analysis of the hydrophobic mismatches between the helical peptide and the bilayer, it has been suggested that the shorter peptides promote the formation of inverse phases with phosphatidylcholine. Beside the formation of channels, the hydrophobic ionophore peptides such as gramicidin and alamethicin, promote inverted phase formation. Basic peptides including (Arg)$_9$ and TAT have the opposite effect compared to hydrophobic peptides, because they generally convert the hexagonal phases to lamellar phases. No isotropic phase has been detected for TAT peptide and (Arg)$_9$ on lipid mixture (PC/PG; molar ratio, 7:3).[48,49] The amphipathic helices of lytic peptides generally facilitate the formation of nonlamellar phases, either micelles, cubic or hexagonal phases. Since the bilayer to H_{II} phase-transition temperature T_H can be either lowered or raised, there is no clear relationship between the lytic action and T_H.[47] Penetratin is not a perfect amphipathic peptide because of its charge distribution. The lipid polymorphism effects have been analyzed with different lipid and organizations such as SUV made of a mixture of PC/PS (4:1).[23] The ^{31}P line-widths of SUV represent a good probe to detect fusogenic (increase of line-width) or mitogenic (decrease of line-width) properties. The addition of penetratin did not affect the line-widths of SUV (PC/PS, 4:1). However, with SUV (PC/PE, 3:1), the addition of penetratin induces two anisotropic resonances typical of bilayers and the hexagonal phase. Thus, the disruption of SUV vesicles by penetratin depends on lipid composition. The shapes of ^{31}P resonance in aqueous dispersions of PC/PS (4:1) exhibited a strong asymmetric shape typical of bilayer systems which remained constant in the presence of 3 mM pAntp analogs.[23] The aqueous dispersion of PC/PE mixture (5:4) corresponds to an equilibrium between bilayer and hexagonal phase which were shifted to the extended bilayers by the addition of penetratin. This behavior is typical of basic peptides. The decrease of high-field peak ^{31}P resonance on a crude lipid dispersion extracted from rat brain embryos emphasizes that other components, such as phospholipids, gangliosides, and cholesterol, can play a major role in the equilibrium between bilayer and hexagonal phase.[23] In liquid NMR, the small difference in chemical shifts between isotropic peak and the low-field peak of

H_{II} phase does not allow discrimination between micelle, inverted micelle, cubic, and rhomboic phases.

6.5.4 BILAYER DEFORMATION

Beside the modifications on the phospholipids structuration of membrane bilayers (Ld, Lo, H_{II}, etc.) observed in micelles, bicelles, and small unilamelar vesicles (SUVs), CPPs and APs are able to induce macroscopic modifications on membranes. Several works using large and giant unilamellar vesicles (LUV and GUV) have been published. LUVs are vesicles of 100 nm usually obtained by extrusion of multilamellar vesicles, and GUVs are vesicles larger than 10 μm diameter that can be obtained by spontaneous formation or preferentially by electroformation. The curvature of the bilayers can be important for membrane behavior, and LUVs could be regarded as models for small, highly curved intercellular vesicles, such as synaptic vesicles and GUVs, as models of the less curved plasma membrane.

It is clear from the literature that there is no internalization in SUV and that the internalization in LUV is slow and depends on the lipid composition and on the trans-bilayer potential.[50] The formation of pores is associated to the calcein leakage from LUVs and can be detected by fluorescence spectroscopy. By using sensitive resonance energy transfer measurements in GUVs, it has been shown that different penetratin analogs are able to cross the membranes.[51,52] Thorén et al.[6] have shown that in GUV stabilized by 1% of glycerol, penetratin internalization is fast and reversible. Other studies have shown that some basic peptides can dramatically change giant vesicle morphology. Indeed, Kinnunen's group has identified two behaviors for four APs on PC/PG (9:1) GUVs: they first detected formation of large endocytosis-like vesicles with temporin B and magainin 2 ($\phi \approx 8 \ \mu$m), whereas indolicidin and temporin L led to a carpet formation followed by small vesicle formation ($\phi \approx 0.9 \ \mu$m).[53,54] Interestingly, Menger et al.[55] have observed on PC/PG (9:1) GUV that the addition of poly-lysine induces major deformations of vesicle membranes. The perpendicular deformation of bilayer induced by poly-lysine on the GUVs, surface forms ropes which could deliver peptide by a concerted opening–closing mechanism.

In order to study peptide specificities, we have analyzed the morphological effects induced by several CPPs on GUVs made of PC/PG in 9:1 molar ratio and generated by electroformation. We have observed no morphological effects by addition of 0.5 mM of substance P (RPKPQQFFGLM-NH$_2$) even after 12 h. On the contrary, as shown in Figure 6.3, the addition of 0.5 μM of penetratin, amphipathic peptide MAP-WR (RRWRRWWRRWWRRWRR-NH$_2$) or (Arg)$_9$ leads to the formation of ropes inside giant vesicles. These three peptides differ in their structural parameters such as charge, conformation, or hydrophobicity, but their effects on GUVs seem very similar in terms of mode of action or in rope formation kinetic (several minutes). Beside, the addition of peptide MAP-LR (RRLRRLLRRLLRRLRR-NH$_2$) on the same giant vesicles, prevents the formation of ropes and induces a fast decrease of vesicle size. This phenomenon may be related to pore formation on GUV bilayers.

6.6   INVERTED LIPID MODELS

The physical properties of many APs and CPPs are so similar that they cannot simply explain the conservation or loss of membrane impermeability and the uptake or nonuptake in mammalian cells. The APs often have high hydrophobic moment (μ) values, but there is no simple correlation between their hydrophobic moments and lytic activities.[56] In Bowes human melanoma cells, leakage of 2-deoxy-D-1[^{3}H]-glucose-6-phosphate increases with the hydrophobic moment of the CPP. In other cell lines, lactate dehydrogenase, hemoglobine, calcein-AM leakage and ethidium homodimer-1 entry have been compared. All CPPs present a very weak cytotoxicity compared to mastoparan.[57] However, the introduction of a fluorophore (rhodamine) on the CPPs (penetratin, TAT, transportan, and poly-arginine) increases their toxicity more or less strongly.[58] The hydrophobic moment along the peptide sequence is strongly modified by the position of the hydrophobic fluorophore. This modification affects the peptide–membrane interactions, inducing the partial loss of membrane impermeability or a change in the internalization efficiency.[6] The four most popular CPPs, penetratin, TAT, $(Arg)_{11}$, and transportan, functionalized by rhodamine become lytic at a concentration of around 10 μM, close to the efficient concentration of most of the APs. The addition of a hydrophobic fluoroprobe may shift the CPP–membrane interaction towards one of the models described to rationalize the AP–membrane interactions (carpets, detergent-like effects, transmembrane helical bundle, etc.). Many data suggest that fluorescent probes could also favor the endocytosis-mediated internalization, resulting in a retention of fluorescence in the endosomic compartments. It has also been shown that the intracellular localization and distribution of fluorescent CPPs depend on the nature of the fluoroprobe.[59] All these results suggest that data obtained from confocal microscopy using fluorescent peptides must be interpreted with caution. In contrast, substitution of a hydrophobic fluorophore by a hydrophilic cargo (hydrophilic peptide or acidic oligonucleotide) seems to lead to direct transport into the cytosolic compartment by an unknown mechanism. The sequence distributions of hydrophobic, hydrophilic, and charged residues determine the three-dimensional structure (α helices, β strands, and random coil), the states of aggregation, the location (membrane surface/interface/hydrophobic phase), and the orientation (parallel, oblique, perpendicular) of peptides. The complementarity between the peptides and phospholipids charges favors the insertion of peptides in mono- or bi-layers and a minimum length of 16 amino acids is required for an α-helix to adopt a transmembrane insertion. The reciprocal or nonreciprocal wedge shapes between peptides and phospho- or glyco- lipids allow the qualitative prediction of their effects on: (1) the gel-to-liquid crystalline transition temperature; (2) the lipid organization by inducing nonlamellar phases (inverted micelles, inverted hexagonal phase, cubic phases, isotropic phases) or by creating discontinuities such as barrel-stave pores (peptide inserted into bilayers with its hydrophobic face lining the water-filled pores); and (3) the transient toroidal holes or fluid isotropic complexes in a detergent-like manner.

^{31}P-NMR data support the formation of lipid-inverted structures for some biconstituents lipidic mixtures (PC/PE) by CPPs, as well as by some APs. This lipid-inverted structure has been invoked to explain both the cytotoxicity of APs

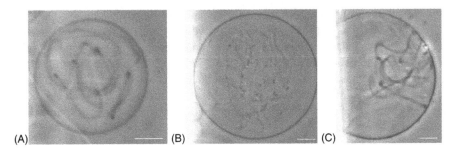

FIGURE 6.3 Rope formation in anionic giant unilamellar vesicles (PC/PG, 90:10%mol) after addition of three different peptides: (A) penetratin; (B) $(Arg)_9$; (C) MAP-RW by phase contrast microscopy (bar 10 μm).

and the translocation of CPPs. The APs and CPPs bind strongly to negatively charged membrane models. The induction of lipid polymorphism and the formation of ropes that evolve into vesicles have been observed for different CPPs when using GUVs made of PC/PG. Different behaviors have been reported for APs. However, since the number of peptides and lipid mixtures analyzed so far is limited, it is difficult to assert that the ropes indeed correspond to an intermediate structure in the cellular internalization.

In living cells, it is impossible to follow the lipid polymorphism of hundreds of molecular species. It is clear that the phospho-, glycero-, glyco-, sphingolipids and cholesterol components are not randomly distributed in the half leaflet of the cellular membranes. This vertical phase separation between the outer and the inner leaflets is amplified by the interactions with cytoskeleton proteins. It is now admitted that lateral phase separations also occur that allow the formation of lipid clusters in different physical states: gel phase (Lβ), lamellar fluid phase (Lα) which can be in liquid-ordered phase (Lo), or liquid disordered phase (Ld). These phase separations form heterogeneous membrane domains such as the transient confinement zones, caveolae, and rafts enriched in cholesterol and sphingolipids. Some of these lipid domains organized in lamellar phase recycle between the plasma membrane and intracellular endocytic compartments that convey exogenous molecules. However, beside this process, we suggest that the membrane permeability towards CPPs and APs should also be due to the instability of lamellar phases in some membrane subdomains leading to the formation of inverted micelles, ropes, or isotropic aggregates. In this hypothesis, every CPP and AP should present different affinities for these various domains. The selectivity for one domain should depend on the sequence of the CPP and AP, the type of fluorophore (generally hydrophobic), the nature of the cargo (peptides, proteins, PNAs, oligonucleotide), the presence of hydrophilic and hydrophobic domains, and the type of charges. This hypothesis rationalizes various apparent contradictory results; however, the challenge is to identify the molecular species associated with the internalization and the cytotoxicity of APs or CPPs. The crosslinking of membrane components (cholesterol, plasmalogen, sphingolipids, glycolipids, and carbohydrates) by APs and CPPs bearing a photo-activable probe represents an interesting way to study the presence, the implications, and the phase behavior of lipid subdomains. On CHO cells, using MALDI-TOF MS analysis, we have not observed a direct relationship between the amounts of membrane-associated and internalized CPP, showing that not all the membrane domains of CPP association to the membrane are able to induce cellular uptake. It will thus be essential in photocrosslinking studies to use an approach that distinguishes between membrane domains only involved in CPP adsorption and domains also promoting internalization.

ACKNOWLEDGMENTS

We thank the FNS and CNRS (ACI Dynamique et Réactivité des Assemblages Biologiques DRAB04/030) and the Conseil Régional d'Ile de France for their financial support. We also thank Solange Lavielle, Odile Convert, Stéphane Balayssac, Fivos Marcopoulos, and Hélène Tran for their contributions. We acknowledge Alain Prochiantz and Alain Joliot for fruitful discussions.

REFERENCES

1. Joliot, A. et al., Antennapedia homeobox peptide regulates neural morphogenesis, *Proc. Natl. Acad. Sci. U.S.A.*, 88, 1864, 1991.
2. Mann, D.A. and Frankel, A.D., Endocytosis and targeting of exogenous HIV-1 Tat protein, *EMBO J.*, 10, 1733, 1991.
3. Derossi, D. et al., The third helix of the Antennapedia homeodomain translocates through biological membranes, *J. Biol. Chem.*, 269, 10444, 1994.
4. Fawell, S. et al., Tat-mediated delivery of heterologous proteins into cells, *Proc. Natl. Acad. Sci. U.S.A.*, 91, 664, 1994.
5. Brooks, H. et al., Tat peptide-mediated cellular delivery: Back to basics, *Adv. Drug Delivery Rev.*, 57, 559, 2005, and references cited herein.
6. Langel, Ü., ed, In *Cell-Penetrating Peptides: Processes and Applications*, CRC Press, Boca Raton, FL, 2002, and references cited herein.
7. Derossi, D. et al., Trojan peptides: The penetratin system for intracellular delivery, *Trends Cell Biol.*, 8, 84, 1998.
8. Dietz, G.P.H. and Bähr, M., Delivery of bioactive molecules into the cell: The Trojan horse approach, *Mol. Cell. Neurosci.*, 27, 85, 2004.
9. Krämer, S.D. et al., No entry for TAT(44–57) into liposomes and intact MDCK cells: Novel approach to study membrane permeation of cell-penetrating peptides, *Biochim. Biophys. Acta*, 1609, 161, 2003.
10. Fischer, R. et al., A quantitative validation of fluorophore-labelled cell-permeable peptide conjugates: Fluorophore and cargo dependence of import, *Biochim. Biophys. Acta*, 1564, 365, 2002.
11. Soughayer, J.S. et al., Characterization of TAT-mediated transport of detachable kinase substrates, *Biochemistry*, 43, 8528, 2004.
12. Derossi, D. et al., Cell internalization of the third helix of the Antennapedia homeodomain is receptor-independent, *J. Biol. Chem.*, 271, 18188, 1996.
13. Girault, S. et al., Coupling of MALDI-TOF mass analysis to the separation of biotinylated peptides by magnetic streptavidin beads, *Anal. Chem.*, 68, 2122, 1996.

14. Burlina, F. et al., Quantification of the cellular uptake of cell-penetrating peptides by MALDI-TOF mass spectrometry, *Angew. Chem.*, 44, 4244, 2005.
15. Ong, S.E. and Mann, M., Mass spectrometry-based proteomics turns quantitative, *Nat. Chem. Biol.*, 1, 252, 2005, and references cited herein.
16. Balayssac, S. et al., Interaction in a membrane-mimicking environment and cellular uptake of cell-penetrating peptides: comparison of penetratin and peptides derived from other homeodomains, *Biochemistry*, 45, 1408, 2006.
17. Theodore, L. et al., Intraneuronal delivery of protein kinase C pseudosubstrate leads to growth cone collapse, *J. Neurosci.*, 15, 7158, 1995.
18. Aussedat, B. et al., Quantification of efficiencies of cargo delivery by peptidic or pseudo-peptidic trojan carriers using maldi-tof mass spectrometry, *Biochim. Biophys. Acta*, 1758, 375, 2006.
19. Zaro, J.L. and Shen, W.C., Quantitative comparison of membrane transduction and endocytosis of oligopeptides, *Biochem. Biophys. Res. Commun.*, 307, 241, 2003.
20. Bechinger, B., Structure and functions of channel-forming peptides: Magainin cecropins melittin and alamethicin, *J. Membr. Biol.*, 156, 197, 1997, and references cited herein.
21. Hariton-Gazal, E. et al., Targeting of nonkaryophilic cell-permeable peptides into the nuclei of intact cells by covalently attached nuclear localization signals, *Biochemistry*, 41, 9208, 2002.
22. Rietveld, A. et al., Studies on the lipid dependency and mechanism of the translocation of the mitochondrial precursor protein apocytochrome c across model membrane, *J. Biol. Chem.*, 261, 3846, 1986.
23. Berlose, J.P. et al., Conformational and associative behaviours of the third helix of antennapedia homeodomain in membrane mimetic environments, *Eur. J. Biochem.*, 242, 372, 1996.
24. Dufourc, E.J. et al., Reversible disc-to-vesicle transition of melittin-DPPC complexes triggered by phospholipid acyl chain melting, *FEBS lett.*, 201, 205, 1986.
25. Siegel, D.P., Energetics of intermediates in membrane fusion: Comparison of stalk and inverted micellar intermediate mechanisms, *Biophys. J.*, 65, 2124, 1993.
26. Hao, M. and Maxfield, F.R., Characterization of rapid membrane internalization and recycling, *J. Biol. Chem.*, 275, 15279, 2000.
27. Conner, S.D. and Schmid, S.L., Regulated portals of entry into the cell, *Nature*, 422, 37, 2003 references cited in here
28. Salamon, Z. et al., Plasmon-waveguide resonance and impedance spectroscopy studies of the interaction between penetratin and supported lipid bilayer membranes, *Biophys. J.*, 84, 1796, 2003.
29. Magzoub, M. et al., Comparison of the interaction, positioning, structure induction and membrane perturbation of cell-penetrating peptides and nontranslocating variants with phospholipid vesicles, *Biophys. Chem.*, 103, 271, 2003.
30. Binder, H. and Lindblom, G.A., Molecular view on the interaction of the Trojan peptide penetratin with the polar interface of lipid bilayers, *Biophys. J.*, 87, 332, 2004.
31. Lindberg, M. et al., Structure and positioning comparison of two variants of penetratin in two different membrane mimicking systems by NMR, *Eur. J. Biochem.*, 270, 3055, 2003.
32. Zhang, W. and Smith, S.O., Mechanism of penetration of Antp(43–58), *Biochemistry*, 44, 10110, 2005.
33. Tassetto, M. et al., Plant and animal homeodomains use convergent mechanisms for intercellular transfer, *EMBO Rep.*, 6, 885, 2005.

34. Lequin, O. et al., Helical structure of dermaseptin B2 in a membrane-mimetic environment, *Biochemistry*, 42, 10311, 2003.
35. Noinville, S. et al., Conformation, orientation, and adsorption kinetics of dermaseptin B2 onto synthetic supports at aqueous/solid interface, *Biophys. J.*, 85, 1196, 2003.
36. Bechinger, B., The structure, dynamics and orientation of antimicrobial peptides in membranes by multidimensional NMR spectroscopy, *Biochim. Biophys. Acta*, 1462, 157, 1999, and references cited herein.
37. Umezawa, N. et al., Translocation of a β-peptide across cell membranes, *J. Am. Chem. Soc.*, 124, 368, 2002.
38. Wender, P.A. et al., Oligocarbamate molecular transporters: Design, synthesis, and biological evaluation of a new class of transporters for drug delivery, *J. Am. Chem. Soc.*, 124, 13382, 2002.
39. Fillon, Y.A. et al., Cell penetrating agents based on a polyproline helix scaffold, *J. Am. Chem. Soc.*, 127, 11798, 2005.
40. Girault, S. et al., Localization of the substance P binding site in the human NK-1 tachykinin receptor with photolabelled peptides, *Eur. J. Biochem.*, 240, 215, 1996.
41. Sachon, S. et al., Met174 side chain is the site of photoinsertion of a competitive peptide antagonist of the substance P receptor photoreactive in position 8, *FEBS Lett.*, 544, 45, 2003.
42. Jullian, V. et al., Carbon–carbon bond ligation between cyclodextrin and peptide by photo-irradiation, *Tetrahedron Lett.*, 44, 6437, 2003.
43. Breslow, R. et al., Photo reactions of charged benzophenone with amphiphiles in micelles and multicomponent aggregates as conformational probes, *J. Am. Chem. Soc.*, 100, 3458, 1978.
44. Delfino, J. et al., Design, synthesis, and properties of a photoactivatable membrane-spanning phospholipidic probe, *J. Am. Chem. Soc.*, 115, 3458, 1993.
45. Deseke, E. et al., Intrinsic reactivities of amino acids towards photoalkylation with benzophenone. A study preliminary to photolabeling of the transmembrane protein glycophorin A, *Eur. J. Org. Chem.*, 2, 243, 1998.
46. Ridder, A.N. et al., Photo-crosslinking analysis of preferential interactions between a transmembrane peptide and matching lipids, *Biochemistry*, 43, 4482, 2004.
47. Epand, R.M., Lipid polymorphism and protein–lipid interactions, *Biochim. Biophys. Acta*, 1376, 353, 1998, and references cited herein.
48. Gonçalves, E. et al., Binding of oligoarginine to membrane lipids and heparan sulfate: Structural and thermodynamic characterization of a cell-penetrating peptide, *Biochemistry*, 44, 2692, 2005.
49. Ziegler, A. et al., Protein transduction domains of HIV-1 and SIV Tat interact with charged lipid vesicles, binding mechanism and thermodynamic analysis, *Biochemistry*, 42, 9185, 2003.
50. Terrone, D. et al., Penetratin and related cell-penetrating cationic peptides can translocate across lipid bilayers in the presence of a transbilayer potential, *Biochemistry*, 42, 13787, 2003.
51. Persson, D. et al., Vesicle size dependent translocation of penetratin analogs across lipid membranes, *Biochim. Biophys. Acta*, 1665, 142, 2004.
52. Thorén, P.E.G. et al., Membrane binding and translocation of cell-penetrating peptides, *Biochemistry*, 43, 3471, 2004.
53. Zhao, H. et al., Comparison of the membrane association of two antimicrobial peptides, magainin 2 and indolicidin, *Biophys. J.*, 81, 2979, 2001.
54. Zhao, H. et al., Interactions of the antimicrobial peptides temporins with model membranes. Comparison of temporins B and L, *Biochemistry*, 41, 4425, 2002.

55. Menger, F.M. et al., Migration of poly-L-lysine through a lipid bilayer, *J. Am. Chem. Soc.*, 125, 2846, 2003.
56. Dathe, M. and Wieprecht, T., Structural features of helical antimicrobial peptides: Their potential to modulate activity on model membranes and biological cells, *Biochim. Biophys. Acta*, 1462, 71, 1999.
57. Saar, K. et al., Cell-penetrating peptides: A comparative membrane toxicity study, *Anal. Biochem.*, 345, 65, 2005.
58. Jones, S.W. et al., Characterisation of cell-penetrating peptide-mediated peptide delivery, *Br. J. Pharmacol.*, 145, 1093, 2005.
59. Szeto, H.H. et al., Fluorescent dyes alter intracellular targeting and function of cell-penetrating tetrapeptides, *FASEB J.*, 19, 118, 2005.

7 Membrane Interactions of Cell-Penetrating Peptides

Elin K. Esbjörner, Astrid Gräslund, and Bengt Nordén

CONTENTS

7.1   INTRODUCTION

While cell studies repeatedly provide evidence that cell-penetrating peptides (CPPs) can deliver biochemically functional cargo into cells, the molecular mechanisms by which the import occurs are still shrouded in mystery. The initial belief that CPP–cargo constructs managed to enter cells via direct lipid interactions has somewhat come to nought after the discovery that studies using fixed cells may have caused artifactual results due to the redistribution of peptide.[1,2] The reevaluation of cell uptake has pointed to the importance of endocytosis in CPP cell entry, especially when CPPs are conjugated to cargoes.[3–6] Still, the concept of peptide translocation through a lipid membrane remains important because subsequent studies on live cells have proven that entry can occur when endocytosis is shut down.[7] Furthermore, CPP–cargo constructs that internalize via endocytosis must indeed be able to escape degradation (hence cross the endosomal membrane) in order to assert biological effects. CPPs that are reported to use direct membrane translocation mechanisms are usually rich in arginines or sufficiently hydrophobic to form pores (e.g., transportan). However, the archetypal CPP penetratin seems to enter cells entirely by endocytosis.

Studying CPPs in model membranes becomes important for understanding CPP–membrane interactions on a molecular level. To date, our studies have mostly focused on peptide–membrane interactions that can explain how CPPs may translocate a membrane without destroying it (i.e., how CPPs can enter cells or exit endosomes without being toxic). Understanding how peptide–membrane inter-actions come into play in shuttling CPPs via different endocytotic routes becomes increasingly important in explaining why certain CPPs are more efficient than others. It is possible that the ability to disturb the membrane, or the ability to bind to and thereby recruit specific membrane components, is important to awaken the endocytotic machinery. Binding to, for example, heparan sulfates on the cell surface rather than to the lipids as such has in fact been found to be important in cell entry.[8]

In this chapter we review our current understanding of CPP conformation in, and interactions with, model membranes. We also try to connect structural and chemical characteristics of CPPs to possible uptake mechanisms in order to obtain a biophysical classification that can predict behavior in cell systems. Further, we describe techniques that have been developed in our laboratories and which are particularly useful for studying CPPs. We also discuss the usefulness of model membrane systems in the context of understanding how CPPs react with live cells. A list of the CPPs discussed in this chapter together with information on their origin and the amino acid sequences is provided in Table 7.1.

7.1.1   MODEL MEMBRANES

Since cell membranes are far too complex to be suitable for most spectroscopic techniques and other biophysical methods, simplified model membranes need to be used in order to obtain any relevant information on CPP–lipid interactions

TABLE 7.1

The Origin and Amino Acid Sequence of CPPs Discussed in This Chapter

Peptide	Origin	Sequence
Penetratin	Antennapedia homeodomain	RQIKIWFQNRRMKWKK
PenArg	Arg-enriched penetratin	RQIRIWFQNRRMRWRR
PenLys	Lys-enriched penetratin	KQIKIWFQNKKMKWKK
Pen2W2F	Tryptophan-substituted penetratin	RQIKIFFQNRRMKFKK
Transportan	(galanin-mastoparan)	GWTLNSAGYLLGKINLKALAALAKKIL
TatP59W	HIV-1 protein (48–60) Pro for Trp substitution at position 59	GRKKRRQRRRPWQ
TatLysP59W	Lys-enriched TatP59W	GKKKKKQKKKPWQ
R₇W	Oligoarginine with a terminal Trp residue	RRRRRRRW
MPrPp	N-terminus of mouse PrP (residues 1–28)	MANLGYWLLALFVTMWTDVGLCKKRPKP
BPrPp	N-terminus of bovine PrP (residues 1–30)	MVKSKIGSWILVLFVAMWSDVGLCKKRPKP

Note: Tryptophan residues are in bold while arginines are underlined, and lysines are in italic.

on a molecular level. The cell membrane is a multifaceted matrix composed of a wide variety of lipids, it embeds fairly large amounts of proteins and is coated with carbohydrates.

In optical spectroscopy, large size is a major problem since membrane vesicles may cause scattering effects that distort spectra in a way that is not possible to correct by just subtracting a blank spectrum. Unilamellar lipid vesicles can be easily prepared from natural and synthetic lipid preparations and are suitable cell models because they encompass a double bilayer and a fairly large inner volume. If light scattering is not a severe problem, large unilamellar vesicles (LUVs) are preferred to smaller liposomes (SUVs, diameter around 30–40 nm) due to their smaller curvature and better stability (LUVs are normally stable for days at room temperature). Still, LUVs have far smaller dimensions than the plasma membrane but their size is comparable to endosomes. Giant unilamellar vesicles (GUVs) can be the size of a cell or even larger ($>1 \mu m$ diameter) and are observable under the microscope, facilitating direct observations of, for example, membrane translocation. A drawback with GUVs is a lack of long-term stability as they tend to collapse on the cover glass surface and show leakage on fairly short timescales (hours).

In nuclear magnetic resonance (NMR), an even smaller membrane than SUVs needs to be used to have sufficiently fast tumbling. Micelles are often used as membrane mimetics, but their high curvature and lack of double bilayer make them rather artificial membrane models. More recently developed are the so-called bicelles, best described as elongated micelles encompassing a small double-bilayer stretch. The stability of bicelles is a problem and the possibility to vary the lipid

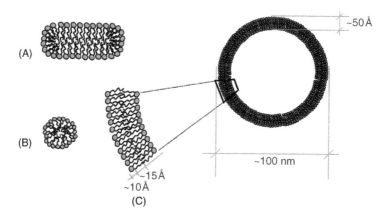

FIGURE 7.1 Schematic cross-sections of model membranes used in spectroscopy and NMR. (A) A phospholipid bicelle. These membrane mimicking solvents are typically composed of both short- and long-chain lipids such as mixtures of DHPC and DPPC. (B) A detergent micelle. For example, SDS micelles are composed of 60 SDS molecules and have a molecular weight of approximately 20 kDa, whereas micelles composed of DHPC contain roughly 35 molecules. (C) A large unilamellar phospholipid vesicle (liposome). Liposomes prepared by the extrusion method are typically 100 nm in diameter. The bilayer thickness is approximately 50 Å if lipids such as DOPC or POPC are used.

composition is limited. Figure 7.1 is a schematic representation of a LUV, a bicelle, and a micelle.

One of the major advantages with model membranes is the possibility to choose a specific lipid composition and concentrate investigations on peptide–membrane interactions to specific components. One difficulty in this experimental approach is how to choose an appropriate lipid composition for the studied effect. On the other hand, lateral diffusion in a membrane is an extremely fast process and peptides may recruit lipids optimal for binding in this way, thus lending a certain tolerance to variations in lipid composition.[9]

7.1.2 The Importance of Arginines in CPP Uptake

Uptake of a series of cationic CPPs has been studied in live adrenal pheochromocytoma cells (PC-12) under normal conditions (37°C), as well as under conditions where endocytosis has been shut down (4°C or where intracellular ATP is depleted).[7] The importance of arginines in efficient uptake has been demonstrated by comparing penetratin to its arginine- and lysine-enriched versions, PenLys and PenArg, and comparing TatP59W to its lysine-enriched analog, TatLysP59W. From Figure 7.2A it is evident that the lysine-enriched peptides cannot enter cells at all. Further, the influence of endocytosis on uptake efficiency has been investigated (Figure 7.2B and C). Penetratin can apparently only enter these cells via endocytosis whereas PenArg is internalized both at low temperature and with ATP depletion, again showing the significance of the arginines. The oligoarginine R_7W enters very efficiently under all studied conditions. Interestingly,

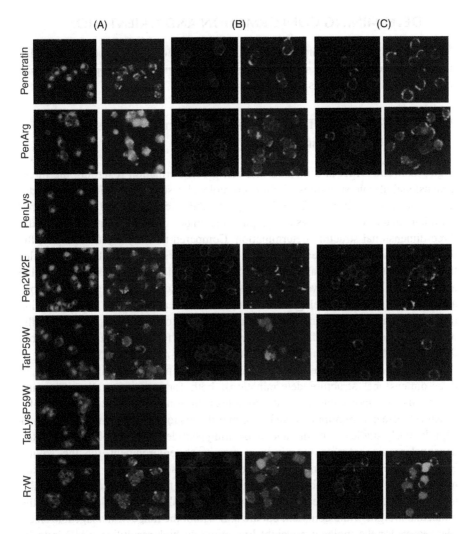

FIGURE 7.2 Confocal laser scanning microscopy images of live unfixed PC-12 cells incubated for 1 h with 1 μM FM 4–64 (red fluorescence) or 5 μM carboxyfluorescein-labeled peptide (green fluorescence). (A) Uptake at 37 °C. (B) Uptake at 4°C. (C) Uptake in cells depleted of intracellular ATP using rotenone. Uptake via endocytosis is abolished in (B) and (C). (Images modified from Thoren, P.E. et al., *Biochem. Biophys. Res. Comm.*, 307, 100–107, 2003. With permission.)

TatP59W can enter cells at low temperature, but not when intracellular ATP is depleted. The possible influence of the two Trp residues in penetratin has been explored by substituting these residues for phenylalanines (Pen2W2F). Whereas earlier reports on fixed cells have claimed that the Trps are crucial for uptake, we conclude that the uptake of the Trp-free Pen2W2F is very similar to that of penetratin under all conditions tested.

7.2 DETERMINING CONFORMATION AND ORIENTATION

Peptide conformation and orientation relative to the membrane can be determined in liposome model membrane systems using polarized light spectroscopy or in simpler membrane mimetic systems using NMR.

7.2.1 NMR STUDIES OF THREE-DIMENSIONAL STRUCTURE AND POSITIONING OF CPPS IN MEMBRANE MODEL SYSTEMS

High-resolution NMR is a method that can give information regarding the three-dimensional atomic structures of biomacromolecules such as proteins and peptides in an aqueous solution. In this respect, the method is often considered complementary to x-ray crystallography, the major method of macromolecular three-dimensional structure determination. Compared to x-ray crystallography, the disadvantages of NMR are to do with the relatively small size of the molecules that can be studied (typically < 50 kD without special techniques). The relative advantages of NMR involve possibilities to study in detail molecular interactions and dynamics, as well as being able to choose the solvent properties in a rather wide range. Often, isotope labeling with ^{15}N or ^{13}C is used to aid in the assignment of the resonances. Solid-state NMR is also coming into use for biomolecular studies, but development has been relatively slower than for high-resolution NMR.

Biomembrane-associated proteins and peptides present special problems in three-dimensional structure determinations, both for x-ray crystallography and NMR. Since the molecular size of the complex is limiting in NMR, it is not generally possible to use phospholipids, vesicles, or multilayers to associate with the protein or peptide. Early studies on membrane-associating peptides by NMR often used mixed aqueous/organic solvents to mimic a membrane environment for the solute. However, it is generally agreed that detergent micelles, or so-called bicelles (mixed micelles), are better membrane mimetics and therefore these are preferentially used.[10,11] Such membrane model systems obviously add considerably to the molecular weight of the complex, but are still within the range of manageable sizes. The reason for the molecular weight limitations in high-resolution NMR is to be found in the relaxation behavior of the studied molecules. The relatively slow motion associated with a large molecule or complex gives rise to rapid relaxation. In turn, this leads to broadened resonances, which in severe cases cannot be resolved or interpreted. A micelle composed of the detergent SDS typically contains approximately 60 SDS molecules. Each headgroup carries a negative charge. Micelles composed of 1,2-Diheptanoyl-sn-Glycero-3-Phosphocholine (DHPC) would contain approximately 35 molecules per micelle, and these have zwitterionic headgroups.[12] Both micelles are relatively small (typically 20 kD). When such micelles are used to mimic a membrane environment for peptides of a few kD size, the NMR spectra of the peptide are generally of high quality and can be used for studies of three-dimensional structure and positioning. A disadvantage is that the curvature of the micelle is considerably higher than one would expect in a biological membrane, and of course the chemical composition is quite different.

The bicelles are composed of two types of phospholipids (typically a mixture of 1,2-Dimyristoyl-*sn*-Glycero-3-Phosphocholine (DMPC) and DHPC) with long and short fatty chains, respectively, and assemble to form disk-like aggregates in an aqueous solution. The bicelle size depends on the relative fractions of the two components. A q-value describes the ratio of long-chained to short-chained phospholipids. Small q-values give rise to relatively isotropic bicelles, which are used in this kind of NMR study. It is possible to replace some of the zwitterionic DMPC with negatively charged DMPG to give a fractional charge to the headgroup surface provided by the long-chained lipids. Whereas detergent micelles are relatively robust regarding their properties under varying conditions, the bicelles are more fragile, and one must take care to keep them under conditions in which they keep their integrity. If isotope-labeled peptides are not available, it is advantageous to use deuterated detergents or phospholipids, which are commercially available.

Positioning studies of peptides in model membrane systems by high-resolution NMR make use of paramagnetic relaxation agents. These can be either external in the solvent or covalently attached to components that insert into the membrane mimetic system. Fatty acid or phospholipid derivatives with doxyl (spin label) groups at different positions have been used. The rationale is that resonances from peptide nuclei located in the vicinity of the relaxation probe will be selectively broadened. In practice, the resolution is somewhat limited because of the dynamic nature of the complexes, but one can at least distinguish between parts of the peptide that are exposed to the external solvent, those in the headgroup layer, and those in the interior of the micelle/bicelle.

7.2.1.1   NMR Studies of Transportan

Transportan is composed of the N-terminal 1–12 residues of galanin and linked by a lysine to mastoparan at its C-terminus. Transportan is a highly efficient CPP with potent membrane interactions. Recent studies suggest that it translocates into cells by different parallel mechanisms: endocytosis in the form of macropinocytosis, as well as direct translocation through the plasma membrane. Structure studies of transportan have been performed in various solvent systems, using circular dichroism (CD) and NMR. In aqueous 50-mM phosphate buffer solution (pH 7), CD spectra show contributions of α-helix (30%) and random coil (70%). An aqueous solvent with 30% HFP, or a solvent with SDS micelles (100 mM SDS in water, pH 3), or with phospholipids vesicles of varying lipid composition in terms of fraction of charged headgroups, increases the contribution of α-helix by approximately a factor of 2 according to CD in all cases, regardless of headgroup charge. These results indicate that the hydrophobicity of the membrane mimetic environment, rather than its charge, dictates the interaction.[13] Shielding from interaction with the aqueous solvent obviously leads to the induced secondary structure.

High-resolution NMR has been used to study the three-dimensional structure and positioning of transportan in SDS micelles.[14,15] With the SDS micelles, the C-terminal mastoparan part has a well-developed α-helix, partly buried in the interior of the micelle, whereas the N-terminal part is less well structured and is located in the headgroup region of the micelle. The middle segment connecting the

FIGURE 7.3 Solution structure of transportan in zwitterionic bicelles composed of DMPC/DHPC in a ratio of 0.33/1. Overlay of 25 calculated low-energy structures, fitted either for (left) residues 3–26 or (right) residues 16–27. (From Bárány-Wallje, E., Andersson, A., Graslund, A., and Maler, L. *FEBS Lett.*, 567, 265–269, 2004. With permission.)

two parts appears to be most exposed to the solvent. Figure 7.3 shows the transportan peptide backbone structure determined in the zwitterionic bicelles. Also, in this solvent, the C-terminal α-helix is well developed. The N-terminus appears better structured than in the SDS micelle and an additional α-helical part can be seen, especially between residues 5 and 8. In this case, the hinge region around residue 15 is not well determined. The paramagnetic probes in this solvent clearly showed that the whole molecule was located at the interphase between membrane and solvent, i.e., in the headgroup region. The difference in localization between the micelle and the bicelle solvents may possibly be found in the higher curvature of the micelle, since its dimensions are close to those of the peptide.

7.2.1.2 NMR Studies of Penetratin

The structure of penetratin has also been determined in SDS micelles and in acidic bicelles.[16] In both cases a relatively well-defined straight α-helix was determined for the central part of the peptide (Figure 7.4). In the bicelle system (DMPC/DMPG/DHPC) in the relative amounts 0.45/0.05/0.5, the penetratin molecule was found to

FIGURE 7.4 Solution structure of penetratin in acidic bicelles composed of DMPC/DMPG/DHPC with relative contents 0.45/0.05/0.5. Overlay of 20 low-energy structures, fitted for residues 3–13. (From Lindberg, M., Biverstahl, H., Graslund, A., and Maler, L., *Eur. J. Biochem.*, 270, 3055–3063, 2003. With permission.)

reside more or less parallel to the bicelle surface with its hydrophobic residues interacting with the interior of the bicelle. The positioning in the SDS micelle was much less well defined, although the available evidence suggested that most of the peptide, except possibly the two most N-terminal residues, was located inside the micelle.

7.2.1.3   NMR Studies of Prion Protein-Derived Peptides

Certain evidence suggests that the prion protein retains its signal sequence, residues 1–22 for mouse PrP, under some conditions in the cell-trafficking process.[17] A peptide corresponding to the signal sequence plus an nuclear localization signal (NLS)-like highly basic segment with 5–6 residues corresponding to the proper protein sequence has great similarities to certain CPPs. Indeed, this peptide translocates into cells and, moreover, it has membrane-perturbing properties similar to the antibacterial melittin peptide, although higher concentrations are needed for mPrP(1–28).[18,19] NMR studies of mouse PrP(1–28) in SDS micelles and the corresponding bovine PrP(1–30) in zwitterionic bicelles have shown induction of α-helix in both peptides interacting with the membrane model system.[18,20] A similar observation in a DHPC micelle environment has also been made for a related peptide, mouse Dpl(1–30), which has a sequence derived from the Doppel protein (a homologue to the PrP). For bovine PrP(1–30) and mouse Dpl(1–30), in which positioning studies were made, the results suggested transmembrane localization of both peptides.[21]

7.2.2   POLARIZED LIGHT SPECTROSCOPY

CD spectroscopy is a routine tool for the assessment of secondary structure of proteins, and relies on the fact that each canonical secondary structure element has a unique CD feature in the peptide chromophore absorption region (190–240 nm). Determining the secondary structure content in small peptides is difficult due to predominant end effects. Measures of the α-helical content obtained in this way should be treated with caution and used only for comparison between the samples acquired and analyzed in similar ways, and not as an absolute measure of the secondary content. The CD contributions from various side-chain chromophores are built up in a complicated way depending on the type of side chains and on the environment. For example, the CD contributions of Trp residues are substantially enhanced by insertion of the chromophore into the lipid bilayer, which complicates the analysis even more.[22]

Linear dichroism (LD) is the differential absorption of light polarized parallel and perpendicular to a macroscopic orientation axis. In contrast to CD, which relates the chirality of the transition moments in a molecule, LD provides information on their orientation. Thus, in LD, a macroscopically oriented sample is required. LUVs may be deformed into ellipsoidal shape by shear flow. Even though the deformation is modest, in many cases it is sufficient to produce an LD signal from chromophores that are aligned relative to the liposome membrane surface. In this way, peptides that bind to the membrane have been analyzed with respect to the preferred orientation of α-helix as well as side-chain chromophores.[23–25]

7.2.2.1 The Influence of Arginines and Lysines in α-Helix Formation of Penetratin Peptides

Figure 7.5 shows typical LD and CD spectra of penetratin and its arginine- and lysine-enriched versions, PenArg and PenLys. The appearance of a negative double peak around 225 nm ($n–\pi^*$ transition of the amide chromophore) and a positive peak around 207 nm ($\pi–\pi^*$ low energy exciton coupling transition) indicates a preferred orientation of the peptide α-helix that is parallel to the membrane surface in all three cases and agrees with the NMR results presented in Section 7.2.1.2. The spectral features in the peptide bond region of the penetratin LD spectrum cannot solely be explained as stemming from an oriented pure α-helix, but in addition require that some peptide residues, probably end residues, adopt a flattened conformation parallel to the membrane surface.[24] The CD spectra shown in Figure 7.5B are typical α-helical and, from the mean molar residue ellipticity at 222 nm, the α-helical content has been estimated to be 48, 56, and 73% for penetratin, PenArg, and PenLys, respectively. Clearly, replacing arginines with lysines has a more pronounced effect on α-helicity than vice versa. In our hands, lysines seem to be more easily adapted to a helical structure than do arginines since substitution of the two central arginines in penetratin for lysines in PenLys stabilizes the α-helical structure. Interestingly, the magnitude of the CD signal opposes that of the LD, showing that, for the penetratin peptides, the peptide with the shortest α-helix (penetratin) has the strongest LD signal and vice versa. This relationship

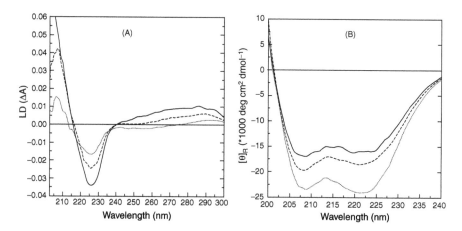

FIGURE 7.5 (A) LD of penetratin (solid line), PenArg (dashed line), and PenLys (dotted line) in LUVs composed of POPC:POPG (80:20 molar ratio). The P/L molar ratio was 1:100 and the peptide concentration 50 μM. Spectra were recorded in a buffer containing 50% (w/v) sucrose to match the refractive index of the lipid vesicles and thereby reducing light scattering, and also to increase the viscous drag (and so the liposome orientation) in the Couette flow cell. Although this buffer decreases scattering it has significant absorption in the short wavelength region (<210 nm) which severely impairs the spectral resolution. (B) CD spectra on penetratin (solid line), PenArg (dashed line) and PenLys (dotted line) recorded under identical conditions (on the same samples) as the LD spectra.

between α-helical content and LD signal in the peptide bond region strengthens the evidence that residues not participating in the α-helix can cause strong LD signals by stretching out along the direction of the membrane surface.

7.2.2.2 Structural Flexibility in Penetratin

The secondary structure of membrane-bound penetratin is flexible and both α-helical and β-sheet forms have been suggested.[22,26–29] At a low peptide-to-lipid molar ratio (typically 1:100) the peptides bind preferentially in α-helical configuration, whereas at high mixing ratios a predominant β-structure can be observed if the surface charge density is increased or if vesicle aggregation occurs. β-Sheet formation has also been observed for peptides associated within SDS micelles and DPPC/DPPS monolayers.[22,30] An interesting observation is that the arginine-for-lysine substituted version of penetratin, PenLys, can undergo an α-helix to β-sheet transition during vesicle aggregation but not in SDS micelles, regardless of the peptide-to-surfactant ratio. PenLys has been found to have a very low affinity for the plasma membrane in PC-12 cells and hence could not be internalized at all.[7] By contrast, penetratin and PenArg were efficiently bound to the plasma membrane, mostly in patches. The initial binding was followed by internalization via endocytosis in the case of penetratin whereas PenArg could enter cells even during conditions preventing endocytosis. Regardless of whether aggregation or only sufficiently high surface charge density is needed to drive β-sheet formation in model membrane systems, the conformational flexibility of penetratin is interesting and, based on the observations with PenLys, we do not find it unlikely that it is an important parameter for efficient cellular uptake.

7.3 THE BINDING OF CATIONIC CPPs TO LIPID BILAYERS CAN BE DESCRIBED BY AN ELECTROSTATIC MODEL

A prerequisite for all spectroscopic peptide–membrane studies is to know to what extent a peptide is bound to the membrane as it is necessary to exclude spectral contributions from unbound peptides free in solution. In addition, the thermodynamic affinity is an important parameter for characterizing the nature of binding. The affinity of peptide binding is normally assessed by titration of a liposome solution with peptide (or vice versa) with the subsequent construction of binding isotherms. Working with cationic peptides, such as penetratin, has been found problematic due to their inherent propensity to adhere to glass and silica surfaces, plastics, etc.[28] This in turn leads to difficulties in determining the exact peptide concentration in the sample. We have developed a method to evaluate more accurately the degree of bound and free peptide in each titration point, based on a multivariate whole-spectra approach with a least-square projection of titration spectra onto reference spectra of free and bound peptide, making use of the emission intensity increase and blueshift that follow membrane binding.[28]

Binding of a hydrophilic peptide to a charged membrane involves a complex interplay between electrostatic attractions, van der Waal, hydrophobic, and dehydration interactions. For peptides such as penetratin, which adopt an ordered secondary structure upon binding, not only peptide–lipid interactions become important but also intrapeptide interactions involving formations of hydrogen bonds. Despite this principal complexity, it has been shown that the binding of cationic CPPs can be quite accurately described using a simple binding model based on the Gouy–Chapman theory combined with a two-state partition equilibrium, as was first described by Beschiaschvili and Seelig.[31] A thorough description of the model, together with detailed information on how to compute binding isotherms from whole fluorescence spectra, has been given elsewhere.[28] However, the essential elements of the model are described below. The concentration immediately above the membrane, C_M, can be determined according to

$$C_M = C_f \exp(z_p F \psi_0 / RT) \tag{7.1}$$

where C_f is the peptide concentration in the bulk solution, z_p is the effective peptide charge, F is the Faraday constant, ψ_0 is the surface potential, R is the gas constant, and T is the absolute temperature. C_M is related to the bound peptide-to-lipid ratio, r, according to

$$r = K_p C_M \tag{7.2}$$

where K_p is the surface partition constant. Binding isotherms for penetratin, PenArg, PenLys TatP59W, TatLysP59W, and R_7W have previously been published, and a summary of binding parameters can be found in Table 7.2.[32,33] The surface partition constant K_p deals with the nonelectrostatic, mainly hydrophobic, contributions to peptide binding. The hydrophobic effect stems from positioning nonpolar peptide residues in a membrane environment and also from forming favorable intramolecular bonds when forming a defined secondary structure in the membrane. In addition, penetratin peptides acquire amphipathicity upon membrane binding. Forming an α-helix allows for positioning of nonpolar residues on one side, giving the peptide a hydrophobic face. The Tat peptides and R_7W do not form an ordered secondary structure upon binding, as evidenced by a random coil signature in CD and no signal at all in LD. Indeed, this is reflected in the magnitude of K_p and hence the hydrophobic contribution to the binding energy, which is in general lower than for the penetratin peptides. An interesting observation is that a high number of arginines seems to increase the hydrophobic contribution since PenArg has a K_p that is almost twice that of penetratin, which in turn is significantly larger than the value for PenLys. Moreover, the oligoarginine analog R_7W has a K_p one order of magnitude larger than TatP59W, even though it contains only one uncharged residue. Comparing TatP59W with its arginine-for-lysine substituted analog TatLysP59W again shows a decrease in K_p. This most likely reflects the superior ability of arginines over lysines to form charge-neutralized connections through hydrogen bonding with negatively charged membrane components, such as phosphate groups. Indeed, the Wimley and White hydrophobicity scale predicts that partitioning of an arginine residue into a neutral membrane is 0.18 kcal/mol more favorable than that of a lysine.[34]

TABLE 7.2
Binding Parameters for CPP Binding to DOPC:DOPG LUVs Containing 40 mol% Anionic Lipids

Peptide	K_p (M^{-1})[a]	z_p[b]	Formal Charge[c]	ΔG_h[d] kcal/mol
Penetratin	907/836[e]	4.3/5.7	+7 (3/4)	−4.0/−4.0
PenArg	1891	4.2	+7 (7/0)	−4.5
PenLys	579	4.7	+7 (0/7)	−3.8
TatP59W	34	5.0	+8 (6/2)	−2.1
TatLysP59W	5	4.5	+8 (0/8)	−1.0
R$_7$W	307	5.1	+7 (7/0)	−3.4

Note: Data in this table have been computed from binding isotherms in Persson et al.[22] and Thoren et al.[33]

[a] Surface partition constant.

[b] Effective peptide charge.

[c] Formal peptide charge. The number of arginines and lysines, respectively (Arg/Lys) are in parentheses.

[d] The hydrophobic contribution to the free energy or binding computed as $\ln(K_p)$.

[e] The two values are from two measurements performed under slightly varying conditions.[22,33] The former data were measured in DOPC:DOPG vesicles containing 2.5 mol% of a PEG–lipid (DSPE-PEG2000) used to prevent vesicle aggregation at elevated peptide to lipid ratios. Binding of PenArg and PenLys was assessed using the same experimental setup. The latter data were measured using liposomes with a PEG-concentration of 5 mol% together with TatP59W, TatLysP59W, and R$_7$W.

The z_p values in Table 7.2, the peptide effective charges, are always lower than the true charges. This has previously been observed in studies of a number of charged peptides and has been attributed mainly to discrete charge effects, separation of charges in the peptide, and effects of associated counter ions.[35–38]

7.4 MEMBRANE TRANSLOCATION IN MODEL SYSTEMS

In order to understand how CPPs can be transported across a lipid bilayer, it is desirable to set up tests for translocation in model membrane systems. Here, the effect of lipid composition and even external stimuli such as ion potentials and pH gradients can be selectively tested.

7.4.1 TRANSLOCATION INTO GIANT UNILAMELLAR VESICLES (GUVs)

An apparent and straightforward method to study peptide translocation is by using confocal laser scanning microscopy (CLSM) to observe transport of fluorescently labeled peptides into GUVs. Our laboratory in Gothenburg uses a dehydration/rehydration technique to produce both multilamellar and unilamellar GUVs in the range of 1–300 μm.[39] In Figure 7.6, translocation of fluorescein-labeled PenArg is compared to a 20-bp oligonucleotide. It is evident from Figure 7.6A that PenArg has a very high propensity to bind to the membrane, but since the line profile in

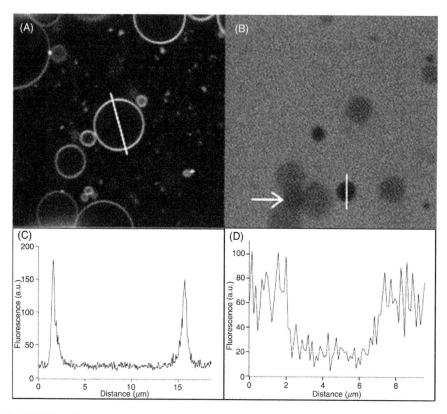

FIGURE 7.6 Comparison of membrane translocating ability of fluorescently labeled PenArg and a fluorescently labeled oligonucleotide. (A) PenArg. (B) Oligonucleotide. (C) Line profile of (A). (D) Line profile of (B). To establish that the fluorescence intensity recorded inside the GUV in (A) is not due to artifactual stray light or out-of focus fluorescence emanating from the intensely fluorescent membranes, bleaching experiments were performed. By selectively exciting a region inside the vesicle with high laser intensity, rapid photobleaching could be achieved. In a few minutes the fluorescence intensity was recovered, reaching the level observed prior to bleaching. The same phenomenon was observed when selectively exciting a region in the external medium. (Images adapted from Persson, D. et al., *Biochim. Biophys. Acta*, 1665(1–2), 142–155, 2004. With permission.)

Figure 7.6C shows that the fluorescence intensity inside the GUV is comparable to that of the surrounding medium, it can be concluded that peptide is able to translocate. Figure 7.6B clearly shows that the oligonucleotide neither binds to the membrane nor translocates to its lumen.

7.4.2 STUDYING PEPTIDE TRANSLOCATION INTO LUVs USING RET

Studying transport across membranes in LUVs can be a cumbersome task due to the difficulty of distinguishing between the inner and outer membrane leaflet. We have developed a method to exclusively label the inner leaflet of LUVs with a methylcoumarin-labeled lyso lipid. This lipid, which was first described by

Wimley and White, has an absorption maximum of 335 nm ($\varepsilon = 14{,}950$ M^{-1} cm^{-1}) and is thus an excellent acceptor of tryptophan fluorescence.[40] The Förster distance, R_0, for the Trp–lysoMC donor acceptor pair is ~ 25 Å, and thus energy transfer across the membrane is, at best, very weak. By labeling vesicles only on the inside we have shown that it is possible to sensitively and selectively assess peptide translocation by monitoring quenching of Trp fluorescence in combination with the enhancement of lysoMC fluorescence.[32,33] The asymmetric labeling of LUVs was achieved by extracting lysoMC from initially symmetrically labeled vesicles by repeatedly "washing" them with large lipid particles, and thereafter purifying the LUVs by centrifugation. The amount of lysoMC remaining in the outer leaflet after each washing step can be assessed by raising the external pH to 11, which causes a shift in lysoMC fluorescence. It was also established that the flip-flop rate in these vesicles is very slow and the vesicles remain asymmetric for hours after preparation, even at room temperature.[32] We have used the lysoMC assay to study the translocation ability of cationic CPPs penetratin, TatP59W, and R$_7$W as well as the analogs PenArg, PenLys, and TatLysP59W in LUVs composed of a variety of different lipids. Tryptophan octyl ester (TOE) was included as a positive reference since it has previously been shown to equilibrate across the membrane of 1-Palmitoyl-2-Oleoyl-*sn*-Glycero-3-[Phospho-*rac*-(1-glycerol)] (POPG) and 1-Palmitoyl-2-Oleoyl-*sn*-Glycero-3-Phosphocholine (POPC) LUVs, whereas the amphipathic α-helical peptide Ac-18A-NH$_2$ was chosen as a negative reference due to its inability to cross lipid bilayers.[40]

Translocation into the lumen of asymmetrically labeled LUVs was assessed by recording fluorescence spectra before and after the addition of peptide to the external solution. The excitation wavelength was set to 260 nm in order to maximize the Trp/lysoMC absorption ratio and hence exclusively excite the donor chromophore. Spectra were recorded between 300 and 500 nm. Resonance energy transfer (RET) is commonly monitored as the degree of quenching of the donor chromophore (here Trp). Due to extensive problems with peptide adsorption to glass and plastics, it is difficult to control the exact concentration of peptide in each sample; therefore, it is more convenient to monitor the enhancement of acceptor fluorescence upon adding a small volume of peptide to a LUV suspension. Figure 7.7 shows spectra acquired in EggPC:DOPG:lysoMC liposomes (molar ratio 60/39/1). As seen in Figure 7.7B, all peptides clearly give rise to RET to lysoMC in the inner leaflet since the enhance lysoMC fluorescence is too large to stem from the few acceptor chromophores remaining in the outer leaflet (1:10,000 lipids compared to 1:100 lipids in the inner leaflet). However, there is a much larger difference in RET when comparing the easily translocating TOE with Ac-18A-NH$_2$. It is evident from Figure 7.7B and from the calculated RET enhancement in Table 7.3 that the investigated CPPs are much closer to the negative reference Ac-18A-NH$_2$ than the positive reference TOE. To give an indication to what extent the peptides may have entered the LUVs, the entry efficiency has been calculated and presented in Table 7.3. In the calculation, it is assumed that TOE readily equilibrates over the membrane and that Ac-18A-NH$_2$ does not translocate at all. In addition, the possibility that different peptides may have slight variations in RET efficiency has been ignored. Keeping these assumptions in mind, the tabulated entry efficiencies in

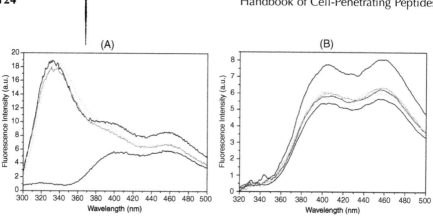

FIGURE 7.7 RET from tryptophan to lysoMC present exclusively in the inner leaflet of LUVs composed of EggPC:DOPG (60/40 molar ratio). (A) Fluorescence emission spectra recorded before and after addition of peptide to 100 μM LUVs. The excitation wavelength was 260 nm and tryptophan concentration $\sim$0.7 μM. The lower black line was recorded before addition of peptide and is hence the lysoMC fluorescence in the absence of donor (Trp). The upper black line is TOE, the solid gray line Ac-18A-NH$_2$, and the dotted line is penetratin. (B) Enhancement of lysoMC fluorescence due to energy transfer from Trp. The spectral contribution from Trp has been subtracted. Energy transfer from TOE (upper black line), Ac-18A-NH$_2$ (thick, solid, dark gray line), Penetratin (dashed gray line), PenArg (dotted gray line), PenLys (short, dotted gray line), and TatP59W (solid gray line).

TABLE 7.3
Summary of Energy Transfer Enhancement from Trp to lysoMC in the Inner Leaflet of LUVs with Two Different Lipid Compositions

Peptide	EggPC:DOPG		DOPC:DOPG		Soybean Lecithin Extract	
	RET[a]	Inside (%)[b]	RET[a]	Inside (%)[b]	RET[a]	Inside (%)[b]
Ac-18A-NH$_2$	1.09	0	1.11	0	1.06	0
TOE	1.43	45	1.45	45	1.38	45
Penetratin	1.10	1	1.10	<0	1.05	<0
PenArg	1.11	2	1.11	1	—	—
PenLys	1.11	3	1.10	<0	—	—
TatP59W	1.13	5	—	—	1.08	3
R$_7$W	—	—	—	—	1.08	3

[a] ET is the enhancement in lysoMC fluorescence due to energy transfer and was calculated by a least-square projection of the spectrum recorded after adding peptide to a suspension of asymmetrically labeled LUVs onto reference spectra of membrane-bound peptide in LUVs without lysoMC and of the acceptor lysoMC prior to addition of peptide as described in the text.

[b] The relative enhancement is calculated as $((RET_{peptide} - RET_{Ac-18A-NH_2})/(RET_{TOE} - RET_{Ac-18A-NH_2})) \times 0.45 \times 100\%$, and is thus an indication of the percentage of peptide that has entered into the lumen. The factor 0.45 is to account for the fact that the inner membrane surface of a 100-nm LUV is 45% of the total surface (inside + outside), assuming that TOE readily equilibrates between the inner and outer leaflet.

Table 7.3 never exceed 5% and are thus most likely not significant but rather an effect of the peptides having somewhat differing abilities to cause transbilayer RET. This is supported by the fact that RET is sometimes even lower for the penetratin peptides than for the negative reference.

7.4.3 Modeling the Endosomal Escape of Penetratin and bPrPp

As discussed in the Introduction, many CPP–cargo constructs enter cells via endocytosis, yet they assert the desired biological effect inside the cell, indicating that CPPs must be able to avoid proteolytic degradation by escaping from the endosome. Little is known about the escape mechanisms of CPPs but it has been indicated that endosomal acidification is important, as is the case in the entry of several viruses. Endosomal escape was modeled in an LUV system by encapsulating peptide in the lumen of liposomes with a composition resembling that of the endosomal membrane (DPPC:cholesterol:sphingomyelin 50:30:20). A trans-membrane pH gradient (ΔpH_{mem}) was created with the help of the K^+/H^+ ionophore nigericin.[41] Under the experimental conditions used, the pH inside the LUVs was lowered to 5.5 compared to 7.4 in the external solution, as evidenced using the pH-sensitive dye phenol red. The obtained ΔpH_{mem} is comparable to in vivo gradients over late endosome membranes. Escape was monitored by adding the aqueous quencher acrylamide to the external medium (total concentration 100 mM) and measuring the degree of Trp quenching upon escape. The maximum level (100% escape) was obtained by lysing the LUV with Triton-X, yielding the fluorescence intensity value F_{max}. The percentage of peptide escape was computed according to

$$\text{escape} = \left(\frac{F_T - F_0}{F_{max} - F_0} \right) \times 100\% \qquad (7.3)$$

where F_T is the intensity at time $t = T$ and F_0 is the intensity at the time of nigericin addition ($t = 0$). Figure 7.8 shows the escape of penetratin (A) and the bovine form of the prior peptide (bPrPp) (B) with time in the absence and presence of ΔpH_{mem}. As can be seen in Figure 7.8A, peptide escape in the absence of ΔpH_{mem} is practically nonexistent for penetratin ($\sim 3\%$), while approximately 25% of the peptide escapes after 2 h in the presence of nigericin. Complexing penetratin with heparan sulfate does not seem to affect the escape rate but decreases the efficiency to approximately 10%. Interestingly, the escape efficiency agrees fairly well with values obtained in a live cell study, showing 20–30% escape from endosomal compartments.[4] bPrPp (Figure 7.8B) is more efficient in escaping the LUVs, reaching almost 50% escape after 60 min incubation with nigericin. It should be emphasized that bPrPp does escape LUVs also in the absence of a potential (10% escape), which is most likely an effect of this peptide being able to form transient pores (see Section 7.5.1). The insets in (A) and (B) show escape at lower peptide concentration (1 μM) in the presence of ΔpH_{mem}. The apparent lower efficiency indicates that a threshold concentration is needed for escape.

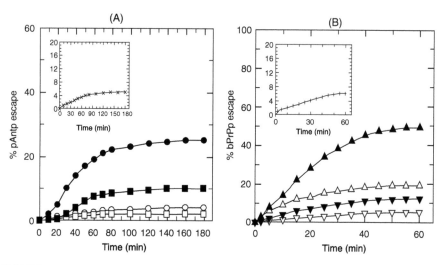

FIGURE 7.8 Escape of nonlabeled peptides from LUVs. (A) 20 μM penetratin (pAntp) alone in the absence ($\bigcirc$) or presence ($\bullet$) of a transmembrane pH gradient ΔpH_{mem}; 20 μM penetratin complexed with 12 μM heparan sulfate (HS) in the absence ($\square$) or presence ($\blacksquare$) of ΔpH_{mem}. (B) 20 μM bPrPp (pAntp) alone in the absence ($\triangle$) or presence ($\blacktriangle$) of a transmembrane pH gradient ΔpH_{mem}; 20 μM bPrPp complexed with 12 μM HS in the absence ($\triangledown$) or presence ($\blacktriangledown$) of ΔpH_{mem}. The LUVs were composed of DPPC:cholesterol:spingho-myelin (50:30:20 molar ratio). The total lipid concentration was 100 μM. Insets show peptide escape at 1 μM peptide concentration. Note that the y-axis scale is not the same in (A) and (B). (Adapted from Magzoub, M. et al., *Biochemistry*, 2005, published online. With permission.)

7.5 MEMBRANE PERTURBATION EFFECTS

The ability to cause membrane perturbation seems to be a prerequisite for CPPs that enter cells via nonendocytotic pathways or translocate across model membranes. CPPs are generally associated with low toxicity and therefore pore formation is unlikely to be an option for entry, whereas other forms of membrane destabilization may play important roles, as suggested in Section 7.5.2.

7.5.1 LEAKAGE STUDIES

Peptide-induced leakage of calcein entrapped in LUVs can be used to monitor membrane perturbations related to toxicity. Leakage caused by penetratin has been compared to the two prion-derived peptides, the murine form (mPrPp) and bPrPp, in Figure 7.9. The toxic pore-forming bee venom peptide melittin causes complete leakage of vesicle contents already at very low peptide-to-lipid molar ratios and is included as a reference. In comparison, penetratin is nontoxic even at extremely elevated peptide-to-lipid ratios, causing no or little leakage. The two prion-derived peptides fall in between melittin and penetratin. Leakage is efficient at low peptide-to-lipid ratios but shows saturation at higher concentrations, which indicates that the formed pores are transient. It can be speculated that the saturation behavior is due to

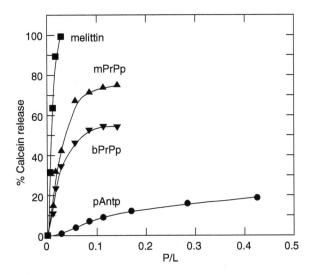

FIGURE 7.9 Peptide-induced leakage of entrapped calcein from LUVs composed of 250 μM POPC:POPG (30:70 molar ratio). Increasing concentrations of peptide were added to the LUV samples while monitoring the dequenching of calcein upon release from the vesicles. 70 mM calcein was entrapped inside the vesicles and the buffer used was 50 mM phosphate (pH 7.4).

increased self-aggregation of the prion-derived peptides resulting in inhibition of its activity.[19]

7.5.2 CPPs TATP59W AND R$_7$W ARE ABLE TO INDUCE MEMBRANE FUSION

Many cationic CPPs are able to cause aggregation of negatively charged lipid vesicles at elevated peptide-to-lipid molar ratios.[42] The process is easily monitored by measuring the change in optical density at 436 nm.[29] Comparing penetratin to TatP59W, TatLysP59W, and R$_7$W reveals that the Tat peptides and R$_7$W are more prone to cause aggregation and that the obtained aggregates do not dissociate as has been shown for penetratin.[42] In order to investigate whether or not the vesicle aggregation event led to vesicle fusion, the extent of lipid mixing between two populations of liposomes was measured using RET. One population of liposomes was labeled with the fluorescent lipids NBD-PE (donor) and Rh-PE (acceptor) and subsequently mixed with unlabeled vesicles. Fusion of vesicles after the addition of peptide would lead to an increased accessible surface area for donor and acceptors, and hence a decrease in RET. Figure 7.10 shows the extent of lipid mixing as a function of the peptide-to-lipid molar ratio for vesicles composed of DOPC:DOPG (60:40 molar ratio) (A) and vesicles composed of DOPC:DPPG (60:40 molar ratio) (B). In Figure 7.10A, it can be seen that lipid mixing, and hence vesicle fusion in DOPC:DOPG vesicles, is very efficient for TatP59W and R$_7$W but does not occur at all for penetratin. Exchanging DOPG for the saturated lipid DPPG decreases membrane fusion substantially, as can be

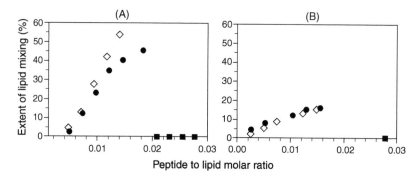

FIGURE 7.10 Extent of lipid mixing 30 min after addition of the CPPs R_7W ($\diamond$), TatP59W ($\bullet$), TatLysP59W ($\triangle$), or penetratin ($\blacksquare$) at different peptide-to-lipid molar ratios. (A) LUVs composed of DOPC:DOPG with and without 0.6 mol% NBD-PE and Rh-PE. (B) LUVs composed of DOPC:DPPG with and without 0.6 mol% NBD-PE and Rh-PE. The extent of lipid mixing was calculated from the intensity increase in donor emission according to %leakage $= 100 \times ((I(t) - I_0)/(I_r - I_0))$, where $I(t)$ is the intensity at time t after peptide addition, I_0 is the intensity before peptide addition, and I_T is the intensity at infinite dilution caused by lysing the vesicles with Triton X-100. (Data adapted from Thorén, P.E. et al., *Biophy. Chem.*, 114(2–3), 169–179, 2005. With permission.)

seen in (B). Both TatP59W and R_7W can enter live cells at a low temperature (4°C) whereas in the same study penetratin was shown to enter only via endocytosis. It is possible that the mechanism of nonendocytotic entry is related to the ability to cause localized destabilization of the plasma membrane. The lipid-mixing assay indicates that penetratin is benign in this context whereas TatP59W and R_7W can evidently cause severe membrane perturbations. The decreased fusion efficiency upon incorporation of saturated lipids could be an indication that nonendocytotic entry does not occur at lipid rafts. It should be noted that membrane destabilization in this context is not synonymous to pore formation because, in the same report, it has been established that none of these peptides causes vesicle leakage.

7.5.3 CPP Binding Has Effects on Lipid Chain Ordering

Membrane perturbation caused by peptide binding can be measured by monitoring changes in lipid chain ordering. This is measured by fluorescence anisotropy of polarization of a lipophilic chromophore incorporated into the membrane. An increase in chain ordering is accompanied by an increased orientation of the chromophore, and hence its anisotropy. Diphenylhexatriene (DPH) was used to measure how CPPs affect chain ordering in vesicles composed of POPC:POPG. The polarization is plotted as a function of the peptide-to-lipid molar ratio in Figure 7.11. Penetratin does not cause any changes in DPH polarization, indicating that this CPP binds only in the headgroup region of the membrane. Melittin, on the contrary, is a pore-forming peptide and its transmembrane orientation causes increased chain ordering, as evidenced by the

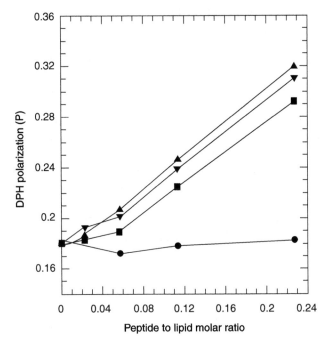

FIGURE 7.11 Peptide-induced changes in fluorescence polarization of membrane incorporated DPH in LUVs composed of POPC:POPG (30:70 molar ratio) for penetratin (●), melittin (■), mPrPp (▲), and bPrPp (▼). (Modified from Magzoub, M. et al., *Biochim. Biophys. Acta*, 1716(2), 126–136, 2005. With permission.)

elevated level of DPH polarization. The two prion peptides are even more effective in causing membrane ordering than is melittin.

7.6 CHARACTERIZING MEMBRANE BINDING USING INTRINSIC TRYPTOPHAN FLUORESCENCE

Tryptophan has optical properties that make it a very useful probe in studying protein conformation and interactions. It has a fairly high quantum yield (0.13) and its extinction coefficient is much higher than the other fluorescent residues tyrosine and phenylalanine.[43] The tryptophan emission is extremely sensitive to solvent polarity, which makes it very useful in studying binding to membranes. A tryptophan residue fully exposed to water exhibits a maximum wavelength (λ_{max}) around 350 nm, whereas a tryptophan inside the hydrocarbon core of a membrane can have a blueshifted emission in which λ_{max} is ~320 nm. In addition, moving from buffer into the membrane usually increases the emission intensity. Tryptophan fluorescence can also be quenched using a wide variety of quenchers ranging from all-purpose substances, such as acrylamide and iodide, to spin-labels and carboxyl groups.

7.6.1 DETERMINING THE INSERTION DEPTH USING BROMINATED LIPIDS

Depth-dependent quenching of tryptophan by quenchers attached at specified positions on the lipid acyl chains has been used to estimate the peptide penetration depth in the membrane. The tryptophan fluorescence intensity, $F(h)$, was measured in three sets of vesicles containing 30 mol% brominated lipid with two bromines on one of the acyl chains in position (6–7), (9–10), or (11–12) along the acyl chains, and thereafter compared to fluorescence intensity in vesicles without quencher, F_0. The quenching ratios were plotted against the quencher position from the bilayer center, h, and the average insertion depth, h_m, was computed using distribution analysis (Equation 7.2) or an extended version of the parallax method (Equation 7.3)[44]:

$$\text{DA:} \quad \ln\frac{F_0}{F(h)} = \frac{S}{\sigma\sqrt{2\pi}} \exp\left[-\frac{(h-h_m)^2}{2\sigma^2}\right] \tag{7.4}$$

$$\text{PM:} \quad \begin{cases} \ln\dfrac{F_0}{F(h)} = \pi C[R_c^2 - (h-h_m)^2] & h-h_m < R_c \\[2mm] \ln\dfrac{F_0}{F(h)} = 0 & h-h_m \geq R_c \end{cases} \tag{7.5}$$

The average insertion depth was computed for penetratin and the analogs PenArg and PenLys, as well as for TatP59W and R_7W. The results are presented in Table 7.4 and the distances presented are an average of the values obtained with the

TABLE 7.4
Membrane Insertion Depths Estimated Using Depth-Dependent Quenching by Brominated Lipids

| | Distance from Bilayer Center (Å)[a] | |
| | Surface Charge | |
Peptide	20%	40%
Penetratin	10.6	10.6
PenArg	10.5	10.4
PenLys	10.4	10.5
TatP59W	10.3	9.9
R_7W	10.7	10.6

Note: Depths are presented as distance from the bilayer center and measured in Å.

[a] The distance from the bilayer center was computed using the distribution analysis (ref) and extended parallax (ref) methods. The presented distance is the average of the values computed with the two respective methods. Distances usually differed substantially ≤ 1.5 Å in all cases.

two methods. Clearly, the surface charge does not affect the insertion depth. However, it is somewhat surprising that the tryptophan residues seem to protrude several Å into the hydrocarbon. Using two fatty acid spin probes, the insertion depth of penetratin has been estimated to be 13 Å in vesicles with a 30% surface charge.[27]

7.6.2 Evaluating Membrane Interactions of Cationic CPPs from Characteristic Tryptophan Fluorescence Parameters

Tryptophan fluorescence parameters describing binding to DOPC:DOPG or POPC:POPG LUVs at molar ratios 80:20 are shown in Table 7.5. The penetratin peptides exhibit a 12-nm blueshift upon binding to membrane, which is normal for moving a tryptophan from the buffer to the headgroup region. TatP59W and R_7W have slightly longer emission maximum wavelengths, both in buffer and when bound to the membrane, which may be related to the fact that the single Trp residue in these peptides is near the end whereas the penetratin peptides have one tryptophan in the middle of the sequence. The Φ_F for peptides in buffer is similar for almost all five CPPs and is approximately 2% lower than for pure Trp, most likely due to more efficient quenching from backbone carbonyls. The ratio between the quantum yield in membrane and in buffer shows the increase in fluorescence upon binding. Unexpectedly, the intensity increase is not the same for all peptides, generally being

TABLE 7.5
Tryptophan Fluorescence Parameters Describing CPP Binding to LUVs Composed of DOPC:DOPG or POPC:POPG at Molar Ratios 80:20

Peptide		λ_{max} (nm)[a]	Φ_F[b]	$\left(\dfrac{\Phi_{F,mem}}{\Phi_{F,buf}}\right)$[c]	$\langle\tau\rangle$ (ns)[d]	$\dfrac{\langle\tau\rangle_{mem}}{\langle\tau\rangle_{buf}}$	k_r (10^7 s^{-1})	$\dfrac{k_{r,mem}}{k_{r,buf}}$
Penetratin	Buffer	349	11.3	1.45	2.08	1.16	5.4	1.2
	Membrane	337	16.4		2.43		6.7	
PenArg	Buffer	349	10.5	1.50	2.10	1.18	5.0	1.3
	Membrane	337	15.8		2.47		6.4	
PenLys	Buffer	349	10.6	1.32	2.07	0.98	5.1	1.3
	Membrane	337	14.0		2.03		6.9	
TatP59W	Buffer	350	11.1	1.53	1.87	1.71	5.9	0.9
	Membrane	342	17.0		3.20		5.3	
R_7W	Buffer	352	11.0	2.04	2.26	1.56	4.9	1.3
	Membrane	341	22.4		3.52		6.4	

[a] Wavelength of the tryptophan emission maximum in buffer and bound to LUVs.
[b] Peptide quantum yield determined using the Φ_F of tryptophan (0.13) as reference.
[c] Relative intensity increase upon binding to membranes determined by comparison of the total emission intensity in a 1 μM peptide solution before and after addition of LUVs (100 μM lipid).
[d] The average fluorescence lifetime was computed according to $\langle\tau\rangle = \sum_i a_i\tau_i$, where a is the normalized preexponential factor and τ is the lifetime. Fluorescence lifetimes were recorded using TCSPC. The decay could not be satisfactorily fitted using less than three lifetimes.

larger for the arginine-enriched versions. The average fluorescence lifetimes also increase when the peptide binds to the membrane, the exception being PenLys where the lifetimes are more or less constant. Evidently, again, the effect is largest for TatP59W and R_7W. The ratio between the radiative rate constants in the membrane and in buffer is similar for almost all peptides (TatP59W being an exception), and hence the Trp chromophore has identical intrinsic properties in all peptides. This is a strong indication that the observed differences in quantum yields and lifetimes between the CPPs are indeed due to differences in interaction of the whole peptide with the membrane.

7.7   CONCLUDING REMARKS

Based on the peptides discussed in this chapter, a classification of CPPs can be made, comprising four distinct, yet interconnected classes of CPPs: (1) arginine-rich peptides; (2) basic/amphipathic peptides; (3) amphipathic peptides; and (4) toxic translocating peptides. To the first class belongs, of course, oligoarginines, in this chapter represented by the Trp-carrying heptaarginine R_7W, but also the TatP59W peptide. These peptides can enter cells via nonendocytotic pathways and have membrane-perturbing properties not related to the direct toxicity caused by membrane leakage. Penetratin and its analogs sort under class (2). There are indications that penetratin is indeed one of the least toxic CPPs, which may be due to its inability to enter cells via nonendocytotic pathways. Arginine residues are extremely important for penetratin, as evidenced by the lack of uptake of the lysine-enriched version PenLys (see Section 7.1.2). Secondary structure induction occurs upon binding to the membrane and there is some gain in amphipathicity accompanied with α-helix formation. Transportan is found in the class of amphipathic peptides (3). Its charged residues and hydrophobic face may, in combination, be important in binding sufficiently strongly to the cell surface and thereby triggering endocytosis, but the amphipathic nature is probably more important in transient pore formation. The prion-derived peptides sort under class (4) and are peptides that cannot be considered as CPPs in the sense that they are far too toxic to function as delivery vectors.

There has been an ongoing discussion about whether or not secondary structure is important for CPP efficiency, and with many reports on induced secondary structures of CPPs and related peptides it has become clear that CPP efficiency is in the broad sense at least not directly correlated with induction of an α-helical structure. Instead, some other interesting correlations appear, for instance that the positioning of an induced peptide α-helix in various membrane model systems seems to be correlated with the potency of causing leakage effects from phospholipid vesicles. This, in turn, may offer hints about the CPP mechanisms in operation for different peptides. Penetratin, with its position in the headgroup layer of a partly negatively charged membrane, causes no vesicle leakage, and even though it has been shown to cross GUV membranes there is no evidence that this CPP can translocate the membranes of smaller vesicles, unless a pH gradient is present across the membrane.[41] This suggests that, for penetratin, endocytosis may be the only pathway of entry into a cell, which is in agreement with uptake studies in live cells (see Section 7.1.2). By comparing penetratin to its analogs PenArg and PenLys, it becomes clear that the arginine residues are extremely important for efficient cell association and subsequent entry. By substituting the lysines in penetratin for arginines, it is even possible to make this peptide (PenArg) enter cells via nonendocytotic pathways. When comparing the secondary structure of the penetratin peptides, it becomes clear that the membrane-induced α-helical structure may even be negatively correlated to uptake because PenLys with its high degree of α-helicity is not internalized at all.[7]

Transportan may be a somewhat different case. The peptide is located in the headgroup region of the bilayer, but preliminary data suggest that it does give rise to some leakage in vesicles, although higher concentrations are needed than for the prion protein-derived peptides.[21,41] For transportan, one could expect that more direct membrane penetration mechanisms might operate in addition to endocytosis, accounting for the observation that transportan can also enter cells when endocytosis has been shut down.[45] Transportan has, indeed, been implicated to be a more efficient CPP than penetratin, but it is also unfortunately more toxic.[46] The ability of transportan to act as a CPP has been attributed to its amphipathicity and this property may be important in the formation of transient pores.

The toxicity of the prion peptides that renders them unacceptable as CPPs is most likely related to their primary amphipathic nature, with a rather long hydrophobic N-terminal stretch and a hydrophilic lysine-rich C-terminus. This is perhaps somewhat surprising since the suspicion that prion-derived peptides might have CPP properties arose from the similarity in secondary structure to a chimeric peptide with documented CPP properties.[18,47] It should be noted that bPrPp has a charged N-terminal part, making it less polarized in terms of hydrophobic/hydrophilic character. This is reflected in induced leakage experiments in which bPrPp causes significantly less leakage than does mPrPp, but also in DPH polarization measurements in which mPrPp has a stronger effect on chain ordering. In the light of these experiments, bPrPp seems to be less toxic than its mPrPp analog. Nevertheless, the lesson to be learnt from the prion proteins is that primary amphipathic character can be associated with membrane toxicity, which is likely to be the case also for transportan, even though substantially higher concentrations may be needed. Penetratin, by contrast, only gains amphipathicity upon forming an α-helix and thus does not possess one end that can penetrate the membrane. This may be the reason why penetratin has always been associated with low toxicity.[46] It is interesting to note that the lysine residues in transportan are beneficial for cell association whereas PenLys seemingly does not bind to the cell surface at all. Hence, it can be hypothesized that positive charge alone is not enough but either needs to be accompanied by a certain level of amphipathicity in the rest of the peptide (as in transportan) or by the hydrogen-bonding capability of arginine (as in penetratin).

Tat-peptides and oligoarginines clearly owe their efficiency to the arginine residues. No secondary structure is induced in these peptides upon binding to membranes, which may in fact facilitate their rather efficient entry via non-endocytotic mechanisms. The CPPs TatP59W and R_7W are able to induce membrane fusion in phospholipid vesicles (see Section 7.5.2), and R_7W shows an outstanding increase in fluorescence quantum yield upon binding to model membranes. Still, these peptides most likely reside in the headgroup region of the membrane and translocation in model systems has only been observed in GUVs. Even though leakage has not been observed in phospholipid vesicles, the ability of R_7W to cause vesicle fusion may be related to the toxicity that has been associated with oligoarginines in some studies.[46,48]

Returning to the classification of CPPs, it seems as if the more efficient the CPP, the higher the risk of toxicity. This relation seems to be true on both sides of the spectrum, since both oligoarginines (efficient entry without pore formation) and amphipathic peptides such as transportan and toxic prion peptides have been associated with toxicity. It can be speculated that penetratin, being in the middle, has beneficial attributes from both arginine-rich peptides and amphipathic peptides. The arginine residues may be helpful in creating charge-neutralized complexes on the membrane surface, ensuring strong and efficient binding, whereas the acquired amphipathicity upon secondary structure induction may be important in anchoring the CPP in the headgroup region, ensuring efficient endocytosis. Even though the number of arginines in penetratin is seemingly too low to allow for entry via nonendocytotic mechanisms, the substitution of the lysine residues for arginines (PenArg) is enough to make this pathway accessible. It would indeed be interesting to investigate further how this exchange might affect toxicity and membrane perturbation. It should also be noted that the modified Tat-peptide TatP59W can enter cells at low temperature, i.e., via nonendocytotic pathways. Tat(48–60) peptides, such as penetratin, are often related to low toxicity and thus this is an interesting case for further studies.[7,48]

It has already been established that entry via endocytosis is the only accessible pathway for the majority of CPP–cargo constructs, at least when cargo sizes exceed that of simple fluorescent markers or very small compounds. It is also possible that this pathway is indeed the most desirable one in terms of reduced toxicity. Clearly, the major challenge for the future development of CPPs is the ability to efficiently escape the endosomes without being degraded. It has already been shown that penetratin can indeed translocate from the lumen of an endosome-mimicking phospholipid vesicle in the presence of a transbilayer pH gradient (see Section 7.4.3). The escape ability of other CPPs needs to be further assessed in model systems, as well as being thoroughly investigated in cell systems. In addition, it may be important to enhance the endosomal escape of CPPs using pH-dependent membrane destabilizing agents. In addition, in the future, the issue of cell specificity needs to be addressed for the practical applications of CPPs. Here, one can envisage the addition of further sequence motifs that will promote interactions with particular components of the specific cell type to be targeted.

REFERENCES

1. Lundberg, M. and Johansson, M., Positively charged DNA-binding proteins cause apparent cell membrane translocation, *Biochem. Biophys. Res. Commun.*, 291(2), 367, 2002.

2. Richard, J.P., Melikov, K., Vives, E., Ramos, C., Verbeure, B., Gait, M.J., Chernomordik, L.V., and Lebleu, B., Cell-penetrating peptides. A reevaluation of the mechanism of cellular uptake, *J. Biol. Chem.*, 278(1), 585, 2003.

3. Fittipaldi, A., Ferrari, A., Zoppe, M., Arcangeli, C., Pellegrini, V., Beltram, F., and Giacca, M., Cell membrane lipid rafts mediate caveolar endocytosis of HIV-1 Tat fusion proteins, *J. Biol. Chem.*, 278(36), 34141, 2003.

4. Nakase, I., Niwa, M., Takeuchi, T., Sonomura, K., Kawabata, N., Koike, Y., Takehashi, M., Tanaka, S., Ueda, K., Simpson, J.C., Jones, A.T., Sugiura, Y., and Futaki, S., Cellular uptake of arginine-rich peptides: Roles for macropinocytosis and actin rearrangement, *Mol. Ther.*, 10(6), 1011, 2004.

5. Saalik, P., Elmquist, A., Hansen, M., Padari, K., Saar, K., Viht, K., Langel, U., and Pooga, M., Protein cargo delivery properties of cell-penetrating peptides. A comparative study, *Bioconjug. Chem.*, 15(6), 1246, 2004.

6. Wadia, J.S., Stan, R.V., and Dowdy, S.F., Transducible TAT-HA fusogenic peptide enhances escape of TAT-fusion proteins after lipid raft macropinocytosis, *Nat. Med.*, 10(3), 310, 2004.

7. Thoren, P.E., Persson, D., Isakson, P., Goksor, M., Onfelt, A., and Norden, B., Uptake of analogs of penetratin, Tat(48–60) and oligoarginine in live cells, *Biochem. Biophys. Res. Commun.*, 307(1), 100, 2003.

8. Goncalves, E., Kitas, E., and Seelig, J., Binding of oligoarginine to membrane lipids and heparan sulfate: Structural and thermodynamic characterization of a cell-penetrating peptide, *Biophys. J.*, 88(1), 234, 2005.

9. Golan, D.E., Alecio, M.R., Veatch, W.R., and Rando, R.R., Lateral mobility of phospholipid and cholesterol in the human erythrocyte membrane: Effects of protein–lipid interactions, *Biochemistry*, 23(2), 332, 1984.

10. Damberg, P., Jarvet, J., and Graslund, A., Micellar systems as solvents in peptide and protein structure determination, *Methods Enzymol.*, 339, 271, 2001.

11. Vold, R.R., Prosser, R.S., and Deese, A.J., Isotropic solutions of phospholipid bicelles: A new membrane mimetic for high-resolution NMR studies of polypeptides, *J. Biomol. NMR*, 9(3), 329, 1997.

12. Hauser, H., Short-chain phospholipids as detergents, *Biochim. Biophys. Acta*, 1508(1–2), 164, 2000.

13. Magzoub, M., Kilk, K., Eriksson, L.E., Langel, U., and Graslund, A., Interaction and structure induction of cell-penetrating peptides in the presence of phospholipid vesicles, *Biochim. Biophys. Acta*, 1512(1), 77, 2001.

14. Barany-Wallje, E., Andersson, A., Graslund, A., and Maler, L., NMR solution structure and position of transportan in neutral phospholipid bicelles, *FEBS Letters*, 567(2–3), 265, 2004.

15. Lindberg, M., Jarvet, J., Langel, U., and Graslund, A., Secondary structure and position of the cell-penetrating peptide transportan in SDS micelles as determined by NMR, *Biochemistry*, 40(10), 3141, 2001.

16. Lindberg, M., Biverstahl, H., Graslund, A., and Maler, L., Structure and positioning comparison of two variants of penetratin in two different membrane mimicking systems by NMR, *Eur. J. Biochem.*, 270(14), 3055, 2003.

17. Stewart, R.S. and Harris, D.A., A transmembrane form of the prion protein is localized in the Golgi apparatus of neurons, *J. Biol. Chem.*, 280(16), 15855, 2005.

18. Lundberg, P., Magzoub, M., Lindberg, M., Hallbrink, M., Jarvet, J., Eriksson, L.E., Langel, U., and Graslund, A., Cell membrane translocation of the N-terminal (1–28) part of the prion protein, *Biochem. Biophys. Res. Commun.*, 299(1), 85, 2002.

19. Magzoub, M., Oglecka, K., Pramanik, A., Goran Eriksson, L.E., and Graslund, A., Membrane perturbation effects of peptides derived from the N-termini of unprocessed prion proteins, *Biochim. Biophys. Acta*, 1716(2), 126, 2005.

20. Biverstahl, H., Andersson, A., Graslund, A., and Maler, L., NMR solution structure and membrane interaction of the N-terminal sequence (1–30) of the bovine prion protein, *Biochemistry*, 43(47), 14940, 2004.

21. Papadopoulos, E., Oglecka, K., Maler, L., Jarvet, J., Wright, P.E., Dyson, H.J., and Graslund, A., NMR solution structure of the peptide fragment 1-30, derived from unprocessed mouse Doppel protein, in DHPC micelles, *Biochemistry*, 45(1), 159–166, 2006.

22. Persson, D., Thoren, P.E.G., Lincoln, P., and Norden, B., Vesicle membrane interactions of penetratin analogues, *Biochemistry*, 43(34), 11045, 2004.

23. Ardhammar, M., Lincoln, P., and Norden, B., Invisible liposomes: Refractive index matching with sucrose enables flow dichroism assessment of peptide orientation in lipid vesicle membrane, *Proc. Natl Acad. Sci.*, 99(24), 15313, 2002.

24. Ardhammar, M., Mikati, N., and Norden, B., Chromophore orientation in liposome membranes probed with flow dichroism, *JACS*, 120(38), 9957, 1998.

25. Brattwall, C.E., Lincoln, P., and Norden, B., Orientation and conformation of cell-penetrating peptide penetratin in phospholipid vesicle membranes determined by polarized-light spectroscopy, *JACS*, 125(47), 14214, 2003.

26. Magzoub, M., Eriksson, L.E., and Graslund, A., Conformational states of the cell-penetrating peptide penetratin when interacting with phospholipid vesicles: Effects of surface charge and peptide concentration, *Biochim. Biophys. Acta*, 1563(1–2), 53, 2002.

27. Magzoub, M., Eriksson, L.E., and Graslund, A., Comparison of the interaction, positioning, structure induction and membrane perturbation of cell-penetrating peptides and nontranslocating variants with phospholipid vesicles, *Biophys. Chem.*, 103(3), 271, 2003.

28. Persson, D., Thoren, P.E., Herner, M., Lincoln, P., and Norden, B., Application of a novel analysis to measure the binding of the membrane-translocating peptide penetratin to negatively charged liposomes, *Biochemistry*, 42(2), 421, 2003.

29. Persson, D., Thoren, P.E., and Norden, B., Penetratin-induced aggregation and subsequent dissociation of negatively charged phospholipid vesicles, *FEBS Letters*, 505(2), 307, 2001.

30. Bellet-Amalric, E., Blaudez, D., Desbat, B., Graner, F., Gauthier, F., and Renault, A., Interaction of the third helix of Antennapedia homeodomain and a phospholipid monolayer, studied by ellipsometry and PM-IRRAS at the air–water interface, *Biochim. Biophys. Acta*, 1467(1), 131, 2000.

31. Beschiaschvili, G. and Seelig, J., Peptide binding to lipid bilayers. Binding isotherms and zeta-potential of a cyclic somatostatin analogue, *Biochemistry*, 29(49), 10995, 1990.

32. Persson, D., Thoren, P.E., Esbjorner, E.K., Goksor, M., Lincoln, P., and Norden, B., Vesicle size-dependent translocation of penetratin analogs across lipid membranes, *Biochim. Biophys. Acta*, 1665(1–2), 142, 2004.

33. Thoren, P.E., Persson, D., Esbjorner, E.K., Goksor, M., Lincoln, P., and Norden, B., Membrane binding and translocation of cell-penetrating peptides, *Biochemistry*, 43(12), 3471, 2004.

34. Wimley, W.C. and White, S.H., Experimentally determined hydrophobicity scale for proteins at membrane interfaces, *Nat. Struct. Biol.*, 3(10), 842, 1996.

35. Carnie, S. and McLaughlin, S., Large divalent cations and electrostatic potentials adjacent to membranes. A theoretical calculation, *Biophys. J.*, 44(3), 325, 1983.
36. Schwarz, G. and Beschiaschvili, G., Thermodynamic and kinetic studies on the association of melittin with a phospholipid bilayer, *Biochim. Biophys. Acta*, 979(1), 82, 1989.
37. Seelig, A. and Macdonald, P.M., Binding of a neuropeptide, substance P, to neutral and negatively charged lipids, *Biochemistry*, 28(6), 2490, 1989.
38. Stankowski, S., Surface charging by large multivalent molecules. Extending the standard Gouy–Chapman treatment, *Biophys. J.*, 60(2), 341, 1991.
39. Karlsson, M., Nolkrantz, K., Davidson, M.J., Stromberg, A., Ryttsen, F., Akerman, B., and Orwar, O., Electroinjection of colloid particles and biopolymers into single unilamellar liposomes and cells for bioanalytical applications, *Anal. Chem.*, 72(23), 5857, 2000.
40. Wimley, W.C. and White, S.H., Determining the membrane topology of peptides by fluorescence quenching, *Biochemistry*, 39(1), 161–170, 2000.
41. Magzoub, M., Pramanik, A., and Graslund, A., Modeling the endosomal escape of cell-penetrating peptides: Transmembrane pH gradient driven translocation across phospholipid bilayers, *Biochemistry*, 44(45), 14890–14897, 2005.
42. Thoren, P.E.G., Persson, D., Lincoln, P., and Norden, B., Membrane destabilizing properties of cell-penetrating peptides, *Biophys. Chem.*, 114(2–3), 169, 2005.
43. Lakowicz, J.R., *Principles of Fluorescence Spectroscopy*, 2nd ed., Kluwer Academic/Plenum Publishers, New York, 1999.
44. London, E. and Ladokhin, A.S., Measuring the depth of amino acid residues in membrane-inserted peptides by fluorescence quenching, In *Peptide–Lipid Interactions*, Academic Press Inc., San Diego, 2002.
45. Chaubey, B., Tripathi, S., Ganguly, S., Harris, D., Casale, R.A., and Pandey, V.N., A PNA-transportan conjugate targeted to the TAR region of the HIV-1 genome exhibits both antiviral and virucidal properties, *Virology*, 331(2), 418, 2005.
46. Jones, S.W., Christison, R., Bundell, K., Voyce, C.J., Brockbank, S.M., Newham, P., and Lindsay, M.A., Characterisation of cell-penetrating peptide-mediated peptide delivery, *Br. J. Pharmacol.*, 145(8), 1093, 2005.
47. Chaloin, L., Vidal, P., Heitz, A., Van Mau, N., Mery, J., Divita, G., and Heitz, F., Conformations of primary amphipathic carrier peptides in membrane mimicking environments, *Biochemistry*, 36(37), 11179, 1997.
48. Saar, K., Lindgren, M., Hansen, M., Eiriksdottir, E., Jiang, Y., Rosenthal-Aizman, K., Sassian, M., and Langel, U., Cell-penetrating peptides: A comparative membrane toxicity study, *Anal. Biochem.*, 345(1), 55, 2005.

8 Interactions of Cell-Penetrating Peptides with Model Membranes

Sébastien Deshayes, May C. Morris, Gilles Divita, and Frédéric Heitz

CONTENTS

8.1 INTRODUCTION

A large number of peptides, such as cell-penetrating peptides (CPPs), which exhibit their biological activities at the cell membrane are hydrosoluble. Because the task of solving the structure of membrane-embedded compounds is still difficult, it is tempting to deduce the membrane form from that identified for the water-soluble one. However, in spite of the recent progress in the identification of peptide and protein structures, the mechanisms and the structural consequences involved in the transfer of peptides from one of these media to the other is still not well understood. Clearly, several different situations are encountered going from a conservation of the overall structure to a major refolding. In fact, this concerns the initial and final steps of the transfer, and in the case of a water to membrane transfer an additional step, although transitory, has to be examined and concerns the adsorption at the water–membrane interface.[1] Another factor which plays a crucial role in the possible lipid-induced conformational changes lies in the chemical structure of the peptides and thus of their hydrophobic profile and flexibility providing the possibility for some sequences to strongly interact with the membrane components.

8.1.1 GENERAL CONSIDERATIONS

Membrane-active peptides are generally built of hydrophilic and hydrophobic sequences which generate amphipathic properties. When in solution in aqueous media, they fold such as to expose their hydrophilic residues toward the solvent and thus will often adopt a globular form which can be roughly compared to a micelle. This form is stabilized by the cohesion forces between the hydrophobic domains, and when they are transferred into an organic medium, the tendency will be reversed such as to anchor the hydrophilic and hydrophobic parts in the polar and nonpolar media, respectively. Hence, the structural versatility of such peptides is a direct consequence of their amphipathic character and the structural changes by interfacial uptake are known as superficial unfolding.

In practice, the situation can be much more complicated due to the complexity of biological membranes. As an example, conformational changes can depend on the nature of the two phases separated by the interface, as revealed by the behavior of hydrophobin.[2] Reconstitution of a natural membrane with all constituents (different types of phospholipids differing by their fatty acids and their headgroups also by the presence of other lipids, such as sphingolipids or cholesterol) and structural heterogeneities (rafts) is, up to now, an unrealistic task. The lipid composition and, therefore, the membrane organization strongly depend on the type of cell.

Here, we will describe the interactions of carrier peptides with membranes and how to estimate them. We will also describe the possible membrane-induced conformational changes of peptides in association with the most frequent methods used for the identification of the conformational changes. These phenomena will be illustrated with several examples characterizing the different types of behaviors observed. We will also provide examples illustrating the necessity of a multi-disciplinary approach in order to avoid possible misinterpretation of data when various methods of investigation are used separately.

8.2 METHODS OF INVESTIGATIONS AND THEIR LIMITATIONS

8.2.1 CIRCULAR DICHROISM

Circular dichroism (CD) is frequently used to identify the secondary structures of proteins or peptides. However, CD has some drastic constraints with regard to the sample. For peptides or proteins in solution, it is strongly recommended to record spectra in the far UV region, down to 180 nm, to obtain confident information related to the conformation. Such a low wavelength precludes the use of most of the common buffers (hepes, acetate, etc.) and also of some salts. A membrane-mimicking situation requires the presence of phospholipids, which will be organized as vesicles. Except for a few synthetic phospholipids, such as DOPG, which leads to transparent vesicles, most phospholipids form vesicles leading to milky solutions or suspensions with high scattering power. One issue to overcome these difficulties lies in the use of either lysophospholipids or detergents such as SDS, which form nonscattering micelles. However, a cautious interpretation of the data is required since micelle has geometrical constraints which do not exist for membranes and, in some cases the results obtained in this medium do not reflect a membrane situation.[3] CD has some advantages such as the determination of the orientation of helical sections of peptides or proteins in membrane. What is required is the preparation of the membranes in a multilayer array and the CD spectra measured at the normal and oblique incident angles with respect to the planes of the layers. Taking into account the artifacts due to dielectric interfaces, linear dichroism, and birefringence, this method allows the detection of orientation modifications when the hydration state is varied.[4]

8.2.2 FOURIER TRANSFORM INFRARED SPECTROSCOPY

Fourier transform infrared spectroscopy (FTIR) spectroscopy appears to be one of the most powerful methods for the identification of protein secondary structures and thus for the determination of conformational changes upon variations of the environment.[5–11] In addition, infrared spectroscopy can provide information for all membrane components. Most data reported so far have been obtained from solution and thus cannot account for the relative orientation of the bound peptide versus the phospholipids. To this end, several models have been constructed. Using polarized attenuated total reflection infrared spectroscopy, determination of the helical order parameter of melittin was revealed and the orientation of the helix was shown to depend on the hydration of the membrane preparation.[12] The first approach was carried out in situ using infrared refection (IRRAS) on phospholipid monolayers on a Langmuir trough. A considerable improvement was then provided by polarization modulation of the incident light (PMIRRAS), which has proved to be an efficient way to greatly increase the surface absorption detectivity while eliminating the intense isotropic absorption occurring in the sample environment.[13,14]

8.2.3 NMR

Beside crystal structure resolution, nuclear magnetic resonance (NMR) is the most precise method for a detailed identification of proteins and peptides structures. However, there are some limitations which lie mainly in the molecular weight of the particle under examination. While the structure of a protein of $\sim$150 residues can be easily solved, an increase of its molecular weight by oligomerization or by incorporation into a lipidic medium (vesicles being an example) precludes the acquisition of resolved spectra. For peptides interacting with membranes, most investigations were carried out in the presence of micelles, which are considered as membrane-mimicking media and give rise to resolved spectra. However, the question of membrane relevance to the data obtained in such a medium has to be raised.[3,15–17] This problem of conformational identification is overcome by the use of solid-state NMR, which has considerable potential for studies of membrane-embedded peptides.[18,19]

8.2.4 MASS SPECTROMETRY

Reports of the use of mass spectrometry to observe noncovalent complexes between lipids and soluble proteins, membrane proteins, or peptides have appeared only very recently. Compared to older mass spectrometric ionization techniques, electrospray mass spectrometry (ES-MS) has the advantage of enabling the preservation of noncovalent associations that exist in solution.[20,21] However, there is often concern that the gas-phase ions representing noncovalent complexes observed by mass spectrometry may not reflect the status of the component molecules in solution.[22–25] Moreover, it has been established that, as solvent molecules escape from the final charged electrospray droplets via evaporation, hydrophobic interactions are weakened, whereas electrostatic interactions are strengthened. In biological systems, noncovalent lipid–peptide or lipid–protein interactions are characterized by both electrostatic and hydrophobic components. Efforts have been made to specifically investigate the detailed binding specificities between selected phospholipids and model carrier peptides.

8.2.5 Atomic Force Microscopy

Atomic force microscopy (AFM) allows the observation of nanometer-sized particles and provides information on the topographical organization of transferred monolayers[26–29] or bilayers.[30–32] AFM observation of monolayers requires transfer to a solid substrate and because of its high vertical resolution, provides details on the domains existing in phase-separated thin films together with the localization of the protein and generally the observations deal with the hydrophobic side of the monolayer.

Bilayers, which are more biologically relevant, are obtained by double transfer or by fusion of small unilamellar vesicles to a glass or mica surface and the observations are made on the hydrophilic face of the bilayer. They allow the identification of membrane domain and their constituents.

8.2.6 Electrophysiological Measurements

These experiments provide information on the ability for the CPPs to induce pore formation when incorporated into lipid bilayers. Two major types of set up can be used: the first one deals with natural membranes such as those of oocytes which allow patch clamp experiments, while the second type of experiment is carried out on reconstituted planar lipid bilayers. The advantage of oocyte experiments lies in the fact that we are dealing with natural membranes while reconstituted lipid bilayers more easily allow single channel recordings and also allow decisions to be made as to whether the transmembrane currents are receptor-mediated or not.[33,34]

8.2.7 Monomolecular Films at the Air–Water Interface

Monomolecular films are good investigative models since the thermodynamic relationship between monolayers and bilayer membranes is direct, and mono-molecular films at the air–water interface overcome limitations such as regulation of lipid lateral-packing and lipid composition which occur in bilayers.[35] Measurements of the peptide–lipid interactions can be achieved by forming a lipid monomolecular film at a given surface pressure, close to the maximum obtained for the peptide at the air–water interface,[36] and by injecting into the subphase aliquots of the peptide solution.[36–39]

Mixed lipid/protein films can also be obtained by spreading on the water surface a mixture of the peptide and lipid dissolved in a volatile solvent[40–42] or by deposition of vesicles containing the desired mixture.[43] The conformational analysis of peptides interacting with lipids in an interfacial situation can be made by FTIR on Langmuir–Blodgett films transferred on to a germanium plate[44,45] or by CD when the transfer is made on to quartz plates.[46,47] Another advantage of monolayer lies in the fact that the lateral pressure can be varied and thus allows the study of the influence of this pressure on the conformational state of the peptide.[48] However, such analyses have to be made with care since the transfer process can induce structural modifications. Thus, in situ methods have been developed accordingly.[49,50]

8.2.8 BILAYERS

Contrary to monolayers, bilayer properties are more restricted but are better models for plasmic membranes and most investigations have been carried out with liposomes. After uptake of the peptide or the protein by the liposomes, conformational analysis can be carried out by CD, x-ray, or infrared spectroscopy.[5,51–54] For helix-rich proteins, insertion into the bilayers induces pore formation,[55] while other peptides or proteins interact only with receptors being membrane-embedded. These receptor–peptide interactions can induce conformational changes and in some cases can promote fusion processes.[56] Analysis of membrane-embedded peptides or proteins can also be carried out on bilayers obtained by the transfer protocol either by the Langmuir–Blodgett or the Langmuir–Schaeffer methods.[57]

8.2.9 FLUORESCENCE

Most proteins contain intrinsic fluorophores which are the aromatic amino acids tryptophan (Trp), tyrosine (Tyr), and phenylalanine (Phe). Protein fluorescence is generally obtained by excitation at the absorption maximum near 280 nm or higher wavelengths. Under these conditions, Phe is not excited and hence this residue cannot be considered as an appropriate probe for fluorescence investigations on proteins. In fact, the most useful residue is Trp since it can be selectively excited from 295 up to 305 nm and is sensitive to both specific and general environment effects.[58] The position of the maximum emission provides information on the Trp environment. A low wavelength maximum (around 330 nm) is indicative of a Trp in a nonpolar environment, i.e., in contact with the phospholipid hydrocarbon chains, while a high wavelength maximum (around 355 nm) reflects a polar environment close to the phospholipid headgroups.

Other types of measurements based on fluorescence properties can be carried out. These are fluorescence transfer and quenching experiments. The latter deal with accessibility of quenchers, such as I^-, Cs^+, trichloroethanol, or acrylamide to the probe,[59] while the former can be carried out using fluorescently labeled lipids or specifically bromine labeled phospholipids which can provide information on the positioning of the fluorophore within the hydrophobic phase of the membrane bilayer.

8.3 EXAMPLES OF ANALYSIS BASED ON TWO PRIMARY AMPHIPATHIC CPPs

These peptides have a common hydrophilic domain, the nuclear localization sequence (NLS) of SV40 large T antigen: PKKKRKV. All bear a WSQ sequence, which acts as a linker between the hydrophilic and hydrophobic domains, thereby maintaining their integrity. For MPG, the selected hydrophobic sequence was chosen from HIV-1 fusion protein gp41. For Pep peptides, the hydrophobic sequence was selected from the dimerization motif at the interface of HIV-1 reverse transcriptase. All of these peptides are acetylated and bear a cysteamine group at their N- and C-termini, respectively:

GALFLGFLGAAGSTMGAWSQPKKKRKV for MPG
KETWWETWWTEWSQPKKKRKV for Pep-1

Several variants of the parental MPG peptide have been tested. The first one (MPG-W) is the result of a W^7 to F substitution and results in efficient nuclear localization in fibroblasts.[60–62] This behavior prompted us to synthesize other variants in order to control the addressed subcellular compartment for nucleic acid cargoes. This was achieved by a K to S substitution in position 23 corresponding to the second lysine of the NLS sequence, generating the vector peptide MPG-Δ^{NLS} which is efficient to transfer RNAs,[63,64] siRNAs,[65,66] double-stranded phosphorothioate oligonucleotides,[67] and more sophisticated oligonucleotides such as $N^{3'} \rightarrow P^{5'}$ thio-phosphoramidates without formation of any covalent linkage.[68,69]

The peptides of the Pep family were designed with the aim of delivering peptides, proteins, or PNAs.[70,71] Pep-1, the leader peptide of the Pep family, forms complexes with proteic cargoes which are rapidly delivered into a great variety of cell lines and applied in cellulo and in vivo.[72,73]

8.3.1 MECHANISMS OF TRANSLOCATION

Identification of the mode of action of CPPs is crucial for the design of future generations of CPPs. Investigation of the mechanism of internalization requires identification of several physicochemical properties of the carrier peptides. First, it is crucial to elucidate the type of interaction that they can elicit in the presence of membrane components, mainly the phospholipids. It is also necessary to identify the structural criteria, mainly the peptide primary and secondary structures, which can influence the internalization process. Finally, three main entry mechanisms can be examined: direct penetration into the membrane, translocation through formation of a transient structure, and an endocytosis-mediated entry.

With respect to the interactions of CPPs with the plasma membrane, some differences can be noticed depending on the nature of the CPP. As an example, penetratin differs from transportan in that the former interacts preferentially with negatively charged membranes, whereas interactions of the latter do not depend on charges.[74] Furthermore, comparison of internalization properties of all-L with those of all-D peptides[75,76] indicate that this process is not receptor-mediated as confirmed by the use of giant unilamellar vesicles.[77] Other investigations point out the importance of the heparan sulfates (HS) present at the cell surface. Indeed, internalization of Tat, penetratin, and polyarginines is inhibited upon degradation of the heparan sulfates, by addition of heparin or sulfated polysaccharides or HS-deficient cell lines.[78–80] However, other authors claim that Tat and penetratin can enter cells independently of the presence of heparan sulfates.[77,81]

A last mechanism has recently been proposed for internalization mediated by peptides belonging to the family of primary amphipathic peptides, namely MPG and Pep-1. On the basis of physicochemical investigations, including CD, FTIR, and NMR spectrometry[82–85] associated with electrophysiological measurements and investigations dealing with the use of systems mimicking model membranes such as monolayers, two very similar models have been proposed. Both are based on the formation of transient pore-like structures. The main difference between the model proposed for MPG and that proposed for Pep-1 is found in the structure, giving rise to the pore structure. For MPG, it is formed by a β-barrel structure,[84] while that of Pep-1 depends on association of helices,[85] which were used and enabled us to the proposal of pore-based models accounting for the cellular internalization process of MPG and Pep-1.

8.3.2 Conformational Investigations and Physicochemical Approaches

8.3.2.1 Pep-1

8.3.2.1.1 *Structural Characterization of Pep-1 in Solution and in the Presence of SDS*

For concentrations ranging between 0.1 and 0.3 mg/mL, the CD spectrum of Pep-1 in solution in water shows a single negative band centered at 202 nm associated with a shoulder around 220 nm, suggestive of a poorly ordered structure. Increasing the concentration of Pep-1 up to 3 mg/mL promotes a dramatic modification and leads to a spectrum representative of a helical conformation.[86] This tendency for Pep-1 to adopt, at least in part, a helical structure is confirmed by observations in media containing SDS micelles, where a spectrum with two minima at 207 and 222 nm and a maximum at 192 nm is obtained, characterizing a helical conformation. The helical folding of Pep-1 is confirmed by NMR investigations. The sequential and medium range NOEs observed for Pep-1 in H_2O and Pep-1 in the presence of SDS. The observed NOEs for Pep-1 in H_2O in the segments 4–13 are consistent with the existence of a helical secondary structure in this part of the sequence. In the presence of SDS, the observed NOEs are characteristic of a helix for the same segment (residues 4 through 13). In addition, several NOEs detected at the beginning of the sequence are indicative of a 3_{10} helix. Interestingly, the NLS moiety of Pep-1 remained unstructured in both media.[85]

8.3.2.1.2 *Structural Characterization of Pep-1 in the Presence of Phospholipids*

The structural states of Pep-1 in the presence of phospholipids have also been investigated. CD shows that addition of vesicles of DOPC/DOPG 80/20 to a dilute solution of Pep-1 in water induced a structural transition with formation of a helical structure. From the FTIR point of view, all spectra showed a complex contour of the Amide I band. In the absence of lipids, two distinct contributions were observed, one at 1625–1630 cm^{-1} and the other centered at 1655–1660 cm^{-1}. The presence of phospholipids generated a significant decrease of the 1625 cm^{-1} contribution associated with a broadening of the 1655–1660 cm^{-1} contribution (Figure 8.1).[5] The finding of a broad band is in full agreement with the NMR data obtained in SDS (see above) and indicates that two helical forms are maintained in a lipidic medium (1655 cm^{-1}, α-helix; 1665 cm^{-1}, helix 3_{10}).[87,88]

8.3.2.1.3 Studies at the Air–Water Interface: Adsorption at Lipid-Containing Interfaces

The ability of Pep-1 to penetrate into lipidic media has been determined by the monolayer approach. This required first the determination of the saturating surface pressure induced by the peptide at a lipid-free air–water interface. The variation of the surface tension as a function of the peptide in the subphase showed a saturation at $5 \times 10^{-7}\, M$ of Pep-1 and that the corresponding surface tension was rather low (4 mN/m) compared to other CPP (>15 mN/m),[3,89] indicating a weak amphipathic character for Pep-1. The understanding of the mechanism through which Pep-1 penetrates into membranes was performed through penetration experiments using phospholipids in the liquid expanded (DOPC and DOPG) or liquid condensed (DPPC and DPPG) states. In the case of liquid-expanded monolayers, a strong increase in the surface pressure is observed. The most important characteristics of these experiments was found in the fact that (1) both DOPC and DOPG yielded identical cpi (45 mN/m) and that (2) extrapolations at zero initial pressure were high

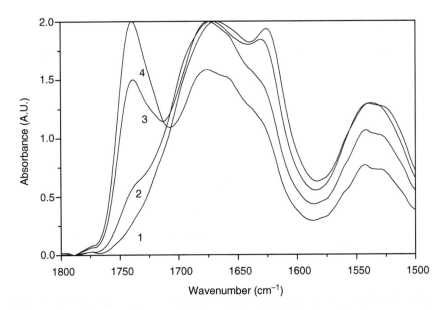

FIGURE 8.1 FTIR spectra (Amide I and II region) of Pep-1/phospholipid mixtures at various peptide/lipid ratios. Spectra 1–4 correspond to peptide/lipid ratios of 1/0, 1/1, 1/10, and 1/20, respectively. (From Deshayes, S. et al., *Biochemistry*, 43, 1449, 2004. With permission.)

(16 and 32 mN/m for DOPC and DOPG, respectively), and significantly different from that measured in the absence of lipid. For the two other phospholipids, DPPC and DPPG, again the cpi is high (33 mN/m) and does not depend on the nature of the headgroups. (Figure 8.2). The high values found for the cpi suggest that the Pep-1 can spontaneously insert into natural membranes.[90] Extrapolations at zero initial pressure provide a good indication that strong peptide–lipid interactions can occur at least in monolayers, with all lipids except DPPC.

8.3.2.1.4 Penetration into Phospholipid Vesicles

The ability of Pep-1 to penetrate into phospholipid bilayers was investigated by fluorescence spectroscopy. Pep-1 contains five Trp residues in its hydrophobic moiety, which are probes for monitoring interactions and changes in its environment upon penetration into phospholipid bilayers by intrinsic fluorescence. Additions of phospholipid vesicles to a solution of Pep-1 induced significant changes in the fluorescence spectrum. A shift of the fluorescence maximum (from 350 to 328 nm) accompanied by quenching of fluorescence occurred when the lipid/peptide ratio increased from 0 to 5 (Figure 8.3). The blue shift of fluorescence is consistent with a change in the environment of the Trp residues from polar to nonpolar in the presence of phospholipid vesicles.[58] This provided a good indication that the Trp residues are embedded in the lipidic core when Pep-1 encounters a membrane. In addition, quenching suggested a transfer between Trp residues which are engaged in clusters when the peptide is in a helical structure.

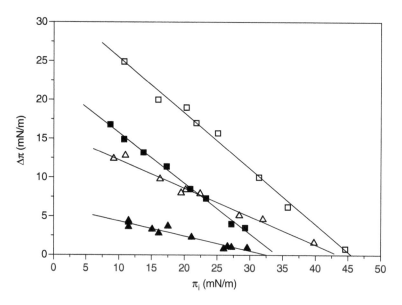

FIGURE 8.2 Pep-1 induced variation of the surface pressure as a function of the initial surface pressure of the phospholipid monolayer. (Δ) DOPC, (□) DOPG, (▲) DPPC, and (■) DPPG. Extrapolation at zero initial pressure gives the critical pressure of insertion. (From Deshayes, S. et al., *Biochemistry*, 43, 1449, 2004. With permission.)

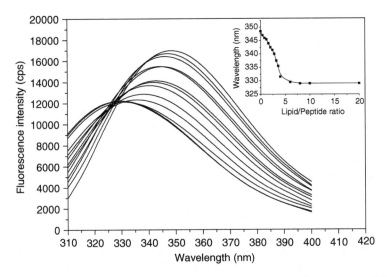

FIGURE 8.3 Effect of the addition of phospholipid vesicles on the fluorescence of the Trp residues of Pep-1. The various spectra correspond to the peptide/lipid ratio quoted on the inset, which shows the variation as a function of the peptide/lipid ratio of the wavelength corresponding to the maximum of the fluorescence emission. (From Deshayes, S. et al., *Biochemistry*, 43, 1449, 2004. With permission.)

8.3.2.1.5 Formation of Carrier/Cargo Complexes and Structural Consequences

The ability of Pep-1 to interact with a cargo peptide was investigated by fluorescence spectroscopy together with the structural consequences of this interaction. Binding of the cargo peptide to Pep-1 induced a marked quenching of the intrinsic tryptophan fluorescence of Pep-1, with a saturating value of 36%. Moreover, a blue shift of the fluorescence emission maximum of 11 nm suggested that the Trp residues of Pep-1 interact directly with the cargo peptide. Saturation was reached at a concentration, fivefold lower than that of the Pep-1, suggesting that the cargo peptide interacts with more than one molecule of Pep-1. As to the conformational consequences of these interactions, CD observations indicate that the formation of a complex does not induce any structural modification of Pep-1. Upon addition of phospholipid vesicles to a preformed Pep-1/cargo complex, the modifications of the CD spectrum show the same trend as those obtained for free Pep-1. This indicated that the particle formed by complexes of Pep-1 and its cargo interacts with lipid bilayers and that this interaction promotes, at least in part, a conformational transition of Pep-1 to an α-helical form.

8.3.2.1.6 Electrophysiological Measurements

When Pep-1 (10 μM) were applied to voltage-clamped oocytes, a marked increase in membrane conductance was noted. This increase was best visualized by an increase in membrane current recorded during voltage ramps applied from -80 mV (the usual holding potential) to $+80$ mV. These observations suggested that the permeabilizing capabilities of Pep-1 are due to the formation of membrane ion channels.

8.3.2.2 MPG

8.3.2.2.1 CD and FTIR Investigations

In water, MPG shows a CD spectrum with a single minimum at 198 nm, characteristic of a nonordered structure. The spectrum remains unchanged either when the concentration of peptide is varied or when phosphate buffer is added. The influence of phospholipids on the conformation of the peptides was examined by FTIR. For peptide MPG, the FTIR spectra show that in the absence and in the presence of phospholipids (DOPC, DOPG, DPPC, and DPPG) at any peptide/lipid ratio, the sheet structure, as characterized by the presence of a major Amide I band component centered around 1625 cm^{-1}, is favored.

8.3.2.2.2 Studies at the Air–Water Interface: Adsorption at Lipid-Containing Interfaces

The ability of peptide MPG to insert into phospholipid monolayers spread at the air–water interface was carried out as shown for Pep-1. The various plots of pressure variation vs. the initial surface pressure showed that, although still high, the critical pressure of insertion increases from 30 to 38 mN/m for DPPC and DPPG, respectively, indicating a better uptake by negatively charged phospholipids. For lipids in the liquid expanded (LE) state, insertion depends on both the initial conformational state of the peptide and the nature of the headgroups. In all cases, the critical pressures of insertion are high. Further examination of the various plots also provided information on the interactions occurring between the peptide and the phospholipids considered here. These are based on the pressure value measured for an air–water interface at low lipid content (i.e., at low initial pressure). The finding of surface pressures higher than these obtained for the pure peptides at saturation indicated strong peptide–lipid interactions.[91]

8.3.2.2.3 Compression Isotherms

The compression isotherms obtained for MPG when mixed with various phospholipids, DOPC, DOPG, DPPC, and DPPG. The isotherms of the pure lipids are in agreement with those already reported.[92–94] The isotherms were analyzed by examining the variation of the mean molecular area (that of the contributions of both the peptide and the lipid) as a function of the peptide/lipid ratio at a given and constant surface pressure. These variations at a pressure of 20 mN/m, which is below that of the collapse, revealed a small but significant and reproducible deviation from linearity and that several situations are encountered depending on the peptide–lipid combination. Interestingly, for lipids in the LE state, namely DOPC and DOPG, positive and negative deviations from linearity occur depending on the peptide conformation. The nonlinear variation of the mean molecular area provides a good argument suggesting that the peptides and lipids are miscible and interact.[95–97] With phospholipids in the LC state (DPPC and DPPG), all peptide–lipid combinations reveal an expansion of the mean molecular area, indicating that the peptides interact with and are miscible with DPPC and DPPG (Figure 8.4).

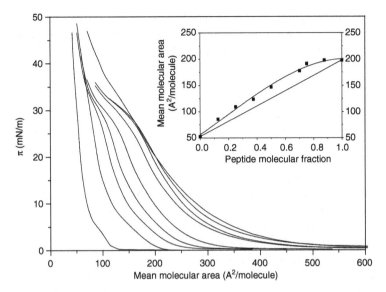

FIGURE 8.4 Compression isotherms of peptide MPG in the presence of DPPC and at various peptide/lipid ratios. The insets show the variations of the mean molecular areas as a function of peptide/lipid ratios. (From Deshayes, S. et al., *Biochemistry*, 43, 7698, 2004. With permission.)

8.3.2.2.4 AFM Observations

Due to the strong possibility of artifactual observations, it appears that some types of experiments require some controls using peptides closely related to the leader peptides, such as MPG, but with different sequences so that they can adopt other conformational states. For this purpose, an analog of MPG was designed to adopt a helical structure in its hydrophobic domain. This peptide will be referred to hereafter as [Pα]. In this section and in that devoted to the analysis by mass spectrometry, the observations carried out on MPG are systematically compared with those of [Pα].

AFM observations were performed in air on LB films transferred at a surface pressure of 26 mN/m on to a cleaved mica surface. Accordingly, the AFM tip imaged the hydrophobic side of the monolayer. As compared to pure DOPC films, the roughness of peptide [Pα]-containing DOPC monolayers was significantly increased. This also applied to peptide [Pα]-containing DOPG films. This roughness corresponded to the presence of round-shaped structures 15–25 nm in diameter. As in the films made of pure [Pα] peptide filamentous structures ~15 nm in diameter, up to 1 μm in length and protruding by ~0.5 nm were also present in [Pα]/DOPC monolayers. However, they were not detected in [Pα]/DOPG films. Addition of [Pα] to the solution spread at the air–water interface had a marked effect for films made of saturated phospholipid species DPPC and DPPG. Thus, instead of a flat surface, with perhaps a few linear defects,[29,98] AFM imaging revealed a complex topography with light domains, corresponding to DPPC liquid condensed phase surrounded by a darker matrix, ~1.1 nm thinner. The size of the LC domains varies from a couple to a few hundred nanometers. The structure of the DPPG films was also disorganized with the presence of both large and small smooth LC domains surrounded by a darker matrix, enriched in peptide. As for DPPC, the lighter domains protruded by ~ 1.1 nm from the darker matrix. On the other hand, no filaments could be visualized in [Pα]/DPPG monolayers. This disruption of gel phase organization induced by the peptide strongly suggests the existence of hydrophobic interactions between [Pα] and the phospholipids. In addition, the absence of filaments for PG species also suggests that the negatively charged PG polar headgroups enhance the lipid–peptide interactions at the expense of peptide–peptide interactions. The images obtained with [MPG] and phosphatidylcholine phospholipids were comparable to those obtained with [Pα] with the presence of both filaments and small aggregates in DOPC films and an important reduction of the DPPC gel phase domains associated with the appearance of filaments in the darker matrix. The presence of filaments in these samples was unexpected because pure [MPG] formed globular aggregates, 20 nm in section, but no filaments. The results of experiments performed with DOPG and DPPG revealed a behavior that markedly differed from that of [Pα]. First, in DOPG the size of peptide-induced structures significantly increased, up to ~ 50 nm for individual globular structures. Furthermore, larger domains of various shapes, likely resulting from the coalescence of individual globular structures, are also formed. Gel phase DPPG domains can no longer be identified in [MPG]–DPPG films that appeared to be constituted of aggregates of different sizes and thicknesses. This drastic reorganization of monolayers likely results from the strongest interactions between [MPG] and the negatively charged phospholipids.

8.3.2.2.5 Mass Spectrometry

Detailed binding specificities between selected phospholipids and carrier peptides have been examined.[99,100] DMPG, which carries a negative charge, exhibits a stronger binding to MPG and Pα than DMPC, which is zwitterionic in nature. The increased electrostatic interaction clearly played a significant role in stabilizing peptide–PG complexes. Methanol addition (known to weaken hydrophobic interactions) disrupted binding between MPG/[Pα] and PC/PG that had been observed from 100% aqueous solutions; a stronger initial interaction required a higher percentage of methanol to destroy binding. These results indicate that detected MPG/[Pα]–lipid complexes were already formed in 100% aqueous solution, with the hydrophobic effect being a primary driving force promoting the interaction. Upon addition of methanol, however, it showed an initial slight decrease in binding to lipids that was followed by an increase in detected binding upon further methanol addition. This behavior contrasts with that observed for the fusion peptides and PG, and thus offers additional evidence that hydrophobic interactions play a key role in allowing the mass spectrometric observation of the latter complexes. Furthermore, with a lowering of pH, detected [Pα]–DMPC complexes exhibited an immediate and steep drop in binding, whereas the detected MPG–DMPC binding was slightly augmented at pH 3.7 and then decreased gradually at even higher acidity. A comparison of lipid–peptide binding as a function of the degree of unsaturation offered the opportunity to test the effect of the hydrophobicity of the lipid on binding. While both [Pα] and MPG exhibited affinities for unsaturated lipid,

(18:1) PC bound slightly more strongly to MPG than the less hydrophobic (18:3) PC. Again, evidence for the importance of an initial solution hydrophobic effect in establishing binding between carrier peptides and cell membrane phospholipids is offered. Study findings corroborate the notion that hydrophobic interactions between fusion proteins and cell membrane phospholipids can serve to initiate membrane perturbation in the early stages of viral fusion. At the same time, the ability of ES-MS to provide information regarding the strength of noncovalent interactions that originated from a hydrophobic effect is established.

8.3.2.2.6 Polynucleotide Binding and Induced Conformational Changes

Fluorescence and gel shift assays were used to monitor formation of a peptide/oligonucleotide complex. A solution of HEX-labeled oligonucleotide (18-mer) was titrated with MPG. Changes in fluorescence intensity were monitored and show an abrupt linear decrease, the extrapolation of which reveals a ratio of about seven peptides per oligonucleotide. In terms of charges, this result indicates that the positive/negative charge ratio is 2 for the complex, in good agreement with previous results obtained with a double-stranded DNA.[101] An identical stoichiometry for the complex was obtained by titration of the peptide by the oligonucleotide. Moreover, determination of the equilibrium constants from both experiments indicates that the peptide exhibits high affinity for the oligonucleotide, respectively, 6 and 5 nM.

In gel shift assays, oligonucleotide was incubated with the peptide at different peptide/oligonucleotide ratios and then submitted to gel migration. For a charge ratio $(+/-)$ of 1 or lower, the free form of the oligonucleotide can be detected, while an increase of this ratio leads to disappearance of this form. Taken together, these data indicate that the carrier peptide MPG strongly interacts with an oligonucleotide cargo to form a complex containing most probably more than seven peptide molecules per oligonucleotide.

Addition of the oligonucleotide to a solution of MPG in water in an oligonucleotide/peptide ratio ranging from 0 to 0.05 promotes changes in the CD spectrum. These spectral variations indicate a decrease in the amount of unfolded peptide to the benefit of sheet structures. Interestingly, the CD spectra obtained for oligonucleotide/peptide ratios ranging from 0 to 0.1 indicate formation of a small but significant amount of β-sheet structure. The mixtures corresponding to the ratios previously used in transfection assays (0.1 or 0.2)[63,64,66] were used for further experiments related to the influence of phospholipids on the structure of the peptide engaged in a complex with oligonucleotide. Typical observed variations of the CD spectrum reveal that addition of DOPG induces spectral modifications that are similar to those observed in the absence of oligonucleotide, thereby indicative of formation of a β-sheet-based structure.

8.3.2.2.7 Peptide-Induced Ionic Leakage

As already shown for Pep-1, when MPG in its free form or engaged in a complex at an oligonucleotide/peptide ratio of 0.1 was applied to voltage-clamped oocytes, marked increases in membrane conductance were detected. The presence of nucleic acids has no effect on the reversal potential, but appeared to reduce the current amplitude. It was impossible to decide whether this decrease in the current arose from an artificial decrease of the peptide concentration because it is engaged in the complex, or to modification of the transmembrane current characteristics. Nevertheless, electrophysiological measurements, which strongly resembled those obtained for Pep-1 and for other channel-forming, CPP suggested that the peptide-induced membrane permeabilization properties were due to formation of an ion channel.

8.3.3 THE MODELS

On the basis of physicochemical investigations, including CD, FTIR, and NMR spectrometry associated with electrophysiological measurements and investigations dealing with the use of systems mimicking model membranes, such as monolayers at the air–water interface and transferred monolayers, two very similar models have been proposed. For MPG, it is formed by a β-barrel structure (Figure 8.5), while that of Pep-1 depends on association of helices forming transient pores. For both peptides the structured domains of the carrier molecule correspond to the hydrophobic domain, while the rest of the molecule remains unstructured.

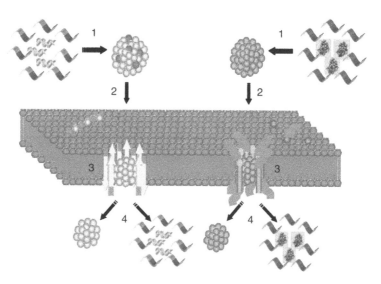

FIGURE 8.5 Model for the translocation process of MPG/cargo (left) and Pep-1/cargo (right) complexes through phospholipid bilayers. For both carriers, the four steps correspond to (1) formation of the complex; (2) membrane uptake of the complex; (3) translocation through the bilayer; and (4) release into the cytoplasmic side (color convention: red corresponds to the NLS of both carriers, yellow and orange to the hydrophobic domain of MPG and Pep-1, respectively). (From Deshayes et al., *Biochim. Biophys. Acta*, 1667, 141, 2004 and Deshayes, S. et al., *Biochemistry*, 43, 1449, 2004. With permission.)

8.4 CONCLUSIONS AND PERSPECTIVES

Although, some of the CPP described above begin to receive applications mainly in research laboratories, some weaknesses are still encountered, in particular due to the lack of specificity toward targets. This implies that some modifications of their sequences or the introduction of nonnatural amino acids bearing side-chains that can improve their specificity is required. However, it is now known that modifications in their sequences, although minor, can strongly modify their ability to act as drug carriers. It is therefore necessary to identify precisely the criteria, which can define an efficient, cell-penetrating peptide with a high degree of drug transfer. This requires the understanding of the interactions of these carrier peptides with all the membrane components in association with the structural consequences of these interactions. These criteria have to be satisfied and will help in the design of the next generation of carrier peptides.

REFERENCES

1. Heitz, F. and Van Mau, N., Protein structural changes induced by their uptake at interfaces, *Biochim. Biophys. Acta*, 1597, 1, 2002.
2. de Vocht, M.L. et al., Structural characterization of the hydrophobin SC3, as a monomer and after self-assembly at hydrophobic/hydrophilic interfaces, *Biophys. J.*, 74, 2059, 1998.
3. Chaloin, L. et al., Conformations of primary amphipathic carrier peptides in membrane mimicking environments, *Biochemistry*, 36, 11179, 1997.
4. Wu, Y., Huang, H.W., and Olah, G.A., Method of oriented circular dichroism, *Biophys. J.*, 57, 797, 1990.
5. Tamm, L.K. and Tatulian, S.A., Infrared spectroscopy of proteins and peptides in lipid bilayers, *Q. Rev. Biophys.*, 30, 365, 1997.
6. Goormaghtigh, E., Cabiaux, V., and Ruysschaert, J.M., Determination of soluble and membrane protein structure by Fourier transform infrared spectroscopy. I. Assignments and model compounds. II. Experimental aspects, side chain structure, and H/D exchange. III. Secondary structure, In *Subcellular Biochemistry*, Hiderson, H.J. and Ralston, G.B., eds., Vol. 23, Physico-chemical Methods in the Study of Biomembranes, New York, Plenum Press, pp. 329–450.
7. Goormaghtigh, E., Raussens, V., and Ruysschaert, J.M., Attenuated total reflection infrared spectroscopy of proteins and lipids in biological membranes, *Biochim. Biophys. Acta*, 1422, 105, 1999.
8. Haris, P.I. and Chapman, D., The conformational analysis of peptides using Fourier transform IR spectroscopy, *Biopolymers*, 37, 251, 1995.
9. Vié, V. et al., Detection of peptide–lipid interactions in mixed monolayers, using isotherms, atomic force microscopy, and Fourier transform infrared analyses, *Biophys. J.*, 78, 846, 2000.
10. Van Mau, N. et al., Lipid induced organization of a primary amphipathic peptide: A coupled AFM-monolayer study, *J. Membr. Biol.*, 167, 241, 1999.
11. Tatulian, S. and Tamm, L., Secondary structure, orientation, oligomerization, and lipid interactions of the transmembrane domain of influenza hemagglutinin, *Biochemistry*, 39, 496, 2000.
12. Frey, S. and Tamm, L.K., Orientation of melittin in phospholipids bilayers. A polarized attenuated total refelection infrared study, *Biophys. J.*, 60, 922, 1991.

13. Blaudez, D. et al., Polarization-modulated FT-IR spectroscopy of spraed monolayers at the air–water interface, *Appl. Spectrosc.*, 47, 869, 1993.

14. Cornut, I. et al., *In situ* study by polarization-modulated Fourier transform infrared spectroscopy of the structure and orientation of lipids and amphipathic peptides at the air–water interface, *Biophys. J.*, 70, 305, 1996.

15. Lindberg, M. et al., Secondary structure and position of the cell-penetrating peptide transportan in SDS micelles as determined by NMR, *Biochemistry*, 40, 3141, 2001.

16. Vidal, P. et al., Conformational analysis of primary amphipathic carrier peptides and origin of the various cellular localizations, *J. Membr. Biol.*, 162, 259, 1998.

17. Magzoub, M. et al., Interaction and structure induction of cell-penetrating peptides in the presence of phospholipids vesicles, *Biochim. Biophys. Acta*, 1512, 77, 2001.

18. Cross, T., Solid-state nuclear magnetic resonance characterization of gramicidin channel structure, *Methods Enzymol.*, 289, 672, 1997.

19. Yang, J., Gabrys, C.M., and Weliky, D.P., Solid-state nuclear magnetic resonance evidence for an extended β strand conformation of the membrane-bound HIV-1 fusion peptide, *Biochemistry*, 40, 8126, 2001.

20. Ganem, B., Li, Y.T., and Henion, J.D., Observation of noncovalent enzyme–substrate and enzyme–product complexes by ion-spray mass spectrometry, *J. Am. Chem. Soc.*, 113, 7818, 1991.

21. Katta, V. and Chait, B.T., Observation of the heme-globin complex in native myoglobin by electrospray-ionization mass spectrometry, *J. Am. Chem. Soc.*, 113, 8534, 1991.

22. Ogorzalek-Loo, R.R. et al., Observation of a noncovalent ribonuclease S-protein/ S-peptide complex by electrospray ionization mass spectrometry, *J. Am. Chem. Soc.*, 115, 4391, 1993.

23. Light-Wahl, K.J., Winger, B.E., and Smith, R.D., Observation of the multimeric forms of concanavalin A by electrospray ionization mass spectrometry, *J. Am. Chem. Soc.*, 115, 5869, 1993.

24. Knight, W.B. et al., Electrospray ionization mass spectrometry as a mechanistic tool: Mass of human leukocyte elastase and a β-lactam-derived E-I complex, *Biochemistry*, 32, 2031, 1993.

25. Fitzgerald, M.C. et al., Probing the oligomeric structure of an enzyme by electrospray ionization time-of-flight mass spectrometry, *Proc. Natl Acad. Sci. U.S.A.*, 93, 6851, 1996.

26. Edidin, M., Lipid microdomains in cell surface membranes, *Curr. Opin. Struct. Biol.*, 7, 528, 1997.

27. Zazadzinski, J.A. et al., Langmuir–Blodgett films, *Science*, 263, 1726, 1994.

28. ten Grotenhuis, E.R. et al., Phase behavior of stratum corneum lipids in mixed Langmuir–Blodgett monolayers, *Biophys. J.*, 71, 1389, 1996.

29. Vié, V. et al., Distribution of ganglioside GM1 between two-components, two phase phosphatidylcholine monolayers, *Langmuir*, 14, 4574, 1998.

30. Mou, J.D., Czalkowsky, M., and Shao, Z., Gramicidin A aggregation in supported gel state phosphatidylcholine bilayers, *Biochemistry*, 34, 3222, 1996.

31. Gliss, C. et al., Direct detection of domains in phospholipids bilayers by grazing incidence diffraction of neutrons and atomic force microscopy, *Biophys. J.*, 74, 2443, 1998.

32. Hollars, C.W. and Dunn, R.C., Submicron structure of L-α-dipalmitoylphosphatidyl-choline monolayers and bilayers probed by confocal, atomic force, and near field microscopy, *Biophys. J.*, 75, 342, 1998.

33. Chaloin, L. et al., Ionic channels formed by primary amphipathic peptides, *Biochim. Biophys. Acta*, 1375, 62, 1998.
34. Dé, E. et al., Conformation and ion channel properties of a five helix bundle protein, *J. Pept. Sci.*, 7, 41, 2001.
35. Brockman, H., Lipid monolayers: Why use half a membrane to characterize protein-membrane interactions?, *Curr. Opin. Struct. Biol.*, 9, 438, 1999.
36. Rafalsky, M., Lear, J.D., and DeGrado, W.F., Phospholipid interactions of synthetic peptides representing the N-terminus of HIV gp41, *Biochemistry*, 29, 7917, 1990.
37. Chiang, C.-M. et al., Conformational change and inactivation of membrane phospholipids-related activity of cardiotoxin V from Taiwan cobra venom at acidic pH, *Biochemistry*, 35, 9167, 1996.
38. Van Mau, N. et al., The SU glycoprotein 120 from HIV-1 penetrates into lipid monolayers mimicking plasma membranes, *J. Membr. Biol.*, 177, 251, 2000.
39. Nordera, P., Dalla Serra, M., and Menestrina, G., The adsorption of *Pseudomonas aeruginosa* exotoxin A to phospholipid monolayers is controlled by pH and surface potential, *Biophys. J.*, 73, 1468, 1997.
40. Vilallonga, F., Altschul, R., and Fernandez, M.S., Lipid–protein interactions at the air–water interface, *Biochim. Biophys. Acta*, 135, 406, 1967.
41. Colacicco, G., Applications of monolayer techniques to biological systems: Symptoms of specific lipid protein interactions, *J. Colloid Interface Sci.*, 29, 345, 1969.
42. Mita, T., Lipid–protein interaction in mixed monolayers from phospholipids and proteins, *Bull. Chem. Soc. Jpn.*, 62, 3114, 1989.
43. Verger, R. and Pattus, F., Lipid–protein interactions in monolayers, *Chem. Phys. Lipids*, 30, 189, 1982.
44. Silvestro, L. and Axelsen, P.H., Infrared spectroscopy of supported lipid monolayer, bilayer, and multilayer membranes, *Chem. Phys. Lipids*, 96, 69, 1998.
45. Silvestro, L. and Axelsen, P.H., Fourier transform infrared linked analysis of conformational changes in annexinV upon membrane binding, *Biochemistry*, 38, 113, 1999.
46. Briggs, M.S. et al., Conformations of signal peptides induced by lipids suggest initial steps in protein export, *Science*, 233, 206, 1986.
47. Burger, K.N.J. et al., The interaction of synthetic analogs of the N-terminal fusion sequence of influenza virus with a lipid monolayer. Comparison of fusion-active and fusion-defective analogs, *Biochim. Biophys. Acta*, 1065, 121, 1991.
48. Lindblom, G. and Quist, P.-O., Protein and peptide interactions with lipids: Structure, membrane function and new methods, *Curr. Opin. Colloid Interface Sci.*, 3, 499, 1998.
49. Taylor, S.E. et al., Structure of a fusion peptide analog at the air–water interface, determined from surface activity, infrared spectroscopy and scanning force microscopy, *Biophys. Chem.*, 87, 63, 2000.
50. Elmore, D.L. and Dluhy, R.A., Application of 2D IR correlation analysis to phase transitions in Langmuir monolayer films, *Colloids Surf. A: Physicochem. Eng. Aspects*, 171, 225, 2000.
51. Cortijo, M. et al., Intrinsic protein-lipid interactions. Infrared spectroscopic studies of gramicidin A, bacteriorhodopsin and Ca^{2+}-ATPase in biomembranes and reconstituted systems, *J. Mol. Biol.*, 157, 597, 1982.
52. Dempsey, C.E., The actions of melittin on membranes, *Biochim. Biophys. Acta*, 1031, 143, 1990.

53. Latal, A. et al., Structural aspects of the interaction of peptidyl–glycylleucine–carboxyamide, a highly potent antimicrobial peptide from frog skin, with lipids, *Eur. J. Biochem.*, 248, 938, 1997.

54. Cummings, C.E. et al., Structural and functional studies of a synthetic peptide mimicking a proposed membrane inserting region of a *Bacillus thuringiensis* δ-endotoxin, *Mol. Membr. Biol.*, 11, 87, 1994.

55. Baty, D. et al., A 136-amino-acid-residue COOH-terminal fragment of colicin A is endowed with ionophoric activity, *Eur. J. Biochem.*, 30, 409, 1990.

56. Pécheur, I. et al., Protein-induced fusion can be modulated by target membrane lipids through a structural switch at the level of the fusion peptide, *J. Biol. Chem.*, 275, 3936, 2000.

57. Vié, V. et al., Lipid-induced pore formation of the *Bacillus thuringiensis* Cry1Aa insecticidal toxin, *J. Membr. Biol.*, 180, 185, 2001.

58. Lakowicz, J.R., *Principles of Fluorecsence Spectroscopy*, Plenum Press, New York, 1986.

59. Lakey, J.H. et al., Membrane insertion of the pore-forming domain of colicin A. A spectroscopic study, *Eur. J. Biochem.*, 196, 599, 1991.

60. Chaloin, L. et al., Synthetic primary amphipathic peptides as tools for the cellular import of drugs and nucleic acids, *Curr. Top. Pept. Protein Res.*, 3, 153, 1999.

61. Chaloin, L. et al., Design of carrier peptide–oligonucleotide conjugates with rapid membrane translocation and nuclear localization properties, *Biochem. Biophys. Res. Commun.*, 243, 601, 1998.

62. Vidal, P. et al., Solid-phase synthesis and cellular localization of a C- and/or N-terminal labelled peptide, *J. Pept. Sci.*, 2, 125, 1996.

63. Morris, M.C. et al., A new peptide vector for efficient delivery of oligonucleotides into mammalian cells, *Nucleic Acids Res.*, 25, 3730, 1997.

64. Morris, M.C. et al., A novel potent strategy for gene delivery using a single peptide vector as a carrier, *Nucleic Acids Res.*, 27, 3510, 1999.

65. Morris, K.V. et al., Small interfering RNA-induced transcriptional gene silencing in human cells, *Science*, 305, 1289, 2004.

66. Simeoni, F. et al., Insight into the mechanism of the peptide-based gene delivery system MPG: implications for delivery of siRNA into mammalian cells, *Nucleic Acids Res.*, 31, 2717, 2003.

67. Labialle, S. et al., New invMED1 element cis-activates human multidrug-related MDR1 and MVP genes, involving the LRP130 protein, *Nucleic Acids Res.*, 32, 3864, 2004.

68. Gryaznov, A. et al., Oligonucleotide N3′ → P5′ thio-phosphoramidate telomerase template antagonists as potential anticancer agents, *Nucleosides Nucleotides Nucleic Acids*, 22, 577, 2003.

69. Asai, A. et al., A novel telomerase template antagonist (GRN163) as a potential anticancer agent, *Cancer Res.*, 63, 3931, 2003.

70. Morris, M.C. et al., A new potent peptide carrier for the delivery of biologically active proteins into mammalian cells, *Nat. Biotechnol.*, 19, 1173, 2001.

71. Morris, M.C. et al., Combination of a new generation of PNAs with a peptide-based carrier enables efficient targeting of cell cycle progression, *Gene Ther.*, 11, 757, 2004.

72. Eum, W.S. et al., In vivo protein transduction: Biologically active intact pep-1-superoxide dismutase fusion protein efficiently protects against ischemic insult, *Free Radic. Biol. Med.*, 37, 1656, 2004.

73. Aoshiba, K. et al., Alveolar wall apoptosis causes lung destruction and emphysematous changes, *Am. J. Respir. Cell. Mol. Biol.*, 28, 555, 2003.

74. Magzoub, M. et al., Interaction and structure induction of cell-penetrating peptides in the presence of phospholipid vesicles, *Biochim. Biophys. Acta*, 1512, 77, 2001.

75. Wender, P.A. et al., The design, synthesis, and evaluation of molecules that enable or enhance cellular uptake: Peptoid molecular transporters, *Proc. Natl Acad. Sci. U.S.A.*, 97, 13003, 2000.

76. Derossi, D. et al., Cell internalization of the third helix of the Antennapedia homeodomain is receptor independent, *J. Biol. Chem.*, 271, 18188, 1996.

77. Thoren, P.E. et al., The antennapedia peptide penetratin translocates across lipid bilayers — the first direct observation, *FEBS Lett.*, 482, 265, 2000.

78. Fuchs, S.M. and Raines, R.T., Pathway for polyarginine entry into mammalian cells, *Biochemistry*, 43, 2438, 2004.

79. Tyagi, M. et al., Internalization of HIV-1 tat requires cell surface heparan sulfate proteoglycans, *J. Biol. Chem.*, 276, 3254, 2001.

80. Console, S. et al., Antennapedia and HIV transactivator of transcription (TAT) 'protein transduction domains' promote endocytosis of high molecular weight cargo upon binding to cell surface glycosaminoglycans, *J. Biol. Chem.*, 278, 35109, 2003.

81. Silhol, M. et al., Different mechanisms for cellular internalization of the HIV-1 Tat-derived cell penetrating peptide and recombinant proteins fused to Tat, *Eur. J. Biochem.*, 269, 494, 2002.

82. Deshayes, S. et al., Primary amphipathic cell-penetrating peptides: Structural requirements and interactions with model membranes, *Biochemistry*, 43, 7698, 2004.

83. Plénat, T. et al., Interaction of primary amphipathic cell-penetrating peptides with phospholipid-supported monolayers, *Langmuir*, 20, 9255, 2004.

84. Deshayes, S. et al., On the mechanism of nonendosomal peptide-mediated cellular delivery of nucleic acids, *Biochim. Biophys. Acta*, 1667, 141, 2004.

85. Deshayes, S. et al., Insight into the mechanism of internalization of the cell-penetrating peptide Pep-1 through conformational analysis, *Biochemistry*, 43, 1449, 2004.

86. Fasman, G.D., ed., *Circular Dichroism and the Conformational Analysis of Biomolecules*, Plenum Press, New York, 1996.

87. Fringeli, U.P. and Fringeli, M., Pore formation in lipid membranes by alamethicin, *Proc. Natl. Acad. Sci. U.S.A.*, 76, 3852, 1979.

88. Kennedy, D.F. et al., Studies of peptides forming 3(10)- and alpha-helices and beta bend ribbon structures in organic solution and in model biomembranes by Fourier transform infrared spectroscopy, *Biochemistry*, 30, 6541, 1991.

89. Vidal, P. et al., Conformational analysis of primary amphipathic carrier peptides and origin of the various cellular localizations, *J. Membr. Biol.*, 162, 259, 1998.

90. Demel, R.A. et al., Relation between various phospholipase actions on human red cell membranes and the interfacial phospholipid pressure in monolayers, *Biochim. Biophys. Acta*, 406, 97, 1975.

91. Maget-Dana, R., The monolayer technique: A potent tool for studying the interfacial properties of antimicrobial and membrane-lytic peptides and their interactions with lipid membranes, *Biochim. Biophys. Acta*, 1462, 109, 1999.

92. Lakhdar-Ghazal, F. and Tocanne, J.F., Phase behaviour in monolayers and in water dispersions of mixtures of dimannosyldiacylglycerol with phosphatidylglycerol, *Biochim. Biophys. Acta*, 644, 284, 1981.

93. Tamm, L.K. and McConnell, H.M., Supported phospholipids bilayers, *Biophys. J.*, 47, 105, 1985.

94. Ruano, M.L. et al., Differential partitioning of pulmonary surfactant protein SP-A into regions of monolayers of dipalmitoylphosphatidylcholine and dipalmitoylphosphatidylcholine/dipalmitoylphosphatidylglycerol, *Biophys. J.*, 74, 1101, 1998.

95. Crisp, D.J., A two-dimensional phase rule. I. Derivation of a two-dimensional phase rule for planar interface. II. Some applications of a two-dimensional phase rule for a single surface, *Surface Chemistry*, Vol. 17. Butterworths, London, 1949.

96. Gaines, G.L., Mixed monolayers, In *Insoluble Monolayers at Liquid–Gas Interfaces*, Prigogine, I. ed., Interscience, New York, p. 281, 1966.

97. Taneva, S. and Keough, K.M.W., Pulmonary surfactant proteins SP-B and SP-C in spread monolayers at the air-water interface. I. Monolayers of pulmonary surfactant protein SP-B and phospholipids, *Biophys. J.*, 66, 1137, 1994.

98. Yuan, C. and Johnston, L.J., Distribution of ganglioside GM1 in L-α-dipalmitoyl-phosphatidylcholine/cholesterol monolayers: a model for lipid rafts, *Biophys. J.*, 79, 2768, 2000.

99. Li, Y. et al., Hydrophobic component in noncovalent binding of fusion peptides to lipids as observed by electrospray mass spectrometry, *Rapid Commun. Mass Spectrom.*, 18, 135, 2004.

100. Li, Y. et al., Lipid–peptide noncovalent interactions observed by nano-electrospray FT-ICR, *Anal. Chem.*, 77, 1556, 2005.

101. Marthinet, E. et al., Modulation of the typical multidrug resistance phenotype by targeting the MED-1 region of human MDR1 promoter, *Gene Ther.*, 7, 1224, 2000.

9 The Import Mechanism of Cationic Cell-Penetrating Peptides and Its Implications for the Delivery of Peptide Inhibitors of Signal Transduction

Mariola Fotin-Mleczek, Söhnke Voss, and Roland Brock

CONTENTS

9.1   INTRODUCTION

The import of peptide-based inhibitors of intracellular signal transduction pathways has been one the earliest applications of cell-penetrating peptides (CPPs). Especially where no low-molecular weight inhibitors are available, peptides that correspond to pseudo-substrates or interaction motifs provide a short-cut to inhibitors by rational design. Applications of such peptides range from tissue culture experiments to therapeutic interventions in whole animals.[1–3]

The efficient cellular uptake of CPP–peptide conjugates by an uptake mechanism that assumed direct permeation of the plasma membrane was a driving force for the widespread application of these tools. The documented success of the CPP-based delivery of inhibitors may serve as a justification for a rather uncritical application of these molecules that little considered the cell biology of CPPs. Still one must be surprised that even though detailed studies about the structure-activity relationship of CPPs had been performed, demonstrating that import efficiency strongly depended on the sequence of the CPP, until 2002 the impact of different peptide cargoes on import efficiency had not been addressed quantitatively.[4,5] It was shown that import efficiency of fluorescein-labeled conjugates of peptides with the CPP penetratin, as measured by intracellular fluorescence, was strongly affected by even slight modifications of the cargo.[6] Therefore, a quantitative control of import efficiency is an absolute requirement for conducting analyses of structure-activity relationships for different peptide conjugates inside the cell.[7]

While such differences in import efficiency can easily be corrected for by incubating cells with different peptide concentrations, two further points raise concerns to which degree CPP–inhibitor conjugates meet the expectations of highly controllable tools for interfering with intracellular signaling: (1) endocytosis-dependent import and (2) proteolytic breakdown.

Over the past years, the initial concept of a direct permeation of the plasma membrane has been confronted with import by endocytosis.[8,9] Endocytosis comprises several distinct pathways. So far, no coherent picture has emerged for the endocytic uptake of CPPs with respect to the preferred route of entry and the dependence of the internalization pathway on the CPP and the cargo. However, it has become clear that CPP–cargo molecules are sequestered in the endolysosomal compartment and that this sequestration severely limits the biological activity of the cargo.[10–13] Moreover, the mechanism by which CPPs are released from the endolysosomal pathway is only poorly defined.[14]

Given that the endolysosomal pathway is a compartment with high proteolytic activity, in addition to simply restraining CPP–cargo constructs from reaching their targets in the cytoplasm or nucleus, at least for peptide cargoes, degradation by proteases should be a further factor severely limiting the biological activity of these molecules. In one of our own studies using fluorescence correlation spectroscopy (FCS) for quantitating the uptake of fluorescently labeled CPPs, we obtained strong evidence that the major part of the intracellular fluorescence originated from proteolytic fragments.[15] Analysis of lysates of cells incubated with fluorescein-labeled penetratin by MALDI-TOF mass spectroscopy, even though semiquantitative in nature, also indicated that the major part of the peptides inside the cell are proteolytically degraded.[16] This view was challenged by a quantitative analysis of peptide uptake using a quantitative MALDI-TOF-based protocol, in which very little digestion products were detected.[17] Still, one cannot ignore a large body of data obtained in the context of the analysis of intracellular peptide processing in antigen presentation to T-lymphocytes, demonstrating that in the cytosol the major part of a peptide is degraded within seconds.[18]

As outlined in Section 9.2, endocytosis provides a variety of portals into the interior of a cell. In this respect, given the cellular uptake of CPPs by endoytosis, the term "Trojan horse approach" seems to accurately reflect the character of CPP-mediated delivery. However, the ability to sneak into the cell undetected, implicated in this reference to the ancient myth, would only apply if CPPs hitchhike constitutive processes. Given the interdependence of endocytosis, cellular signal transduction, and the cytoskeleton, any active induction of endocytosis would be expected to interfere with, or at least modulate, cellular function. Data have been presented that CPPs do in fact induce endocytosis and CPPs, therefore, are not mere modulators of import by passive mechanisms but exert a cellular activity by themselves, i.e., CPPs do not only modify the pharmacokinetics of their respective cargo but possess their own pharmacodynamic properties.[11,19] These activities go beyond the induction of cytotoxicity, an experimental control that is routinely included in studies using CPP constructs. Instead, at nontoxic concentrations as well, CPPs may interfere with cellular signal transduction in a way that may either mimic or mask the intended biological activity of a cargo.[11] Therefore, the comparison with the Trojan horse may not be well chosen.

In our own work we could show that the cationic CPPs penetratin, nona-arginine, and the HIV-1 Tat protein-derived Tat peptide induced the internalization of TNF receptors 1 and 2 and of the EGFR receptor from the plasma membrane of HeLa cells in a concentration-dependent manner.[11,20,21] For the TNF receptors, we confirmed that this internalization occurred without receptor activation. Internalization of the receptors diminished the capability of the cells to respond to their ligands.

This chapter will summarize the present knowledge on the endocytic import of CPPs, including a short introduction to endocytosis itself. In the following sections, we will (1) outline the implications of endocytosis on the bioavailability of CPPs and their cargoes inside the cytoplasm and nucleus and point out strategies to overcome the problems associated with endosomal capture; (2) present the available data on the interference of CPPs with cellular signal transduction; and finally, (3) discuss practical issues of CPP experiments. While the review of data will focus on the three cationic CPPs penetratin, nona-arginine, and the Tat peptide, the experimental guidelines will be useful for the application of CPPs for mediating the cellular import of inhibitors of signal transduction in general.

9.2 ENDOCYTIC IMPORT OF CATIONIC CPPS

9.2.1 ENDOCYTOSIS

Especially over the last 3 years, with new data on the "cell biology" of CPPs, the answer to the question "How do CPPs cross the plasma membrane?" became rather more elusive than clear. At present it seems very likely that there will not be a single answer just as a variety of different CPPs do not constitute a single question. Still, even though our understanding of the mechanism of cell entry is still fragmentary, there is solid data from a considerable number of laboratories supporting a key role of endocytosis in the internalization process. Endocytosis, i.e., the pinching off of plasma membrane-enclosed vesicles containing extracellular fluid and its constituents, represents an essential mode of communication and material exchange between the cell and its environment.

Endocytosis comprises distinct pathways, which can be subdivided into two groups, phagocytic and pinocytic (Figure 9.1).[22,23] Pinocytosis occurs in all cells and encompasses a variety of processes leading to the uptake of fluids, solutes, and membrane components. Phagocytosis, on the other hand, relates to the uptake of large particles and is restricted to cells such as macrophages, monocytes, and neutrophils that are specialized for the elimination of pathogens and infected or apoptotic cells. The regulation of these processes is highly complex and, despite the enormous progress in the analysis of the endocytic machinery, many details are still poorly understood. At least four different pinocytic pathways can be distinguished: caveolae/lipid raft-mediated endocytosis, clathrin-mediated endocytosis (CME), macropinocytosis, and clathrin- and caveolin-independent endocytosis. These pathways differ with regard to the size of endocytic vesicles, the nature of the cargo, and the mechanism of vesicle formation.

CME ensures the continuous uptake of nutrients such as cholesterol-laden low-density lipoprotein (LDL) and iron-laden transferrin (Tfn) and occurs in all mammalian cells in a constitutive and an activation-dependent manner.[24,25] Internalization begins with the formation of so-called coated pits, cytosolic structures with clathrin as the key structural protein. For activation-dependent CME, internalization is initiated by the binding of molecules to their corresponding receptors at the plasma membrane that in turn leads to the enrichment of ligand–receptor complexes in coated pits. Adaptor proteins actively support this process.[26] Coated pits invaginate, pinch off, and form endocytic vesicles. This process is supported by the GTPase dynamin. By controlling the levels of receptors at the cell surface, e.g., by rapid downregulation of activated receptors, CME represents a key mechanism for regulating cellular signal transduction.[22]

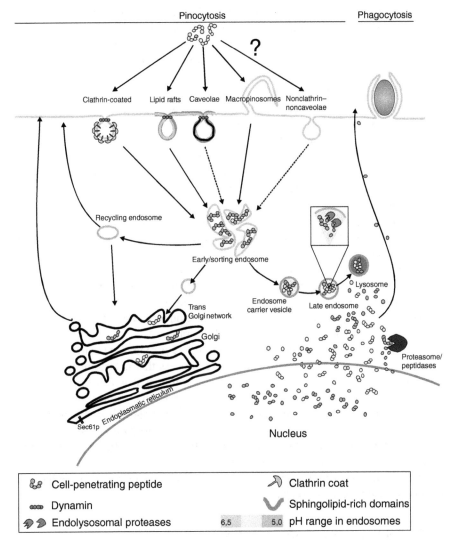

FIGURE 9.1 Endocytosis-dependent uptake of cationic CPPs. There is experimental evidence that, of the four pinocytic pathways, i.e., clathrin-mediated endocytosis, caveolae/lipid raft-mediated endocytosis, macropinocytosis, and clathrin-/caveolae-indepen-dent endocytosis, at least the first three are involved in the uptake of cationic CPPs. Endosomal acidification has been shown to contribute to the release of peptides from the vesicular compartment into the cytoplasm. In endocytic vesicles and in the cytoplasm, CPPs encounter proteases and undergo degradation. Digested peptide fragments can be released from the cell. The biological activity of imported peptides demonstrates that intact peptides reach the cytoplasm and nucleus. The impact of Golgi-disrupting drugs on uptake suggests a role of retrograde transport in the intracellular trafficking.

Caveolae were discovered 50 years ago and were described as flask-shaped pits, with a smooth surface and a diameter of typically 55–65 nm, that cover the surface of many mammalian cells.[27,28] Very similar to lipid rafts, caveolae are detergent insoluble, low-density membrane domains rich in cholesterol and sphingolipids.[29] On the molecular level, caveolae are characterized by the presence of caveolin-1.[30]

In contrast to CME, the role of caveolae in endocytosis is still unclear.[31] Caveolar endocytosis is supposed to provide at least in part a nonacidic, nondigestive route for cellular internalization.[31,32] The kinase-dependent disruption of the actin cytoskeleton and recruitment of dynamin-2 to the site of internalization facilitate internalization of caveolae.[33] The absence of caveolae from some cell types (lymphocytes, many neuronal cells) together with the mild phenotype of caveolin knockouts raise questions on the physiological role of caveolae-mediated endocytosis.[34]

It has been proposed that caveolae and lipid rafts mediate a common endocytic pathway that is defined by clathrin independence, dynamin dependence, sensitivity to cholesterol depletion, and the morphology and lipid composition of the vesicular intermediate.[32] This caveolae/lipid raft-dependent endocytosis mediates the internalization of sphingolipids, sphingolipid-binding toxins (cholera toxin and shiga toxin), GPI-anchored proteins, the growth hormone receptor, viruses (SV40), and bacteria.[35,36] However, there has also been evidence supporting that caveolae form domains that are in fact distinct from lipid rafts.[37,38]

Macropinosomes are rather large vesicles (0.5–2 μm) that are heterogeneous in size and originate from membrane ruffles in a process that is accompanied by the rearrangement of microfilaments. Although macropinocytosis can occur constitutively in some cells, the major role of this process is seen in the intracellular delivery of solutes and membranes in response to growth factors and mitogenic agents.[39]

This type of endocytosis is only poorly understood. The pathway is more-or-less defined by not being any of the previous types than by a specific set of characteristics. Nonetheless, there is evidence that clathrin- and caveolae-independent endocytosis utilizes a unique pool of endocytic vesicles in order to mediate the delivery of distinct cargo molecules.[40]

9.2.2 ENDOCYTOSIS OF CATIONIC CPPs

CPPs were found to be delivered by CME, macropinocytosis, and caveolae/lipid raft-mediated endocytosis. So far, it cannot be excluded that clathrin- and caveolae-independent endocytosis is also involved in the uptake of CPPs. The lack of such evidence is the consequence of the lack of well-established tools for the study of this little-defined pathway. The reasons for the absence of a coherent picture are multifaceted and comprise: (1) the use of different cell lines and different peptides; (2) the application of peptides at different concentrations; (3) different time windows chosen for the analysis of the uptake; (4) the application of different inhibitors as well as the application of different cargoes coupled to different CPPs.

The present data strongly indicate that indeed different endocytic pathways are involved in the internalization of CPPs. Moreover, no correlation between one specific CPP and one specific endocytic route can be defined so far, if such correlation exists at all. Nevertheless, independent of the endocytic pathway contributing to the internalization of a CPP, sequestration and proteolytic breakdown represent major challenges for the application of CPPs (Figure 9.1). To this point, it cannot be excluded that the individual endolysosomal pathways differ with respect to the release of peptides. Moreover, one can only speculate that, due to different proteolytic environments within vesicles of different origin, the uptake route may directly affect the integrity of peptides. Unfortunately, so far no data exist with respect to this interesting topic.[8]

9.2.2.1 The Tat Peptide

For the Tat peptide, a partial colocalization with transferrin, the tracer for clathrin-mediated endocytosis, was shown.[9,41] Furthermore, in our own work, we demonstrated a weak colocalization of penetratin and the Tat peptide with dextran, a marker for macropinocytosis, even though this finding is difficult to reconcile with the postulated similarities of CPP and toxin trafficking along the retrograde pathway.[16] Further contributions demonstrated that fluorescently labeled Tat peptide and Tat–PNA constructs accumulate in endocytic vesicles originating from clathrin-dependent endocytosis.[9,42] However, a clathrin-independent but lipid raft-dependent endocytosis has also been reported.[43] In contrast to the rather slow, endocytosis-mediated uptake of the Tat peptide, in another study, a formation of dense aggregates on the plasma membrane and a rapid increase of fluorescence in the cytoplasm and the nucleus within seconds was observed using time-lapse microscopy of mouse fibroblasts incubated with a fluorescein-labeled Tat peptide.[44] Interestingly, even though microscopy was performed in living cells, only few endocytic vesicles could be observed. The cellular basis for these apparently conflicting results still needs to be resolved.

Tat peptide-mediated protein import was reported to be (1) clathrin- and lipid raft-dependent; (2) dependent on a lipid raft-dependent macropinocytosis; and (3) purely raft-dependent.[13,45,46] For the Tat peptide-protein conjugates and the Tat peptide alone, Dowdy's group obtained congruent results on an uptake by macropinocytosis that, however, was also sensitive to the lipid raft-disrupting agent methyl-β-cyclodextrin.[13,19] However, it remains to be established to what degree the full-length Tat protein, the Tat peptide, and Tat peptide-protein conjugates share common import pathways.

9.2.2.2 Polyarginines

For polyarginines, Futaki and coworkers provided evidence for a role of macropinocytosis.[47] However, the impact of the inhibition of macropinocytosis on the reduction of uptake depended on the chain length of the oligo-arginine peptides, indicating that additional pathways differentially contribute to the uptake of these molecules. The inability of nona-arginine to enter living cells deficient in heparan sulfate furthermore suggested that binding to heparan sulfate is necessary for internalization.[48]

9.2.2.3   Penetratin

For penetratin also, inhibitors of metabolism or endocytosis impaired cellular uptake and there is evidence for a clathrin-independent and lipid raft-dependent endocytosis.[43,49] Moreover, similar to the Tat peptide, penetratin promotes the endocytosis of high molecular weight cargo upon binding to cell surface glycosaminoglycans and endosomal acidification is involved in the release of the peptide into the cytosol.[16]

9.2.3   Intracellular Trafficking of Cationic CPPs

The contribution of endocytosis to the cellular uptake of CPPs has added two new layers of complexity to the CPP-mediated delivery of cargoes into the nucleus and cytoplasm: (1) intracellular trafficking in endosomal compartments, and (2) release from the endosomal compartment.

The intracellular trafficking of CPPs has been studied with a repertoire of inhibitors that interfere with the later steps of the endocytic pathway. Drin et al. investigated the effect of vinblastine, nocodazole (both microtubule-disrupting agents), cytochalasine D, ammonium-chloride, chloroquine, bafilomycin A1, and nystatin on the cellular uptake of tetramethylrhodamine-conjugated penetratin and SynB peptides.[49] The highest inhibition of cellular delivery (70%) was achieved by bafilomycin A1. This observation was extended by our own data showing that, in addition to penetratin, the cellular uptake of a fluorescein-labeled analog of poly-arginine was also significantly inhibited by this compound.[16] In addition, bafilomycin A1 affected the intracellular localization of these CPPs, blocking the release of fluorescence from vesicular compartments into the cytoplasm (Figure 9.2). Ammonium chloride also reduced the cellular delivery of Antp and blocked the nuclear delivery of the fluorescein-conjugated Tat peptide and a fluorescein-labeled $(VRR)_4$ peptide consisting of beta amino acids.[41,49] Finally, chloroquine also reduced uptake of fluorescein-labeled peptide analogs as measured by a reduction in cellular fluorescence — a phenotype that is somewhat difficult to reconcile, however, with the chloroquine-induced increases in transfection efficiency, cross-presentation, and cytoplasmic delivery of proteins (see below).[10,16,49–51]

Taken together, these reports demonstrate a crucial role of endosomal acidification for the release of peptides into the cytoplasm. However, while one may easily understand that the inhibition of endosomal acidification affects the subcellular distribution of peptides, it is more difficult to understand why peptide uptake should be reduced as well.

Interesting phenotypes were obtained by incubation of cells with nordihy-droguaiaretic acid (NDGA).[16] Preincubation of cells with this Golgi-disrupting agent reduced the uptake of cationic CPPs into HeLa cells. In contrast, when cells were treated with this compound following a 2-h pulse with a fluorescein-labeled Tat peptide, (1) the cytoplasmic fluorescence increased, while (2) the vesicular fluorescence of an Alexa647-labeled dextran remained unaffected. These data provided strong evidence that the Tat peptide and dextran traffic along different

Penetratin

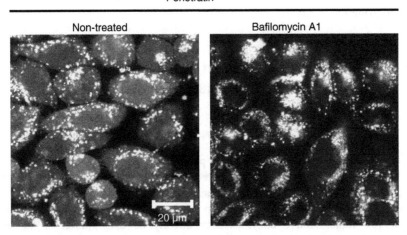

Non-treated   Bafilomycin A1

20 µm

FIGURE 9.2 Endosomal acidification is required for the release of peptides from vesicles into the cytoplasm. HeLa cells were treated with bafilomycin A1, an inhibitor of endosomal acidification (300 nM) for 30 min or left untreated, washed and incubated with fluorescein-labeled penetratin (10 μM) for a further 30 min at 37°C before image acquisition by confocal microscopy at RT.

endosomal pathways and that Tat peptides reach the trans Golgi network and/or Golgi complex.

A series of recent papers provided more mechanistic studies on endosomal release. It was shown that the guanidinium group has a high propensity to form complexes with negatively charged molecules. Formation of such complexes with negatively charged lipid-head groups at the plasma membrane promotes the partitioning of guanidinium group-rich oligomers into the lipid bilayer.[52,53] Terrone et al. demonstrated the importance of a transbilayer electric potential as a driving force for membrane permeation.[54] Diverse lipid compositions enabled a substantial potential-dependent (inside negative) uptake of different cationic CPPs (penetratin, R6-Gly-Cys, and K6-Gly-Cys) into large unilamellar vesicles.

Considering the necessity of heparan sulfate for the efficient internalization of nona-arginines and the observation that at high peptide to lipid ratios these peptides induce endosomal leakage, for these peptides a four-stage model of cytoplasmic delivery was proposed: (1) binding to cell surface heparan sulfate; (2) uptake by endocytosis; (3) release upon heparan sulfate degradation; and finally, (4) leakage from endocytic vesicles.[55] This model was complemented by a further in vitro study using large unilamellar vesicles demonstrating that for penetratin, vesicular exit occurs only in the presence of a transmembrane proton gradient and does not involve rupture of vesicles. Furthermore, it was concluded that endosomal exit requires a minimum threshold concentration and dissociation from heparan sulfates.[56]

Finally, endosomal release has been associated with the molecular composition of endosomal vesicles. For example, the unconventional phospholipid lysobisphosphatidic acid (LBPA) is abundant in late endosomes and has not been detected elsewhere in the cell. The lipid is involved in protein and lipid trafficking through late endosomes.[57]

9.3   RELEVANCE OF THE IMPORT MECHANISM FOR THE CELLULAR UPTAKE OF INHIBITORS OF SIGNAL TRANSDUCTION

9.3.1   SEQUESTRATION AND BREAKDOWN IN THE ENDOSOMAL COMPARTMENT

The endosome is a compartment of high proteolytic activity. Cathepsins, a diverse set of proteases, constitute the major proteolytic activities encountered by a protein along the endolysosomal pathway.[58] These proteases are present throughout the endolysosomal compartment. The activity of these enzymes is strongly dependent on the pH value. Active enzymes may already be encountered in the early endosome; however, the major part of the activity is localized in late endosomes and lysosomes. Most research on the relevance of individual endolysosomal proteases comes from the analysis of antigen processing in the MHC class II pathway.[59–61] However, in this pathway, degradation starts from folded proteins, so little is known about the turnover of peptides.

Evidence that endolysosomal capture is a limiting step in the release of peptides into the cytoplasm comes from the analysis of cross-presentation.[50] Cross-presentation refers to the presentation by MHC class I molecules of peptides entering cells via endocytosis. For MHC class I molecules the cytosol is the source of presented peptide.[62,63] Incubation of cells with chloroquine increased the endosomal release of an antigen and the efficiency of cross-presentation. Nevertheless, inhibitors of endolysosomal proteolysis apparently did not affect the biological activity of CPP–MHC I epitope constructs.[64,65]

9.3.2   PROTEOLYTIC BREAKDOWN IN THE CYTOSOL

Apart from a general notion that inside the cytosol peptides should be subject to rapid proteolytic degradation, rather little concrete information exists on the kinetics and extent of degradation of peptides in this compartment. The cytosol is rich in proteases that contribute to the breakdown of peptides. The tripeptidyl-peptidase II constitutes a major proteolytic activity inside the cytoplasm, as does the thimet oligopeptidase and aminopeptidases.[66–68] In contrast, the ubiquitin-dependent proteasomal breakdown is rather considered the major proteolytic activity for the degradation of proteins. Still, the intracellular processing of penetratin–MHC I epitope constructs was sensitive to the proteasome inhibitor lactacystin.[64]

With respect to the turnover of peptides by cytosolic proteolytic activities, again most information comes from the analysis of antigen processing and presentation, this time in the MHC class I pathway. Using an elegant combination of fluorescence-based and biochemical approaches it was demonstrated that peptides introduced into the cell by microinjection distributed rapidly in the cytoplasm and nucleus by free diffusion inside each compartment and across the nuclear pore.[18] Inside the nucleus, some peptide bound to DNA-associated proteins. By following the dequenching of fluorescence of a nonapeptide carrying two different fluorophores, these researchers determined a half-life of 7 sec, a time sufficient to cross the diameter of a cell only once by free diffusion. However, such a rapid degradation was only observed for peptides with a free N-terminus. Once the N-terminus was blocked, no dequenching of fluorescence was observed over a period of 50 sec. Later, it was shown that peptide length has an impact on half-life. For 10-mer and 12-mer peptides, a half-life of 20–30 sec was reported.[69]

Still, a large number of applications exist in which peptides have been used successfully as inhibitors of intracellular signal transduction.[1] In addition, the rapid breakdown of peptides contrasts with our own work for the proapoptotic Smac-derived peptide AVPIAQK.[11] For this peptide the free N-terminus is stringently required for its bioactivity as the N-terminal amino acids AVPI are directly involved in the binding to the target protein.[70] Nevertheless, the AVPIAQK peptide alone, introduced into cells via electroporation, still exerted a strong proapoptotic effect several hours after the cell entry. For AVPIAQK C-terminally conjugated to the N-terminus of nonaargine, biological activity was observed 12 h after incubation of cells with this molecule. It is conceivable that the interaction of cargo peptide with its target protein protects the peptide from the proteolytic degradation.

9.3.3 STRATEGIES TO INDUCE EFFICIENT ENDOSOMAL RELEASE

Endosomal capture has been recognized as a problem in the cellular delivery of a variety of molecules intended to exert their activity inside the cytoplasm or nucleus. For that reason, different strategies have been presented to promote the endosomal release of molecules. Given the growing awareness of sequestration of CPPs inside endosomes as a factor limiting the bioactivity of these molecules, it is to be expected that several of these strategies will also find their application in the field of CPPs, at least in tissue culture experiments. In one case (see below) the benefit of combining a membrane fusogenic peptide with CPP-based cell targeting has already been demonstrated.[13] However, destabilization of endosolysomal membranes is a delicate issue as release of endolysosomal proteases acts as a trigger for programmed cell death.[71]

The antimalaria drug chloroquine has long been known to promote the cellular activity of substances taken up via endocytosis. Chloroquine is a lysosomotropic agent that exerts a dual effect on the endolysosomal compartment. First, endosomal acidification is abolished, thereby inhibiting endolysosomal proteases. Second, at least at not too high concentrations, chloroquine promotes the release of molecules into the cytoplasm without a detectable endosome rupture.[50] However, it is not clear to what degree the second effect is merely a consequence of the first one. Using a fusion construct of the Cre recombinase with the Tat peptide and expression of a reporter gene as a functional read out, Wadia et al. demonstrated that treatment of 3T3 fibroblasts with chloroquine increased the number of cells expressing the reporter gene in a concentration-dependent manner.[13] However, concentrations of this drug required to strongly enhance the activity of the CPP construct were highly cytotoxic, possibly due to endosome swelling and membrane rupture.

Coadministration of the Tat–Cre construct together with a conjugate consisting of the Tat peptide and the 20 N-terminal amino acids of the influenza virus hemagglutinin protein that acts as a pH-dependent membrane fusogenic moiety also increased the efficiency of Cre delivery, however, without showing toxic side effects.

A nonphysiological pH-sensitive, membrane-disruptive targeting peptide for the cellular delivery of macromolecules was generated by acetylating lysines of the membrane disruptive peptide melittin with a maleic acid anhydride. At low pH, the masking groups dissociated. The toxicity of the melittin was greatly reduced by confining its activity to the endosomal lumen.[72]

Light has been employed as a further agent for endosome disruption.[73] The endosomal pathway is loaded with photosensitizers, which, upon illumination with light, induce rupture of the endosomal membrane. For fluorescently labeled oligo-arginine peptides, laser illumination in the confocal microscope in the absence of an additional photosensitizer yielded a redistribution from vesicular structures into the cytosol without induction of toxicity.[74]

9.3.4 INTERFERENCE WITH CELLULAR SIGNAL TRANSDUCTION

Most CPPs are at least rich in basic amino acids if not purely polycationic. Considering that for polycationic substances a range of cell biological activities has been reported, it is, therefore, surprising that biological tests for side effects of CPPs have been restricted to a mere assessment of cytotoxicity.

Polycationic substances, including neuropeptides, the venom peptide mastoparan, or the synthetic polyamine compound 48/80 have been described to induce degranulation of mast cells.[75–77] It has been proposed that the polycations bind to the negatively charged membrane of mast cells and directly activate heterotrimeric G-proteins.[78] In vitro experiments revealed that the cationic peptides stimulate purified and reconstituted G-proteins from calf brain and from rat peritoneal mast cells.[79,80]

Furthermore, cationic antimicrobial peptides represent a large and abundant group of cationic peptides in nature. As some of these peptides also act as CPPs, it should be worth considering documented effects on cellular signaling of antimicrobial peptides for obtaining clues in which ways CPPs may interfere with cellular signaling.[81] In addition to the plasma membrane, cationic antimicrobial peptides are known to address targets, such as receptors and enzymes. For instance, magainin 2, indolicidin, and temporins B and L enhance the activity of mammalian secretory phospholipase A2.[82] For indolicidin, calmodulin-binding activity has also been described.[83]

A further immune modulatory property of polycationic substances (arginine- and, to a lesser extent, lysine-rich molecules) is their ability to synergistically enhance the LPS-induced activation of monocytes. The polyamines are thought to bind to the negatively charged LPS and thereby enhance the agonistic activity of the TLR4 ligand.[84]

The interference with signaling is not limited to cationic peptides. Cationic transfection agents led to the activation and downregulation of insulin receptors.[85] It would be interesting to investigate whether the same was observed for cationic CPPs.

In our own work, the relevance of the import mechanism for the bioavailability of functional peptides and interference with cellular signaling was demonstrated using the proapoptotic AVPIAQK peptide that enhances TNF receptor-dependent apoptosis.[11] The peptide was introduced into cells either by electroporation or by conjugation to penetratin. Although both strategies were able to deliver cargo into the cells with similar efficiencies, the cellular destination reached by the peptides was different. Peptide introduced into the cells via transient membrane permeation was homogenously distributed within the cytoplasm and nucleus. The same peptide delivered via penetratin remained predominantly enriched within endocytic vesicles. Only a little amount of peptide was able to reach the cytoplasm, a stringent requirement for exerting its biological effect. While for the electroporated peptide the intended enhancement of TNF-induced apoptosis was observed, very surprisingly, for the penetratin conjugate at low concentrations, caspase activity was reduced and cells were protected against TNF-mediated cell death. Further analyses revealed that penetratin, nona-arginine, and the Tat peptide induced the internalization of TNF receptors in a concentration-dependent manner (Figure 9.3). This induction of internalization was clathrin-dependent and completely decoupled from receptor activation. The critical step in the initiation of TNF signaling is the ligand-induced crosslinking of receptors in the plasma membrane. Very likely, the cationic CPPs failed to induce receptor clustering and, therefore, the internalized receptor remained signaling incompetent. As the cationic CPPs already interfered with the cellular signal transduction at the very early stage of the signaling cascade,

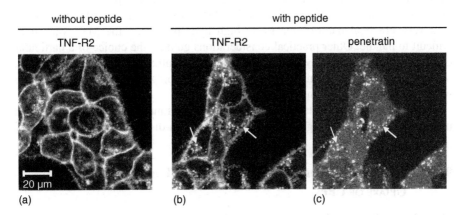

FIGURE 9.3 Cationic CPPs induce internalization of TNF receptor 2. After immunostaining of TNF-R2 with TNF-R2-specific antibody-Zenon Alexa Fluor 647 conjugates HeLa-TNF-R2 cells were washed and incubated with Antp (10 μM) for 1 h at 37°C. Cells were analyzed by multichannel confocal microscopy; (a, b) receptor staining; (c) distribution of fluorescently labeled peptide. White arrows indicate the colocalization between peptide and TNF receptor 2 in internalized vesicles.

their effect on TNF-mediated signaling was general and not restricted to only one of the possible TNF-mediated responses. In contrast, no internalization was observed for the receptor Fas. In this case, the peptides did not interfere with signaling and for the cargo peptide the intended biological effect was observed in a concentration-dependent manner. These differences may relate to the association of Fas and the TNF receptors with distinct membrane microdomains, which may be differentially involved in the uptake of CPPs. These results demonstrated that CPPs may be used for the cellular import of inhibitor peptides. However, detailed knowledge on the import mechanism is required to implement the necessary experimental controls to ascertain the validity of the experimental data.

9.4   EXPERIMENTAL DESIGN

9.4.1   THE NATURE OF THE CPP

A total of about 30 different CPPs has been described so far, not including truncated versions and peptides with single amino acid substitutions. These peptides vary in length from 8 to about 30 amino acids and in their physicochemical characteristics, with some being purely cationic, such as the oligoarginines, others being amphiphilic with a high fraction of positively charged residues, such as penetratin, and again others being purely hydrophobic, such as the basic fibroblast growth factor-derived membrane translocating peptide (MTS).[86]

For some CPPs a clear preference for a specific type of cargo, such as nucleic acids or peptides/proteins, is given.[87,88] However, within the group of CPPs considered suitable for the cellular delivery of peptides, also large differences in import efficiency were observed. We compared the cellular uptake of the purely hydrophobic membrane translocating sequence (MTS) and of penetratin.[15] Uptake of penetratin was more efficient than of the MTS. However, further differences were of a very practical nature. Due to its high hydrophobicity, the MTS was more difficult to handle. Other practical considerations guiding the choice of a particular CPP may affect the ease by which a CPP may be synthesized. For oligoarginines, incomplete couplings during peptide synthesis will only lead to a shorter peptide with basically the same functional characteristics.

Kelemen et al. demonstrated that the success of a functional peptide to act as an inhibitor of intracellular signaling greatly depends on the CPP used for delivery.[89]

9.4.2   EXPERIMENTAL APPROACHES TO ANALYZE THE ENDOCYTIC
UPTAKE OF CPPs

The CPP-mediated induction of internalization of TNF receptors (see above) demonstrates that the mechanism of endocytosis may have major implications for the CPP-mediated import of functional molecules. A detailed understanding of the uptake mechanism of CPPs may guide the design of CPPs that do not interfere with the localization of molecules in the plasma membrane and avoid compartments with high proteolytic activity. Questions in this context relate to (1) the preferences of a specific CPP and a CPP–cargo conjugate for an endocytic pathway; (2) the impact of cargo on the endocytic trafficking of CPP–cargo conjugates; (3) the induction of endocytosis by the CPP; and (4) the release of peptides from vesicles. For points 1–3, researchers can resort to a panel of biological and pharmacological tools (Table 9.1).

These tools include fluorescently or radioactively labeled tracers, predominantly internalized via one of the endocytic pathways, as well as a large number of pharmacological inhibitors interfering with a specific step of the endocytic machinery.[8] Although the selectivity of both tracers and inhibitors with respect to one specific endocytic pathway is not always absolute, the combination of different approaches enables a reliable analysis of the internalization process. Furthermore, cell lines lacking a protein required for a specific endocytic pathway or cells expressing a dominant-negative mutant form of such a protein represent highly specific tools for the analysis of endocytic pathways.

TABLE 9.1
Strategies and Tools for Determining the Role of an Endocytic Pathway in the Cellular Uptake of a CPP

	Tracers	Inhibitors	Genetically Modified Cell Lines
Clathrin-mediated endocytosis	Transferrin	Chlorpromazine	DN dynamin (dynK44A)
	Low-density lipoprotein	Wortmannin	DN epsin 15
	Epidermal growth factor	Hyperosmotic sucrose media	
		Phenylarsine oxide	
Caveolae-/lipid raft-mediated endocytosis	Cholera toxin β subunit	Methyl-β-cyclodextrin	Caveolin-1 knock out
	GPI-anchored proteins	Filipin	
	Growth hormone receptor	Nystatin	
		Cholesterol oxidase	
		Cytochalasin D	
Macropinocytosis	Dextran	Amiloride	–
		Wortmannin	
Clathrin- and caveolae-independent endocytosis	Interleukin-2	–	–

In particular, the specificity of the inhibitors is frequently questioned. Therefore, consistent results obtained for different tools and independent experimental strategies are considered necessary in order to conclude on any of these pathways. DN, dominant negative.

9.4.3 Monitoring Import and Distribution of CPPs

For most experiments investigating the cellular uptake of CPPs, fluorescently labeled analogs of CPPs have been employed. For penetratin, different fluorophores had little impact on the relative uptake efficiencies of analogs labeled at different positions within the peptide. However, differences in the cellular distribution of fluorescence were observed.[5] Moreover, even though the understanding of the cellular trafficking of CPPs has benefited enormously from the use of fluorescently tagged CPPs in live cell microscopy, one should be aware that the subcellular distribution of the fluorescent dye may not represent that of the CPP, but rather that of a proteolytic fragment, especially for longer incubation times. Moreover, for longer incubation times, the peptide may equilibrate in different endocytic compartments, thereby compromising the interpretation of results on endocytic trafficking. If cells are washed insufficiently before analysis by flow cytometry, peptides associated with the plasma membrane may be mistaken for peptides taken up into the cells.[9] Fixation may strongly affect the distribution of molecules inside the cell. However, it is obvious that a CPP conjugated to a high molecular weight protein will behave differently towards fixation than a CPP, alone. Moreover, too long incubation times after removal of peptide from the incubation medium may lead to an exit of peptides and proteolytic fragments from the cells. If this leakage occurs preferentially for peptides in the cytoplasm, then erroneous results on peptide distribution will also be obtained. Finally, the results will strongly depend on the read-out. Functional read-outs such as the determination of the activity of the Cre-recombinase will only detect intact protein reaching the nucleus.

9.4.4 Cellular Factors

For the intracellular peptide distribution, a cell-type dependence has been observed.[16] For cationic CPPs, HeLa cells showed a mostly vesicular staining while in MC57 fibrosarcoma cells, fluorescence was prominent in the cytoplasm and in the nucleus. For a given cell line, the cell cycle may also play a role. In the G1 phase uptake of a conjugate of the Cre-recombinase with the signal sequence hydrophobic region (SSHR) was only half as efficient as uptake during other phases of the cell cycle.[90]

Furthermore, toxicity of cationic peptides is also cell-specific. Several cationic antimicrobial peptides selectively killed tumor cells.[91] This antitumor activity is presumably due to a slightly higher amount of negatively charged phospholipids in the cell membrane of cancer cells in comparison to normal eukaryotic cells.

Moreover, cell density is an important factor to ensure reproducibility of results. It was demonstrated that rather than peptide concentration, the peptide to cell ratio strongly affects the loading efficiency.[92]

9.5 CONCLUSIONS

It has become clear that CPPs are not mere pharmacokinetic modifiers of their respective cargoes. Instead, the CPPs themselves constitute pharmacodynamically active entities. We do not know how many studies using a CPP-based delivery strategy were discontinued because they provided unexpected or nonexplicable data. The side effects connected with CPPs do not mean to question the use of these peptides as delivery tools for studying cellular signal transduction. On the contrary, understanding of the molecular basis of these activities enables the inclusion of stringent controls that will guarantee the successful and effective application of CPPs.

For the application of CPPs in the import of inhibitors of cellular signal transduction, at present it is advisable to include at least the following controls:

- Instead of using only the CPP without cargo, a CPP with a cargo carrying only minor structural modification, e.g., single amino-acid exchanges known to abolish the biological activity, should be tested.
- All CPP–cargo constructs should be fluorescently labeled. The relative cellular uptake of all constructs should be quantitated and the subcellular distribution of fluorescence analyzed by fluorescence microscopy.
- In addition to standard cell viability tests, the integrity and morphology of cells should be investigated by microscopy.
- For CPP–cargo constructs interfering with cellular responses induced by the activation of cell surface receptors, the effect of each individual CPP–cargo construct on surface expression of the receptor has to be determined.
- Controls also have to include cells treated with CPP–cargo constructs in the absence of a stimulus.

A recent comparison of import efficiencies versus toxicity for the Tat peptide, poly-arginine, penetratin, and transportan revealed that, because of higher toxicity, the peptide that shows the most efficient import is not necessarily the best import mediator.[43] These results demonstrate that a one-dimensional judgment fails to identify a CPP best suited for a cell biological application. One may therefore expect that, once further tests are included, there will be several "best" CPPs for different applications.

ACKNOWLEDGMENTS

R.B. gratefully acknowledges financial support from the Volkswagen Foundation (Nachwuchsgruppen an Universitäten, I/77 472).

REFERENCES

1. Dietz, G.P.H. and Bähr, M., Delivery of bioactive molecules into the cell: The Trojan horse approach, *Mol. Cell. Neurosci.*, 27, 85, 2004.
2. Fischer, P.M., Krausz, E., and Lane, D.P., Cellular delivery of impermeable effector molecules in the form of conjugates with peptides capable of mediating membrane translocation, *Bioconjugate Chem.*, 12, 825, 2001.
3. Joliot, A. and Prochiantz, A., Transduction peptides: From technology to physiology, *Nat. Cell Biol.*, 6, 189, 2004.

4. Fischer, P.M. et al., Structure–activity relationship of truncated and substituted analogues of the intracellular delivery vector penetratin, *J. Pept. Res.*, 55, 163, 2000.

5. Fischer, R. et al., A quantitatitive validation of fluorophore-labelled cell-permeable peptide conjugates: Fluorophore and cargo dependence of import, *Biochim. Biophys. Acta-Biomembr.*, 1564, 365, 2002.

6. Derossi, D. et al., The third helix of the Antennapedia homeodomain translocates through biological membranes, *J. Biol. Chem.*, 269, 10444, 1994.

7. Liu, X.-Y. et al., Identification of a functionally important sequence in the cytoplasmic tail of integrin β3 by using cell-permeable peptide analogs, *Proc. Acad. Natl Sci. U.S.A*, 93, 11819, 1996.

8. Fotin-Mleczek, M., Fischer, R., and Brock, R., Endocytosis and cationic cell-penetrating peptides — a merger of concepts and methods, *Curr. Pharm. Des.*, 11, 3613, 2005.

9. Richard, J.P. et al., Cell-penetrating peptides: A reevaluation of the mechanism of cellular uptake, *J. Biol. Chem.*, 278, 585, 2003.

10. Caron, N.J., Quenneville, S.P., and Tremblay, J.P., Endosome disruption enhances the functional nuclear delivery of Tat-fusion proteins, *Biochem. Biophys. Res. Commun.*, 319, 12, 2004.

11. Fotin-Mleczek, M. et al., Cationic cell-penetrating peptides interfere with TNF signaling by induction of TNF receptor internalization, *J. Cell Sci.*, 118, 3339, 2005.

12. Sengoku, T. et al., Tat-calpastatin fusion proteins transduce primary rat cortical neurons but do not inhibit cellular calpain activity, *Exp. Neurol.*, 188, 161, 2004.

13. Wadia, J.S., Stan, R.V., and Dowdy, S.F., Transducible TAT-HA fusogenic peptide enhances escape of TAT-fusion proteins after lipid raft macropinocytosis, *Nat. Med.*, 10, 310, 2004.

14. Fischer, R. et al., Break on through to the other side — biophysics and cell biology shed light on cationic cell-penetrating peptides, *ChemBioChem*, 6, 2126, 2005.

15. Waizenegger, T., Fischer, R., and Brock, R., Intracellular concentration measurements in adherent cells: A comparison of import efficiencies of cell-permeable peptides, *Biol. Chem.*, 383, 291, 2002.

16. Fischer, R. et al., A stepwise dissection of the intracellular fate of cationic cell-penetrating peptides, *J. Biol. Chem.*, 279, 12625, 2004.

17. Burlina, F. et al., Quantification of the cellular uptake of cell-penetrating peptides by MALDI-TOF mass spectrometry, *Angew. Chem., Int. Ed. Engl.*, 44, 4244, 2005.

18. Reits, E. et al., Peptide diffusion, protection, and degradation in nuclear and cytoplasmic compartments before antigen presentation by MHC class I, *Immunity*, 18, 97, 2003.

19. Kaplan, I.M., Wadia, J.S., and Dowdy, S.F., Cationic TAT peptide transduction domain enters cells by macropinocytosis, *J. Control. Release*, 102, 247, 2005.

20. Mitchell, D.J. et al., Polyarginine enters cells more efficiently than other polycationic homopolymers, *J. Pept. Res.*, 56, 318, 2000.

21. Vives, E., Brodin, P., and Lebleu, B., A truncated HIV-1 Tat protein basic domain rapidly translocates through the plasma membrane and accumulates in the cell nucleus, *J. Biol. Chem.*, 272, 16010, 1997.

22. Conner, S.D. and Schmid, S.L., Regulated portals of entry into the cell, *Nature*, 422, 37, 2003.

23. Johannes, L. and Goud, B., Facing inward from compartment shores: How many pathways were we looking for?, *Traffic*, 1, 119, 2000.

24. Brodsky, F.M. et al., Biological basket weaving: Formation and function of clathrin-coated vesicles, *Annu. Rev. Cell Dev. Biol.*, 17, 517, 2001.

25. Schmid, S.L., Clathrin-coated vesicle formation and protein sorting: An integrated process, *Annu. Rev. Biochem.*, 66, 511, 1997.

26. Robinson, M.S. and Bonifacino, J.S., Adaptor-related proteins, *Curr. Opin. Cell Biol.*, 13, 444, 2001.

27. Palade, G.E., Fine structure of blood capillaries, *J. Appl. Phys.*, 24, 1424, 1953.

28. Yamada, E., The fine structure of the renal glomerulus of the mouse, *J. Biophys. Biochem. Cytol.*, 1, 551, 1955.

29. Simons, K. and Ikonen, E., Functional rafts in cell membranes, *Nature*, 387, 569, 1997.

30. Rothberg, G.E. et al., Caveolin, a protein component of caveolae membrane coats, *Cell*, 68, 673, 1992.

31. Parton, R.G. and Richards, A.A., Lipid rafts and caveolae as portals for endocytosis: New insights and common mechanisms, *Traffic*, 4, 724, 2003.

32. Nabi, I.R. and Le, P.U., Caveolae/raft-dependent endocytosis, *J. Cell Biol.*, 161, 673, 2003.

33. Pelkmans, L., Puntener, D., and Helenius, A., Local actin polymerization and dynamin recruitment in SV40-induced internalization of caveolae, *Science*, 296, 535, 2002.

34. Drab, M. et al., Loss of caveolae, vascular dysfunction, and pulmonary defects in caveolin-1 gene-disrupted mice, *Science*, 293, 2449, 2001.

35. Duncan, M.J., Shin, J.S., and Abraham, S.N., Microbial entry through caveolae: Variations on a theme, *Cell Microbiol.*, 4, 783, 2002.

36. Nichols, B.J. and Lippincott-Schwartz, J., Endocytosis without clathrin coats, *Trends Cell Biol.*, 11, 406, 2001.

37. Liu, J. et al., Organized endothelial cell surface signal transduction in caveolae distinct from glycosylphosphatidylinositol-anchored protein microdomains, *J Biol. Chem.*, 272, 7211, 1997.

38. Sowa, G., Pypaert, M., and Sessa, W.C., Distinction between signaling mechanisms in lipid rafts vs. caveolae, *Proc. Natl Acad. Sci. U.S.A*, 98, 14072, 2001.

39. Amyere, M. et al., Constitutive macropinocytosis in oncogene-transformed fibroblasts depends on sequential permanent activation of phosphoinositide 3-kinase and phospholipase C, *Mol. Biol. Cell*, 11, 3453, 2000.

40. Lamaze, C. et al., Interleukin 2 receptors and detergent-resistant membrane domains define a clathrin-independent endocytic pathway, *Mol. Cell*, 7, 661, 2001.

41. Potocky, T.B., Menon, A.K., and Gellman, S.H., Cytoplasmic and nuclear delivery of a TAT-derived peptide and a beta-peptide after endocytic uptake into HeLa cells, *J Biol. Chem.*, 278, 50188, 2003.

42. Richard, J.P. et al., Cellular uptake of unconjugated TAT peptide involves clathrin-dependent endocytosis and heparan sulfate receptors, *J. Biol. Chem.*, 280, 15300, 2005.

43. Jones, S.W. et al., Characterisation of cell-penetrating peptide-mediated peptide delivery, *Br. J. Pharmacol.*, 145, 1093, 2005.

44. Ziegler, A. et al., The cationic cell-penetrating peptide CPP(TAT) derived from the HIV-1 protein TAT is rapidly transported into living fibroblasts: Optical, biophysical, and metabolic evidence, *Biochemistry*, 44, 138, 2005.

45. Ferrari, A. et al., Caveolae-mediated internalization of extracellular HIV-1 tat fusion proteins visualized in real time, *Mol. Ther.*, 8, 284, 2003.

46. Saalik, P. et al., Protein cargo delivery properties of cell-penetrating peptides. A comparative study, *Bioconjug. Chem.*, 15, 1246, 2004.

47. Nakase, I. et al., Cellular uptake of arginine-rich peptides: Roles for macropinocytosis and actin rearrangement, *Mol. Ther.*, 10, 1011, 2004.

48. Fuchs, P. et al., Primary structure and functional scFv antibody expression of an antibody against the human protooncogen c-myc, *Hybridoma*, 16, 227, 1997.

49. Drin, G. et al., Studies on the internalization mechanism of cationic cell-penetrating peptides, *J. Biol. Chem.*, 278, 31192, 2003.

50. Accapezzato, D. et al., Chloroquine enhances human CD8 + T cell responses against soluble antigens in vivo, *J. Exp. Med.*, 202, 817, 2005.

51. Erbacher, P. et al., Putative role of chloroquine in gene transfer into a human hepatoma cell line by DNA/lactosylated polylysine complexes, *Exp. Cell Res.*, 225, 186, 1996.

52. Sakai, N. and Matile, S., Anion-mediated transfer of polyarginine across liquid and bilayer membranes, *J. Am. Chem. Soc.*, 125, 14348, 2003.

53. Sakai, N. et al., Direct observation of anion-mediated translocation of fluorescent oligoarginine carriers into and across bulk liquid and anionic bilayer membranes, *ChemBioChem*, 6, 114, 2005.

54. Terrone, D. et al., Penetratin and related cell-penetrating cationic peptides can translocate across lipid bilayers in the presence of a transbilayer potential, *Biochemistry*, 42, 13787, 2003.

55. Fuchs, S.M. and Raines, R.T., Pathway for polyarginine entry into mammalian cells, *Biochemistry*, 43, 2438, 2004.

56. Magzoub, M., Pramanik, A., and Graslund, A., Modeling the endosomal escape of cell-penetrating peptides: Transmembrane pH gradient driven translocation across phospholipid bilayers, *Biochemistry*, 44, 14890, 2005.

57. Matsuo, H. et al., Role of LBPA and Alix in multivesicular liposome formation and endosome organization, *Science*, 303, 531, 2004.

58. Pillay, C.S., Elliott, E., and Dennison, C., Endolysosomal proteolysis and its regulation, *Biochem. J.*, 363, 417, 2002.

59. Chapman, H.A., Endosomal proteolysis and MHC class II function, *Curr. Opin. Immunol.*, 10, 93, 1998.

60. Villadangos, J.A. and Ploegh, H.L., Proteolysis in MHC class II antigen presentation: Who's in charge?, *Immunity*, 12, 233, 2000.

61. Villadangos, J.A. et al., Early endosomal maturation of MHC class II molecules independently of cysteine proteases and H-2DM, *EMBO J.*, 19, 882, 2000.

62. Ackerman, A.L. and Cresswell, P., Cellular mechanisms governing cross-presentation of exogenous antigens, *Nat. Immunol.*, 5, 678, 2004.

63. Heath, W.R. and Carbone, F.R., Cross-presentation in viral immunity and self-tolerance, *Nat. Rev. Immunol.*, 1, 126, 2001.

64. Pietersz, G.A., Li, W., and Apostolopoulos, V., A 16-mer peptide (RQIKIWFQNRR-MKWKK) from antennapedia preferentially targets the class I pathway, *Vaccine*, 19, 1397, 2001.

65. Lu, J. et al., TAP-independent presentation of CTL epitopes by trojan antigens, *J. Immunol.*, 166, 7063, 2001.

66. Geier, E. et al., A giant protease with potential to substitute for some functions of the proteasome, *Science*, 283, 978, 1999.

67. Saric, T. et al., Major histocompatibility complex class I-presented antigenic peptides are degraded in cytosolic extracts primarily by thimet oligopeptidase, *J. Biol. Chem.*, 276, 36474, 2001.

68. Taylor, A., Aminopeptidases: Structure and function, *FASEB J.*, 7, 290, 1993.

69. Neijssen, J. et al., Cross-presentation by intercellular peptide transfer through gap junctions, *Nature*, 434, 83, 2005.

70. Chai, J. et al., Structural and biochemical basis of apoptotic activation by Smac/DIABLO, *Nature*, 406, 855, 2000.

71. Ferri, K.F. and Kroemer, G., Organelle-specific initiation of cell death pathways, *Nat. Cell Biol.*, 3, E255–E263, 2001.

72. Rozema, D.B. et al., Endosomolysis by masking of a membrane-active agent (EMMA) for cytoplasmic release of macromolecules, *Bioconjug. Chem.*, 14, 51, 2003.

73. Hogset, A. et al., Photochemical internalisation in drug and gene delivery, *Adv. Drug Deliv. Rev.*, 56, 95, 2004.

74. Maiolo, J.R. III, Ottinger, E.A., and Ferrer, M., Specific redistribution of cell-penetrating peptides from endosomes to the cytoplasm and nucleus upon laser illumination, *J. Am. Chem. Soc.*, 126, 15376, 2004.

75. Mousli, M. et al., Activation of rat peritoneal mast cells by substance P and mastoparan, *J. Pharmacol. Exp. Ther.*, 250, 329, 1989.

76. Paton, W.D., Compound 48/80: A potent histamine liberator, *Br. J. Pharmacol. Chemother.*, 6, 499, 1951.

77. Repke, H. et al., Histamine release induced by Arg-Pro-Lys-Pro(CH2)11CH3 from rat peritoneal mast cells, *J. Pharmacol. Exp. Ther.*, 243, 317, 1987.

78. Mousli, M. et al., G protein activation: A receptor-independent mode of action for cationic amphiphilic neuropeptides and venom peptides, *Trends Pharmacol. Sci.*, 11, 358, 1990.

79. Fischer, T. et al., The mechanism of inhibition of alkylamines on the mast-cell peptidergic pathway, *Biochim. Biophys. Acta*, 1176, 305, 1993.

80. Higashijima, T., Burnier, J., and Ross, E.M., Regulation of Gi and Go by mastoparan, related amphiphilic peptides, and hydrophobic amines. Mechanism and structural determinants of activity, *J. Biol. Chem.*, 265, 14176, 1990.

81. Takeshima, K. et al., Translocation of analogues of the antimicrobial peptides Magainin and Buforin across human cell membranes, *J. Biol. Chem.*, 278, 1310, 2003.

82. Zhao, H. and Kinnunen, K.J., Modulation of the activity of secretory phospholipase A2 by antimicrobial peptides, *Antimicrob. Agents Chemother.*, 47, 965, 2003.

83. Sitaram, N., Subbalakshmi, C., and Nagaraj, R., Indolicidin, a 13-residue basic antimicrobial peptide rich in tryptophan and proline, interacts with Ca(2+)-calmodulin, *Biochem. Biophys. Res. Commun.*, 309, 879, 2003.

84. Bosshart, H. and Heinzelmann, M., Arginine-rich cationic polypeptides amplify lipopolysaccharide-induced monocyte activation, *Infect. Immun.*, 70, 6904, 2002.

85. Pramfalk, C. et al., Insulin receptor activation and down-regulation by cationic lipid transfection reagents, *BMC Cell Biol.*, 5, 7, 2004.

86. Hawiger, J., Noninvasive intracellular delivery of functional peptides and proteins, *Curr. Opin. Chem. Biol.*, 3, 89, 1999.

87. Morris, M.C. et al., A peptide carrier for the delivery of biologically active proteins into mammalian cells, *Nat. Biotechnol.*, 19, 1173, 2001.

88. Simeoni, F. et al., Insight into the mechanism of the peptide-based gene delivery system MPG: Implications for delivery of siRNA into mammalian cells, *Nucleic Acids Res.*, 31, 2717, 2003.

89. Kelemen, B.R., Hsiao, K., and Goueli, S.A., Selective in vivo inhibition of mitogen-activated protein kinase activation using cell-permeable peptides, *J. Biol. Chem.*, 277, 8741, 2002.

90. Jo, D. et al., Cell cycle-dependent transduction of cell-permeant Cre recombinase proteins, *J. Cell. Biochem.*, 89, 674, 2003.

91. Papo, N. and Shai, Y., New lytic peptides based on the D,L-amphipathic helix motif preferentially kill tumor cells compared to normal cells, *Biochemistry*, 42, 9346, 2003.

92. Hällbrink, M. et al., Uptake of cell-penetrating peptides is dependent on peptide-to-cell ratio rather than on peptide concentration, *Biochim. Biophys. Acta*, 1667, 222, 2004.

10 Mechanisms of Cargo Delivery by Transportans

Margus Pooga, Kärt Padari, and Ülo Langel

CONTENTS

10.1 INTRODUCTION

During the past 10 years of research in the field of cell-penetrating peptides (CPPs), more than 200 sequences of different origin and properties have been described.

A comprehensive list of applications of CPP-mediated transport has been given in an excellent review by Dietz and Bähr.[1] The common ability of CPPs to cross cellular membranes and to introduce hydrophilic bioactive cargoes into cells has led to the next course of studies directed towards the search for cell type-specific sequences and most effective combinations of CPP and cargo. The cell-penetrating peptide transportan, discussed in detail in this chapter, is a membrane-interacting peptide which has proven to be an efficient transporter for different cargoes in various cell lines. Transportan as a CPP was discovered serendipitously in the search for novel galanin receptor ligands. The coupling of the galanin (1–13) fragment to the N-terminus of mastoparan yielded a 27 amino acid long galparan,[2] which at high concentration activated G-proteins in a receptor-independent manner.[3] Replacement of [13]Pro, the last amino acid from the galanin part, by Lys provided a suitable attachment point for reporter groups like biotin and fluorophores and allowed the characterization of its cellular uptake.[4] Surprisingly, the biotinylated peptide accumulated in cultured cells and showed the ability to deliver other molecules into cells.

10.2   TRANSPORTANS

Based on experiments with fixed cells, the cellular uptake of transportan was suggested to be temperature- and energy-independent. In Bowes melanoma and other cells, transportan did not distribute uniformly in the cell cytoplasm and nucleus, but concentrated in membranous structures, showing high affinity of transportan to membranes.[5] Later series of deletion analogs were designed in order to define the regions responsible for the membrane translocation properties and to find a peptide with no affinity for galanin receptors and no interference with G-proteins.[6] The analog, TP10, which lacked six amino acid residues from the N-terminus of transportan, retained its cell translocation ability, but did not bind to galanin receptors nor modulated the activity of G-proteins. The C-terminal part of transportan and its analogs is necessary for membrane translocation,[6] and is later shown to fold into an α-helical structure upon interaction with lipid vesicles and in the aqueous environment.[7] The C-terminus of transportan interacts with phospholipids using the hydrophobic face of the α-helix[8] and its positive charges are important for cell entry.[6]

10.3   CELLULAR DELIVERY OF CARGOES

Different cargoes coupled to transportan and its shorter analog, TP10, have been efficiently transduced into cells. Transportan has been used for delivery of medium-sized hydrophilic biomolecules — peptides and PNA oligomers[5,9–13] and proteins like GFP, antibodies, and large complexes into different cells in culture.[14] Different strategies could be used for attaching the carrier peptide to the cargo. The labile disulfide bond has been used mostly for small cargo molecules.[14] The disulfide bridge is quickly cleaved in the reducing cytosol, allowing the dissociation of the cargo molecule from the carrier peptide after cellular internalization. TP–cargo conjugates have also been prepared by using a bifunctional cross-linker, which couples the cysteine of peptide with amino groups of the cargo protein. However, the covalent bond is not necessary and cargo can be delivered into cells after forming a stable noncovalent complex, like biotin-modified CPP with (strept)avidin.[14]

10.3.1 Delivery of Peptide Nucleic Acids

Transportan and TP10 have been used mostly for transducing *peptide nucleic acids* (PNA) oligomers into cells. PNAs are oligonucleotides in which the negatively charged sugar-phosphate skeleton has been replaced by a neutral polyamide backbone.[15] PNA oligomers bind to complementary RNA and DNA in a parallel or antiparallel orientation forming highly stable PNA–DNA or PNA–RNA duplexes.[16] PNA oligomers are promising reagents for antisense technology due to very high resistance to proteases and nucleases.[17] Unfortunately, contrary to deoxynucleotide and phosphorothioate oligomers, PNA molecules are very poorly taken up by cells.

PNA oligomers can efficiently be delivered into cells both in vitro and in vivo, when conjugated to transportan or TP10. The human galanin receptor type 1 (GalR1) antisense PNA conjugated to transportan or penetratin targeting the nucleotides of coding sequences 1–21 and 18–38 of GalR1 mRNA showed strong antisense effects in Bowes melanoma cells in culture and blocked specifically the expression of galanin receptor. Intrathecal injection of the PNA sequence complementary to region 18–38 of the rat GalR1 mRNA with carrier peptide caused the downregulation of galanin receptor type 1 in rat spinal cord in vivo and modulated the pain transmission.[18] The role of subregions of the targeted sequence 18–38 in GalR1 was studied in more detail by using transportan and TP10 for cellular transduction of respective shorter PNA oligomers in order to define the regulatory region(s) of the targeted mRNA.[19]

Kaushik and collaborators used transportan as the delivery vector for internalization of PNA oligomers targeted to the HIV transactivation response region (PNA$_{TAR}$) in CEM cells. They coupled PNA oligomers to transportan via a disulfide linkage and examined the functional efficacy of conjugate, detecting a significant inhibition of Tat-mediated transactivation of HIV-1 TAR element in the highly conserved 5' long terminal repeat (LTR). PNA$_{TAR}$ conjugated with transportan retained its specific binding affinity for its target sequence on the TAR RNA and revealed functional activity by inhibiting Tat-mediated transactivation of the HIV-1 LTR, as judged from the expression levels of luciferase reporter gene in CEM and Jurkat cells. The ability of the conjugate to block the Tat–TAR interaction at relatively low concentrations and to inhibit transcription of HIV-1 mRNA in chronically HIV-1-infected H9 cells was also demonstrated.[10]

Recently, Chaubey et al. demonstrated that the uptake of PNA$_{TAR}$–transportan (TP) conjugate by CEM cells is not dependent on temperature. Moreover, trypsination of CEM cells treated with fluorescein-labeled PNA$_{TAR}$–TP reduced the fluorescence signal from cells by less than 20%, suggesting that most of the conjugate entered cells and only a small fraction of it was on the surface, accessible to trypsin. The kinetics analysis of the uptake of PNA_{TAR}–TP resulted in a sigmoidal curve with a cooperativity index of 6, indicating very rapid cellular uptake. Importantly, the PNA_{TAR}–TP conjugate inserted into HIV-1 virions reduced viral infectivity by 60% at 100 nM concentration by blocking endogenous reverse transcription. Analogous conjugate with the scrambled PNA sequence was not able to associate with HIV-1 virions and did not have an effect on the viral infectivity.[20] The anti-HIV-1 virucidal effect of CPP conjugates with anti-TAR–PNA was also detected with TP10, penetratin, and Tat peptides with IC_{50} in the range of 28–37 nM. The IC_{50} for inhibition of HIV-1 replication, however, was the lowest for PNA_{TAR}–TP, at 0.4 μM concentration.[21]

Östenson and coworkers demonstrated the efficient downregulation of phosphotyrosine phosphatase σ level in isolated pancreatic islets of spontaneously type 2 diabetic Goto-Kakizaki rats by delivering antisense PNA into cells as a disulfide-linked construct with transportan.[13]

Beletskii et al. used TP–PNA conjugates to study whether the structure of inactivated X-chromosome could be disrupted in living cells by the administration of sequence-specific PNAs against the *Xist* transcript. The nontranslated RNA *Xist* has been shown to be necessary and sufficient to initiate the silencing pathway of X-chromosome. Transportan coupled to 19-mer antisense PNA targeted against the repetitive C-region in the first exon of *Xist* RNA abolished binding of *Xist* to the X-chromosome, thereby preventing the formation of Xi and its association with macrohistone H2A.[9] The 6- and 9-mer PNA conjugated to transportan or TP10 have also been used for transduction of NFκB decoy double-stranded oligonucleotide into Rinm5F cells. Treatment with 1 μM decoy oligonucleotide for 1 h blocked the IL-1β-induced NFκB binding activity and suppressed the IL-6 mRNA expression.[12]

10.3.2 Delivery of Peptides and Proteins

Various cell-penetrating peptides have been used to deliver small peptides, which mimic regulatory domains of intracellular proteins, to eukaryotic cells.[22,23] Transportan and MAP peptide showed higher efficiency in delivering a small model peptide cargo into the cytosol of Bowes melanoma cells than Tat or penetratin peptide.[24] TP10 was recently used for delivery of bioactive mimetic peptides into the RBL-2H3 mast cells. The peptides derived from protein kinase C (PKC) and CB_1 cannabinoid receptor conjugated to TP10 promoted the exocytosis of β-hexosaminidase indicating that the TP10-mediated delivery system is suitable for studying the activities of intracellular proteins that regulate cell signaling and membrane trafficking.[11]

Mostly the model proteins, like GFP, antibodies, and streptavidin–gold, have been used for demonstrating the potential of transportan in protein cellular transduction.[14,25] The cellular delivery of a cargo protein can be achieved by using different linkages to transportan — stable and labile covalent bonds and noncovalent complexation.[14]

10.3.3 ELECTRON MICROSCOPY IN DETECTION OF CPPS AND DELIVERED CARGOES IN CELLS

The unrivaled resolution and sensitivity of electron microscopy has seldom been used for assessing the cellular whereabouts of CPPs and their cargoes. First, penetration of biotin-derivatized pAntp (43–58) and D-(43–58) into cortical-striatal cells of E15 rat embryos was demonstrated by postembedding cryo-electron microscopy. Both peptides localized to the cell cytoplasm and nucleus, preferentially in the heterochromatin region.[26] The localization of Engrailed homeoprotein, which is known to travel from cell to cell by a nonendocytic process,[27] in caveolae-like structures of COS-7 cells was demonstrated by electron microscopy.[28]

We demonstrated the transduction of gold-labeled streptavidin into vesicular structures and cytosol of Bowes melanoma and COS-7 cells after complexing it with biotin–transportan.[14] Electron microscopy was used to examine the condensation of plasmid DNA with Vpr peptide derived from viral protein R of HIV-1 (and -2) and the cellular uptake of formed complexes.[29] Very recently, de la Fuente and Berry demonstrated the translocation of Tat-peptide-derivatized gold nanoparticles with diameter of about 3 nm into the cytoplasm and the nucleus of immortalized human primary fibroblasts by using transmission electron microscopy.[30]

10.4 CELL TRANSDUCTION MECHANISM OF TRANSPORTAN

In order to characterize the protein cellular delivery by transportan, we used noncovalent complexes of biotinylated peptide with labeled avidin or streptavidin. Transportan translocated covalently and noncovalently coupled tetramethylrhodamine-labeled avidin efficiently into COS-7 cells yielding a mostly granular localization pattern in cells, reminiscent of endocytosis, though some diffuse staining of the cytoplasm was also detected.[31]

10.4.1 DELIVERY WITH TRANSPORTAN IS CELLULAR-ENERGY-DEPENDENT

The endocytic pathways used by CPPs to facilitate the cellular uptake of proteins can be distinguished by specifically inhibiting a particular pathway. The blockage of a certain pathway could in principle be compensated by other uptake mechanisms. However, the application of inhibitors has yielded valuable information about the uptake and intracellular routing of CPPs.[32] Interference with cellular uptake mechanisms affects the cargo delivery efficiency of different CPPs to a different extent.[25] The inhibition of the most efficient pathway, clathrin-dependent endocytosis, by the hyperosmolar medium decreased the cellular uptake of FITC-labeled avidin complexed with biotinylated transportan only 1.5-fold compared to the uptake of complexes under normal isotonic medium conditions as revealed by FACS analysis. The minor inhibition of the internalization of the avidin–transportan complexes in the hyperosmolar medium indicates that the clathrin-dependent endocytosis is neither the only nor the major uptake mechanism used by CPPs and other clathrin-independent pathways are also involved. The involvement of other energy-dependent pathways in the uptake of transportan–avidin complexes was corroborated by the depletion of cellular energy by inhibiting oxidative phosphorylation and glycolysis with the inhibitors of cellular metabolism, sodium azide, and deoxyglucose. Blockage of oxidative phosphorylation with sodium azide diminished the internalization of transportan–avidin complexes, while deoxyglucose affected the uptake less. Although the coapplication of inhibitors had an additive inhibitory effect on the cellular translocation of transportan–avidin complexes leading to about fivefold decrease, the stable association of CPPs with the cells was not completely blocked, suggesting that at least the step of interaction of peptide–protein complexes with cells probably contains a component not dependent on the cellular energy.[25]

10.4.2 Protein Transduction by Transportan is Blocked at Low Temperature

Lowering of the incubation temperature below 18°C has been a popular method for blocking endocytosis in cells. However, some studies have demonstrated that even at 10°C, the cellular uptake is not completely blocked but dramatically reduced.[33,34] The first studies in the CPP field suggested that these peptides enter cultured cells at 4°C,[26,35,36] proposing that currently known endocytic pathways do not dominate in translocation. Based mostly on the indirect immunofluorescence data from fixed cells, the CPPs were defined as peptides, which enter cells in an energy- and receptor-independent manner and translocate across membranes in a nonendocytic fashion. The independence from the cellular energy and endosomal pathways of the translocation of cationic peptides into eukaryotic cells was later questioned and disproved.[37–40] Lowering of the temperature decreases the fluidity of the plasma membrane and partially depolarizes cells, leading to a dramatically reduced internalization of CPPs.[38] Assuming that endocytosis is abolished at low temperatures, we assessed the transportan-induced uptake of fluorescent Texas Red-labeled streptavidin at 0°C.[14] After 2 h of incubation of COS-7 cells with peptide–protein complexes at 0°C, the fluorescence signal was detectable by fluorescence microscopy mostly at the plasma membrane, but also weakly in the peripheral cytoplasm of the cells. Quantification of the association of CPP–protein complexes with plasma membrane by using flow cytometry showed that lowering the incubation temperature from 37 to 4°C led to about a 10-fold drop in fluorescence intensity, confirming dramatically decreased uptake of biotinyl–transportan–avidin complexes by cells.[25] Additionally, we can assume that at low temperature the majority of complexes are not internalized, but remain associated with the plasma membrane, because most of the FITC-labeled avidin associated with HeLa cells was dissociated from cells after extensive treatment with trypsin or pronase.[25]

More detailed information about the internalization of transportan–protein complexes at low temperature was obtained by electron microscopy using the complexes of biotinylated transportan with streptavidin–gold conjugates. Corroborating earlier results, only a negligible amount of gold particles had associated with the plasma membrane after incubation of cells with the transportan–streptavidin complexes on ice for 1 h, and no complexes were detected in the cells.[41] Surprisingly, at temperatures of 8–10°C, the cellular translocation of the transportan–streptavidin complexes was not abolished completely, but the amount of internalized complexes had markedly decreased and their vesicular transport toward the cell center was inhibited as revealed by electron microscopy. The absence of uptake at 4°C suggests the prevalence of energy-dependent processes in the internalization of CPP–protein complexes and conjugates. The decrease of the fluidity of the plasma membrane could play an essential role in blocking the cellular uptake of transportan–streptavidin complexes at 4°C. On the other hand, the fact that transportan and TP10 are able to deliver protein into cells even at 10°C supports the idea that the uptake pathways are still active at low temperatures.[42]

10.4.3 THE CHOLESTEROL-DEPENDENT UPTAKE

Recent studies have provided evidence that cell translocation of CPPs and their cargoes is dependent on the organization and composition of membrane lipids, especially cholesterol. The fusion protein pTat–Cre was shown to internalize via a lipid raft mediated macropinocytosis,[43] while pTat–EGFP fusion proteins were shown to internalize via caveolae-mediated uptake.[44,45] Both these endocytic uptake mechanisms are raft-mediated and cholesterol-dependent. Removal of cholesterol from the plasma membrane of cells by extraction with methyl-β-cyclodextrin leads to the disappearance of rafts and caveolae and inhibits their reformation.[46] Cholesterol-depleted HeLa cells take up transportan–avidin complexes less efficiently than the untreated cells as shown by the flow cytometry,[25] suggesting that the organization of the membrane lipids is essential for transportan-mediated protein delivery. The results obtained by FACS analysis are supported by the electron and fluorescence microscopical observations. Transportan–protein complexes associate preferentially with cholesterol-rich subdomains of the plasma membrane of HeLa cells colocalizing with the raft marker cholera toxin B subunit as demonstrated by confocal microscopy.

10.4.4 VESICLES MEDIATING THE CELLULAR UPTAKE OF TRANSPORTAN AND ITS CARGOES

Mammalian cells take up nutrients and extracellular material from the surrounding medium by endocytosis using different types of vesicles to deliver internalized substances to specific destinations. Fluorescently labeled avidin delivered into COS-7 cells by transportan either as a covalently coupled construct or in a noncovalent complex gave rise to the granular staining of the cytoplasm.[14] Confocal microscopy confirmed earlier results demonstrating that complexes of biotinyl–transportan or biotinyl–TP10 with Alexa Fluor 488-labeled avidin localized in a punctuate manner in the cortical and perinuclear cytoplasm after delivery into HeLa cells, suggesting an accumulation within vesicular structures, probably endosomes.[25] More detailed visualization by electron microscopy studies confirmed that transportan and TP10 translocated the streptavidin–gold conjugates into HeLa and

Bowes cells mostly within vesicles of different size and morphology. In order to specify the intracellular trajectories and structures of endocytic pathways by which the CPP–protein complexes translocate into cells, we used double-labeling by visualizing organelles involved in endocytosis and subsequent cellular trafficking of CPP–protein complexes simultaneously.[41] The localization of the transportan–avidin complexes overlapped only partially with the classical endocytosis marker, fluorescently labeled transferrin in HeLa cells, as demonstrated by confocal microscopy, suggesting that the internalization of proteins by transportan and TP10 is not predominantly mediated by the clathrin-dependent endocytic pathway of the transferrin receptor. An analogous result was obtained by electron microscopy, when following the uptake of the gold-labeled streptavidin complexed with transportan along with transferrin by HeLa cells. Only a partial colocalization of the transportan–streptavidin complexes with transferrin was detected at the plasma membrane and in cells in multivesicular bodies. However, at high concentration of both, transferrin often associated with regions of cell surface, where transportan–streptavidin complexes were located.[41]

Recently, Giacca's group demonstrated that the green fluorescent protein coupled to Tat peptide or Tat protein is taken up by cells by a clathrin-independent caveolae-mediated process.[44,45] To assess whether the caveolar internalization is involved in the transportan-mediated protein transduction, we tested the colocalization of transportan–avidin complexes with the fluorescently labeled B-subunit of cholera-toxin, which is known to label rafts and use caveolar endocytosis to enter the cells.[47] The localization of transportan–avidin complexes overlapped well with the cholera toxin B-subunit at the plasma membrane and in the cortical cytoplasm of HeLa cells. The colocalization with cholera toxin was more extensive than with transferrin, suggesting that lipid-raft dependent pathway is prevailing in the transportan-mediated cellular delivery of proteins. The ultrastructural analysis by electron microscopy confirmed that the TP–streptavidin complexes localize often to cholesterol-rich domains and caveosome-like structures.

The recruitment of cargo molecules transduced into cells is of the highest importance for application of CPPs as delivery vectors. The molecules taken up by endocytosis are carefully sorted in cells and routed to the structures of the degradative pathway, Golgi complex, ER, or recruited to the plasma membrane. The transportan–avidin complexes accumulated in large vesicles in the perinuclear area after longer incubations. The localization of the complexes to some extent overlapped with the Golgi network visualized with fluorescent ceramide derivatives as seen in confocal microscopy (Figure 10.1B). Staining of lysosomes with the antibody against lysosome-associated membrane protein-2 (LAMP2) or Lyso Tracker showed that the transportan–protein complexes were concentrating in late endosomes and lysosomal structures after 20 h (Figure 10.1C), which was also confirmed by ultrastructural analysis. However, no specific accumulation of the transportan–protein complexes to endoplasmic reticulum or Golgi network was observed by fluorescence microscopy or electron microscopy.[41] These results suggest the involvement of different endocytic pathways in the transportan-mediated cellular delivery of proteins.

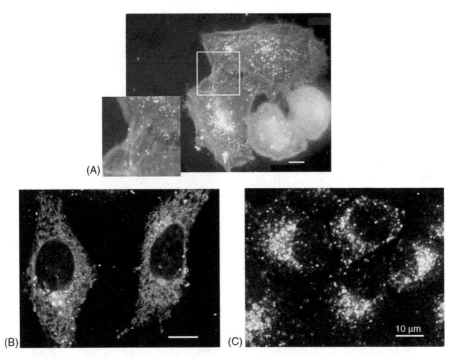

FIGURE 10.1 Localization of transportan–avidin complexes in relation to the actin cytoskeleton and organelles. HeLa cells incubated with complexes of 1 μM (A) or 0.5 μM (B, C) biotinyl–transportan and 0.15 μM Alexa Fluor 488-labeled avidin for 1 h (green). Actin fibers were stained after fixation of cells with 30 nM phalloidin-Texas Red (A). Golgi apparatus was visualized by coincubating the live cells with transportan–avidin complexes and 0.5 μM BODIPY-TR-C$_5$ ceramide (B). Cells were incubated with transportan–avidin complexes for 1 h and after 24 h chase period lysosomes were visualized by staining of cells with LAMP2 antibody and Alexa Fluor 594-labeled anti mouse antibody (red) (C). Colocalized structures are visible as yellow. Fluorescence microphoto of HeLa cells (A) and confocal sections at equatorial level of nuclei (B, C). (From Padari, K. et al., *Bioconjug. Chem.*, 16, 1399, 2005. With permission.)

10.4.5 ASSOCIATION OF TRANSPORTAN–PROTEIN COMPLEXES WITH PLASMA MEMBRANE, FILOPODIA, AND CYTOSKELETON

Electron microscopy provides detailed information about the interaction of CPP–protein complexes with the plasma membrane, internalization, and intracellular localization as compared to fluorescence microscopy. Therefore, we used electron microscopy to map the transportan-mediated protein transduction pathways by using the gold-labeled streptavidin complexed with biotinylated transportan. Based on ultrastructural observations, we could distinguish at least two morphologically different modes of association of transportan–streptavidin complexes. The aggregates of peptide–protein complexes associated preferentially with filopodia and extensions of plasma membrane in HeLa and Bowes cells (Figure 10.2A–C).

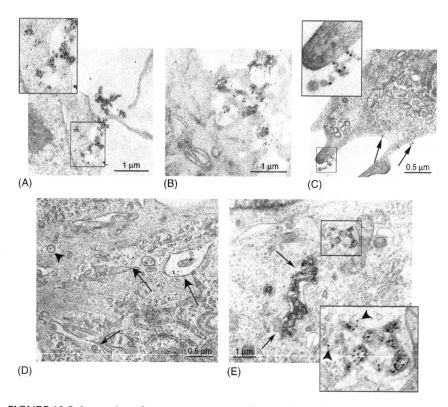

FIGURE 10.2 Interaction of transportan–streptavidin complexes with the plasma membrane of cells and localization in vesicular structures. Human Bowes melanoma cells (A, E) incubated with the complexes of gold-labeled streptavidin (10 nm, dilution 1:50) and 5 μM biotinyl–transportan for 30 min (A) or 2 h (E) at 37°C. HeLa cells (B, C, D) incubated with the complexes of gold-labeled streptavidin (10 nm, dilution 1:50) and 2.5 μM biotin–transportan for 1 h. (From Padari, K. et al., *Bioconjug. Chem.*, 16, 1399, 2005. With permission.)

The complexes of biotinyl–transportan with gold-labeled streptavidin were also detectable between the cells mostly as elongated structures of 0.4–2 μm length. The electron-dense background under the complexes also suggests an association with longer filopodia. The smaller fraction of complexes associated with the flat areas (arrows in Figure 10.2C), apparently being in close contact with the plasma membrane (enlarged section in Figure 10.2C). Large transportan–streptavidin aggregates induce morphological changes in the plasma membrane upon association, leading to the formation of membrane invaginations as demonstrated in the Figure 10.2A and B. Later, the complexes shift deeper into the cytoplasm, being finally fully engulfed by the cell into endocytic vacuoles of 0.5–1.0 μm, suggesting internalization by macropinocytosis. Moreover, the association of the transportan–streptavidin complexes with the extensions of cell surface and the induction of invaginations of the plasma membrane are similar to changes preceding the macropinocytosis of some pathogenic bacteria.[48,49] On the other hand, the transportan–streptavidin complexes are also detectable in cells in other types of vesicular structures — in multivesicular bodies (large arrows in Figure 10.2D), small caveosome-like vesicles (arrowhead in Figure 10.2D), and long tubulovesicular structures (small arrow in Figure 10.2D and Figure 10.2E). The localization of transportan–streptavidin complexes in the vesicular structures with heterogeneous size and shape suggests that different endocytic pathways are used by transportan to gain entry to the cell. In Bowes melanoma cells, the membrane of some intracellular vesicles was not continuous (arrows in Figure 10.2E), implying that a fraction of the vesicles had lost intactness and the complexes had escaped into the cytosol (enlarged section in Figure 10.2E).[41]

The second mode of internalization, which is characteristic for smaller complexes containing a single or a few gold particles, was also detected in addition to vesicular uptake. Smaller complexes did not induce detectable changes in the morphology of the plasma membrane upon interaction. The transportan–streptavidin complexes with a single gold particle were present rarely in the cortical cytoplasm of Bowes and HeLa cells. The number of complexes detected in the cytosol was very small and could be an artifact of fixation[37,50] or embedding of cells for EM. However, we cannot completely exclude the translocation of the complexes using short-living, pore-like structures as hypothesized for carrier peptide MPG and Pep-1[51] or the passive uptake characterized for some CPPs.[42,52,53] The transportan–streptavidin complexes associated mainly with membrane extensions and protrusions supported by actin cytoskeleton as visualized with Texas Red-labeled phalloidin (Figure 10.1A). The complexes aligned with the actin fibers at the intercellular filopodia, as well as in the cortical cytoplasm of the cells, suggesting that the intracellular translocation of the peptide–protein complexes might be mediated by the actin cytoskeleton.[41]

We have detected transportan–protein complexes in vesicles of different size and morphology with some overlap with the tested markers of endocytic structures, like transferrin, cholera toxin B-subunit, Golgi network, lysosomes, etc. Therefore, the involvement of several endocytic pathways in parallel for transportan-mediated protein transduction is very likely. The efficiency of delivering a cargo molecule into cells by CPP is dependent on a multitude of factors, such as the character of the used cargo and carrier peptide, cell type, and uptake conditions.

10.5 CONCLUSIONS

Transportan and its analogs facilitate the cellular uptake of various cargo molecules by using several endocytic pathways. Transportan-mediated delivery of cargoes into cells is dependent on cellular energy and the composition of the plasma membrane, and seems to be assisted by filopodia and actin cytoskeleton. Deliverable cargo molecules are mainly present in large vesicles with irregular shape reminiscent of macropinosomes. The respective vesicles are not present in the control cells incubated with the cargo molecules only, implying that the carrier peptide induces the formation of macropinosome-like vesicles and the subsequent cellular uptake. Some of the complex-containing structures are surrounded by a discontinuous membrane suggesting the destabilization of vesicles by transportan and the liberation of cargo molecules to the cytosol. However, the amount of cargo molecules in the cytosol is very small compared to the complexes found in vesicles. The facilitation of escape of the cargo molecules from the endosomal pathway structures is one of the main challenges in CPP-mediated cellular delivery in the future. Better understanding of the translocation mechanisms of CPPs would also help in designing the delivery system for targeting functional molecules to specific sites of action in vivo.

10.6 EXPERIMENTAL DETECTION OF CPP–PROTEIN COMPLEXES IN CELLS BY ELECTRON MICROSCOPY

10.6.1 COMPLEX FORMATION BETWEEN GOLD-LABELED STREPTAVIDIN AND BIOTINYL–TRANSPORTAN

Streptavidin adsorbed on colloidal gold can be complexed with biotinyl derivatives of CPPs and used as a simple model for the electron microscopy studies of the cellular delivery of protein by CPPs.

Reagents:

1. Biotinyl–CPP, 1 mM stock solution in Milli-Q quality water. We used (N, ε)-Lys^{13}-biotinyl–transportan, and (N, ε)-Lys^{7}-biotinyl–TP10.
2. L-Leucine, 20 mg/mL in distilled water.
3. Gold-labeled streptavidin (particle size 10 nm; Amersham Biosciences, UK).
4. Iscove's modified Dulbecco's culture medium (IMDM) supplemented with 10% fetal calf serum (FCS), 100 IU/mL penicillin and 100 μg/mL streptomycin.

Protocol:

1. Dissociate the multimers of biotinyl–transportan. Mix 2.5 μL biotinyl–transportan stock solution with 2.5 μL L-leucine solution and incubate for 5 min at room temperature for each well of the 24-well tissue culture plate at 5 μM peptide concentration. This step is only necessary for transportan and not needed for other biotinyl–CPPs.
2. Add 10 μL of streptavidin–gold (final dilution 1:50) and incubate for 3 min at room temperature.
3. Dilute the formed complexes of streptavidin–gold with biotinyl–transportan by adding 500 μL fresh culture medium and apply directly to cells.

10.6.2 TREATMENT OF CELLS WITH TRANSPORTAN–STREPTAVIDIN COMPLEXES

Protocol:

1. Grow HeLa or Bowes melanoma cells on the round plastic tissue culture coverslips (Thermanox, diameter 13 mm, EMS, Electron Microscopy

Science, Fort Washington, PA) in 24-well plates in IMDM for 2 days to 80–100% confluence.

2. Remove the culture medium from wells by aspiration and add the solution of streptavidin–gold complexed with biotinyl–transportan on to the cells (500 μL per well).

3. Incubate at 37°C under standard condition for 30 min to 24 h depending on the type of experiment.

10.6.3   PREPARATION OF CELLS TREATED WITH TRANSPORTAN–PROTEIN COMPLEXES FOR TRANSMISSION ELECTRON MICROSCOPY

Reagents:

1. Isoosmotic 0.1 M sodium cacodylate buffer containing 0.05 M sucrose, pH adjusted to 7.3 with HCl.

2. Fixative: 3% glutaraldehyde (EMS) in 0.1 M sodium cacodylate–HCl buffer (pH 7.3) containing 0.05 M sucrose.

3. Osmium tetraoxide solution for postfixation. Dilute 0.014 g potassium ferrocyanate in 0.5 mL cacodylate buffer (final concentration 0.7%) and add 1.5 mL of 2% osmium tetraoxide stock solution (EMS) to prepare 2 mL of working solution.

4. 50, 60, 70, 80, 90, and 96% solution of ethanol in Milli-Q water.

5. Mixture of epoxy resin embedding kit components (Fluka Chemie AG, Germany). To make 10 mL, mix the following components:

4.6 mL	EPON 812.
4 mL	DDSA (dodecenyl succinic anhydride).
1.4 mL	(N)MNA (methyl-nadic-anhydride).
0.18 mL	DMP-30 [2,4,6-Tris(dimethylaminomethyl)phenol]

6. Solution of 2% uranyl acetate (Serva, Germany) in 50% ethanol for the staining of epoxy resin sections.

7. Solution of lead citrate. Dissolve 0.03 g of lead citrate (EMS) in 10 mL of distilled water (CO_2 free) in a centrifuge tube. Add 0.1 mL of 10 N carbonate-free sodium hydroxide and seal the tube. Shake vigorously until the lead citrate is dissolved.

Protocol:

1. Remove the complexes-containing incubation medium by aspiration. Wash the cells on Thermanox coverslips twice for 5 min with cacodylate buffer.

2. Fix the cells with 3% glutaraldehyde in cacodylate buffer for 1 h at 4°C.

3. Wash three times for 20 min with cacodylate buffer.

4. Postfix cells with $OsO_4/K_3Fe(CN)_6$ in cacodylate buffer for 30 min at 4°C.

5. Wash the cells with cacodylate buffer three times for 10 min.

6. Dehydrate the cells in ethanol series (50, 60, 70, 80, 90, and 96%) for 10 min each step. For final dehydration step, transfer the Thermanox coverslip with cells from 24-plate wells into glass vial with acetone for 5 min.

7. Infiltrate the epoxy resin into cells by incubating in the following mixtures of 1:3, 1:1, and 3:1 (v/v) epoxy resin and acetone, respectively, at room temperature for 2 h each step.

8. Incubate the cells in the epoxy resin for 2 h.

9. Cut the plastic coverslips with cells into smaller pieces with scissors. Transfer the pieces with cells on upper side of coverslips into the flat bottom polyethylene embedding capsules (EMS).

10. Fill the capsules with freshly prepared epoxy resin, polymerize at 37°C for overnight and then at 60°C for additional 24 h.

11. Remove the plastic coverslips from the top of polymerized resin blocks and cut cells embedded in resin into ultrathin sections (40–60 nm).

12. Stain the sections collected on the copper grids (EMS) with solution of 2% uranyl acetate for 1 min. Wash twice in 50% ethanol and dry.

13. To achieve better contrast, stain the epoxy resin sections with lead citrate for 2 min. Wash the sections on grids with 0.01 N NaOH for 1 min before staining with solution of lead citrate. Wash again with 0.01 N NaOH, rinse with H_2O for 2 min, and dry in air.

14. Examine the preparates with electron microscope (JEM-100S, JEOL, Japan).

ACKNOWLEDGMENTS

This work was supported by grants from EC Framework 5 (QLK3-CT-2002-01989), Estonian Science Foundation (ESF 5588), Swedish Royal Academy of Sciences, and Swedish Research Councils NT and Med. We thank Helin Räägel and Pille Säälik for help with the manuscript and all our colleagues for their contribution.

REFERENCES

1. Dietz, G.P. and Bähr, M., Delivery of bioactive molecules into the cell: The Trojan horse approach, *Mol. Cell. Neurosci.*, 27, 85, 2004.

2. Langel, Ü. et al., A galanin-mastoparan chimeric peptide activates the Na+,K(+)-ATPase and reverses its inhibition by ouabain, *Regul. Pept.*, 62, 47, 1996.

3. Zorko, M. et al., Differential regulation of GTPase activity by mastoparan and galparan, *Arch. Biochem. Biophys.*, 349, 321, 1998.

4. Pooga, M. et al., Novel galanin receptor ligands, *J. Pept. Res.*, 51, 65, 1998.

5. Pooga, M. et al., Cell penetration by transportan, *FASEB J.*, 12, 67, 1998.

6. Soomets, U. et al., Deletion analogues of transportan, *Biochim. Biophys. Acta*, 1467, 165, 2000.

7. Magzoub, M. et al., Interaction and structure induction of cell-penetrating peptides in the presence of phospholipid vesicles, *Biochim. Biophys. Acta*, 1512, 77, 2001.
8. Barany-Wallje, E. et al., NMR solution structure and position of transportan in neutral phospholipid bicelles, *FEBS Lett.*, 567, 265, 2004.
9. Beletskii, A. et al., PNA interference mapping demonstrates functional domains in the noncoding RNA Xist, *Proc. Natl. Acad. Sci. U.S.A.*, 98, 9215, 2001.
10. Kaushik, N. et al., Anti-TAR polyamide nucleotide analog conjugated with a membrane-permeating peptide inhibits human immunodeficiency virus type 1 production, *J. Virol.*, 76, 3881, 2002.
11. Howl, J. et al., Intracellular delivery of bioactive peptides to RBL-2H3 cells induces beta-hexosaminidase secretion and phospholipase D activation, *Chembiochem.*, 4, 1312, 2003.
12. Fisher, L. et al., Cellular delivery of a double-stranded oligonucleotide NFkappaB decoy by hybridization to complementary PNA linked to a cell-penetrating peptide, *Gene Ther.*, 11, 1264, 2004.
13. Östenson, C.G. et al., Overexpression of protein-tyrosine phosphatase PTP sigma is linked to impaired glucose-induced insulin secretion in hereditary diabetic Goto-Kakizaki rats, *Biochem. Biophys. Res. Commun.*, 291, 945, 2002.
14. Pooga, M. et al., Cellular translocation of proteins by transportan, *FASEB J.*, 15, 1451, 2001.
15. Nielsen, P.E. et al., Sequence-selective recognition of DNA by strand displacement with a thymine-substituted polyamide, *Science*, 254, 1497, 1991.
16. Knudsen, H. and Nielsen, P.E., Antisense properties of duplex- and triplex-forming PNAs, *Nucleic Acids Res.*, 24, 494, 1996.
17. Demidov, V.V. et al., Stability of peptide nucleic acids in human serum and cellular extracts, *Biochem. Pharmacol.*, 48, 1310, 1994.
18. Pooga, M. et al., Cell penetrating PNA constructs regulate galanin receptor levels and modify pain transmission in vivo, *Nat. Biotechnol.*, 16, 857, 1998.
19. Kilk, K. et al., Targeting of antisense PNA oligomers to human galanin receptor type 1 mRNA, *Neuropeptides*, 38, 316, 2004.
20. Chaubey, B. et al., A PNA-transportan conjugate targeted to the TAR region of the HIV-1 genome exhibits both antiviral and virucidal properties, *Virology*, 331, 418, 2005.
21. Tripathi, S. et al., Anti-HIV-1 activity of anti-TAR polyamide nucleic acid conjugated with various membrane transducing peptides, *Nucleic Acids Res.*, 33, 4345, 2005.
22. Oehlke, J. et al., Cellular uptake of an alpha-helical amphipathic model peptide with the potential to deliver polar compounds into the cell interior non endocytically, *Biochim. Biophys. Acta*, 1414, 127, 1998.
23. Begley, R. et al., Biodistribution of intracellularly acting peptides conjugated reversibly to Tat, *Biochem. Biophys. Res. Commun.*, 318, 949, 2004.
24. Hällbrink, M. et al., Cargo delivery kinetics of cell-penetrating peptides, *Biochim. Biophys. Acta*, 1515, 101, 2001.
25. Säälik, P. et al., Protein cargo delivery properties of cell-penetrating peptides. A comparative study, *Bioconjug. Chem.*, 15, 1246, 2004.
26. Derossi, D. et al., Cell internalization of the third helix of the Antennapedia homeodomain is receptor-independent, *J. Biol. Chem.*, 271, 18188, 1996.
27. Maizel, A. et al., A short region of its homeodomain is necessary for engrailed nuclear export and secretion, *Development*, 126, 3183, 1999.

28. Joliot, A. et al., Association of Engrailed homeoproteins with vesicles presenting caveolae-like properties, *Development*, 124, 1865, 1997.
29. Coeytaux, E. et al., The cationic amphipathic alpha-helix of HIV-1 viral protein R (Vpr) binds to nucleic acids, permeabilizes membranes, and efficiently transfects cells, *J. Biol. Chem.*, 278, 18110, 2003.
30. de la Fuente, J.M. and Berry, C.C., Tat peptide as an efficient molecule to translocate gold nanoparticles into the cell nucleus, *Bioconjug. Chem.*, 16, 1176, 2005.
31. Pooga, M. and Langel, Ü., Targeting of cancer-related proteins with PNA oligomers, *Curr. Cancer Drug Targets*, 1, 231, 2001.
32. Fischer, R. et al., A stepwise dissection of the intracellular fate of cationic cell-penetrating peptides, *J. Biol. Chem.*, 279, 12625, 2004.
33. Silverstein, S.C. et al., Endocytosis, *Annu. Rev. Biochem.*, 46, 669, 1977.
34. Steinman, R.M. et al., Pinocytosis in fibroblasts. Quantitative studies in vitro, *J. Cell. Biol.*, 63, 949, 1974.
35. Derossi, D. et al., The third helix of the Antennapedia homeodomain translocates through biological membranes, *J. Biol. Chem.*, 269, 10444, 1994.
36. Vivés, E. et al., A truncated HIV-1 Tat protein basic domain rapidly translocates through the plasma membrane and accumulates in the cell nucleus, *J. Biol. Chem.*, 272, 16010, 1997.
37. Richard, J.P. et al., Cell penetrating peptides: A reevaluation of the mechanism of cellular uptake, *J. Biol. Chem.*, 278, 585, 2003.
38. Drin, G. et al., Studies on the internalization mechanism of cationic cell-penetrating peptides, *J. Biol. Chem.*, 278, 31192, 2003.
39. Console, S. et al., Antennapedia and HIV transactivator of transcription (TAT) 'protein transduction domains' promote endocytosis of high molecular weight cargo upon binding to cell surface glycosaminoglycans, *J. Biol. Chem.*, 278, 35109, 2003.
40. Koppelhus, U. et al., Cell-dependent differential cellular uptake of PNA, peptides, and PNA–peptide conjugates, *Antisense Nucleic Acid Drug Dev.*, 12, 51, 2002.
41. Padari, K. et al., Cell transduction pathways of transportans, *Bioconjug. Chem*, 16, 1399, 2005.
42. Ziegler, A. et al., The cationic cell-penetrating peptide CPP(TAT) derived from the HIV-1 protein TAT is rapidly transported into living fibroblasts: Optical, biophysical, and metabolic evidence, *Biochemistry*, 44, 138, 2005.
43. Wadia, J.S. et al., Transducible TAT-HA fusogenic peptide enhances escape of TAT-fusion proteins after lipid raft macropinocytosis, *Nat. Med.*, 10, 310, 2004.
44. Ferrari, A. et al., Caveolae-mediated internalization of extracellular HIV-1 tat fusion proteins visualized in real time, *Mol. Ther.*, 8, 284, 2003.
45. Fittipaldi, A. et al., Cell membrane lipid rafts mediate caveolar endocytosis of HIV-1 Tat fusion proteins, *J. Biol. Chem.*, 278, 34141, 2003.
46. Thyberg, J., Caveolae and cholesterol distribution in vascular smooth muscle cells of different phenotypes, *J. Histochem. Cytochem.*, 50, 185, 2002.
47. Montesano, R. et al., Noncoated membrane invaginations are involved in binding and internalization of cholera and tetanus toxins, *Nature*, 296, 651, 1982.
48. Ammendolia, M.G. et al., A Sphingomonas bacterium interacting with epithelial cells, *Res. Microbiol.*, 155, 636, 2004.
49. Garcia-Perez, B.E. et al., Internalization of *Mycobacterium tuberculosis* by macropinocytosis in nonphagocytic cells, *Microb. Pathog.*, 35, 49, 2003.
50. Lundberg, M. and Johansson, M., Positively charged DNA-binding proteins cause apparent cell membrane translocation, *Biochem. Biophys. Res. Commun.*, 291, 367, 2002.

51. Deshayes, S. et al., On the mechanism of nonendosomial peptide-mediated cellular delivery of nucleic acids, *Biochim. Biophys. Acta*, 1667, 141, 2004.
52. Thorén, P.E. et al., Membrane binding and translocation of cell-penetrating peptides, *Biochemistry*, 43, 3471, 2004.
53. Maiolo, J.R. et al., Effects of cargo molecules on the cellular uptake of arginine-rich cell-penetrating peptides, *Biochim. Biophys. Acta*, 1712, 161, 2005.

11 Tat-Mediated Peptide/Protein Transduction In Vivo

Wenge Shi and Steven F. Dowdy

CONTENTS

11.1 INTRODUCTION

The direct intracellular delivery of large hydrophilic therapeutic agents such as proteins, active peptide domains, or nucleic acids has, until recently, been difficult to achieve due primarily to the bioavailability barrier of the plasma membrane, which effectively prevents the uptake of such macromolecules by limiting their passive entry. Traditional approaches to modulate protein function have largely relied on the serendipitous discovery of specific drugs and small molecules, which could be delivered easily into the cell. However, the usefulness of these pharmacological agents is limited by their tissue distribution and, unlike "information-rich" macromolecules, they often suffer from poor target specificity, unwanted side effects, and toxicity. Likewise, over the past several decades, the development of molecular techniques for gene delivery and expression of proteins has provided for tremendous advances in our understanding of cellular processes but has been of surprisingly little benefit for the management of genetic disorders. Apart from these gains, however, the transfer of genetic material into eukaryotic cells in vivo either using viral vectors or by nonviral methods, such as microinjection, electroporation, or chemical transfection, and the use of liposomes, remains problematic. Moreover, in vivo, gene therapy approaches relying on adenoviral vectors are associated with significant difficulties relating to toxicity and immunogenicity which have contributed to poor performance in several clinical trials.[1]

Remarkably, the identification of a particular group of proteins with the enhanced ability to cross the plasma membrane in a receptor-independent fashion has led to the discovery of a class of protein domains with cell membrane penetrating properties, or so-called protein transduction domain (PTD) and cell penetrating peptides (CPP).[2,3] The first example of protein transduction was observed when the full length HIV Tat protein was found to be capable of entering mammalian cells and activating transcription from an HIV long terminal repeat promoter construct.[4,5] Subsequent studies defined the specific region of the protein necessary for cellular uptake.[6] In a similar fashion, the Antp or penetratin peptide was derived from a Drosophila Antennapedia homeodomain protein.[7] In addition, VP22 derived from a herpes simplex viral protein and even short peptides consisting entirely of arginines[8] or lysines[9] show membrane penetrating ability. In fact, all PTD are rich in positively charged amino acids such as lysines and arginines, the basic nature of which appears to be essential for their cell-permeation properties.[9] A fusion of these PTD peptide sequences with heterologous proteins is sufficient to cause their rapid transduction into a variety of different cells in a rapid, concentration-dependent manner. Moreover, this novel technique for protein and peptide delivery appears to circumvent many problems associated with DNA and drug-based methods. This technique may represent the next paradigm in our ability to modulate cell function and offers a unique avenue for the treatment of disease.

Further interest in these peptides was stimulated by the observation that PTDs can also facilitate systemic delivery of recombinant proteins to a large number of tissues in a living mouse.[10] Among various PTDs used as carriers for relatively small cargo such as peptides and oligonucleotides, Tat has been the predominant PTD used in the delivery of large molecules such as full length proteins.[11] Since their initial description,[12] many in-frame Tat fusion proteins produced in bacteria have been shown to enter mammalian cells and carry out intracellular functions ranging from cytoskeletal reorganization to recombination of genomic DNA.[13] Tat peptides have even been used to transfer much larger molecules, such as 45 nm iron beads, lambda phage, adenovirus, liposomes complexed with plasmid DNA, and nanoparticles.[14] These results raised the possibility that large intracellular proteins linked to PTDs might be used therapeutically just as extracellular proteins (i.e., insulin, monoclonal antibodies) are employed in clinical practice today.

In this review, we will discuss the mechanisms of PTD-mediated protein transduction and also focus on recent studies that have demonstrated the ability of PTD conjugates in treating small animal models of cancer and other diseases.

11.2 MECHANSISMS OF PTD-MEDIATED TRANSDUCTION

11.2.1 EARLIER CLAIMS OF ENERGY-INDEPENDENT UPTAKE

As PTD-mediated protein transduction gained popularity over the last decade, there is no shortage of clear demonstrations that the PTD was able to cross the cell membrane and the cargo covalently attached to the PTD was able to elicit intracellular events. However, simultaneous studies aimed at determining the mechanism of uptake suggested disparate mechanisms.

Despite the initial suggestion of an involvement of an endocytic pathway,[15,16] mechanistic studies during the late 1990s and early 21st century claimed that PTDs penetrated cells directly across the cell membrane in a temperature- and energy-independent process. The transduction occurs rapidly within minutes and is not dependent on energy, as the import occurs at both 37°C and 4°C.[7,17–19] They are not inhibited by inhibitors of endocytosis and a decrease of cellular ATP pool.[20] These observations were based on direct visualization using fluorescence microscopy on fixed and permeabilized cells. In addition, neither sequence inversion nor synthesis with D-amino acids ablates the function of Tat and Antp, suggesting that the cell entry is either not mediated by a chiral receptor or specific protein transporter[7,8] or mediated by some ubiquitously expressed "receptors." Earlier, other ways commonly used to assess protein transduction were based on fluorescence activated cell sorter (FACS) analysis of transduced cells or on assays for enzymatic activity. Although these methods avoid artifacts caused by cell fixation, they do not necessarily distinguish if transduced cargoes are internal or cell surface-bound despite extensive washing with buffered saline. Results were strongly influenced by the cell surface binding of these PTD domains. Therefore, a few models were proposed as the potential mechanisms of protein transduction. One such model suggested a direct penetration of transduced proteins through the membrane lipid bilayer.[17] Another model suggested an inverted micelle intermediate within the membrane involved in Tat transduction, resulting from the interaction of positively charged CPP dimers with negatively charged phospholipids.[7] However, all these models asked for a radical rethinking of membrane properties, while at the same time they seem to have difficulty explaining the lack of toxicity of CPPs to cells[21] and the CPPs ability to deliver large cargoes.

11.2.2 INVOLVEMENT OF ENDOCYTOSIS IN UPTAKE

The earlier claim of an energy-independent protein transduction was challenged. Richard et al.[22] demonstrated that cell fixation, even in mild conditions, leads to the artifactual redistribution of the Tat and R9 nonconjugated peptides into the nucleus, while fluorescence microscopy on live unfixed cells shows characteristic endosomal distribution of peptides. They showed that flow cytometry analysis cannot be used validly to evaluate cellular uptake unless a step of trypsin digestion of the cell membrane-adsorbed peptide is included in the protocol. After all, flow cytometry analysis indicates that the kinetics of uptake is similar to the kinetics of endocytosis. Furthermore, peptide uptake is inhibited by incubation at low temperature and cellular ATP pool depletion. Therefore, they suggested an involvement of endocytosis in the cellular internalization of cell-penetrating peptides.

Similarly, Lundberg et al.[23,24] reported that translocation of PTDs (VP22, Tat, polyarginine, and polylysine) fused to the green fluorescent protein (GFP) is due to a fixation artifact by visualizing PTDs in living cells and fixed cells. They showed that the investigated PTDs strongly adhered to the surface of living cells and were internalized by constitutive endocytosis. No cytosolic or nuclear import of the proteins was detected. In contrast, the PTD–GFP fusion proteins were redistributed to the cytosol and nucleus directly after fixation. However, limited by low sensitivity, the direct visualization technique does not definitely exclude the minute transduced GFP into the nucleus.

Endocytosis is an essential cellular process for the uptake of a wide variety of extracellular factors.[25] There are a few functionally distinct mechanisms: phagoctosis restricted to specialized cell types, or endocytosis (clathrin-dependent, raft/caveolin-dependent endocytosis, clathrin- and caveolin-independent endocytosis, and macropinocytosis). Ferrari et al.[26] studied the uptake of Tat–eGFP into live HeLa and CHO cells and suggested that the uptake of Tat-fusion protein is mediated by caveolin-dependent endocytosis based on a colocalization of caveolin-1 and inhibition of internalization by cytochalasin D. However, the caveolin-dependent pathway is not essential for the uptake of unconjugated Tat peptide, as evidenced by the efficient internalization of Tat in the presence of the known inhibitors of the raft/caveolin-dependent pathway and for cells lacking or deficient in caveolin-1 expression.[27,28] In addition, cytochalasin D, a so-called caveolar trafficking inhibitor used in this study, in fact inhibits more general F-actin elongation.[29]

In order to determine the true mechanism of Tat-mediated transduction while avoiding these potential pitfalls, Wadia et al.[28] used a phenotypic assay for cellular uptake based on a LoxP-Stop-LoxP GFP gene reporter assay and the transduction of Cre recombinase. In this system exogenous Tat-Cre protein must enter the cell, translocate to the nucleus, and excise the transcriptional STOP DNA segment in live cells in a nontoxic fashion before scoring positive for eGFP expression. Treatment of reporter T cells for as little as 5 min was sufficient to induce recombination, confirming that the cellular uptake was a rapid process. Authors also demonstrated a colocalization of fluorescently labeled Tat-Cre with endocytosis markers in live cells. They went on to show that Tat-fusion proteins are rapidly internalized by lipid raft-dependent macropinocytosis and transduction was independent of interleukin-2 receptor/raft-, caveolar-, and clathrin-mediated endocytosis and phagocytosis. Due to the ultrasensitive nature of assay, a remaining question is how many molecules are sufficient to cause Cre mediated-DNA cleavage. Similarly, Tat peptides enter mammalian cells by macropinocytosis.[27] Subsequently, more groups confirmed the involvement of endocytosis in the cellular uptake of CPP proteins,[30] antennapedia- and Tat-mediated peptide delivery,[31] and of Tat PTD peptide–gold nanoparticle conjugates.[32] Transduction's requirement for active endocytosis may provide an explanation for the earlier observation that well-differentiated epithelial cells in the urinary bladder of living rats in vivo form a permeability barrier to Tat basic domain.[33]

So far, clathrin-dependent endocytosis,[34] raft/caveolin-dependent endocytosis,[26] and macropinocytosis[27,28] are all implicated in the uptake of CPPs and their cargoes into cells. The uptake mechanisms may be different dependent on the kind of CPPs, size of cargoes, conjugated versus nonconjugated peptides, read-out, or cell types.

11.2.3 THE STEPS OF PROTEIN TRANSDUCTION

All PTDs share a common feature of cationic charged amino acids and form a tight and rapid interaction with the ubiquitous presence of extracellular glycosaminogly-cans, such as heparan sulfate, heparin, and chondroitin sulfate B. Externally added heparin inhibits the cellular uptake of Tat-GFP or Tat peptides, as does treatment of heparinase III.[20,35] Whereas a significant part of Tat peptide uptake involves heparan sulfate receptors, efficient internalization of peptide is observed even in their absence,[35] indicating the involvement of other receptors, perhaps a negative charged portion of phospholipids. All these make the binding of the highly cationic CPP peptide to various anionic membrane components electrostatically a strong candidate as the primary step of protein translocation.[36] After internalization by endocytic pathway, a portion of CPPs may then be released from heparan sulfate due to hydrolytic degradation of heparan sulfate.

The rate-limiting step of protein transduction lies in the release of CPP cargoes into the cell cytoplasm and nucleus and reaching the desired sites before they are degraded in endocytic vesicles. Although macropinosomes are thought to be inherently leaky vesicles compared to other types of endosomes and do not fuse into lysosomes,[37,38] the majority of the Tat proteins remain trapped within these intracellular compartments, functioning as internal reservoirs up to 24 h following treatment. Therefore, developing ways to improve the release of transduced molecules from endosomes or macropinosomes is critical to making intracellular delivery more efficient (Figure 11.1).

Virus relies on host cell machinery to replicate itself and develop different strategies to infect a host cell, including taking advantage of endocytic routes. Influenza virus evolves to use its hemagglutinin (HA2) protein to destabilize endosomal lipid membranes at low pH and escape from endosomes after internalization. Knowledge of Tat-mediated macropinocytosis invites the develop-ment of a new strategy to enhance cargo release. Tat-HA2, a transducible peptide derived from this protein, markedly enhanced Tat-Cre escape from macropinosomes and significantly enhanced intracellular DNA recombination by a Tat-Cre fusion protein in living cells. It is likely that such enhancers of transduction, either in the form of peptides or small molecules, will also be used to facilitate protein transduction in living organisms.[28]

Another interesting way to enhance the release of endocytosed proteins trapped in the endosome in a controlled manner was demonstrated by Matsushita et al.[39] They showed that exposure to light at 480 nm stimulated endosomal release of FITC-labeled transduced R11-PTD, Tat-PTD, Antp-PTD, and R11-p53 protein. The use of such a photo-acceleration strategy seems to be efficient in vitro with minimum

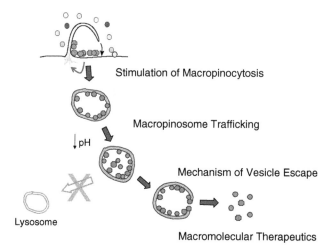

FIGURE 11.1 TAT-mediated transduction occurs by macropinocytosis. Cationic peptides bind to cell surface proteoglycans and stimulate macropinocytosis, a specialized form of fluid-phase endocytosis. Macropinosomes decrease their pH; however, they do not appear to traffic to lysomes, thus avoiding rapid degradation of peptides and proteins. By an unknown mechanism, the cationic TAT domain destabilizes the macropinosome lipid bilayer, resulting in escape of a small amount of the TAT cargo into the cytoplasm. Thus, escape into the cytoplasm is the rate-limiting step and use of membrane destabilizing or membrane fusion domains, such as influenza's HA2 pH sensitive fusion domain, results in dramatic increases in transduction into the cytoplasm (see Wadia, J.S., Stan, R.V., and Dowdy, S.F., *Nat. Med.,* 10, 310, 2004.)

cytotoxicity which, albeit, can be limited in vivo by light accessibility. On the other hand, this strategy can be used to design site-specific therapy.

11.2.4 TARGETING DIFFERENT COMPARTMENTS

Without additional targeting signals, the intracellular localization of the proteins delivered by PTD appears to be cytoplasmic[40] or nuclear.[28] Further, targeting into different cellular compartments can be facilitated by the addition of selective targeting signals. For example, Shokolenko et al.[41] efficiently directed Tat-GFP and Tat-exonuclease III into mitochondria of breast cancer cells and modulated mitochondrial function and cell survival utilizing the mitochondrial targeting signal (MTS) from hMnSOD on the N-terminus of Tat-fusion proteins.

11.3 IN VIVO APPLICATIONS OF PROTEIN TRANSDUCTION

PTDs have proven their ability to manipulate the biology of cultured mammalian cells by delivering cargoes such as peptides and recombinant proteins. However, the ability to deliver large, biologically active molecules to the interior of cells in a living organism would also be of tremendous benefit. To determine whether PTDs might facilitate such delivery, a Tat-β-galactosidase fusion protein was delivered to mice by intraperitoneal (IP) injection.[10] Analysis of tissue sections revealed delivery of the fusion protein to many, if not most, tissues of the mouse, including the brain. Importantly, the X-gal assay used in this study demonstrated that the Tat-β-galactosidase, a 120 kD protein enzyme, retained its activity in vivo. These observations stimulated a number of groups to examine the ability of PTD peptides and proteins to modulate the biology of cells and tissues in vivo. In the last 6 years, PTDs have been successfully used to treat small animal models of cancer and other diseases.

11.3.1 ANTICANCER

A number of studies have examined the ability of transducible peptides and proteins to inhibit tumor growth in vivo. Given the lack of specificity of current cancer therapy and the limitations of gene therapy, PTDs appear to be an attractive means by which to introduce tumor suppressors or other pro-apoptotic proteins directly into the cancer cells that make up tumors in vivo.

Nearly every human cancer has mutation of genes in the p53 pathway,[42] so restoration of p53 function in cancer cells would be a useful application of in vivo protein transduction. Our laboratory attempted to do this by linking Tat to a peptide derived from the C-terminus of p53.[43] This p53C′ peptide was previously shown to activate wildtype and certain p53 mutants in cancer cells, leading to apoptosis.[44–46] One potential pitfall of using peptides in living organisms is the short half-life of many peptides secondary to degradation. To address this, we synthesized a retro-inverso (RI) version of the Tat-p53C′ peptide by inverting the peptide sequence and using D-amino acids. This double inversion of peptide structure often leaves the surface topology of the side chains intact and has been used extensively to stabilize biologically active peptides for in vivo applications. After confirming that the RI-Tatp53C′ peptide retained the p53-activating function of its parental peptide, we tested its efficacy in three different models of human cancer. In the first model, we found that IP injection led to a delivery of peptide to subcutaneous tumors and caused significant inhibition of their growth. In a second model, peptide administration resulted in a sevenfold increase in lifespan in a model of terminal peritoneal carcinomatosis. Finally, peptide-treated mice harboring a model of B-cell lymphoma achieved a 50% long-term survival (>200 days), whereas control mice died an average of 35 days after lymphoma cell inoculation.[43] Notably, two of these models utilized immune-competent mice, showing that the immune system is not an absolute barrier, at least in the short term, to successful in vivo protein transduction. The fact that these studies utilized models of terminal malignancy is also relevant because anticancer therapeutics are defined as clinically successful by their ability to alleviate pathology and extend survival and not simply by their ability to reduce tumor size.

Another highly successful anticancer protein transduction strategy has involved the use of a peptide derived from the Smac protein, a mitochondrial protein that can inactivate the inhibitor of apoptosis (IAP) proteins. Multiple groups have found that the N-terminus of Smac can be linked to either Tat or Antp to facilitate cellular uptake[47,48] and that Smac-Tat or Smac-Antp sensitized cells to proapoptotic stimuli. One group took these results a step further and tested the function of the Smac-Tat peptide in an intracranial glioblastoma xenograft model.[47] The authors first showed that local administration of TRAIL, a death receptor ligand with specificity for tumor cells, reduced tumor volume and moderately extended the life of nude mice bearing established intracranial U87MG tumors. By contrast, local treatment with Smac-Tat alone had no effect on tumor growth. When TRAIL and Smac-Tat were co-administered, however, there was a synergistic effect on tumor volume and mouse survival. Whereas control mice all died of tumor burden by 35 days after tumor cell injection, mice treated with Smac-Tat and the highest doses of TRAIL survived beyond 70 days after the start of the experiment. Furthermore, histological analysis revealed no evidence of remaining tumor in the brain. The general applicability of this approach was confirmed by an independent group, who used nude mice harboring subcutaneous lung cancer xenografts to show that intratumoral injection of a transducible Smac peptide sensitized the cancer cells to systemic cisplatin therapy.[49]

Some groups have sought to take advantage of the tumor microenvironment when designing anticancer protein transduction strategies. Tumor hypoxia in a solid tumor mass has long been recognized as a cause of resistance to current cancer therapies. In a clever series of experiments, Harada et al.[50] fused the oxygen-dependent degradation domain (ODD) of HIF-1α to the Tat-β-gal and the Tat-caspase-3 fusion proteins. The ODD domain stimulates degradation of HIF-1α under normoxic, but not under hypoxic, conditions. Thus, the authors hypothesized that a Tat-ODD fusion protein would be stable, specifically in the hypoxic core of tumors, but would be degraded and nonfunctional in normal tissue. Indeed, after IP injection of Tat-ODD-β-gal into tumor-bearing nude mice, only the hypoxic regions of the tumors showed evidence of Tat-ODD-β-gal protein. By contrast, Tat-β-galactosidase protein could be detected throughout tumors after IP delivery. Furthermore, Tat-ODD-β-gal was undetectable in normal mouse liver after IP injection, again unlike the parental Tat-β-gal protein. Next, tumor-bearing mice were given IP injections of a Tat-ODD-caspase-3 protein. Tat-ODD-caspase-3 reduced tumor growth without causing the toxic side effects that would be expected from delivering active caspase-3 to an entire mouse. This group recently extended their results by testing the Tat-ODD-caspase-3 protein in a rodent model of malignant ascites in which cancer cells grow in the hypoxic environment of the peritoneum.[51] They found that administration of Tat-ODD-caspase-3 induced a 60% cure rate in an otherwise lethal model of cancer. Similarly, Willam et al.[52] used Tat delivery of ODD peptides to modulate angiogenesis in vivo. These studies show that functional domains such as ODD can be used to modulate the type of tissue in which Tat-fusion proteins are active and, in this way, increase their specificity for cancer cells.

Using protein transduction to reconstitute tumor suppressor function has also been reported in several mouse tumor models in vivo. For example, delivery of p27(Kip) tumor suppressor protein leads to partial responses.[53] Other examples of systemic delivery of CPP include the fusion of Tat to a peptide derived from the von Hippel–Lindau (VHL) tumor suppressor that inhibits insulin-like growth factor-1 receptor signaling in renal cell carcinomas.[54] IP administration of Tat-VHL peptide slowed the growth of subcutaneous renal cell carcinoma tumors in nude mice, primarily through inhibition of cell proliferation rather than by induction of apoptosis. Peptide treatment also appeared to reduce tumor invasion into the underlying tissue. This study also provides strong immunohistochemical evidence that the Tat-VHL peptide is homogeneously delivered to the tumors after IP injection. Another report found that IP delivery of an Antp-p16 fusion peptide moderately inhibited the growth of pancreatic cancer cells growing as intraperitoneal and as subcutaneous tumors in nude mice.[55] Although p16 functions primarily as an inhibitor of cell-cycle progression, the authors found that Antp-p16 slowed tumor growth by inducing apoptosis of cancer cells in vivo. In neither study was any toxicity to normal tissue observed,[54,55] indicating that cancer cells may be much more sensitive to the effects of transducible tumor suppressor peptides than nontransformed cells. These observations serve as a starting point to further develop the delivery of peptide and proteins to specifically target tumors while sparing normal tissue.

Several other reports have shown that PTDs can be used to attack cancer cells in vivo by targeting a number of the signal transduction pathways known to be dysregulated during tumorigenesis. Local administration of CPPs that block Cdk2/cyclin A activity,[56] sequester mdm2,[57] or induce apoptosis by mitochondrial disruption[58] has been shown to reduce tumor burden in living animals.

11.3.2 TREATMENT FOR ASTHMA

Asthma is caused by immunological hyperresponsiveness in airways. Myou et al.[59] showed that IP administration of a dominant negative Ras fused to Tat PTD (Tat-dnRas) inhibits the airway inflammatory response by cytokine blockage in a mouse asthma model. Uptake of Tat-dnRas was demonstrated in leukocytes. Treatment of animals with 3–10 mg/kg Tat-dnRas blocked the inflammatory cell infiltration (largely eosinophils and mononuclear cells) and mucin production around the airways caused by OVA. Tat-dnRas also blocked Ag-induced Th2 cytokine production in lung tissue.

11.3.3 PROTECTING CELLS FROM PATHOLOGICAL APOPTOSIS

Apoptosis has to be tightly regulated for maintaining normal tissue homeostasis. Dysregulated cell apoptosis is associated with several pathologic processes, for example, transient increase in apoptosis during acute hepatitis B, autoimmune diseases, ischemia–reperfusion injury, sepsis, or allograft rejection. Given the high clinical relevance, it is of importance to develop new tools to modulate cell apoptosis.

To specifically inhibit the initiation of death receptor-mediated apoptosis, Krautwald et al.[60] designed a Tat-FLIP (FLICE inhibitory protein, a caspase-8 inhibitor) fusion protein and demonstrated its ability to transduce cross the cell membrane and interfere with the activation of Fas-inducing signaling complex inside the lymphocytic cell and prevent cells from undergoing apoptosis in vitro.

Systemic application of Tat-FLIPS prolongs survival and reduces multi-organ failure due to Fas-receptor-mediated lethal apoptosis in mice.

The Bcl-2 family of proteins regulates apoptosis chiefly by controlling mitochondrial membrane permeability. The Bcl2-homology domain 4 (BH4) of Bcl-2/Bcl-xL has previously been shown essential for the prevention of apoptotic mitochondrial changes, including the release of cytochrome c and apoptotic cell death. Sugioka et al.[61] showed that Tat-BH4 was cytoprotective in ex vivo and in vivo rodent models. IP injection of Tat-BH4 peptide greatly inhibited x-ray-induced apoptosis in the small intestine of mice and partially suppressed Fas-induced fulminant hepatitis. In addition, this peptide markedly suppressed heart failure after ischemia–reperfusion injury in isolated rat heart, probably by preventing mitochondrial dysfunction. These findings demonstrate that Tat-BH4 peptide exerts anti-apoptotic activity both in vivo and ex vivo, and imply that it may be a useful therapeutic agent for diseases involving mitochondrial dysfunction and apoptosis.

11.3.4 Ischemia

Cerebral ischemia (stroke) remains one of the leading causes of morbidity and mortality. Cerebral ischemia and the restoration of blood flow to the ischemic tissue (reperfusion) lead to neuronal necrosis and apoptosis possibly by a variety of mechanisms, including energy depletion, peri-infarct depolarization, and excito-toxicity secondary to N-methyl-D-aspartate receptors (NMDAR) activation. So far, only one medical therapy for cerebral ischemia has been approved (recombinant tissue plasminogen activator) and it must be administered within 3 h of the onset of ischemia to be effective, a time point at which most patients have not obtained medical care.[62] Because of the ability of Tat PTD to cross cell membranes and the blood–brain barrier, even when coupled with larger peptides, protein transduction appears to be promising for the development of new stroke therapies to target multiple levels of the neuronal response to ischemia in order to block cell death and preserve neurological function after stroke.

First, NMDARs mediate ischemic brain damage by interacting with the intracellular postsynaptic density protein PSD-95 and lead to the production of reactive oxygen species (ROS), such as nitric oxide and cell death, but also mediate essential neuronal excitation. Aarts et al.[63] found that an NMDAR-derived peptide could be linked to Tat and delivered to cultured neurons. This Tat-NMR2 peptide then sequestered the PSD-95, thereby blocking NMDAR-mediated apoptosis. Intravenous administration of Tat-NMR2 within 1 h after cerebral artery occlusion also reduced cerebral infarction volume and led to better neurological scores in rats. Importantly, this peptide had no effect on NMDAR-mediated currents as in vivo NMDAR blockade is too deleterious to neurons to be used as a stroke therapy.

Another target for stoke treatment is ROS, implicated in reperfusion injury after transient focal cerebral ischemia. The antioxidant enzyme, Cu,Zn-superoxide dismutase (SOD), is one of the major means by which cells counteract the deleterious effects of ROS after ischemia. Therefore, Tat-SOD fusion protein was tested for neuronal protection against cell death and ischemic insults. Immunohisto-chemical analysis revealed that Tat-SOD injected intraperitoneally into mice has access to various tissues including brain neurons. When intraperitoneally injected into gerbils, Tat-SOD prevented neuronal cell death in the hippocampus in response to transient forebrain ischemia.[64]

Previous data had shown that JNK activity increased after NMDAR activation and was at least partly responsible for inducing neuronal cell death through phosphorylation of its multiple effectors. Recently, inhibition of the c-Jun N-terminal kinase (JNK) has been shown to block neuronal cell death after ischemia in vivo.[62,65] The authors therefore linked Tat to a 20 amino acid peptide derived from JNK-interacting protein in order to competitively block interactions between JNK and its substrates. The authors also generated a stable, retro-inverso version of this peptide, which they termed D-JNKI-1. They found that D-JNKI-1 inhibited JNK activity in vitro and in vivo. They also showed that administration of the peptide was effective in reducing infarct size in rodent models of both transient and permanent cerebral artery occlusion. For example, intraventricular injection of D-JNKI-1 up to 6 h after transient middle cerebral artery occlusion led to a ~90% reduction in infarct size at time points up to 14 days postocclusion. Peptide-treated ischemic mice also displayed no decline in locomotor performance at up to 14 days postocclusion when compared to ischemic mice given no treatment. This is an important experimental endpoint because new medications for stroke victims will ultimately be judged on their ability to improve neurological outcomes rather than to reduce absolute infarct size.

PKCδ is another protein kinase implicated in mediating cerebral repurfusion injury in vivo. Bright et al.[66] demonstrated that Tat-δV-1, a selective PKCδ inhibitor peptide, can be detected in pyramidal cells in the cortex through the brain–blood barrier 30 min after being injected into internal carotid artery. More importantly, delivery of Tat-δV-1 during reperfusion reduced apoptotic cell death and enhanced prosurvival signals, in part by blocking intrinsic cell death pathways that contribute to secondary damages.

Kilic et al.[67] evaluated the protective property of Tat-glial line-derived neurotrophic factor (GDNF) fusion protein in focal cerebral ischemia using mouse ischima–reperfusion model. Tat-GDNF was intravenously applied over 10 min immediately after reperfusion. Tat-GDNF significantly reduced cell death and increased the number of viable neurons in the striatum, where disseminated tissue injury was observed. This approach may be of clinical interest because such fusion proteins can be intravenously applied and reach the ischemic brain regions when delivered both before and after an ischemic insult.

More examples of the therapeutic potential of Tat-fusion proteins have been provided by studies using a recombinant Tat-Bcl-xL protein. A member of the Bcl-2 family of proteins, Bcl-xL can act at the mitochondria to suppress apoptosis in multiple cell types. In one recent study,[68] Tat-Bcl-xL (but not Bcl-xL) was shown to inhibit apoptosis in cultured neurons at concentrations (30–100 nM) much lower than those used with traditional small molecule apoptosis inhibitors. The authors also demonstrated the in vivo efficacy of Tat-Bcl-xL by using a murine model of stroke in which cerebral artery occlusion leads to focal ischemia followed by neuronal apoptosis. The authors found that IP administration of 9 mg/kg Tat-Bcl-xL protein could significantly decrease the size of the cerebral infarction ($>60\%$) and reduce neuronal caspase-3 activity. The protein was effective even if it was administered 45 min after the ischemic episode. In contrast, nontransducible Bcl-xL protein had no effect on neuronal apoptosis in vivo. These results were corroborated by another report in which intravenous Tat-Bcl-xL administration reduced neuronal apoptosis and infarct volume in a murine stroke model.[67] In a separate paper, this group also showed that Tat-Bcl-xL was able to suppress axotomy-induced apoptosis in retinal ganglion cells after local delivery into the vitreous space of the eye.[69] Furthermore, the Tat-Bcl-xL protein could still be detected in cultured cerebellar granule cells 10 days after protein addition, indicating that not all proteins are so short-lived as to be poor candidates for therapeutics.

Other reports have provided more evidence for the therapeutic utility of Tat-Bcl-xL. In one example, a gain of function Bcl-xL mutant (termed FNK, which refers to the amino acid substitutions Y22F, Q26N, and R165K) was linked to the Tat PTD.[70] The Tat-FNK protein was cytoprotective in cultured neurons and was detected in the brain after IP administration to mice. More importantly, IP administration of Tat-FNK to gerbils (5 mg/kg) protected hippocampal CA1 neurons from cell death after transient global ischemia.

All these approaches may therefore offer new perspectives for future strategies in stroke therapy. In the future, combination of above strategies with existing therapy may greatly improve the ischemia response with minimum side effects.

11.3.5 Diabetes

Excess reactive-oxygen species are linked to the several pathological processes of human diseases, including carcinogenesis, ischemia, radiation injury, and inflammation/immune injury. Among the key cellular enzymes by which cells detoxify free radicals and protect themselves from oxidative damage, Cu,Zn-SOD draws extensive research interest for its therapeutic potential. It has been shown that exogenous Cu,Zn-SOD fused with Tat protein can be directly transduced into the cells, and the delivered enzymatically active Tat-SOD exhibits a cellular protective function against oxidative stress.[71] Naturally, the Tat-SOD is tested for cell death prevention in disease models, such as insulin-dependent diabetes mellitus, in which ROS are considered an important mediator in insulin-producing pancreatic β cell destruction and triggering the disease development. Eum et al.[72] tested the idea by injecting Tat-SOD in streptozotocin-induced diabetic mouse model. A single IP injection of Tat-SOD resulted in the delivery of this biologically active enzyme to the pancreas. Moreover, increased radical scavenging activity in the pancreas was induced by multiple injections of Tat-SOD, and this enhanced the tolerance of pancreatic β cells to oxidative stress. These results suggest that the transduction of Tat-SOD offers a new strategy for protecting pancreatic β cells from destruction by relieving oxidative stress in ROS-implicated diabetes.

11.4 FUTURE DIRECTION

Numerous studies have revealed the ability of PTDs to modulate the biology of living organisms. These studies have often taken place in the context of animal models of cancer, stroke, and other diseases that are responsible for a substantial amount of morbidity and mortality. Importantly, many of these reports have demonstrated not only the modulation of intracellular biology in vivo, but also an improvement in relevant clinical endpoints (e.g., survival for cancer, neurological function for stroke).

The pharmacokinetics and potential immunogenicity of PTD-linked molecules need to be examined in more detail. Such studies may enable the design of "next-generation" PTDs that have even more favorable biodistribution or immunogenic properties than current PTDs that were designed primarily for tissue culture experiments. However, the intrinsic heterogeneity of polypeptides may make it difficult to generalize the results of pharmacokinetic and immunogenicity studies from one Tat-fusion protein to another. In addition, the results reviewed here already show that at least some Tat-fusions have adequate biodistribution to ameliorate models of disease and that the immune system is not an absolute barrier to in vivo efficacy in these cases.

The quantity and quality of reports of successful in vivo transduction during the past 6 years has been exciting and encouraging. There is still preclinical work to be done before the design of phase I clinical trials using some of the most efficacious peptides and full-length proteins linked to PTDs thus far. Nevertheless, in the field of medical oncology, PTDs have tremendous potential for translating what has been learned about the molecular basis of cancer into viable clinical treatments. Cancer is a disease of multiple genetic alterations, and so it is unlikely that any one Tat-fusion protein will yield a cure of any particular class of cancer. However, many types of cancer are so refractory to treatment that new agents which provide even modest gains in survival are welcomed. Even more importantly, we envision a near future in which molecular analysis will reveal the specific genes and pathways that are mutated for each patient's tumor. In such a case, a combination of Tat-fusion proteins could be used to restore tumor-suppressor pathways that are missing and block the oncogenic pathways that are upregulated in that particular cancer.

In addition to the initiation of clinical trials of PTDs, there remains much preclinical work to be carried out in the field of protein transduction. For example, it is clear that delivery of Tat-fusion proteins to the cellular interior can be enhanced,[28] but the enhancing agents described so far likely represent only the tip of the iceberg. More enhancers of transduction must be sought, both by rational, mechanism-driven approaches and by screens of large chemical libraries. After discovery and characterization in cultured cells, these molecules must then be tested for their ability to enhance protein transduction in living animals with a minimum of toxicity.

Finally, the relative lack of specificity of protein transduction must be addressed.[11] The ability to theoretically target any cell is advantageous in some respects. However, in some cases it may also hinder the delivery of adequate levels of PTDs to the tissue of interest without delivering so much to other tissues that side effects ensue. The ability to target CPPs to specific cell types or selective activation in specific cells[50,73] has the potential to widen their therapeutic index. Targeted CPPs could theoretically both reduce the side effects of delivery to undesired tissue and lower the total amount of administered polypeptide needed to achieve a therapeutic effect.

Therapeutic use of PTDs is unlikely to be a panacea for any human disease. However, the preclinical studies published in the last 6 years have shown that they have tremendous potential to become the basis of an entirely new class of therapeutic agents, the intracellular biologicals.

REFERENCES

1. Nathwani, A.C. et al., Current status and prospects for gene therapy, *Vox Sang.*, 87(2), 73, 2004.
2. Joliot, A. and Prochiantz, A., Transduction peptides: From technology to physiology, *Nat. Cell Biol.*, 6(3), 189, 2004.
3. Trehin, R. and Merkle, H.P., Chances and pitfalls of cell penetrating peptides for cellular drug delivery, *Eur. J. Pharm. Biopharm.*, 58, 209, 2004.
4. Frankel, A.D. and Pabo, C.O., Cellular uptake of the tat protein from human immunodeficiency virus, *Cell*, 55, 1189, 1988.
5. Green, M. and Loewenstein, P.M., Autonomous functional domains of chemically synthesized human immunodeficiency virus tat trans-activator protein, *Cell*, 55, 1179, 1988.
6. Brooks, H., Lebleu, B., and Vives, E., Tat peptide-mediated cellular delivery: Back to basics, *Adv. Drug Deliv. Rev.*, 57, 559, 2005.
7. Derossi, D. et al., Cell internalization of the third helix of the Antennapedia homeodomain is receptor-independent, *J. Biol. Chem.*, 271, 18188, 1996.
8. Wender, P.A. et al., The design, synthesis, and evaluation of molecules that enable or enhance cellular uptake: Peptoid molecular transporters, *Proc. Natl Acad. Sci. U.S.A.*, 97, 13003, 2000.
9. Bullok, K.E. et al., Characterization of novel histidine-tagged Tat–peptide complexes dual-labeled with (99m)Tc-tricarbonyl and fluorescein for scintigraphy and fluorescence microscopy, *Bioconjug. Chem.*, 13, 1226, 2002.
10. Schwarze, S.R. et al., In vivo protein transduction: Delivery of a biologically active protein into the mouse, *Science*, 285, 1569, 1999.
11. Vives, E., Present and future of cell-penetrating peptide mediated delivery systems: Is the Trojan horse too wild to go only to Troy?, *J. Controlled Release*, 2005.
12. Nagahara, H. et al., Transduction of full-length TAT fusion proteins into mammalian cells: TAT-p27Kip1 induces cell migration, *Nat. Med.*, 4, 1449, 1998.
13. Wadia, J.S. and Dowdy, S.F., Modulation of cellular function by TAT mediated transduction of full length proteins, *Curr. Protein Pept. Sci.*, 4, 97, 2003.
14. Dietz, G.P. and Bahr, M., Delivery of bioactive molecules into the cell: The Trojan horse approach, *Mol. Cell Neurosci.*, 27, 85, 2004.
15. Mann, D.A. and Frankel, A.D., Endocytosis and targeting of exogenous HIV-1 Tat protein, *EMBO J.*, 10, 1733, 1991.
16. Fawell, S. et al., Tat-mediated delivery of heterologous proteins into cells, *Proc. Natl Acad. Sci. U.S.A.*, 91, 664, 1994.
17. Vives, E., Brodin, P., and Lebleu, B., A truncated HIV-1 Tat protein basic domain rapidly translocates through the plasma membrane and accumulates in the cell nucleus, *J. Biol. Chem.*, 272, 16010, 1997.
18. Futaki, S. et al., Arginine-rich peptides. An abundant source of membrane-permeable peptides having potential as carriers for intracellular protein delivery, *J. Biol. Chem.*, 276, 5836, 2001.

19. Derossi, D. et al., The third helix of the Antennapedia homeodomain translocates through biological membranes, *J. Biol. Chem.*, 269, 10444, 1994.
20. Suzuki, T. et al., Possible existence of common internalization mechanisms among arginine-rich peptides, *J. Biol. Chem.*, 277, 2437, 2002.
21. Hallbrink, M. et al., Cargo delivery kinetics of cell-penetrating peptides, *Biochim. Biophys. Acta*, 1515, 101, 2001.
22. Richard, J.P. et al., Cell-penetrating peptides. A reevaluation of the mechanism of cellular uptake, *J. Biol. Chem.*, 278, 585, 2003.
23. Lundberg, M. and Johansson, M., Positively charged DNA-binding proteins cause apparent cell membrane translocation, *Biochem. Biophys. Res. Commun.*, 291, 367, 2002.
24. Lundberg, M., Wikstrom, S., and Johansson, M., Cell surface adherence and endocytosis of protein transduction domains, *Mol. Ther.*, 8, 143, 2003.
25. Conner, S.D. and Schmid, S.L., Regulated portals of entry into the cell, *Nature*, 422(6927), 37–44, 2003.
26. Ferrari, A. et al., Caveolae-mediated internalization of extracellular HIV-1 tat fusion proteins visualized in real time, *Mol. Ther.*, 8, 284, 2003.
27. Kaplan, I.M., Wadia, J.S., and Dowdy, S.F., Cationic TAT peptide transduction domain enters cells by macropinocytosis, *J. Controlled Release*, 102, 247, 2005.
28. Wadia, J.S., Stan, R.V., and Dowdy, S.F., Transducible TAT-HA fusogenic peptide enhances escape of TAT-fusion proteins after lipid raft macropinocytosis, *Nat. Med.*, 10, 310, 2004.
29. Sampath, P. and Pollard, T.D., Effects of cytochalasin, phalloidin, and pH on the elongation of actin filaments, *Biochemistry*, 30, 1973, 1991.
30. Soane, L. and Fiskum, G., TAT-mediated endocytotic delivery of the loop deletion Bcl-2 protein protects neurons against cell death, *J. Neurochem.*, 95, 230, 2005.
31. Jones, S.W. et al., Characterisation of cell-penetrating peptide-mediated peptide delivery, *Br. J. Pharmacol.*, 145, 1093, 2005.
32. Tkachenko, A.G. et al., Cellular trajectories of peptide-modified gold particle complexes: Comparison of nuclear localization signals and peptide transduction domains, *Bioconjug. Chem.*, 15, 482, 2004.
33. Violini, S. et al., Evidence for a plasma membrane-mediated permeability barrier to Tat basic domain in well-differentiated epithelial cells: Lack of correlation with heparan sulfate, *Biochemistry*, 41, 12652, 2002.
34. Vendeville, A. et al., HIV-1 Tat enters T cells using coated pits before translocating from acidified endosomes and eliciting biological responses, *Mol. Biol. Cell*, 15, 2347, 2004.
35. Richard, J.P. et al., Cellular uptake of unconjugated TAT peptide involves clathrin-dependent endocytosis and heparan sulfate receptors, *J. Biol. Chem.*, 280, 15300, 2005.
36. Ziegler, A. and Seelig, J., Interaction of the protein transduction domain of HIV-1 TAT with heparan sulfate: Binding mechanism and thermodynamic parameters, *Biophys. J.*, 86, 254, 2004.
37. Meier, O. et al., Adenovirus triggers macropinocytosis and endosomal leakage together with its clathrin-mediated uptake, *J. Cell Biol.*, 158, 1119, 2002.
38. Norbury, C.C. et al., Class I MHC presentation of exogenous soluble antigen via macropinocytosis in bone marrow macrophages, *Immunity*, 3, 783, 1995.
39. Matsushita, M. et al., Photo-acceleration of protein release from endosome in the protein transduction system, *FEBS Lett.*, 572, 221, 2004.

40. Vocero-Akbani, A.M. et al., Killing HIV-infected cells by transduction with an HIV protease-activated caspase-3 protein, *Nat. Med.*, 5, 29, 1999.

41. Shokolenko, I.N. et al., TAT-mediated protein transduction and targeted delivery of fusion proteins into mitochondria of breast cancer cells, *DNA Repair (Amst)*, 4, 511, 2005.

42. Vousden, K.H., Activation of the p53 tumor suppressor protein, *Biochim. Biophys. Acta*, 1602, 47, 2002.

43. Snyder, E.L. et al., Treatment of terminal peritoneal carcinomatosis by a transducible p53-activating peptide, *PLoS Biol.*, 2, E36, 2004.

44. Selivanova, G. et al., Restoration of the growth suppression function of mutant p53 by a synthetic peptide derived from the p53 C-terminal domain, *Nat. Med.*, 3, 632, 1997.

45. Selivanova, G. et al., Reactivation of mutant p53: A new strategy for cancer therapy, *Semin. Cancer Biol.*, 8, 369, 1998.

46. Kim, A.L. et al., Conformational and molecular basis for induction of apoptosis by a p53 C-terminal peptide in human cancer cells, *J. Biol. Chem.*, 274, 34924, 1999.

47. Fulda, S. et al., Smac agonists sensitize for Apo2L/TRAIL- or anticancer drug-induced apoptosis and induce regression of malignant glioma in vivo, *Nat. Med.*, 8, 808, 2002.

48. Vucic, D. et al., SMAC negatively regulates the anti-apoptotic activity of melanoma inhibitor of apoptosis (ML-IAP), *J. Biol. Chem.*, 277, 12275, 2002.

49. Yang, L. et al., Predominant suppression of apoptosome by inhibitor of apoptosis protein in nonsmall cell lung cancer H460 cells: Therapeutic effect of a novel polyarginine-conjugated Smac peptide, *Cancer Res.*, 63, 831, 2003.

50. Harada, H., Hiraoka, M., and Kizaka-Kondoh, S., Antitumor effect of TAT-oxygen-dependent degradation-caspase-3 fusion protein specifically stabilized and activated in hypoxic tumor cells, *Cancer Res.*, 62, 2013, 2002.

51. Inoue, M. et al., Targeting hypoxic cancer cells with a protein prodrug is effective in experimental malignant ascites, *Int. J. Oncol.*, 25, 713, 2004.

52. Willam, C. et al., Peptide blockade of HIFalpha degradation modulates cellular metabolism and angiogenesis, *Proc. Natl Acad. Sci. U.S.A.*, 99, 10423, 2002.

53. Snyder, E.L., Meade, B.R., and Dowdy, S.F., Anti-cancer protein transduction strategies: Reconstitution of p27 tumor suppressor function, *J. Controlled Release*, 91, 45, 2003.

54. Datta, K. et al., The 104-123 amino acid sequence of the beta-domain of von Hippel–Lindau gene product is sufficient to inhibit renal tumor growth and invasion, *Cancer Res.*, 61, 1768, 2001.

55. Hosotani, R. et al., Trojan p16 peptide suppresses pancreatic cancer growth and prolongs survival in mice, *Clin. Cancer Res.*, 8, 1271, 2002.

56. Mendoza, N. et al., Selective cyclin-dependent kinase 2/cyclin A antagonists that differ from ATP site inhibitors block tumor growth, *Cancer Res.*, 63, 1020, 2003.

57. Harbour, J.W. et al., Transducible peptide therapy for uveal melanoma and retinoblastoma, *Arch. Ophthalmol.*, 120, 1341, 2002.

58. Mai, J.C. et al., A proapoptotic peptide for the treatment of solid tumors, *Cancer Res.*, 61, 7709, 2001.

59. Myou, S. et al., Blockade of airway inflammation and hyperresponsiveness by HIV-TAT-dominant negative Ras, *J. Immunol.*, 171, 4379, 2003.

60. Krautwald, S. et al., Transduction of the TAT-FLIP fusion protein results in transient resistance to Fas-induced apoptosis in vivo, *J. Biol. Chem.*, 279, 44005, 2004.

61. Sugioka, R. et al., BH4-domain peptide from Bcl-xL exerts anti-apoptotic activity in vivo, *Oncogene*, 22, 8432, 2003.

62. Borsello, T. et al., A peptide inhibitor of c-Jun N-terminal kinase protects against excitotoxicity and cerebral ischemia, *Nat. Med.*, 9, 1180, 2003.
63. Aarts, M. et al., Treatment of ischemic brain damage by perturbing NMDA receptor-PSD-95 protein interactions, *Science*, 298, 846, 2002.
64. Kim, D.W. et al., Transduced Tat-SOD fusion protein protects against ischemic brain injury, *Mol. Cells*, 19, 88, 2005.
65. Hirt, L. et al., D-JNKI1 a cell-penetrating c-Jun-N-terminal kinase inhibitor, protects against cell death in severe cerebral ischemia, *Stroke*, 35, 1738, 2004.
66. Bright, R. et al., Protein kinase C delta mediates cerebral reperfusion injury in vivo, *J. Neurosci.*, 24, 6880, 2004.
67. Kilic, U. et al., Intravenous TAT-GDNF is protective after focal cerebral ischemia in mice, *Stroke*, 34, 1304, 2003.
68. Cao, G. et al., In vivo delivery of a Bcl-xL fusion protein containing the TAT protein transduction domain protects against ischemic brain injury and neuronal apoptosis, *J. Neurosci.*, 22, 5423, 2002.
69. Dietz, G.P., Kilic, E., and Bahr, M., Inhibition of neuronal apoptosis in vitro and in vivo using TAT-mediated protein transduction, *Mol. Cell Neurosci.*, 21, 29, 2002.
70. Asoh, S. et al., Protection against ischemic brain injury by protein therapeutics, *Proc. Natl Acad. Sci. U.S.A.*, 99, 17107, 2002.
71. Kwon, H.Y. et al., Transduction of Cu,Zn-superoxide dismutase mediated by an HIV-1 Tat protein basic domain into mammalian cells, *FEBS Lett.*, 485, 163, 2000.
72. Eum, W.S. et al., HIV-1 Tat-mediated protein transduction of Cu,Zn-superoxide dismutase into pancreatic beta cells in vitro and in vivo, *Free Radic. Biol. Med.*, 37, 339, 2004.
73. Jiang, T. et al., Tumor imaging by means of proteolytic activation of cell-penetrating peptides, *Proc. Natl Acad. Sci. U.S.A.*, 101, 17867, 2004.

12 Proteoglycans as Endocytosis Receptors for CPPs

Mattias Belting, Anders Wittrup, and Staffan Sandgren

CONTENTS

12.1 INTRODUCTION

Cells continuously export, import, and recycle molecules over the plasma membrane by fundamental processes that can be capitalized upon for efficient intracellular delivery of bioactive macromolecules, such as proteins and nucleic acids. Proteoglycans emerge as key players in membrane transport, e.g., in the uptake of polyamines, peptides, peptide–nucleic acid polyplexes, and microbes. This chapter discusses proteoglycans in the context of endocytosis and CPP internalization.

12.2 WHAT IS A PROTEOGLYCAN?

Proteoglycans (PGs) are a superfamily of molecules with a multitude of functions spanning from cartilage shock absorption to regulation of morphogenesis. Their importance in physiology is reflected by the severe and often complex phenotypes of various PG transgenic animals (Table 12.1). By definition, PGs are proteins that

TABLE 12.1
Phenotypes of PG-Related Knockout Animals

Species	Genotype	Phenotype	Refs.
Caenorhabditis elegans	HS 2-OST −/−	Defective cell migration	1
C. elegans	Syndecan-1 −/−	Defective egg laying	2
C. elegans	Xyl-T −/−; Gal-TII−/−	Defective vulval morphogenesis	3
C. elegans	CS synthase −/−	Defective early embryogenesis and vulval development; defective embryonic cytokinesis and cell division	4,5
Drosophila	HS 6-OST	Defective tracheal branching	6
Drosophila	Dally/Dally like −/−	Impaired morphogen distribution	7,8
Mouse	HS 2-OST	Renal agenesis, defective eye and skeleton development; defective cerebral cortex development	9,10
Mouse	Syndecan-3 −/−	Resistance to diet-induced obesity; muscular dystrophy; enhanced level of long-term potentiation	11–13
Mouse	Syndecan-1 −/−	Resistance to mammary tumor development; defective wound healing	14–16
Mouse	HS polymerase −/−, conditional	Defect midline axon guidance	17
Mouse	Perlecan HS −/−	Defective lens capsule; defective acetylcholine esterase localization; impaired angiogenesis, wound healing and tumor growth	18–20
Mouse	Perlecan −/−	Defective cardiac morphogenesis; defective cartilage development	21,22
Mouse	Glypican-3 −/−	SGBS; Impaired hematopoietic differentiation	23,24
Mouse	NDST-1 −/−	Neonatal mortality due to respiratory distress syndrome; Defective cerebral and craniofacial development	9,25
Mouse	NDST-2 −/−	Abnormal mast cells	9
Mouse	NDST-1 −/−, conditional	Impaired neutrophil infiltration in inflammation	26
Mouse	HS polymerase (EXT1) −/−	Defective gastrulation and early embryonic development	9,27
Mouse	HS epimerase −/−	Defective renal, lung, and skeletal development	28

(continued)

Table 12.1   **(Continued)**

Species	Genotype	Phenotype	Refs.
Mouse	HS 3-OST-1 $-/-$	Normal hemostasis, intrauterine growth-retardation	29
Zebra fish	HS 6-OST $-/-$	Defective muscle development; defective angiogenesis	30,31
Zebra fish	HS polymerase $-/-$	Defective axon sorting in optic tract	32
Zebra fish	UDP-Glc dehydrogenase $-/-$	Defective cardiac valve formation	33

SGBS, Simpson–Golabi–Behmel syndrome; 2-, 3- and 6-OST, respective O-sulfotransferase; Xyl-T, xylosyl transferase; Gal-TII, galactosyltransferase II; NDST, N-deacetylase-sulfotransferase; EXT, hereditary multiple exostoses; UDP, uridine-di-phosphate; Glc, glucose.

have been substituted with one to several glycosaminoglycan (GAG) polysaccharide chains. GAGs are subgrouped according to the type of saccharides that compose the basic polymeric structure.[1-3] Here, we will focus on the GAG heparan sulfate (HS), which is a polymer of repeating N-acetyl glucosamine (GlcNAc)-D-glucuronic acid (GlcA) disaccharide units. To assemble a fully modified HS chain, at least a dozen genes (not taking into account the various enzyme isoforms) need to be expressed. General modifications occurring in the synthesis of an HS chain are: N-deacetylation/N-sulfation of GlcNAc residues; epimerization of GlcA to L-iduronic acid (IdoA); sulfation at the 2-O position of IdoA; and sulfation at the 6-O and 3-O positions of GlcNAc. An overview of the HS assembly is presented in Figure 12.1.

Due to its polyanionic nature provided by the sulfate and carboxyl groups, HS contains natural binding sites for a wide variety of polybasic structures. From a physico-chemical point of view, HS may thus be regarded as an "extracellular equivalent to DNA". Accordingly, the monoclonal anti-HS antibody HEPSS-1 (Seikagaku, Japan), which displays no cross-reactivity with other GAGs, e.g., chondroitin sulfate and keratan sulfate, recognizes single-stranded DNA, and anti-DNA antibodies that occur in, for example, systemic lupus erythematosus patients, and display cross-reactivity with HS epitopes.[4] Along the same line, heparin (a highly sulfated variant of HS) affinity chromatography is widely used for the purification of DNA-binding proteins, e.g., transcription factors, steroid hormone receptors, and histones. In the case of proteins, consensus sequences for efficient HS binding, such as xBBBxxBx, xBBxBx, BBxBB, and TxxBxxTBxxxTBB (where B is a basic amino acid, x is a hydropathic amino acid, and T is a turn) have been proposed.[5] Nonpeptide ligands, such as the polyamine spermine, have also been reported to bind avidly to highly sulfated HS.[6] A general problem when studying the function of HS is thus the great number of potential binding partners; importantly, a growing number of proteins and peptides have been shown to depend on HS binding for proper biological activity in vitro and in vivo.[1-3]

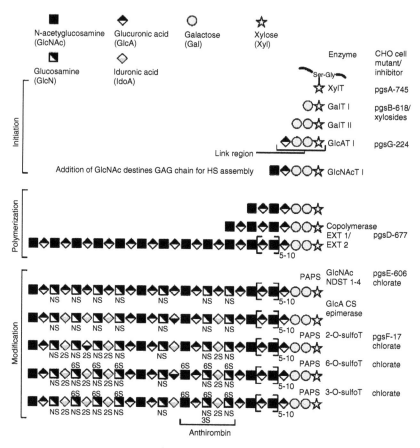

FIGURE 12.1 Biosynthesis of heparan sulfate. HS generally displays a patched appearance, with highly modified regions interspersed between regions of lower modification grade. The sequence responsible for antithrombin III binding is depicted. Esko and co-workers[2] have developed a CHO cell system with mutants deficient in various steps of GAG synthesis. Mutants deficient in enzymes involving link region formation are deficient in all GAGs, i.e., pgsA-745, pgsB-618, pgsG-224. The EXT1 mutant (pgsD-677) is deficient in the copolymerase responsible for HS chain polymerization and is thus HS deficient. Two mutants that produce under sulfated HS are available, i.e., the NDST-1 mutant (pgsE-606) and the 2-*O*-sulfotransferase mutant (pgsF-17). Monosaccharide symbols in accordance with Nomenclature Committee, Consortium for Functional Glycomics. T, transferase; sulfoT, sulfo-transferase; NS, *N*-sulfate; (2, 3 or 6)S, (2, 3 or 6)-O-sulfates; PAPS, 3'-phosphoadenosine 5'-phosphosulfate (sulfate donor).

12.2.1 PROTEOGLYCANS FORM A POLYANIONIC ENVELOPE AT THE PLASMA MEMBRANE

PGs are present in the extracellular matrix, in intracellular vesicles, and at the cellsurface. Here, we will focus on membrane-associated HSPGs. The two major families of cell surface HSPGs are the syndecans and the glypicans (Figure 12.2).[1–3]

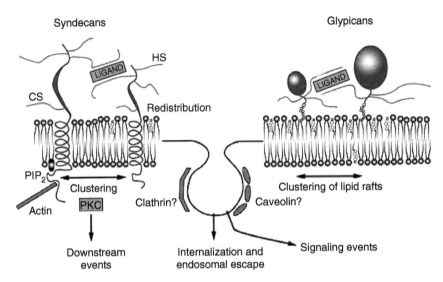

FIGURE 12.2 General characteristics of syndecan and glypican proteoglycans using syndecan-4 and glypican-1 as models. Clustering of syndecan molecules induces intracellular signaling events; redistribution into the lipid raft partition of the plasma membrane; and internalization via endocytosis. Ligand binding to HS on glypican molecules induces clustering of lipid rafts, which may trigger endocytosis. Once internalized, further destiny may include signaling events or endosomal escape. PIP_2, phophoinositol (4,5) bisphosphate; PKC, protein kinase C.

The syndecan family comprises four members in mammals with varying tissue distribution; syndecan-1 is expressed predominantly in epithelial and plasma cells; syndecan-2 in epithelial, fibroblast, and neuronal cells; syndecan-3 almost exclusively in neuronal and musculoskeletal tissue; and syndecan-4 in virtually every cell type studied. Syndecans are structurally similar with a cytoplasmic, extracellular, and transmembrane domain (depicted in Figure 12.2). The transmembrane region is highly conserved, whereas the extracellular and the cytoplasmic domains are varied, yielding core proteins with molecular weights ranging from 22 to 43 kDa. The core proteins all have N-terminal GAG attachment domains that are generally substituted with HS. Other GAG attachment sites, generally substituted with chondroitin sulfate, can be located near the trans-membrane region. The short cytoplasmic domain contains three important regions; conserved regions one and two (C1 and C2) and a variable region (V). The V domain displays a high degree of heterogeneity and has been studied particularly well in syndecan-4. In syndecan-4, the V domain contains a phosphatidylinositol-4,5-bisphosphate (PIP_2) binding site. PIP_2 binding to the V domain allows syndecan-4 to interact with, and activate, protein kinase Cα, initiating intracellular events involved in the regulation of tyrosine kinase receptor signaling.[7] Syndecans are important regulators of cell–matrix interactions and cell–cell interactions, especially so during development. Further expanding the role of syndecans in signaling and cell–cell interaction, it has been demonstrated that protein kinase Cγ mediates phosphorylation of the cytoplasmic domain of syndecan-2 in right side, but not left side, ectodermal cells, as an instructive function in *Xenopus* left–right development.[8]

The glypican family has six members in mammals. Mature glypican core proteins have a molecular weight of 60–70 kDa. The bulk of the protein is thought to form a tight globular domain due to the presence of 14 conserved cystein residues, presumably engaged in disulfide bond formation.[9] The globular domain has been suggested to regulate HS assembly onto glypican core proteins. Glypicans have three GAG attachment sites in direct succession located between the globular domain and the C-terminal glycosylphosphatidylinositol (GPI) anchor attachment sequence. The GPI anchor attaches glypicans to membranes and directs glypicans to regions of the plasma membrane with specific lipid characteristics, i.e., lipid rafts and caveolae (see further below), which is thought to be a general behavior of GPI-anchored proteins.[10] Several glypican members are primarily found in the CNS and neurons. The expression pattern of glypicans is dramatically different in adult mammals as compared to embryos, e.g., in adults, only glypican-1 is ubiquitously distributed. Glypican-1 has been ascribed a role in growth factor signaling; glypican-2 is believed to be important for axonal guidance; and glypican-3 has been suggested to be important for insulin growth factor signaling. The functions of glypicans 4–6 are essentially unknown. Glypican homologues in *Drosophila*, division abnormally delayed (*Dally*) and *Dally like*, are involved in morphogenesis and are associated with gradient formation of *wingless*, *hedgehog*, and *decapentaplegic*.[8,10] Altogether, both glypicans and syndecans are involved in the formation of morphogen gradients and developmental patterning, having partly overlapping and partly distinct effects on developmental processes.[11] In fact, the redundancy between the different cell-surface HSPGs described above has not been thoroughly investigated, and it remains to be established to what degree the core protein carrying the HS chain determines the biological function of the HSPG moiety.

12.3 ENDOCYTOSIS MECHANISMS

Traditionally, endocytosis is divided into phagocytosis (cell eating) and pinocytosis (cell drinking). Pinocytosis occurs by a number of ill-defined pathways, including clathrin-mediated (classical) endocytosis, caveolin-mediated endocytosis, clathrin and caveolin independent endocytosis, and macropinocytosis.[12] Clathrin-mediated endocytosis was first discovered due to the distinct appearance of the clathrin coat in electron micrographs. The bulk constituents of the polygonal coat are two oligomeric proteins: the clathrin triskelion and the adaptor protein complexes. The adaptor protein complexes have multiple roles and come in four heterotetrameric forms, AP1–4. APs trigger clathrin assembly and link the clathrin coat to membranes by interacting with the cytoplasmic portion of transmembrane proteins and with PIP$_2$. AP1 and AP2 localize clathrin coat formation to the *trans* Golgi network and the plasma membrane, respectively, making AP2 responsible for clathrin-dependent endocytosis. APs are also thought to concentrate cargo into forming clathrin-coated pits by interacting with and recruiting transmembrane receptors.

Another key player in clathrin-mediated endocytosis, as well as in other types of vesicle formation, is the guanosine triphosphatase dynamin. Its role in endocytosis was identified in the *Drosophila* mutant *shibire*, expressing a temperature-sensitive dynamin homologue. The fly exhibited a rapid and reversible paralysis upon exposure to nonpermissive temperature due to blocked endocytosis affecting synaptic signaling. Dynamin is capable of generating force and polymerizes into rings when purified. It is thought to polymerize around the neck of invaginating buds and to pinch off coated vesicles, with the exact mechanism still being inexplicable. There are three isoforms of dynamin in mammals (1–3), with several splice variants supposedly involved in membrane fission at distinct cellular locations. Dynamin-1 is expressed exclusively in neurons, dynamin-2 is expressed in all tissues, and dynamin-3 is restricted to the testis, the brain, the lung, and the heart.[13] Several mutant dominant negative dynamin constructs have been used by researchers to investigate the role of dynamin in different aspects of endocytosis.

Studies on the action of actin inhibitors in clathrin-mediated endocytosis show variable, and sometimes contradictory, results. Two recent studies indicate an essential role for actin dynamics in clathrin-mediated endocytosis. Zhu et al.[14] report that cortactin (an F-actin binding protein) and dynamin drive the fission of clathrin-coated pits in an actin polymerization dependent manner, whereas Yarar et al.[15] state that actin dynamics is essential for the processes of coated pit formation as well as maturation into coated vesicles and translocation into the cytosol.

No drugs are available that target the clathrin-mediated pathway in a specific manner. Overexpression of fragments or mutants of proteins associated with the formation of clathrin-coated vesicles, e.g., mutants of Eps15, the μ2 subunit of AP2, and dynamin, inhibit endocytosis dependent on clathrin. A recent approach is RNA interference directed against the same subset of proteins, e.g., AP2 or the clathrin heavy chain. In the case of RNA interference, effects have been very modest. This has been explained by the long half-life of most proteins involved in clathrin-mediated endocytosis and by functional redundancy.[16]

Caveolae appear as 50–100 nm bulb-like invaginations of the plasma membrane.[17] The caveolin family comprises three members in mammals (1–3) that differ in their tissue distribution. Caveolin-1 and -2, having two and three isoforms, respectively, are coexpressed in most cell types, and are abundant in endothelial cells and fibroblasts. The expression of caveolin-3 is restricted to muscle cells, astrocytes, and chondrocytes. Caveolin-1 is known to homo-oligomerize or hetero-oligomerize with caveolin-2 to form complexes of 14–16 units. These complexes are thought to be the assembly units for caveolar coat formation. Introduction of the caveolin-1 gene in lymphocytes devoid of caveolin induced formation of caveolae. In the absence of caveolin-1, caveolae formation is abolished and caveolin-2 seems to be completely retained at the level of the Golgi complex. Upon reintroduction of caveolin-1, caveolin-2 transport is restored, resulting in the formation of hetero-oligomers and caveolae, indicating that caveolin-2 is functionally dependent on caveolin-1. Surprisingly, caveolin knockout mice are viable and show only mild phenotypes.[18] Aberrations in caveolin-1 null mice are related to lung function and vasculature while caveolin-3 null mice display mild-to-moderately dystrophic muscle tissue. Investigations on the endocytosis and transcytosis of albumin, which are thought to be caveolae-dependent events, have produced contradictory results. Embryonic fibroblasts from caveolin-1-null mice failed to show surface labeling or internalization of fluorophor-labeled albumin, while transfection of caveolin-1 into the null cells restored uptake. Likewise, injected albumin conjugated with 5 nm gold is readily internalized by pulmonary endothelium in wildtype mice, whereas no uptake can be seen in caveolin-1 null mice, as shown by electron microscopy. However, both albumin concentration in cerebrospinal fluid, which is presumed to depend on caveolar transcytosis, and albumin-dependent extravascular osmotic pressure are normal in caveolin-1 null mice.

It has recently been suggested that caveolae are a subpopulation of lipid rafts. Rafts are detergent-insoluble, low-density membrane fractions rich in cholesterol and sphingolipids, whereas caveolae are cholesterol- and sphingolipid-rich invaginations of the plasma membrane that partition into raft fractions and whose expression is associated with caveolin-1.[19] The fact that caveolae share many characteristics with membrane lipid rafts suggests that caveolae and rafts share a common endocytic pathway, which is defined by clathrin independence, dynamin dependence, sensitivity to cholesterol depletion, and similar membrane lipid composition. This definition, although debated, unites caveolar endocytosis with noncaveolar raft endocytosis and is referred to as caveolae-/raft-dependent endocytosis. Still, the exact roles of caveolin in caveolar/raft formation and transport remain unknown. Caveolae have been shown to anchor to the actin cytoskeleton. The interaction is mediated by the actin binding protein filamin, which associates with the N-terminus of caveolin-1. As a consequence, disruption of actin filaments results in increased lateral movement of caveolae. SV40 internalization and transport were inhibited by actin disruption. However, a recent study indicates that actin depolymerization can induce caveolae-mediated endocytosis, making the story more complex.[20] Thus, the exact role of actin in caveolae-mediated endocytosis remains to be elucidated.

A role for macropinocytosis has been suggested in processes, such as satisfying nutritional requirements, sampling of soluble antigens in dendritic cells, and membrane turnover in motile cells.[21] Macropinocytosis is, in many ways, mechanistically similar to phagocytosis. The process starts with formation of ruffles, i.e., actin-driven plasma membrane protrusions. The ruffles organize to form large folds and cups that are then either relaxed and flattened again, or progressively more invaginated to form a flask-like appearance similar to those seen in micropinocytosis. Actin dynamics have a profound role during all phases of macropinocytosis. Macropinocytosis differs from other types of endocytosis in that it is, supposedly, receptor independent. Phagocytosis as well as clathrin- and caveolae-mediated endocytosis all involve receptors and transmembrane proteins that offer selectivity in cargo, whereas macropinocytosis is thought to mediate unspecific uptake of surrounding solutes.[21] The membrane of the forming macropinosome most certainly contains a multitude of receptors, but these are thought not to be specifically concentrated. Since the volume to surface area ratio is high, macropinocytosis may be an efficient way to internalize nutrients.

12.4   ENDOCYTOSIS OF CPPs

In 1988, it was reported that HIV-1-trans activator of transcription (Tat) was secreted from infected cells and internalized by surrounding cells in a paracrine fashion. Once inside the nucleus of an HIV-1 infected cell, Tat functions as an antitermination factor allowing viral replication. A few years later, Derossi et al.[22] reported that the Drosophila Antennapedia homeodomain peptide was internalized by nerve cells and transported to the nucleus, inducing morphological differentiation. In both cases, the ability to enter cells was found to arise from a short stretch of amino acids. In the case of HIV-Tat, the sequence is GRKKRRQRRRPPQC (herein referred to as Tat), and in the case of Antennapedia RQI-KIWFQNRRMKWKK (named penetratin). The common name protein transduction domain (PTD) was adopted to indicate that the active fragment is part of a native protein.[23] However, one of the first reports on peptide-mediated delivery of proteins into cells is that of Ryser and Shen[24] from 1978. They provided evidence that poly lysine (PLL) conjugated to albumin or horseradish peroxidase greatly enhances protein uptake into cultured fibroblasts.

As the family of peptides with similar properties expanded to include artificially designed and chimerical peptides, the wider term cell-penetrating peptide (CPP) was introduced.[25] CPP conjugates have also been used to deliver biologically active proteins in vivo.[23] The list of CPPs is rapidly expanding, and a general theme amongst the most commonly used CPPs is a positive net charge. However, not all peptides with the capability to enter cells exhibit positive net charges. Here, emphasis will be on basic CPPs (Table 12.2). The term CPP and the concept of CPPs, as they were originally proposed, contain implications on the mechanism of internalization and may thus be inappropriate, as is discussed in more detail in other chapters. The CPP field has been forced to reevaluate several of its findings since it has become apparent that fixation may induce artifacts[26] and that flow cytometry fails to discriminate between cell attachment and true internalization, unless stringent trypsination and rinsing protocols are employed.[27] It is, however, beyond any doubt that cargo internalization by CPPs is not an artifact, because delivery of CPP–protein and CPP–oligonucleotide conjugates elicits biological effects that require intracellular localization, and because delivery of CPP–DNA plasmid polyplexes results in plasmid expression. An elaborate method to prove that CPP–protein conjugates do indeed reach the nuclei of target cells is the Cre-loxP system, which has been used to provide evidence that Tat-Cre and penetratin-Cre do in fact reach cell nuclei when administered to cells in culture.[28] In the case of Tat-Cre uptake, it has been suggested that uptake occurs via macropinocytosis since: (1) uptake occurs via endocytosis as evidenced by colocalization with an unspecific marker for endocytosis, FM4-64; (2) cholesterol depletion using β-methylcyclodextrin (β-MCD) and nystatin inhibited uptake and activity, speaking in favor of lipid raft/caveolae-mediated uptake or possibly macropinocytosis; (3) fluorescent Tat-Cre did not colocalize with caveolin-1-red fluorescent protein, excluding caveolae

TABLE 12.2
Examples of Basic CPPs

Name	Sequence	Source	Net Charge
Tat	GRKKRRQRRRPPQCa	aa 48–60 of the HIV-1 Tat protein	+8
Penetratin	RQIKIWFQNRRMKWKK	aa 43–58 of the third helix of the Antennapedia protein	+7
LL-37	LLGDFFRKSKEKIGKEFK-RIVQRIKDFLRNLVPR-TESCa	Human antimicrobial peptide	+6
Prion peptide	MVKSKIGSWILVLF-VAMWSDVGLCKKRPKP	aa 1–28 of the bovine prion protein	+5
LMW protamine	VSRRRRRRGGRRRR	Degraded salmon protamine	+10
Protamine	PRRRRSSSRPVRRRRRPRV-SRRRRRRGGRRRR	Intact peptide, expressed in spermatides	+21
MAP	KLALKLALKALKAALKLA	Model amphipathic peptide	+5
Homopolymers of basic aa	$R_n{}^a$ and K_n	Synthetic	+n
Transportan	GWTLNSAGYLLGKINKA-LAALAKISILa	Chimera of active part of galanin (aa 1–12) and mastoparan	+3

[a] In the case of poly-arg, number of residues seems important for efficient internalization. Optimal size is reportedly ~8 arg residues. However, this result was obtained after fixation of cells. LMW, low molecular weight; a, amide; aa, amino acids.

mediated uptake; (4) expression of a dominant negative dynamin construct did not hamper Tat-Cre activity whilst uptake of fluorescent transferrin was inhibited, speaking against clathrin-mediated uptake.[29] Other investigators have recently reported that clathrin-mediated pathways are crucial for the uptake of fluorophor-conjugated Tat.[30] In this study, nystatin had no effect, whereas chlorpromazine and potassium-depletion markedly reduced Tat internalization. The notion that clathrin-mediated endocytosis is the internalization route for Tat conjugated to low molecular weight cargo was supported by an elaborate report by Vendeville and coworkers.[31] Not only do these authors present electron micrographs of Tat-5 nm gold conjugates localized to coated pits; using dominant negative constructs of Eps15 and intersectin, the importance of these clathrin associated proteins is demonstrated. Caveolae/lipid raft mediated uptake has likewise been reported as the mode for CPP internalization.[32,33] In these reports, Tat-GFP administered to cells displayed colocalization with caveolin-1 positive vesicular structures and CTxB, whereas no colocalization with transferrin was observed. Moreover, cholesterol depletion by β-MCD severely reduced uptake in these studies.

To conclude, the route of entry for Tat protein and Tat peptide seems to elude concise definition, which may reflect a pan-endocytic route of entry.

12.4.1 Role of Cell Surface Proteoglycans in Endocytosis

PGs emerge as multifunctional carriers for the internalization of extracellular ligands,[3] as exemplified in Table 12.3. Several viruses and intracellular bacteria are thus dependent on PGs for efficient internalization and infectivity. In the case of the FGF family of growth factors, evidence has accumulated that internalization is a requirement for FGF activity, a process that has been shown to depend on HSPG.[34] Recently, it was established that syndecan-4-FGF complexes are cointernalized following FGF-mediated clustering of syndecan-4 molecules.[35] These results were obtained using an Fc-receptor-syndecan-4 chimera previously shown to form heterodimers with coexpressed native syndecan-4, a system that according to the authors reproduces the behavior of native syndecan-4.[36] According to this model, FGF binding induces oligomerization of syndecan-4 and endocytosis of the complexes (Figure 12.2). Macropinocytosis was suggested as the entry pathway since internalization appeared lipid raft-dependent, clathrin- and dynamin-independent, and amiloride-sensitive.[35] Interestingly, unclustered syndecan-4 was shown to reside predominantly in the nonraft membrane partition, whereas clustering induced redistribution of syndecan-4 into the noncaveolae raft partition.[36]

Polyamines represent a family of small, polybasic, growth-promoting substances that are essential for cell proliferation and differentiation in both prokaryotes and in eukaryotic cells. Polyamine homeostasis is maintained through tightly regulated pathways for biosynthesis and degradation, as well as transporters for the utilization of extracellular polyamines. Whereas several polyamine transport proteins have been characterized in bacteria and yeast cells, no mammalian polyamine transporter

---

TABLE 12.3
Examples of Ligands and Pathogens that Are Internalized via HSPG

Ligand	Function In	HS-Binding Motif
Apolipoprotein E[1]	Lipid metabolism	SHLRKLRKRLLRDADD
Basic fibroblast growth factor[2]	Cell-growth	GHFKDPKRLYCKNGGF
TFPI–Xa complex[3,4]	Regulation of coagulation	GGLIKTKRKRKKQRVKIAY
Spermine[5]	Cell growth and differentiation	

Pathogen	Associated disease	
Neisseria gonorrhoeae[6]	Acute gonorrhoea	
Enterococcus faecalis[7]	Nosocomial bacteremia and endocarditis	
Herpes simplex virus[8]	Cold sores; genital herpes. encephalitis.	
Adeno associated virus-2[9]	Possible role in miscarriage and trophoblastic disease	
HIV-1[10,11]	Acquired immunodeficiency syndrome	

TFPI, tissue factor pathway inhibitor; Xa, activated coagulation factor X.

has been identified. Our group has demonstrated a link between glypican turnover and polyamine transport in mammalian cells. Under conditions where cells are dependent on extracellular polyamines, i.e., inhibition of endogenous polyamine synthesis, HSPGs in general and glypicans in particular exhibit increased affinity for spermine,[37,38] suggesting that the HS synthesis machinery can be modulated according to the need to internalize specific extracellular ligands. As mentioned above, several GPI-anchored proteins have been reported to recycle between the plasma membrane and intracellular compartments. The existence of a recycling pathway has also been suggested for glypican-1,[9] and may be a means by which HSPGs function to shuttle cargo from the exterior of the cell to the correct intracellular compartment for processing or signal induction.[3,39] During glypican recycling, the HS chains are thought to be degraded by both enzymatic and nonenzymatic processes. The nonenzymatic process involves a deaminative cleavage of HS catalyzed by nitric oxide (NO). NO is produced by NO synthases and may be sequestered by the glypican core protein as SNO groups, which may occur at any of the 14 conserved cystein residues of the glypican globular domain. Upon arrival at the Golgi compartment, HS chains are resynthesized onto the remaining glypican core, which then arrives at the plasma membrane to initiate another cycle. Interestingly, glypicans contain the nuclear localization sequence KRRR(G/A)K, and endogenous glypican-1 has been located to the nuclei of neural cells.[40] Ligand binding to HS chains of glypican may induce assembly of glypicans at the plasma membrane, followed by cointernalization and subsequent nuclear targeting of the ligand–HSPG complex. Indeed, several studies have demonstrated the presence of endogenous HS in the nucleus.[3]

12.4.2 ROLE OF PROTEOGLYCANS IN THE ENDOCYTOSIS OF CPPS

Early on, negatively charged species of phospholipids and gangliosides were pointed out as the most probable binding partners in CPP interaction with the plasma membrane.[22] However, accumulating data recognized cell surface HSPG as another group of negatively charged plasma membrane-associated molecules involved in CPP internalization. Since the initial findings that efficient transfection using PLL–DNA polyplexes and cationic lipid–DNA lipoplexes requires cell surface HSPG, and that the polyamine spermine is dependent on cell surface HSPG for efficient internalization, the role of cell surface HSPG as a plasma membrane carrier has been firmly cemented.[3] The fact is that HSPG mediates internalization of spermine stimulated investigations on the internalization of spermine-based cationic lipids, e.g., lipofectamine. Regarding CPP internalization, it has convincingly been reported that cell surface HSPGs are essential for the uptake of fluorophor-conjugated Tat, Tat–protein constructs, and Tat–DNA polyplexes. Indeed, HSPG has proven essential for entry of a penetratin–protein conjugate, as well as poly arginine peptides.[3,39]

It may be speculated that binding of polybasic compounds, such as CPPs and their conjugates, induces cell surface HSPG clustering by forming bridges between HS chains attached onto different PG core protein monomers. As mentioned above, in the case of syndecan-4 this would result in intracellular kinase activity and downstream events leading to internalization, whereas in the case of glypican clustering would induce lipid raft mediated endocytosis (see Figure 12.2). In the case of polyplexes with a positive net charge, clustering of cell surface HSPG appears even more likely due to the sheer size of polyplex components. This model would explain the common internalization mechanism for arginine-rich or polybasic peptides. It also provides a plausible explanation for the observed induction of endocytosis and uptake of nonconjugated and noncomplexed proteins in the presence of polybasic proteins and peptides.[41,42]

The fact that CPPs and CPP containing formulations enter cells via cell surface HSPG may have important implications when it comes to target cell specificity. HS synthesis is cell-type specific rather than core protein specific, i.e., HS on glypicans are structurally similar to HS on syndecans. However, the topographical localization of HS chains on these two PG species differs. Glypican HS chains are attached to the core protein in close vicinity to the plasma membrane, and syndecan HS chains are attached at more peripheral sites. HS binding ligands may therefore be exposed to syndecan HS prior to glypican HS. Unless this or other unknown facts are of vital importance, administered CPPs should be distributed to the HS chains irrespective of core protein. If so, what does that imply in terms of internalization route? GPI-anchored glypicans are supposed to reside in membrane lipid rafts and are thought to be internalized via lipid raft- and caveolae-mediated endocytosis. Syndecans normally reside in the nonlipid raft membrane fraction but (at least in the case of syndecan-4) seem to redistribute to the lipid raft fraction upon ligand-induced clustering. Both species thus potentially utilize the raft/caveolae pathway. Macropinocytosis has also been suggested as a way of entrance for bFGF–syndecan-4 complexes.[35] Moreover, a recent study suggests that internalization of Tat involves both PGs and clathrin-dependent endocytosis.[30] It may thus be speculated that CPP internalization via cell surface HSPG will be distributed across the entire range of endocytic routes depending on: (1) the subset of PG core proteins present; and (2) the preponderance of these core proteins for specific endocytic pathways. Accordingly, clathrin-mediated endocytosis, caveolar/lipid raft-mediated endocytosis, and macropinocytosis have all been suggested as the PG-dependent route of entry for Tat. Moreover, Ignatovich et al.[43] demonstrated that the mode of internalization of Tat–DNA polyplexes varies depending on the cell line used.

12.5 FUTURE PERSPECTIVES

The implications for CPP-mediated drug and gene delivery are still to be fully investigated and will likely result in the production of new diagnostics and treatment strategies to the benefit of the patient. Although the field has shifted so as to generally accept the notion that polybasic CPPs enter cells by HSPG-dependent endocytosis, the exact mechanism and the specific players involved in, for example, endosomal escape remain to be determined. Another key question that needs to be addressed is to what extent cell/tissue specific targeting via HSPG-dependent entry pathways can be achieved. As an example, do tumor cells express unique HS epitopes that allow specific CPP-cargo delivery to tumor tissue in vivo?

Approaches that may finally answer these questions are currently being developed in our laboratory.

Research on the CPP mode of internalization has brought into focus the physiological aspects of macromolecular delivery. Especially intriguing is the possible existence of intercellular communication via regulated transfer of transcription factors, morphogens, and growth factors,[28] and the possible role for HSPG in these processes.

ACKNOWLEDGMENTS

We would like to thank our colleagues at the Section of Oncology, Lund University, and The Swedish Research Council, The Swedish Cancer Fund, The AICR, and FLÄK for generous support. We apologize to all authors whose original work, due to space limitations, could not be cited in this review.

REFERENCES

1. Kjellén, L. and Lindahl, U., Proteoglycans: Structures and interactions, *Annu. Rev. Biochem.*, 60, 443, 1991.
2. Esko, J.D. and Selleck, S.B., Order out of chaos: Assembly of ligand binding sites in heparan sulfate, *Annu. Rev. Biochem.*, 71, 435, 2002.
3. Belting, M., Heparan sulfate proteoglycan as a plasma membrane carrier, *Trends Biochem. Sci.*, 28, 145, 2003.
4. Faaber, P. et al., Cross-reactivity of human and murine anti-DNA antibodies with heparan sulfate. The major glycosaminoglycan in glomerular basement membranes, *J. Clin. Invest.*, 77, 1824, 1986.
5. Cardin, A.D. and Weintraub, H.J., Molecular modeling of protein-glycosaminoglycan interactions, *Arteriosclerosis*, 9, 21, 1989.
6. Belting, M. et al., Heparan sulfate/heparin glycosaminoglycans with strong affinity for the growth-promoter spermine have high anti-proliferative activity, *Glycobiology*, 6, 121, 1996.
7. Tkachenko, E., Rhodes, J.M., and Simons, M., Syndecans: New kids on the signaling block, *Circ. Res.*, 96, 488, 2005.
8. Kramer, K.L., Barnette, J.E., and Yost, H.J., PKCgamma regulates syndecan-2 inside-out signaling during xenopus left-right development, *Cell*, 111, 981, 2002.
9. Fransson, L.Å. et al., Novel aspects of glypican glycobiology, *Cell Mol. Life Sci.*, 61, 1016, 2004.
10. Ikezawa, H., Glycosylphosphatidylinositol (GPI)-anchored proteins, *Biol. Pharm. Bull.*, 25, 409, 2002.
11. Rawson, J.M. et al., The heparan sulfate proteoglycans Dally-like and Syndecan have distinct functions in axon guidance and visual-system assembly in Drosophila, *Curr. Biol.*, 15, 833, 2005.
12. Conner, S.D. and Schmid, S.L., Regulated portals of entry into the cell, *Nature*, 422, 37, 2003.
13. Henley, J.R., Krueger, E.W., Oswald, B.J., and McNiven, M.A., Dynamin-mediated internalization of caveolae, *J. Cell Biol.*, 141, 85, 1998.
14. Zhu, J. et al., Regulation of cortactin/dynamin interaction by actin polymerization during the fission of clathrin-coated pits, *J. Cell Sci.*, 118, 807, 2005.

15. Yarar, D., Waterman-Storer, C.M., and Schmid, S.L., A dynamic actin cytoskeleton functions at multiple stages of clathrin-mediated endocytosis, *Mol. Biol. Cell*, 16, 964, 2005.
16. Sorkin, A., Cargo recognition during clathrin-mediated endocytosis: A team effort, *Curr. Opin. Cell Biol.*, 16, 392, 2004.
17. Rothberg, K.G. et al., Caveolin, a protein component of caveolae membrane coats, *Cell*, 68, 673, 1992.
18. Williams, T.M. and Lisanti, M.P., The caveolin genes: From cell biology to medicine, *Ann. Med.*, 36, 584, 2004.
19. Nabi, I.R. and Le, P.U., Caveolae/raft-dependent endocytosis, *J. Cell Biol.*, 161, 673, 2003.
20. Shen, L. and Turner, J.R., Actin depolymerization disrupts tight junctions via caveolae-mediated endocytosis, *Mol. Biol. Cell*, 16, 3919, 2005.
21. Swanson, J.A. and Watts, C., Macropinocytosis, *Trends Cell Biol.*, 5, 424, 1995.
22. Derossi, D., Chassaing, G., and Prochiantz, A., Trojan peptides: The penetratin system for intracellular delivery, *Trends Cell Biol.*, 8, 84, 1998.
23. Schwarze, S.R., Ho, A., Vocero-Akbani, A., and Dowdy, S.F., In vivo protein transduction: Delivery of a biologically active protein into the mouse, *Science*, 285, 1569, 1999.
24. Ryser, H.J. and Shen, W.C., Conjugation of methotrexate to poly(L-lysine) increases drug transport and overcomes drug resistance in cultured cells, *Proc. Natl Acad. Sci. U.S.A.*, 75, 3867, 1978.
25. Lindgren, M., Hällbrink, M., Prochiantz, A., and Langel, Ü., Cell-penetrating peptides, *Trends Pharmacol. Sci.*, 21, 99, 2000.
26. Lundberg, M. and Johansson, M., Is VP22 nuclear homing an artifact?, *Nat. Biotechnol.*, 19, 713, 2001.
27. Richard, J.P., Cell-penetrating peptides: A re-evaluation of the mechanism of cellular uptake, *J. Biol. Chem.*, 278, 585, 2003.
28. Joliot, A. and Prochiantz, A., Transduction peptides: From technology to physiology, *Nat. Cell Biol.*, 6, 189, 2004.
29. Wadia, J.S., Stan, R.V., and Dowdy, S.F., Transducible TAT-HA fusogenic peptide enhances escape of TAT-fusion proteins after lipid raft macropinocytosis, *Nat. Med.*, 10, 310, 2004.
30. Richard, J.P. et al., Cellular uptake of unconjugated TAT peptide involves clathrin-dependent endocytosis and heparan sulfate receptors, *J. Biol. Chem.*, 280, 15300, 2005.
31. Vendeville, A. et al., HIV-1 Tat enters T cells using coated pits before translocating from acidified endosomes and eliciting biological responses, *Mol. Biol. Cell*, 15, 2347, 2004.
32. Ferrari, A. et al., Caveolae-mediated internalization of extracellular HIV-1 tat fusion proteins visualized in real time, *Mol. Ther.*, 8, 284, 2003.
33. Fittipaldi, A. et al., Cell membrane lipid rafts mediate caveolar endocytosis of HIV-1 Tat fusion proteins, *J. Biol. Chem.*, 278, 34141, 2003.
34. Colin, S. et al., In vivo involvement of heparan sulfate proteoglycan in the bioavailability, internalization, and catabolism of exogenous basic fibroblast growth factor, *Mol. Pharmacol.*, 55, 74, 1999.
35. Tkachenko, E., Lutgens, E., Stan, R.V., and Simons, M., Fibroblast growth factor 2 endocytosis in endothelial cells proceed via syndecan-4-dependent activation of Rac1 and a Cdc42-dependent macropinocytic pathway, *J. Cell Sci.*, 117, 3189, 2004.

36. Tkachenko, E. and Simons, M., Clustering induces redistribution of syndecan-4 core protein into raft membrane domains, *J. Biol. Chem.*, 277, 19946, 2002.
37. Belting, M., Persson, S., and Fransson, L.Å., Proteoglycan involvement in polyamine uptake, *Biochem. J.*, 338, 317, 1999.
38. Belting, M. et al., Glypican-1 is a vehicle for polyamine uptake in mammalian cells: A pivital role for nitrosothiol-derived nitric oxide, *J. Biol. Chem.*, 278, 47181, 2003.
39. Belting, M., Sandgren, S., and Wittrup, A., Nuclear delivery of macromolecules: Barriers and carriers, *Adv. Drug Deliv. Rev.*, 57, 505, 2005.
40. Liang, Y. et al., Glypican and biglycan in the nuclei of neurons and glioma cells: Presence of functional nuclear localization signals and dynamic changes in glypican during the cell cycle, *J. Cell Biol.*, 139, 851, 1997.
41. Ryser, H.J. and Hancock, R., Histones and basic polyamino acids stimulate the uptake of albumin by tumor cells in culture, *Science*, 150, 501, 1965.
42. Kaplan, I.M., Wadia, J.S., and Dowdy, S.F., Cationic TAT peptide transduction domain enters cells by macropinocytosis, *J. Controlled Release*, 102, 247, 2005.
43. Ignatovich, I.A. et al., Complexes of plasmid DNA with basic domain 47–57 of the HIV-1 Tat protein are transferred to mammalian cells by endocytosis-mediated pathways, *J. Biol. Chem.*, 278, 42625, 2003.

Part III

Functionality of Shuttling Proteins and Protein-Derived Peptides

Alain Prochiantz

It was almost 20 years ago, in 1988, that Frankel and Pabo reported the capture of HIV-TAT by cells and its transport to the nucleus.[1] This finding, although not really pursued at the time, suggested that transcription factors could be transported between cells. The underlying concept of messenger proteins was then entirely developed on the basis of a line of research initiated in 1991 in my laboratory and that of Joliot.[2–4] Indeed, we have demonstrated that several homeoprotein transcription factors are exchanged between cells and started to identify some functions associated with this newly discovered mode of signal transduction.[5] This point will be alluded to later and fully developed in chapter 13 of this book. In the context of this introduction, it is important to identify the two peptidic domains involved in cell internalization and secretion.[6] The internalization domain corresponds to the 16 amino acid-long third helix of the homeodomain, called penetratin. This peptide is internalized through an endocytosis-independent mechanism and is capable of carrying attached cargoes of various sizes and compositions into the cell interior.[7–9]

Since then, the family of cell-penetrating peptides (CPPs), also known as transduction peptides, has increased at a rapid pace. The most emblematic ones are the objects of specific chapters. Their use as new tools for basic research and as innovative therapeutic agents is also described in this book. I have no doubt that many other compounds, peptidic, lipopeptidic, or nonpeptidic, will be described in the near future and will find original and important applications. Among these applications, the ability of getting across tight junction epitheliums, in particular at the level of the intestinal barrier and of the blood–brain barrier, is most exciting. However, before we can get to the point of rationally designing such vectors, we must fully understand the mechanisms of internalization of most of these peptides and of the strategies that they use to gain access to the cell cytoplasm or to distinct organelles. It is my bias that all these peptides differ and thus use different internalization strategies.

For example, there is a huge difference between a poly-Arg or Tat peptide entirely composed of charged amino acids and penetratin, which shows a specific distribution of hydrophilic and hydrophobic residues, allowing it to form an amphipathic α-helix or β-sheet. It is thus not surprising, quite the opposite, that peptides with different structures are internalized through different mechanisms. For example, penetratin is internalized in the absence of endocytosis, possibly through the induction of inverted micelles,[8,10] whereas TAT-derived peptides, rich in arginine residues, are first endocytosed into a vesicular compartment and secondarily released into the cytoplasm following vesicle disruption.[11] In addition, subtle chemical properties can intervene, as demonstrated by the fact that, within penetratin, changing a tryptophan into a phenylanin blocks internalization even though the two amino acids are hydrophobic.[8] To this diversity of sequences and mechanisms, it must be added that the same peptide may use two pathways (or more) in parallel, and that specific pathways could be selected depending on the attached cargoes, the membrane composition, or the physiological state of the target cell.[6]

In a way, it is surprising that, despite the fact that transduction was discovered more that 10 years ago and given the medical importance of the topic, so little is known for certain regarding entry mechanisms. An explanation could be found, among several factors, in opinion biases and in immediate commercial interests. The success of the "how to cross a biological membrane" program will now depend on the ability of involved groups to adopt a more rational, collaborative, and systematic approach. This will necessitate strong experimental interactions between the biologists — often satisfied by the pragmatic result "it works" — and the chemists and physicists able to dissect the structural constraints and the sequence of interactions between peptides and lipids that explain why some peptides can traverse a lipid bilayer, be it artificial or natural. I also strongly believe that we must put more effort on the physiology behind CPPs and that the functions associated with transcription factor exchange must be unraveled. It is only if all these conditions are made possible that the scientific community will eventually produce several compounds, not necessarily peptides, with high and safe transduction properties.

In spite of the hope raised by the ability to address hydrophilic compounds into live cells in vivo, including, possibly, across tight junction epitheliums, one must be fully aware of the difficulties ahead of us. Many of these new compounds can be toxic or mutagenic, and we know very little of their biodistribution. In addition, it is necessary to enter the correct cells — for example, the sick ones — and, once in the cells, to reach the appropriate compartment: nucleus, mitochondria, various regions of the cytoplasm, etc. When it comes to crossing tight junction epitheliums, we must ensure that the peptides and their cargoes are not entrapped in the endothelial cells, but truly travel across the cells. It is, for example, striking that, although penetratin has been reported to gain access to the brain in vivo,[12] our own unpublished results demonstrate that it cannot cross a tight junction epithelium unless a true export sequence has been added to the construct (Dupont et al., unpublished results). The take-home message would thus be that before we can get the Golden Fleece home, there will be plenty of problems to solve, plenty of work to do, and plenty of information to share; short cuts will not help.

Indeed, because specific chapters will develop the theme of homeoprotein transfer physiological functions, the recent findings on this original mode of signal transduction have not been discussed. However, it is important to realize that several diseases are genetically linked to mutations in homeogene or to homeogene-perturbed expression. This is why many homeoproteins, because they include within their sequence domains that allow internalization and secretion, can be considered as "natural cargoes" and used, at least theoretically, as therapeutic proteins. Amsellem et al.[13] have used this strategy to amplify CD34-positive stem cells, in vitro and in vivo. This is certainly a very important result as it demonstrates that the discovery of CPPs opens the way to new developments in protein therapy, complementary to, and as important as, gene therapy. Indeed, we all know that many pharmaceutical companies prefer small compounds, but the success of therapeutic antibodies and of genetically engineered and produced proteins, for example insulin, should incite us to explore this path. In this context, it is certainly useful not to let the practical consequences of the phenomenon of protein transduction, in term of applications, impede the study of its physiological meaning. Indeed, it is a clear lesson of scientific history that deciphering a new physiological pathway can only shed light on mechanisms at work in many pathologies.

REFERENCES

1. Frankel, A.D., Bredt, D.S., and Pabo, C.O., Tat protein from human immuno-deficiency virus forms a metal-linked dimer, *Science*, 240(4848), 70, 1988.
2. Joliot, A. et al., Antennapedia homeobox peptide regulates neural morphogenesis, *Proc. Natl Acad. Sci. U.S.A.*, 88, 1991, 1864.
3. Prochiantz, A., Messenger proteins: Homeoproteins, TAT and others, *Curr. Opin. Cell Biol.*, 12(4), 400, 2000.
4. Prochiantz, A. and Joliot, A., Can transcription factors function as cell–cell signalling molecules?, *Nat. Rev. Mol. Cell Biol.*, 4(10), 814, 2003.
5. Brunet, I. et al., The transcription factor Engrailed-2 guides retinal axons, *Nature*, 438, 98, 2005.
6. Joliot, A. and Prochiantz, A., Transduction peptides: From technology to physiology, *Nat. Cell Biol.*, 6(3), 189, 2004.
7. Derossi, D. et al., Cell internalization of the third helix of the Antennapedia homeodomain is receptor-independent, *J. Biol. Chem.*, 271(30), 18188, 1996.
8. Derossi, D., Chassaing, G., and Prochiantz, A., Trojan peptides: The penetratin system for intracellular delivery, *Trends Cell Biol.*, 8(2), 84–87, 1998.

9. Derossi, D. et al., The third helix of the Antennapedia homeodomain translocates through biological membranes, *J. Biol. Chem.*, 269(14), 10444, 1994.

10. Berlose, J.P. et al., Conformational and associative behaviours of the third helix of Antennapedia homeodomain in membrane-mimetic environments, *Eur. J. Biochem.*, 242, 372, 1996.

11. Snyder, E.L. and Dowdy, S.F., Cell penetrating peptides in drug delivery, *Pharm. Res.*, 21(3), 389, 2004.

12. Rousselle, C. et al., New advances in the transport of doxorubicin through the blood–brain barrier by a peptide vector mediated strategy, *Mol. Pharmacol.*, 57, 679, 2000.

13. Amsellem, S. et al., Ex vivo expansion of human hematopoietic stem cells by direct delivery of the HOXB4 homeoprotein, *Nat. Med.*, 9(11), 1423, 2003.

13 Cell Permeable Peptides and Messenger Proteins, from a Serendipitous Observation to a New Signaling Mechanism

Alain Prochiantz

CONTENTS

13.1 PLACE AND SHAPE: THE PROBLEM AT THE ORIGIN OF AN UNEXPECTED OBSERVATION

In the mid-1980s, we observed that brain neurons in culture adopted different polarity patterns depending on the origin of the astrocytes on which they were plated.[1,2] It was particularly striking that dendrites would only develop when neurons and astrocytes were derived from the same structure. This allowed us to established a theoretical link between developmental morphogenetic programs and positional information.

This was the first known indication of astrocyte heterogeneity, but it also pointed toward a specific class of genes that regulate shape as a function of place. At the time, these genes, primarily homeogenes, were not seen as cell-shape or cell-polarity (a subprogram of shape) regulators, but as organ-shape regulators. Nevertheless, it seemed to us that morphological programs acting at the cellular and multicellular levels may share similar mechanisms, for example making use of the intensity and position of forces internal to the cells or applied at their surfaces.

The idea did not meet with frank success since the dogma was that homeogenes were purely for cell lineage and organ shape. Nevertheless, we wanted to test our hypothesis and we decided to do so by injecting a homeodomain within live neurons. The logic was that the injected homeodomain would gain access to the nucleus and displace endogenous homeoproteins away from their cognate sites, thus revealing their morphological function at the single-cell level. We used the homeodomain of Antennapedia for practical reasons and on the basis of the strong sequence conservation between homeodomains.

The results were as expected with a strong change in cell shape,[3] but the surprise was total when, adding the homeodomain into the culture medium for a control, we observed the same phenotype. This suggested that either the effect of the injected homeodomain was due to its leakage outside of the cells—an artifact—or that the homeodomain was internalized. We verified the latter possibility and observed, much to our surprise, that the 60 amino acid-long polypeptide was captured by the cells and addressed to their nuclei.[3]

13.2   A DISTURBING OBSERVATION UNDER RATIONAL SCRUTINY

This observation was so unexpected and disturbing that we decided to identify the mechanism involved in the Antennapedia homeodomain (AntpHD) capture. Interestingly, the intracellular distribution, showing uniform cytoplasmic staining and nuclear accumulation, was at odds with endocytosis. Indeed, and I want to insist on this point, uptake was observed at 4°C, with the same uniform cytoplasmic staining. To preclude that this diffusion was due to pAntpHD redistribution following fixation, the same experiments carried out with a FITC-tagged homeodomain on live cells and with the help of confocal microscopy gave identical results.[3] Finally, it was verified that the pAntpHD retrieved from the cells was intact, demonstrating very limited degradation, even after several days.[4]

These results were published in the early 1990s and replicated recently. There is thus no reason to question the conclusion that pAntpHD is captured at 4°C and 37°C through an endocytosis-independent mechanism. Indeed, this does not preclude that pAntpHD can also take another route. More importantly, the results obtained with pAntpHD and peptides of the penetratin family (see below) are not universal. It is very possible, in fact extremely likely, that other cell-penetrating peptides (CPPs) use endocytosis for internalization. It is also possible that attaching a cargo will modify CPP mode of entry, and this also holds true for homeodomain-derived CPPs.

However, what seems secure today was very controversial 15 years ago and we really needed to have a mechanism of entry based on the different properties of peptide variants. This is why we developed a site-directed mutagenesis approach that allowed us to demonstrate that the third helix of the homeodomain, 16 amino acid-long, was necessary and sufficient for internalization.[5,6] This was quite an important observation because it opened the way to chemical approaches on the penetratin family of peptides, all derived from the third helix of pAntpHD. This led us to demonstrate the absence of a chiral receptor, the capture of variants not capable of adopting an α-helical structure in a hydrophobic environment, and the importance of a tryptophan residue in position 48 of the homeodomain (W48).[7,8] The fact that changing this W48 for a phenylalinine, although not modifying its amphiphilic properties, nearly abolishes penetratin internalization separates the penetratin family from other CPPs, Tat in particular, in which hydrophobic residues are absent.

13.3 HOMEOPROTEIN SHUTTLING

Following penetratin discovery, many CPPs were developed which were used with increasing success, both in vitro and in vivo, to address biological cargoes into the cell interior. Many of the molecules targeted by this strategy are either cytoplasmic or nuclear, demonstrating true access to these compartments. This means that, even in the case of CPPs internalized by endocytosis, enough of the product escapes the endocytic compartment to exert cytoplasmic and nuclear pharmacological activities.[9–11]

However, this overwhelming success had the counterproductive effect of distracting the attention of the researchers from investigating the transduction field from a physiological viewpoint. This is particularly clear in the case of TAT transactivation factor. Indeed, although TAT protein was the first identified transducing transcription factor, few studies have been devoted to this aspect of HIV physiology.[12]

Our laboratory, in contrast, has concentrated on the identification of physiological functions associated with full-length homeoprotein intercellular transfer. We have first demonstrated the internalization of full length homeoprotein, such as Hoxa5, Engrailed, Otx2, Pax6, and HoxB4 (unpublished results and reviewed in Ref. [13]). In fact, given the structural conservation of homeodomain third helices, it is extremely likely that most homeoproteins are internalized.

However, for intercellular signaling, homeoproteins also need to be secreted and, in the absence of a classical secretion signal, the mechanism had to be unconventional. We have shown, in collaboration with the group of Alain Joliot, that secretion requires a small sequence, upstream of the penetratin sequence and slightly overlapping with it, and thus within the homeodomain.[14] This secretion signal also acts as a nuclear export signal and takes the protein out of the nucleus and into a secretion compartment enriched in cholesterol and glycoshingolipids.[15] It is interesting to note that the two mechanisms for entry and secretion require different signals and are based on different mechanisms. Again, the conservation of the two peptidic domains suggests that the intercellular transfer is a property shared by many homeoproteins.

13.4 FUNCTIONAL STUDIES

Once the mechanisms and signals responsible for intercellular exchange were identified, it became obvious, based on the conservation of the import and export domains within most homeoproteins, that this biological phenomenon had to serve a function. An even stronger argument in favor of a physiological function was provided by the work of Joliot and colleagues on plants.[16] In plants, protein intercellular passage takes place due to intercellular bridges named plasmodesmata. It was thus extremely interesting to observe that the homeodomain of Knotted-1, a plant homeoprotein, translocates between animal cells, and that all transfer mutants and revertants have exactly the same behavior in plant and in animal models. This conservation of transfer mechanisms between the two phyla is probably the best indication that this phenomenon has important physiological functions, but what are they?

My laboratory is presently working on three functions.[13] The first function is the establishment of borders in the developing neuroepithelium, the second is axonal guidance, and the last one concerns the regulation of critical period in the visual cortex. Here, I will only describe the case of axonal guidance because it is the most advanced and also because it demonstrates a new function for homeoprotein transcription factors.

A popular paradigm in the axonal guidance field is provided by the retino-tectal map. On the tectum, a midbrain structure that receives visual and auditory information, axons in provenance from the retina navigate and form connections in an order that projects the temporal/nasal retina axis onto the anterior/posterior axis of the tectum. Accordingly, the posterior part of the tectum, where the homeoprotein engrailed is highly expressed, attracts and repels nasal and temporal axons, respectively. In a collaborative study with the group of Christine Holt in Cambridge, we showed that, placed in an engrailed gradient, temporal axons are repelled and nasal axons are attracted by the source of the protein. This morphogenetic effect of Engrailed requires both its internalization and its ability to locally regulate translation.[17]

This raises the issue of homeoproteins as translation regulators. Interestingly, regulation of translation is accompanied by Engrailed-induced phosphorylation of eIF4E, a key initiation factor for protein synthesis.[17] It is noteworthy that many homeoproteins, including Engrailed, Otx2, Bicoid, HoxA9, and Proline-rich homeoprotein, directly bind eIF4E through a small domain conserved among more than 200 members of the family.[18–23] It can thus be proposed that homeoproteins are transcription and translation regulators and that regulation at the two levels is both cell-autonomous and noncell-autonomous.

We cannot end this list of considerations without addressing the consequences of homeoprotein transduction in therapeutics. It is now well established that many homeoproteins are involved in adult physiological functions and that several diseases can be genetically linked to mutations in homeogenes. Homeoproteins, because they include within their sequence a transduction peptide, can thus be considered as natural cargoes and produced as therapeutic proteins. This approach has been recently developed by Fischelson and colleagues,[24] who have used HOXB4 to stimulate the proliferation of CD34-positive stem cells.

REFERENCES

1. Denis-Donini, S., Glowinski, J., and Prochiantz, A., Glial heterogeneity may define the three-dimensional shape of mouse mesencephalic dopaminergic neurones, *Nature*, 307(5952), 641, 1984.

2. Chamak, B. et al., MAP2 expression and neuritic outgrowth and branching are coregulated through region-specific neuro-astroglial interactions, *J. Neurosci.*, 7(10), 3163, 1987.

3. Joliot, A. et al., Antennapedia homeobox peptide regulates neural morphogenesis, *Proc. Natl Acad. Sci. U.S.A.*, 88, 1864, 1991.

4. Joliot, A.H. et al., alpha-2,8-Polysialic acid is the neuronal surface receptor of antennapedia homeobox peptide, *New Biol.*, 3(11), 1121, 1991.

5. Bloch-Gallego, E. et al., Antennapedia homeobox peptide enhances growth and branching of embryonic chicken motoneurons in vitro, *J. Cell. Biol.*, 120(2), 485, 1993.

6. Le Roux, I. et al., Neurotrophic activity of the Antennapedia homeodomain depends on its specific DNA-binding properties, *Proc. Natl Acad. Sci. U.S.A.*, 90(19), 9120, 1993.

7. Derossi, D. et al., Cell internalization of the third helix of the Antennapedia homeodomain is receptor-independent, *J. Biol. Chem.*, 271(30), 18188, 1996.

8. Derossi, D., Chassaing, G., and Prochiantz, A., Trojan peptides: The penetratin system for intracellular delivery, *Trends Cell Biol.*, 8(2), 84, 1998.

9. Lindgren, M. et al., Cell-penetrating peptides, *Trends Pharmacol. Sci.*, 21, 99, 2000.

10. Schwarze, S.R. and Dowdy, S.F., In vivo protein transduction: Intracellular delivery of biologically active protein, compounds and DNA, *Trends Pharmacol. Sci.*, 21, 45, 2000.

11. Joliot, A. and Prochiantz, A., Transduction peptides: From technology to physiology, *Nat. Cell Biol.*, 6(3), 189, 2004.

12. Frankel, A.D., Bredt, D.S., and Pabo, C.O., Tat protein from human immuno-deficiency virus forms a metal-linked dimer, *Science*, 240(4848), 70, 1988.

13. Prochiantz, A. and Joliot, A., Can transcription factors function as cell-cell signalling molecules?, *Nat. Rev. Mol. Cell Biol.*, 4(10), 814, 2003.

14. Maizel, A. et al., A short region of its homeodomain is necessary for Engrailed nuclear export and secretion, *Development*, 126, 3183, 1999.

15. Joliot, A. et al., Association of engrailed homeoproteins with vesicles presenting caveolae-like properties, *Development*, 124(10), 1865, 1997.

16. Tassetto, M. et al., Plant and animal homeodomains use convergent mechanisms for intercellular transfer, *EMBO Rep.*, 6(9), 885, 2005.

17. Brunet, I. et al., The transcription factor engrailed-2 guides retinal axons, *Nature*, 438, 94, 2005.

18. Dubnau, J. and Struhl, G., RNA recognition and translational regulation by a homeodomain protein, *Nature*, 379(6567), 694, 1996.

19. Nedelec, S. et al., Emx2 homeodomain transcription factor interacts with eukaryotic translation initiation factor 4E (eIF4E) in the axons of olfactory sensory neurons, *Proc. Natl Acad. Sci. U.S.A.*, 101(29), 10815, 2004.

20. Topisirovic, I. and Borden, K.L., Homeodomain proteins and eukaryotic translation initiation factor 4E (eIF4E): An unexpected relationship, *Histol. Histopathol.*, 20(4), 1275, 2005.

21. Topisirovic, I. et al., The proline-rich homeodomain protein, PRH, is a tissue-specific inhibitor of eIF4E-dependent cyclin D1 mRNA transport and growth, *EMBO J.*, 22(3), 689, 2003.

22. Topisirovic, I. et al., Aberrant eukaryotic translation initiation factor 4E-dependent mRNA transport impedes hematopoietic differentiation and contributes to leukemo-genesis, *Mol. Cell Biol.*, 23(24), 8992, 2003.

23. Topisirovic, I. et al., Eukaryotic translation initiation factor 4E activity is modulated by HOXA9 at multiple levels, *Mol. Cell Biol.*, 25(3), 1100, 2005.
24. Amsellem, S. et al., Ex vivo expansion of human hematopoietic stem cells by direct delivery of the HOXB4 homeoprotein, *Nat. Med.*, 9(11), 1423, 2003.

14 Pharmacology and In Vivo Efficacy of Pepducins in Hemostasis and Arterial Thrombosis

Lidija Covic, Boris Tchernychev, Suzanne Jacques, and Athan Kuliopulos

CONTENTS

14.1 INTRODUCTION

Although remarkably diverse in sequence and function, all G protein-coupled receptors (GPCRs) share a highly conserved topological arrangement of a seven-transmembrane helical core domain joined by three intracellular loops, three extra-cellular loops, and amino- and carboxyl-terminal domains.[1] Overwhelming evidence suggests that most, if not all, GPCRs have a strong propensity to dimerize and/or form large oligomeric structures[2] that can lead to cooperative binding of ligands to the dimeric unit. A key event for the switch from inactive to active receptor is ligand-induced conformational changes of transmembrane helices 6 (TM6), TM2, and TM7.[3] In turn, these helical movements alter the conformation of the intracellular loops of the receptor to promote activation of associated heterotrimeric G proteins. Mutagenesis studies[4–6] demonstrated that the third intracellular loop (i3) mediates a large part of the coupling between receptor and G protein. I3 loops expressed as minigenes have also been shown to directly compete with receptors for G_q binding,[7] and can activate G proteins as soluble peptides in cell-free conditions.[8] GPCRs couple to single or multiple G proteins through G_α and/or G_α-associated ($\beta\gamma$) subunits.

A unique mechanism of activation of a subclass of GPCRs, known as protease-activated receptors (PARs),[9–13] is by proteolytic cleavage of their N-terminal domains which exposes a tethered ligand. The PAR family consists of four receptors, PAR1–4, which can be cleaved by thrombin (PAR1, 3, and 4) or trypsin (PAR2).[9,12–14] A larger repertoire of proteases have been identified that can activate PARs, such as the activated factor X, cathepsin G, plasmin, and, most recently, matrix metalloprotease-1.[12,13,15–18] PARs can also be selectively activated by short peptides that mimic their tethered ligands.[19]

PAR1 is a promiscuous GPCR and couples to G_i to activate PI3K, MAPK, JNK, and Akt involved in proliferation and cell survival; $G_{12/13}$ to stimulate Rho-dependent cell transformation, invasion, and metastasis, and G_q to stimulate phospholipase C-β and PKC and tyrosine kinase pathways. PAR1, together with PAR4, play a prominent role in human platelet activation.[20–24] Mice that lack PAR4 manifest severe bleeding.[25] PAR1 and PAR4 have also been also implicated in inflammation.[26] Thus, therapeutics targeted against PAR1 and PAR4 could be beneficial in the treatment of patients with a high risk of thrombotic events, and may have wider application in other types of disease. Here, we present a novel technology and identify cell-penetrating lipopeptide inhibitors based on the intracellular loops of GPCRs, as exemplified by PAR1 and PAR4.

14.2 PEPDUCIN TECHNOLOGY

14.2.1 Design

Using sequence and structural information of any GPCR target ($n = 700$–$1,000$), the peptidic portion of the pepducin is first designed and then appended to a panel of lipids. The array of lipid–peptide variants is screened to identify the best combination. PD/PK properties of a pepducin may be modulated by changing the lipid tag and keeping the peptide constant. To date, 11 different lipids (e.g., C8, C10, C12, C14, C16, C18 acyl chains, steroids) have been tested and in vitro data can predict how a certain lipid will affect the solubility, cell-penetrating ability, efficacy, and half-life of a given pepducin in animals.

Palmitoylated peptides were synthesized by standard fmoc solid-phase synthetic methods with C-terminal amides. Palmitic acid was dissolved in 50% *N*-methyl pyrolidone/50% methylene chloride and coupled overnight to the deprotected N-terminal amine of the peptide. After cleavage from the resin, palmitoylated peptides were purified to >95% purity by C_{18} or C_4 reverse-phase chromatography.[18,26–28]

14.2.2 Pepducin™ Platform

Our approach has been to systematically develop pepducin compounds that target the receptor–G protein interface, a completely unexplored area in drug discovery (Figure 14.1A,B). We have invented a lipopeptide production platform that can rapidly generate effective and potent drug candidates that essentially target any GPCR at its intracellular surface. Since the pepducin is derived from intracellular GPCR loops, a proposed mechanism of pepducin modulation of GPCR signaling is by mimicking and competing with contacts made by GPCR and G protein. Thus, pepducins may also be useful for the examination of mechanism(s) of G protein activation by receptors. Lipidated peptides have previously been used as cell-penetrating antiprotease alkylating agents[30] and as inhibitors of platelet integrins.[31] More than 100 different pepducin compounds have been successfully tested in vitro against a wide variety of receptors, including PAR1, PAR2, PAR4, CCKA, CCKB, SSTR2, MC4, GLP-1, P2Y12 ADP, CXCR1, CXCR2, CXCR4, CCR1, CCR2, CCR5, and in vivo for three indications (inflammation, cancer, and vascular disease).[18,27–29,32] After delivering its cargo across the plasma membrane, the lipid moiety anchors the compound to the lipid bilayer, thereby concentrating the pepducin to the target receptor–G protein interface and increasing the effective molarity. The lipid tag also protects the pepducin from degradation and clearance leading to high plasma levels and long half-lives.

14.2.3 Lipid Moiety and Cell-Penetrating Properties

We previously reported that the short seven amino acid (aa) hydrophobic membrane-tether in P1-i3-26 conferred no activity, whereas attachment of > 14 aa hydrophobic residues in P1-i3-33 gave full agonist activity.[27] Hence, membrane tethering alone

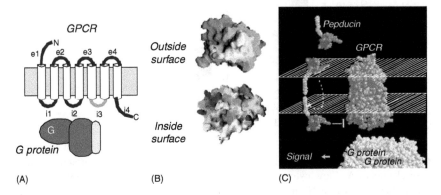

FIGURE 14.1 Pepducin drug platform. (A) The structural models of seven-transmembrane GPCR receptor and i3-loop peptides are based on the x-ray structure of (B) rhodopsin.[1] (C) Proposed mechanism of inhibition of GPCR by its cognate pepducin at the intracellular surface of the plasma membrane. Cell-penetrating pepducins derived from the sequences of the intracellular third loops (i3) of PAR1 (P1pal-12, red) are shown with a covalently attached palmitate lipid (green) and the heterotrimeric G protein is shown in white.

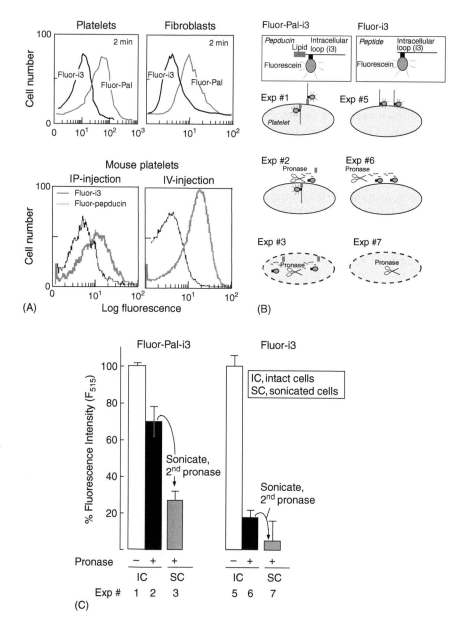

FIGURE 14.2 IP- and IV-delivered palmitoylated i3-loop peptide penetrates circulating mouse platelets. (A) Flow cytometry was conducted on platelets or Rat1 fibroblasts[28] that were incubated with 10 mM Fluor-Pal-i3 or Fluor-i3 for 2 min in PBS/0.1% fetal calf serum and then treated with 2 U pronase for 15 min at 37°C and washed prior to FACS. Mice were injected with 17 ng/g of palmitoylated (Fluor-Pal-i3) or nonpalmitoylated (Fluor-i3) fluorescein-labeled i3-loop peptides. Whole blood was collected after 15 min and platelets were isolated by centrifugation and treated with 2 U of pronase for 15 min prior to flow cytometry as described.[28] (B) Palmitoylated i3-loop peptides penetrate intact cells. Human platelets were incubated with either 1 mM Fluor-Pal-i3 or Fluor-i3 for 2 min at 25°C in PBS.

is not likely to be sufficient for activity. Likewise, previous studies have demonstrated that attachment of lipids[30] or >11–12 hydrophobic amino acid tags confers cell-penetrating abilities to a wide range of peptides and proteins.[33] Dowdy's group[34] used this strategy to efficiently deliver over 50 proteins into a variety of human and mouse cells.

To determine whether lipidation confers cell-penetrating abilities, palmitoylated and biotinylated i3 peptides were labeled with fluorescein (Fluor) and incubated with platelets and PAR1-Rat1 fibroblasts. The cells were then treated with pronase to digest extracellularly bound peptides and analyzed by flow cytometry. As shown in Figure 14.2A, both platelets and fibroblasts remained strongly fluorescent when treated with Fluor-Pal-i3, as compared to the nonpalmitoylated Fluor-i3. Disruption of the cell membrane abrogates protection against pronase digestion only with Fluor-Pal-i3 and not Fluor-i3, consistent with the palmitoylated i3 peptide being membrane-permeant (Figure 14.2B).

Palmitoylated and nonpalmitoylated i3 loop peptides tagged with fluorescein[28] were delivered into the mouse via a catheter placed in the internal jugular vein (IV) or into the intraperitoneal (IP) cavity. The fluorescently labeled i3-loop peptides were allowed to circulate for 15 or 30 min, respectively, then blood was collected and the platelets purified. After treatment with pronase to remove peripherally bound peptides, flow cytometry revealed that circulating mouse platelets exposed to the palmitoylated Fluor-Pal-i3 acquired significantly higher fluorescence relative to platelets exposed to the nonpalmitoylated Fluor-i3 (Figure 14.2A).

14.3 PEPDUCIN PHARMACOLOGY

14.3.1 PHARMACOKINETIC STUDIES

To determine pepducin bioavailability, we radioactively labeled P4pal-10[28] with ^{14}C-acetamide (^{14}C-P4pal-10). The P4pal-10 sequence is shown in Figure 14.3A. We then determined the following PK properties shown in Figure 14.3B: (1) high and prolonged plateau phase; (2) large volume of distribution (46 mL in 23 g mouse with approximately 2 mL blood volume); and (3) large area under the curve (AUC).

Platelets were washed three times in PBS buffer and subjected to spectrofluorimetry. The excitation spectra was set at 480 nm, and the emission spectra was collected between 490 and 600 nm. The emission maximum intensity for fluorescein-5-EX is at 515 nm. Fluorescence intensity for intact cells (IC) labeled with Fluor-Pal-i3 (Exp #1) or Fluor-i3 (Exp #5) were set as 100%, and all further treatments in Exp #2–3 and #6–7 are expressed relative to these values. Cells from Exp #1 and Exp #5 were treated with 5 U of pronase for 15 min at 37°C in PBS. Cells were washed three times in PBS and their relative fluorescence is shown in Exp #2 and Exp #6. These intact cells were then sonicated for 20 sec to disrupt the membranes and treated for a second time with 5 U of pronase for 15 min at 37°C. The crude membranes from these samples were precipitated with 0.5 M NaCl (final concentration), and centrifuged for 30 min at 14,000 g at 4°C. The pellets were resuspended in an equal volume and relative fluorescence intensity is shown in Exp #3 and Exp #7.

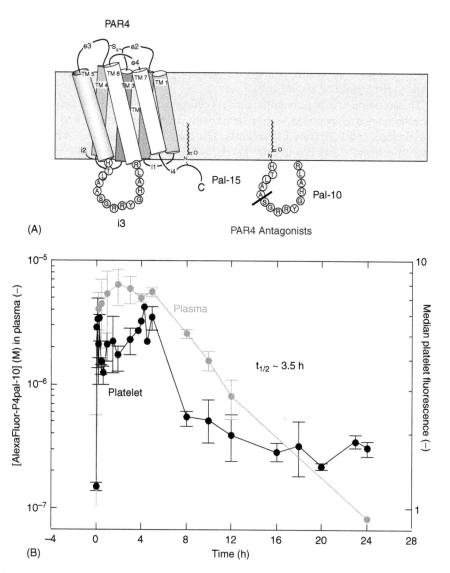

FIGURE 14.3 Pharmacokinetic (PK) properties of pepducins. (A) Cell-penetrating pepducins derived from the sequences of the intracellular third loops (i3) of PAR4 (P4pal-15 and P4pal-10) are shown with a covalently attached palmitate lipid. (B) Pharmacokinetics of IV (0.93 mg/kg) administered pepducin P4pal-10-AlexaFluor.

Intravenous single bolus dosing of mice with Alexa Fluor-labeled P4pal-10 (0.93 mg/kg) revealed that high drug levels in platelets and plasma (5 μM) were maintained for 5 h followed by elimination with a half-life of 3.5 h. P4pal-10 weakly binds human serum albumin (Kd = 0.3–0.7 mM); therefore, pepducin activity in blood should not be substantially reduced by the presence of albumin.

14.3.2   PHARMACODYNAMIC STUDIES

Next, we carried out PD studies on mice with the PAR4 P4pal-10 pepducin because mice lack platelet PAR1. Previously, we showed[28] (Figure 14.4A) that P4pal-10 blocks platelet function by targeting its cognate receptor, PAR4. Low-dose bolus IV infusion of P4pal-10 (0.37 mg/kg) extended bleeding time by three- to five-fold at 5 min and substantial platelet protective effects were maintained during the first hour time period (Figure 14.4B). By 4 h, half of the protective effect remained and, by 24 h, bleeding time had returned to baseline (Figure 14.4A,B). Subcutaneous injection of P4pal-10 gave a more prolonged half-life of ~ 14 h (Figure 14.4A). Subcutaneous (SC) injections of P4pal-10 (3.2 mg/kg) gave a six-fold prolongation of bleeding time at 4 h (Figure 14.4B). Similar to PK studies, PD studies indicate that pepducin compounds have a high biological activity and high efficacy because they are not easily eliminated, are systemically distributed, and penetrate tissues very rapidly.

We also carried out a preliminary PD evaluation of the P4pal-10 pepducin after bolus IV administration (0.37 mg/kg) in 10 kg baboons (conducted by Dr. Andras

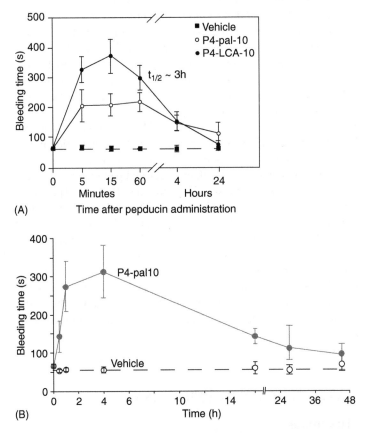

FIGURE 14.4 Pharmacodynamic (PD) properties of pepducins. (A) Pharmacodynamis of acyl versus steroid pepducins P4pal-10 and P4LCA-10 administered IV using 0.37 mg/kg. (B) The effect of subQ administration (3 mg/kg) of P4pal-10 on mouse tail bleeding time.

Gruber). There was complete (100%) blockade of PAR4 platelet function 25 h after administration with no cross-inhibition of PAR1 or ADP receptors, suggesting that even after 25 h there was still substantial active drug levels in the baboon (data not shown). Although preliminary, these data suggest that palmitoylated pepducins may have long-lived PK and PD parameters in larger animals.

In order to explore whether varying the lipid could be used to tailor pepducin properties, we tested the effect of different lipid moieties with the identical peptide sequence moiety on efficacy and PD. Lithocholic-modified P4-LCA-10 (0.37 mg/kg IV) doubled the bleeding time in mice relative to palmitate for the first hour but was cleared at a similar rate as the palmitate version (Figure 14.4A). These data suggest that the lipid moiety may provide an effective way of manipulating drug half-life, bioavailability, and efficacy.

14.3.3 SPECIFICITY

As previously described,[28] we tested the selectivity of the P1pal-12 pepducin for PAR1 versus other human platelet GPCRs including PAR4, thromboxane A2 (TXA_2), $P2Y_1$ and $P2Y_{12}$ ADP receptors, and non-GPCR receptors for collagen ($\alpha_2\beta_1$, GPVI/FcγII) and von Willebrand factor (GPIb/IX/V).[35–38] P1pal-12 was selective for PAR1 and did not block aggregation induced by the extracellular ligands of the other platelet receptors. It is noteworthy that the anti-PAR1 pepducin, P1pal-12, does not inhibit PAR4-dependent platelet aggregation. Conversely, the P4pal-10 PAR4 antagonist could also partially inhibit activation of PAR1 by SFLLRN, but did not appreciably block aggregation in response to agonists for TXA_2, ADP, collagen, or GPIb/IX/V receptors. Thus, P4pal-10 can inhibit signaling from both PAR4 and PAR1 thrombin receptors with higher selectivity for PAR4 over PAR1. We now show that pepducins derived from i1 loops of GPCRs may in some cases exhibit higher specificity than i3-derived pepducins.[29] Kaneider et al.[29] made CXCR1 and CXCR2-i1 versus -i3 pepducins. Receptor specificity of these pepducins was validated to be specific to CXCR1 and CXCR2 and inhibition was not observed with CXCR4, CCR1, CCR2, CCR5, PAR1, or PAR4 receptors. The i1-loop pepducin of CXCR1/2 was selective for IL-8 receptors. Because the i1 loop is on the opposite side of the GPCR relative to the i3 loop, we tested whether CXCR1/2 i1 as compared to i3 pepducins could identify potential heterodimeric[39,40] GPCR interactions. CXCR4 was expressed on human embryonic kidney (HEK) cells in combination with either CXCR1 or CXCR2. We demonstrated that only the CX1/2pal-i3 pepducin could inhibit CXCR4-dependent migration if CXCR1 was also present.

14.4   IN VIVO EFFICACY OF PEPDUCINS

14.4.1 HEMOSTASIS

In order to determine the potential therapeutic value of blockade of a particular signaling pathway, we used pepducins that inhibit PAR receptors in mouse animal models. Since mouse platelets lack PAR1 and use dual PAR3/PAR4 receptors, we tested blockade of PAR4 on primary hemostasis. We showed[28] that IV preinfusion of P4pal-10 or SC administration prolonged bleeding time (Figure 14.4A,B) by inhibiting systemic platelet activation. Interestingly, 50% of the mice injected with P4pal-10 formed unstable thrombi at the amputated tail tips.[28] Conversely, rebleeding was not observed in any of the control mice injected with vehicle alone or PAR-1 pepducin P1pal-12. Our previous studies[20,21] with PAR4 predicted that the physiological role of PAR4 in humans is to control the stability of platelet–platelet aggregates. Thus, the in vivo P4pal-10 effects in the mouse are consistent with the predicted role of PAR4. Furthermore, our P4pal-10 and P1pal-12 pepducin in vivo effects are in close agreement with the bleeding phenotypes of PAR1$(-/-)$[41] and PAR4$(-/-)$[25] mice. Together, our initial in vivo mouse pepducin experiments have demonstrated "proof-of-principle" for rapid determination of the potential therapeutic blockade of a particular signaling pathway.

14.4.2 ARTERIAL THROMBOSIS

Thrombosis is a leading cause of death worldwide and platelets play a primary role in the initiation of arterial thrombosis. In order to evaluate PAR4 as a therapeutic target for antithrombotic therapy, we examined the effect of PAR4 protection on the large vessel (carotid) artherial thrombosis. Mouse carotid artery thrombus formation in FeCl$_3$-induced carotid artery injury in mice treated with P4pal-10 (0.28 mg/kg) was compared to vehicle alone using a Doppler probe in order to monitor and record arterial blood flow. Administration of P4pal-10 significantly ($p < .001$) delayed time of occlusion over a 30 min period (Figure 14.5). In contrast to vehicle-treated mice whose carotid artery occluded by 90% ($n = 16$), the majority (70%) of P4pal-10 treated mice were protected over the 30 min period. These results identify PAR4 as a potential therapeutic target for antithrombotic therapy.

14.4.3 OTHER POTENTIAL INDICATIONS

We have also tested pepducins for efficacy in animal models of cancer and systemic inflammation.[18,27–29,32] In the cancer studies, nude mice were inoculated in their mammary fat pads with the invasive breast cancer cell line, MCF7-PAR1/N55, and the mice treated with either vehicle or PAR1 pepducin, P1pal-7 (10 mg/kg SC every other day) for 6 weeks.[18] By the 6-week time point, P1pal-7 significantly inhibited (62%, $p < .01$) the growth of the MCF7-PAR1/N55 xenografts. Tumors were excised from the mammary pads 610 weeks postinoculation, and hematoxylin and eosin staining of the MCF7-PAR1/N55 breast tumors revealed extensive replacement of normal mammary tissue.[18] P1pal-7 treatment caused a significant (75%, $p < .002$) reduction in blood vessel density at the center of the tumors.

   Several other laboratories have reported the use of PAR pepducins.[26,41–43] For example, Houle et al.[26] evaluated the role of PAR4 in a rodent paw inflammation model. Administration of PAR4 agonist to the mouse hind paw caused an inflammatory reaction. These effects were blocked in mice pretreated with P4pal-10. Furthermore, Keuren et al.[42] used PAR-1 and PAR-4 selective intracellular

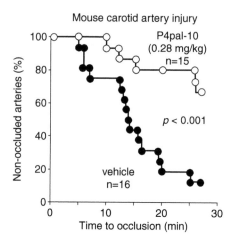

FIGURE 14.5 In vivo efficacy of pepducins in arterial thrombosis. Changes in the blood flow were recorded using Doppler probe after artery was fitted into a transverse cylindrical groove (0.55 mm in diameter). Once a stable signal was obtained, thrombosis was initiated with application of filter strip saturated with $FeCl_3$ for 2 min; after removal, carotid artery occlusion was monitored over 30 min.

inhibitors (P1pal-12 and P4pal-10) to demonstrate that PAR-1 was required for sustained Ca^{2+} flux and procoagulant activity rather than PAR-4 in collagen plus thrombin-stimulated platelets.

To further illustrate that the pepducin approach could be generally applied to GPCR families other than PARs, pepducins were tested for efficacy against CXCR1 and CXCR2[29] in systemic inflammatory response syndrome. Administration of x1/2pal-i3 or x1/2LCA-i1 after induction of sepsis by cecal ligation and puncture (CLP) resulted in >85% survival even after pepducins were administered 8 h post-CLP. This increase in survival was accompanied with reversal of shock, disseminated intravascular coagulation, and multiorgan failure.

14.5 CONCLUSION

GPCRs constitute the largest family of cell surface receptors with more than 1000 members, yet only a few have been successfully targeted. Here, we present pepducins as a new class of highly stable lipidated peptides that act on the inside surface of the cell membrane and block signaling between the receptor and its effectors. Inhibitors targeted against GPCRs represent a new approach for blocking GPCR-mediated cellular signaling. We have extensively tested pepducins for efficacy in animal models of cancer, thrombosis, and systemic inflammation.[18,27–29,32] From pharmacodynamic, pharmacokinetic, biodistribution, and toxicity studies in animals, pepducins appear to possess appropriate drug-like properties. These studies demonstrate that PD activity and PK drug levels are strongly correlated. Moreover, pepducins are simple to synthesize and purify in large quantities and may provide a useful complement to mouse knockout technology to identify GPCRs that may be important therapeutic targets in a variety of disorders.

ACKNOWLEDGMENTS

This research was supported by NIH grants HL64701 and HL57905 (A.K.) and CA104406 (L.C.).

REFERENCES

1. Palczewski, K. et al., Crystal structure of rhodopsin: A G protein-coupled receptor, *Science*, 289, 739, 2000.
2. Milligan, G., G protein-coupled receptor dimerization: Function and ligand pharmacology, *Mol. Pharmacol.*, 66, 1, 2004.
3. Gether, U. and Kolbilka, B.K., G Protein-coupled receptors II. Mechanism of agonist activation, *J. Biol. Chem.*, 273, 17979, 1998.
4. Cotecchia, S., Ostrowski, J., Kjelsberg, M.A., Caron, M.G., and Lefkowitz, R.J., Discrete amino acid sequences of the alpha 1-adrenergic receptor determine the selectivity of coupling to phosphatidylinositol hydrolysis, *J. Biol. Chem.*, 267, 1633, 1992.
5. Kostenis, E., Conklin, B.R., and Wess, J., Molecular basis of receptor/G protein coupling selectivity studied by coexpression of wild type and mutant m2 muscarinic receptors with mutant $G\alpha_q$ subunits, *Biochemistry*, 36, 1487, 1997.
6. Kjelsberg, M.A., Cotecchia, S., Ostrowski, J., Caron, M.G., and Lefkowitz, R.J., Constitutive activation of the α_{1B}-adrenergic receptor by all amino acid substitutions at a single site, *J. Biol. Chem.*, 267, 1430, 1992.
7. Luttrell, L.M., Ostrowski, J., Cotecchia, S., Kendal, H., and Lefkowitz, R.J., Antagonism of catecholamine receptor signaling by expression of cytoplasmic domains of the receptors, *Science*, 259, 1453, 1993.
8. Okamoto, T. et al., Identification of a Gs activator region of the β2-adrenergic receptor that is autoregulated via protein kinase A-dependent phosphorylation, *Cell*, 67, 723, 1991.
9. Rasmussen, U.B. et al., cDNA cloning and expression of a hamster α-thrombin receptor coupled to Ca_{2+} mobilization, *FEBS Lett.*, 288, 123, 1991.
10. Vu, T.-K.H., Hung, D.T., Wheaton, V.I., and Coughlin, S.R., Molecular cloning of a functional thrombin receptor reveals a novel proteolytic mechanism of receptor action, *Cell*, 64, 1057, 1991.
11. Nystedt, S., Emilsson, K., Wahlestedt, C., and Sundelin, J., Molecular cloning of a potential proteinase activated receptor, *Proc. Natl Acad. Sci. U.S.A.*, 91, 9208, 1994.
12. Kahn, M.L. et al., A dual thrombin receptor system for platelet activation, *Nature*, 394, 690, 1998.
13. Xu, W.-F. et al., Cloning and characterization of human protease-activated receptor 4, *Proc. Natl Acad. Sci. U.S.A.*, 95, 6642, 1998.
14. Vu, T.-K.H., Wheaton, V.I., Hung, D.T., Charo, I., and Coughlin, S.R., Domains specifying thrombin–receptor interaction, *Nature*, 353, 674, 1991.
15. Camerer, E., Huang, W., and Coughlin, S.R., Tissue factor- and factor X-dependent activation of protease-activated receptor 2 by factor VIIa, *Proc. Natl Acad. Sci. U.S.A.*, 97, 5255, 2000.

16. Sambrano, G.R., Weiss, E.J., Zheng, Y.W., Huang, W., and Coughlin, S., Cathepsin G activates protease-activated receptor-4 in human platelets, *J. Biol. Chem.*, 275, 6819, 2000.

17. Kuliopulos, A. et al., Plasmin desensitization of the PAR1 thrombin receptor: Kinetics, sites of truncation, and implications for thrombolytic therapy, *Biochemistry*, 38, 4572, 1999.

18. Boire, A. et al., PAR1 is a matrix metalloprotease-1 receptor that promotes invasion and tumorigenesis of breast cancer cells, *Cell*, 120, 303, 2005.

19. Hollenberg, M.D. and Compton, S.J., Proteinase-activated receptors, *Pharmacol. Rev.*, 54, 203, 2002.

20. Covic, L., Gresser, A.L., and Kuliopulos, A., Biphasic kinetics of activation and signaling for PAR1 and PAR4 thrombin receptors in platelets, *Biochemistry*, 39, 5458, 2000.

21. Covic, L., Singh, C., Smith, H., and Kuliopulos, A., Role of the PAR4 thrombin receptor in stabilizing platelet–platelet aggregates as revealed by a patient with Hermansky–Pudlak syndrome, *Thromb. Haemost.*, 87, 722, 2002.

22. Jacques, S., LeMasurier, M., Sheridan, P.J., Seeley, S.K., and Kuliopulos, A., Substrate-assisted catalysis of the PAR1 thrombin receptor: Enhancement of macromolecular association and cleavage, *J. Biol. Chem.*, 275, 40671, 2000.

23. Seeley, S. et al., Structural basis for thrombin activation of a protease-activated receptor: Inhibition of intramolecular liganding, *Chem. Biol.*, 10, 1033, 2003.

24. Gerszten, R.E. et al., Specificity of the thrombin receptor for agonist peptide is defined by its extracellular surface, *Nature*, 368, 648, 1994.

25. Sambrano, G.R., Weiss, E.J., Zheng, Y.W., Huang, W., and Coughlin, S.R., Role of thrombin signalling in platelets in haemostasis and thrombosis, *Nature*, 413, 74, 2001.

26. Houle, S., Papez, M.D., Ferazzini, M., Hollenberg, M.D., and Vergnolle, N. Neutrophils and the kallikrein–kinin system in proteinase-activated receptor 4-mediated inflammation in rodents. *Br. J. Pharmacol.*, 146, 670, 2005.

27. Covic, L., Gresser, A.L., Talavera, J., Swift, S., and Kuliopulos, A., Activation and inhibition of G protein-coupled receptors by cell-penetrating membrane-tethered peptides, *Proc. Natl Acad. Sci. U.S.A.*, 99, 643, 2002.

28. Covic, L., Misra, M., Badar, J., Singh, C., and Kuliopulos, A., Pepducin-based intervention of thrombin receptor signaling and systemic platelet activation, *Nat. Med.*, 8, 1161, 2002.

29. Kaneider, N., Agarwal, A., Leger, A.J., and Kuliopulos, A., Reversing systemic inflammatory response syndrome with chemokine receptor pepducins, *Nat. Med.*, 11, 661, 2005.

30. Wikstrom, P., Kirschke, H., Stone, S., and Shaw, E., The properties of peptidyl diazoethanes and chloroethanes as protease inactivators, *Arch. Biochem. Biophys.*, 270, 286, 1989.

31. Stephans, G. et al., A sequence within the cytoplasmic tail of GpIIb independently activates platelet aggregation and thromboxane synthesis, *J. Biol. Chem.*, 273, 20317, 1998.

32. Kuliopulos, A. and Covic, L., Blocking receptors on the inside: Pepducin-based intervention of PAR signaling and thrombosis, *Life Sci.*, 74, 255, 2003.

33. Rojas, M., Donahue, J.P., Tan, Z., and Lin, Y.-Z., Genetic engineering of proteins with cell membrane permeability, *Nat. Biotech.*, 16, 370, 1998.

34. Schwarze, S.R., Ho, A., Vocero-Akbani, A., and Dowdy, S.F., In vivo protein transduction: Delivery of a biologically active protein into the mouse, *Science*, 285, 156, 1999.

35. Murray, R. and FitzGerald, G.A., Regulation of thromboxane receptor activation in human platelets, *Proc. Natl Acad. Sci. U.S.A.*, 86, 124, 1989.
36. Woulfe, D., Yang, J., and Brass, L., ADP and platelets: The end of the beginning, *J. Clin. Invest.*, 107, 1503, 2001.
37. Moroi, M., Jung, S.M., Okuma, M., and Shinmyozu, K., A patient with platelets deficient in glycoprotein VI that lack both collagen-induced aggregation and adhesion, *J. Clin. Invest.*, 84, 1440, 1989.
38. Savage, B., Almus-Jacobs, F., and Ruggeri, Z.M., Specific synergy of multiple substrate–receptor interactions in platelet thrombus formation under flow, *Cell*, 94, 657, 1998.
39. Stampfuss, J.J., Inhibition of platelet thromboxane receptor function by a thrombin receptor-targeted pepducin, *Nat. Med.*, 9, 1447, 2003.
40. Breitwieser, G.E., G protein-coupled receptor oligomerization, *Circ. Res.*, 94, 17, 2004.
41. Connolly, A.J., Ishihara, H., Kahn, M.L., Farese, R.V., and Coughlin, S.R., Role of the thrombin receptor in development and evidence for a second receptor, *Nature*, 381, 516, 1996.
42. Keuren, J.F.W. et al., Synergistic effect of thrombin on collagen-induced platelet procoagulant activity is mediated through protease-activated receptor-1, *Arteriosler. Thromb. Vasc. Biol.*, 25, 1499, 2005.
43. Hollenberg, M.D., Saifeddine, M., Sandhu, S., Houle, S., and Vergnolle, N., Proteinase-activated receptor-4: Evaluation of tethered ligand-derived peptides as probes for receptor function and as inflammatory agonists in vivo, *Br. J. Pharmacol.*, 143, 443, 2004.

15 HOXB4 Homeoprotein Transfer Promotes the Expansion of Hematopoietic Stem Cells

Sophie Amsellem and Serge Fichelson

CONTENTS

15.1 INTRODUCTION

All blood cells derive from multipotent hematopoietic stem cells (HSCs) through lineage-specific committed progenitors that undergo terminal differentiation following a single pathway. As in most stem cells, HSCs represent a very rare population and have the capacity to both long-term selfrenew and differentiate, according to body needs.[1,2] Hematopoiesis is a process tightly regulated by a series of signals, either endogenous (intrinsic regulation through transcription factors) or exogenous provided by the stem cell environment, mainly in so-called "bone marrow hematopoietic niches" (extrinsic regulation) (Figure 15.1). Difficulties

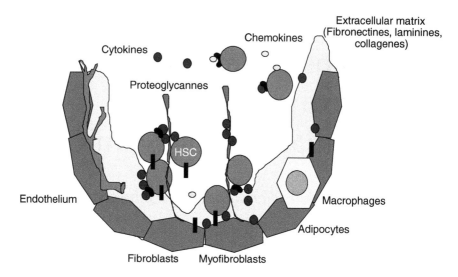

FIGURE 15.1 Schematic representation of the so-called "hematopoietic niche."

in identifying and isolating HSCs are not only related to their scarcity, but also to their heterogeneity in both their phenotype and potentialities and to the absence of a precise localization inside the bone marrow. Therefore, HSC-enrichment techniques in humans are based on their phenotypic characteristics (cell surface antigens such as CD34 and CD38) and metabolic characteristics (dye efflux of Hoechst 33342). The presence of stem cells in such a cell population is further identified by functional tests performed in vivo for the most immature ones or in vitro for progenitors (Figure 15.2).

Currently, a major challenge in the field of bone marrow transplantation (BMT), as well as in the definition of new protocols of cell or gene therapy, is to obtain a sufficient number of HSCs. Actually, patients suffering from hematopoietic diseases or cancers often require a BMT as a result of a chemotherapy- and/or radiotherapy-induced irreversible aplastic anemia. Sources of human HSCs are: (1) the bone marrow that can be taken out from anesthetized patients or normal donors by quite invasive multipuncture procedures; (2) the peripheral blood after HSC mobilization by several sets of cytokine injection, then cytapheresis collection(s); and (3) the umbilical cord blood from which low cell numbers just make available the transplantation of small infants. As a consequence, HSC use in therapeutic protocols, including cell engineering for transplantation or gene therapy, requires methods that allow ex vivo expansion of these cells without loss of their pluripotentiality.

So far, most ex vivo human primitive hematopoietic cell expansion protocols involve the use of cytokine mixtures whose value in the maintenance of biological properties of these cells remains controversial, because the absence of an irreversible differentiation process during culture cannot be ruled out. Other factors have been identified recently, such as Sonic Hedgehog and Wnt molecules.[3–6] The efficacy of these factors on the expansion of human HSCs seems promising but requires more

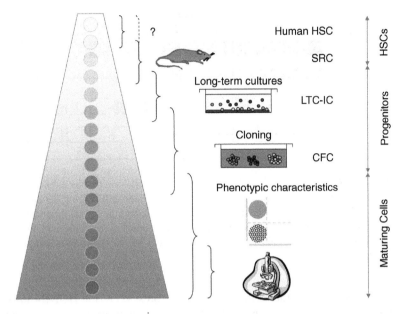

FIGURE 15.2 Functional tests used to study human hematopoiesis. Hematopoiesis can be represented as a pyramidal system whose top comprises the HSCs and base comprises the most mature blood cells. This system can be considered as a continuum that can be explored at several levels by specific functional tests. These tests are based on a retrospective study of the progeny of cells to be studied or numbered. Schematically, the upper compartment of HSCs is explored by in vivo xenogenic transplantation into immunodeficient NOD-SCID mice. Such human cells are called SCID repopulating cells (SRCs). The compartment of immature progenitors is tested by long-term liquid cultures to determine the presence of LTC-ICs able to generate more mature clonogenic cells after 5 to 8 weeks. Clonogenic cells (or colony forming cells [CFCs]) can give rise to morphologically identifiable cell colonies after 15 days in semisolid culture conditions. Final steps of differentiation are composed of cells that can be identified on either specific surface markers or morphological characteristics.

investigation. An alternative approach consists of using transcription factors that can lead to cell expansion. A set of transcription factors has been shown to get involved in stem cell selfrenewal and expansion. Among them, homeoproteins emerged as particularly attractive candidates. The importance of homeogenes had been experimentally established by experiments of overexpression or, conversely, inhibition of expression in mouse models. In humans, a series of such genes are frequently overexpressed or involved in recurrent translocations in leukemias. Moreover, numerous homeogenes, particularly class I homeogenes (HOX genes), are normally expressed during human hematopoiesis (Figure 15.3) but, when overexpressed, most of them lead to either immortalization or leukemic transformation of hematopoietic cells. HOXB4, whose physiological expression is restricted to very immature hematopoietic cells, was chosen because of its apparent safety.

Experiments were first performed with murine HSCs by using gene transfer mediated by recombinant retroviruses that comprised the HOXB4 coding sequence.

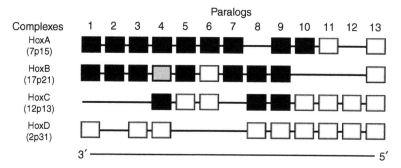

FIGURE 15.3 HOX gene expression in developing hematopoietic tissues in humans. Black boxes correspond to genes that are expressed in hematopoietic cells during development or regeneration of the hematopoietic tissues. The class I homeobox genes are organized in four complexes located on different chromosomes (between brackets are indicated the human chromosome localizations). Genes located on homologous loci of the various complexes are called paralogous genes.

Interestingly, such experiments that led to overexpression of the HOXB4 factor in bone marrow cells selectively entailed strong expansion of primitive hematopoietic cell populations.[7] This effect was reproducible both in vitro and in vivo and restricted to HSCs and progenitors: the peripheral blood cell populations remained balanced and no leukemogenic process has been observed so far in long-term experiments.[8–10] Thus, the HOXB4 homeoprotein appeared as the key factor to be used for primitive hematopoietic cell amplification. However, even if no leukemia had been detected in mice, a permanent retrovirus-related genetic alteration of human HSCs, along with constitutive expression of HOXB4, could be regarded as potentially deleterious or harmful in the prospect of a therapeutic use. Consequently, the expansion of human HSCs required an alternative approach. In that way, we established an in vitro model that allowed us to succeed in expanding HSCs and immature progenitors with the HOXB4 protein.

15.2 METHODOLOGY USED FOR HOXB4 PROTEIN TRANSFER INTO HUMAN HSCS

Our aim was so to expand human primitive hematopoietic cells exclusive of the addition of exogenous cytokines or retrovirus gene transfer. Translocation/transfer of homeoproteins from cell to cell, particularly HOX proteins, had been firmly established by the team of Prochiantz as a passive and reversible process.[11–13] This process is triggered by the homeodomain of the proteins that behaves as a cell-penetrating structure. The biological activities of such transferred homeoproteins were shown to remain unchanged.[14] Therefore, we designed a protocol that involved the secretion of human HOXB4 protein (hHOXB4) by a stromal feeder cell layer, leading to the passive transfer of this protein into human target cells (Figure 15.4). The murine MS-5 stromal cell line supports human hematopoiesis in long-term

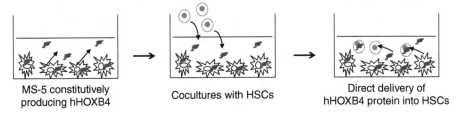

MS-5 constitutively producing hHOXB4 → Cocultures with HSCs → Direct delivery of hHOXB4 protein into HSCs

FIGURE 15.4 General design for the HOXB4 protein transduction into HSCs.

culture.[15] To establish MS-5 cells that constitutively produce hHOXB4 homeo-protein, we transduced these cells with several hHOXB4 cDNA constructs cloned into the TRIPΔU3-EF1α lentiviral vector.[16] These constructs contained either the wildtype hHOXB4 cDNA alone (HOXB4 construct) or the hHOXB4 cDNA preceded by a sequence encoding the signal peptide (SP) of the immunoglobulin κ light chain (SP-HOXB4 construct). This latter construct was designed to enhance the secretion of hHOXB4, considering that the signal peptide is cleaved during protein secretion. As a control, we used a construct containing the enhanced green fluorescent protein (EGFP) cDNA (EGFP construct). We detected the hHOXB4 protein expression by Western blot in these engineered MS-5 cells transduced with the various hHOXB4 constructs, and 100% of these cells stably expressed the hHOXB4 protein. Cells that had integrated the SP-HOXB4 construct (MS-5/SP-HOXB4 cells) exhibited strong and specific cytoplasmic labeling, while cells transduced with the wildtype HOXB4 construct (MS-5/HOXB4 cells) displayed intense nuclear labeling. ELISA tests performed on supernatants of the MS-5 cells further used for coculture experiments allowed us to detect hHOXB4 in the supernatant from MS-5/SP-HOXB4, but not in the supernatants from MS-5/HOXB4 or MS-5/EGFP cells. Altogether, our results indicated that we had established stromal cells able to stably produce and export hHOXB4, thereby supplying a continuous source of this homeoprotein in culture supernatant. We further ascertained that functional hHOXB4 protein could efficiently enter human hematopoietic cells when cocultured with the producing cells and retained its specific DNA binding activity as assessed in electrophoretic mobility shift assay (EMSA).[17]

15.3 DEMONSTRATION OF HUMAN HEMATOPOIETIC CELL EXPANSION AFTER HOXB4 PROTEIN TRANSFER[17]

15.3.1 AMPLIFICATION OF THE PRODUCTION OF TOTAL CELLS AND CLONOGENIC PROGENITORS (CFCs) IN LONG-TERM COCULTURE OF CD34[+] CELLS WITH hHOXB4-PRODUCING MS-5 CELLS

To assess whether the passive transfer of hHOXB4 could result in the amplification of the human immature hematopoietic cells, we first performed long-term cocultures of HSC-enriched cell suspensions (CD34[+] cells) with the various MS-5 cells described above. As the untransduced MS-5 cells and MS-5/EGFP cells displayed sustain long-term cultures of human hematopoietic cells (unpublished data), we used the MS-5/EGFP as control cells in our experiments. Every week over 5 weeks we determined: (1) the total cell numbers; (2) the percentage and absolute numbers of CD34$^+$ cells; and (3) the number of clonogenic progenitors. After 4–5 weeks, cocultures with any of MS-5/HOXB4 or MS-5/SP-HOXB4 displayed numbers of total cells and CD34$^+$ cells 2.5 to 3 times higher than cocultures with MS-5/EGFP cells (Figure 15.5A). Moreover, absolute numbers of CFCs present at the end of the cocultures with MS-5/HOXB4 or MS-5/SP-HOXB4 cells were four times higher than those found in cocultures with MS-5/EGFP cells (Figure 15.5B). These data demonstrated that hHOXB4-producing MS-5 cells favored the amplification of human CD34$^+$ and progenitors cells in coculture, even when no active secretion of HOXB4 was induced by a signal peptide, resulting in low HOXB4 concentration in the culture medium.

15.3.2 Amplification of CFCs and LTC-ICs in Long-Term Cocultures of CD34$^+$CD38^{low} Cells with hHOXB4-Producing MS-5 Cells

To determine the expansion rate of more primitive cells, we performed long-term cocultures of CD34$^+$CD38^{low} cells with HOXB4-producing MS-5 cells. Expansion rates of total cells, CD34$^+$ cells, and CFCs numbers were, respectively, 2.2 ± 0.9, 2 ± 0.9, and 3 ± 1 after 4–5 weeks of cocultures with MS-5/SP-HOXB4 when compared to cocultures with MS-5/EGFP cells. Interestingly, such amplifications could not be obtained when these immature CD34$^+$CD38^{low} cells were cocultured with MS-5 cells transduced by the HOXB4 construct devoid of an SP (Table 15.1).

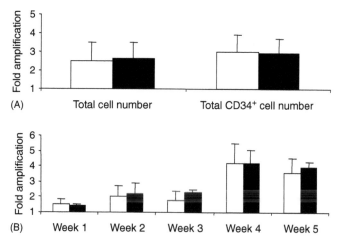

FIGURE 15.5 HOXB4-induced expansion of human hematopoietic progenitors. Expansion of (A) total cells and (B) clonogenic progenitors (CFCs) in long-term cocultures of CD34$^+$ hematopoietic cells with: □ MS-5/HOXB4 cells, ■ MS-5/SP-HOXB4 cells.

TABLE 15.1
Human Hematopoietic Cell Expansion after Long-Term Coculture with the Various Engineered MS-5 Stromal Cell Lines

	Stromal Cell Lines Used		
Human Cells Numbered	**MS-5/EGFP**	**MS-5/HoxB4**	**MS-5/SP-HoxB4**
Total cells	×237	×145 (0.61)	×530 (2.2)
CD34$^+$ cells	×60	×35 (0.58)	×122 (2)
CFCs	×13.8	×16.5 (1.2)	×41.3 (3)
LTCICs	×3.5–8.5	×6–14 (1.6)	×18–44 (5)

Human cord blood-derived CD34+/CD38 cells were plated with MS-5 cells transduced by hHoxB4, hHoxB4 preceded by a signal peptide, or EGFP as control. After 4 to 5 weeks of coculture, the various hematopoietic cell populations were evaluated using in vitro techniques. Results presented here correspond to net amplifications of these cells, relatively to the cell inputs at day 0. Numbers in parentheses correspond to relative amplifications as compared to cocultures made under similar conditions on MS-5/EGFP control cells.

To study whether the most primitive hematopoietic cells had been efficiently amplified in cocultures with HOXB4-producing MS-5 cells, we established secondary long-term cultures. CD34$^+$ cells derived from 5-week primary cocultures were sorted by flow cytometry, then plated at limiting dilutions on untransduced MS-5 cells for 5 additional weeks in LTC conditions, and the number of LTC-ICs was thus determined. Cocultures that had been performed with either MS-5/HOXB4 or MS-5/SP-HOXB4 cells showed LTC-IC frequency increases of 1.6 ± 0.3 and 5 ± 1, respectively (Table 15.1). The number, size, and nature of CFCs (BFU-E vs. CFU-GM) derived from a single LTC-IC remained unaffected whether they were from cocultures with HOXB4-producing or nonproducing MS-5 cells (unpublished data). In conclusion, assuming that the frequency of LTC-ICs in a cord blood-derived CD34$^+$CD38^{low} cell population is 2–5%,[15] after 5 weeks of primary cultures, the overall enrichment in total LTC-ICs was 18–44 times with MS-5/SP-HOXB4 and 6–14 times with MS-5/HOXB4 cells versus 3.5–8.5 times with MS-5/EGFP cells (Table 15.1). These results represented an ex vivo demonstration that the presence of HOXB4 protein leads to significant expansion of human primitive hematopoietic cells.

15.3.3 ENHANCED NOD-SCID MICE ENGRAFTMENT AND REPOPULATING ACTIVITY OF HOXB4-EXPANDED HUMAN PRIMITIVE HEMATOPOIETIC CELLS

Human hematopoietic cells that can engraft nonobese diabetic-severe immuno-deficient (NOD-SCID) mice, the SRCs are considered as the most primitive hematopoietic cells experimentally identifiable in humans.[18,19] To quantitatively ascertain the effects of the hHOXB4 protein on SRCs, we determined, by limiting dilution analysis, both frequency and repopulating activity of human CD34$^+$ cord blood cells at day 0 before culture and after 2-week and 4 to 5-week cocultures with MS-5/SP-HOXB4 or MS-5/EGFP cells.

At day 0, the SRC frequency was 1 in $\sim$19,100 cells whereas after 2 weeks of coculture with MS-5/SP-HOXB4, this frequency increased to 1 in $\sim$7,500 cells, demonstrating that the repopulating cells expanded by $\sim$2.5-fold over their input values. This expansion remained quite stable because, after 5 weeks of coculture, we still found 1 SRC in 9,000 cells. Moreover, the ability of the hHOXB4 protein to increase the repopulating capacity of human SRCs was shown by the enhanced hematopoietic cell chimerism in mice transplanted with human cells cocultured with MS-5/SP-HOXB4 (10.5% CD45$^+$ human cells, range 0.2–41.5%) compared to cells cocultured with MS-5/EGFP (1.5% human CD45$^+$ cells, range 0.6–2.6%), after 4–5 weeks of coculture. These data demonstrated that HOXB4 treatment was able to ex vivo amplify the number of human SRCs initially seeded in cultures while maintaining their ability of long-term reconstitution in animals.

Finally, phenotypic analysis revealed that HOXB4-induced expansion of human SRCs did not change their normal differentiation program in vivo: proportions of human myeloid cells (CD14$^+$/CD15$^+$), B-lymphoid cells (CD19$^+$) as well as CD34$^+$ cells were not altered at any time point when compared to controls. Thus, after HOXB4-induced primary long-term ex vivo expansion, SRCs could engraft NOD-SCID mice without alteration in their differentiation capacity, demonstrating that hHOXB4 induces straightforward amplification of human HSCs.

15.4   DISCUSSION AND CONCLUSIONS

We established a new protocol of ex vivo human primitive hematopoietic cell expansion that does not involve their genetic alteration. We showed that the continuous presence of the hHOXB4 protein resulted in an absolute number increase of the most immature hematopoietic cells identifiable in human, LTC-ICs, and SRCs, as well as of more mature progenitors.

Expansion of human HSCs has concentrated many efforts, and several strategies towards this aim have arisen, particularly the use of cytokine mixtures and/or additional molecules, and the modifications of the HSC genome by way of gene transduction. As it was not known if the use of cytokines might result in the loss of some properties of the amplified stem cells, a strategy that consisted of the use of transcription factors seemed more suitable. On the basis of the experiments reported by Sauvageau and his team in the murine model,[7,8,10] HOXB4 appeared to be an attractive candidate. Moreover, HOXB4 was demonstrated as a key factor in the determination of definitive hematopoiesis as it could confer definitive lympho-myeloid potential to embryonic stem cells and to yolk sac-derived HSCs in mice.[20] Similar technology that consisted of the transduction of human stem cells with recombinant retroviruses containing HOXB4 also led to successful expansion of human HSCs.[21] Expansion rates thus obtained with human cells were similar to the ones reported in our study. However, retrovirus-mediated constitutive expression of HOXB4 in human HSCs, as well as potential risks of insertional mutagenesis, might be hazardous for further therapeutic applications.[22]

To circumvent these drawbacks, the new protocol that we designed involved hHOXB4 protein transfer in HSCs. Homeodomain-containing proteins are known for their spontaneous ability to get across cell membranes through specific structures related to the canonical penetratin, a cell penetrating peptide derived from the third helix of Antennapedia homeodomain.[23] A number of messenger proteins or peptides such as the synthetic peptide carrier Pep-1,[24] as well as a peptide derived from the HIV transactivation factor Tat,[25,26] and Transportan 10,[27,28] are currently available as cell-penetrating molecules. Herpes simplex VP22 protein and the penetrating peptide pVEC derived from murine vascular endothelial cadherin also directly translocate into cells.[29,30] All these molecules can thus reach the cytoplasm or nucleus and steer coupled molecules into cells. The translocation of these proteins or peptides follows a nonclassical pathway as their interaction with the cell membrane is believed to induce the formation of inverted micelles in which the molecules are trapped.[31–34] Micelles then open and release the molecules inside the cell.[33] Up to our experiments, most studies on homeoprotein passive translocation had focused on neuronal and adherent cells.[11–14] Our study extended these observations to human hematopoietic cells. We showed that the translocated HOXB4 protein favored the expansion of most primitive hematopoietic cell populations: long-term culture initiating cells (LTC-ICs) and NOD-SCID mouse repopulating cells (SRCs), as well as myeloid progenitors.[17] Such HOXB4-amplified HSCs could not be distinguished from nonamplified ones since no difference appeared in the phenotype of cells derived from the various coculture conditions.

Whereas $CD34^+$ cells exhibited equal responsiveness to amplifying activities of both MS-5/SP-HOXB4 and MS-5/HOXB4 cells, a more immature $CD34^+CD38^{low}$ cell population appeared much more insensible to MS-5/HOXB4 cells. This difference in the behavior of the two cell populations could be related to dose-dependent variations in their HOXB4 requirement for expansion; among $CD34^+$ cells, $CD34^+CD38^{low}$ cells (which correspond to 10–15% of the total $CD34^+$ cell population) seemed less sensitive than $CD34^+CD38^{high}$ cells to the action of MS-5/HOXB4 cells that produced low amounts of hHOXB4 protein, undetectable in culture supernatants by ELISA tests. Moreover, we also performed direct transduction of hHOXB4 into human $CD34^+$ cells with the recombinant retroviruses used to transduce the MS-5 cells abovementioned. Instead of an expansion, we observed a clear deleterious effect of such transduced HOXB4 on the growth of human HSCs and progenitors (unpublished data). A dose–response effect had been previously suggested when HOXB4-related myeloid differentiation of HSCs was studied in mice and in humans.[35,36] Actually, strong HOXB4 expression in HSCs would direct these cells toward myeloid differentiation rather than increased proliferation.[36] However, this statement remains controversial since it has been suggested that HOXB4-induced stem cell expansion would be related to the HOXB4 modulation of c-Myc gene expression: the increased expression of HOXB4 would down-regulate the level of c-Myc, resulting in an enhancement of HSC selfrenewal[37,38] (see Klump et al.[39] for a review).

Concerning our experimental model, the question arose whether HOXB4-dependent cell expansion might be related to either hHOXB4 transfer into hematopoietic cells (direct effect) or putative changes in HOXB4-expressing MS-5 cells (indirect effect). Several observations argued against the latter mechanism: (1) similar effects on CD34$^+$ cells were obtained with MS-5 cells expressing HOXB4 or SP-HOXB4 while HOXB4 subcellular localization in the MS-5 cells was completely different; (2) phenotypic characteristics (aspect, growth rate) of MS-5 cells were not modified whatever the transduction performed; (3) CD34$^+$CD38^{low} cells were much less sensitive to the action of nonactively secreted hHOXB4, thus providing an internal control against an indirect effect of HOXB4 through MS-5 cell changes; and finally, (4) similar results had been reported after retroviral transduction of HOXB4 into human HSCs.[21] These observations strongly argued in favor of a direct involvement of HOXB4 in the HSC amplification we obtained. This point will be definitely confirmed by the use of purified recombinant HOXB4 protein.

Our results in the human model were corroborated by the work of Krosl et al.[40] which consisted of the utilization of a recombinant Tat–HOXB4 fusion protein to expand murine HSCs in culture. Such a soluble factor was also able to reach its subcellular nuclear targets and lead to an expansion of stem cells without alteration of their further in vivo differentiation and proliferation potentials. However, the usefulness of the presence of the Tat peptide was not proved in that report, because HOXB4 by itself is able to enter cells and induce their expansion.[17]

In conclusion, our results provide the basis for the development of new therapeutic strategies that include HOXB4-mediated expansion of human HSCs and possibly of other stem cells.

ACKNOWLEDGMENTS

This work was supported by the Ligue Nationale contre le Cancer (LNCC).

REFERENCES

1. Orkin, S.H., Diversification of haematopoietic stem cells to specific lineages, *Nat. Rev. Genet.*, 1, 57, 2000.
2. Weissman, I.L., Anderson, D.J., and Gage, F., Stem and progenitor cells: Origins, phenotypes, lineage commitments, and transdifferentiations, *Annu. Rev. Cell. Dev. Biol.*, 17, 387, 2001.
3. Bhardwaj, G., Murdoch, B., Wu, D., Baker, D.P., Williams, K.P., Chadwick, K., Ling, L.E., Karanu, F.N., and Bhatia, M., Sonic hedgehog induces the proliferation of primitive human hematopoietic cells via BMP regulation, *Nat. Immunol.*, 2, 172, 2001.
4. Murdoch, B., Chadwick, K., Martin, M., Shojaei, F., Shah, K.V., Gallacher, L., Moon, R.T., and Bhatia, M., Wnt-5A augments repopulating capacity and primitive hematopoietic development of human blood stem cells in vivo, *Proc. Natl Acad. Sci. U.S.A.*, 100, 3422, 2003.

5. Willert, K., Brown, J.D., Danenberg, E., Duncan, A.W., Weissman, I.L., Reya, T., Yates, J.R., 3rd, and Nusse, R., Wnt proteins are lipid-modified and can act as stem cell growth factors, *Nature*, 423, 448, 2003.
6. Reya, T. and Clevers, H., Wnt signalling in stem cells and cancer, *Nature*, 434, 843, 2005.
7. Sauvageau, G., Thorsteinsdottir, U., Eaves, C.J., Lawrence, H.J., Largman, C., Lansdorp, P.M., and Humphries, R.K., Overexpression of HOXB4 in hematopoietic cells causes the selective expansion of more primitive populations in vitro and in vivo, *Genes Dev.*, 9, 1753, 1995.
8. Antonchuk, J., Sauvageau, G., and Humphries, R.K., HOXB4 overexpression mediates very rapid stem cell regeneration and competitive hematopoietic repopulation, *Exp. Hematol.*, 29, 1125, 2001.
9. Antonchuk, J., Sauvageau, G., and Humphries, R.K., HOXB4-induced expansion of adult hematopoietic stem cells ex vivo, *Cell*, 109, 39, 2002.
10. Thorsteinsdottir, U., Sauvageau, G., and Humphries, R.K., Enhanced in vivo regenerative potential of HOXB4-transduced hematopoietic stem cells with regulation of their pool size, *Blood*, 94, 2605, 1999.
11. Derossi, D., Calvet, S., Trembleau, A., Brunissen, A., Chassaing, G., and Prochiantz, A., Cell internalization of the third helix of the Antennapedia homeodomain is receptor-independent, *J. Biol. Chem.*, 271, 18188, 1996.
12. Derossi, D., Joliot, A.H., Chassaing, G., and Prochiantz, A., The third helix of the Antennapedia homeodomain translocates through biological membranes, *J. Biol. Chem.*, 269, 10444, 1994.
13. Maizel, A., Bensaude, O., Prochiantz, A., and Joliot, A., A short region of its homeodomain is necessary for engrailed nuclear export and secretion, *Development*, 126, 3183, 1999.
14. Joliot, A., Trembleau, A., Raposo, G., Calvet, S., Volovitch, M., and Prochiantz, A., Association of Engrailed homeoproteins with vesicles presenting caveolae-like properties, *Development*, 124, 1865, 1997.
15. Issaad, C., Croisille, L., Katz, A., Vainchenker, W., and Coulombel, L., A murine stromal cell line allows the proliferation of very primitive human CD34+ +/ CD38− progenitor cells in long-term cultures and semisolid assays, *Blood*, 81, 2916, 1993.
16. Sirven, A., Ravet, E., Charneau, P., Zennou, V., Coulombel, L., Guetard, D., Pflumio, F., and Dubart-Kupperschmitt, A., Enhanced transgene expression in cord blood CD34(+)-derived hematopoietic cells, including developing T cells and NOD/SCID mouse repopulating cells, following transduction with modified trip lentiviral vectors, *Mol. Ther.*, 3, 438, 2001.
17. Amsellem, S., Pflumio, F., Bardinet, D., Izac, B., Charneau, P., Romeo, P.H., Dubart-Kupperschmitt, A., and Fichelson, S., Ex vivo expansion of human hematopoietic stem cells by direct delivery of the HOXB4 homeoprotein, *Nat. Med.*, 9, 1423, 2003.
18. Hogan, C.J., Shpall, E.J., McNiece, I., and Keller, G., Multilineage engraftment in NOD/LtSz-scid/scid mice from mobilized human CD34+ peripheral blood progenitor cells, *Biol. Blood Marrow Transplant.*, 3, 236, 1997.
19. Kobari, L., Pflumio, F., Giarratana, M., Li, X., Titeux, M., Izac, B., Leteurtre, F., Coulombel, L., and Douay, L., In vitro and in vivo evidence for the long-term multilineage (myeloid, B, NK, and T) reconstitution capacity of ex vivo expanded human CD34(+) cord blood cells, *Exp. Hematol.*, 28, 1470, 2000.

20. Kyba, M., Perlingeiro, R.C., and Daley, G.Q., HoxB4 confers definitive lymphoid-myeloid engraftment potential on embryonic stem cell and yolk sac hematopoietic progenitors, *Cell*, 109, 29, 2002.

21. Buske, C., Feuring-Buske, M., Abramovich, C., Spiekermann, K., Eaves, C.J., Coulombel, L., Sauvageau, G., Hogge, D.E., and Humphries, R.K., Deregulated expression of HOXB4 enhances the primitive growth activity of human hematopoietic cells, *Blood*, 100, 862, 2002.

22. Li, Z., Dullmann, J., Schiedlmeier, B., Schmidt, M., von Kalle, C., Meyer, J., Forster, M., Stocking, C., Wahlers, A., Frank, O., Ostertag, W., Kuhlcke, K., Eckert, H.G., Fehse, B., and Baum, C., Murine leukemia induced by retroviral gene marking, *Science*, 296, 497, 2002.

23. Perez, F., Joliot, A., Bloch-Gallego, E., Zahraoui, A., Triller, A., and Prochiantz, A., Antennapedia homeobox as a signal for the cellular internalization and nuclear addressing of a small exogenous peptide, *J. Cell. Sci.*, 102, 717, 1992.

24. Morris, M.C., Depollier, J., Mery, J., Heitz, F., and Divita, G., A peptide carrier for the delivery of biologically active proteins into mammalian cells, *Nat. Biotechnol.*, 19, 1173, 2001.

25. Fawell, S., Seery, J., Daikh, Y., Moore, C., Chen, L.L., Pepinsky, B., and Barsoum, J., Tat-mediated delivery of heterologous proteins into cells, *Proc. Natl Acad. Sci. U.S.A.*, 91, 664, 1994.

26. Schwarze, S.R., Ho, A., Vocero-Akbani, A., and Dowdy, S.F., In vivo protein transduction: Delivery of a biologically active protein into the mouse, *Science*, 285(5433), 1569–1572, 1999.

27. Lindberg, M., Jarvet, J., Langel, U., and Graslund, A., Secondary structure and position of the cell-penetrating peptide transportan in SDS micelles as determined by NMR, *Biochemistry*, 40, 3141, 2001.

28. Zorko, M. and Langel, U., Cell-penetrating peptides: Mechanism and kinetics of cargo delivery, *Adv. Drug Deliv. Rev.*, 57, 529, 2005.

29. Elliott, G. and O'Hare, P., Intercellular trafficking and protein delivery by a herpesvirus structural protein, *Cell*, 88, 223, 1997.

30. Elmquist, A., Lindgren, M., Bartfai, T., and Langel, U., VE-cadherin-derived cell-penetrating peptide, pVEC, with carrier functions, *Exp. Cell Res.*, 269, 237, 2001.

31. Derossi, D., Chassaing, G., and Prochiantz, A., Trojan peptides: The penetratin system for intracellular delivery, *Trends Cell Biol.*, 8, 84, 1998.

32. Lindgren, M., Hallbrink, M., Prochiantz, A., and Langel, U., Cell-penetrating peptides, *Trends Pharmacol. Sci.*, 21, 99, 2000.

33. Prochiantz, A., Messenger proteins: Homeoproteins, TAT and others, *Curr. Opin. Cell Biol.*, 12, 400, 2000.

34. Schwarze, S.R. and Dowdy, S.F., In vivo protein transduction: Intracellular delivery of biologically active proteins, compounds and DNA, *Trends Pharmacol. Sci.*, 21, 45, 2000.

35. Rideout, W.M., 3rd, Hochedlinger, K., Kyba, M., Daley, G.Q., and Jaenisch, R., Correction of a genetic defect by nuclear transplantation and combined cell and gene therapy, *Cell*, 109, 17, 2002.

36. Brun, A.C., Fan, X., Bjornsson, J.M., Humphries, R.K., and Karlsson, S., Enforced adenoviral vector-mediated expression of HOXB4 in human umbilical cord blood CD34+ cells promotes myeloid differentiation but not proliferation, *Mol. Ther.*, 8, 618, 2003.

37. Pan, Q. and Simpson, R.U., Antisense knockout of HOXB4 blocks 1,25-dihydroxyvitamin D3 inhibition of c-myc expression, *J. Endocrinol.*, 169, 153, 2001.
38. Wilson, A., Murphy, M.J., Oskarsson, T., Kaloulis, K., Bettess, M.D., Oser, G.M., Pasche, A.C., Knabenhans, C., Macdonald, H.R., and Trumpp, A., c-Myc controls the balance between hematopoietic stem cell self-renewal and differentiation, *Genes Dev.*, 18, 2747, 2004.
39. Klump, H., Schiedlmeier, B., and Baum, C., Control of self-renewal and differentiation of hematopoietic stem cells: HOXB4 on the threshold, *Ann. N.Y. Acad. Sci.*, 1044, 6, 2005.
40. Krosl, J., Austin, P., Beslu, N., Kroon, E., Humphries, R.K., and Sauvageau, G., In vitro expansion of hematopoietic stem cells by recombinant TAT–HOXB4 protein, *Nat. Med.*, 9, 1428, 2003.

16 Applications of Cell-Penetrating Peptides as Signal Transduction Modulators

Sarah Jones and John Howl

CONTENTS

16.1   INTRODUCTION

16.1.1   BACKGROUND

A fundamental paradigm of intracellular signal transduction is that of protein–protein interactions. The molecular interfaces and regulatory domains that mediate these protein–protein interactions are therefore potential targets for mimetic peptides that can selectively modulate signal transduction. Moreover, attachment of peptide fragments, corresponding to these molecular interfaces, to cell-penetrating peptide (CPP) vectors not only provides a vehicle for the introduction of peptidomimetic probes into the cytosolic environment, but offers the distinct advantage of intracellular delivery of peptide cargoes into intact and living cellular systems.[1,2] Furthermore, the bioassays used herein represent membrane translocation, protein binding, and activation/inhibition of biological responses and are therefore perhaps a more rigorous assessment of the utility of our TP10 chimerae to act as signal transduction modulators (STMs).

CPPs are now firmly established as a proven vehicle for the intracellular delivery of bioactive cargoes. In the past decade, numerous CPPs have been employed as relatively inert delivery vectors to deliver bioactive moieties that include peptides,[1,3–5] proteins,[6–8] various nucleic acids,[9] and antisense oligonucleotides.[10,11]

Transportan (galanin(1–12)-Lys-MP) is a chimeric amino-terminal extended analog of MP that demonstrates efficient cell-penetrating properties.[12] Moreover, the deletion analog transportan 10 (TP10; galanin(7–12)-Lys-MP) maintains efficient translocating properties, but lacks the inhibitory action on GTPases, a feature of most transportan analogs.[13] Furthermore, one mechanistic advantage of the TP10 system is that, following translocation, peptide cargoes are liberated by the intracellular reduction of cystine.[12,13] Hence, in this study we utilized TP10 as a suitably inert peptide vector to effect the intracellular delivery of peptide cargoes.

To assess the utility of our TP10 chimerae as STM, our studies have utilized peptide cargoes that mimic functional domains of signal-transducing proteins and included mimetic peptide cargoes derived from (1) the C-terminal juxtamembrane region of the central cannabinoid receptor (CB$_1$); (2) a sequence from the regulatory domain of protein kinase C; and (3) the carboxyl terminal of G$_i$α3. These rationally designed mimetic peptide cargoes included sequences designed to target and regulate the activity of G$_{i/o}$α and PKC, proteins widely reported to modulate secretory pathways.

16.1.2 BIOACTIVE PEPTIDOMIMETIC CARGOES

Bioactive peptide cargoes were translocated into rat basophilic cells RBL-2H3, a prototypic cell line for the study of mast cell secretion, and secretion of the pharmacological mediator β-hexosaminidase was measured. The rationale for using the proposed peptide modulators of G_i/G_o protein signaling and for assessing subsequent secretory events in the RBL-2H3 cell line is supported by the following contentions: (1) $G_i\alpha3$ is considered to be an integral component of the mast cell secretory response[14]; (2) secretory events in this cell line are largely dependent on the activation of PTX-sensitive G-proteins, G_i and G_o[14,15]; (3) the mast cell degranulating tetradecapeptide mastoparan (MP), a potent β-hexoseaminidase secretagog of RBL-2H3 cells,[16,17] is also a known activator of heterotrimeric G-proteins, particularly G_i and G_o; (4) $G_i\alpha3$ has been identified as a specific target for MP in other mast cell lines.[14]

16.1.2.1 $CB_1^{401-417}$

A peptide mimetic of the C-terminal juxtamembrane region of the rat CB_1 receptor (RSKDLRHAFRSMFPSSE) autonomously activates G_i/G_o proteins and competitively disrupts interaction of the CB_1 receptor with $G_o\alpha$ and $G_i\alpha3$.[18–20] This peptide contains a serine substitution for the natural Cys^{416}, a strategy that was initially intended to reduce spontaneous disulfide-bridge formation.[19] Therefore, a Cys^0-extended analog of this peptide Cys^0-$[Ser^{416}]CB_1^{401-417}$ and the natural sequence of the same segment ($CB_1^{401-417}$; H-RSKDLRHAFRSMFPSCE-NH$_2$) were synthesized and conjugated to TP10 ($[Lys^7N^{\epsilon Cys(Npys)}]$TP10). Amino-terminal extension of $[Ser^{416}]CB_1^{401-417}$ with Cys^0 allowed for conjugation via disulfide linkage to $[Lys^7N^{\epsilon Cys(Npys)}]$TP10 (Figure 16.1). Figure 16.2A illustrates the regulatory domain of the CB_1 receptor from which these mimetic peptides were derived.

16.1.2.2 $G_i\alpha3^{346-355}$

Cargo $G_i\alpha3^{346-355}$ (KNNLKECGLY) is a peptide mimetic of the carboxyl-terminal of the secretory G-protein subunit $G_i\alpha3$ and is conserved both within human and rodent genomes.[21] A corresponding peptide has been used to confirm a role of plasma membrane-bound $G_i\alpha3$ in the exocytotic response of melanotrophs.[22] Significantly, the $G_i\alpha3$ peptide, but not a corresponding sequence from $G_i\alpha1/2$, ablated MP-induced changes in membrane capacitance when dialyzed into rat pituitary melanotrophs using a patch clamp pipette.[22] Related peptides, introduced into permeabilized cells, have provided further evidence for a specific role of $G_i\alpha3$ in mediating MP-induced secretory events in both chromaffin cells[23] and peritoneal mast cells.[14] Collectively, these studies indicate that mimetic peptides of the carboxyl-terminal of $G_i\alpha3$ can modulate the activities of heterotrimeric G-proteins.

16.1.2.3 $Cys^0PKC^{238-249}$

Cargo $Cys^0PKC^{238-249}$ is a sequence from the middle of the C2 region of the regulatory domain of human PKCβ1 ($PKC^{238-249}$). This peptide has alternatively

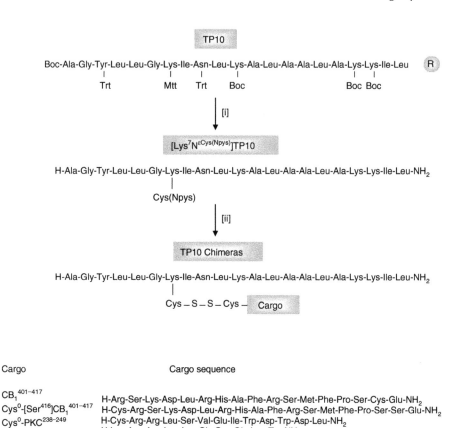

FIGURE 16.1 Schematic representation of the synthetic route to TP10 chimerae. TP10 and peptide cargoes were manually synthesized (0.1 mmol scale) on Rink amide methylbenzhydrylamine resin (R) by employing an N-α-Fmoc protection strategy with HBTU/HOBt activation. Step 1 indicates selective removal of the 4-methyltrityl group of Lys[7] with trifluoracetic acid (TFA) (3% (v/v) in dichloromethane (DCM; 2×10 min), acylation with Boc-Cys(Npys) (two equivalents) and cleavage with TFA/H_2O/triisopropylsilane (95:2.5:2.5%) to yield the fully deprotected [Lys^{7}N$^{\varepsilon Cys(Npys)}$]TP10. Disulfide bond formation, to generate TP10 chimeras, was achieved by dissolving [Lys^{7}N$^{\varepsilon Cys(Npys)}$]TP10 and individual cargoes (twofold molar ratio) in a minimum volume of DMF/DMSO/$C_2H_3O_2$Na (0.1 M) pH 5 (3:1:1) and mixing overnight (step 2). Cargo sequences are shown and cysteine residues indicated in bold represent the point of attachment of peptide cargoes to [Lys^{7}N$^{\varepsilon Cys(Npys)}$]TP10. (From Howl, J., Jones, S., and Farquhar, M., *Chembiochem.*, 4, 1312, 2003. With permission.)

been named pseudo-RACK1 (receptor for activated C kinase), since homologous sequences have been found in four WD40 repeat domains of RACK-1,[24] a scaffolding protein that binds a variety of signaling proteins including PKC. The pseudo RACK-1 peptide directly binds PKC and acts as an agonist of PKC in vivo[24] (Figure 16.2B). For this study, the original sequence of pseudo-RACK1 (SVEIWD) was extended to include an additional β-PKC-derived sequence including one further WD motif and an amino-terminal Cys for conjugation to TP10 (Figure 16.2B).[1]

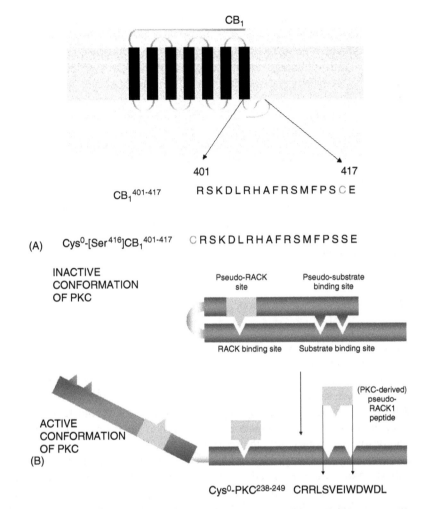

FIGURE 16.2 (A) Mimetic peptides derived from the C-terminal juxtamembrane region of the central cannabinoid G protein-coupled receptor (CB_1). (B) Origins of the bioactive cargo $Cys^0PKC^{238-249}$ (pseudo RACK-1; H-CRRLSVEIWDWDL-NH_2) from the C2 region of the regulatory domain of PKC. (Part (B) modified from Ron, D. and Mochly-Rosen, D., *Proc. Natl Acad. Sci. USA*, 92, 492, 1995 (24) and from Souroujon, M.C. and Mochly-Rosen, D., *Nat. Biotech.*, 16, 919, 1998 (25).)

16.2 METHODS TO STUDY THE INTRACELLULAR TRANSLOCATION OF BIOACTIVE PEPTIDES

16.2.1 Peptide Synthesis

All peptide cargoes were synthesized with an amidated carboxyl terminal to increase resistance to proteolysis. TP10 and peptide cargoes were manually synthesized (0.1 mmol scale) on Rink amide methylbenzhydrylamine (MBHA) resin employing an N-α-Fmoc protection strategy with 2-(1-H-benzotriazole-1-yl)-1,1,3,3-tetra-methyluronium hexafluorophosphate/N-hydroxybenzotriazole) (HBTU/HOBt) activation. The synthetic route to TP10-conjugated peptides is illustrated and described in Figure 16.1. TP10 chimerae and individual cargoes were purified to apparent homogeneity by semi-preparative scale HPLC.[26] The predicted masses of all peptides used (average M + H$^+$) were confirmed to an accuracy of +1 by matrix-assisted laser desorption ionization (MALDI) time-of-flight mass spectrometry (Kratos Kompact Probe operated in positive ion mode).[1]

16.2.2 Cell Culture

U373MG and RBL-2H3 cell lines were maintained in DMEM in a humidified atmosphere of 5% CO_2 at 37°C. Medium contained L-glutamine (0.1 mg/mL) and was supplemented with 10% (w/v) fetal bovine serum (FBS), penicillin (100 U/mL) and streptomycin (100 μg/mL).

16.2.3 Measurement of Translocation Kinetics

Analysis of the translocation kinetics of a fluoresceindiacetate [CFDA]-conjugated analogue of $G_i\alpha3^{346-355}$ was used to confirm effective intracellular delivery of our TP10 chimerae. Following intracellular delivery, CFDA is cleaved by intracellular esterases to generate a fluorescein chromophore that can be quantitatively detected by spectrofluorimetry. Fluorescent labeling of $G_i\alpha3^{346-355}$ to generate [CFDA]$G_i\alpha3^{346-355}$ was attained by acylation of the lysine amino terminal with CFDA succinimide ester.[29] Conjugation of this cargo to TP10 was achieved as with other cargoes in accordance with the scheme outlined in Figure 16.1.

RBL-2H3 cells were suspended in balanced salt solution (BSS) and transferred to a stirred, temperature-regulated quartz cuvette in a fluorescence spectropho-tometer at 37°C. The CFDA fluorescein chromophore was detected using excitation/emission wavelengths of 495 and 525 nm, respectively. The fluorescent signal was calculated at six time points by subtracting background fluorescence from the composite signal.[29]

16.2.4 Measurement of β-Hexoseaminidase from RBL-2H3 Cells

RBL-2H3 cells were cultured as above in 24-well plates and transferred to labeling medium consisting of Ham's medium supplemented with 0.1% (w/v) BSA and 10 mM HEPES, pH 7.4. Cells were incubated with peptides for the designated time periods at 37°C. Secreted β-hexoseaminidase was assayed by incubating samples of medium with 1 mM p-nitrophenyl N-acetyl-β-D-glucosamide in 0.1 M sodium citrate buffer for 1 h at 37°C. Addition of 0.1 M Na_2CO_3/$NaHCO_3$ buffer (pH 10.5) allowed determination of β-hexoseaminidase activity by colorimetric analysis at 405 nm using a microtiter plate reader.[1] Secretion of β-hexoseaminidase was calculated as fold over basal stimulation.

16.2.5 Measurement of Phospholipase D (PLD) Activity

PLD activity was determined as previously reported[16] by using a transphos-phatidylation assay in the presence of 0.3% (v/v) butan-1-ol.[27] Measurement of the accumulation of metabolically stable phosphatidylbutanol (PBut) was used as an index of PLD activity. RBL-2H3 cells were cultured as above and grown to a confluent monolayer in 24-well plates. Cells were labeled with [³H]palmitic acid (1 μCi) for 24 h, incubated with peptides for the designated time periods and PBut was separated from other lipids by TLC.[27] The radioactivity in PBut fractions was determined by liquid scintillation spectroscopy.

16.2.6 Detection of p42/44 MAPK Phosphorylation

Dual phosphorylation of p42/44 MAPK was detected by Western immuno-blotting. U373MG cells were cultured as above in six-well plates. Cells were stimulated in serum-free media containing peptide or vehicle (medium alone) for the designated time periods at 37°C. Cell lysates were prepared[29] and subjected to SDS-PAGE in 10% gels. Proteins were electrotransferred to PVDF membrane and blocked overnight at 4°C. Membranes were probed with primary (rabbit Anti-Active MAPK, Promega) and secondary (donkey anti-rabbit, conjugated to horseradish peroxidase, Promega) antibodies, according to the manufacturer's instructions and visualized by enhanced chemiluminescence. Total p42/44 MAPK, specific for dually, mono, and nonphosphorylated forms of p42/44 MAPK, were obtained by stripping blots in stripping buffer (100 mM β-mercaptoethanol, 2% (w/v) SDS and 62.5 mM Tris–HCl, pH 7.6) at 50°C for 30 min, blocking overnight, then probing with rabbit anti-ERK1/2 primary antibody rabbit (Promega) and treated as above. Where necessary, signal intensities were quantified by measuring the mean pixel intensitiy of bands of interest using Scion Image Beta 4.02 software. Signal intensity values are expressed as fold stimulation over basal.

16.2.7 MTT Assay of Cellular Viability

Cell viability was measured by the 3-(4,5-dimethylthazol-2-yl)-2,5-diphenyl tetrazolium bromide (MTT) conversion assay.[28] Cells were cultured as above in 96-well plates and treated with medium (without serum) containing peptide or vehicle (medium alone) for the designated time periods at 37°C. Stimulation medium was removed and the cells were incubated in medium containing MTT (0.5 mg/mL) for 3 h at 37°C. Medium was aspirated and the insoluble formazan product was solubilized with DMSO. MTT conversion was determined by colorimetric analysis at 540 nm. Cell viability was expressed as a percentage of cells treated with vehicle (medium) alone.

16.3   RESULTS

16.3.1   Translocation Kinetics of TP10-$G_i\alpha3^{346-355}$

The translocation kinetics of TP10-mediated cargo delivery was determined using the TP10-$G_i\alpha3^{346-355}$ construct labeled at the amino-terminus with CFDA (Figure 16.3). Intracellular esterases rapidly generate the fluorescein chromophore allowing the kinetics of penetration to be monitored in real time. These spectrofluorimetric analyses indicated that the TP10-$G_i\alpha3^{346-355}$ construct translocated into RBL-2H3 cells with first order, saturable kinetics with a half-life of 3 min at a concentration of 3 μM.[29]

16.3.2   TP10 Chimerae-Induced Secretion of β-Hexoseaminidase

Intracellular translocation of $Cys^0PKC^{238-249}$ and $CB_1^{401-417}$ as [$Lys^7N^{\varepsilon Cys}$]TP10 chimeric constructs induced a concentration-dependent exocytosis of β-hexose-aminidase, producing an 11- and sixfold increase at 3 μM, respectively (Figure 16.4A). However, TP10-Cys^0[Ser^{416}]$CB_1^{401-417}$ was a very weak secretagog, only producing a 2.5-fold increase in β-hexoseaminidase secretion at 3 μM and TP10-$G_i\alpha3^{346-355}$ was without effect. Figure 16.4A further indicates that the TP10 vector ([$Lys^7N^{\varepsilon Cys}$]TP10) was a relatively weak secretagog active only at a concentration (3 μM) above those required for efficient cellular translocation. When tested alone, peptide cargoes $Cys^0PKC^{238-249}$, $CB_1^{401-417}$, [Ser^{416}]$CB_1^{401-417}$, and $G_i\alpha3^{346-355}$ were inactive (data not shown).

16.3.3   Mechanisms of $Cys^0PKC^{238-249}$-Induced Secretion of β-Hexoseaminidase

We have previously reported that β-hexoseaminidase secretory efficacies are positively correlated to the activation of phospholipase D (PLD) in RBL-2H3 cells.[16] Therefore, we anticipated that translocation of $Cys^0PKC^{238-249}$ into RBL-2H3 cells would also stimulate PLD activity. Intracellular delivery of $Cys^0PKC^{238-249}$ activated PLD over the same concentration range required to promote β-hexoseaminidase secretion, whilst [$Lys^7N^{\varepsilon Cys}$]TP10 and the unconjugated cargo $Cys^0PKC^{238-249}$ were inactive (Figure 16.4B).

16.3.4   Intracellular Delivery of $G_i\alpha3^{346-355}$ Does Not Modulate β-Hexoseaminidase Secretion from RBL-2H3 Cells

Intracellular delivery of $G_i\alpha3^{346-355}$ did not directly influence the biochemical pathway leading to the exocytosis of β-hexoseaminidase. We have previously demonstrated that a range of MP analogs promotes concentration-dependent secretion of β-hexoseaminidase from RBL-2H3 cells.[16,17] MP binds the carboxyl-terminal of heterotrimeric α subunits.[30] Moreover, the conserved decapeptide sequence KNNLKECGLY ($G_i\alpha3^{346-355}$), corresponding to the carboxyl terminal domain of $G_i\alpha3$, has successfully been used to perturb MP-mediated secretory

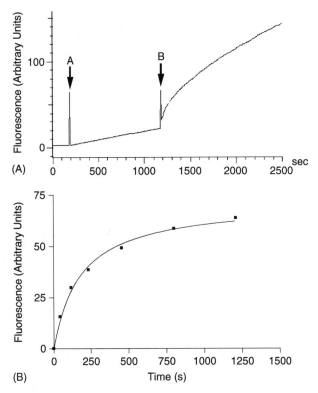

FIGURE 16.3 (A) Translocation and fluorimetric analysis of TP10-[CFDA]$G_i\alpha3^{346-355}$. This figure demonstrates the metabolic generation of the fluorescein chromophore following internalization of TP10-[CFDA]$G_i\alpha3^{346-355}$. Arrow A indicates the addition of unconjugated [CFDA]$G_i\alpha3^{346-355}$ (3 μM) to RBL-2H3 cells and arrow B indicates the subsequent addition of TP10-[CFDA]$G_i\alpha3^{346-355}$ (3 μM). Following addition of the nonpermeable and nonfluorescent peptide [CFDA]$G_i\alpha3^{346-355}$ (3 μM) (A), a gradual linear increase in fluorescence over a time period of 100 sec was observed and is likely to reflect the activity of extracellular esterases. This background change in signal intensity was calculated at Δ 0.019 units/sec. Ordinate: fluorescence intensity (arbitrary units Ex/Em 495/525 nm); Abscissa: time (sec). (B) Kinetic analysis of TP10-mediated translocation. To establish the specific fluorescent signal (F) generated by translocated [CFDA]$G_i\alpha3^{346-355}$, background fluorescence activity, representative of extracellular esterases, was subtracted from the composite signal at six time points. Background fluorescence was calculated at Δ 0.038 units/sec from the time of addition of TP10-[CFDA]$G_i\alpha3^{346-355}$ to cells, as indicated by arrow B in panel A. Data were analyzed using Figure P software (Biosoft) and demonstrate that internalization occurred with first-order kinetics ($F = F_{max} \times t/t_{0.5} + t$) with a half-life ($t_{0.5}$) of 180.8 sec and a maximum signal intensity (F_{max}) of 71.5 units. (Panels A and B were reprinted from Jones, S. et al., *Biochim. Biophys. Acta*, 1745, 207, 2005. With permission.)

events.[14,22,31] Therefore, we anticipated that TP10-$G_i\alpha3^{346-355}$ would reduce the magnitude of MP-induced secretion of β-hexoseaminidase (Figure 16.5, columns A and B). Addition of $G_i\alpha3^{346-355}$ alone had no effect on β-hexoseaminidase secretion (Figure 16.5, column C; $p > 0.5$, Student's t-test). All other secretory responses

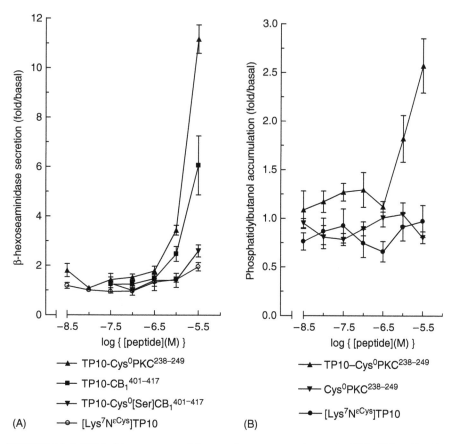

FIGURE 16.4 TP10 chimerae as modulators of secretory pathways in RBL-2H3 cells. (A) Comparative secretory efficacies of TP10 chimerae. The figure compares the concentration-dependent secretion of β-hexoseaminidase by TP10 chimerae (TP10-$CB_1^{401-417}$, TP10-Cys^0[Ser^{416}]$CB_1^{401-417}$ and TP10-$Cys^0PKC^{238-249}$) and the TP10 vector alone ([$Lys^{7N\varepsilon Cys}$]TP10). RBL-2H3 cells were treated with peptides for 60 min, prior to measurement of β-hexoseaminidase secretion. Data points are mean ± s.e.m. from at least three experiments performed in triplicate. (B) Stimulation of PLD by TP10-$Cys^0PKC^{238-249}$. RBL-2H3 cells were incubated with peptides for 60 min and phosphatidylbutanol accumulation determined as an index of PLD activity. Data points are mean ± s.e.m. from three independent experiments each performed in triplicate. (Panels A and B from Howl, J. et al., *Chembiochem.*, 4, 1312, 2003. With permission.)

(Figure 16.5, columns D–H) were significant ($p < 0.05$) compared with basal levels of secretion. Both the vector [$Lys^{7N\varepsilon Cys}$]TP10 (Figure 16.5, column D) and TP10-$G_i\alpha3^{346-355}$ (Figure 16.5, column E) were weak secretagogs. Addition of TP10-$G_i\alpha3^{346-355}$ followed by MP produced an additive effect on the magnitude of β-hexoseaminidase secretion (Figure 16.5, column F). Figure 16.5 includes comparative data demonstrating the magnitude of secretory responses to

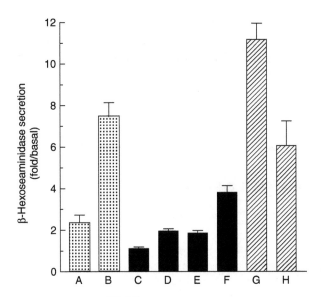

FIGURE 16.5 Translocated $G_i\alpha3^{346-355}$ does not modulate β-hexoseaminidase secretion from RBL-2H3 cells. This figure compares the fold/basal secretory activities of peptides and peptide combinations (A–H). All data are means ± s.e.m. from at least three experiments each performed in triplicate. (A) MP (10 μM, 15 min); (B) MP (30 μM, 15 min); (C) $G_i\alpha3^{346-355}$ (3 μM, 60 min); (D) [Lys$^{7N\varepsilon Cys}$]TP10 (3 μM, 60 min); (E) TP10-$G_i\alpha3^{346-355}$ (3 μM, 60 min); (F) TP10-$G_i\alpha3^{346-355}$ (3 μM, 45 min)+MP (10 μM, 15 min); (G) TP10-Cys^{0}PKC$^{238-249}$ (3 μM, 60 min); (H) TP10-CB$_1^{401-417}$ (3 μM, 60 min). (From Jones, S. et al., *Biochim. Biophys. Acta* 1745, 207, 2005. With permission.)

TP10-Cys^{0}PKC$^{238-249}$ (Figure 16.5, column G) and TP10-CB$_1^{401-417}$ (Figure 16.5, column H) as previously described. The secretory efficacies of TP10-$G_i\alpha3^{346-355}$ and the unconjugated vector [Lys$^{7N\varepsilon Cys}$]TP10 were comparable and clearly a consequence of the MP-containing sequence of the carboxyl terminal of TP10. These observations indicate that $G_i\alpha3^{346-355}$ does not directly influence the biochemical pathway leading to the exocytosis of β-hexoseaminidase or attenuate the secretory activity of MP.

16.3.5 INTRACELLULAR TRANSLOCATION OF $G_i\alpha3^{346-355}$ ACTIVATES A SIGNAL TRANSDUCTION CASCADE THAT TERMINATES IN THE DUAL PHOSPHORYLATION OF P42/44 MAPK

Intracellular delivery of $G_i\alpha3^{346-355}$ did not directly influence the biochemical pathway leading to the exocytosis of β-hexoseaminidase or abrogate the secretory activity of MP. We have recently reported that MP and a range of structural homologues induce the dual phosphorylation of p42/44 MAPK in U373MG cells.[17] As secretory events complicate the temporal analysis of MAPK activation in RBL-2H3,[17] the studies presented here also used the U373MG astrocytoma line. Therefore, and with reference to previous observations using peptides derived from

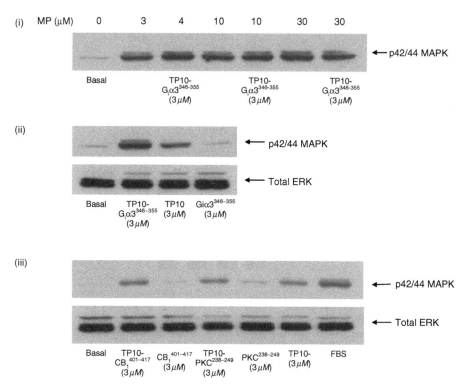

FIGURE 16.6 Intracellular delivery of $G_i\alpha3^{346-355}$, a peptide corresponding to the MP-G_i interface, using the cell penetrant vector [Lys^{7}N$^{\varepsilon Cys}$]TP10. U373MG cells were pre-incubated with peptides, vector conjugated to cargo, cargo alone, and the [Lys^{7}N$^{\varepsilon Cys}$]TP10 vector alone (designated here as TP10) for 30 min at a concentration of 3 μM followed by MP or vehicle alone for a further 30 min. Panel (i): MP at the concentrations indicated produced a marked increase in levels of dually phosphorylated p42/44 MAPK which was unaffected by treatment with TP10-$G_i\alpha3^{346-355}$ (3 μM). Panel (ii): Treatment of unstimulated U373MG cells with TP10-$G_i\alpha3^{346-355}$ increased the dual phosphorylation of p42/44 MAPK sixfold and was comparable to signal intensites mediated by 30 μM MP (5.8-fold) (panel [i]) and FBS (7.7-fold) (panel [iii]). The cell penetrant peptide vector [Lys^{7}N$^{\varepsilon Cys}$]TP10 produced a minor increase in levels of dually phosphorylated p42/44 MAPK (2.7-fold), whereas $G_i\alpha3^{346-355}$ was inactive and gave a 0.8-fold stimulation. Panel (iii): Peptides previously shown to be biologically active secretagogs, following intracellular delivery by the [Lys^{7}N$^{\varepsilon Cys}$]TP10 vector (TP10-Cys^{0}PKC$^{238-249}$ and CB$_1^{401-417}$),[1] had no effect on levels of dually phosphorylated p42/44 MAPK. [Lys^{7}N$^{\varepsilon Cys}$]TP10 cell penetrant chimeras produced a minor enhancement in signal intensity comparable to that of [Lys^{7}N$^{\varepsilon Cys}$]TP10 alone, whilst the cargoes alone were inactive. Total levels of p42/44 MAPK, designated here as total ERK, remained unaltered (panels [ii] and [iii]). (From Jones, S. et al., *Biochim. Biophys. Acta* 1745, 207, 2005. With permission.)

the $G_i\alpha3$ carboxyl-terminal,[14,22,31] we anticipated that the intracellular delivery of the predicted peptide sequence corresponding to the MP-G_i interface ($G_i\alpha3^{346-355}$) may influence MP-induced p42/44 MAPK activation in U373MG cells. Figure 16.6 clearly shows that intracellular delivery of $G_i\alpha3^{346-355}$ as a [Lys^{7}N$^{\varepsilon Cys}$]TP10 chimeric construct (TP10-$G_i\alpha3^{346-355}$) did not abrogate MP-induced phosphorylation of p42/44 MAPK. However, an enhanced signal intensity was evident following the introduction of the TP10-$G_i\alpha3^{346-355}$ chimera to unstimulated cells, giving a sixfold stimulation. This result clearly demonstrates that intracellular delivery of $G_i\alpha3^{346-355}$ promotes dual phosphorylation of p42/44 MAPK. Control experiments indicated that the [Lys^{7}N$^{\varepsilon Cys}$]TP10 vector weakly activated p42/44 MAPK phosphorylation (2.7-fold), whilst the peptide cargo $G_i\alpha3^{346-355}$ alone had no effect on basal levels of phosphorylation, giving a 0.8-fold stimulation. Likewise, p42/44 MAPK phosphorylation was not increased by intracellular delivery of control peptides derived from the regulatory domain of protein kinase C (Cys^{0}PKC$^{238-249}$)[1,24] and the third intracellular loop of the CB$_1$ receptor (CB$_1^{401-417}$),[1,20] observations that confirmed the specificity of the $G_i\alpha3^{346-355}$ peptide.

16.3.6 TP10 AND TP10 CHIMERAE DO NOT REDUCE CELLULAR VIABILITY

We have reported that MP and a range of structural analogs induce cell death of U373MG cells.[17] These cytotoxic actions of MP are not a result of random pore formation, but are most probably mediated by the ability of MP to target and modulate the activity of specific intracellular signaling proteins in this cell line.[32] As previously mentioned, one such protein target is p42/44 MAPK. As an effective peptidyl delivery system, TP10 should confer effective translocation properties without altering cellular viability. Given that TP10-$G_i\alpha3^{346-355}$ promotes dual phosphorylation of p42/44 MAPK and concentrations of MP that activate p42/44 MAPK (3–30 μM) significantly reduce cellular viability,[17] in addition to MP being a component of TP10, we determined whether exposure of U373MG to TP10-$G_i\alpha3^{346-355}$ and its components, might also reduce cellular viability. Similarly, we measured the viability of RBL-2H3 cells following treatment with [Lys^{7}N$^{\varepsilon Cys}$]TP10. U373MG cell viability readings were measured following treatment with peptides for 4 h, a time point sufficient to observe the cytotoxic effects of MP on U373MG cells. RBL-2H3 cell viability was measured following 60 min of treatment with [Lys^{7}N$^{\varepsilon Cys}$]TP10 (3 μM), a time period and concentration that were used for effective cellular penetration. Figure 16.7 clearly demonstrates that [Lys^{7}N$^{\varepsilon Cys}$]TP10 and TP10-$G_i\alpha3^{346-355}$ did not alter the viability of both cell lines tested and only MP produced a significant change in cellular viability (p = .0022, Mann–Whitney test).

16.4 DISCUSSION

16.4.1 TP10 AS AN EFFICIENT VECTOR FOR THE INTRACELLULAR DELIVERY OF PEPTIDES

The aim of this study was to develop a cell penetrant, inert, nontoxic system to effect the intracellular delivery of bioactive peptide cargoes that modulate signal transduction. Kinetic analysis data presented here provide an estimate of the rate of internalization of TP10 ($t_{0.5}$) of approximately 3 min, which compares to previous

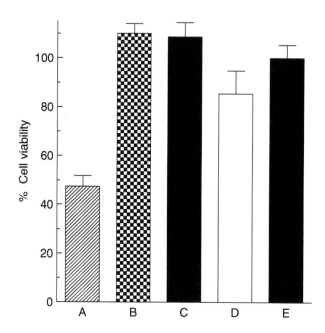

FIGURE 16.7 TP10 and TP10 chimerae do not alter the viability of U373MG and RBL-2H3 cells. U373MG cells were incubated with peptides in accordance with the following conditions: (A) MP (30 μM, 4 h); (B) TP10-G$_i$3$\alpha^{346-355}$ (3 μM, 4 h); (C) [Lys$^{7N\epsilon Cys}$]TP10 (3 μM, 4 h); (D) G$_i$3$\alpha^{346-355}$ (3 μM, 4 h). RBL-2H3 cells were exposed to [Lys$^{7N\epsilon Cys}$]TP10 (3 μM) for 60 min (E). Viability data (mean $\pm$ s.e.m., n=6) are expressed as a percentage of viable cells treated with vehicle alone.

estimates of 8 min in Bowes melanoma,[13] and is faster than the endocytotic internalization of the Tat CPP.[33]

The [CFDA]G$_i$α3$^{346-355}$ construct was shown to be an effective tool to determine the translocation kinetics of TP10. CFDA is nonfluorescent when exogenously added to cells, but is rapidly metabolized by intracellular esterases to unmask the fluorescein chromophore.[34] This feature is appropriate for the labeling of both CPPs and cargoes that are differentially localized following intracellular delivery, since metabolism of CFDA occurs within the cytosol, organelles, and other intracellular compartments.[29]

An ideal CPP should possess high translocation efficiency at concentrations that have little or no influence on the biology or viability of eukaryotic cells. MP, a component of TP10, reduces U373MG cell viability in a concentration-dependent manner and promotes dual phosphorylation of p42/44 MAPK in this cell line.[17] Activation of p42/44 MAPK can influence cell fate to promote proliferation, differentiation, or apoptosis; disparate outcomes that are seemingly dependent upon intensity and duration of intra- and extracellular signals.[35–37] Thus, it was important to determine whether TP10 and the MAPK activator TP10-G$_i$3α$^{346-355}$ influenced cellular viability. Our data indicate that neither TP10-G$_i$α3$^{346-355}$ nor [Lys$^{7N\epsilon Cys}$]TP10 reduced the viability of U373MG cells at a concentration (3 μM)

that provides effective cellular delivery. Moreover, cellular viability was maintained for a minimum period of 4 h of exposure to peptides, a time course much longer than that required to observe the biological effects of TP10-translocated peptide cargoes.

16.4.2 Biological Activities of Peptide Cargoes

Translocated cargoes displayed sequence-dependent biological activities. Table 16.1 summarizes the biological activities of each peptide tested. Intracellular translocation of $CB_1^{401-417}$ into RBL-2H3 cells promoted a concentration-dependent exocytosis of β-hexoseaminidase, whilst translocation of the Ser^{416}-substituted analog Cys^0-[Ser^{416}]$CB_1^{401-417}$, previously reported to exhibit intrinsic biological activity,[18–20] demonstrated only weak secretory properties. It is possible that Cys^{416} is an essential pharmacophore for the interaction of $CB_1^{401-417}$ with its intracellular target in RBL-2H3 cells. However, it is also probable that the Cys^0-extension and the amidated carboxyl terminus of Cys^0-[Ser^{416}]$CB_1^{401-417}$ reduces interaction with potential intracellular targets that modulate secretion, such as heterotrimeric G-proteins.

Similarly, intracellular delivery of $Cys^0PKC^{238-249}$ into RBL-2H3 cells as a [$Lys^{7NεCys}$]TP10 chimeric construct promoted a concentration-dependent secretion of β-hexoseaminidase, whilst being the most efficacious secretagog tested. Additional experiments carried out in our laboratory attempted to establish the intracellular protein targets for TP10-$Cys^0PKC^{238-249}$; namely, whether the secretory properties of TP10-$Cys^0PKC^{238-249}$ were a consequence of PKC activation. However, neither downregulation of Ca^{2+}-sensitive isoforms of PKC with phorbol 12-myristate 13-acetate (PMA, 100 nM) nor selective inhibition of the catalytic activity of Ca^{2+}-insensitive isoforms of PKC with R0-32-0432 (1 μM) attenuated the secretory efficacy of TP10-$Cys^0PKC^{238-249}$.[1] Thus, it may be concluded that the intracellular target for $Cys^0PKC^{238-249}$ is not a PKC isoform. Interestingly, a database search (SIB BLAST network service) identified a $Cys^0PKC^{238-249}$ homologue (SPLRVELWDWDM) that is located within the C2 domain of rasGAP (GTPase activating protein)-related protein.[38] It may therefore be postulated that TP10-$Cys^0PKC^{238-249}$ modulates the normal function of C2 domains, which is to promote calcium-dependent phospholipid binding.

TABLE 16.1
Summary of the Biological Activities of Peptide Cargoes

	Biological Activity	
Peptide Cargo	Secretion	p42/44 MAPK Stimulation
$Cys^0PKC^{238-249}$ CRRLSVEIWDWDL	Active	Inactive
$CB_1^{401-417}$ RSKDLRHAFRSMFPSCE	Active	Inactive
Cys^0[Ser^{416}]$CB_1^{401-417}$ CRSKDLRHAFRSMFPSSE	Inactive	Not determined
$G_i3\alpha^{346-355}$ KNNLKECGLY	Inactive	Active

It is likely that translocated $Cys^0PKC^{238-249}$ targets an alternative component of the regulated secretory response in RBL-2H3 cells that is not PKC. Evidence to support this contention is demonstrated by the observation that translocated $Cys^0PKC^{238-249}$ activated PLD, an enzyme intimately involved in the antigen-stimulated secretory response of this cell line.[39] PLD hydrolyzes phosphatidylcholine to generate phosphatidate (PA). Data from our laboratory[1] have indicated that inhibition of the synthesis of PA by PLD does not attenuate TP10-$Cys^0PKC^{238-249}$-induced secretion. Thus, the exocytotic response induced by TP10-$Cys^0PKC^{238-249}$ is not dependent upon de novo synthesis of PA. Therefore, PLD may mediate secretion of β-hexoseaminidase from RBL-2H3 cells by enzymatic modification of lipid membranes rather than as a consequence of PA synthesis. Finally, it is probable that TP10-$Cys^0PKC^{238-249}$ either directly stimulates PLD or interacts with known modulators of PLD, such as small monomeric GTPases and ADP-ribosylation factor 6 (ARF6).[40]

Translocated $G_i\alpha3^{346-355}$ was without effect on β-hexoseaminidase exocytosis and did not influence the secretory activity of MP. These findings are in contrast to previous reports that have used similar peptides to block MP-induced secretory events.[14,22,23] One explanation of these data could be that translocated $G_i\alpha3^{346-355}$ does not inhibit MP binding to the $G_i\alpha3$ protein. Alternatively, the lack of secretory activity in response to translocated $G_i\alpha3^{346-355}$ could indicate that the G_i3 protein α subunit (or at least its carboxyl terminal) does not contribute to the exocytosis of secretory lysosomes in RBL-2H3 cells. Given the reported specificity and selectivity of action of peptides derived from the extreme carboxyl-termini of G-protein α subunits[22] in other mast cell models, this interpretation would suggest that the secretion of β-hexoseaminidase from RBL-2H3 is not a $G_i\alpha3$-dependent phenomenon. Moreover, recent observations have indicated that $G_i\alpha3$ is not the elusive exocytotic G-protein (G_E) responsible for regulated secretory events in mast cells.[41]

Intracellular delivery of $G_i\alpha3^{346-355}$ did not abrogate MP-mediated phosphorylation of p42/44 MAPK in U373MG cells. However, translocated $G_i\alpha3^{346-355}$ appeared to display intrinsic biological activity in U373MG astrocytic tumor cells by promoting a kinase cascade that terminated in the phosphorylation of p42/44 MAPK. These results indicate that translocated $G_i\alpha3^{346-355}$ binds and activates at least one intracellular target protein. Support for this contention is provided by previous studies indicating that a peptide derived from the carboxyl domain of transducin ($G_t\alpha^{293-314}$) binds and activates cGMP-phosphodiesterase, a novel intracellular effector of rhodopsin-transducin signaling.[42,43]

Our findings provide direct evidence that a component of the MAPK signaling pathway is a molecular target for activated $G_i\alpha3$ in U373MG cells. This interpretation is in accordance with extensive reports that both α and βγ subunits of heterotrimeric G-proteins can activate signaling events that lead to the dual phosphorylation of p42/p44 MAPK.[44] The G-protein activator, MP, also induces dual phosphorylation of p42/44 MAPK so it is plausible that both the translocated $G_i\alpha3^{346-355}$ peptide and MP activate a common signaling target upstream of p42/44 MAPK phosphorylation. Further support for this contention is strengthened by reports that MP binds both carboxyl and amino termini of G-protein alpha subunits[30,45] and Western blot analysis confirms that $G_i\alpha3$ is expressed in U373MG cells.[32] However, translocated $G_i\alpha3^{346-355}$ does not mimic other biological actions of MP, such as induction of U373MG cell death.[17] It is speculated that MP could mediate its cytotoxic effects in U373MG cells by binding to mitochondrial membranes to promote the release of apoptotic proteins such as cytochrome C.[32,46]

Data presented herein indicate that TP10-mediated translocation is an effective methodology for the delivery of bioactive cargoes that modulate intracellular signal transduction. Such an approach provides a general strategy to both study and manipulate cell signaling pathways, transduced via G-proteins and divergent kinase cascades, that represent potential drug targets. Our current investigations are focused on the identification of intracellular protein targets that bind bioactive peptide cargoes by adopting a combination of affinity purification and mass spectrometry.

ACKNOWLEDGMENTS

The authors would like to thank Michelle Farquhar for her considerable contribution to the presented work and Wiley-VCH Verlag GmbH & Co. and Elsevier for permission granted to reproduce published figures.

REFERENCES

1. Howl, J., Jones, S., and Farquhar, M., Intracellular delivery of bioactive peptides to RBL-2H3 cells induces beta-hexosaminidase secretion and phospholipase D activation, *Chembiochem.*, 4, 1312, 2003.
2. Lindgren, M. et al., Cell-penetrating peptides, *Trends Pharmacol. Sci.*, 21, 99, 2000.
3. Shibagaki, N. and Udey, M.C., Dendritic cells transduced with protein antigens induce cytotoxic lymphocytes and elicit antitumor immunity, *J. Immunol.*, 168, 2393, 2002.
4. Begley, R. et al., Biodistribution of intracellularly acting peptides conjugated reversibly to Tat, *Biochem. Biophys. Res. Commun.*, 318, 949, 2004.
5. Dunican, D.J. and Doherty, P., Designing cell-permeant phosphopeptides to modulate intracellular signaling pathways, *Biopolymers*, 60, 45, 2001.
6. Fawell, S. et al., Tat-mediated delivery of heterologous proteins into cells, *Proc. Natl Acad. Sci. U.S.A.*, 91, 664, 1994.
7. Nagahara, H. et al., Transduction of full-length TAT fusion proteins into mammalian cells: TAT-p27Kip1 induces cell migration, *Nat. Med.*, 4, 1449, 1998.
8. Schwarze, S.R. et al., In vivo protein transduction: Delivery of a biologically active protein into the mouse, *Science*, 285, 1569, 1999.
9. Pooga, M. et al., Cell penetrating PNA constructs regulate galanin receptor levels and modify pain transmission in vivo, *Nat. Biotech.*, 16, 857, 1998.
10. Astriab-Fisher, A. et al., Conjugates of antisense oligonucleotides with the Tat and antennapedia cell-penetrating peptides: Effects on cellular uptake, binding to target sequences, and biologic actions, *Pharm. Res.*, 19, 744, 2002.
11. Fisher, L. et al., Cellular delivery of a double-stranded oligonucleotide NfkappaB decoy by hybrization to complementary PNA linked to a cell-penetrating peptide, *Gene Ther.*, 11, 1264, 2004.

12. Pooga, M. et al., Cell penetration by transportan, *FASEB J.*, 12, 67, 1998.

13. Soomets, U. et al., Deletion analogues of transportan, *Biochim. Biophys. Acta.*, 1467, 165, 2000.

14. Aridor, M. et al., Activation of exocytosis by the heterotrimeric G protein G_i3, *Science*, 262, 1569, 1993.

15. Higashijima, T. et al., Mastoparan, a peptide toxin from wasp venom, mimics receptors by activating GTP-binding regulatory proteins (G proteins), *J. Biol. Chem.*, 263, 6491, 1988.

16. Farquhar, M. et al., Novel mastoparan analogs induce differential secretion from mast cells, *Chem. Biol.*, 9, 63, 2002.

17. Jones, S. and Howl, J., Charge delocalisation and the design of novel mastoparan analogues: Enhanced cytotoxicity and secretory efficacy of [Lys^5, Lys^8, Aib^{10}]MP, *Regul. Pept.*, 121, 121, 2004.

18. Howlett, A.C. et al., Characterization of CB_1 cannabinoid receptors using receptor peptide fragments and site-directed antibodies, *Mol. Pharmacol.*, 53, 504, 1998.

19. Mukhopadhyay, S. et al., Regulation of G_i by the CB_1 cannabinoid receptor C-terminal juxtamembrane region: Structural requirements determined by peptide analysis, *Biochemistry*, 38, 3447, 1999.

20. Mukhopadhyay, S. et al., The CB_1 cannabinoid receptor juxtamembrane C-terminal peptide confers activation to specific G proteins in brain, *Mol. Pharmacol.*, 57, 162, 2000.

21. Itoh, H. et al., Presence of three distinct molecular species of G_i protein α subunit. Structure of rat cDNAs and human genomic DNAs, *J. Biol. Chem.*, 263, 6656, 1988.

22. Kreft, M. et al., The heterotrimeric G_i3 protein acts in slow but not in fast exocytosis of rat melanotrophs, *J. Cell. Sci.*, 112, 4143, 1999.

23. Vitale, N. et al., Trimeric G proteins control regulated exocytosis in bovine chromaffin cells: Sequential involvement of G_o associated with secretory granules and G_i3 bound to the plasma membrane, *Eur. J. Neurosci.*, 8, 1275, 1996.

24. Ron, D. and Mochly-Rosen, D., An autoregulatory region in protein kinase C: The pseudoanchoring site, *Proc. Natl Acad. Sci. U.S.A.*, 92, 492, 1995.

25. Souroujon, M.C. and Mochly-Rosen, D., Peptide modulators of protein–protein interactions in intracellular signaling, *Nat. Biotech.*, 16, 919, 1998.

26. Howl, J. and Wheatley, M., V1a vasopressin receptors: Selective biotinylated probes, *Methods Neurosci.*, 13, 281, 1993.

27. Wakelam, M.J.O., Hodgkin, M.A., and Martin, A., The measurement of phospholipase D-linked signaling in cells, *Methods Mol. Biol.*, 41, 271, 1995.

28. Carmichael, J. et al., Evaluation of a tetrazolium-based semiautomated colorimetric assay: Assessment of chemosensitivity testing, *Cancer Res.*, 47, 936, 1987.

29. Jones, S. et al., Intracellular translocation of the decapeptide carboxyl terminal of $G_i3\alpha$ induces the dual phosphorylation of p42/p44 MAP kinases, *Biochim. Biophys. Acta*, 1745, 207, 2005.

30. Weingarten, R. et al., Mastoparan interacts with the carboxyl terminus of the alpha subunit of G_i, *J. Biol. Chem.*, 265, 11044, 1990.

31. Ferry, X. et al., Activation of $\beta\gamma$ subunits of G_i2 and G_i3 proteins by basic secretagogues induces exocytosis through phospholipase C β and arachidonate release through phospholipase C γ in mast cells, *J. Immuno.*, 167, 4805, 2001.

32. Jones, S., *Design, Synthesis and Evaluation of Receptor Mimetic Peptides as Modulators of Signal Transduction and Cytotoxic Agents*, Ph.D. thesis, University of Wolverhampton, UK, 2005.

33. Richard, J.P. et al., Cell-penetrating peptides. A reevaluation of the mechanism of cellular uptake, *J.Biol.Chem.*, 278, 585, 2003.
34. Kao, J.P., Practical aspects of measuring $[Ca^{2+}]$ with fluorescent indicators, *Methods Cell Biol.*, 40, 155, 1994.
35. Sewing, A. et al., High-intensity Raf signal causes cell cycle arrest mediated by p21Cip1, *Mol. Cell. Biol.*, 17, 5588, 1997.
36. Pumiglia, K.M. and Decker, S.J., Cell cycle arrest mediated by the MEK/mitogen-activated protein kinase pathway, *Proc. Natl Acad. Sci. U.S.A.*, 94, 448, 1997.
37. Murphy, L.O. et al., Molecular interpretation of ERK signal duration by immediate early gene products, *Nat. Cell Biol.*, 4, 556, 2002.
38. Allen, M. et al., Restricted tissue expression pattern of a novel human rasGAP-related gene and its murine ortholog, *Gene*, 218, 17, 1998.
39. Brown, F.D. et al., Phospholipase D1 localises to secretory granules and lysosomes and is plasma-membrane translocated on cellular stimulation, *Curr. Biol.*, 8, 835, 1998.
40. Powner, D.J., Hodgkin, M.N., and Wakelam, M.J.O., Antigen-stimulated activation of phospholipase D1β by Rac1, ARF6 and PKCα in RBL-2H3 cells, *Mol. Biol. Cell*, 13, 1252, 2002.
41. Moqbel, R. and Lacey, P., Exocytotic events in eosinphils and mast cells, *Clin. Exp. Allergy*, 29, 1017, 1999.
42. Rarick, H.M., Artemyev, N.O., and Hamm, H.E., Site of effector activation on the α subunit of transducin, *Science*, 256, 1031, 1992.
43. Artemyev, N.O. et al., Sites of interaction between rod G-protein α-subunit and cGMP phosphodiesterase γ-subunit, *J. Biol. Chem.*, 267, 25067, 1992.
44. Liebmann, C., Regulation of MAP kinase activity by peptide receptor signalling pathway: Paradigms of multiplicity, *Cell Signal.*, 13, 777, 2001.
45. Higashijima, T., Burnier, J., and Ross, E., Regulation of G_i and G_o by mastoparan, related amphiphilic peptides, and hydrophobic amines. Mechanism and structural determinants of activity, *J. Biol. Chem.*, 265, 14176, 1990.
46. Ellerby, H. et al., Establishment of a cell-free system of neuronal apoptosis: Comparison of premitochondrial, mitochondrial, and postmitochondrial phases, *J. Neurosci.*, 17, 6165, 1997.

17 SynB Vectors as Drug Carriers

Jamal Temsamani

CONTENTS

17.1 INTRODUCTION

A large number of hydrophilic molecules, such as peptides, proteins, and oligonucleotides, are poorly taken up by cells since they do not efficiently cross the lipid bilayer of the plasma membrane. This is considered to be a major limitation for their use as therapeutic agents in biomedical research and in the pharmaceutical industry. In particular, it has been widely accepted that peptide neuromodulators fail to significantly affect their target cells within the brain when administered peripherally. This is likely to be due to the existence of the blood–brain barrier (BBB), a complex biological interface, which prevents transport of most drugs from the vasculature into the brain parenchyma.

During the last decade, several cell-penetrating peptides (CPPs) have been described, such as SynB vectors, penetratin, transportan, and Tat, that allow the intracellular delivery of polar, biologically active compounds in vitro and in vivo.[1–3] These peptides, belonging to various families, are heterogeneous in size (10–25 amino acids) and sequence. However, all these peptides possess multiple positive charges and some of them share common features such as important theoretical hydrophobicity and helical moment (reflecting the peptide amphipathicity), and the ability to interact with lipid membrane and to adopt a significant secondary structure upon binding to lipids. The facility with which they cross the membrane into the cytoplasm, even when carrying hydrophilic molecules, has provided a new and powerful tool in biomedical research.[4,5] An even more difficult task was to use these peptide vectors to deliver drugs across the BBB. This chapter will emphasize the use of SynB vectors for intracellular and brain delivery.

17.2   SYNB VECTORS

SynB vectors are a new family of vectors derived from the antimicrobial peptide protegrin 1 (PG-1), an 18 amino acid peptide originally isolated from porcine leucocytes (Figure 17.1).[6–8] As previously reported, the PG-1 peptide interacts with, and forms pores in, the lipid matrix of bacterial membranes.[9,10] Since it has been shown that the pore formation capability of PG-1 depends on its cyclization, various linear analogs of PG-1, lacking the cysteine residues, were designed (Figure 17.1).[10] These linear peptides (SynB vectors) are able to interact with the cell surface and cross the plasma membrane while their membrane-disrupting activity has been lost.[11] Furthermore, the internalization of these peptides into cells does not appear to be dependent on a chiral receptor since the D-enantiomer form penetrates as efficiently as the parent peptide (L-form), and retro-inverso sequences exhibit identical penetrating activity. To explore the mechanism further, we first determined the influence of low temperature and metabolic inhibition on the cellular uptake of SynB peptides.[12] We observed that the intracellular accumulation of these peptides is abolished at low temperature or in ATP-deprived cells. Further mechanistic insights were provided by an examination of the influence of endocytosis inhibitors on the internalization of the SynB peptides. In control cells, the vesicular accumulation of peptides and Lucifer Yellow (LY) resulted in fluorescence staining in the perinuclear region, characteristic of late endosomes and lysosomes.[13] In the presence of microtubule-disrupting reagents, the cellular accumulation of peptides was reduced and, as found with LY, peptides remained inside vesicles localized to the cell periphery rather than the perinuclear region. Further support for this

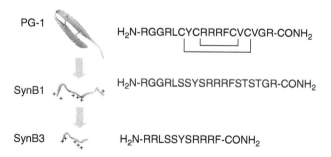

FIGURE 17.1 Primary sequence of protegrin (PG-1), SynB1, and SynB3.

localization came from experiments where cells were treated either with NH_4Cl or chloroquine; a marked change in the morphology of fluorescently labeled vesicles was seen, confirming that the peptides were within an acidic compartment.[12] In addition, these two treatments inhibited peptide accumulation into the cells, as measured by flow cytometry. Because the formation of endosome carrier vesicles is dependent on the acidification of their luminal interiors, it is likely that the neutralization of the endosomal pH blocks the transfer of peptide to late endosomes, as observed elsewhere for fluid-phase markers.[14] In conclusion, these results suggest that SynB peptides enter via a common adsorptive-mediated endocytosis pathway rather than a translocation mechanism.

These linear analogs were the starting point for developing a new potent strategy for drug delivery into complex biological membranes, such as the BBB. Further optimizations have led to the development of peptides that have improved properties.

17.3 ENHANCEMENT OF CELL DELIVERY

17.3.1 INTRODUCTION TO INTRACELLULAR DELIVERY

The ability to deliver large hydrophilic molecules, such as peptides, proteins, nucleic acids, and large particles, into cells is a big challenge due to the bioavailability restriction imposed by the cell membrane. The plasma membrane of the cell forms an effective barrier that limits the intracellular uptake to those sufficiently nonpolar and smaller in size (less than 600 Da). The discovery of CPPs that can translocate efficiently across the plasma membrane of cells has, hence, opened up fascinating perspectives for the development of cell delivery. Several reviews have described the successful transport of cargoes using CPPs both in vitro and in vivo.[4,15] We will therefore concentrate on a few examples that highlight the application of SynB vectors.

17.3.2 DELIVERY OF SMALL MOLECULES

Doxorubicin, an anthracycline, is one of the most widely used anticancer chemotherapeutic agents. It has a strong antiproliferative effect on many solid tumors and it is currently used for the treatment of breast, ovarian, and lung cancers.[16] However, its clinical use is limited by clinical drug resistance. Limited uptake or increased drug excretion out of tumor cells is often noted when tumors acquire resistance to anticancer drugs. The mechanism underlying chemoresistance has been partly associated with overproduction of the drug-efflux pump, called gp 170 or P-gp.[17] To overcome this problem, we tested the capability of SynB vectors that are able to cross cellular membranes to deliver doxorubicin in P-gp-expressing cells.[18] The cell uptake and antitumor effect of peptide-conjugated doxorubicin were tested in the human erythroleukemic sensitive (K562) and resistant (K562/ADR) cells. In resistant cells, doxorubicin had a lower uptake than in sensitive cells (Figure 17.2). Interestingly, the uptake of the conjugated doxorubicin was similar

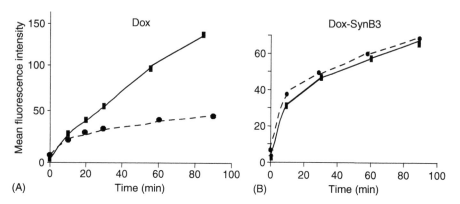

FIGURE 17.2 Cell uptake of free and vectorized doxorubicin in sensitive cells (K562, ■) and resistant cells (K562/ADR, ●). The uptake was measured at different times using flow cytometry.

in both resistant and sensitive cells. The conjugate showed potent dose-dependent inhibition of cell growth against K562/ADR cells, as compared with doxorubicin alone. Doxorubicin exhibited IC_{50} concentrations in the resistant cells that were significantly higher than those observed for vectorized doxorubicin. After treatment of the resistant cells with verapamil, the intracellular levels of doxorubicin were markedly increased and consequent cytotoxicity was improved. In contrast, treatment of resistant cells with verapamil did not cause any further enhancement in the cell uptake nor in the cytotoxic effect of the conjugated doxorubicin, indicating that the conjugate bypasses the P-gp.[18] These results indicate that vectorization of doxorubicin with peptide vectors is effective in overcoming multidrug resistance.

We have also coupled camptothecin and assessed its cell uptake in K562 cells. Camptothecin is an anticancer agent that kills cells by converting DNA topoisomerase I into a DNA-damaging agent. Although camptothecin and its derivatives are now being used to treat tumors in a variety of clinical protocols, the low water solubility of the drug and its unique pharmacodynamics and reactivity in vivo limit its delivery to cancer cells. Camptothecin was coupled to a SynB vector and its cell uptake was measured using fluorescent microscopy and flow cytometry. Figure 17.3 shows that SynB vector significantly enhances the cell uptake of camptothecin.

17.3.3 DELIVERY OF BIOLOGICALLY ACTIVE PEPTIDES AND PROTEINS

The discovery of CPPs allows the design of constructs reaching the interior of cells and interacting with intracellular proteins.

An interesting application of the CPPs is the design of cytotoxic T lymphocytes-inducing vaccines for the treatment or prevention of infectious and malignant diseases. The cytotoxic T lymphocytes (CTL) are probably the most effective mechanism that the immune system has to eliminate abnormal cells expressing viral

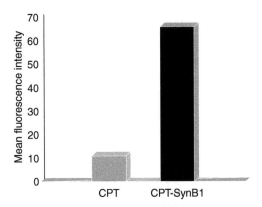

FIGURE 17.3 Cell uptake of free (CPT) and vectorized (CPT-SynB1) camptothecin following 60-min incubation in K562 cells. The uptake was measured using flow cytometry.

protein of tumor-associated antigen. The use of conventional vaccines, such as killed pathogens or recombinant proteins, suffers from the inability of antigen-presenting cells (APCs) to trigger a CTL response via the MHC classes I pathway. This problem has been addressed by conjugating peptide and protein antigens to SynB vectors.

We have investigated the ability of these peptide vectors to deliver antigens into target cells.[19] The antigens used consisted of a T cell epitope from a *Mycobacterium tuberculosis* antigen (Mtb39), $Mtb_{250-258}$ (Mtbp) and also a peptide from influenza nucleoprotein (flu $NP_{147-155}$) (NPp) and a recombinant protein, also from *M. tuberculosis*, Mtb8.4.[20-22]

In vitro cell culture studies indicated that the levels of cell uptake of the two 9-mer peptide antigens, $NP_{147-155}$ and $Mtb_{250-258}$, and the *M. tuberculosis* antigen, Mtb8.4 protein, into K562 was low. However, when these peptide and protein antigens were covalently linked to the SynB vectors, the uptake increased significantly (Figure 17.4A). The enhancement of maximal internalization was about two- to eightfold, depending on the peptide vector used.[19]

Furthermore, selected SynB vectors, when conjugated to these same antigens, were used as immunogens. Mice were immunized with the various formulations of SynB conjugates and the CTL response was measured.[19] The free peptides or protein antigens did not induce a significant CTL activity compared to naïve mice. However, conjugation of these antigens with SynB vectors led to a significant enhancement in the CTL response (Figure 17.4B).

An important consideration when delivering peptides, formulated in any way, is whether the induced T cell is capable of recognizing cells expressing and presenting the naturally processed epitope. This is a necessary requirement for induction of antitumor or antipathogen responses in vivo. For the Mtb39 system, we addressed this issue because we were able to immunize mice with the SynB4/Mtbp, then restimulate in vitro with EL4 tumor cells expressing the entire Mtb39 antigen (E1 cell line). After two rounds of in vitro stimulation, CTL assays were performed using E1 cells (and EL4 cells as negative control) as targets. Our data indicate

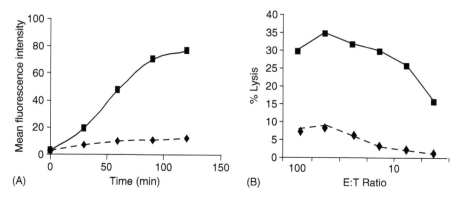

FIGURE 17.4 Cellular delivery of a Mtb8.4 protein antigen (8.4 kDa). (A) Cell uptake of free and vectorized fluorescently labeled protein in K562 cells. The cell uptake was measured using flow cytometry. (B) Identification of Mtb8.4-specific CTL after immunization with 1 μg of vectorized or free protein (intradermal). Mice were immunized on days 0 and 21 and then on day 42 spleens were harvested and responses were measured as described previously. (From Day, F.H. et al., *J. Immunol.* 170, 1498, 2003.) ◆, Mtb8.4; ■, Mtb8.4-SynB4.

that CTL were identified that could indeed lyse these Mtb39-expressing targets.[19] Similar results were obtained when other CPPs, such as penetratin and Tat, were used.[23–25]

17.4 ENHANCEMENT OF BRAIN DELIVERY

The BBB poses a formidable obstacle to drug therapy for the central nervous system (CNS). It consists of a monolayer of polarized endothelial cells connected by complex tight junctions that separates the blood compartment from the extracellular fluid compartment of the brain parenchyma.[26] The most important factors determining the extent to which a molecule will be delivered from the blood into the CNS are lipid solubility, molecular mass, and charge. Therefore, based simply on lipid solubility and molecular mass, about 95% of the CNS drugs will almost certainly be impeded by the BBB.[27]

17.4.1 Strategies for Overcoming the BBB

To overcome the limited access of drugs to the brain, at least three strategies have been developed that achieve BBB penetration:[28–30]

1. Neurosurgery-based strategies, which bypass the BBB by means of intraventricular drug infusion, intracerebral infusion, or disruption of the BBB.
2. Pharmacology-based strategies, some examples of which employ lipidation or chemical modification of the drug to improve its ability to diffuse across the BBB.

3. Physiology-based strategies, which take advantage of BBB nutrient carriers or specific receptors, mediating transport via these transporter systems. One example has been to conjugate the therapeutic drug with a protein or a monoclonal antibody that gains access to the brain by either receptor or adsorptive-mediated transcytosis (e.g., transferrin receptor-mediated transfer).

However, problems have been encountered with many of these approaches. For example, the risk of infection and neurosurgical costs limit the use of neurosurgery-based procedures, whereas increasing the lipophilicity of a drug decreases its solubility in serum and may enhance its accumulation in other nontarget sites. Therefore, new and noninvasive methods are urgently needed. An alternative approach, which overcomes many of the drawbacks of the existing methods, is the use of CPPs. We will focus here on a few examples that highlight the application of SynB vectors. Few other reviews have described the successful brain transport of cargoes using other CPPs.[15,31]

17.4.2 Delivery of Small Molecules

The efficacy of SynB vectors to enhance the brain uptake of the anticancer agent doxorubicin was assessed using in situ cerebral perfusion in rats and mice.[32,33] This vectorization of doxorubicin to SynB vectors via a succinate linker significantly enhanced its brain uptake without compromising BBB integrity (Figure 17.5). Interestingly, the SynB-vectorized doxorubicin bypasses the P-glycoprotein, which has been shown to be present in the luminal membrane of the BBB endothelial cells and restricts the brain entrance of a broad number of therapeutic compounds, including cytotoxic drugs.[18] Similar results were obtained after intravenous

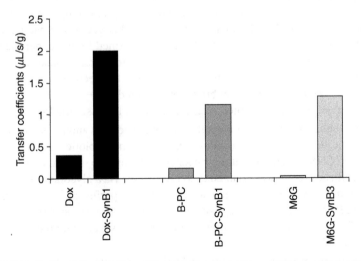

FIGURE 17.5 Brain delivery of small molecules. Transfer coefficients (Kin) of free and vectorized drugs uptake in the right hemisphere of rodents after 60 sec of perfusion in buffer.

administration. The brain concentrations were higher for vectorized doxorubicin compared to that of free doxorubicin.[32] Interestingly, vectorized doxorubicin shows significantly lower levels in the heart, strongly suggesting that cardiotoxicity — the main side effect of doxorubicin — could be reduced using this strategy. The blood clearance of the vectorized doxorubicin was reduced during the first 180 min, allowing the compound to be more exposed to brain and other tissues. The tissue-to-plasma-partition coefficients were calculated and compared with those of free doxorubicin. The calculated tissue distribution advantage (TDA) was found to be > 1 in brain, indicating a more important brain uptake for the vectorized doxorubicin than would have been expected from the observed increase in plasma levels.

In another study, we have coupled paclitaxel with SynB3 vector via a succinate linker in order to improve its brain uptake and chemical properties.[34] The study demonstrated that coupling of paclitaxel with SynB3 significantly enhances its solubility. Paclitaxel is commercially available formulated with polyethoxylated castor oil (Cremophor EL). This excipient is believed to play a major role in the hypersensitivity reactions observed in patients while on paclitaxel therapy.[35] Therefore, using this peptide vector strategy, large amounts of solubilizing agents such as Cremophor EL, polysorbate, and ethanol may not be required, so that side effects typically associated with these solubilizing agents, such as anaphylaxis, hypotension, and flushing, could be reduced. We then evaluated the brain uptake of free and coupled paclitaxel using the in situ mouse brain perfusion method. Although paclitaxel is very lipophilic, we only observed a low brain uptake of paclitaxel ($K_{in} = 0.53 \pm 0.2$ µL/g/s). This corroborates with studies showing that administration of paclitaxel by intravenous administration resulted in low concentrations in the CNS.[36] This low brain permeability could be explained by the efflux activity of P-gp at the BBB. Paclitaxel is actually transported by P-gp expressed at the brain capillaries in the physiological state.[37,38] A significant increase in paclitaxel brain uptake was obtained for the conjugated drug compared to free paclitaxel.[34] This increase in brain uptake obtained for the vectorized paclitaxel might be explained by the translocation properties of the SynB3 vector and also by the fact that vectorized paclitaxel is not recognized by the P-gp. This was furthermore confirmed by the experiment using P-gp deficient mice (mdr1a($-/-$)). We observed that brain uptake of paclitaxel was increased in P-gp-deficient mice, confirming the recognition by the P-gp. Interestingly, vectorized paclitaxel brain uptake in mdr1a($-/-$) mice was identical to that observed in wild-type mice, suggesting that it bypasses the P-gp.[34]

Similar enhancement in brain uptake was obtained with another small molecule: the antibiotic benzyl-penicillin (B-Pc).[39] Beta-lactam antibiotics are often used for treatment of CNS infections, but their poor penetration into the brain does not allow sufficient efficacy. B-Pc was coupled to SynB1 vector via a glycolamidic ester linker and the brain uptake was measured using in situ brain perfusion.[39] The brain uptake of coupled B-Pc showed an average of eightfold increase in comparison to free B-Pc (Figure 17.5). This increase was quite similar for the seven explored gray areas of the rat brain.

Recently, we have shown that vectorization of morphine-6-glucuronide (M6G) by SynB3 can enhance its brain transport.[40] The main metabolism pathway of morphine includes liver glucuronidation to M6G and morphine-3-glucuronide (M3G). M6G is thought to contribute to the pharmacological effects of the parent drug.[41–43] Various clinical trials have used M6G as the therapeutic drug in preference to morphine.[44–46] Antinociception studies in experimental animals have demonstrated that, although M6G and morphine are almost equally potent after systemic administration, the analgesic potency of M6G is more than 100-fold higher than morphine after intracerebroventricular injection, a route of administration that bypasses the BBB in vivo.[41–43] These pharmacological data suggest that the brain penetration of M6G is significantly attenuated relative to that of morphine, probably due to the presence of the glucuronide moiety of M6G, conferring a higher hydrophilic character.

Vectorization of M6G with the SynB3 vector (Syn1001) resulted in a significant improvement in the brain uptake (Figure 17.5). Interestingly, this increase in brain uptake was accompanied by an enhancement in the antinociceptive activity of M6G.[40] We showed by different tests of nociception in animals that the effect of Syn1001 is dose-dependent and was more potent than free M6G or morphine. The ratio of the antinociceptive ED50 of M6G over Syn1001 was approximately 4 on a molar basis. This enhancement was due to the vectorization of M6G since free peptide (SynB3) had no antinociceptive effect and no affinity for the opioid receptors. The fact that vectorized M6G binds to the *mu* receptors in vitro indicates that free M6G does not need to be cleaved from the vector in order to have a pharmacological effect. In this study, M6G was conjugated to the SynB3 vector via a linker containing a disulfide bond. The disulfide-based linker system has been shown to be stable in plasma for several hours though labile in the brain.[47] It is not clear yet in which form Syn1001 binds to its opioid receptors in vivo. Further studies are needed to assess the mechanism and rate of cleavage of vectorized M6G within the brain.

Respiratory depression is one of the most disturbing side effects associated with opioid drugs. Case reports have implicated M6G in respiratory depression.[41,48,49] We have observed that vectorization of M6G with SynB3 reduces its effect on respiratory depression significantly.[40] This suggests that vectorization of M6G does not only lead to an enhancement of its in vivo activity, but also to an improvement of its side-effects profile.

17.4.3 DELIVERY OF PEPTIDES AND PROTEINS

In a pharmacological application focused on pain management, the brain uptake of an enkephalin analog, dalargin, was enhanced significantly after vectorization.[50] Dalargin is a hexapeptide analog of leu-enkephalin containing d-Ala in the second position and an additional C-terminal arginine. While the intracerebroventricular injection of this peptide induces analgesic action, its systemic administration shows no activity in central analgesic mechanisms.[50] This is probably due to the poor brain uptake of dalargin. We have shown by in situ brain perfusion that vectorization with SynB vectors markedly enhances the brain uptake of dalargin (Figure 17.6). Free or conjugated dalargin was also administered intravenously to mice and anti-nociception was determined using the hot plate test, an assay known to be mediated

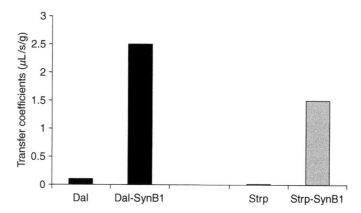

FIGURE 17.6 Brain delivery of peptides and proteins. Transfer coefficients (Kin) of free and vectorized molecules uptake in the right hemisphere of mice after 60 sec of perfusion in buffer. Dal, dalargin; Strp, streptavidin.

by central receptors. The results show that intravenous administration of dalargin to mice exhibited no analgesic activity. In contrast, conjugation of dalargin to SynB vectors led to a considerable enhancement of analgesic activity immediately after the intravenous injection.[50]

In another application we have shown that SynB vectors are able to transport large molecules, such as the protein streptavidin (MW approximately 60,000 Da), across the BBB. Radiolabeled streptavidin [^{125}I] was attached to a biotin moiety linked to the peptide vector. The biotinylated vector was added in large excess to allow sufficient binding of the radiolabeled strepatvidin. In situ brain perfusion studies in mice showed that attachment of SynB vector to streptavidin resulted in a significant enhancement uptake into the brain (Figure 17.6). As expected, free or biotin complexed streptavidin were unable to cross the BBB. Although this approach was successful in enhancing the brain uptake of proteins, such as streptavidin, we were unable to repeat this success for larger molecules such as IgGs. Vectorization of IgGs with SynB vectors did not lead to a significant enhancement of brain transport using the in situ brain perfusion method or by intravenous administration. This suggests that there is a size limit to this approach. However, one cannot rule out that coupling of very large molecules might make the peptide inaccessible to the membrane surface and therefore unable to perform its duty.

17.4.4 MECHANISM OF BRAIN UPTAKE

We have shown by in situ perfusion studies, a technique allowing a first-pass exposure, that the internalization of vectorized doxorubicin is a saturable process.[33] The measured Km, which was in the range of 4 to 9 μM, compares well with the values observed for substrates reported to be taken up by adsorptive-mediated endocytosis. Furthermore, no difference in brain uptake was seen between doxorubicin linked to L-SynB or D-SynB vectors, indicating that a stereospecific receptor is not a requirement for its brain transport. In addition, we have reported that the passage of peptides can be inhibited in a competitive manner by polycationic molecules such as poly(L)lysine or protamine, which act as endocytosis inhibitors. These observations suggest that the crossing of BBB by SynB vectors is via an energy-dependent adsorptive-mediated endocytosis mechanism.[33]

It is known that at physiological pH values the luminal surface of the brain endothelium presents an overall negative charge (due in part to significant sialylation) and thus creates an environment more selective to positively charged substances.[51,52] The SynB peptides are positively charged. This net positive charge is likely to play a key role in the adsorptive-mediated endocytosis process wherein electrostatic interactions of the peptide vector with the surface of endothelial cells may mediate surface binding and subsequent internalization of the peptide vectors into the brain capillaries.

For the transcytosis of peptides through the BBB, three steps have been proposed: (1) binding and internalization at the luminal side of endothelial cell membrane; (2) diffusion through the cytoplasm of endothelial cells; and (3) externalization at the basolateral side of endothelial cells.[53] The main components of the basal membrane are type IV collagen, fibronectin, laminins, chondroitin, and heparan sulfate from glycosaminoglycans. The most abundant component, type IV collagen, polymerizes with laminin and fibronectin proteins via protein-binding domains, such as integrin and lectin receptors.[54] These components not only provide a mechanical supporting structure for the capillary wall, they are also important as a negatively charged barrier arising from the chondroitin and heparan sulfate residues, in addition to the anionic properties of the luminal and abluminal membranes of the endothelial cells.[55] Our results suggest that adsorptive-mediated endocytosis occurs at least at the luminal side of brain capillaries. The similarity in behavior observed for the peptide vectors studied suggests that the externalization at the abluminal side of endothelial cells may also be via a receptor-independent mechanism. However, since endocytosis inhibitors have only been tested at the luminal side of the endothelial cells, it is formally possible that a different mechanism may be involved in the externalization step.

17.5 PHARMACOLOGICAL CONSIDERATIONS

The development of the peptide-conjugated cargo system is based on the premise that enhancing cell or brain uptake of a drug will result in a therapeutic benefit. However, this development will require careful evaluation to determine if the supposed benefit gained by modifying the drug or adding a vector will not be offset by anticipated problems created by the peptide vector. Potential toxicity and immunogenicity of the peptide vectors must be considered before clinical application. A preliminary study showed that no immunogenic effect was obtained with SynB1 vector. Conjugation of a drug with a transport vector may often lead to a loss of biological activity, requiring the development of a linker strategy that will allow the drug molecule to be cleaved off the drug transporter once it reaches its site of action. We have observed that some therapeutic molecules are still active when they are in the vectorized form. This is the case of dalargin and M6G. In this latter case, binding to the *mu* opioid receptor has been increased. However, in many cases, such as doxorubicin and paclitaxel, the drug needed to be released from the vector in order to be active. In many studies, the cargo was coupled to the vector via a chemical linker, such as a disulfide bond. This enables the reduction of the bond in the cell and the release of the cargo. When dealing with protein cargoes, the chemistry of linking becomes more complicated because many functional groups are present within a protein. Therefore, in most CPPs, the conjugates were prepared by fusion proteins and recombination.

CPPs have proven their ability to deliver various molecules such as small molecules and peptides in cultured cells. However, the ability to deliver these cargoes in vivo would be of tremendous benefit. Therefore, the development of vectors will not only depend on the efficiency with which these SynB vectors and other CPPs can cross the cell membrane, but also on their pharmacokinetic and biodistribution profiles. Ideally, an efficient peptide vector should have the dual effects of enhancing the uptake and increasing the systemic bioavailability of the drug in plasma. A conjugate with a poor pharmacokinetic profile and a reduced plasma concentration will result in low delivery, irrespective of the extent to which the cell membrane permeability has been increased due to the peptide vector. SynB vectors and other CPPs are mostly in L-amino acids form and therefore susceptible to degradation by proteases. The half-life of some of the conjugates in plasma, such as vectorized doxorubicin, was about 10 min. We have tried to enhance the stability of the SynB vectors by using D-amino acids. However, administration of an all D-amino acid peptide resulted in an accumulation of the compound in liver and other highly perfused tissues, suggesting a potential toxicity of these compounds. Nevertheless, one can imagine making a mixture of L- and D-amino acid peptides and thereby increasing the stability of the vector without having the potential side effects and accumulation in certain tissues.

The fact that others as well as ourselves have observed in vivo activity with various conjugated drugs and for different indications suggests that these conjugates reach their site of action in animals. However, little is available in the literature regarding the fate of these conjugates in vivo.

Finally, in order to develop this strategy at an industrial scale for clinical use in humans, the conjugated drug should be easy to manufacture on a large scale and should be cost-effective.

17.6 CONCLUSION

The ability to deliver large hydrophilic molecules is a great challenge due to the bioavailability restriction imposed by the cell membrane. As rapid advances in cell and molecular biology lead to a proliferation of potent molecules that cannot be effectively delivered to cells and brain by conventional means, continuing refinement of the new delivery methods will be essential for realizing the potential of these molecular drugs. The use of SynB vectors and other CPPs for drug delivery represents a novel and promising approach. Several significant points emerge from this review. First, the cell culture and animal experiments have provided evidence that SynB-conjugated cargoes can successfully transport drugs across the membranes of many different cell types, and furthermore can traverse the more demanding BBB. Secondly, the approach is broadly applicable to many diseases. Finally, the insights gained during the last decade have led to the design of more efficient peptide vectors. The optimization of these vectors will also be aided by a greater understanding of the transport mechanisms operating at the cell membranes and the BBB.

The success of the peptide-mediated strategy for clinical use will depend not only on their efficiency and safety, but also on the ultimate cost. Large-scale applications and new methodologies are being implanted to increase the yield and reduce the cost. Because of the progress made to date and the tremendous potential of this approach, it is reasonable to state that we can expect to see the beneficial effect of these peptide-conjugated cargoes in humans in the coming years.

ACKNOWLEDGMENTS

I would like to thank my colleagues at Synt:em and at INSERM U26 for their work and helpful discussions.

REFERENCES

1. Temsamani, J. and Scherrmann, J.M., Peptide vectors as drug carriers, *Prog. Drug Res.*, 61, 221, 2003.
2. Lindgren, M. et al., Cell-penetrating peptides, *Trends Pharmacol. Sci.*, 21, 99, 2000.
3. Joliot, A. and Prochiantz, A., Transduction peptides: From technology to physiology, *Nat. Cell Biol.*, 6, 189, 2004.
4. Langel, U., ed., *Cell Penetrating Peptides Processes and Applications*, CRC Press, Boca Raton, FL, 2002.
5. Drin, G. et al., Peptide delivery to the brain via adsorptive-mediated endocytosis: Advances with SynB vectors, *AAPS PharmSci.*, 4, 26, 2006.
6. Kokryakov, V.N. et al., Protegrins: Leukocyte antimicrobial peptides that combine features of corticostatic defensins and tachyplesins, *FEBS Lett.*, 327, 231, 1993.
7. Harwig, S.S. et al., Determination of disulphide bridges in PG-2, an antimicrobial peptide from porcine leukocytes, *J. Pept. Sci.*, 1, 207, 1995.
8. Aumelas, A. et al., Synthesis and solution structure of the antimicrobial peptide protegrin-1, *Eur. J. Biochem.*, 237, 575, 1996.
9. Mangoni, M.E. et al., Change in membrane permeability induced by protegrin 1: Implication of disulphide bridges for pore formation, *FEBS Lett.*, 383, 93, 1996.
10. Sokolov, Y. et al., Membrane channel formation by antimicrobial protegrins, *Biochim. Biophys. Acta*, 1420, 23, 1999.
11. Drin, G. and Temsamani, J., Translocation of protegrin I through phospholipid membranes: role of peptide folding, *Biochim. Biophys. Acta*, 1559, 160, 2002.
12. Drin, G. et al., Studies on the internalization mechanism of cationic cell-penetrating peptides, *J. Biol. Chem.*, 278, 31192, 2003.
13. Gruenberg, J., Griffiths, G., and Howell, K.E., Characterization of the early endosome and putative endocytic carrier vesicles in vivo and with an assay of vesicle fusion in vitro, *J. Cell Biol.*, 108, 1301, 1989.
14. Clague, M. et al., Vacuolar ATPase activity is required for endosomal carrier vesicle formation, *J. Biol. Chem.*, 269, 21, 1994.

15. Snyder, E.L. and Dowdy, S.F., Cell penetrating peptides in drug delivery, *Pharm. Res.*, 21, 389, 2004.
16. Monneret, C., Recent developments in the field of antitumour anthracyclines, *Eur. J. Med. Chem.*, 36, 483, 2001.
17. Gottesman, M.M. and Pastan, I., Biochemistry of multidrug resistance mediated by the multidrug transporter, *Annu. Rev. Biochem.*, 62, 385, 1993.
18. Mazel, M. et al., Doxorubicin–peptide conjugates overcome multidrug resistance, *Anti-Cancer Drugs*, 12, 107, 2001.
19. Day, F.H. et al., Induction of antigen-specific CTL responses using antigens conjugated to short peptide vectors, *J. Immunol.*, 170, 1498, 2003.
20. Dillon, D.C. et al., Molecular characterization and human T-cell responses to a member of a novel *Mycobacterium tuberculosis* mtb39 gene family, *Infect. Immun.*, 67, 2941, 1999.
21. Falk, K. et al., Allele-specific motifs revealed by sequencing of self-peptides eluted from MHC molecules, *Nature*, 351, 290, 1991.
22. Corr, M. et al., Gene vaccination with naked plasmid DNA: Mechanism of CTL priming, *J. Exp. Med.*, 184, 1555, 1996.
23. Lu, J. et al., TAP-independent presentation of CTL epitopes by Trojan antigens, *J. Immunol.*, 166, 7063, 2001.
24. Kim, D.T. et al., Introduction of soluble proteins into the MHC class I pathway by conjugation to an HIV tat peptide, *J. Immunol.*, 159, 1666, 1997.
25. Pietersz, G.A. et al., A 16-mer peptide (RQIKIWFQNRRMKWKK) from antenna-pedia preferentially targets the Class I pathway, *Vaccine*, 19, 1397, 2001.
26. Brightman, M.W., Morphology of blood–brain interfaces, *Exp. Eye Res.*, 25, 1, 1977.
27. Pardridge, W.M., Why is the global CNS pharmaceutical market so under penetrated?, *Drug Discov. Today*, 7, 5, 2002.
28. Pardridge, W.M., New approaches to drug delivery through the blood–brain barrier, *Tibtech*, 12, 239, 1994.
29. Jolliet-Riant, P. and Tillement, J.P., Drug transfer across the blood–brain barrier and improvement of brain delivery, *Fundam. Clin. Pharmacol.*, 13, 16, 1999.
30. Temsamani, J. et al., Vector-mediated drug delivery to the brain, *Expert Opin. Biol. Ther.*, 1, 773, 2001.
31. Temsamani, J. and Vidal, P., The use of cell-penetrating peptides for drug delivery, *Drug Discov. Today*, 9, 1012, 2004.
32. Rousselle, C. et al., New advances in the transport of doxorubicin through the blood–brain barrier by a peptide vector-mediated strategy, *Mol. Pharmacol.*, 57, 679, 2000.
33. Rousselle, C. et al., Enhanced delivery of doxorubicin into the brain via a peptide-vector-mediated strategy: Saturation kinetics and specificity, *J. Pharmacol. Exp. Ther.*, 296, 124, 2001.
34. Blanc, E. et al., Peptide-vector strategy bypasses P-glycoprotein efflux, and enhances brain transport and solubility of paclitaxel, *Anticancer Drugs*, 15, 947, 2004.
35. Gelderblom, H. et al., The drawbacks and advantages of vehicle selection for drug formulation, *Eur. J. Cancer*, 37, 1590, 2001.
36. Eiseman, J.L. et al., Plasma pharmacokinetics and tissue distribution of paclitaxel in CD2F1 mice, *Cancer Chemother. Pharmacol.*, 34, 465, 1994.
37. Fellner, S. et al., Transport of paclitaxel (Taxol) across the blood–brain barrier in vitro and in vivo, *J. Clin. Invest.*, 110, 1309, 2002.
38. Gallo, J.M. et al., The effect of P-glycoprotein on paclitaxel brain and brain tumor distribution in mice, *Cancer Res.*, 63, 5114, 2003.

39. Rousselle, C. et al., Improved brain delivery of benzylpenicillin with a peptide-vector-mediated strategy, *J. Drug Target*, 10, 309, 2002.

40. Temsamani, J. et al., Improved brain uptake and pharmacological activity profile of morphine-6-glucuronide using a peptide vector-mediated strategy, *J. Pharmacol. Exp. Ther.*, 313, 712, 2005.

41. Abbott, F.V. and Palmour, R.M., Morphine-6-glucuronide: Analgesic effects and receptor binding profile in rats, *Life Sci.*, 43, 1685, 1988.

42. Paul, D. et al., Pharmacological characterization of morphine-6 beta-glucuronide, a very potent morphine metabolite, *J. Pharmacol. Exp. Ther.*, 251, 477, 1989.

43. Frances, B. et al., Further evidence that morphine-6 beta-glucuronide is a more potent opioid agonist than morphine, *J. Pharmacol. Exp. Ther.*, 262, 25, 1992.

44. Thompson, P.I. et al., Respiratory depression following morphine and morphine-6-glucuronide in normal subjects, *Br. J. Clin. Pharmacol.*, 40, 145, 1995.

45. Grace, D. and Fee, J.P., A comparison of intrathecal morphine-6-glucuronide and intrathecal morphine sulfate as analgesics for total hip replacement, *Anesth. Analg.*, 83, 1055, 1996.

46. Lotsch, J. and Geisslinger, G., Morphine-6-glucuronide: An analgesic of the future?, *Clin. Pharmacokinet.*, 40, 485, 2001.

47. Letvin, N.L. et al., In vivo administration of lymphocyte-specific monoclonal antibodies in nonhuman primates. In vivo stability of disulfide-linked immunotoxin conjugates, *J. Clin. Invest.*, 77, 977, 1986.

48. Pelligrino, D.A. et al., Comparative ventilatory effects of intravenous versus fourth cerebroventricular infusions of morphine sulfate in the unanesthetized dog, *Anesthesiology*, 71, 250, 1989.

49. Peat, S.J. et al., Morphine-6-glucuronide: Effects on ventilation in normal volunteers, *Pain*, 45, 101, 1991.

50. Rousselle, C. et al., Improved brain uptake and pharmacological activity of dalargin using a peptide-vector-mediated strategy, *J. Pharmacol. Exp. Ther.*, 306, 371, 2003.

51. Hardebo, J.E. and Kahrstrom, J., Endothelial negative surface charge areas, and blood–brain barrier function, *Acta Physiol. Scand.*, 125, 495, 1985.

52. Vorbrodt, A.W., Ultrastructural cytochemistry of blood–brain barrier endothelia, *Prog. Histochem. Cytochem.*, 18, 1, 1988.

53. Bar, R.S. et al., Insulin binding to microvascular endothelium of intact heart: A kinetic and morphometric analysis, *Am. J. Physiol.*, 244, 477, 1983.

54. Virgintino, D. et al., An immunohistochemical and morphometric study on astrocytes and microvasculature in the human cerebral cortex, *Histochem. J.*, 29, 655, 1997.

55. Bertossi, M., Virgintino, D., Maiorano, E., Occhiogrosso, M., Roncali, L. et al., Ultrastructural and morphometric investigation of human brain capillaries in normal and peritumoral tissues, *Ultrastruct. Pathol.*, 21, 41, 1997.

Part IV

Applications of Cell-Penetrating Peptides in Gene Modulation and Protein Transport

Steven F. Dowdy

Although various earlier observations had shown that large polymers of lysine could be used to deliver molecules into cells, the watershed moment for cell penetrating peptides (CPPs)/protein–peptide transduction domains (PTDs) came in 1988 when independent observations by Green and Loewenstein,[1] and Frankel and Pabo[2] showed that HIV1 TAT protein could enter cells. Although it took over 10 years to go from a few papers published each year to well over a 100 papers each year, these observations opened the door for formalizing macromolecular delivery. Conse-quently, today we have many PTDs, many different types of cargo being delivered, and many different ideas on the delivery mechanism across the cell membrane. Although we are blessed in the richness and depth of our data, we have many unanswered questions and have clearly not yet realized the full potential of macromolecular delivery for clinically relevant therapeutics.

Based on mere publication numbers, the cationic PTDs, namely polyArg, TAT, and Antp, have risen to the top of the heap for cargo delivery into cells. Although there are many other cationic PTDs that have been identified or synthesized with similar amounts of cationic charge, these three have clearly become the Xerox of PTDs and have been used to deliver a wide variety of cargo including: small-molecules, peptides, oligonucleotides and PNAs, and full-length proteins and protein domains. Intracellular delivery is dependent either on direct conjugation to the cargo or on incredibly high avidity for the cargo, such as biotin:streptavidin. Importantly, one must also consider the potential negative consequence of having the PTD attached to the cargo once inside the cell. While many cargoes seem unaffected, clearly there are those that are negatively affected by continued conjugation to the PTD. Redox- and esterase-sensitive linkers are one avenue that has been exploited to avoid this potential issue and will likely be further optimized in the future. Surprisingly, there has been a relatively minimal published effort on the generation of synthetic cationic PTDs. While we have taken advantage of naturally occurring cationic PTDs, synthetically derived cationic delivery PTDs will be an area of intense investigation as a necessary area to improve compound stability clinically. The use of D-isomer residues to avoid peptidase degradation is also a significant step in this direction.

In contrast to the significant limitations of information-poor small-molecule intracellular therapeutics, where the vast majority are inhibitors, cationic PTD-mediated delivery opens entirely new avenues of therapeutic intervention, especially in the area of introducing information-rich macromolecular cargo that has the potential to amplify the signal. The ability to deliver biologically active enzymes, transcription factors and repair proteins is the single greatest therapeutic potential of PTD-mediated delivery. Although this is over the horizon in the face of the tried and true small-molecule pharmaceutical industry, Genentech, Amgen, and Genzyme have been extremely successful in developing protein therapeutics. Indeed, small-molecules therapeutics were chosen, not because of their specificity compared to macromolecules, but because they can enter cells and macromolecules could not. In fact, gene therapy was invented, not to introduce DNA into cells, but to express biologically active proteins to specifically treat human diseases in ways that small-molecules will never be able to. However, PTDs now change all of that thinking and begin to allow for testing the potential benefit to deliver protein therapeutics directly into cells. Given the rapid development over the last 5 years, it will be interesting to see how far we can go in the next 5 years.

Although there is universal consent that fixation artifacts, first identified in Sweden by Lundberg and Johansson[3] in 2002, led to artificial redistribution of PTD-delivered cargo and misinterpretations of mechanistic experiments, the mechanism of how PTDs enter cells remains somewhat controversial. Endocytosis remains at the top of the mechanistic list. My laboratory has proposed macropinocytosis for both TAT-proteins and TAT-peptides,[4] and others have proposed caveolin-mediated endocytosis and some nonendocytotic, direct cell membrane-penetrating approaches. While there may very well be different mechanisms of entry for each and every PTD, given the dependency on cationic charge, this seems highly unlikely for polyArg, TAT, and Antp PTDs. However, the proof is always in the pudding. To define the entry mechanism(s) accurately, we need to stop relying solely on visualization methods and focus on approaches that generate clear and defined phenotypes on live cells. Anything less only results in the publication of more confusion, rather than mechanistic clarity.

As discussed in the following chapters, the future of PTD-mediated macromolecular delivery lies in both understanding and thereby improving the mechanism of delivery and improving the activity of the cargo. While there are a number of clinical trials currently underway using PTD-mediated delivery, both of these areas of research will require significant investments of time and resources before we even begin to fully exploit PTD-mediated delivery of therapeutics in the clinics.

REFERENCES

1. Green, M. and Loewenstein, P.M., Autonomous functional domains of chemically synthesized human immunodeficiency virus tat trans-activator protein, *Cell*, 55, 1179, 1988.
2. Frankel, A.D. and Pabo, C.O., Cellular uptake of the tat protein from human immunodeficiency virus, *Cell*, 55, 1189, 1988.
3. Lundberg, M. and Johansson, M., Positively charged DNA-binding proteins cause apparent cell membrane translocation, *Biochem. Biophys. Res. Commun.*, 291, 367, 2002.
4. Wadia, J.S., Stan, R.V., and Dowdy, S.F., Transducible TAT-HA fusogenic peptide enhances escape of TAT-fusion proteins after lipid raft macropinocytosis, *Nat. Med.*, 10, 310, 2004.

18 Peptide Conjugates of Oligonucleotide Analogs and siRNA for Gene Expression Modulation

*John J. Turner, Andrey A. Arzumanov,
Gabriela Ivanova, Martin Fabani,
and Michael J. Gait*

CONTENTS

18.1 THE CHALLENGE OF DELIVERY OF OLIGONUCLEOTIDE ANALOGS AND siRNA INTO CELLS IN CULTURE AND IN VIVO

Unlike many small molecule drugs, oligonucleotides and short interference RNA (siRNA) are of too high mass (4000–12,000 Da) to permeate freely into mammalian cells. In addition, natural DNA and RNA, as well as many oligonucleotide analog types, carry substantial numbers of negative charges due to their phosphate backbones, and so are repelled by anionic charges, such as in sulfated polysaccharides on cell surfaces. Thus, they are not ideal agents to enter cells either in culture or in vivo. Surprisingly, perhaps, some modified antisense oligonucleotides, notably those containing phosphorothioate linkages, have been taken to clinical trials.[1,2] The mechanisms by which such oligonucleotides may enter cells in vivo are not well understood, but their uptake may be mediated by binding to various serum proteins. By contrast, in cell culture, in order to be useful in gene silencing or modulation, antisense and siRNA reagents usually require formulation in some way to enhance their cell uptake and delivery into the cytosol or nucleus.[3]

Common laboratory cells in culture, such as cancer-derived cells, are usually transfected by complexation of an oligonucleotide or siRNA with a cationic lipid,[4] of which there are now a very large number available commercially. Some more sophisticated versions are even being used for in vivo delivery, for example of siRNA.[5] However, cationic lipid transfection is not effective for all cell types in culture, especially for primary cells, and commonly shows dose-limiting cell toxicity. Thus, alternative, chemically simple cell-delivery techniques that do not require formulation have been sought for some years (for a recent review, see Ref. [6]).

One approach designed to increase cell entry was to attach a cationic poly-L-lysine.[7] Later, Prochiantz and colleagues suggested the conjugation of oligonucleotides or their analogs with certain cell-penetrating peptides (CPPs),[8] also known as protein transduction domains (PTDs). A range of such CPPs has been proposed and many such peptides have been found to translocate into cells efficiently (see other chapters in this book). Despite the initial promise of this idea, examples of real success in covalently attached CPPs delivering active, negatively charged antisense oligonucleotides and siRNA cargoes into cells in culture have been disappointingly few. By contrast, there are a significant number of good examples of biological activities obtained for peptide conjugates of electrically neutral peptide nucleic acids (PNAs) or phosphorodiamidate morpholino oligo-nucleotides (PMOs) analogs. In addition, certain CPPs designed to be used as oligonucleotide or siRNA complexes have proved promising as transfection agents,[9] such as the peptide MPG used for siRNA delivery.[10]

There have been a number of recent reviews on the synthesis and gene silencing activity of peptide–oligonucleotide conjugates,[11–14] as well as general reviews on the various methods of gene expression modulation.[1,15,16] This chapter describes some of the main methods of chemical synthesis of CPP conjugation to oligonucleotide types that we and others have used in gene expression modulation (RNase H-inducing and steric block), as well as to siRNA, and highlights the successes and problems in applying such conjugates to cell culture systems involving RNA targeting.

18.2 CHEMICAL SYNTHESIS OF PEPTIDE–OLIGONUCLEOTIDE CONJUGATES

Single-stranded oligonucleotides are usually chemically modified by incorporation of various nucleoside derivatives to protect against degradation by serum and cellular nucleases. Double-stranded siRNA is generally sufficiently stable in cell culture to be used without modification, but are usually heavily modified with nucleoside analogs for in vivo use.[5] In each case, it is now possible to synthesize a range of peptide–oligonucleotide conjugates through a variety of different conjugation chemistries that include either stable or cleavable linkages. A number of reviews on the synthesis of oligonucleotide–peptide conjugates are available.[14,17–19] In this chapter, we cite some of these chemistries and methods, especially those of which we have had experience. Since there is as yet no general consensus on the rules of conjugation chemistry that can lead to biological activity, the field is insufficiently mature to make firm recommendations between different linkage chemistries, type of spacer, or orientation of the oligonucleotide and peptide components. However, it is true to say that a majority of publications involving biological activity of such conjugates have utilized cleavable disulfide linkages.

There are two main synthetic strategies for conjugate synthesis: (1) total stepwise solid-phase synthesis and (2) solution-phase or solid-phase fragment coupling of peptides with oligonucleotides that have each been prepared separately on their own solid supports. Whereas stable linkages can be prepared by either routes (1) or (2), only the latter is possible for cleavable disulfide linkages.

18.2.1   SYNTHESIS OF CPP CONJUGATES TO PHOSPHATE-CONTAINING OLIGONUCLEOTIDES

No fully reliable procedure has been found for total stepwise solid-phase synthesis of phosphate-containing oligonucleotides stably conjugated to peptides, despite numerous suggestions.[20–24] This is because it is hard to find entirely compatible sets of protecting groups simultaneously for both base exocyclic amino groups and amino acid side chains. Further, internucleotide and interamino acid coupling reaction chemistries were developed for different scales and have different reaction rates. We developed a good method for solid-phase synthesis of 3′-conjugates of oligonucleotides to peptides and other small molecules,[25] and later adapted this for a total stepwise synthesis of oligonucleotide-(3′→N) conjugates of peptides.[21] However, we experienced difficulties with peptides containing arginine for which no compatible side-chain protecting group was available.

In our experience, the most generally applicable conjugation chemistries have involved fragment coupling in aqueous solution. Peptide and oligonucleotide fragments are first prepared on their own solid supports by conventional automated solid-phase synthesis procedures. Each fragment requires a masked functional group to be incorporated during the assembly, which can be released chemoselectively during a simple deprotection step. A specific chemical reaction is then initiated in aqueous solution between the functional group on the oligonucleotide and that on the peptide to produce the desired conjugate. An advantage of the fragment-coupling route is that both components can be purified (usually by HPLC) before conjugation, making it easier to identify the conjugation product in the event of a low yield. A disadvantage of the route is that the aqueous conditions of conjugation sometimes need to be adjusted to maintain the solubility of both components, as well as of the conjugate product.

We have used the fragment coupling approach in aqueous solution to generate several new types of linkage between oligonucleotide and peptide components. For example, reaction of a cysteine-substituted oligonucleotide with a thioester-substituted peptide produces a stable amide linkage through the mechanism of "native ligation" (Figure 18.1A).[26] Alternatively, reaction of an aldehyde-containing oligonucleotide with a cysteine-containing peptide forms a thiazolidine linkage, or with an aminooxy peptide forms an oxime linkage, or with a hydrazinopeptide forms a hydrazide (Figure 18.1B).[27] An aldehyde oligonucleotide is unstable and must be generated by periodate treatment of a *cis*-diol substituted oligonucleotide just prior to conjugation.[28]

However, the most convenient and simple methods of fragment conjugation rely on synthesis of the oligonucleotide component with an alkyl thiol linker on either the 5'- or 3'-end (Figure 18.2). Such thiol-functionalized oligonucleotides are readily

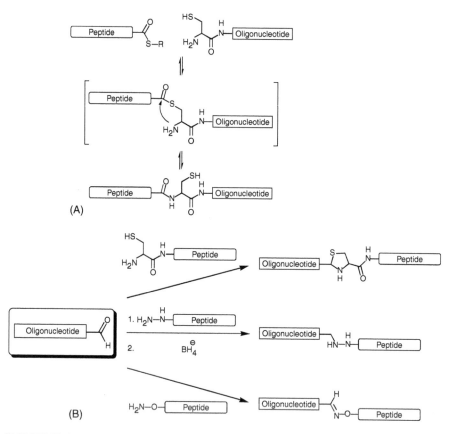

FIGURE 18.1 (A) Chemical reaction involved in "native ligation" conjugation of a peptide thioester and an oligonucleotide functionalized with a 5'-cysteine to form a stable amide linkage. (B) Chemical reactions involved in conjugation of an oligonucleotide functionalized with an aldehyde moiety and a peptide functionalized with a cysteine, hydrazide, or amino-oxy group to form thiazolidine, hydrazine, and oxime linkages, respectively.

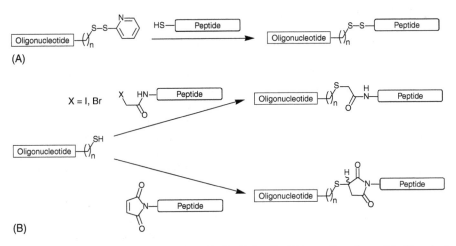

FIGURE 18.2 (A) Chemical reaction involved in disulfide conjugation. (B) Chemical reactions involved in conjugation of a thiol-functionalized oligonucleotide to form thioether or thiomaleimide linkages with a peptide.

obtained commercially. The thiol moiety can be reacted with a cysteine-containing peptide to form a cleavable disulfide linkage (Figure 18.2A) or with a bromoacetyl-substituted peptide to form a stable thioether linkage (Figure 18.2B). Similarly, a maleimide derivative of the peptide can be prepared for reaction with a thiol-substituted oligonucleotide via a Michael-type addition reaction (Figure 18.2B).

In recent studies, we have utilized predominantly disulfide conjugation to prepare conjugates of 2′-O-methyloligoribonucleotides, as well as their mixmers with locked nucleic acids (LNA) to a range of CPPs.[29] One of the two thiol-containing components is first activated with a pyridylsulfide or nitropyridylsulfide group before addition of the second thiol component. This allows specific heterodisulfide formation without concomitant homodisulfide formation of the peptide or oligonucleotide component. The reaction is very rapid and is complete within a few minutes. In the case of cationic CPPs, there are severe problems in maintaining solubility of the components and the resultant conjugate in aqueous solution because of aggregation and precipitation. We found that a high concentration of formamide in both conjugation and HPLC purification steps allows such disulfide conjugates to be readily isolated in pure form. The product of the conjugation reaction is loaded on to a Resource Q anion exchange column and excess peptide is eluted rapidly whilst the conjugate is retained and eluted later using a higher salt wash. We have recently provided full protocols for such disulfide conjugations for typical cases.[30]

We have applied the same principles to the synthesis of conjugates between one of the two siRNA strands and a CPP. Here, the RNA strand is functionalized with a 3′-thiopropyl or 5′-thiohexyl linker, activated with pyridylsulfide, and coupled to a cysteine-containing peptide.[30]

18.2.2 Synthesis of PNA–Peptide Conjugates

Total stepwise solid-phase synthesis of PNA stably conjugated to peptides is straightforward. The coupling of a PNA monomer unit, using either tBoc/benzyl-oxycarbonyl[31] or Fmoc/Bhoc chemistry,[32,33] is very similar to the coupling of amino acids to form a peptide chain. Thus, it is possible to use either manual total synthesis or a commercial peptide synthesis machine to assemble PNA–peptides containing all amide linkages. The advantage of such total synthesis methodology is its simplicity and the ability to place amino acids at either or both ends of the PNA oligomer and to add a commercially available polyether spacer (known as an O-linker) if desired. We have used a robotic APEX 396 peptide synthesizer for assembly of PNA–peptides by the Fmoc/Bhoc method.[34]

Alternatively, fragment coupling by the disulfide route has been very popular.[35–37] This is entirely analogous to the methods described for disulfide conjugation between peptides and negatively charged oligonucleotides (Figure 18.2), except that, in the case of PNA, a thiol group is introduced on to the C- or N-terminus merely by the addition of a cysteine residue during solid-phase synthesis (Figure 18.3). A cysteine residue is also placed at the C- or N-terminus of the CPP, or an internal cysteine residue may be utilized. In contrast to fragment conjugation of CPPs with negatively charged oligonucleotides, there is less likelihood of precipitation of the conjugate and we have found that most disulfide-linked CPP–PNA conjugates can be purified by reversed-phase HPLC by careful use of shallow gradients of acetonitrile and a water-bath heated column.[30,34]

For tracking of the PNA component of disulfide-linked PNA–peptide conjugates, it is necessary to incorporate a fluorescent label (such as fluorescein) during chemical synthesis. It is possible to place such a label on the N-terminus by a condensation reaction with, for example carboxyfluorescein diacetate, either directly or via an O-linker.[37] Alternatively, a fluorescein label can be incorporated on the side chain of a lysine residue through use of a commercially available Fmoc-Lys(FAM)-COOH derivative, although this can only be used at the N-terminus of the PNA–peptide.

18.3 GENE EXPRESSION MODULATION BY PEPTIDE CONJUGATES OF ANTISENSE OR STERIC BLOCK OLIGONUCLEOTIDES

18.3.1 Anionic Oligonucleotide Analog Conjugates with CPPs

We recently reviewed the literature available on antisense activity and cell uptake studies of conjugates of various peptides with oligodeoxyribonucleotides or their

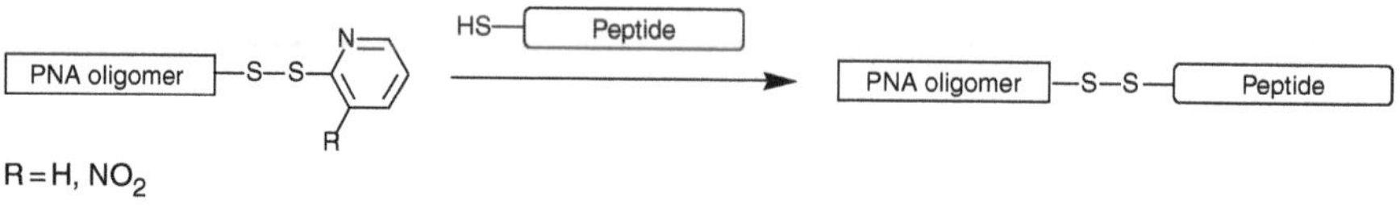

FIGURE 18.3 Chemical reaction in conjugation of a PNA with a peptide to form a disulfide linkage.

phosphorothioate derivatives capable of inducing RNase H cleavage when bound to a complementary RNA target.[14] Many of the early confocal microscopy studies of cell uptake of such materials were carried out using cell fixatives, conditions that were later shown to result in significant redistribution of peptide-containing materials, and therefore such results must be viewed with care.[38] There are several examples of protein expression reduction using conjugates containing the CPP penetratin[39–41] or in one case an HIV-1 Tat peptide.[41] Another example is of 15-mer phosphorothioate oligodeoxynucleotide stably conjugated to a LALLAK peptide that reduced expression of the GLUT-1 mRNA and accordingly reduced cell proliferation.[42] However, there are also examples of the lack of activity. Eighteen disulfide-linked conjugates of peptides with phosphorothioate oligonucleotides at either the C- or N-terminus were synthesized that contained either a hydrophobic signal sequence from Kaposi fibroblast growth factor or the same domain extended at the C-terminus by a NLS from transcription factor κB.[43] Fluorescent labeling and confocal microscopy showed predominant localization in cytosolic vesicles of monkey kidney fibroblast cells CV1-P. All conjugates failed to show antisense activity, but were readily delivered into the nucleus by cationic liposomes. Interestingly, no further successful examples of RNase-H-dependent activity of such CPP–oligonucleotide conjugates have been published in the last few years.

One study described the delivery of a steric block oligonucleotide into the cell nucleus to control RNA splicing. In this case, an 18-mer $2'$-O-methyl oligoribonucleotide (OMe) phosphorothioate was disulfide-conjugated to the Tat or penetratin peptide and the conjugate was reported in each case to localize in the nucleus, as well as in large cytosolic vesicles in HeLa cells.[44] Uptake was shown to be endocytotic and energy-dependent, the initial entry being into cytosolic vesicles. The conjugated oligonucleotides showed dose-dependent upregulation of a splicing-defective luciferase gene that was stably integrated into the HeLa cells. However, the activity level was significantly less than that of cationic lipid-delivered oligonucleotide.[44]

We recently tested a number of CPPs (Tat, penetratin, transportan, or R_9F_2) disulfide-linked conjugates to steric block oligonucleotides composed of either OMe phosphorothioate or mixmer phosphodiesters of locked nucleic acid (LNA) and OMe residues. We found that most of these CPPs were effective at enhancing delivery of the oligonucleotides into cytosolic compartments of HeLa and fibroblast cells, but they were not subsequently seen to be released into the cytosol or cell nuclei.[29] As a result, these CPP conjugates were unable to inhibit expression of firefly luciferase in a stably integrated plasmid system in HeLa cells where expression is under HIV-1 Tat/*trans*-activation responsive region (TAR)-dependent *trans*-activation control (Figure 18.4A). By contrast, these oligonucleotides and their CPP conjugates were mostly effective when delivered by cationic lipids.

Thus, we conclude that the main effect of conjugation of these types of cationic CPP to anionic oligonucleotides is to obtain sufficient charge neutralization to promote binding to the cell surface in order to allow cell uptake by endocytosis, but the amount of release from cytosolic vesicles is too poor to obtain sufficient steric block activity of the HIV-1 TAR RNA target in the nucleus (the site of action of the HIV-1 Tat protein) using a highly expressed plasmid-encoded reporter. It is not clear

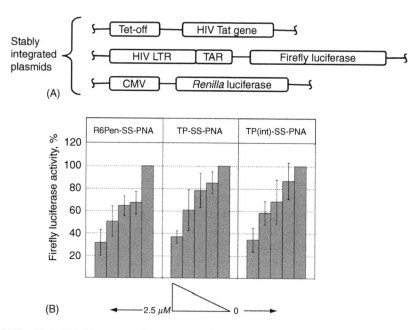

FIGURE 18.4 (A) Three plasmid constructs that are stably integrated into HeLa cells for use in assaying of inhibition of GL3 luciferase expression through interference with Tat-dependent *trans*-activation by the HIV-Tat protein of the TAR RNA target, together with control *Renilla* luciferase expression from a constitutive CMV promoter. (B) Dose-dependent inhibition of firefly luciferase expression exhibited by three disulfide-linked PNA–peptide conjugates after treatment for 24 h of the HeLa cell line as shown in (A). The PNA is a 16-mer directed to the apical stem-loop of the HIV-1 TAR RNA and the peptides are R_6-penetratin and transportan (linked in two different ways to the PNA). (From Turner, J.J. et al., *Nucleic Acids Res.*, 33, 6837, 2005.)

at this stage if it is possible to improve the release of such conjugates from endocytotic vesicles by use of different CPP conjugates. Unfortunately, the addition of further cationic charges on to the CPP (such as an additional six arginine residues on to the end of penetratin) merely caused the conjugate to become very insoluble and highly aggregated.[29] Thus, further improvements to vesicle release may be extremely hard to achieve by mere covalent attachment of peptides.

18.3.2 CHARGE-NEUTRAL PNA AND PMO OLIGONUCLEOTIDE CONJUGATES WITH CPPs

The charge neutrality of PNA gives this oligonucleotide analog a significant advantage over anionically charged varieties. Although "naked" PNA is not taken up by cells any better than unmodified oligodeoxyribonucleotides,[45] it does not require more than a few lysine residues to be added to the PNA (during chemical synthesis) to bestow sufficient positive charges to enhance cellular uptake substantially (presumably due to improved binding to the cell surface). By the use of fluorescent labeling, uptake of cationic conjugated PNA, whether through stable or disulfide linkages, has been shown to be primarily endocytotic into cytosolic vesicles.[36,45,46]

There are now several examples of antisense activity of PNA conjugated to a variety of peptides, either through disulfide linkages or through stable amide bonds (reviewed in Ref. [14]). There is some controversy, however, as to what extent stably linked conjugates can exert activity as compared to those linked through potentially cleavable disulfide linkages. The first example of antisense activity within cells was shown for a 21-mer PNA disulfide coupled to the CPPs penetratin or transportan (a hybrid of a section of the neuropeptide galanin and the wasp venom peptide mastoparan), which blocked expression of the galanin receptor in human Bowes cells.[35] When injected intrathecally into rats, the conjugates showed pain-modifying biological activity and were nontoxic at the concentration used. It was thought that the disulfide bond might be cleaved quickly upon cell entry, allowing PNA to enter the cytosol because there was concern that more stably conjugated PNA–peptides might be shunted to other cellular destinations. Amongst several other similar studies, transportan disulfide-linked to a 16-mer PNA complementary to the HIV-1 TAR RNA sequence blocked Tat-dependent *trans*-activation in transiently transfected CEM and Jurkat T-cells and also exhibited antiviral activity in chronically HIV-infected H9 cells at micromolar concentrations.[47]

However, there are now examples of PNA–peptide conjugates that contain stable amide linkages and that also exert antisense activity. A PNA stably conjugated to a hydrophobic peptide FLFLFL showed increased uptake into mouse macrophages and dose-dependent inhibition of nitric oxide synthase expression.[48] A 17-mer anti-c-myc PNA conjugated to NLS peptide PKKKRKV downregulated c-myc oncogene expression and showed nuclear localization in BL cells.[32] In this case, the concentration of PNA–peptide used was relatively high (10 μM). Recently, a PNA oligomer stably linked to the same SV40 NLS or to R_7 through an O-linker and targeted to a stem-loop RNA encapsidation signal of duck heptatitis B virus (HBV) was shown to decrease HBV DNA synthesis by 65% in primary duck hepatocytes.[49] Further, a PNA conjugated at the N-terminus to eight lysine residues inhibited CD40 gene expression in murine B-cell lymphoma cells by redirection of splicing at exon 6 splice junction.[50] In primary murine macrophages, induction of IL-12 was attenuated when cells were treated with micromolar concentrations of the PNA–peptide conjugate. Thus, it seems that even rather simple cationic peptides stably conjugated to PNA can help deliver PNA into the nucleus, at least to some extent and in some cell lines. However, the concentrations needed for a biological effect tend to be relatively high (micromolar range).

Our own studies of PNA–peptides have concentrated on steric blocking of the HIV-1 TAR RNA target.[34] Our HeLa cell line carries stably integrated luciferase reporters[51–53] and has significant advantages for the study of Tat-dependent *trans*-activation inhibition (Figure 18.4A). In the stably integrated system, HIV-1 Tat is produced to control production of GL3-firefly luciferase from the HIV-1 LTR. Control *Renilla* luciferase is expressed under constitutive CMV promoter direction. In order to see significant steric block inhibition of the TAR RNA target, PNA must be delivered efficiently to the nucleus of almost all cells. The aim is to obtain reduction in firefly luciferase activity without concomitant *Renilla* luciferase inhibition. Thus, the system would be expected to mimic an integrated HIV-1 provirus better than transient transfection and in addition is useful to assess to what extent *trans*-activation inhibition by TAR-targeting reagents can be used as an antiviral mechanism.

We synthesized a number of CPP conjugates of a 16-mer PNA targeted to the apical stem-loop of TAR.[34] These included a $(Lys)_8$ peptide stably linked to the N-terminus of the PNA, as well as a number of other stably linked conjugates of the PNA with Tat, transportan, or the SV40 NLS in various orientations. None of the stably linked conjugates was able to show inhibition of Tat-dependent *trans*-activation in the HeLa cell assay when incubated for up to 24 h. Several disulfide-linked PNA conjugates were also synthesized. Whereas Tat, penetratin, and NLS conjugates were inactive, two types of CPP (transportan or a novel chimeric peptide R_6-penetratin) were found to exhibit dose-dependent *trans*-activation inhibition activity when incubated for 24 h (Figure 18.4B). When 100 μM of the lysosomotropic reagent chloroquine was co-administered, inhibition was seen within 6 h for both types of disulfide-linked construct, which suggested that release from endosomal compartments is a limiting factor.

In collaborative studies with Lebleu and colleagues in Montpellier, we showed previously that a conjugate of PNA stably conjugated with the cationic peptide Tat via an O-linker and N-terminally labeled with fluorescein was taken up primarily into cytosolic vesicles in HeLa cells.[38] Interestingly, when the same Tat–PNA conjugate was coadministered with 100 μM chloroquine with our HeLa test cell line, dose-dependent *trans*-activation inhibition was observed, even though the same conjugate was inactive in the absence of chloroquine.[34] There was confocal microscopy evidence also for release of the fluorescent Tat–PNA conjugate from punctate cytosolic vesicles into a more general fluorescence in the cytosol. Stably linked transportan–PNA conjugates that were fluorescein-labeled showed most fluorescence nearer the cell surface within or close to the cell membrane and fewer punctate vesicles. Coadministration of chloroquine did not result in general fluorescence release into the cytosol, but instead some cells showed accumulation of fluorescence on the inside of the nucleus.[34] Thus, there may be different classes of CPP that have different effects on PNA localization. Nevertheless, activity is clearly limited by trapping of the conjugates in cytosolic vesicles or membrane-bound compartments.

In our *trans*-activation assay system, inhibition of gene expression was only seen in the absence of chloroquine for certain disulfide-linked CPP–PNA conjugates when administered for 24 h.[34] This seems to confirm that a cleavable linker between PNA and CPP is optimal. However, it is unclear if the need for extended incubation times is due to slow endosomal release or slow cleavage of the disulfide linkage or a combination of both. A previous study suggested that disulfide conjugation of a stably linked, NLS-modified PNA with another CPP, such as penetratin, was required to confer both release into the cytosol and entry of the PNA into the nucleus, but no antisense activity results were shown to confirm that such localization corresponds with antisense activity.[36] It is now necessary for careful structure-activity relationships to be set up where both nuclear antisense activity and cell localization are monitored as a function of CPP–PNA structure and linkage to see if some better rules can be established as to the optimal requirements.

Analogous uncharged analogs are phosphorodiamidate morpholino oligo-nucleotides (PMO), which have been synthesized as stably linked conjugates with cationic peptides, notably R_9F_2. For example, the α-amino group of the arginine-rich peptide was coupled to a 5'-thiomaleimide-substituted PMO prepared by use of one of a number of commercially available cross-linking agents.[54,55] Interestingly, no difference was found in steric block antisense activity in upregulation of luciferase expression (concentrations required in the low micromolar range) compared to disulfide-linked conjugates, even though the uptake and distribution patterns of fluorescently labeled PMO–peptides appeared dissimilar.[55] Stably linked PMO–peptides have been used recently to target coronavirus RNA and to attenuate growth.[56,57]

18.4 GENE EXPRESSION INHIBITION BY PEPTIDE CONJUGATES OF siRNA

In contrast to antisense activity which mostly requires nuclear delivery, siRNA is active predominantly or wholly within cytosolic compartments due to the requirement for the action of the RISC complex (for reviews, see Refs [58–60]). Synthetic siRNA usually consists of two complementary RNA strands of 19-residues, sometimes extended by two single-stranded dT or dU residues. Just as for antisense, the standard cellular delivery agents are cationic lipid preparations. Unfortunately, too few studies have yet been published with peptide conjugates to generate structure-activity relationships. Linkage to the 5'-end of the antisense strand would appear undesirable, because this strand requires phosphorylation in order to observe RNA interference activity, but conjugation to either end of the sense strand looks reasonable in principle. In one study, transportan and penetratin conjugates disulfide-linked to the 5'-end of the sense strands of siRNA were prepared for targeting of luciferase or EGFP mRNA reporters.[61] However, the disulfide linkages in these conjugates were prepared by a nonspecific oxidation method and the conjugates were added without purification directly to mouse fibroblasts cells. Thus, it is not clear what proportion of gene expression inhibition activity observed was due to the conjugate itself or instead to excess peptide complexing the siRNA or its conjugate.

In a second study, Tat peptide was conjugated to the 3'-end of the antisense strand of siRNA through a stable thiomaleimide linkage and the sense strand carried a 5'-fluorescent label.[62] In this case, it was not clear if the native gel electrophoresis method of purification was fully effective in removal of Tat–peptide complexes, but siRNA cell uptake did correlate dose-dependently with EGFP gene silencing (IC_{50} in the 200–400 nM range) in a HeLa cell transient transfection reporter system.

So far, we have failed to see inhibition of firefly luciferase activity within a stably integrated HeLa cell test system, such as shown in Figure 18.4A, with siRNA 5'-disulfide conjugated on the sense strand to the Tat peptide, to penetratin or to transportan (Fabani et al., unpublished). However, further more detailed structure-function studies are in progress with various peptide–siRNA conjugates using an alternative cell assay targeting the mRNA of an endogenous gene, where gene expression interference may be easier to achieve.

18.5   FUTURE PERSPECTIVES

It is true to say that the literature to date is very confusing on the use of covalently attached CPPs to synthetic oligonucleotides, their analogs, and siRNA to modulate gene expression. Some good examples of gene expression inhibition and redirection of splicing have been published, especially in the case of charge-neutral PNA and PMO. However, there are also examples where CPP conjugation has not led to sufficient cellular delivery to achieve a biological effect. Our experience has tended more towards the latter, especially for anionic oligonucleotides conjugated to CPPs. This may be because we have used largely a demanding test system (Figure 18.4A) that requires significant nuclear delivery to most cells in order to achieve a measurable level of expression inhibition. Further, some claims of success are based on activity levels only in the high micromolar range. Our view is that activity levels below $1 \mu M$ are mostly needed if the results are likely to have potential therapeutic extension, particularly because of cost and possible toxicities of conjugates.

Clearly the type of oligonucleotide backbone chemistry has dramatic effects on how an attached CPP can influence cell surface recognition and entry into endosomal pathways. However, little is still known as to which uptake pathways predominate for conjugates in different cell types and under what conditions. Further, the type of CPP (cationic, hydrophobic) may help direct the oligonucleotide into different uptake modes and there is some controversy remaining as to whether endosomal uptake pathways can indeed be avoided by certain CPPs (such as transportan). Another problem is that there is no consensus as to the types of linkage or spacer that are optimal between CPP and oligonucleotide, the orientation of each component, or indeed whether or not the linkage should be cleavable (such as disulfide). Faced with these many uncertainties and the considerable number of parameter variables, it is surprising perhaps that there have been successes at all!

The best perspective is perhaps in limiting the expectations of what conjugation of a CPP to an oligonucleotide or siRNA might be able to achieve. Already, CPPs have been able to achieve the first aim: that is, to overcome the reluctance of an oligonucleotide to attach to the cell surface and thus be available to enter an endosomal or other membrane-involved uptake pathway. There are now numerous literature examples of cells amply filled with fluorescent oligonucleotides, and improvements in confocal microscopy techniques can now distinguish more clearly the types of vesicles wherein most conjugates are initially sequestered. However, it is unrealistic to expect a single peptide attachment to be able to impart (on an anionic oligonucleotide, for example) as much endosomal release (through membrane destabilization and pH alteration) as can be obtained with a cationic lipid reagent or a CPP used as a packaging agent, which stoichiometrically is present in far larger quantities.

The aim now must surely be to concentrate efforts on oligonucleotide backbone types that are more readily affected by CPPs in uptake and to carry out wider ranging structure-function studies that aim to correlate a biological phenotype with the biology of cell uptake and trafficking. We believe our current studies to be a small step in this direction, but much remains to be done. Our experience to date is that none of the currently developed CPPs has proved sufficiently powerful to obtain strong endosomal or membrane release for attached oligonucleotides, PNA, or siRNA, although some have promise for further development.

Looking towards clinical potential, if the type of biological activity needed does not require complete gene expression modulation (for example, in splicing redirection) then that may be easier to achieve with a single CPP attachment, especially for PNA or PMO. For siRNA or related reagents, these appear to be intrinsically active at lower concentrations and, in addition, targeting of cytosolic compartments where the RISC complex is believed to act may be more readily achievable than nuclear antisense activities. Further, some applications may not even need the CPPs to direct the oligonucleotide analog into cells, but into other biological entities such as virions, which may be easier to penetrate (e.g., the recent studies on CPP-PNA as HIV virucidal agents).[63,64]

Finally, the challenge for CPP delivery of oligonucleotides and siRNA will be to what extent any rules generated in cell lines can be applied in vivo. Clearly, peptides will need to be stabilized from proteolysis, in the same way as oligonucleotides require protection against nuclease attack and the use of D-peptides or other peptidomimetics will surely be essential. Further, toxicology and biodistribution studies of PNA–peptides in animals have only recently started to be published (e.g., Ref. [65]) and in vivo studies have yet to be reported on siRNA–peptide conjugates. Such experiences with conjugates in vivo will clearly shape the thinking of those research groups actively pursuing covalently attached CPPs as oligonucleotide and siRNA cell-delivery agents.

ACKNOWLEDGMENTS

This work is funded in part by a grant from EC Framework 5 (contract QLK3-CT-2002-01989). We thank David Owen, Donna Williams, and Matthew Watson for their research support. We also thank the many collaborating groups throughout the world that have helped us towards the goal of improving cell delivery of oligonucleotides and their analogs.

REFERENCES

1. Kurreck, J., Antisense technologies. Improvement through novel chemical modifications, *Eur. J. Biochem.*, 270, 1628, 2003.
2. Gleave, M.E. and Monia, B.P., Antisense therapy for cancer, *Nat. Rev. Cancer*, 5, 468, 2005.
3. Thierry, A.R. et al., Cellular uptake and intracellular fate of antisense oligonucleotides, *Curr. Opin. Mol. Ther.*, 5, 133, 2003.
4. Bennett, C.F. et al., Cationic lipids enhance cellular uptake and activity of phosphorothioate antisense oligonucleotides, *Mol. Pharmacol.*, 41, 1023, 1992.
5. Morrissey, D.V. et al., Potent and persistent in vivo anti-HBV activity of chemically modified siRNAs, *Nat. Biotechnol.*, 23, 1002, 2005.

6. Shi, F. and Hoekstra, D., Effective intracellular delivery of oligonucleotides in order to make sense of antisense, *J. Controlled Release*, 97, 189, 2004.

7. Lemaitre, M., Bayard, B., and Lebleu, B., Specific antiviral activity of poly (L-lysine)-conjugated oligodeoxyribonucleotide sequence complementary to vesicular stomatitis virus N protein mRNA initiation site, *Proc. Natl Acad. Sci. U.S.A.*, 84, 648, 1987.

8. Derossi, D. et al., The third helix of the Antennapedia homeodomain translocates through biological membranes, *J. Biol. Chem.*, 269, 10444, 1994.

9. Lochmann, D., Jauk, E., and Zimmer, A., Drug delivery of oligonucleotides by peptides, *Eur. J. Pharm. Biopharm.*, 58, 237, 2004.

10. Simeoni, F. et al., Insight into the mechanism of the peptide-based gene delivery system MPG: Implications for delivery of siRNA into mammalian cells, *Nucleic Acids Res.*, 31, 2717, 2003.

11. Stetsenko, D.A. et al., Peptide conjugates of oligonucleotides as enhanced antisense agents: A review, *Mol. Biol.*, 34, 852, 2000 (in Russian).

12. Gait, M.J., Peptide-mediated cellular delivery of antisense oligonucleotides and their analogues, *Cell. Mol. Life Sci.*, 60, 1, 2003.

13. Juliano, R.L., Peptide–oligonucleotide conjugates for the delivery of antisense and siRNA, *Curr. Opin. Mol. Ther.*, 7, 132, 2005.

14. Zatsepin, T.S. et al., Conjugates of oligonucleotides and analogues with cell penetrating peptides as gene silencing agents, *Curr. Pharm. Des.*, 11, 3639, 2005.

15. Opalinska, J.B. and Gewirtz, A.M., Nucleic-acid therapeutics: Basic principles and recent applications, *Nat. Rev.*, 1, 503, 2002.

16. Scherer, L.J. and Rossi, J.J., Approaches for the sequence-specific knockdown of mRNA, *Nat. Biotechnol.*, 21, 1457, 2003.

17. Zubin, E.M. et al., Oligonucleotide–peptide conjugates as potential antisense agents, *FEBS Lett.*, 456, 59, 1999.

18. Corey, D.R., Synthesis of oligonucleotide–peptide and oligonucleotide–protein conjugates, In *Methods of Molecular Biology. Bioconjugation Protocols: Strategies and Methods*, Niemeyer, C.M. ed., Vol. 283, Humana Press, Totowa, NJ, p. 197, 2004.

19. Stetsenko, D.A. and Gait, M.J., Chemical methods for peptide–oligonucleotide conjugate synthesis, In *Methods of Molecular Biology. Oligonucleotide Synthesis. Methods and Applications*, Herdewijn, P. ed., Vol. 288, Humana Press, Totowa, NJ, p. 205, 2004.

20. Haralambidis, J. et al., The synthesis of polyamide oligonucleotide conjugate molecules, *Nucleic Acids Res.*, 18, 493, 1990.

21. Stetsenko, D.A., Malakhov, A.D., and Gait, M.J., Total stepwise solid-phase synthesis of oligonucleotide (3'-N)–peptide conjugates, *Org. Lett.*, 4, 3259, 2002.

22. Basu, S. and Wickstrom, E., Solid-phase synthesis of a D-peptide–phosphorothioate oligodeoxynucleotide conjugate from two arms of a polyethylene glycol-polystyrene support, *Tetrahedron Lett.*, 36, 4943, 1995.

23. Antopolsky, M. et al., Towards a general method for the stepwise solid-phase synthesis of peptide–oligonucleotide conjugates, *Tetrahedron Lett.*, 43, 527, 2002.

24. De Napoli, L. et al., A new solid-phase synthesis of oligonucleotides 3'-conjugated with peptides, *Bioorg. Med. Chem.*, 7, 395, 1999.

25. Stetsenko, D.A. and Gait, M.J., A convenient solid-phase method for synthesis of 3'-conjugates of oligonucleotides, *Bioconjug. Chem.*, 12, 576, 2001.

26. Stetsenko, D.A. and Gait, M.J., Efficient conjugation of peptides to oligonucleotides by 'native ligation', *J. Org. Chem.*, 65, 4900, 2000.

27. Zatsepin, T.S. et al., Synthesis of peptide–oligonucleotide conjugates with single and multiple peptides attached to 2'-aldehydes through thiazolidine, oxime and hydrazine linkages, *Bioconjug. Chem.*, 13, 822, 2002.

28. Zatsepin, T.S. et al., Synthesis and properties of modified oligodeoxyribonucleotides containing 2'-O-(2,3-dihydroxypropyl)uridine and 2'-O-(2-oxoethyl)uridine, *Russ. J. Bioorg. Chem.*, 27, 45, 2001.

29. Turner, J.J., Arzumanov, A.A., and Gait, M.J., Synthesis, cellular uptake and HIV-1 Tat-dependent trans-activation inhibition activity of oligonucleotide analogues disulphide-conjugated to cell-penetrating peptides, *Nucleic Acids Res.*, 33, 27, 2005.

30. Turner, D.H. et al., Disulfide conjugation of peptides to oligonucleotides and their analogues, *Curr. Protoc. Nucleic Acids Chem.*, 428, 1, 2006.

31. Mier, W. et al., Peptide–PNA conjugates: Targeted transport of antisense therapeutics into tumors, *Angewandte Chem. Int. Ed.*, 2003, 1968, 2003.

32. Cutrona, G. et al., Effects in live cells of a c-myc anti-gene PNA linked to a nuclear localization signal, *Nat. Biotechnol.*, 18, 300, 2000.

33. Sazani, P. et al., Systemically delivered antisense oligomers upregulate gene expression in mouse tissues, *Nat. Biotechnol.*, 20, 1228, 2002.

34. Turner, J.J. et al., Cell-penetrating peptide conjugates of peptide nucleic acids (PNA) as inhibitors of HIV-1 Tat-dependent trans-activation in cells, *Nucleic Acids Res.*, 33, 6837, 2005.

35. Pooga, M. et al., Cell penetrating PNA constructs regulate galanin receptor levels and modify pain transmission in vivo, *Nat. Biotechnol.*, 16, 857, 1998.

36. Braun, K. et al., A biological transporter for the delivery of peptide nucleic acids (PNAs) to the nuclear compartment of living cells, *J. Mol. Biol.*, 318, 237, 2002.

37. Kaushik, N. et al., Anti-TAR polyamide nucleotide analog conjugated with a membrane-permeating peptide inhibits Human Immunodeficiency Virus Type I production, *J. Virol.*, 76, 3881, 2002.

38. Richard, J.-P. et al., Cell-penetrating peptides. A re-evaluation of the mechanism of cellular uptake, *J. Biol. Chem.*, 278, 585, 2003.

39. Allinquant, B. et al., Downregulation of amyloid precursor protein inhibits neurite outgrowth in vitro, *J. Cell Biol.*, 128, 919, 1995.

40. Troy, C.M. et al., Downregulation of Cu/Zn superoxide dismutase leads to cell death via the nitric oxide-peroxynitrite pathway, *J. Neurosci.*, 16, 253, 1996.

41. Astriab-Fisher, A. et al., Antisense inhibition of P-glycoprotein expression using peptide–oligonucleotide conjugates, *Biochem. Pharmacol.*, 60, 83, 2000.

42. Chen, C.P. et al., Synthesis of antisense oligonucleotide-peptide conjugate targeting to GLUT-1 in HepG-2 and MCF-7 cells, *Bioconjug. Chem.*, 13, 525, 2002.

43. Antopolsky, M. et al., Peptide–oligonucleotide phosphorothioate conjugates with membrane translocation and nuclear localization properties, *Bioconjug. Chem.*, 10, 598, 1999.

44. Astriab-Fisher, A. et al., Conjugates of antisense oligonucleotides with the Tat and Antennapedia cell-penetrating peptides: Effect on cellular uptake, binding to target sequences, and biologic actions, *Pharm. Res.*, 19, 744, 2002.

45. Koppelhus, U. et al., Cell-dependent differential cellular uptake of PNA, peptides and PNA–peptide conjugates, *Antisense Nucleic Acid Drug Dev.*, 12, 51, 2002.

46. Kaihatsu, K., Huffman, K.E., and Corey, D.R., Intracellular uptake and inhibition of gene expression by PNAs and PNA–peptide conjugates, *Biochemistry*, 43, 14340, 2004.

47. Kaushik, M.J., Basu, A., and Pandey, P.K., Inhibition of HIV-1 replication by anti-transactivation responsive polyamide nucleotide analog, *Antiviral Res.*, 56, 13, 2002.

48. Scarfi, S. et al., Modified peptide nucleic acids are internalized in mouse macrophages RAW 264.7 and inhibit inducible nitric oxide synthase, *FEBS Lett.*, 451, 264, 1999.

49. Robaczewska, M. et al., Sequence-specific inhibition of duck hepatitis B virus reverse transcription by peptide nucleic acids (PNA), *J. Hepatol.*, 42, 180, 2005.

50. Siwkowski, A.M. et al., Identification and functional validation of PNAs that inhibit murine CD40 expression by redirection of splicing, *Nucleic Acids Res.*, 32, 2695, 2004.

51. Arzumanov, A. et al., Inhibition of HIV-1 Tat-dependent *trans*-activation by steric block chimeric 2′-O-methyl/LNA oligoribonucleotides, *Biochemistry*, 40, 14645, 2001.

52. Arzumanov, A. et al., A structure-activity study of the inhibition of HIV-1 Tat-dependent *trans*-activation by mixmer 2′-O-methyl oligoribonucleotides containing locked nucleic acid (LNA), α-LNA or 2′-thio-LNA residues, *Oligonucleotides*, 13, 435, 2003.

53. Holmes, S.C., Arzumanov, A., and Gait, M.J., Steric inhibition of human immunodeficiency virus type-1 Tat-dependent *trans*-activation *in vitro* and in cells by oligonucleotides containing 2′-O-methyl G-clamp ribonucleoside analogues, *Nucleic Acids Res.*, 31, 2759, 2003.

54. Moulton, H.M. et al., HIV Tat peptide enhances cellular delivery of antisense morpholino oligomers, *Antisense Nucleic Acid Drug Dev.*, 13, 31, 2003.

55. Moulton, H.M. et al., Cellular uptake of antisense morpholino oligomers conjugated to arginine-rich peptides, *Bioconjug. Chem.*, 15, 290, 2004.

56. Neuman, B.W. et al., Antisense morpholino-oligomers directed against the 5′-end of the genome inhibit coronavirus proliferation and growth, *J. Virol.*, 78, 5891, 2004.

57. Neuman, B.W. et al., Inhibition, escape, and attenuated growth of severe acute respiratory syndrome coronavirus treated with antisense morpholino oligomers, *J. Virol.*, 79, 9665, 2005.

58. Jones, S.W., de Souza, P.M., and Lindsay, M.A., siRNA for gene silencing: A route to drug target discovery, *Curr. Opin. Pharmacol.*, 4, 522, 2004.

59. Hannon, G.J. and Rossi, J.J., Unlocking the potential of the human genome with RNA interference, *Nature*, 431, 371, 2004.

60. Manoharan, M., RNA interference and chemically modified small interfering RNAs, *Curr. Opin. Chem. Biol.*, 8, 1, 2004.

61. Muratovska, A. and Eccles, M.R., Conjugate for efficient delivery of short interfering RNA (siRNA) into mamalian cells, *FEBS Lett.*, 558, 63, 2004.

62. Chiu, Y.-L. et al., Visualizing a correlation between siRNA, localization, cellular uptake and RNAi in living cells, *Chem. Biol.*, 11, 1165, 2004.

63. Chaubey, B. et al., A PNA-Transportan conjugate targeted to the TAR region of the HIV-1 genome exhibits both antiviral and virucidal properties, *Virology*, 331, 418, 2005.

64. Tripathi, S. et al., Anti-HIV-1 activity of anti-TAR polyamide nucleic acid conjugated with various membrane transducing peptides, *Nucleic Acids Res.*, 33, 4345, 2005.

65. Boffa, L.C. et al., Therapeutically promising PNA complementary to a regulatory sequence for c-myc: Pharmacokinetics in an animal model of human Burkitt's lymphoma, *Oligonucleotides*, 15, 85, 2005.

19 Lego-Like Strategy for Assembling Multivalent Peptide-Based Vehicles Able to Deliver Molecular Cargoes into Cells

Stephen A. Waschuk, Gregory M.K. Poon, and Jean Gariépy

CONTENTS

19.1   INTRODUCTION

In protein architectures such as immunoglobulins, valency refers to the number of copies of a given antigen-binding domain. Any valency number higher than one will affect the avidity of an antibody for a given antigen. More recently, valency in terms of the number of protein transduction domains (PTDs) or nuclear localization signals (NLSs) integrated into a molecular cargo or a peptide scaffold has been shown to play a major role in enhancing the import and nuclear localization of peptide-based vehicles in cells.[1-5]

More precisely, peptide dendrimers containing several polycationic peptide transduction domains,[2,3,6-9] or the introduction of multiple PTDs within the scaffold of a cargo, can dramatically enhance the cellular uptake of the resulting conjugates.[5] Similarly, the rate of nuclear import is also dependent on the number of nuclear localization sequences associated with a cargo.[1,10]

Peptide oligomerization domains have been used to create multivalent molecules displaying high avidity for a ligand.[11-13] This review focuses on the design of peptide-based delivery vehicles and, in particular, on peptide scaffolds that can incorporate valency effects in terms of delivering molecules into cells.

19.1.1   THE MULTIVALENT DISPLAY OF PEPTIDE IMPORT SIGNALS LEADS TO THE EFFICIENT CELLULAR UPTAKE AND ROUTING OF MACROMOLECULES INTO CELLS

Multivalent interactions abound in biological systems. Examples include viral and bacterial adhesion to host cell surfaces,[14,15] host cell–cell interactions,[16] and many aspects of the humoral immune response.[17] In terms of receptor–ligand interactions, a multivalent ligand may interact with a multivalent receptor or a cluster of monovalent receptors.[18] Antibodies probably represent the most widely described examples of multivalent protein complexes. For instance, dimeric IgA molecules contain two antigen-binding domains while IgM pentamers display 10 such sites.[17,19] In addition to antibody–antigen interactions, mannose sugars present on the Fc domain of antibodies interact with mannose receptors on the surface on macrophages. This additional ligand allows for the simultaneous binding of multiple antibodies decorating the surface of foreign particles to macrophages and the activation of phagocytosis by such cells.[17]

One of the most promising areas of research for the potential treatment of diseases such as cancer rests upon the specific delivery of drugs and therapeutic macromolecules into target cells. For this objective to be realized, vectors that enable the efficient internalization of macromolecules into cells must first be developed. The intracellular routing of macromolecules, in particular peptide-based therapeutics, into specific organelles also represents a desirable feature that needs to be encoded by delivery vectors. Of the many approaches that have been investigated to date, peptide-based vehicles offer the broadest range of options in terms of guiding the delivery of macromolecules to and into cells. Molecular assemblies incorporating one or more such signals have been constructed and shown to act as vehicles to modulate the localization, processing, and duration of a macromolecule inside cells,[2] all of which are important parameters in the targeted delivery of therapeutic agents.

With reference to cellular import of cargo macromolecules, polycationic peptide sequences such as polylysine, polyarginine, and polyornithine sequences are known for their cell-penetrating properties.[20–25] These PTDs have commonly been used to shuttle macromolecules such as DNA or proteins across cellular membranes. Very short, arginine-rich peptides have recently been identified as PTDs. They include a short region of the HIV-1 transactivator of transcription domain (Tat peptide; GRKKRRQRRRAP, residues 48–60), the third helix of the Drosophila Antennapedia homeodomain (Antp peptide; SGRQIKIWFQNRRMKWKKC, residues 43–58), and part of the herpesvirus protein VP22 (DAATATRGRSAASRPTER-PRAPARSASRPRRPVE, residues 267–301).[26–33] Similarly, lysine- and arginine-rich sequences, such as the SV40 large T-antigen nuclear localization sequence (NLS; TPPKKKRKVEDP) and its synthetic analogs, represent efficient signals for intracellular routing of cargoes into the nucleus.[1,10,34] While the effectiveness of these cell-penetrating and routing signal peptides have been found to depend on the identity of the peptide and the cell type under study, the presentation of such sequences in association with cargoes universally increases their import and routing into cells.[4]

19.1.2 THERMODYNAMIC AND MECHANISTIC BASES OF MULTIVALENCY

It is widely acknowledged that nature exploits multivalency to compensate for weak binding by monovalent interactions. Without regard to mechanism, a general treatment of the thermodynamics of avidity enhancements found in oligovalent ligands can be formulated as follows and the observed stability of a multivalent ligand-receptor complex, ΔG_{obs}, is derived from the relationship:

$$\Delta G_{obs} = n\Delta G_{mono} + \Delta G_{int}, \tag{19.1}$$

where n is the valency of the ligand, ΔG_{mono} is the stability for the monovalent ligand, and ΔG_{int} captures the aggregate positive and negative contributions of polyvalency to ΔG_{obs}. Recently, Kitov and Bundle[35] have provided a microscopic treatment of the interaction free energy, focusing on the highly favorable entropy contribution from the purely statistical degeneracy of the bound states Ω:

$$\Delta S_{avidity} = R \ln \Omega \tag{19.2}$$

(where R is the gas constant). This "avidity entropy" can become the dominant contributor to ΔG_{int} with increasing valency of the ligand and/or the binding sites of the receptor.

Mechanistically, avidity enhancement due to multivalency is primarily considered an entropic phenomenon, where the binding of the first element dramatically reduces the loss in translational and rotational entropy associated with binding of successive elements of the ligand. This initial binding event effectively results in a large increase in the local concentration (>1 M) of the recognition elements in the environment of the receptor (the chelate effect).[19,36,37] In other words, a multivalent ligand exhibits increased affinity for a receptor even if the receptor is only monovalent, because of the greatly increased local concentration of the recognition elements near the receptor.[18] Experimentally, however, the potential of the chelate effect is rarely fully realized, and although a significant enhancement in avidity is observed, binding of successive equivalents has generally been found to be negatively cooperative,[38] even with designer ligands.[39] This phenomenon is attributable to the lack of spatial complementarity between the ligand's recognition elements and the receptor,[17] and manifests itself thermodynamically as an increasingly unfavorable interaction entropy for successive equivalents while enthalpy is favorable or remains unchanged.[38,40,41]

Thus, thermodynamic investigations have highlighted the importance of the complementary geometry and conformational rigidity in designing a ligand or vehicle for multivalent display of recognition elements. Such considerations are relevant to the import of NLS-containing vehicles and their cargoes since such sequences are recognized by NLS-binding proteins.[42,43] In the case of cationic PTDs, their import into cells may occur through more than one mechanism.[44–50] Nevertheless, anionic glycosaminoglycans such as heparan sulfate have been implicated in their binding to cells.[51,52] The description of valency effects upon binding of PTDs to such cell surface molecules is expected to be complex, however, in view of the abundance of cationic and anionic groups present on PTDs and glycosaminoglycans and the stability of such ionic interactions under physiological conditions (0.15 M NaCl).

19.1.3 THE MULTIVALENT DISPLAY OF PTD AND NLS SEQUENCES ON CARGO MOLECULES

A number of approaches are available for incorporating more than one PTD molecule onto a cargo. These approaches can rely on simple strategies such as the aggregation/condensation/complexation of PTD peptides with DNA (plasmid) or the oligomerization of a PTD-containing protein.[4] Polycationic peptides (PTD as well as NLS sequences) have also been chemically crosslinked to polymers or related macromolecules.[23,27,29,31] The extent of crosslinking, and therefore valency in such cases, is determined by the availability of reactive groups on the solvent-accessible surface of the cargo. These coupling approaches can result in a relatively uniform distribution of conjugates in terms of their molecular weight range and expected valency. Direct coupling methods continue to be employed as exemplified by the recent production of nanoparticles decorated with protein transduction domains.[5]

Using solid-phase peptide synthesis methods, it is possible to construct well-defined constructs displaying multivalent protein transduction domains. Two major advantages of such methods are the ability to introduce nonnaturally occurring residues and the rapid creation of peptide dendrimers using α- or ε-amino groups of lysine. This peptide branching strategy, developed by Tam,[53–55] enables one to synthesize dendrimeric peptide-based vehicles displaying multiple PTD and NLS peptides. Specifically, our laboratory has exploited this synthesis method to generate

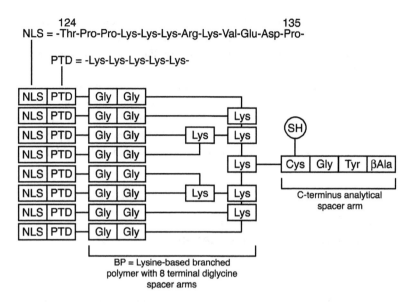

FIGURE 19.1 Structure of a synthetic peptide dendrimer (loligomer) that is able to penetrate eukaryotic cells and self-localize to the nucleus. It consists of a five-amino acid, C-terminal region (analytical arm) and eight identical N-terminal arms linked together through a branched lysine polymer. Each N-terminal arm contains a domain representing the nuclear localization signal of the SV40 large T-antigen residues 124–135 (NLS, Thr-Pro-Pro-Lys-Lys-Lys-Arg-Lys-Val-Glu-Asp-Pro) and a lysine pentapeptide acting as a protein transduction domain. The thiol side chain on the cysteine residue incorporated in the analytical arm can be derivatized with either a fluorescent group or a biotin moiety.

branched peptides called loligomers (Figure 19.1). The term loligomer is derived from the Latin root *loligo* referring to members of the squid family, thereby emphasizing the branched or squid-like nature of these peptides. In particular, loligomers have been designed to contain eight SV40 large T-antigen NLS and eight pentalysine stretches acting as PTD sequences, which allow such peptide dendrimers to penetrate cells and relocate to their nucleus.[3] Loligomers also include a C-terminal arm to introduce reporter groups into their structures.[9] These peptide dendrimers typically outperform their linear, monomeric homologues (one arm of the peptide dendrimer incorporating the PTD and NLS sequences) in terms of cellular uptake and nuclear localization into CHO cells.[3] Loligomers harboring antigenic epitopes have also been shown to generate cytotoxic T-cell responses.[7] In addition, loligomers are proven to be effective at importing plasmids into cells as the cationic arms associate via electrostatic interactions with the phosphate backbone of DNA.[8] As in the case of essentially all cationic PTDs-based delivery methods, a significant fraction of internalized complexes remain trapped in vesicular compartments, and are degraded or recycled out of the cell.[56] Approaches are thus needed to maximize their release to the cytosol. Such strategies will require the introduction of additional peptide (or polymer) domains into such delivery constructs.[50,57,58] Unfortunately, loligomers are assembled by solid-phase peptide synthesis, a chemical strategy that places limits on the size and homogeneity of the final constructs.

An alternative route for creating peptide dendrimers is to incorporate an oligomerization domain within linear peptide constructs. Such constructs would self-assemble into organized structures based on their programmed quaternary structure. This strategy can be readily realized by recombinant approaches, and offers a more defined, systematic approach for incorporating PTD and NLS peptides into fusion constructs.[59]

19.2 SEARCHING FOR USEFUL OLIGOMERIZATION DOMAINS

Having established that valency plays a role in enhancing the properties of PTD and NLS sequences, one needs to consider the features of oligomerization domains which may best be suited to display peptide routing domains. Two important features distinguish selfassembling peptides from synthetic dendrimers. First, oligomerization domains are held together by noncovalent interactions. Second, such domains spontaneously self-assemble according to the conformation encoded by the primary sequence of the oligomerization motif. This feature may be perturbed by the addition of peptide domains. A set of criteria was thus developed (Table 19.1) to guide the selection of selfassembling domains that may serve as appropriate scaffolds for integrating peptide routing domains such as PTDs into multitasking peptide delivery vehicles. A nonexhaustive list of oligomerization domains fitting these criteria was then drafted (Table 19.2), and the merits of the most promising candidates discussed in the context of designing peptide vehicles. One such motif was selected and linear peptides incorporating an oligomerization domain, as well as cell routing sequences, were synthesized and evaluated in cell-based assays (Section 19.2.8).

TABLE 19.1
Selection Criteria for Oligomerization Domains

1. Known mechanism of oligomerization and a defined high resolution structure
2. Length restriction for each monomer (less than 100 residues) with a preference for short peptide domains
3. Thermal stability
4. Monomer N- and C-termini located away from the oligomer core, with a uniform distribution surrounding the structure
5. Facile expression and purification forming a stable oligomeric structure in solution

---

TABLE 19.2
A Nonexhaustive List of Potential Self-Assembling Peptide Domains

Oligomerization State	Protein	Source	Oligomerization Domain	PDB ID
Dimer	HIV-1 capsid	Human immunodeficiency virus 1	C-terminus, residues 146–231	1A43[61]
	T-cell surface antigen CD2	*Rattus norvegicus*	N-terminus, residues 1–99	1A64[62]
	GAL4	*Saccharomyces cerevisiae*	Residues 50–106	1HBW[63]
	GCN4 leucine zipper	*Saccharomyces cerevisiae*	Residues 249–281	2ZTA[64]
	Human EB1	*Homo sapiens*	C-terminus, residues 185–255	1YIB[65]
Trimer	HLA-DR antigen associated invariant chain	*Homo sapiens*	Residues 118–192	1IIE[66]
	Pulmonary surfactant associated protein-D	*Homo sapiens*	Residues 223–257 with an additional seven collagen triplets	1M7L[67]
	Whisker antigen control protein	Bacteriophage T4	C-terminus, residues 457–483	1RFO[68]
Tetramer	Human tumor-suppressor protein p53	*Homo sapiens*	Residues 326–356	1AIE[69]
	Regulatory protein Mnt	*Salmonella* bacteriophage P22	C-terminus, residues 52–82	1QEY[70]
	Theoretical right-handed coiled-coil	Synthetic construct	35 Residues	1RH4[71]
	Heterogeneous nuclear ribonucleoprotein C	*Homo sapiens*	Residues 180–207	1TXP[72]
	Vasodilator-stimulated phosphoprotein	*Homo sapiens*	Residues 336–380	1USE[73]
Higher	Verotoxin 1 (pentamer)	*Escherichia coli*	69 Residues	1BOV[74]
	Small nuclear ribonucleoprotein homologue (SM-LIKE) (heptamer)	*Pyrobaculum aerophilum*	81 Residues	1LNX[75]

19.2.1 CRITERIA FOR SELECTING A SELF-ASSEMBLING PEPTIDE SCAFFOLD

From a practical standpoint, one of the most critical criteria in selecting an oligomerization domain is the knowledge of its mechanism of oligomerization. For this reason, the search was restricted to those proteins or peptides whose structure and oligomerization properties had been carefully defined, and moreover, had been expressed or synthesized as stand-alone protein domains, with structures that had been solved by NMR spectroscopy or by x-ray crystallography. Furthermore, an ideal oligomerization domain should yield a stable quaternary structure encoded by a minimal number of residues. The self-assembling domain should be encoded by less than 100 residues, with shorter domains being preferable. More specifically, by restricting the length of the self-assembling scaffold, one can control the size for the final constructs, which will include import and intracellular routing signals as well as a cargo.

One challenge in choosing a small oligomerization domain is that a short peptide may assemble into multimers that lack stability as a result of fewer possible intermonomeric interactions in comparison to larger self-assembling modules. Stability data for many of the potential oligomeric structures remain scarce. For this reason, a third criterion in the selection of self-assembling monomers is to infer the stability of an oligomeric structure by monitoring the nature and area of the interface between monomers in terms of defined hydrophobic and hydrogen bonding interactions.

An additional criterion is the location of the N- and C-termini of each monomer within its oligomeric structure. Exposed terminal ends represent the most practical sites to insert PTDs, as well as other domains. It is thus essential that residues located at either termini be exposed and not involved in destabilizing the quaternary structure. There may thus be a need to incorporate a spacer region to separate an import or routing signal from the oligomerization domain.

Finally, one should avoid sequences containing cysteine residues since they may lead to the formation of unplanned disulfide bridges between monomeric units and may potentially lead to aggregation.

19.2.2 PRACTICAL CONSIDERATIONS IN SELECTION OF SELF-ASSEMBLING OLIGOMERIZATION DOMAINS

A search was performed for candidate oligomerization domains by applying the criteria outlined in Section 19.2.1 to the structures of known domains deposited in the RCSB Protein Databank.[60] A nonexhaustive list of domains meeting such criteria is presented in Table 19.2. These domains have been produced in isolation of the native protein and their quaternary structure deduced from multidimensional NMR and x-ray diffraction studies. While it may be intuitively more logical to only consider peptide elements yielding the highest possible valency, this strategy may actually be counterproductive since the resulting increase in mass of the final peptide oligomers would make them more immunogenic and lower their ability to penetrate into solid tumors (in the case of designing cancer therapeutics).

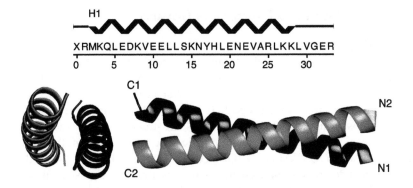

FIGURE 19.2 Secondary structure element and ribbon representation of the GCN4 leucine zipper dimerization domain from *Saccharomyces cerevisiae*. The ribbon representations (front and side views) were constructed using MacPyMOL (From DeLano, W.L., *The PyMOL Molecular Graphics System*, DeLano Scientific, San Carlos, CA, 2002.) and the available set of coordinates for the PDB structure 2ZTA. The linear depiction of secondary structure elements was obtained from the EBI PDBsum web resource. (From Laskowski, R.A., Chistyakov, V.V., Thornton, J.M., *Nucleic Acids Res* 33 (Database issue), D266–D268, 2005.)

19.2.3 SELF-ASSEMBLING DIMERIZATION DOMAINS

Dimerization domains form the largest category of elements identified in our search. Only five motifs were retained based on the criteria listed in Table 19.1. These motifs are mostly based on a coiled-coil (leucine zipper) structural element, exemplified by the dimerization domain of the yeast transcriptional activator GCN4 (sequence RMKQLEDKVEELLSKNYHLENEVARLKKLVGER, Figure 19.2).[64] The coiled-coil motif has been used as a dimerization element by numerous groups and offers great potential in terms of generating homo- and heterodimeric constructs.[64,76–79] Significantly, coiled-coils are made of tandem repeats of seven residues (heptad) forming two turns of a coiled-coil helix, and the consequences of altering heptad number and composition on the stability of the dimer have been well characterized.[80,81] The fact that several versions of this motif exist in nature suggest that it does represent an optimized scaffold for creating peptides incorporating up to four import/routing sequences (two C- and N-termini).

The N-terminus of the rat T-cell surface antigen CD2 (residues 1–99)[62] represents an alternate template to the coiled-coil motif for forming a bivalent scaffold. This peptide assembles into a dimer when its monomers exchange domains within its hydrophilic core, resulting in a markedly intertwined arrangement of mostly β strands. Significantly, while the wildtype protein exists in an equilibrium between monomers and metastable dimers, a small mutation can alter the monomer/dimer equilibrium towards the preferential formation of dimers (R87A), and a double deletion mutant leading to formation of a tetramer (ΔM46ΔK47).[62] The details of this interconversion will be discussed in greater detail in Section 19.2.7.

19.2.4 SELF-ASSEMBLING TRIMERIZATION DOMAINS

An interesting trimerization domain is a 64-residue long peptide derived from the coiled-coil domain of pulmonary surfactant associated protein D (GSPGLKGDKGIPGDKGAKGESGLPDVASLRQQVEALQGQVQHLQAAFS-QYKKVELPNGGIPHRD),[67] which forms a symmetric coiled-coil trimer with each coil comprising of four heptad repeats (Figure 19.3A; the ribbon representation is limited to those residues that appear in the NMR structure) with some stabilization resulting from interactions of the aromatic side chains within the coiled-coil segment (VASLRQQVEALQGQVQHLQAAFSQYKK). The full 64-residue peptide also includes seven collagen triplet repeats fused to the N-terminus of the coiled-coil domain, as well as several additional residues at the C-terminus. This peptide had previously been shown to form a stable trimer over a pH range of

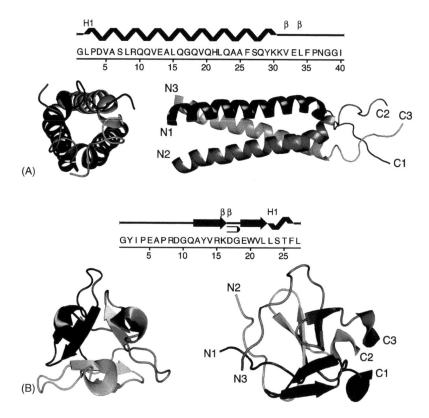

FIGURE 19.3 Secondary structure elements and ribbon representations of (A) the trimerization domain of the human pulmonary surfactant associated protein-D and (B) the trimeric foldon domain from the whisker antigen control protein fibritin trimerization domain from bacteriophage T4. The ribbon representations were constructed using MacPyMOL and the available sets of coordinates for the PDB structures 1M7L and 1RFO, respectively. The linear depictions of secondary structure elements were obtained from the EBI PDBsum web resource.

$3.0–8.5.$[82] Notably, much of the series of collagen triplets are absent in the crystal structure, suggesting they are unstructured following oligomerization and had previously been found to not be necessary for stable trimerization.[82] Furthermore, the additional C-terminal residues exhibit a random conformation in the oligomer, suggesting it might be possible to form a similarly stable trimer with fewer than 50 residues per monomer. The C-terminus of this domain is distal to the oligomerization motif and thus appears to be a site well suited for introducing PTD and NLS sequences.

The 27-residue foldon domain of bacteriophage T4 fibritin, representing the C-terminal residues 457–483 of the protein (GYIPEAPRDGQAYVRKD-GEWVLLSTFL), is viewed as critical in fibritin trimerization both in vitro and in vivo.[83,84] When isolated, the foldon domain trimerizes in solution, forming a largely β strand structure[68] unique to the trimers fitting our criteria (Figure 19.3B). Its rapid folding and stability[68] suggest that this protein could make a promising scaffold for building peptide trimers. PTDs and NLS modules could be positioned at the N-terminus of each chain within the trimeric motif.

19.2.5   SELF-ASSEMBLING TETRAMERIZATION DOMAINS

All of the tetramerization domains that meet our selection criteria are relatively short peptides of less than 50 residues in length.

The heterogeneous nuclear ribonucleoprotein C (hnRNP C) has been found to form a symmetric, coiled-coil tetramer in vivo.[85] The oligomerization of hnRNP C is

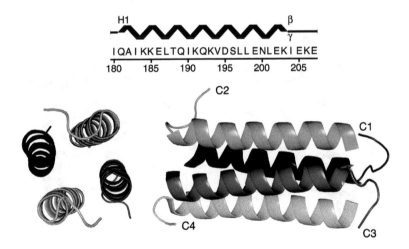

FIGURE 19.4 Secondary structure elements and ribbon representation of the tetramerization domain of human heterogeneous nuclear ribonucleoprotein C. The ribbon representations (front and side views) were constructed using MacPyMOL and the available set of coordinates for the PDB structure 1TXP. The linear depiction of secondary structure elements was obtained from the EBI PDBsum web resource.

encoded by a leucine zipper-like domain: a 28-amino acid long segment (IQAIKKELT QIKQKVDSLLENLEKIEKE) spanning residues 180–207 of its sequence.[72] This domain consists of more than three heptads spanning 22 residues, and results in an antiparallel arrangement of four α-helices, exposing four C-termini equally distributed at both ends of the tetramer (Figure 19.4). Interestingly, a high-affinity RNA-binding domain present within the sequence of hnRNP C comprises of a 40 residue basic sequence adjacent to the N-terminus of its tetramerization domain,[86] suggesting that this segment may potentially serve as a natural PTD domain. The same domain could possibly be re-engineered to bind nucleic acid-based agents (siRNA, for example).

The vasodilator-stimulated phosphoprotein contains a 45-residue tetramerization domain (residues 336–380, PSSSDYSDLQRVKQELLEEVKKELQKV-KEEIIEAFVQELRK RGSP) made of two consecutive 15-residue repeats (residues 344–358 and 359–373).[73] The structure forms a right-handed coiled-coil, with a characteristic pattern of hydrophobic interactions facing both the core and adjacent monomers. This tetramerization domain is highly stable, remaining fully folded over a wide range of temperatures (4–80°C at pH 6.0) and pHs (1–10 at 24°C),[87] which would be extremely useful if multivalent peptides were required to maintain their oligomeric state in such acidic compartments as the endosome or lysosome.

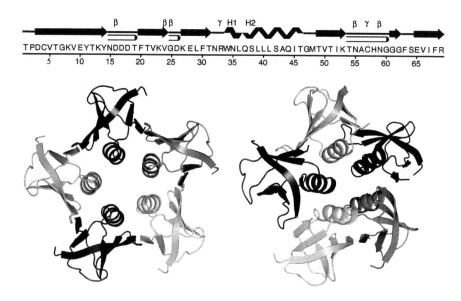

FIGURE 19.5 Secondary structure elements and three-dimensional ribbon representations of the pentameric B subunit of Shiga-like toxin 1 (verotoxin 1) from *Escherichia coli*. The ribbon representations (above and tilted views) were constructed using MacPyMOL and the available set of coordinates for the PDB structure 1BOV. The linear depiction of secondary structure elements was obtained from the EBI PDBsum web resource.

One of the most appealing oligomerization motifs, however, is the 31-residue long tetramerization domain of human p53 (residues 325–355). This motif has been used in the past to design multivalent scFv structures[11,12] and was recently used by our group as a proof of concept in designing self-assembling peptide-based vehicles (as discussed in Section 19.2.8).

19.2.6 SELF-ASSEMBLING HIGHER-ORDER OLIGOMERIZATION DOMAINS

A common, conserved motif in self-assembling pentamers is that found in AB_5 protein toxins, as exemplified by Shiga and Shiga-like toxin 1 (SLT-1; verotoxin 1). Each SLT-1 B-subunit monomer (69 amino acids) is structurally composed of two three-stranded antiparallel β-sheets and an α-helix.[74,88] Each monomer further forms a six-stranded antiparallel β-sheet with each of its two adjacent neighbors, forming a pentamer with the helices lining a pore in the center of the complex (Figure 19.5). Peptide domains have been introduced at the C-terminus of this B subunit without loss of pentamerization, suggesting that this element could serve as an excellent scaffold for designing high-valency peptide vehicles.[13]

19.2.7 INTERCONVERSION OF VALENCY

An important facet to consider in the use of self-assembling oligomers in the design of peptide scaffolds for cell import or intracellular routing is that, in many cases, simple substitutions or alterations within an oligomerization domain can destabilize or reorganize its quaternary structure. While the majority of these mutations leads to monomeric structures, there are examples where such mutations in fact increase the valency. A prime example that was previously mentioned is the N-terminal dimerization domain of the rat T-cell surface antigen CD2. The introduction of an alanine within the dimer core (R87A) caused an approximately 3.5-fold shift towards dimer formation (with the relative proportion of dimer to monomer shifting from 15% forming dimers in the wildtype sequence to approximately 55% in the R87A mutant, as calculated by gel permeation chromatography) (Figure 19.6A). More importantly, the removal of two residues within this domain (Δ46Δ47 mutant, removing Met46 and Lys47) induced two dimers to interact with each other, creating a tetramer (Figure 19.6B).[62]

Another example of mutations altering the valency of a self-assembling domain was observed in the immunoglobulin-binding domain B1 of streptococcal protein G. In these studies, the introduction of four mutations (L5V/F30V/Y33F/A34F) into the monomer caused the formation of a domain-swapped dimer.[89] A subsequent additional mutation within one residue in the dimer (L5V/A26F/F30V/Y33F/A34F) resulted in the formation of a tetramer.[90]

Finally, an Sm-like archaeal protein found in *Pyrobaculum aerophilum* reveals the possibility that, in addition to mutations either abolishing oligo-merization altogether, or increasing the valency of the oligomer, it is also possible to downgrade such valency. More precisely, the *P. aerophilum* Sm-like archaeal protein is 81 residues in length, forming a ring-like heptamer with a high

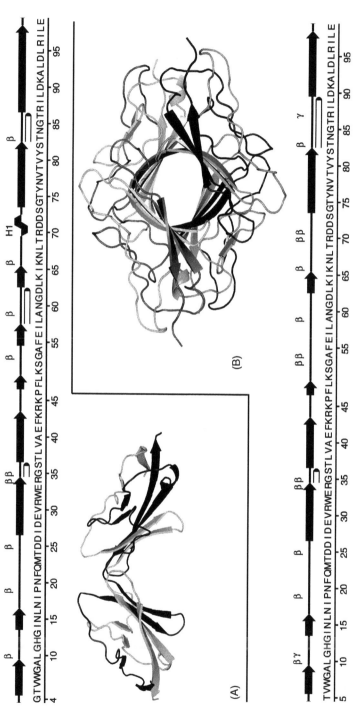

FIGURE 19.6 Secondary structure elements and ribbon representations of the oligomerization domain of the rat T-cell surface antigen CD2 (residues 1–99). Panel A represents the metastable, dimeric wildtype form of this domain, while Panel B depicts the tetrameric double deletion mutant (Δ46Δ47) form of the same motif. The deleted methionine and lysine residues of the Δ46Δ47 mutant do not appear in the secondary structure analyses of either the wildtype or mutant forms of the domain as they have been excluded from the original PDB file. The ribbon representations (top and side views) were constructed using MacPyMOL and the available sets of coordinates for the PDB structures 1A64 and 1A7B, respectively. The linear depictions of secondary structure elements were obtained from the EBI PDBsum web resource.

β-sheet content.[75] Interestingly, while there is no interaction in the crystal structure between the single cysteine residue in each monomer, its substitution to serine (C8S) results in the formation of pentamers, as judged by equilibrium sedimentation analysis.[75]

All these possibilities point to the fact that, as our knowledge and understanding of oligomerization domains increase, it might be possible to take an oligomer which most closely possesses desirable biochemical properties and alter its valency to optimize the multivalent display of routing sequences and PTDs.

19.2.8 A Practical Example: Peptide-Based Vehicles Incorporating the Oligomerization Domain of Human p53

Recent investigations in our laboratory have focused on dendrimers that incorporate the minimal tetramerization domain of human tumor suppressor p53 within their sequence. This tetramerization domain is a 31-amino acid long peptide representing residues 325–355 of human p53. It exists as a stable tetramer with a melting temperature near 80°C under physiological conditions.[91] The crystal and NMR structures of the tetramerization domain of p53 indicate that residues 325–255 forms a "dimer of dimers" in which each monomer adopts a β strand-turn-α helix fold (Figure 19.7A).[92] The p53tet domain offers the advantage of forming peptide dendrimers that approximate the eight-branch scaffolding of loligomers (tetramers with 4 N- and 4 C-terminal sites for introducing functional domains). In order to design peptide dendrimers that would exploit the valency effects previously observed for loligomers, we have recently constructed peptides (by synthetic or recombinant approaches) where PTDs, as well as other functional motifs, were directly fused to the N-terminus of the p53 tetramerization sequence. The resulting constructs were shown to assemble into peptide oligomers (tetramers). More specifically, a deca-arginyl (10R), deca-lysyl (10K) or the HIV-1 Tat sequence (GRKKRRQRRRAP; residues 48–60) was inserted at the N-terminus of the p53[tet]

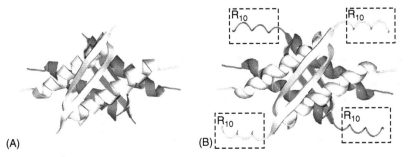

(A)   (B)

FIGURE 19.7 Ribbon structures of (A) the minimal p53 tetramerization domain (325–355) solved by NMR spectroscopy (PDB structure 1PES[92]) and (B) a deca-arginyl-p53tet fusion construct (10R-p53[tet]) where a sequence of 10 arginine residues (dashed box) was fused to each N-terminus of the minimal p53 tetramerization domain. This model was constructed using Swiss-Prot PDB Viewer.

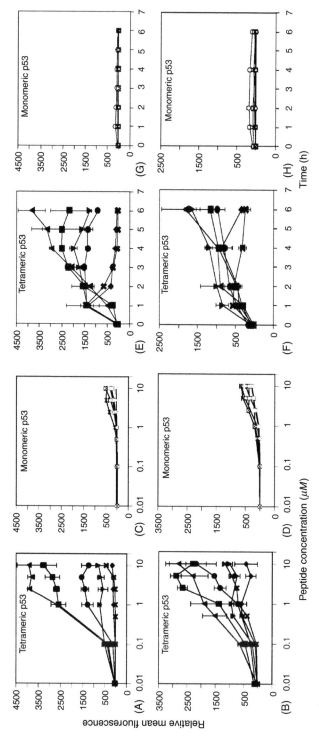

FIGURE 19.8 Internalization of fluorescent p53$^{tet/mono}$-based constructs by CHO (top row) and Vero (bottom row) cells as measured by flow cytometry. (A–D) Dependence of cellular import on peptide concentration. CHO and Vero cells were incubated for 4 h with increasing concentrations of synthetic p53 constructs labeled with fluorescein (A, B) Cellular uptake curves for all tetrameric p53 constructs: NLS-10K- p53^{tet} (●), 10K-p53^{tet} (■), NLS-10R-p53^{tet} (×), 10R-p53^{tet} (7), NLS-TAT-p53^{tet} (B), TAT-p53^{tet} (◆). (C, D) Cellular uptake curves for all monomeric p53 constructs: NLS-10K- p53^{mono} (○), 10K-p53^{mono} (□), NLS-10R-p53^{mono} (×), 10R-p53^{mono} (△), NLS-TAT-p53^{mono} (▽), TAT-p53^{mono} (◇). (E–H) Time dependence of cellular import. Fluorescein-labeled p53^{tet} constructs (0.5 µM; symbols are identical to those cited in panels A through D) were incubated with CHO or Vero cells and the level of cellular uptake measured by flow cytometry at different time intervals following their exposure to cells. Each data point represents the averaged relative mean fluorescence signal from experiments performed in triplicate.

domain. A ribbon model of the structure of the deca-arginyl p53tet peptide is shown in Figure 19.7B. Peptides containing both the PTD as well as the NLS sequence from the SV40 large T-antigen were also constructed. Monomeric versions (p53^{mono}) of these tetrameric peptide vehicles were generated by inserting an L334P mutation within the p53 tetrameric domain.[93] Routing functions, such as the cellular uptake and nuclear localization of the tetrameric constructs, were shown to be dramatically enhanced relative to their monomeric analogues and are strongly correlated with the valency of the signal sequences (Figure 19.8). These constructs were not toxic to cells. Flow cytometry results (Figure 19.8) and transfection assays indicated that tetravalent deca-arginyl peptides (10R-p53tet and NLS-10R-p53tet) were the most efficiently routed constructs into cells.[94]

19.3 CONCLUSIONS

Self-assembling oligomeric peptides are emerging as excellent scaffolds for properly displaying multiple import and intracellular routing peptide sequences. The resulting lego-like, small proteins can now be tailored to exploit valency effects in terms of more effectively carrying out multiple cell targeting tasks.

ACKNOWLEDGMENTS

Financial assistance from the Canadian Breast Cancer Research Alliance in association with the Canadian Cancer Society (J.G.), as well as the Natural Sciences and Engineering Research Council (S.A.W.) is gratefully acknowledged.

REFERENCES

1. Dingwall, C., Sharnick, S.V., and Laskey, R.A., A polypeptide domain that specifies migration of nucleoplasmin into the nucleus, *Cell*, 30(2), 449–458, 1982.
2. Gariepy, J. and Kawamura, K., Vectorial delivery of macromolecules into cells using peptide-based vehicles, *Trends Biotechnol.*, 19(1), 21–28, 2001.
3. Sheldon, K., Liu, D., Ferguson, J., and Gariepy, J., Loligomers: Design of de novo peptide-based intracellular vehicles, *Proc. Natl Acad. Sci. U.S.A.*, 92(6), 2056–2060, 1995.
4. Sung, M., Poon, G.M.K., Gariépy, J., The importance of valency in enhancing the import and cell routing potential of protein transduction domain-containing molecules, *Biochim. Biophys. Acta Biomem.*, 2006 (epub).
5. Zhao, M., Kircher, M.F., Josephson, L., and Weissleder, R., Differential conjugation of tat peptide to superparamagnetic nanoparticles and its effect on cellular uptake, *Bioconjug. Chem.*, 13(4), 840–844, 2002.
6. Bisland, S.K., Singh, D., and Gariepy, J., Potentiation of chlorin e6 photodynamic activity in vitro with peptide-based intracellular vehicles, *Bioconjug. Chem.*, 10(6), 982–992, 1999.
7. Kawamura, K.S., Su, R.C., Nguyen, L.T., Elford, A.R., Ohashi, P.S., and Gariepy, J., In vivo generation of cytotoxic T cells from epitopes displayed on peptide-based delivery vehicles, *J. Immunol.*, 168(11), 5709–5715, 2002.

8. Singh, D., Bisland, S.K., Kawamura, K., and Gariepy, J., Peptide-based intracellular shuttle able to facilitate gene transfer in mammalian cells, *Bioconjug. Chem.*, 10(5), 745–754, 1999.

9. Singh, D., Kiarash, R., Kawamura, K., LaCasse, E.C., and Gariepy, J., Penetration and intracellular routing of nucleus-directed peptide-based shuttles (loligomers) in eukaryotic cells, *Biochemistry*, 37(17), 5798–5809, 1998.

10. Dworetzky, S.I., Lanford, R.E., and Feldherr, C.M., The effects of variations in the number and sequence of targeting signals on nuclear uptake, *J. Cell. Biol.*, 107(4), 1279–1287, 1988.

11. Pluckthun, A. and Pack, P., New protein engineering approaches to multivalent and bispecific antibody fragments, *Immunotechnology*, 3(2), 83–105, 1997.

12. Rheinnecker, M., Hardt, C., Ilag, L.L., Kufer, P., Gruber, R., Hoess, A., Lupas, A., Rottenberger, C., Pluckthun, A., and Pack, P., Multivalent antibody fragments with high functional affinity for a tumor-associated carbohydrate antigen, *J. Immunol.*, 157(7), 2989–2997, 1996.

13. Zhang, J., Tanha, J., Hirama, T., Khieu, N.H., To, R., Tong-Sevinc, H., Stone, E., Brisson, J.R., and MacKenzie, C.R., Pentamerization of single-domain antibodies from phage libraries: A novel strategy for the rapid generation of high-avidity antibody reagents, *J. Mol. Biol.*, 335(1), 49–56, 2004.

14. Karlsson, K.A., Animal glycosphingolipids as membrane attachment sites for bacteria, *Annu. Rev. Biochem*, 58, 309–350, 1989.

15. Lineberger, J.E., Danzeisen, R., Hazuda, D.J., Simon, A.J., and Miller, M.D., Altering expression levels of human immunodeficiency virus type 1 gp120-gp41 affects efficiency but not kinetics of cell-cell fusion, *J. Virol.*, 76(7), 3522–3533, 2002.

16. Denda-Nagai, K. and Irimura, T., MUC1 in carcinoma-host interactions, *Glycoconj. J.*, 17(7–9), 649–658, 2000.

17. Mammen, M., Choi, S.-K., and Whitesides, G.M., Polyvalent interactions in biological systems: Implications for design and use of multivalent ligands and inhibitors, *Angew. Chem. Int. Ed.*, 37(20), 2754–2794, 1998.

18. Kiessling, L.L., Gestwicki, J.E., and Strong, L.E., Synthetic multivalent ligands in the exploration of cell-surface interactions, *Curr. Opin. Chem. Biol.*, 4(6), 696–703, 2000.

19. Crothers, D.M. and Metzger, H., The influence of polyvalency on the binding properties of antibodies, *Immunochemistry*, 9(3), 341–357, 1972.

20. Kollen, W.J., Schembri, F.M., Gerwig, G.J., Vliegenthart, J.F., Glick, M.C., and Scanlin, T.F., Enhanced efficiency of lactosylated poly-L-lysine-mediated gene transfer into cystic fibrosis airway epithelial cells, *Am. J. Respir. Cell. Mol. Biol.*, 20(5), 1081–1086, 1999.

21. Midoux, P., Mendes, C., Legrand, A., Raimond, J., Mayer, R., Monsigny, M., and Roche, A.C., Specific gene transfer mediated by lactosylated poly-L-lysine into hepatoma cells, *Nucleic Acids Res.*, 21(4), 871–878, 1993.

22. Mitchell, D.J., Kim, D.T., Steinman, L., Fathman, C.G., and Rothbard, J.B., Polyarginine enters cells more efficiently than other polycationic homopolymers, *J. Pept. Res.*, 56(5), 318–325, 2000.

23. Ryser, H.J., A membrane effect of basic polymers dependent on molecular size, *Nature*, 215(104), 934–936, 1967.

24. Ryser, H.J., Uptake of protein by mammalian cells: An underdeveloped area. The penetration of foreign proteins into mammalian cells can be measured and their functions explored, *Science*, 159(813), 390–396, 1968.

25. Ryser, H.J. and Hancock, R., Histones and basic polyamino acids stimulate the uptake of albumin by tumor cells in culture, *Science*, 150(695), 501–503, 1965.

26. Akhlynina, T.V., Jans, D.A., Rosenkranz, A.A., Statsyuk, N.V., Balashova, I.Y., Toth, G., Pavo, I., Rubin, A.B., and Sobolev, A.S., Nuclear targeting of chlorin e6 enhances its photosensitizing activity, *J. Biol. Chem.*, 272(33), 20328–20331, 1997.

27. Derossi, D., Joliot, A.H., Chassaing, G., and Prochiantz, A., The third helix of the Antennapedia homeodomain translocates through biological membranes, *J. Biol. Chem.*, 269(14), 10444–10450, 1994.

28. Elliott, G. and O'Hare, P., Intercellular trafficking and protein delivery by a herpesvirus structural protein, *Cell*, 88(2), 223–233, 1997.

29. Ivanova, M.M., Rosenkranz, A.A., Smirnova, O.A., Nikitin, V.A., Sobolev, A.S., Landa, V., Naroditsky, B.S., and Ernst, L.K., Receptor-mediated transport of foreign DNA into preimplantation mammalian embryos, *Mol. Reprod. Dev.*, 54(2), 112–120, 1999.

30. Nori, A., Jensen, K.D., Tijerina, M., Kopeckova, P., and Kopecek, J., Tat-conjugated synthetic macromolecules facilitate cytoplasmic drug delivery to human ovarian carcinoma cells, *Bioconjug. Chem.*, 14(1), 44–50, 2003.

31. Phelan, A., Elliott, G., and O'Hare, P., Intercellular delivery of functional p53 by the herpesvirus protein VP22, *Nat. Biotechnol.*, 16(5), 440–443, 1998.

32. Sobolev, A.S., Rosenkranz, A.A., Smirnova, O.A., Nikitin, V.A., Neugodova, G.L., Naroditsky, B.S., Shilov, I.N., Shatski, I.N., and Ernst, L.K., Receptor-mediated transfection of murine and ovine mammary glands in vivo, *J. Biol. Chem.*, 273(14), 7928–7933, 1998.

33. Vives, E., Brodin, P., and Lebleu, B., A truncated HIV-1 Tat protein basic domain rapidly translocates through the plasma membrane and accumulates in the cell nucleus, *J. Biol. Chem.*, 272(25), 16010–16017, 1997.

34. Goldfarb, D.S., Gariepy, J., Schoolnik, G., and Kornberg, R.D., Synthetic peptides as nuclear localization signals, *Nature*, 322(6080), 641–644, 1986.

35. Kitov, P.I. and Bundle, D.R., On the nature of the multivalency effect: A thermodynamic model, *J. Am. Chem. Soc.*, 125(52), 16271–16284, 2003.

36. Page, M.I. and Jencks, W.P., Entropic contributions to rate accelerations in enzymic and intramolecular reactions and the chelate effect, *Proc. Natl Acad. Sci. U.S.A.*, 68(8), 1678–1683, 1971.

37. Schwarzenbach, G., Der Chelateffekt, *Helv. Chim. Acta*, 35(7), 2344–2363, 1952.

38. Margossian, S.S. and Lowey, S., Interaction of myosin subfragments with F-actin, *Biochemistry*, 17(25), 5431–5439, 1978.

39. Dam, T.K., Roy, R., Page, D., and Brewer, C.F., Negative cooperativity associated with binding of multivalent carbohydrates to lectins. Thermodynamic analysis of the "multivalency effect", *Biochemistry*, 41(4), 1351–1358, 2002.

40. Christensen, T., Gooden, D.M., Kung, J.E., and Toone, E.J., Additivity and the physical basis of multivalency effects: A thermodynamic investigation of the calcium EDTA interaction, *J. Am. Chem. Soc.*, 125(24), 7357–7366, 2003.

41. Dam, T.K., Roy, R., Page, D., and Brewer, C.F., Thermodynamic binding parameters of individual epitopes of multivalent carbohydrates to concanavalin a as determined by "reverse" isothermal titration microcalorimetry, *Biochemistry*, 41(4), 1359–1363, 2002.

42. Kaffman, A. and O'Shea, E.K., Regulation of nuclear localization: A key to a door, *Annu. Rev. Cell. Dev. Biol.*, 15, 291–339, 1999.

43. Weis, K., Importins exportins: How to get in and out of the nucleus, *Trends Biochem. Sci.*, 23(5), 185–189, 1998.

44. Ferrari, A., Pellegrini, V., Arcangeli, C., Fittipaldi, A., Giacca, M., and Beltram, F., Caveolae-mediated internalization of extracellular HIV-1 tat fusion proteins visualized in real time, *Mol. Ther.*, 8(2), 284–294, 2003.

45. Fittipaldi, A., Ferrari, A., Zoppe, M., Arcangeli, C., Pellegrini, V., Beltram, F., and Giacca, M., Cell membrane lipid rafts mediate caveolar endocytosis of HIV-1 Tat fusion proteins, *J. Biol. Chem.*, 278(36), 34141–34149, 2003.

46. Fuchs, S.M. and Raines, R.T., Pathway for polyarginine entry into mammalian cells, *Biochemistry*, 43(9), 2438–2444, 2004.

47. Kaplan, I.M., Wadia, J.S., Dowdy, S.F., and Cationic, T.A.T., Cationic TAT peptide transduction domain enters cells by macropinocytosis, *J. Control Release*, 102(1), 247–253, 2005.

48. Richard, J.P., Melikov, K., Brooks, H., Prevot, P., Lebleu, B., and Chernomordik, L.V., Cellular uptake of unconjugated TAT peptide involves clathrin-dependent endocytosis and heparan sulfate receptors, *J. Biol. Chem.*, 280(15), 15300–15306, 2005.

49. Richard, J.P., Melikov, K., Vives, E., Ramos, C., Verbeure, B., Gait, M.J., Chernomordik, L.V., and Lebleu, B., Cell-penetrating peptides. A reevaluation of the mechanism of cellular uptake, *J. Biol. Chem.*, 278(1), 585–590, 2003.

50. Wadia, J.S., Stan, R.V., and Dowdy, S.F., Transducible TAT-HA fusogenic peptide enhances escape of TAT-fusion proteins after lipid raft macropinocytosis, *Nat. Med.*, 10(3), 310–315, 2004.

51. Console, S., Marty, C., Garcia-Echeverria, C., Schwendener, R., and Ballmer-Hofer, K., Antennapedia and HIV transactivator of transcription (TAT) "protein transduction domains" promote endocytosis of high molecular weight cargo upon binding to cell surface glycosaminoglycans, *J. Biol. Chem.*, 278(37), 35109–35114, 2003.

52. Suzuki, T., Futaki, S., Niwa, M., Tanaka, S., Ueda, K., and Sugiura, Y., Possible existence of common internalization mechanisms among arginine-rich peptides, *J. Biol. Chem.*, 277(4), 2437–2443, 2002.

53. Nardelli, B. and Tam, J.P., Cellular immune responses induced by in vivo priming with a lipid-conjugated multimeric antigen peptide, *Immunology*, 79(3), 355–361, 1993.

54. Tam, J.P., Synthetic peptide vaccine design: Synthesis and properties of a high-density multiple antigenic peptide system, *Proc. Natl Acad. Sci. U.S.A.*, 85(15), 5409–5413, 1988.

55. Tam, J.P. and Spetzler, J.C., Multiple antigen peptide system, *Methods Enzymol.*, 289, 612–637, 1997.

56. Fischer, R., Kohler, K., Fotin-Mleczek, M., and Brock, R., A stepwise dissection of the intracellular fate of cationic cell-penetrating peptides, *J. Biol. Chem.*, 279(13), 12625–12635, 2004.

57. Pack, D.W., Putnam, D., and Langer, R., Design of imidazole-containing endosomolytic biopolymers for gene delivery, *Biotechnol. Bioeng.*, 67(2), 217–223, 2000.

58. Putnam, D., Gentry, C.A., Pack, D.W., and Langer, R., Polymer-based gene delivery with low cytotoxicity by a unique balance of side-chain termini, *Proc. Natl Acad. Sci. U.S.A.*, 98(3), 1200–1205, 2001.

59. Kopecek, J., Smart and genetically engineered biomaterials and drug delivery systems, *Eur. J. Pharm. Sci.*, 20(1), 1–16, 2003.

60. Berman, H.M., Westbrook, J., Feng, Z., Gilliland, G., Bhat, T.N., Weissig, H., Shindyalov, I.N., and Bourne, P.E., The Protein Data Bank, *Nucleic Acids Res.*, 28(1), 235–242, 2000.
61. Worthylake, D.K., Wang, H., Yoo, S., Sundquist, W.I., and Hill, C.P., Structures of the HIV-1 capsid protein dimerization domain at 2.6 A resolution, *Acta Crystallogr. D. Biol. Crystallogr*, 55(Pt 1), 85–92, 1999.
62. Murray, A.J., Head, J.G., Barker, J.J., and Brady, R.L., Engineering an intertwined form of CD2 for stability and assembly, *Nat. Struct. Biol.*, 5(9), 778–782, 1998.
63. Hidalgo, P., Ansari, A.Z., Schmidt, P., Hare, B., Simkovich, N., Farrell, S., Shin, E.J., Ptashne, M., and Wagner, G., Recruitment of the transcriptional machinery through GAL11P: Structure and interactions of the GAL4 dimerization domain, *Genes Dev.*, 15(8), 1007–1020, 2001.
64. O'Shea, E.K., Klemm, J.D., Kim, P.S., and Alber, T., X-ray structure of the GCN4 leucine zipper, a two-stranded, parallel coiled coil, *Science*, 254(5031), 539–544, 1991.
65. Slep, K.C., Rogers, S.L., Elliott, S.L., Ohkura, H., Kolodziej, P.A., and Vale, R.D., Structural determinants for EB1-mediated recruitment of APC and spectraplakins to the microtubule plus end, *J. Cell. Biol.*, 168(4), 587–598, 2005.
66. Jasanoff, A., Wagner, G., and Wiley, D.C., Structure of a trimeric domain of the MHC class II-associated chaperonin and targeting protein Ii, *EMBO J.*, 17(23), 6812–6818, 1998.
67. Kovacs, H., O'Donoghue, S.I., Hoppe, H.J., Comfort, D., Reid, K.B., Campbell, D., and Nilges, M., Solution structure of the coiled-coil trimerization domain from lung surfactant protein D, *J. Biomol. NMR*, 24(2), 89–102, 2002.
68. Guthe, S., Kapinos, L., Moglich, A., Meier, S., Grzesiek, S., and Kiefhaber, T., Very fast folding and association of a trimerization domain from bacteriophage T4 fibritin, *J. Mol. Biol.*, 337(4), 905–915, 2004.
69. Mittl, P.R., Chene, P., and Grutter, M.G., Crystallization and structure solution of p53 (residues 326–356) by molecular replacement using an NMR model as template, *Acta Crystallogr. D. Biol. Crystallogr.*, 54(Pt 1), 86–89, 1998.
70. Nooren, I.M., Kaptein, R., Sauer, R.T., and Boelens, R., The tetramerization domain of the Mnt repressor consists of two right-handed coiled coils, *Nat. Struct. Biol.*, 6(8), 755–759, 1999.
71. Harbury, P.B., Plecs, J.J., Tidor, B., Alber, T., and Kim, P.S., High-resolution protein design with backbone freedom, *Science*, 282(5393), 1462–1467, 1998.
72. Whitson, S.R., LeStourgeon, W.M., and Krezel, A.M., Solution structure of the symmetric coiled coil tetramer formed by the oligomerization domain of hnRNP C: Implications for biological function, *J. Mol. Biol.*, 350(2), 319–337, 2005.
73. Kuhnel, K., Jarchau, T., Wolf, E., Schlichting, I., Walter, U., Wittinghofer, A., and Strelkov, S.V., The VASP tetramerization domain is a right-handed coiled coil based on a 15-residue repeat, *Proc. Natl Acad. Sci. U.S.A.*, 101(49), 17027–17032, 2004.
74. Stein, P.E., Boodhoo, A., Tyrrell, G.J., Brunton, J.L., and Read, R.J., Crystal structure of the cell-binding B oligomer of verotoxin-1 from *E. coli*, *Nature*, 355(6362), 748–750, 1992.
75. Mura, C., Kozhukhovsky, A., Gingery, M., Phillips, M., and Eisenberg, D., The oligomerization and ligand-binding properties of Sm-like archaeal proteins (SmAPs), *Protein Sci.*, 12(4), 832–847, 2003.
76. O'Shea, E.K., Rutkowski, R., and Kim, P.S., Evidence that the leucine zipper is a coiled coil, *Science*, 243(4890), 538–542, 1989.

77. Graddis, T.J., Myszka, D.G., and Chaiken, I.M., Controlled formation of model homo- and heterodimer coiled coil polypeptides, *Biochemistry*, 32(47), 12664–12671, 1993.

78. Lavigne, P., Kondejewski, L.H., Houston, M.E. Jr, Sonnichsen, F.D., Lix, B., Sykes, B.D., Hodges, R.S., and Kay, C.M., Preferential heterodimeric parallel coiled-coil formation by synthetic Max and c-Myc leucine zippers: A description of putative electrostatic interactions responsible for the specificity of heterodimerization, *J. Mol. Biol.*, 254(3), 505–520, 1995.

79. Fairman, R., Chao, H.G., Lavoie, T.B., Villafranca, J.J., Matsueda, G.R., and Novotny, J., Design of heterotetrameric coiled coils: Evidence for increased stabilization by Glu($-$)–Lys($+$) ion pair interactions, *Biochemistry*, 35(9), 2824–2829, 1996.

80. Zhou, N.E., Kay, C.M., and Hodges, R.S., The net energetic contribution of interhelical electrostatic attractions to coiled-coil stability, *Protein Eng.*, 7(11), 1365–1372, 1994.

81. Kohn, W.D., Kay, C.M., and Hodges, R.S., Protein destabilization by electrostatic repulsions in the two-stranded alpha-helical coiled-coil/leucine zipper, *Protein Sci.*, 4(2), 237–250, 1995.

82. Hoppe, H.J., Barlow, P.N., and Reid, K.B., A parallel three stranded alpha-helical bundle at the nucleation site of collagen triple-helix formation, *FEBS Lett.*, 344(2–3), 191–195, 1994.

83. Tao, Y., Strelkov, S.V., Mesyanzhinov, V.V., and Rossmann, M.G., Structure of bacteriophage T4 fibritin: A segmented coiled coil and the role of the C-terminal domain, *Structure*, 5(6), 789–798, 1997.

84. Boudko, S., Frank, S., Kammerer, R.A., Stetefeld, J., Schulthess, T., Landwehr, R., Lustig, A., Bachinger, H.P., and Engel, J., Nucleation and propagation of the collagen triple helix in single-chain and trimerized peptides: Transition from third to first order kinetics, *J. Mol. Biol.*, 317(3), 459–470, 2002.

85. Tan, J.H., Kajiwara, Y., Shahied, L., Li, J., McAfee, J.G., and LeStourgeon, W.M., The bZIP-like motif of hnRNP C directs the nuclear accumulation of pre-mRNA and lethality in yeast, *J. Mol. Biol.*, 305(4), 829–838, 2001.

86. Shahied-Milam, L., Soltaninassab, S.R., Iyer, G.V., and LeStourgeon, W.M., The heterogeneous nuclear ribonucleoprotein C protein tetramer binds U1, U2, and U6 snRNAs through its high affinity RNA binding domain (the bZIP-like motif), *J. Biol. Chem.*, 273(33), 21359–21367, 1998.

87. Zimmermann, J., Labudde, D., Jarchau, T., Walter, U., Oschkinat, H., and Ball, L.J., Relaxation, equilibrium oligomerization, and molecular symmetry of the VASP (336–380) EVH2 tetramer, *Biochemistry*, 41(37), 11143–11151, 2002.

88. Fraser, M.E., Chernaia, M.M., Kozlov, Y.V., and James, M.N., Crystal structure of the holotoxin from Shigella dysenteriae at 2.5 A resolution, *Nat. Struct. Biol.*, 1(1), 59–64, 1994.

89. Byeon, I.J., Louis, J.M., and Gronenborn, A.M., A protein contortionist: Core mutations of GB1 that induce dimerization and domain swapping, *J. Mol. Biol.*, 333(1), 141–152, 2003.

90. Kirsten Frank, M., Dyda, F., Dobrodumov, A., and Gronenborn, A.M., Core mutations switch monomeric protein GB1 into an intertwined tetramer, *Nat. Struct. Biol.*, 9(11), 877–885, 2002.

91. Johnson, C.R., Morin, P.E., Arrowsmith, C.H., and Freire, E., Thermodynamic analysis of the structural stability of the tetrameric oligomerization domain of p53 tumor suppressor, *Biochemistry*, 34(16), 5309–5316, 1995.

92. Lee, W., Harvey, T.S., Yin, Y., Yau, P., Litchfield, D., and Arrowsmith, C.H., Solution structure of the tetrameric minimum transforming domain of p53, *Nat. Struct. Biol.*, 1(12), 877–890, 1994.

93. Davison, T.S., Yin, P., Nie, E., Kay, C., and Arrowsmith, C.H., Characterization of the oligomerization defects of two p53 mutants found in families with Li-Fraumeni and Li-Fraumeni-like syndrome, *Oncogene*, 17(5), 651–656, 1998.

94. Kawamura, K., Sung, M., Bolewska-Pedyczak, E., and Gariépy, J., Probing the impact of valency on the routing of arginine-rich peptides into eukaryotic cells, *Biochemistry*, 45(4), 1116–1127, 2006.

95. DeLano, W.L., The PyMOL Molecular Graphics System,. DeLano Scientific, San Carlos, CA, 2002.

96. Laskowski, R.A., Chistyakov, V.V., and Thornton, J.M., PDBsum more: New summaries and analyses of the known 3D structures of proteins and nucleic acids, *Nucleic Acids Res*, 33(Database issue), D266–D268, 2005.

20 Mechanistic Insights into Guanidinium-Rich Transporters Entering Cells

Jonathan B. Rothbard, Theodore C. Jessop,
Lisa Jones, Elena Goun, Rajesh Shinde,
Christopher H. Contag, and Paul A. Wender

CONTENTS

20.1 INTRODUCTION

Biological membranes have evolved in part to prevent biopolymers and polar molecules from passively entering cells.[1] Numerous organisms have developed proteins, many of which are transcription factors, that breach these biological barriers through a variety of mechanisms.[2] The protein HIV Tat, for example, when used in vitro, rapidly enters the cytosol (and nucleus) of a wide spectrum of cells after endocytosis.[3] However, the nine amino acid peptide required for the uptake of HIV Tat, residues 49–57 (RKKRRQRRR), appears itself to utilize an additional mechanism as evident from its uptake, even at 4°C, by a route differentiated from the intact protein.[4] We have found that guanidinium-rich oligomers enter suspension cells more effectively than the Tat nonamer,[5] often without the production of observable endocytotic vesicles.[6,7] We describe herein studies on the cellular uptake mechanism of guanidinium-rich transporters conjugated to small molecules (MW ca. <3000).

Pertinent to the formulation of a mechanism, our previous studies demonstrated that the guanidinium head groups of Tat 49–57 are critical for its uptake into cells. Replacement of any of the arginine residues with alanine diminished uptake. Conversely, replacement of all nonarginine residues in the Tat nonamer with arginines provided transporters that exhibit superior rates of uptake. Charge itself is necessary, but not sufficient, as is evident from the comparatively poor uptake of lysine nonamers.[5,6] The number of arginines is also important, with optimal uptake for oligomers of 7–15 residues.[6,8] Backbone chirality is not critical for uptake. Even the position of attachment and length of the side chains can be altered, as shown with guanidinium-rich peptoids that exhibit highly efficient uptake. Changes in the backbone composition and in the side chain spacing can also increase uptake.[6,9–12] Even highly branched guanidinium-rich oligosaccharides and dendrimers are efficient transporters.[7,13–15] In contrast to receptor-mediated uptake, an increase in conformational flexibility generally favors uptake.

20.2 ADAPTIVE PARTITIONING: THE CONVERSION OF WATER SOLUBLE TO MEMBRANE SOLUBLE CONJUGATES THROUGH ION-PAIRING OF GUANIDINIUM IONS AND FATTY ACIDS

Several mechanisms could accommodate the above structure-function relationships for guanidinium-rich transporters and some could operate concurrently. A receptor-mediated process is inconsistent with the broad range of structural modifications that promote uptake and especially the observation that more flexible systems work better. Conventional passive diffusion across the nonpolar interior of the plasma membrane is difficult to reconcile with the polarity of the arginine oligomers and the dependency of uptake on the number of charges. In contrast to passive diffusion in which a migrating conjugate maintains its polarity, the polar, positively charged guanidinium oligomers could adaptively diffuse into the nonpolar membrane by recruiting negatively charged cell surface constituents to transiently produce a less polar, ion pair complex. Indeed, the polarity and bioavailability of many proteins and peptides can be changed by the intentional preformation of a noncovalent complex with anionic agents such as sodium dodecyl sulfate (SDS).[16] To test whether a highly water-soluble guanidinium rich oligomer could be rendered lipid soluble through ion pair formation, a fluoresceinated arginine octamer (Fl-aca-D-Arg$_8$-CONH$_2$) was added to a bilayer of octanol and water. Not surprisingly, the highly polar charged system partitioned almost exclusively ($>95\%$) into the water layer (Figure 20.1). However, when a surrogate for a membrane-bound fatty acid salt, namely sodium laurate, was added to this mixture, the transporter partitioned

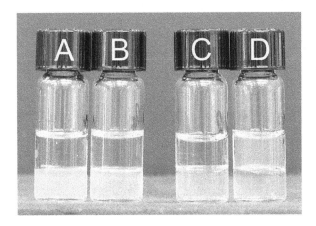

FIGURE 20.1 Octanol/water (upper and lower phase, respectively) partitioning of Fl-aca-Orn$_8$-CONH$_2$ and Fl-aca-D-Arg$_8$-CONH$_2$ alone (A, B) and after addition of sodium laurate (C, D).

completely (99%) into the octanol layer.[17] The relative partitioning was quantified by separation of the layers and analysis of the dissolved agents.

While other polycations, such as short ornithine oligomers, might partition through a similar ion pair mechanism, they are observed to be significantly less effective in cellular uptake than the arginine oligomers. This difference could arise, in part, from the more effective bidentate hydrogen bonding possible for guanidinium groups versus monodentate hydrogen bonding for the ammonium groups. Consistent with this analysis, when ornithine oligomers were submitted to the above two-phase partitioning experiments, they preferentially ($>$95%) stayed in the aqueous layer even with added sodium laurate (A and C).

When varying amounts of sodium laurate were added to fluorescently labeled octamers of arginine, lysine, ornithine, or residues 49–57 of Tat, the arginine octamers were found to require the least equivalents of laurate for uptake into octanol (Figure 20.2A). The Tat peptide (containing, six arginines, two lysines, and a glutamine) completely partitioned into octanol with 5 M equiv. of laurate. Under the same conditions (five equivalents) octamers of lysine and ornithine only partially entered the octanol, with marginally greater partitioning of the lysine derivative. Significantly, the rank hierarchy of the propensity of the peptides to form a lipophilic ion pair and enter octanol was identical to their ability to cross phospholipid membranes.[5,6] The ability of shorter chain fatty acids to mediate the formation of lipophilic ion pairs was also investigated. Addition of salts of acetic, hexanoic, and octanoic acid had little effect on the partitioning of a fluorescently labeled arginine oligomer, whereas the addition of sodium decanoate and laurate to the arginine octamer resulted in a high percentage ($>$80%) of the materials being partitioned into the octanol layer (Figure 20.2B). Collectively, these experiments indicate that guanidinium-rich transporters can change polarity through the adaptive formation of noncovalent complexes with membrane-embedded constituents.

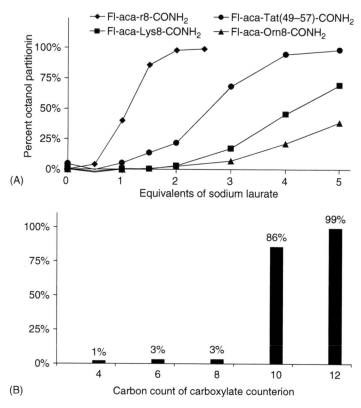

FIGURE 20.2 Role of the stoichiometry and the structure of fatty acids on partitioning of peptide–lipid ion pairs. Effect of equivalents of laurate on octanol partitioning of cationic peptides (A) and the octanol partitioning of Fl-aca-D-Arg$_8$-CONH$_2$ induced by the sodium salts of a variety of fatty acids (two equivalents per guanidinium) (B).

20.3 THE IMPORTANCE OF BIDENTATE HYDROGEN BONDING IN ADAPTIVE PARTITIONING AND CELLULAR UPTAKE

Critical to the adaptive diffusion mechanism is the special ability of guanidinium groups to transiently form bidentate hydrogen bonds with cell surface hydrogen bond acceptors. While incorporating the high basicity and dispersed cationic charge of a guanidinium group, alkylated guanidiniums would have an attenuated ability to form hydrogen bonds with phosphates, carboxylates, or sulfates and thus would be expected to be less effective transporters. This proved to be the case. When octamers of mono- and dimethylated arginine (Fl-aca-Arg^m$_8$-CONH$_2$, Fl-aca-Arg^{mm}$_8$-CONH$_2$), synthesized from the corresponding ornithine octamer, were assayed for cell entry, uptake of the former was reduced by 80% and the latter by greater than 95% when compared with an unalkylated arginine octamer (Figure 20.3).

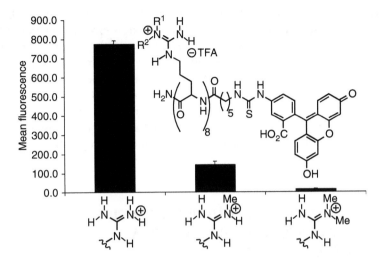

FIGURE 20.3 Reduced uptake of arginine oligomers with mono- and dimethylated guanidiniums. Mean fluorescence of Jurkat cells after treatment (5 min, 50 μM) with Fl-aca-D-Arg$_8$-CONH$_2$ (left), Fl-aca-Arg$^m{}_8$-CONH$_2$ (center), and Fl-aca-Arg$^{mm}{}_8$-CONH$_2$ (right); Fl = Fluorescein-HNC(S)-, aca = aminocaproic acid, Arg^m = N^G-methylarginine, Arg^{mm} = N^G, N^G-dimethylarginine.

The differential uptake by mono- and dimethylated arginines can be rationalized by considering the possible isomers of the alkylated guanidiniums and their ability to form bidentate hydrogen bonds. Molecular modeling of methylated guanidiniums[18] reveals the approximate energies of the various alkylated arginine isomers (Figure 20.4). Monomethylated arginines can assume four possible isomeric forms. Eclipsing of the methyl group with the side chain (configuration D) results in a steric repulsion of approximately 4 kcal/mol,

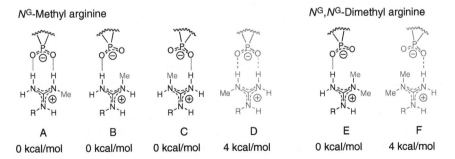

FIGURE 20.4 Configurations and bidentate hydrogen bonding of alkylated guanidiniums. Isomers A, B, and C are calculated to be of approximately equal energy; isomer D is calculated to be 4 kcal/mol higher in energy; isomer F is calculated to be of approximately 4 kcal/mol greater energy than isomer E.

significantly decreasing the population and contribution of this isomer. Only one (A) of the remaining three isomers (A–C) can form a bidentate hydrogen bond, collectively providing a statistical disadvantage relative to the unalkylated arginine. Similar analysis of the dimethylated arginine transporter reveals two possible isomers of unequal energy. The one with lower energy is unable to form bidentate hydrogen bonds, while the other could form a bidentate ion pair, but with a penalty of 4 kcal/mol due to steric interactions. Again, the observed reduction of cellular uptake for the dimethylated arginine transporters correlates with the ability of the head groups to form bidentate hydrogen bonds.

20.4   COUNTER IONS OF GUANIDINIUM-RICH TRANSPORTERS ARE EXCHANGED AT THE CELL SURFACE

If the guanidiniums of a transporter form bidendate ion pairs with anions on the cell surface, then the counter ions associated with the guanidinium groups must be exchanged on the cell surface and not transported into the cell. To visualize the proposed ion exchange, fluorescein was used as a counter ion by adding a molar equivalent of fluorescein to the free base of an octamer of arginine, formed by eluting from a hydroxide anion exchange column, to afford Ac-D-Arg$_8$-CONH$_2$ • 1x fluorescein salt. When human T-cells were treated with this salt for 5 min, washed, and analyzed by flow cytometry, fluorescent microscopy, and fluorescent confocal microscopy, a distinctly different pattern was seen from that observed when the cells were treated with the transporter covalently attached to the fluorescein. In contrast to the bimodal distribution of fluorescence, ranging over three orders of magnitude, exhibited by cells treated with Fl-aca-D-Arg$_8$-CONH$_2$ when analyzed by flow cytometry (Figure 20.5A), cells treated with Ac-D-Arg$_8$-CONH$_2$ • 1x fluorescein salt exhibited a relatively uniform distribution of fluorescence (Figure 20.5C). When the highly stained population resulting from treatment with the covalent conjugate was isolated by fluorescence activated cell sorting, and visualized by fluorescent microscopy and fluorescent confocal microscopy, the peptide was localized in the nuclei, the cytosol, and, most prominently, the nucleoli (Figure 20.5B). In contrast, microscopic analysis of cells treated with the salt revealed solely cell surface staining with a small degree of patching and capping (Figure 20.5D). Moreover, unlike cells treated with the covalent conjugate Fl-aca-D-Arg$_8$-CONH$_2$, the fluorescence of cells treated with Ac-D-Arg$_8$-CONH$_2$ • 1x fluorescein salt was not diminished by pretreatment with either 1% sodium azide or phosphate-buffered saline (PBS) with all sodium salts replaced with equimolar amounts of potassium salts (K+PBS). These data support the hypothesis that the polyguanidinyl transporter binds to anions on the cell surface with a concomitant ion exchange. If the counter ion of the guanidinium is polar and water soluble, it will diffuse into the medium. If the counter ion is hydrophobic, such as fluorescein, it will partition into the membrane.

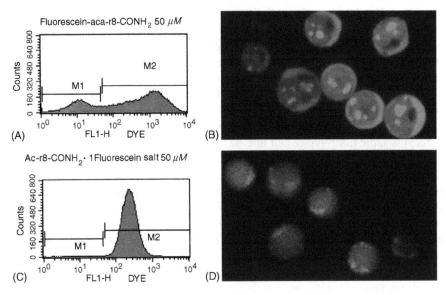

FIGURE 20.5 The cellular staining of lymphocytes differs when treated with either fluorescein covalently conjugated, or complexed as a salt, to guanidinium-rich transporters. Histogram of cellular fluorescence after treatment of the human T-lymphocyte cell line, Jurkat, after being incubated with Fl-aca-D-Arg$_8$-CONH$_2$ (A) or Ac-HN-D-Arg$_8$-CONH$_2$ • 1x fluorescein salt (C) (50 μM, for 5 min). Representative fluorescent micrographs of the treated lymphocytes isolated using gate M2 in (A) and (C) shown in (B) and (D), respectively.

20.5 ROLE OF MEMBRANE POTENTIAL IN TRANSPORTER TRANSLOCATION

While charge complementation with endogenous membrane constituents allows for adaptive entry into the membrane, it does not explain the driving force for passage through the membrane and the energy dependency of uptake observed in some studies. Given that phospholipid membranes in viable cells exhibit a membrane potential, the maintenance of which requires ATP, we reasoned that uptake of cationic-rich transporters might be driven by the voltage potential across most cell membranes.[19] To test this hypothesis, the membrane potential was reduced to close to zero by incubating the cells with an isotonic buffer with potassium concentrations equivalent to that found intracellularly. The intracellular concentration of K^+ in lymphocytes is ~ 140 mM, the extracellular concentration is ~ 5 mM, and it is primarily this concentration gradient that maintains the transmembrane potential.[19] Replacement of the sodium salts in PBS with equimolar amounts of the equivalent potassium salts afforded what was called K+PBS. To test whether the membrane potential in lymphocytes was a factor in transport of guanidinium-rich transporters, fluorescently labeled Tat 49–57 (Fl-aca-Tat$_{49-57}$-CONH$_2$) and an octamer of D-arginine (Fl-aca-D-Arg$_8$-CONH$_2$) were incubated individually with Jurkat cells for 5 min in either PBS or K+PBS. The cells were washed and analyzed by flow cytometry. Uptake was reduced by greater than 90% at all concentrations when the assay was carried out in the presence of a buffer with a high concentration of potassium (Figure 20.6A). The observed inhibition of uptake was equivalent to that seen when the cells were pretreated with sodium azide. The effects of high concentrations of potassium are not limited to lymphocytes as the uptake of Fl-aca-D-Arg$_8$-CONH$_2$ into two macrophage cell lines, J774 and RAW 264 (Figure 20.6B and C), was also inhibited.

To determine whether the reduction of membrane potential could affect endocytosis, the macrophage cell line, J774, was treated with Texas Red transferrin under conditions that permitted endocytosis and in the presence or absence of K+PBS. Visualization of the cellular fluorescence under these conditions revealed no significant difference in the amount of transferrin endocytosed (Figure 20.7). The ability to significantly limit cellular penetration of guanidinium-rich transporters by preincubation with K + PBS is in sharp contrast with its failure to block the uptake of transferrin, providing further evidence that an alternative mechanism to endocytosis is involved in the uptake of guanidinium-rich transporters.

To determine whether peptide translocation required a minimum voltage or varied continuously with the membrane potential, the uptake experiment of the

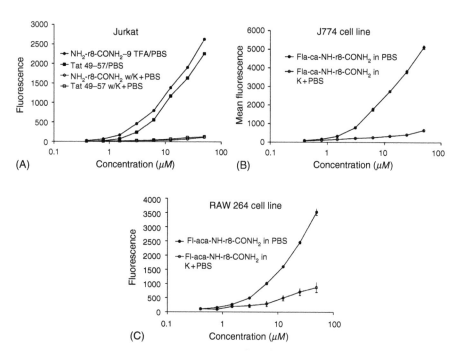

FIGURE 20.6 Cellular uptake of Fl-aca-D-Arg$_8$-CONH$_2$ (circles) and Fl-aca-tat$_{49-57}$-CONH$_2$ (squares) in PBS (solid markers) and K + PBS (outlined markers) in a lymphocyte (A) and two macrophage cell lines (B and C). The macrophage cell lines were trypsinized for 5 min; lymphocytes and macrophages were washed three times with either K+PBS or PBS and incubated with varying concentrations of the fluorescent peptides for 5 min. The cells were washed and analyzed by flow cytometry.

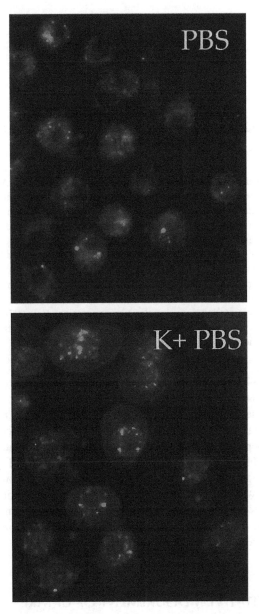

FIGURE 20.7 Demonstration that incubation with high potassium buffers does not inhibit endocytosis in a macrophage cell line. Fluorescent micrographs of J774 cells incubated with Texas Red transferrin at 37°C for 30 min in either PBS or K+PBS.

labeled arginine oligomer was repeated using a series of buffers whose potassium ion concentration varied between 140 mM and zero. The results demonstrated that uptake decreased with an increase in the external concentration of potassium (Figure 20.8A). The uptake, as measured by cellular fluorescence, appeared to vary

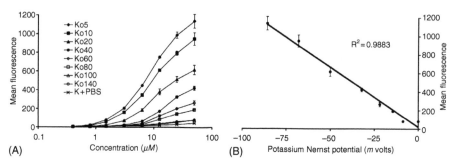

FIGURE 20.8 Differential inhibition of cellular uptake of Fl-aca-D-Arg$_8$-CONH$_2$ in the human lymphocyte cell line, Jurkat, by incubation with isotonic buffers containing increasing amounts of potassium ions. The concentration (mM) of the potassium ions in each of the buffers correspond to the K_o values shown in panel A. Cells were washed three times with the listed buffers, incubated with varying concentrations of Fl-aca-D-Arg$_8$-CONH$_2$ for 5 min in the same buffer. The cells were washed with and kept in the appropriate buffer and analyzed by flow cytometry. The uptake as measured by cellular fluorescence appeared to vary linearly with the K$^+$ Nernst potential calculated across the range of extracellular K$^+$ concentrations (B). Calculation of the K$^+$ Nernst potential (E_K) in cells incubated with buffers with varying K$^+$ concentrations was carried out using the Nernst equation, $E_K = (RT/F) \ln (K_o/K_i)$ where R = the gas constant, T = temperature, F = Faraday's constant, K_o = extracellular potassium ion concentration, and K_i = intracellular potassium ion concentration.

linearly with the potassium Nernst potential calculated across the range of extracellular potassium ion concentrations (Figure 20.8B).[20,21]

To explore whether high potassium buffers inhibited uptake by modulating the membrane potential, or by an alternative effect, lymphocytes were pretreated with gramicidin A, a pore-forming peptide known to reduce membrane potential,[22] prior to the addition of Fl-aca-D-Arg$_8$-CONH$_2$. This procedure reduced cellular uptake by more than 90% (Figure 20.9). The reciprocal experiment, hyperpolarizing the cell to increase uptake, was accomplished with valinomycin, a peptide antibiotic that selectively shuttles potassium ions across the membrane.[23] When Jurkat cells were preincubated with 50 μM valinomycin, the uptake of Fl-aca-D-Arg$_8$-CONH$_2$ was significantly increased (Figure 20.9).

20.6 QUANTITATIVE MEASUREMENT OF UPTAKE AND RELEASE OF A LUCIFERIN PRODRUG IN CELLS

A major focus of our group has been to use arginine-rich peptides to enhance delivery of small molecular weight therapeutics. Conjugation of the transport peptide to molecules such as cyclosporin A[24] and taxol[25] inhibits the drug from binding its receptor. Activity is restored only when the drug is released from the conjugate. Consequently, enhanced uptake of the conjugate will not be therapeutically useful unless the drug is released at the appropriate site and in a timely manner. This additional layer of complexity in drug design has led us to

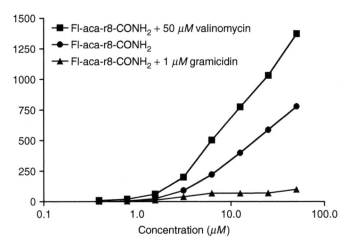

FIGURE 20.9 Cellular uptake of Fl-aca-D-Arg$_8$-CONH$_2$ into the human T-lymphocyte cell line, Jurkat, alone or with preincubation with valinomycin or gramicidin A.

develop assays that allow the measurement of both real-time uptake into cells and animals, and also the release of a drug/probe cargo after cell entry. We have shown that intracellular cargo release is possible and measurable when an oligoarginine transporter is attached through a cysteine disulfide bond to an otherwise cell-impermeable peptide cargo.[26] The resultant conjugate enters cells and is then cleaved to the free bioactive peptide, as determined in a functional assay for ischemic damage. This assay, however, does not lend itself to rapidly evaluating new transporters, linkers, or release systems as it is labor intensive, time consuming, difficult to quantify, and only indirectly measures release of the active cargo. To address these problems we have developed a new releasable linker system that is compatible with a range of transporters and that releases its cargo (luciferin) only after cell entry.[27] The selection of luciferin as a representative cargo that allows measurement of linker–transporter uptake and release in real time in luciferase transfected cells and potentially transgenic animals through the emission of light, a system that emulates drug uptake, intracellular release, and receptor interaction.

A major obstacle in implementing the above strategy initially proved to be the synthetic difficulty of making luciferin conjugates. While luciferin itself has figured prominently as a research tool for decades, little is known about its modification and no information is available on its attachment to a transporter through a releasable linker. After much experimentation, a concise solution to this synthetic problem was developed, as illustrated in Scheme 20.1. In this route, hydroxy thiol **1** is transformed with Aldrithiol to an activated disulfide **2**. The chloroformate **3** is then formed by reaction of disulfide **2** in dichloromethane or methylene chloride (DCM) with phosgene in toluene (20%). Due to the limited solubility of D-luciferin (**6**) in organic solvents, and the desire to avoid protecting groups, the organic solvent is removed in vacuo and the potassium salt of luciferin is added with aqueous base to the chloroformate **3** to form upon acidic workup the carbonate **4**. This carbonate serves as a reagent for conjugation to a variety of transporters. For the purposes of this

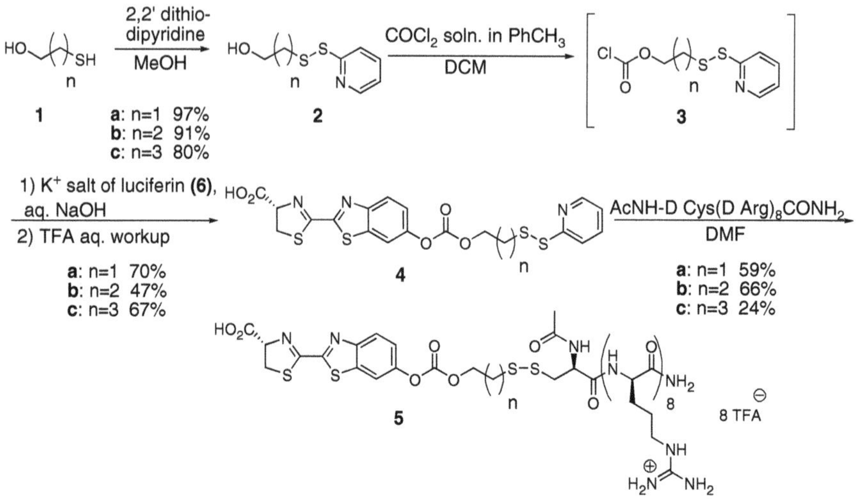

SCHEME 20.1

inaugural study, the thiopyridyl moiety of **4** was displaced with acylated D-cysteine D-octaarginine (AcNHcr8CONH$_2$) to give the transporter–linker conjugate **5**. The avoidance of protecting groups in this sequence provides a flexible, step economical route to these densely functionalized transporter conjugates and bodes well for the use of this system for the synthesis and study of other transporter–linker conjugates.

The stability of conjugate **5** was assayed by measuring its decomposition when incubated in Hepes-buffered saline (HBS, pH 7.4) at 37°C using analytical high performance liquid chromatography (HPLC). The half-lives of the conjugate was shown to be 11 h. The decomposition products were luciferin, alcohol **7**, and CO$_2$ as expected from slow hydrolysis of the carbonate (Scheme 20.2).

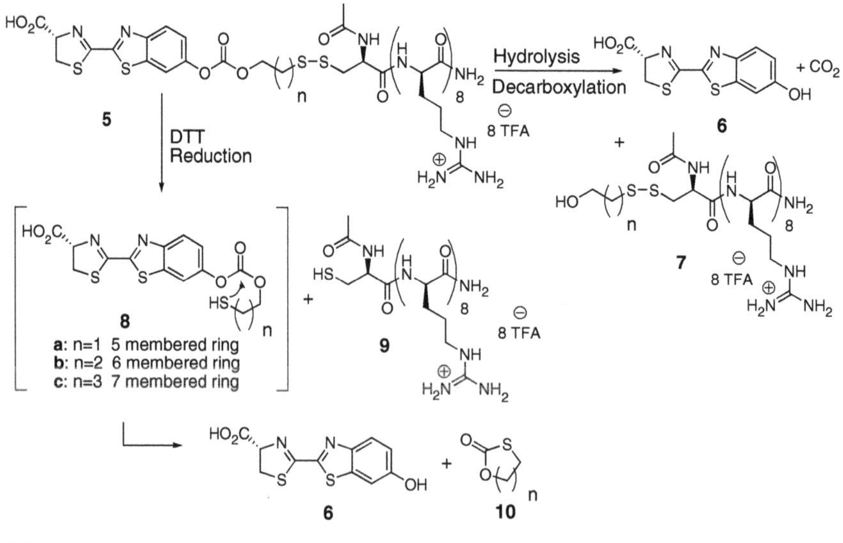

SCHEME 20.2

To determine whether release of luciferin from the conjugates could be induced by disulfide cleavage, as desired for intracellular release, each conjugate was incubated with HBS pH 7.4 with the addition of 10 mM dithiothreitol. The conjugate was cleaved in minutes. Free luciferin was formed without any observable intermediates.

In assays measuring uptake and release, luciferin, known to enter cells by passive diffusion, was used as a positive control. Varying concentrations of luciferin and carbonate **5b** were incubated separately with a prostate tumor cell line stably transfected with luciferase, PC3M-luc.[28]

The cells were plated at 60,000 cells per well in three separate 96-well, flat-bottomed plates 12 h prior to the assay. The plates were pulsed for 1 min with 50 and 25 μM of the potassium salt of luciferin (Xenogen, Alameda, CA) in either HBS pH 7.4 or HBS with all sodium salts replaced with equimolar amounts of potassium salts (K+HBS) in triplicate. The cells were washed once and the resultant fluorescence was measured using a charged coupled device camera and Living Image software ((IVIS200, Xenogen, Corp., Alameda, CA).[29] In both buffers, a significant amount of light was seen at the earliest times (approximately 90 sec after exposure to the drug), which rapidly decayed to background after approximately 300 sec (Figure 20.10A). The decay was sufficiently rapid that pulse times longer than 1 min do not result in greater observed luminescence (data not shown). Incubation of the cells treated with identical concentrations of luciferin in K+HBS did not inhibit luminescence. In fact, the presence of K+PBS resulted in approximately 20% more light than those incubated in HBS, demonstrating that uptake of luciferin was not dependent on the membrane potential (Figure 20.10B).

A distinctly different pattern of bioluminescence was observed when the PC3M-luc cells were exposed to 50 and 25 μM solutions of carbonate **5b** in either HBS or K+HBS. The cells were plated into 96-well plates and were pulsed for 1, 2, and 5 min with 50 and 25 μM releasable luciferin conjugate in the two buffers (Figure 20.11). At these concentrations and pulse times, the bioluminescence increased in the first few seconds and then decayed by 1800 sec (30 min) (Figure 20.11A). The pattern was consistent with the intracellular concentrations of luciferin increasing after the end of the pulse, after which the luciferin is steadily converted into photons. In contrast to luciferin, greater than 95% of the bioluminescence from cells pulsed with the octa D-arginine luciferin conjugate addition was inhibited when the assay was carried out in K+PBS. These experiments provide additional evidence that the transport of the majority of the octaarginine luciferin complex was dependent on the membrane potential.

These experiments emphasize that relatively large amounts of the octa D-arginine-luciferin conjugate enter cells in minutes (Figure 20.12). A comparison of the integrated areas under the curves of luminescence, which reflect the amount of luciferin being released intracellularly, after the different incubation times reveal that the signal approximately doubles when the pulse time increased from 1 to 2 min. However, while the signal doubles again after 5 min, the signal is no longer linear with time after 5 min. These data are very similar to quantitative measurements of the cellular uptake technetium–Tat chelates by Piwnica-Worms and colleagues.[30]

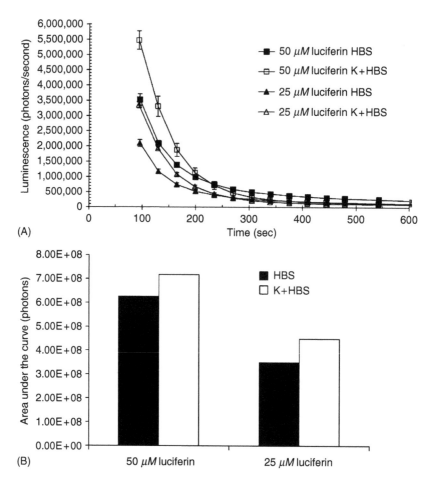

FIGURE 20.10 Observed bioluminescence from a transformed prostate cell line transfected with luciferase, PC3M-luc, when pulsed with luciferin for 1 min in HBS pH 7.4 or K+HBS pH 7.4. The real-time measurement of bioluminescence is shown in (A), while the total amount of photons as measured by integrating the area under the curves is shown in (B).

They demonstrated their radiolabeled peptide chelate approached high picomolar cytosolic concentrations in less than 10 min.

The synthesis and performance of these releasable luciferin conjugates establish an operationally facile method to quantify in real-time uptake and release of new or established transporters and linkers in a cellular assay that emulates drug-conjugate delivery into a cell, drug release, and drug turnover by an intracellular target. The newly introduced disulfide–carbonate linker system can be tuned for hydrolytic stability (from hours to days) without affecting its rapid rate of cargo release (minutes) in cells through disulfide cleavage. The linker–luciferin system was designed to allow quantification of uptake of various transporters in both transfected cells and transgenic animals, thereby providing a facile method to measure and correlate in vitro and in vivo activities in real time.

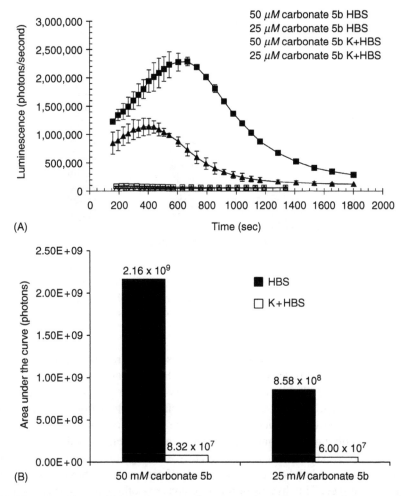

FIGURE 20.11 Observed bioluminescence from a transformed prostate cell line transfected with luciferase, PC3M-luc, when pulsed with carbonate **5b** for 2 min in HBS pH 7.4 or K + HBS pH 7.4. The real-time measurement of bioluminescence is shown in (A), while the total amount of photons as measured by integrating the area under the curves is shown in (B).

20.7 DISCUSSION AND CONCLUSIONS

The development of strategies to increase the bioavailability of drugs and biopolymers is a major challenge as science moves to exploit the opportunities arising from genomic studies, proteomics, systems biology, combinatorial chemistry, and rational drug design. Most therapeutic agents and probes with intracellular targets enter cells through passive diffusion and, as such, they generally must conform to a rather restricted log P range, allowing for solubility in both the polar extracellular milieu and nonpolar membrane. Notwithstanding the attention given to structural diversity in drug discovery, the relative lack of physical property

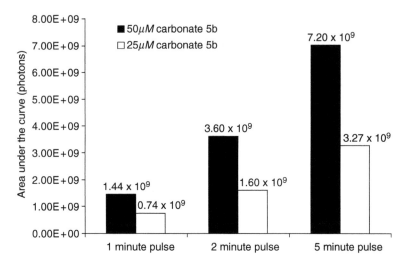

FIGURE 20.12 Observed bioluminescence from a transformed prostate cell line stably transfected with luciferase, PC3M-luc, when pulsed with carbonate **5b** in HBS pH 7.4 for 1, 2, and 5 min. The histograms represent the total amount of photons emitted over 40 min, as measured by integrating the area under the curves of luminescence.

diversity limits the range of agents that can be used as therapeutics and probes. The studies discussed in this review provide a mechanistic rationale for how a freely water-soluble, polar transporter can change its effective polarity, enter nonpolar membranes through adaptive noncovalent association with membrane constituents, and move into the cytosol under the influence of a membrane potential. In this process, guanidinium-rich transporters only weakly associated with their counterions in the polar extracellular milieu[31] readily exchange their counterions to form bidendate hydrogen bonds with phosphates, sulfates and/or carboxylates anchored on the cell surface. This hydrogen bonding converts otherwise polar cationic and anionic functionalities into lipophilic ion pairs whose hydrogen bond associations strengthen as they enter the nonpolar lipid bilayer.[31] The directionality and rate of passage of these ion pairs through the bilayer is dictated by the membrane potential. The ion pair complexes dissociate on the inner leaf of the membrane and the transporters enter the cytosol in the microscopic reverse of the cell surface association. Too few guanidinium groups diminish cell surface adherence and therefore entry; too many could inhibit escape from the inner leaf of the membrane. A range of approximately 7–20 guanidinium groups works well for most cargos with a preference for eight dictated by the cost of goods and a novel method of preparation of octamers.[32]

This mechanism is consistent with the diverse spectrum of guanidinium-rich structures that rapidly enter cells, emphasizing that the key feature for entry into the cell is the formation of the bidentate hydrogen bonds between the guanidinium head groups and anionic structures on the cell surface, and not the structural and conformational details of the transporter. Consequently, the stereochemistry, the composition of the backbone, and the spacing of the guanidiniums extending either from or along the backbone can all be modified with only modest changes in transporter activity. The essential structural requirement appears to be a minimum number of guanidiniums, which explains why even nonlinear dendrimers and oligosaccharides decorated with a sufficient number of guanidiniums are transported effectively.

The importance of the membrane potential suggests that the transporter complex has an overall positive charge, which could arise either from incomplete ion pairing with anionic cellular components or by ion pair formation with zwitterrionic species. Incomplete ion pair formation is entropically favored as the number of guanidiniums increases and would be consistent with the requirement of a minimum of guanidinium head groups for effective transport. In the proposed mechanism, the large number of potential binding sites on the cell surface also explains why inhibition of uptake at sensible molar excesses of a competitive inhibitor is not observed, and also why neither prokaryotic nor eukaryotic transport negative mutants have been generated (unpublished data). Inhibition of uptake by sodium azide is better explained by the dissipation of the membrane potential or inhibition of ion export rather than inhibition of an energy-dependent receptor-mediated pathway of uptake.

An attractive aspect of the mechanism proposed above is that each of the central steps has precedent. Modification of the polarity of highly basic peptides by formation of ion pair complexes with hydrophobic anions is fundamental to ion pair chromatography and has been used to increase the stability of biologically active peptide.[16,33–35] Furthermore, the importance of ion pair formation in the cellular uptake of cationic peptides has been considered by others.[17] Movement of cations, as well as arginines, across biological membranes in the direction of the membrane potential also has significant precedent. Wilson and colleagues[36,37] established that the movement of a cationic cyanine dye into or across the membrane was proportional to membrane potential and is sufficiently quantitative that the dyes are used in flow cytometry to measure the membrane potential of cells.

At least two groups have argued that guanidinium-rich transporters enter cells exclusively by endocytosis.[38,39] Clearly, however, multiple mechanisms of entry are possible given the variety of cargo sizes, cell types, and experimental conditions. For example, for larger cargoes an endocytotic pathway would be favored[4,40] as diffusion of an ion pair complex would be expected to decrease with size. However, even here, endocytotic entry results in an endosomal bound conjugate that has not traversed a barrier but must do so to become freely available in the cytosol. In contrast, the mechanism discussed above allows for adaptive diffusion across the membrane to produce a freely available, cytosolic conjugate. Interestingly, endosomes retain a membrane potential; therefore, the mechanism discussed above could also be operative in the translocation of guanidinium-rich molecules through endosomes into the cytosol. This premise is supported by two recent papers demonstrating that the endosomal exit of FGF-1 and FGF-2, which both contain arginine-rich domains, required the endosomal membrane potential.[41,42] Consequently, the voltage-dependent, lipid-mediated uptake mechanism can be invoked in crossing of both the endosome and the plasma membrane lipid bilayers.

There are other important differences between the designs of our experiments and those reporting endocytotic vesicles in cells treated with various transporter–peptide conjugates. As previously mentioned, the most obvious difference is the molecular weight of the cargo. We have observed endosomal staining of cells treated for 5 min with fluorescent hepta-D-arginine-PNA and octa-D-arginine PEG conjugates as small as 5000 Daltons, not including the counterions, which are not present in smaller conjugates (unpublished results). Most importantly, there is significant cytosolic staining in these cells, which is absent when cargo sizes increase over 10,000 Daltons. These data support the hypothesis that multiple mechanisms can operate simultaneously, with different proportions of the material entering by different pathways at different times. The experiments in this review emphasize that, at early times, with low molecular weight cargo, a membrane potential is necessary for a high percentage of the conjugate to enter the cytosol. However, at these early times, not all cells are uniformly stained (Figure 20.5). The reports that argue that endocytosis is the principal pathway of entry have used longer incubation times that result in a population of cells that are more uniformly stained and are consequently easier to analyze microscopically.[38,39] That there might be more endocytosed material at those times is expected. However, what is not clear is what pathway results in the majority of the conjugates in the cytosol. Another important aspect of comparing the results in the literature is the aggregation state of the different transporters. Piwnica-Worms[30] has published data that support the hypothesis that cell entry has a cooperative component in the mechanism. If so, the dose range of the conjugate used in the experiments might also favor one mechanism over another. These experiments use concentration ranges that are higher than those reports arguing that endocytosis is important.[38,39]

In summary, the data in this paper summarize evidence supporting a voltage-dependent, adaptive translocation mechanism for the cellular uptake of guanidinium-rich molecular transporters attached to small molecular weight cargos. The proposed mechanism emphasizes the importance of ion pair formation based on bidentate hydrogen bonding and electrostatics that allow for adaptive translocation of a polar transporter into a nonpolar membrane and transmembrane migration under the influence of the membrane voltage potential. This mechanism provides an improved molecular understanding of the translocation of guanidinium-rich transporters across phospholipid membranes. In vivo studies have established that arginine-rich peptides are widely distributed in tissue, which indicates that these transporters are capable of crossing a variety of phospholipid and nonphospholipid barriers.[24]

ACKNOWLEDGMENTS

Support of this work by grants from the National Institutes of Health (CA31841, CA31845) and the NSF Center for Biophotonics Science and Technology (002865-SU) are gratefully acknowledged. L.R.J. was supported by the American Chemical Society.

REFERENCES

1. Alberts, B., Lewis, J., Raff, M., Roberts, K., and Watson, J., *Molecular Biology of the Cell*, Garland, New York, 1994.
2. Joliot, A. and Prochiantz, A., Transduction peptides: From technology to physiology, *Nat. Cell Biol.*, 6, 189, 2004.
3. Mann, D.A. and Frankel, A.D., Endocytosis and targeting of exogenous HIV-1 tat protein, *EMBO J.*, 10, 1733, 1991.
4. Silhol, M., Tyagi, M., Giacca, M., Lebleu, B., and Vives, E., Different mechanisms for cellular internalization of the HIV-1 tat-derived cell penetrating peptide and recombinant proteins fused to tat, *Eur. J. Biochem.*, 269, 494, 2002.
5. Wender, P.A., Mitchell, D.J., Pattabiraman, K., Pelkey, E.T., Steinman, L., and Rothbard, J.B., The design, synthesis, and evaluation of molecules that enable or enhance cellular uptake: Peptoid molecular transporters, *Proc. Natl Acad. Sci.*, 97, 13003, 2000.
6. Mitchell, D.J., Kim, D.T., Steinman, L., Fathman, C.G., and Rothbard, J.B., Polyarginine enters cells more efficiently than other polycationic homopolymers, *J. Pept. Res.*, 56, 318, 2000.
7. Luedtke, N.W., Carmichael, P., and Tor, Y., Cellular uptake of aminoglycosides, guanidinoglycosides, and poly-arginine, *J. Am. Chem. Soc.*, 125, 12374, 2003.
8. Futaki, S., Suzuki, T., Ohashi, W., Yagami, T., Tanaka, S., Ueda, K., and Sugiura, Y., Arginine-rich peptides: An abundant source of membrane-permeable peptides having potential as carriers for intracellular protein delivery, *J. Biol. Chem.*, 276, 5836, 2001.
9. Wender, P.A., Rothbard, J.B., Jessop, T.C., Kreider, E.L., and Wylie, B.L., Oligocarbamate molecular transporters: Design, synthesis, and biological evaluation of a new class of transporters for drug delivery, *J. Am. Chem. Soc.*, 124, 13382, 2002.
10. Rothbard, J.B., Kreider, E., Vandeusen, C.L., Wright, L., Wylie, B.L., and Wender, P.A., Arginine-rich molecular transporters for drug delivery: Role of backbone spacing in cellular uptake, *J. Med. Chem.*, 45, 3612, 2002.
11. Umezawa, N., Gelman, M.A., Haigis, M.C., Raines, R.T., and Gellman, S.H., Translocation of a beta-peptide across cell membranes, *J. Am. Chem. Soc.*, 124, 368, 2002.
12. Rueping, M., Mahajan, Y., Sauer, M., and Seebach, D., Cellular uptake studies with beta-peptides, *Chembiochem*, 3, 257, 2002.
13. Wender, P.A., Kreider, E., Pelkey, E.T., Rothbard, J., and Vandeusen, C.L., Dendrimeric molecular transporters: Synthesis and evaluation of tunable polyguanidino dendrimers that facilitate cellular uptake, *Org. Lett.*, 7, 4815, 2005.
14. Futaki, S., Nakase, I., Suzuki, T., Youjun, Z., and Sugiura, Y., Translocation of branched-chain arginine peptides through cell membranes: Flexibility in the spatial disposition of positive charges in membrane-permeable peptides, *Biochemistry*, 41, 7925, 2002.
15. Chung, H.H., Harms, G., Seong, C.M., Choi, B.H., Min, C., Taulane, J.P., and Goodman, M., Dendritic oligoguanidines as intracellular translocators, *Biopolymers*, 76, 83, 2004.
16. Powers, M.E., Matsuura, J., Brassell, J., Manning, M.C., and Shefter, E., Enhanced solubility of proteins and peptides in nonpolar-solvents through hydrophobic ion-pairing, *Biopolymers*, 33, 927, 1993.
17. Sakai, N. and Matile, S., Anion-mediated transfer of polyarginine across liquid and bilayer membranes, *J. Am. Chem. Soc.*, 125, 14348, 2003.

18. MOPAC, AM1, closed shell. Energetics of the various guanidinium configurations of mono- and dimethylated arginine was approximated by modeling N,N'-dimethylguanidinium and N,N,N'-trimethylguanidinium.

19. Aidley, D. and Stanfield, P., *Ion Channels*, Cambridge University Press, Cambridge, UK, 1996.

20. The membrane potential is dominated by the K^+ Nernst potential at high extracellular concentrations of potassium, but contributions of other ionic conductances makes the membrane potential more positive than the potassium Nernst potential at low extracellular concentrations of potassium. Thus, although the uptake rate of Fl-aca-D-Arg$_8$-CONH$_2$ is linear with the potassium Nernst potential (Figure 20.8B), the curve describing the relationship between rate of uptake and membrane potential would lie increasingly to the right of this line at more negative membrane potentials. The addition of valinomycin serves to clamp the membrane potential at the theoretical potassium Nernst potential.

21. Weiss, T., *Cellular Biophysics*, MIT Press, Boston, MA, 1996.

22. Urban, B.W., Hladky, S.B., and Haydon, D.A., Ion movements in gramicidin pores. An example of single-file transport, *Biochim. Biophys. Acta*, 602, 331, 1980.

23. Harada, H., Morita, M., and Suketa, Y., K+ ionophores inhibit nerve growth factor-induced neuronal differentiation in rat adrenal pheochromocytoma PC12 cells, *Biochim. Biophys. Acta*, 1220, 310, 1994.

24. Rothbard, J.B., Garlington, S., Lin, Q., Kirschberg, T., Kreider, E., McGrane, P.L., Wender, P.A., and Khavari, P.A., Conjugation of arginine oligomers to cyclosporin a facilitates topical delivery and inhibition of inflammation, *Nat. Med.*, 6, 1253, 2000.

25. Kirschberg, T.A., VanDeusen, C.L., Rothbard, J.B., Yang, M., and Wender, P.A., Arginine-based molecular transporters: The synthesis and chemical evaluation of releasable taxol-transporter conjugates, *Org. Lett.*, 5, 3459, 2003.

26. Chen, L., Wright, L.R., Chen, C.H., Oliver, S.F., Wender, P.A., and Mochly-Rosen, D., Molecular transporters for peptides: Delivery of a cardioprotective epsilon PKC agonist peptide into cells and intact ischemic heart using a transport system, R(7), *Chem. Biol.*, 8, 1123, 2001.

27. Jones, L.R., Goun E., Shinde, R., Rothbard, J.B., Contag, C., Wender, PA., Releasable luciferin-transporter conjugates: Tools for the real time analysis of cellular uptake and release. *J. Am. Chem. Soc.*, 128, 6526–6527, 2006.

28. Jenkins, D.E., Oei, Y., Hornig, Y.S., Yu, S.F., Dusich, J., Purchio, T., and Contag, C.H., Bioluminescent imaging (BLI) to improve and refine traditional murine models of tumor growth and metastasis, *Clin. Exp. Metastasis*, 20, 733, 2003.

29. Cao, Y., Wagers, A.J., Beilhack, A., Dusich, J., Bachmann, M.H., Negrin, R.S., Weissman, I.L., and Contag, C.H., Shifting foci of hematopoiesis during reconstitution from single stem cells, *Proc. Natl Acad. Sci. U.S.A.*, 101, 221, 2004.

30. Polyakov, V., Sharma, V., Dahlheimer, J., Pica, C., Luker, G., and Piwnica-Worms, Novel tat-peptide chelates for direct transduction of technetium-99m and rhenium into human cells for imaging and radiotherapy, *Bioconjug. Chem.*, 11, 762, 2000.

31. Onda, M., Yoshihara, K., Koyano, H., Ariga, K., and Kunitake, T., Molecular recognition of nucleotides by the guanidinium unit at the surface of aqueous micelles and bilayers. A comparison of microscopic and macroscopic interfaces, *J. Am. Chem. Soc.*, 118, 8524, 1996.

32. Wender, P.A., Jessop, T.C., Pattabiraman, K., Pelkey, E.T., and VanDeusen, C.L., An efficient, scalable synthesis of the molecular transporter octaarginine via a segment doubling strategy, *Org. Lett.*, 3, 3229, 2001.

33. Hancock, W.S., Bishop, C.A., Prestidge, R.L., Harding, D.R., and Hearn, M.T., Reversed-phase, high-pressure liquid chromatography of peptides and proteins with ion-pairing reagents, *Science*, 200, 1168, 1978.

34. Gennaro, M.C., Reversed-phase ion-pair and ion-interaction chromatography, *Adv. Chrom.*, 35, 343, 1995.

35. Schill, G., High-performance ion-pair chromatography, *J. Biochem. Biophys. Met.*, 18, 249, 1989.

36. Wilson, H.A., Seligmann, B.E., and Chused, T.M., Voltage-sensitive cyanine dye fluorescence signals in lymphocytes — plasma-membrane and mitochondrial components, *J. Cell. Phys.*, 125, 61, 1985.

37. Wilson, H.A. and Chused, T.M., Lymphocyte membrane-potential and Ca^{2+}-sensitive potassium channels described by oxonol dye fluorescence measurements, *J. Cell. Phys.*, 125, 72, 1985.

38. Vives, E., Cellular uptake of the tat peptide: An endocytosis mechanism following ionic interactions, *J. Mol. Recognit.*, 16, 265, 2003.

39. Fuchs, S.M. and Raines, R.T., Pathway for polyarginine entry into mammalian cell, *Biochemistry*, 43, 2438, 2003.

40. Wadia, J.S., Stan, R.V., and Dowdy, S.F., Transducible tat-HA fusogenic peptide enhances escape of tat-fusion proteins after lipid raft macropinocytosis, *Nat. Med.*, 10, 310, 2004.

41. Malecki, J., Wiedlocha, A., Wesche, J., and Olsnes, S., Vesicle transmembrane potential is required for translocation to the cytosol of externally added FGF-1, *EMBO J.*, 21, 4480, 2002.

42. Malecki, J., Wesche, J., Skjerpen, C.S., Wiedlocha, A., and Olsnes, S., Translocation of FGF-1 and FGF-2 across vesicular membranes occurs during G(1)-phase by a common mechanism, *Mol. Biol. Cell*, 15, 801, 2004.

21 Cell-Penetrating Short Interfering RNAs and Decoy Oligonucleotides

Samir El-Andaloussi, Henrik Johansson,
Pontus Lundberg, and Ülo Langel

CONTENTS

21.1 INTRODUCTION

As mentioned previously in this book, one of the major obstacles in drug delivery today is the low bioavailability of hydrophilic compounds, such as oligonucleotides (ON). The main reason for this is the inability of these substances to transverse the lipid membrane of cells. ONs are interesting therapeutical tools that can be used as decoys and sequester transcription factors, as well as to initiate degradation of RNA in the form of short interfering RNAs (siRNA). To overcome the bioavailability problem and problems associated with delivery of these macromolecules, such as nuclease degradation, endosomal trapping, and poor nuclear targeting, several carrier systems have been developed, as reviewed by Gardlik et al.[1] These delivery systems mainly include viruses and nonviral vectors such as polycations and cationic liposomes. Viruses are generally considered to be more efficient than nonviral vectors but are potentially immunogenic in vivo. Although polycations and cationic liposomes are widely used on a routine basis today, they are still not efficient enough or are too toxic for in vivo applications.

In the past, several peptides have been utilized for ON delivery. These peptides include various ligands for receptors and growth factors that confer cellular targeting and concomitant receptor-mediated endocytosis of the cargo molecule.[2,3] Other peptide-based vectors include nuclear localization signal sequences (NLS peptides) that mediate cellular uptake and nuclear delivery,[4] or peptides with endosomolytic activity that promote endosomal escape of cargoes.[5] However, most of these peptides with cargoes have a limited internalization frequency.

The discovery of cell-penetrating peptides (CPPs) has led the way to a new path in drug delivery, making it possible to deliver virtually any cargo in various sizes in a relatively nontoxic fashion with high efficiency.[6] This chapter focuses on the delivery of small ONs, in particular siRNAs and decoy ONs, using different CPPs (Table 21.1). For a complete overview of peptides used in ON delivery, see Lochman et al.[7] In order to enhance ON uptake by CPPs, two different strategies are mainly used: either ONs are simply mixed with CPPs to form stable, noncovalent complexes through electrostatic interactions, or the peptides are covalently linked to ONs, i.e., via disulfide bridges (Figure 21.1).

21.2   DELIVERY OF siRNA

21.2.1   VECTORS FOR siRNA DELIVERY

siRNAs are small, double-stranded RNAs, typically 21–23 base pairs in length, that are involved in gene silencing through degradation of mRNA and compacting DNA, thereby blocking transcription. This mechanism is similar to the antisense-mediated

---

TABLE 21.1
CPPs Used in the Delivery of siRNA and Decoy ONs

	Peptide	Target	Conjugation Strategy	Refs.
siRNA				
	MPG, MPGΔNLS	Luciferase, GAPDH	Coincubation	16
	Penetratin, transportan	GFP, luciferase	Disulfide	18
	Tat	EGFP, CDK9	Disulfide	20
	Penetratin	SOD1, Casp3	Disulfide	19
Decoy				
	MPG	MDR-1	Coincubation	47
	Transportan, TP10	NFκB	PNA hybridization	50
	TP10	Myc	PNA hybridization, coincubation	49

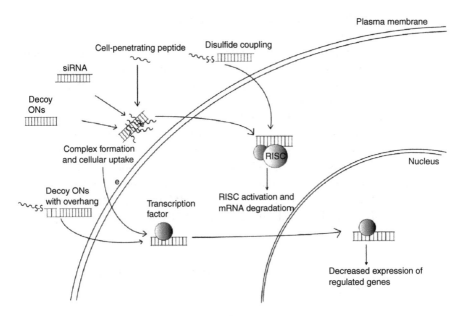

FIGURE 21.1 Different delivery strategies of siRNAs and decoy ONs.

degradation of mRNA except that RNA interference (RNAi) is an endogenous process and therefore involves other enzymes. The RNAi pathway is activated by long, double-stranded RNAs that trigger an enzyme, Dicer, which processes these RNAs into siRNAs. These siRNAs are then incorporated into an RNA-induced silencing complex (RISC) that targets complementary mRNAs for degradation (Figure 21.1). RNAi is an important and highly specific process, mainly active during embryonic development and as a defense against viruses utilizing dsRNA.[8]

Recently, an increasing number of studies have suggested the potential use of siRNAs in basic research and as a therapeutic tool to knock down protein expression both in vitro and in vivo.[8] Due to the hydrophilic nature of siRNAs, they will not be readily internalized by cells. For the majority of eukaryotic cells, the siRNA has to be actively delivered over the plasma membrane by a carrier. The effectors of RNAi can be delivered in the form of oligoribonucleotides, 21–23 base pairs in length, with a two-nucleotide overhang at 3' end, or by vectors expressing two complementary siRNA strands or a short hairpin RNA (shRNA).

In dividing cells, transfection of siRNA has a maximum effect 2–3 days after transfection, with a knock down lasting for approximately 1 week. In nondividing cells, such as neurons, the effect can last for several weeks. For prolonged silencing, vectors expressing siRNA or shRNA under the control of strong or inducible promoters are useful. The vector systems are often based on adenoviruses, adeno-associated viruses, oncoretroviruses, or lentiviruses.[9] The most commonly used method to deliver siRNA in vitro is based on cationic lipids, such as the commercially available oligofectamine™ or lipofectamine™. Other delivery strategies include invasive systems, such as electroporation, calcium phosphate coprecipitation, and microinjection.

A number of alternatives to cationic lipids have been developed recently, in which the main strategy is to connect different functional domains, e.g., a nucleotide-binding domain with a transducing domain and a targeting domain. An example of this is where the RGD peptide domain targeting the integrin receptor is conjugated to PEI via a PEG linker, resulting in effective siRNA delivery.[10] There are also examples where the siRNA delivery has been enhanced by using different nonpeptide vehicles, such as cholesterol[11,12] or atelocollagen.[13,14] A more specific method is to take advantage of the nucleotide binding properties of protamine, in which the protamine part is fused to either a Fab antibody directed against the HIV envelope glycoprotein gp160 targeting HIV-infected cells or an anti-ErbB2, which specifically delivers siRNA to ErB2-expressing breast cancer cells.[15]

21.2.2   COVALENT AND NONCOVALENT STRATEGIES TO DELIVER siRNA USING CPPs

In general, delivery systems based on CPPs can be divided into two groups: noncovalent complex formation, in which the siRNA is simply coincubated with the CPP, or covalent coupling, in which the siRNA and CPP are connected via, for example, a disulfide bridge (Figure 21.1). The noncovalent complex formation strategy has so far worked well in delivering siRNA in vitro (Ref. [16] and our own observations). The advantage of using this strategy is its simplicity, as only mixing of CPP and siRNA is required, followed by addition of the mixture to the cells. This strategy supposedly relies on electrostatic interactions to form a noncovalent complex in which the positively charged CPPs surround the siRNA, masking its negative charges. Since the CPP is typically used in a charge ratio compared to siRNA to neutralize negative charges, there is a molar excess of CPP as compared to the siRNA, yielding complexes of various sizes. This could be a disadvantage in conditions in which a defined cargo/transporter complex size is required, such as in therapeutic applications. If defined properties for siRNA complexes are required the complexes can be purified, but are still under equilibrium between on-bound and nonbound CPPs; thus, a constant exchange will occur changing the size of the complexes.

Another aspect of the complex formation strategy is the fact that uptake can be observed by means of confocal microscopy, but no phenotype, suggesting that the siRNA is trapped in the endosomal compartments and thus unable to exert its effect, or that the CPPs bind too strongly to the siRNA, thus rendering the complex in a state where the CPPs remain bound to the siRNA complex and the siRNA cannot exert its effect. A solution to the first scenario, in which the complex is stuck in the endosomal compartments, is the use of endosomolytic peptides, which have the ability to change conformation when a drop in pH occurs, leading to an insertion of the lysosomotropic peptide into the endosomal membrane and a concomitant disruption. An example of this is where the endosomolytic peptide HA2, fused to Tat, increases Cre-mediated recombination, presumably by increasing the endosomal escape of the Tat-Cre fusion protein.[17]

For covalent coupling between siRNA and CPP, three different CPPs have been used to date: transportan,[18] penetratin,[18,19] and Tat.[20] In this case, there is a defined molecule in which one CPP is coupled to one siRNA molecule. The advantages of using this system are many. First of all there is a defined molecule, which is desirable in therapeutic applications. Also, less peptide is needed in this setting as compared to the complex formation strategy, which might be more cost efficient, and a lower peptide concentration can be advantageous as some of the CPPs show toxicity at higher concentrations. The disadvantage of the method is that it is more expensive as an activated peptide is needed, and some peptide/siRNA could be lost in the purification process. Also, some of the peptides might not be suitable for this kind of approach as they might bind electrostatically to the siRNA, which could be difficult to detect using mass spectrometry, because electrostatic interactions are usually broken upon laser excitation. These CPPs, which are electrostatically bound to the siRNA, may help in delivery by neutralizing negative charge but will render a somewhat undefined complex with possible aggregation during the disulfide formation.

21.2.3 CELL-PENETRATING siRNAs

The first CPP used to introduce siRNA into cells was the MPG peptide, a fusion peptide between a hydrophobic domain from HIV gp41 and a hydrophilic domain from the nuclear localization sequence of SV40 T-antigen,[21] which forms an electrostatic complex with siRNA. This system is extensively covered elsewhere in this book and is therefore not discussed further.

Next, Muratovska and Eccles[18] described the formation of a disulfide bridge between the CPPs, transportan and penetratin, to the 5′ end of one strand of siRNA by using the thiol oxidizing agent diamide. The siRNA with a free thiol, CPP with N-terminal cysteine, was mixed with diamide and incubated for 1 h and then applied on cells. Although the article states that the yield was 80% using this method, no gel or mass spectrometry data were shown that could support the data.

Another study by Davidson et al.[19] also utilized a disulfide bridge formed between penetratin and siRNA, although in this study the peptide moiety was modified with a pyridylthiol for higher reactivity with the 5′ thiol on the siRNA. Before the disulfide reaction was performed, the siRNA solution was pretreated with equimolar amounts of tris(2-carboxyethyl)phosphine to reduce the 5′ thiol on the siRNA, after which penetratin with pyridylthiol was added to form the disulfide bridge. Using this protocol resulted in an estimated yield of 90% by SDS–PAGE. Protein levels were downregulated in primary neurons, which interestingly preceded the degradation of mRNA, where the conventional understanding is that protein levels decrease after mRNA degradation. These results suggest that, at least in this case, the mRNA degradation is preceded by a translational arrest due to an antisense steric blocking effect, followed by degradation of the target mRNA.

Other peptide vectors reported to deliver siRNA effectively consist of a combination of histidine and lysine stretches. Read et al.[22] downregulated the neurotrophin receptor p75NTR by two different peptides using histidine and lysine repeats with cysteine in both ends (Cys–His$_3$–Lys$_3$–His$_3$–Cys and Cys–His$_6$–Lys$_3$–His$_6$–Cys). Also, highly branched histidine and lysine polymers introduced siRNA and downregulated β-galactosidase and luciferase activity.[23] Here, the addition of the RGD peptide ligand targeting the integrin receptor slightly increased the siRNA efficiency.

To date, no CPP has been used for delivery of siRNA in vivo, although transportan,[24] penetratin,[24] Tat,[25] ppTG1, and ppTG20[26] have been successfully used in vivo. siRNA delivery in vivo is probably where CPPs can prove to be of real significance as the conventional methods with which in vitro transfection efficiency is often compared, e.g., oligofectamine, has, due to toxicity, very limited use.

21.3 DELIVERY OF DECOY ONs

21.3.1 THE DECOY CONCEPT

Another interesting biomedical tool, based on ONs and used to interfere with protein activity, is the decoy strategy. This approach is designed to operate on transcription factors and to regulate their activity. Transcription factors are generally nuclear proteins that play a critical role in gene regulation, either exerting a positive or negative effect on gene expression. These regulatory proteins bind specific consensus sequences found in promotor regions of target genes. The consensus-binding sequences are generally 6–10 base pairs in length and are occasionally found in multiple iterations upstream or downstream of transcription initiation sites. Binding of transcription factors, and subsequent interactions of these proteins with each other, as well as with RNA polymerases or their cofactors, yield a complex set of factors that determine the relative transcriptional activity.

The decoy strategy was first used as a tool for investigating transcription factor activity in cell culture systems.[27] Similar decoys can also be devised as therapeutic agents, either to inhibit the expression of genes that are transactivated by the factor in question, or to upregulate genes that are transcriptionally suppressed by the binding of a factor. Decoy ONs have also been applied in vivo to downregulate the activity of E2F, a transcription factor activator involved in cell-cycle proliferation, leading to blocked smooth muscle proliferation and neointimal hyperplasia in injured vessels.[28] In addition to the initial in vivo application, transcription factor decoys have been used to block a negative regulatory element in the promotor of the *Renin* gene in the mouse submandibular gland, demonstrating that decoys can be used to enhance as well as to suppress gene activity in vivo.[29,30] Also, decoy ONs have been introduced into HIV-infected cells, inhibiting transcription of viral genes.[31,32]

Promising therapeutics based on this approach include inhibitors of Stat6, which may be useful in reducing IL-4-induced proliferation of Th cells in allergic diseases,[32] and on the cAMP-response element-binding protein (CREB) in tumor cells.[33] CREB transcription factor decoys have yielded a particularly intriguing effect on the in vitro and in vivo growth of a number of tumor cell lines, in that growth inhibition appears to be tumor specific and has not been observed in any of the noncancerous cells tested.

21.3.2 STABILITY OF DECOY ONs

Although the decoy strategy offers great potential in basic research and in clinical settings, there are still drawbacks associated with the use of ONs as therapeutic tools. First, the phosphodiester backbone of DNA and RNA is readily degraded by nucleases. Therefore, much effort has been made to circumvent this drawback by introducing chemical modifications. Modification of the phosphodiester linkages by various methods confers resistance to degradation by nucleases in vivo and in cell culture. Several analogs of the backbone have been prepared, including phosphorothioates, phosphoroamidates, and methylphosphonates. In decoy procedures, phosphorothioate ONs have been most commonly used due to their relatively stable nature and high solubility even though they exhibit lower affinity for the target.[34] Recently, several groups have reported a novel type of decoy ON chimera bearing two strands with mixed oligodeoxynucleotides and neutrally charged peptide nucleic acid (PNA)–ONs.[35,36] These chimeras have proved to be as efficient as double-stranded DNA decoys and have been used to downregulate the activity of several transcription factors.[37] A major advantage with these decoys is that they can easily be further engineered by adding short peptides covalently to the PNA, which facilitates cellular and nuclear entry.

21.3.3 STRATEGIES TO IMPROVE DECOY DELIVERY

The usefulness of ONs is also limited by their polyanionic nature, which confers poor cellular internalization. Since decoy ONs are poorly taken up by cells, several different delivery vectors have been applied to facilitate delivery. Viral vectors, especially in vivo using lentiviral or adenoviral vectors, and cationic liposomes have been the most frequently used vehicles for the delivery of decoy ONs. Due to the drawbacks with these vectors, discussed previously in this chapter, development of novel vectors in the last years has been a hot topic. One approach has been to combine parts of viruses with liposomes. Hemagglutinating virus of Japan liposomes was used to deliver decoy phosphorothioates targeting E2F in vivo in rats.[38] Other groups have utilized biodegradable polymer microparticles (or microspheres) such as poly(DL-lactic-co-glycolic acid) (PLGA) microspheres for efficient delivery of decoy ONs, targeting NFκB in glioblastoma cells and macrophages.[39,40] Also, anionic carboxylate polystyrene microspheres coated with an ornithine/histidine-based cationic peptide (O10H6) have been used to target NFκB decoys ex vivo.[41] The main advantages with using microspheres in vivo include ease of production and processing, the possibility of injecting directly into the site of action, protection against enzymatic digestion, and mediation of a slow release of ONs that give rise to higher cellular uptake and a more sustained effect. By introducing ornithines and histidines in microspheres, enhanced ON condensation and increased endosomal escape is achieved, respectively.[42,43]

Although microspheres offer promising potential for ex vivo use, there is a need to obtain more efficient vectors for in vivo delivery and nuclear targeting, since most transcription factors are active in the nucleus. Recently, Sakaguchi et al.[44] developed a protein-based delivery vehicle for decoy ONs targeting p53.

The vehicle is composed of glutathione-S-transferase, seven arginine residues, the DNA-binding domain of GAL4, and an NLS, all spaced via flexible glycine linker sequences. Two tandem p53 elements (p53 decoy) are linked to the yeast UAS sequence, which interacts with GAL4 in the vehicle protein. The polyarginine and the NLS facilitate cellular and nuclear delivery, respectively, while the GST sequence is included mainly to simplify purification of the recombinant protein. The ONs are efficiently targeted to the nuclei of cells with concomitant sequestering of p53. In contrast to liposomal transfections, which mainly target decoy ONs to the perinuclear cytoplasm,[45,46] this vehicle protein efficiently translocates p53 decoys into the nuclei of various cell types. Although this is a highly efficient vector, it requires extensive recombinant work and generates a large vehicle (45.8 kDa). Another concern with this method is how stable the interaction between GAL4 and UAS is in vivo in extracellular fluids.

21.3.4 DELIVERY OF DECOY BY CPPs

Because viruses are by far the most efficient vectors, although potentially immunogenic, an optimal vector would be of nonviral nature but still have the essential features of a virus for cellular translocation of cargoes. An example of this design is the MPG peptide. By simply mixing this peptide in a 25:1 molar ratio over decoy ONs containing the MED-1 *cis* element, efficient downregulation of protein activity was achieved.[47] Also, as discussed elsewhere in this book, MPG has been used for delivery of antisense ONs, siRNAs, and plasmids, with convincing results.

We recently reported improved polyethyleneimine (PEI)-mediated plasmid delivery upon addition of TP10 to the transfection cocktail.[48] However, TP10 is incapable of conferring plasmid translocation as a single peptide vector. On the other hand, the same strategy using TP10 as a vector peptide for delivery of decoy ONs targeting the oncogenic protein Myc was shown to be highly efficient. At a charge ratio of $+1.5$ (a molar ratio of 12:1 of TP10 over decoy ONs), uptake of double-stranded oligodeoxynucleotides was improved 100-fold. Introduction of the Myc decoy into human breast cancer cells suppressed cellular proliferation in a dose-dependent manner, showing the strongest effects when complexed with TP10. However, there are some drawbacks for coincubating cationic peptides with ONs. First, CPPs form strong complexes with ONs,[21,49] which is convenient in the sense that peptides protect ONs from nuclease degradation in serum, but also inconvenient because most of the internalized decoys remain inactive when bound to peptides. Also, large complexes that are readily internalized in cell culture may show completely different properties in vivo. In pharmacology, the main concern about using this strategy is the lack of a defined complex size.

Finally, successful coupling of decoy ONs to a CPP with a nonamer PNA as a linker has been reported for the first time.[50] By using a double-stranded decoy ON sequence with one strand containing a flanking nonanucleotide as compared to the other strand, a 9-mer PNA, complementary to the nine nucleotides, can hybridize to that particular sequence. By introducing a cysteine N terminally of PNA, thiolated TP or TP10 forms a disulfide bridge with the PNA yielding a cell-penetrating decoy construct. The decoy ONs, delivered by TP or TP10 in this study, hybridized specifically to NFκB in the cytoplasm and thereby prevented its further binding to DNA and induction of IL-6 transcription.

We recently utilized the same strategy but, in order to enhance further cellular uptake, an NLS sequence was directly synthesized on the N-terminus of PNA (not included in the construct used by Fisher et al.), which, in turn, was coupled to TP10 via a disulfide bridge. Interestingly, when comparing this strategy with the previously mentioned coincubation approach, uptake levels were significantly lower using this method. However, when analyzing the biological response of Myc decoys, i.e., proliferation in this case, we observed 70% growth inhibition when applying Myc decoy in complex with TP10 compared to 40% inhibition using the PNA conjugate as a delivery vector. Quantification of the number of decoy molecules that are internalized per cell reveals that there are approximately 400,000 Myc decoy molecules per cell when delivered by the TP10–PNA conjugate and 20 million decoys per cell when coincubating with TP10. This should be compared with an average of approximately 61,000 Myc proteins per cell for MCF-7 cells in the cytoplasm and nucleus.[51] These data clearly illustrate that, even though the uptake is 20-fold greater when coincubating TP10 with decoy, the biological response is only enhanced twofold as compared to using the TP10–PNA conjugate as a vector. One plausible explanation for this is that not all internalized Myc decoy ON are accessible for the Myc protein in the nucleus and cytoplasm due to the strong interaction between TP10 and DNA. Since the quantitative uptake assay is unable to determine the cellular localization of internalized material, another reasonable explanation could be that a large fraction of TP10 in complex with decoy ONs remains trapped in endosomes during translocation. This is, however, not likely since experiments with endocytosis inhibitors suggest a nonendocytotic uptake mechanism.

In conclusion, the delivery efficiency of TP10 and strong biological effect of Myc decoy in combination with TP10 suggest a novel strategy to downregulate proliferation in tumor cells. Although the exact mechanism(s) of uptake for CPPs remains unclear, their relatively nontoxic nature and potential to deliver virtually any cargo make them an attractive vector for future ON therapeutics.

21.4 SUMMARY

Using CPPs in the delivery of siRNA and decoy ONs is still in its infancy. The methods used to date to connect peptides to ONs depend either on noncovalent interactions and formation of complexes based on electrostatic interactions or on covalent bonds, e.g., disulfide bridges. The main challenge in the future will be the development of peptides that form serum-resistant, stable complexes, which mediate internalization, still having the ability to dissociate inside the cell and concomitantly release ONs. Covalent coupling between CPPs and ONs provides a defined conjugate size but conjugation and purification strategies need to be improved. To date, most conjugates are not purified; thus, there are nonconjugated CPPs and ONs together with the conjugate, possibly changing the properties of the conjugate. Also, cellular localization of CPPs and ONs affects the efficiency of effector molecules, which should be taken into account when designing and comparing different strategies for internalization. Finally, using various strategies, present experimental achievements are most probably paving the way for future therapeutics based on ONs conjugated to CPPs.

ACKNOWLEDGMENTS

We thank Peter Järver for help with Figure 21.1.

REFERENCES

1. Gardlik, R. et al., Vectors and delivery systems in gene therapy, *Med. Sci. Monit.*, 11, RA110, 2005.
2. Wagner, E. et al., Transferrin–polycation conjugates as carriers for DNA uptake into cells, *Proc. Natl Acad. Sci. U.S.A.*, 87, 3410, 1990.
3. Citro, G., Ginobbi, P., and Szczylik, C., Receptor-mediated oligodeoxynucleotides delivery by estradiol and folic acid polylysine conjugates, *Cytotechnology*, 11(Suppl. 1), S30, 1993.
4. Brandén, L.J., Mohamed, A.J., and Smith, C.I., A peptide nucleic acid–nuclear localization signal fusion that mediates nuclear transport of DNA, *Nat. Biotechnol.*, 17, 784, 1999.
5. Pichon, C. et al., Histidylated oligolysines increase the transmembrane passage and the biological activity of antisense oligonucleotides, *Nucleic Acids Res.*, 28, 504, 2000.
6. Järver, P. and Langel, Ü., The use of cell-penetrating peptides as a tool for gene regulation, *Drug Discov. Today*, 9, 395, 2004.
7. Lochmann, D., Jauk, E., and Zimmer, A., Drug delivery of oligonucleotides by peptides, *Eur. J. Pharm. Biopharm.*, 58, 237, 2004.
8. Dykxhoorn, D.M. and Lieberman, J., The silent revolution: RNA interference as basic biology, research tool, and therapeutic, *Annu. Rev. Med.*, 56, 401, 2005.
9. Dorsett, Y. and Tuschl, T., siRNAs: Applications in functional genomics and potential as therapeutics, *Nat. Rev. Drug Discov.*, 3, 318, 2004.
10. Schiffelers, R.M. et al., Cancer siRNA therapy by tumor selective delivery with ligand-targeted sterically stabilized nanoparticle, *Nucleic Acids Res.*, 32, e149, 2004.
11. Lorenz, C. et al., Steroid and lipid conjugates of siRNAs to enhance cellular uptake and gene silencing in liver cells, *Bioorg. Med. Chem. Lett.*, 14, 4975, 2004.
12. Soutschek, J. et al., Therapeutic silencing of an endogenous gene by systemic administration of modified siRNAs, *Nature*, 432, 173, 2004.
13. Takei, Y. et al., A small interfering RNA targeting vascular endothelial growth factor as cancer therapeutics, *Cancer Res.*, 64, 3365, 2004.
14. Minakuchi, Y. et al., Atelocollagen-mediated synthetic small interfering RNA delivery for effective gene silencing in vitro and in vivo, *Nucleic Acids Res.*, 32, e109, 2004.
15. Song, E. et al., Antibody mediated in vivo delivery of small interfering RNAs via cell-surface receptors, *Nat. Biotechnol.*, 23, 709, 2005.
16. Simeoni, F. et al., Insight into the mechanism of the peptide-based gene delivery system MPG: Implications for delivery of siRNA into mammalian cells, *Nucleic Acids Res.*, 31, 2717, 2003.

17. Wadia, J.S., Stan, R.V., and Dowdy, S.F., Transducible TAT-HA fusogenic peptide enhances escape of TAT-fusion proteins after lipid raft macropinocytosis, *Nat. Med.*, 10, 310, 2004.
18. Muratovska, A. and Eccles, M.R., Conjugate for efficient delivery of short interfering RNA (siRNA) into mammalian cells, *FEBS Lett.*, 558, 63, 2004.
19. Davidson, T.J. et al., Highly efficient small interfering RNA delivery to primary mammalian neurons induces microRNA-like effects before mRNA degradation, *J. Neurosci.*, 24, 10040, 2004.
20. Chiu, Y.L. et al., Visualizing a correlation between siRNA localization, cellular uptake, and RNAi in living cells, *Chem. Biol.*, 11, 1165, 2004.
21. Morris, M.C. et al., A new peptide vector for efficient delivery of oligonucleotides into mammalian cells, *Nucleic Acids Res.*, 25, 2730, 1997.
22. Read, M.L. et al., A versatile reducible polycation-based system for efficient delivery of a broad range of nucleic acids, *Nucleic Acids Res.*, 33, e86, 2005.
23. Leng, Q. et al., Highly branched HK peptides are effective carriers of siRNA, *J. Gene Med.*, 7, 977, 2005.
24. Pooga, M. et al., Cell penetrating PNA constructs regulate galanin receptor levels and modify pain transmission in vivo, *Nat. Biotechnol.*, 16, 857, 1998.
25. Schwarze, S.R. et al., In vivo protein transduction: Delivery of a biologically active protein into the mouse, *Science*, 285, 1569, 1999.
26. Rittner, K. et al., New basic membrane-destabilizing peptides for plasmid-based gene delivery in vitro and in vivo, *Mol. Ther.*, 5, 104, 2002.
27. Bielinska, A. et al., Regulation of gene expression with double-stranded phosphorothioate oligonucleotides, *Science*, 250, 997, 1990.
28. Morishita, R. et al., A gene therapy strategy using a transcription factor decoy of the E2F binding site inhibits smooth muscle proliferation in vivo, *Proc. Natl Acad. Sci. U.S.A.*, 92, 5855, 1995.
29. Yamada, T. et al., In vivo identification of a negative regulatory element in the mouse renin gene using direct gene transfer, *J. Clin. Invest.*, 96, 1230, 1995.
30. Tomita, S. et al., Transcription factor decoy to study the molecular mechanism of negative regulation of renin gene expression in the liver in vivo, *Circ. Res.*, 84, 1059, 1999.
31. Li, M.J. et al., Long-term inhibition of HIV-1 infection in primary hematopoietic cells by lentiviral vector delivery of a triple combination of anti-HIV shRNA, anti-CCR5 Ribozyme, and a nucleolar-localizing TAR decoy, *Mol. Ther.*, 2005.
32. Wang, L.H. et al., Targeted disruption of stat6 DNA binding activity by an oligonucleotide decoy blocks IL-4-driven T(H)2 cell response, *Blood*, 95, 1249, 2000.
33. Park, Y.G. et al., Dual blockade of cyclic AMP response element- (CRE) and AP-1-directed transcription by CRE-transcription factor decoy oligonucleotide. gene-specific inhibition of tumor growth, *J. Biol. Chem.*, 274, 1573, 1999.
34. Varma, R.S., Synthesis of oligonucleotide analogues with modified backbones, *Synlett*,1993 621, 1993.
35. Gambari, R., Biological activity and delivery of peptide nucleic acids (PNA)–DNA chimeras for transcription factor decoy (TFD) pharmacotherapy, *Curr. Med. Chem.*, 11, 1253, 2004.
36. Mischiati, C. et al., Complexation to cationic microspheres of double-stranded peptide nucleic acid–DNA chimeras exhibiting decoy activity, *J. Biomed. Sci.*, 11, 697, 2004.

37. Penolazzi, L. et al., Peptide nucleic acid-DNA decoy chimeras targeting NF-kappaB transcription factors: Induction of apoptosis in human primary osteoclasts, *Int. J. Mol. Med.*, 14, 145, 2004.

38. Tomita, N. et al., Gene therapy with an E2F transcription factor decoy inhibits cell cycle progression in rat anti-Thy 1 glomerulonephritis, *Int. J. Mol. Med.*, 13, 629, 2004.

39. De Rosa, G. et al., Enhanced intracellular uptake and inhibition of NF-kappaB activation by decoy oligonucleotide released from PLGA microspheres, *J. Gene Med.*, 7, 771, 2005.

40. Gill, J.S. et al., Effects of NFkappaB decoy oligonucleotides released from biodegradable polymer microparticles on a glioblastoma cell line, *Biomaterials*, 23, 2773, 2002.

41. Kovacs, J.R. et al., Polymeric microspheres as stabilizing anchors for oligonucleotide delivery to dendritic cells, *Biomaterials*, 26, 6754, 2005.

42. Midoux, P. et al., Membrane permeabilization and efficient gene transfer by a peptide containing several histidines, *Bioconjug. Chem.*, 9, 260, 1998.

43. Chamarthy, S.P. et al., A cationic peptide consists of ornithine and histidine repeats augments gene transfer in dendritic cells, *Mol. Immunol.*, 40, 483, 2003.

44. Sakaguchi, M. et al., Targeted disruption of transcriptional regulatory function of p53 by a novel efficient method for introducing a decoy oligonucleotide into nuclei, *Nucleic Acids Res.*, 33, e88, 2005.

45. Griesenbach, U. et al., Cytoplasmic deposition of NFkappaB decoy oligonucleotides is insufficient to inhibit bleomycin-induced pulmonary inflammation, *Gene Ther.*, 9, 1109, 2002.

46. Bene, A., Kurten, R.C., and Chambers, T.C., Subcellular localization as a limiting factor for utilization of decoy oligonucleotides, *Nucleic Acids Res.*, 32, e142, 2004.

47. Marthinet, E. et al., Modulation of the typical multidrug resistance phenotype by targeting the MED-1 region of human MDR1 promoter, *Gene Ther.*, 7, 1224, 2000.

48. Kilk, K. et al., Evaluation of transportan 10 in PEI mediated plasmid delivery assay, *J. Controlled Release*, 103, 511, 2005.

49. El-Andaloussi, S. et al., TP10, a delivery vector for decoy oligonucleotides targeting the Myc protein, *J. Controlled Release*, 110, 189, 2005.

50. Fisher, L. et al., Cellular delivery of a double-stranded oligonucleotide NFkappaB decoy by hybridization to complementary PNA linked to a cell-penetrating peptide, *Gene Ther.*, 11, 1264, 2004.

51. Nieddu, E. et al., Sequence specific peptidomimetic molecules inhibitors of a protein–protein interaction at the helix 1 level of c-Myc, *FASEB J.*, 19, 632, 2005.

22 A Noncovalent Peptide-Based Strategy for Peptide and Short Interfering RNA Delivery

May C. Morris, Sébastien Deshayes,
Federica Simeoni, Gudrun Aldrian-Herrada,
Frédéric Heitz, and Gilles Divita

CONTENTS

22.1   INTRODUCTION

Over the past 10 years, substantial progress has been made in the design of new technologies to improve cellular uptake of therapeutic compounds.[1–6] This evolution has been directly correlated with the dramatic acceleration in the production of new therapeutic molecules, as cell delivery systems described until then were restricted by very specific issues. However, only a few nonviral technologies are efficiently applied in vivo at either preclinical or clinical states.[1,5,6] Their major limitations are the poor stability of the formulations, the rapid degradation of the cargo, as well as its insufficient capability to reach its target. Cell-penetrating peptides (CPPs) constitute one of the most promising generations of tools for delivering biologically active molecules into cells and can thereby have a major impact on the future of treatments.[4,7,8] CPPs have been shown to improve intracellular delivery of various biomolecules efficiently, including plasmid DNA, oligonucleotides, short interfering RNA (siRNA), PNA, proteins, peptides, as well as liposomes into cells both in vivo and in vitro.[7–12] Short synthetic CPPs able to overcome both extracellular and intracellular limitations have been designed. These peptides can trigger the movement of a cargo across the cell membrane into the cytoplasm of the cells and improve its intracellular routing, thereby facilitating the interaction with the target.[7–13] Most of the technologies described to date require the attachment of a CPP to the target cargo, which is achieved by either chemical cross-linking or cloning and expression of a protein fused to the CPP.[14–17] Conjugation methods offer several advantages for in vivo applications, including rationalization, reproducibility of the procedure, together with the control of the stoichiometry of the protein transduction domain (PTD) cargo. However, the covalent PTD technology is limited from the chemical point of view as it is mainly performed via a synthetic disulfide linkage and risks altering the biological activity of the cargoes. In order to offer an alternative to covalent strategies we have proposed a new potent strategy for the delivery of biomolecules into mammalian cells, based on the short amphipathic peptide carriers, MPG and Pep-1.[18–23] MPG and Pep-1 form stable nanoparticles with cargoes without the need for cross-linking or chemical modifications. MPG efficiently delivers nucleic acids (plasmid DNA, oligonucleotides, siRNA) and Pep-1 improves the delivery of proteins and peptides in a fully biologically active form into a variety of cell lines and in vivo.[21–23] The mechanism through which MPG or Pep-1 delivers active macromolecules does not involve the endosomal pathway and therefore allows control of the release of the cargo in the appropriate target subcellular compartment.[24,25] In this chapter we will describe the characteristics of the noncovalent MPG and Pep-1 strategies, and their applications for nucleic acid and peptide transduction, both in vitro and in vivo.

22.2 PEPTIDE-BASED STRATEGY FOR MACROMOLECULE DELIVERY

22.2.1 PEPTIDE-BASED STRATEGY FOR NUCLEIC ACID DELIVERY

The poor permeability of the plasma membrane of eukaryotic cells to drugs or DNA, together with the low efficiency of DNA or oligonucleotides to reach their target within the cells, constitute the two major barriers for the development of therapeutic molecules. As such, the development of multifunctional nonviral-based delivery systems is a major challenge in gene and antisense-based therapeutics.[1,2] A number of nonviral strategies has been proposed, including lipid, polycationic, nanoparticle, and peptide-based formulations.[1,2,5,11,26,27] Although they exhibit several advantages over viral systems, nonviral strategies for therapeutic applications remain limited by their inefficiency to release nucleic acids from the endosomal compartment following cellular uptake, and to translocate DNA into the nucleus. In the last decade, a number of CPP-based gene-delivery systems have been proposed to overcome both extracellular and intracellular limitations.[4,8,11,14]

Several strategies have been designed either to avoid the endosomal pathway or to facilitate the escape of cargoes from early endosomes to prevent their degradation. Peptide carriers that combine DNA binding and membrane destabilizing properties have been developed to facilitate gene transfer into cultured cells and living animals.[1,2,5,11] A number of peptides with pH-dependent fusogenic and endosomolytic activities has been described to improve lysosomal degradation and consequently increase transfection efficiency.[28,29] Amphipathic peptides containing a periodicity of hydrophobic and polar residues, such as the fusion peptide of the HA_2 subunit of influenza hemagglutinin,[28,29] or synthetic analogs GALA, KALA, JTS1,[30–33] histine-rich peptides,[34,35] and Hel 11.7 interact with the phospholipid membrane and induce fusion and lysis. They have been shown to increase transfection efficiency when associated with poly-L-lysine/DNA, condensing peptide/DNA, cationic lipids, polyethyleneimine, or polyamidoamine cascade polymers.[28,29]

The second major barrier to nonviral gene-delivery systems is their poor nuclear translocation, which is, however, essential for transfection of nondividing cells and gene therapy. In order to improve nuclear delivery of cargoes (drugs, DNA), synthetic peptides containing nuclear localization sequences (NLS) have been used extensively.[36–38] Most of these studies were performed with the sequence derived from the SV40 large T antigen NLS PKKKRKV. This sequence is associated with either membrane-penetrating peptides or cationic peptides, but is also directly linked to cargoes or combined with other transfection methods to facilitate delivery into the nucleus.[11] CPPs harboring a NLS have been applied for drug and cargo delivery.[36–38] Moreover, NLS sequences have been associated with different hydrophobic CPPs in order to favor membrane crossing of nuclear targeting cargoes, but also to facilitate DNA binding and compaction for gene delivery.[11]

A number of synthetic multifunctional peptide sequences that integrate multiple targeting and routing signals have been designed to overcome both cytoplasmic and nuclear entry barriers. Peptides comprise either a linear or a branched organization of cytoplasmic-translocating sequence signals and NLS.[11,12] PTDs have also been used to improve the delivery of DNA[40–43] or oligonucleotides,[8,14,15] and have been combined with other lipid-based nonviral methods.[3] Several CPPs have been successfully applied for the delivery of small oligonucleotides in vivo via covalent coupling.[4,8,11,14,15] In contrast, only a few CPPs have been validated in vivo for gene delivery.[2] To date, the secondary amphipathic peptide, PPTG1, constitutes one of the only examples reporting a significant in vivo gene expression response following intravenous injections.[39]

22.2.2 Peptide-Based Strategy for Peptide and Protein Delivery

In order to circumvent the problems of gene therapy technology, an increasing interest is being taken in designing novel strategies that allow the delivery of peptides and full-length proteins into a large number of cells.[4,7,8] However, the development of therapeutic peptides and proteins remains limited by the poor permeability and the selectivity of the cell membrane for large molecules. Recently, the discovery of PTDs has given new hope for the administration of large proteins and peptides in vivo and has proved that "protein therapy" can have a major impact on the future of therapies in a variety of viral diseases and cancers.[7–12] The CPP family includes several peptide sequences: synthetic and natural cell-permeable peptides, PTDs, and membrane-translocating sequences, which are all capable of translocating the cell membrane independently of transporters or specific receptors.[7,12] PTDs have been shown to cross biological membranes efficiently, and to promote the delivery of peptides and proteins into both cultured cells and animals. Peptides derived from the transactivating regulatory protein (TAT) of the human immunodeficiency virus (HIV),[44–47] the third α-helix of the Antennapedia homeodomain protein,[10,48] VP22 protein from the herpes simplex virus,[49] the polyarginine peptide sequence,[50,51] peptides derived from calcitonin[52,53] or from antimicrobial peptides Buforin I and SynB,[54,55] as well as transportan and derivates[4,56] have been successfully used to improve the delivery of covalently linked peptides or proteins into cells and have been shown to be of considerable interest for protein therapeutics.[7–11]

22.3 MPG/PEP-1 NONCOVALENT STRATEGY FOR MACROMOLECULE DELIVERY

22.3.1 MPG and Pep-1 Peptide Families

MPG (27 residues: GALFLGFLGAAGSTMGAWSQPKKKRKV) and Pep-1 (21 residues: KETWWETWWTEWSQPKKKRKV) are primary amphiphatic peptides (Table 22.1), consisting of three domains, two of which are a variable N-terminal hydrophobic motif and a hydrophilic lysine-rich domain which, in both peptide families, is derived from the NLS of SV40 large T antigen (KKKRKV), and is required for the main interactions with nucleic acids, intracellular trafficking of the cargo, and solubility of the peptide vector. A linker domain (WSQP) separates the two domains mentioned above and contains a proline residue, which improves the flexibility and integrity of both the hydrophobic and the hydrophilic domains.[18–23] The two peptide families differ mainly in their hydrophobic domain. The hydrophobic

TABLE 22.1
Peptide Sequences

Peptides	Sequences
MPG	ac-GALFLGFLGAAGSTMGAWSQPKKKRKV-cya
MPG$^{\Delta NLS}$	ac-GALFLGFLGAAGSTMGAWSQPKSKRKV-cya
Pep-1	ac-KETWWETWWTEWSQPKKKRKV-cya
Pep-2	ac-KETWFETWFTEWSQPKKKRKV-cya

motif of MPG (GALFLGFLGAAGSTMGA) derived from the fusion sequence of the HIV protein gp41 is required for efficient targeting to the cell membrane and cellular uptake. The hydrophobic motif of Pep-1 corresponds to a tryptophan-rich cluster (KETWWETWWTEW), which is also required for efficient targeting to the cell membrane and for forming hydrophobic interactions with proteins. Structural and mechanistic investigations have revealed that the flexibility between the two domains of MPG and Pep-1 is crucial for macromolecule delivery, which is maintained by the linker sequence between the fusion and the NLS motifs.[18–26] Both peptide sequences are acetylated at their N-terminus and carry a cysteamide group at their C-terminus, both of which are essential for the stability of the peptides and their transduction mechanism. The cysteamide function has been introduced into several peptide carriers as its offers the advantage of being compatible with Fmoc synthesis and avoids the use of cysteine, thereby protecting its side chain.[57] The cysteamide function is also essential for the cellular uptake mechanism and is required for stabilization of the carrier–cargo particles.[19,20] The sequence of MPG has been modified in order to facilitate rapid release of the cargo in the cytoplasm and to limit its nuclear translocation. A single mutation on the second lysine residue of the NLS sequence to a serine (MPG$^{\Delta NLS}$ GALFLGFLGAAGSTMGAWSQPKSKRKV) abolishes the nuclear translocation property.[20] Several modifications of Pep-1 sequences have also been proposed to stabilize the cargo–carrier complex or to extend the potency of this strategy to other cargo molecules (Table 22.1). A new peptide carrier, Pep-2 (KETWFETWFTEWSQPKKKRKV), has been demonstrated to interact with and facilitate the cellular uptake of PNAs and analogs. The sequence of Pep-2 essentially differs from that of Pep-1 by two Phe residues at positions 5 and 9, which replace Trp residues in the hydrophobic domain.[58]

22.3.2 FORMATION OF CARRIER–CARGO COMPLEXES

MPG and Pep-1 associate rapidly in solution with their respective cargo (oligonucleotide or protein/peptide) through noncovalent electrostatic or hydrophobic interactions and form stable complexes independently of specific sequences.[18,19,22] The formation of carrier–cargo complexes can be easily monitored by fluorescence spectroscopy using the Trp residue located in the central linker sequence of the peptide, or in the hydrophobic domain of Pep-1, as a sensitive sensor of the interaction (Figure 22.1A). MPG exhibits high affinity in the nanomolar range

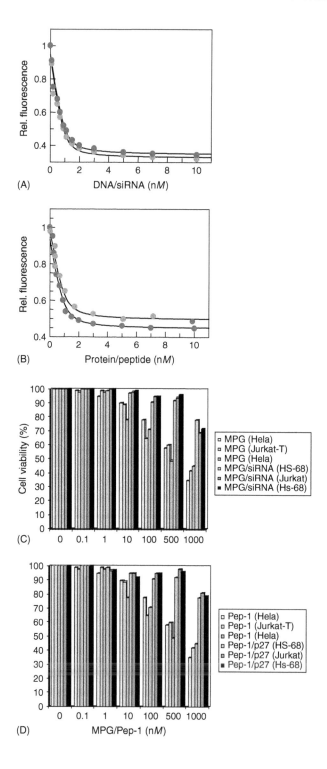

for small single- and double-stranded oligonucleotides as well as for large plasmid DNA.[18,19,22,25] The interaction with nucleic acids is mediated by the NLS domain, followed by peptide–peptide interactions through the gp41 hydrophobic domain, thus generating a peptide cage around the plasmid. The MPG-based nanostructure includes several peptide molecules, from 20 to 40 depending on the size of the nucleic acid.[21,25] Peptide-based nanoparticles have been identified both by light scattering (particle size 200 nm diameter) and gel shift assays,[18,59] and have been shown to improve dramatically the stability of the oligonucleotide inside the cell and significantly protect the nucleic acid from degradation (Figure 22.1A,B). Pep-1 exhibits high affinity for its cargoes in the nanomolar range, and forms stable peptide-based nanoparticles around the cargo, of a size estimated between 100 and 200 nm diameter. The nanoparticles correspond to peptide/cargo ratios of about 10/1 to 15/1 depending on the nature of the cargo (Figure 22.1).

22.3.3 Toxicity of Carrier–Cargo Complexes

An important criterion for a CPP is the balance between efficiency and toxicity of the peptide, as most peptides that interact with the membrane are toxic to a certain extent. Evaluation of the toxicity of MPG and Pep-1 on different cell lines has revealed that neither peptide exhibits any toxicity up to a concentration of 100 μM, and interestingly, when the peptide carriers are associated with their cargoes (MPG: nucleic acids; Pep-1: proteins and peptides), their toxicity is significantly reduced (Figure 22.2C,D). This decrease in toxicity is associated with the stabilization of the structure of the particle, which involves the cysteamide group.[20] Moreover, no in vivo toxicity has been reported for either peptide up to a millimolar concentration.

22.4 CELLULAR UPTAKE MECHANISM OF MPG AND PEP-1

22.4.1 Cellular Uptake Mechanism: A Structural Point of View

Recently, the cellular uptake mechanism of PTDs has been completely revised and shown to be essentially associated with the endosomal pathway.[61–65] However, for most CPPs, the cellular uptake mechanism still needs to be confirmed and there is evidence for several routes of cellular uptake of CPPs, some of which are independent of the endosomal pathway and involve transmembrane potential.[13,66,68,69] An important criterion to be considered is the structural requirement for cellular uptake of CPPs (see Chapter 8). A variety of physical and spectroscopic approaches

---

FIGURE 22.1 Formation and toxicity of Pep-1–cargo complexes. The binding of cargoes to Pep-1 and MPG was monitored by fluorescence spectroscopy in phosphate buffer. A fixed concentration of carrier (0.1 μM) was titrated by increasing concentration of cargoes. (A) MPG was titrated with a short double-stranded siRNA (red circle) and plasmid DNA (blue circle). (B) Pep-1 was titrated with GFP-protein (red circle) and a 32-mer peptide (blue circle), as previously described.[22] The toxicity of MPG (C) and Pep-1 (D) was investigated in different cell lines, including HeLa, HS-68, and Jurkat-T. Cells were incubated in the presence of increasing concentrations, from 0.1 μM to 1 mM of carrier associated or not associated with a cargo molecule at a molar ratio of 20/1. The cell viability was estimated using MTT staining after 24 h incubation.

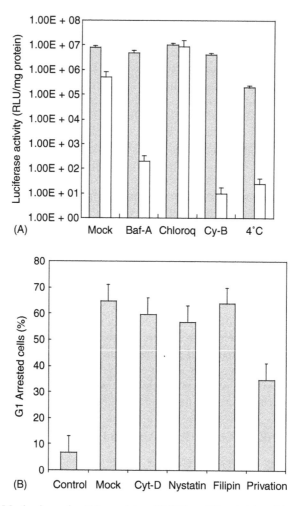

FIGURE 22.2 Mechanism of cellular uptake of MPG and Pep-1. Plasmid encoding for the luciferase and cell cycle inhibitory protein 27^{kip1} were formulated with MPG and Pep-1, at a molar ratio of 1/20, respectively. Experiments were performed on synchronized HS-68 cells in the absence (Mock) or presence of different inhibitors of the endosomal pathway. Cells were incubated in the presence of the inhibitor 1 h prior to addition of carrier–cargo complexes and inhibitors were maintained for 1 h after transfection. The level of expression of luciferase and cell arrested in G1 was analyzed 48 or 24 h after release.

can be combined to gain insight into the structure(s) involved in the interactions of carrier–cargo complexes with lipids, and thus to characterize their mechanism of cellular internalization.[13,70] Both the MPG and Pep-1 peptide families have been shown to interact strongly with membrane lipids, mainly through their hydrophobic domain, either the fusion sequence of MPG or the Trp-rich motif of Pep-1, which are crucial for insertion of the peptide into the membrane. The direct interaction of peptides with lipids limits their association with proteoglycans at the surface of the cell as well as risking uptake through the endosomal pathway.[24,25] The conformation of both MPG and Pep-1 are not significantly affected upon formation of a particle with their cargo. In contrast, NMR, circular dichroism, and FTIR analysis have revealed that interaction of the peptide carrier or of carrier–cargo complexes with phospholipids results in folding of the carrier; MPG and Pep-1 fold into a β-structure and an α-helix, respectively (see Chapter 8).[24,25] The outer part of the "carrier-based nanoparticle" with the cargo plays a key role in the interactions with the membrane and forms transient transmembrane helical or β-structures, depending on the carriers that temporarily affect the cell membrane organization, without associated leakage or toxicity, thereby facilitating insertion into the membrane and initiation of the translocation process. Based on both structural and biophysical investigations, a four-step mechanism has been proposed:

1. Formation of the MPG–cargo or Pep-1–cargo complexes involving hydrophobic and electrostatic interactions depending on the nature of the cargo
2. Interaction of the complex with the external side of the cell involving electrostatic contacts with the phospholipid head groups
3. Insertion of the complex into the membrane, associated with conformational changes which induce membrane structure perturbations
4. Nuclear targeting of the MPG–cargo complex or release of the Pep-1–cargo complex into the cytoplasm with partial "de-caging" of the cargo[13,24,25]

22.4.2 MECHANISM OF CELLULAR UPTAKE: A BIOLOGICAL POINT OF VIEW

Artifacts can be associated with fixation methods, as described by several groups.[61,62] In addition, the use of fluorescent probes attached to CPPs can modify their cellular behavior. As several routes may exist it is essential to identify the one leading to a biological response. Therefore, an essential rule when investigating the uptake mechanism is to correlate the uptake pathway with a biological response associated with the cargo.[65]

22.4.2.1 MPG: Cellular Uptake Mechanism

There are two important steps in the mechanism of MPG-mediated delivery: its cellular uptake mechanism and its impact on the nuclear translocation of the nucleic acid. Experiments based on the ability of MPG to transfect plasmids encoding for the luciferase reporter gene were performed in order to identify the different steps in the MPG uptake mechanism (Figure 22.2B). The level of reporter gene expression was used to investigate the uptake mechanism of MPG–DNA particles in cellulo and the role of the NLS sequence in nuclear import. Uptake experiments performed in the presence of several inhibitors of the endosomal pathway (bafilomycine A, cytochalasin D) and at a low temperature for energy depletion (Figure 22.2A), demonstrated that none of the inhibitors affected the efficiency of MPG, suggesting that the uptake of the MPG–DNA complex associated with a biological response is independent of the endosomal pathway and mediated by the membrane potential. The NLS motif of MPG is required for both electrostatic interactions with DNA and nuclear targeting. MPG–DNA particles are able to interact with the nuclear import machinery, and a single mutation in the NLS abolishes the translocation of DNA into the nucleus and the ability of MPG to transfect quiescent or arrested cell in G1. After crossing the cell membrane, the presence of the NLS domain promotes rapid delivery of the plasmid into the nucleus.[18,20]

22.4.2.2 Pep-1 Cellular Uptake Mechanism

A model experiment based on the ability of Pep-1 to improve the delivery of the cell cycle inhibitor protein p27[Kip1] was performed to understand its uptake mechanism. The cell-cycle inhibitor p27[Kip1] binds to and inhibits Cdk–cyclin complexes involved in the G1/S transition such as Cdk2/cyclin E and Cdk2/cyclin A.[71] A covalent strategy based on a PTD peptide has been successfully used for the delivery of p27[kip1] or derived inhibitory peptides into cultured cells and ex vivo.[17,72] Pep-1 can efficiently deliver p27[kip1] in a biologically active form into nontransformed and cancer cells without the need for cross-linking, which results in a cell-cycle arrest in G1 in more than 70% of cells (Figure 22.2B). The p27[kip1]-associated response was used to investigate the uptake mechanism of Pep-1–cargo particles in cellulo. Uptake experiments were performed in the presence of several inhibitors of the endosomal pathway or an energy-depleted system (Figure 22.2B). Results demonstrated that none of the inhibitors affected the efficiency of Pep-1, with the exception of energy deprivation, which reduced by 50% the biological response associated with Pep-1-mediated p27[kip1] delivery. This result can be directly correlated to modification of membrane potential, known to be required for the uptake of CPPs.[67] We therefore proposed that the uptake of the Pep-1–cargo complex leading to the biological response is independent of the endosomal pathway and is directly correlated to the size of particles and the nature of cargoes.

22.5 APPLICATION TO IN VITRO AND IN VIVO DELIVERY OF THERAPEUTIC MOLECULES

22.5.1 MPG: PEPTIDE-BASED NUCLEIC ACID DELIVERY SYSTEM

22.5.1.1 Gene and Oligonucleotide Delivery

MPG technology has been applied to both plasmid DNA and oligonucleotide delivery, with high efficiency (50–90%), into a large number of cells in suspension and adherent cell lines (Table 22.2). The ability of MPG to improve the nuclear translocation of nucleic acids without requiring nuclear membrane breakdown during mitosis has been reported in several protocols for gene and oligonucleotide delivery on primary cell lines and nondividing cells (Table 22.2).[20,21] As the cellular uptake mechanism and, therefore, the efficiency of MPG are directly correlated with the MPG–cargo particle size, the procedure to obtain the MPG–cargo complexes and the molar ratio are crucial parameters. Although MPG efficiency is not affected by the presence of serum, complexes should be prepared in the absence of serum and the carrier/cargo ratio maintained between 20/1 and 40/1 to avoid any aggregation or endosomal uptake, and to obtain optimal associated biological response (Figure 22.3D).

TABLE 22.2
Pep-1- and MPG-Mediated Delivery of Biomolecules

Cell Lines	Macromolecules	Refs.
MPG family:		
Hela	siRNA/oligonucleotide/plasmid	18–21,59,60,93–95
SW620	siRNA	21,22,95
MEF	siRNA	22,95
HEK	siRNA/oligonucleotide	18–21,59,60,93–95
HUVEC	siRNA/oligonucleotide PS	95
C2C12	siRNA/plasmid	18,21
HS-68	siRNA/oligonucleotide/plasmid	18–21,95
CEM/macrophage	siRNA/oligonucleotide/plasmid	59,60
CAKI-1 renal carcinoma	Thiophosphoramidate	73,74
DV-145 prostate adenocarcinoma	Thiophosphoramidate	73,74
MCF-7	Oligonucleotide PS	59,95
HepG2/primary hepatocytes	siRNA oligonucleotide PS	21,59,95
Pep-1 family:		
3T3/L1—mouse fibroblast	Antibody, protein, peptide	120
HeLa—human cervix carcinoma	Protein, peptide, antibody, PNA	22,23,121,122
HEK/293—human embryonic kidney	Protein/peptide	119
HS-68/WI-38 fibroblast	Protein, peptide, antibody	22,23,97,103,121
PC-12 rat pheochromocytoma	Protein, peptide	122,130
A549 human lung carcinoma	Antibody	113
CV-1 monkey kidney	Antibody	125
C2C12—myotubes	Antibody, protein	132
Primary plant protoplasts Arabidopsis	Protein	114
COS-7 monkey kidney	Protein, oligopeptide	22,23
WISH human placenta carcinoma	Protein	135
Primary mouse embryonic cells	Protein	136
Cardiomyocites	Protein	125
Primary human mesenchymal stem cells	Peptide	98
Macrophage/Jurkat human T-cell leukemia	Antibody	22,23,98,119,124
MCF-7 human breast carcinoma	Protein/PNA	58,100
Primary human monocyte	Protein	127
Primary mouse hepatocytes	Protein	101
Primary rate type II alveolar epithelial	Protein	113
Primary neurons neural retina cells	Antibody, protein, peptide	97,98,99,102
NRK normal rat kidney	Antibody	128,131
Pancreatic cells primary human	Protein	104
Saos-2 human osteosarcoma cells	Antibody	136
Bovine and monkey zygote	Protein, antibody	117,118

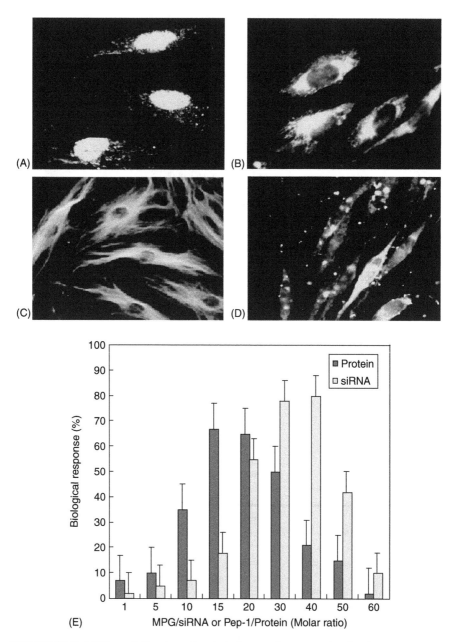

FIGURE 22.3 MPG- and Pep-1-mediated delivery of macromolecules into mammalian cells. 0.1 μM of FITC-labeled siRNA was associated with MPG (A) or MPG$^{\Delta NLS}$ (B) at a molar ratio of 20/1. Monoclonal anti-β-actin (C) was formulated with Pep-1 at a molar ratio of 20/1. A concentration of 0.1 μM of FITC-labeled PNA (D) was associated with Pep-2 at a molar ratio of 10/1. Cells were incubated with complexes for 1 h and then extensively washed prior to observation by confocal microscopy. (Adapted from Simeoni et al., *Nucleic Acids Res.*, 31, 2717, 2003 and Morris et al., *Nat. Biotechnol.*, 19, 1173, 2001.)

MPG technology has been reported to be a potent method for the delivery of unmodified antisense oligonucleotides and full-length antisense construct targeting the cell-cycle regulatory protein Cdc25C,[18,19] locked nucleic acid,[59] phosphorothioate oligonucleotides targeting the promoter of MDR-1 promoter in human CEM leukemia cells,[59,60] and thiophosphoramidate telomerase template antagonist in cancer cells.[73,74] MPG strategy has been extended to the delivery of small hydrophobic peptides and demonstrated to improve the delivery of peptide drugs targeting the maturation of HIV reverse transcriptase in primary T cells and macrophages, associated with strong antiviral activity.[75]

22.5.1.2 siRNA Delivery

siRNAs constitute a powerful tool to silence gene expression posttranscriptionally.[76–79] However, the major limitation of siRNA applications, similar to most antisense or nucleic acid-based strategies, remains their poor cellular uptake associated with the poor permeability of the cell membrane to nucleic acids. Several viral[79–83] and nonviral strategies[84–89] have been proposed to improve the delivery of either siRNAs expressing vectors or synthetic siRNAs, both in cultured cells and in vivo. PTD-based strategies have been developed to improve the delivery of oligonucleotides both in vitro and in vivo. PTD strategy was extended to the delivery of siRNA into cultured cells. Similarly, siRNAs covalently linked to transportan[91] and penetratin[92] have been associated with a silencing response.

MPG forms stable noncovalent complexes with siRNAs, increases their stability, promotes their cellular uptake without the need for prior chemical covalent coupling, and enables robust downregulation of target mRNA (Figure 22.3).[20,21] A variant of MPG harboring a single mutation in the NLS (MPG$^{\Delta NLS}$) has been designed to favor a rapid release of the siRNA into the cytoplasm, which correlates with a more significant biological response. MPG strategy has been used for the delivery of siRNA into a large panel of cell lines including adherent cell lines, cells in suspension, and cancer and primary cell lines that cannot be transfected using other nonviral approaches.[20,21,93–95] MPG$^{\Delta NLS}$ is manufactured under the name of Express^{si} by Genospectra Inc. (http://www.genespectra.com) and updated protocols are available on the company's web site. siRNA targets localize to both the cytoplasm and the nucleus, depending on the mechanism involved. Tampering with the NLS sequence of MPG allows discrimination between delivery to the nucleus and the cytoplasm, and controls the release of the siRNA into the appropriate subcellular compartment (cytoplasm or nucleus). Morris et al. demonstrate that in contrast to other nonviral methods, MPG can efficiently deliver promoter-directed siRNA into the nucleus and, therefore, inhibits transcription.[93,94]

(E) Concentration-dependent MPG- and Pep-1-mediated cellular delivery of siRNA and protein. Fixed concentrations of siRNA targeting GAPDH (50 nM) and of p27^{kip1} (0.5 μM) were associated with either MPG or Pep-1, at different molar ratios (from 1/1 to 60/1). The complexes were overlaid onto cultured HS-68 cells and biological response associated with the cargoes was monitored 24 h after transduction: The level of GAPDH protein and of G1-arrested cell induced by p27^{kip1}.

The delivery of siRNA in vivo constitutes a major challenge. Several viral and nonviral strategies have been proposed including liposome, polycationic vector, antibody, and nanoparticle, all of which are limited, and to date there is no universal approach. The most efficient method for in vivo applications is the nonviral "hydrodynamic" tail-vein injection of mice with high doses of natural and modified siRNA.[85–90] MPG has been successfully applied for the delivery of siRNA in vivo upon intratumoral or intravenous injections[96] as well as for the delivery of siRNA into mouse oocytes.[96] This technology has been applied to target an essential cell cycle protein, cyclin B1; intravenous injections of MPG/siRNA particles have been shown to block tumor growth efficiently.[96]

22.5.2 APPLICATION OF PEP-1 STRATEGY TO MACROMOLECULE TRANSDUCTION

22.5.2.1 Pep-1-Mediated Transduction of Peptides and Large Proteins

Pep-1 technology has been applied to both basic research as well as to the delivery of therapeutic peptides and proteins, with high efficiency (60–80%), into a large number of mammalian cell lines, including nontransformed, cancer, neuronal, and primary cell lines.[22,23,97] This carrier promotes the cellular uptake of small peptides and of large proteins, independently of the size and nature of the polypeptide and of the cell types (Table 22.2).[100–137] It has been shown to deliver antibodies into cells while preserving the ability to recognize their target antigens, which constitutes a major interest for therapeutic application (Figure 22.3). Improving the uptake of proteins and peptides into primary cell lines remains an essential challenge; several protocols have been described for the application of Pep-1 into different primary cell lines including macrophages,[100] hepatocytes,[101] neural retinal,[102] human stem,[103] and pancreatic cells.[104] Gallo and colleagues[98,99] have optimized Pep-1 protocol for the transduction of peptides and proteins into primary neuronal cells. As for MPG, the pep-1/cargo ratio is a critical parameter for the formation of Pep-1–macromolecule complexes, as it is directly associated with the size of the particle and therefore with the cellular uptake mechanism. This ratio should be maintained between 15/1 and 20/1 to avoid aggregation/precipitation and to obtain an optimal associated biological response (Figure 22.3D). Pep-1 is manufactured under the name Chariot™ by Active Motif Inc. (http://www.activemotif.com), and an update of the published work on the Chariot strategy is available on the company's web site.

22.5.2.2 Pep-1-Mediated Transduction of Biologically Active Molecules

Pep-1 strategy has been extended to the delivery of other uncharged and charged cargoes, including siRNA,[105] DNA–protein complexes,[24] replication-deficient viruses,[106] PNAs,[58] and semiconductor quantum dots.[107] PNAs have been efficiently delivered using covalent PTD-based strategies.[108,109] Pep-2 technology has been used to improve significantly the delivery of both uncharged PNAs and derivatives such as HypNA-*p*PNAs (trans-4-hydroxyl-L-proline/phosphonate–PNA)[110] into several cell lines, and this technology has been applied to target an essential cell cycle protein, cyclin B1, and efficiently block cancer cell proliferation.[58]

22.5.2.3 Application of Pep-1-Mediated PNA and Protein Transduction In Vivo

Several Pep-1-based formulations for in vivo applications have been described including intravenous, intratumoral, and intratracheal injections as well as transduction into oocytes or sprays for skin or nasal delivery, suggesting that Pep-1 may be a potent technology for the delivery of therapeutic proteins in vivo.[24,98,111–118] Pep-1 has be shown to be an excellent technology for protein and antibody transduction in primary neurons and to be able to cross the blood–brain barrier.[98–99] Recently, Pep-1 strategy has been applied in vivo for the delivery of caspase 3 into the lungs of mice to produce alveolar wall apoptosis,[113] or to repair a defective step in a cellular signaling pathway in vivo. Maron et al.[115] used Pep-1 for the delivery of PKA into distal lung epithelial cells of rats in order to correct the defect in that protein kinase. Pep-1 strategy was also applied for the evaluation of the antitumoral activity of peptide inhibitors of protein kinase or of antisense PNAs targeting cyclin B1 via intratumoral injections,[116] as well as for the delivery of protein into bovine and mouse oocytes, which offers a potent tool for studying early embryonic development.[117,118]

22.6 CONCLUSIONS AND PERSPECTIVES

In order to offer an alternative to covalent strategies, we have proposed a new potent strategy for the delivery of cargoes into mammalian cells, based on a short amphipathic peptide carrier, which can form stable nanoparticles with cargoes without the need for cross-linking or chemical modifications.[18–23] MPG and Pep-1 technologies have been successfully applied to the delivery of different cargoes (siRNA and peptides) in primary cell lines and in vivo (Table 22.2). These peptide-based strategies present several advantages, including rapid delivery of cargoes into cells with very high efficiency, stability in physiological buffers, and lack of toxicity and of sensitivity to serum. Moreover, the lack of a prerequisite for covalent coupling to formation of carrier–macromolecule particles favors the intracellular routing of the cargo and control of its release in the target cellular compartment. The final localization of the delivered macromolecule is then determined by its inherent intracellular targeting properties.[18–24] A major concern with the cellular uptake of CPPs is to avoid the endosomal pathway or to favor escape of the cargo from early endosomes. MPG and Pep-1 behave significantly differently from other similarly designed CPPs. Although we cannot exclude that MPG or Pep-1 uptake follows several routes, the major cell-translocation mechanism is independent of the endosomal pathway and involves transient membrane disorganization associated with folding of the carrier into either a helical or β-structure within the phospholipid membrane.[24,25]

In conclusion, although it is apparent than MPG and Pep-1 technologies are still in their infancy and need to be optimized for systematic in vivo applications, they are already powerful tools for basic research and for targeting specific cellular events both in vitro and in vivo, as well as in a therapeutic context for screening potential therapeutic molecules. Both technologies constitute an important alternative to covalent strategies and can have a major impact on the use of siRNA, proteins, and peptides for future therapies.

REFERENCES

1. Luo, D. and Saltzman M.W., Synthetic, DNA, delivery system, *Nat. Biotechnol.*, 18, 33, 2000.
2. Niidome, T. and Huang, L., Gene therapy progress and prospects: Non viral vectors, *Gene Ther.*, 10, 991, 2002.
3. Torchilin, V.P., Recent advances with liposomes as pharmaceutical carriers, *Nat. Rev. Drug Discov.*, 4, 145, 2005.
4. Javer, P. and Langel, Ü., The use of cell-penetrating peptides as a toll for gene regulation, *Drug Discovery Today*, 9, 395, 2004.
5. Glover, D.J., Lipps, H.J., and Jans, D.A., Towards safe, nonviral therapeutic gene expression in humans, *Nat. Rev. Genet.*, 6, 299, 2005.
6. Opalinska, J.B. and Gewirtz, A.M., Nucleic-acid therapeutics: Basic principles and recent applications, *Nat. Rev. Drug Discov*, 1, 503, 2002.
7. Langel, Ü., *Cell Penetrating Peptides: Processes and Applications*, Pharmacology & Toxicology Series, CRC Press, Boca Raton, FL, 2002.
8. El-Andaloussi, S., Holm, T., and Langel, Ü., Cell-penetrating peptides: Mechanism and applications, *Curr. Pharma. Design*, 11, 3597, 2005.
9. Wadia, J.S. and Dowdy, S.F., Protein transduction technology, *Curr. Opin. Biotechnol.*, 13, 52, 2002.
10. Joliot, A. and Prochiantz, A., Transduction peptides: From technology to physiology, *Nat. Cell. Biol.*, 6, 189, 2004.
11. Morris, M.C. et al., Translocating peptides and proteins and their use for gene delivery, *Curr. Opin. Biotechnol.*, 11, 461, 2000.
12. Gariepy, J. and Kawamura, K., Vectorial delivery of macromolecules into cells using peptide-based vehicles, *Trends Biotechnol.*, 19, 21, 2000.
13. Deshayes, D. et al., Cell-penetrating peptides: Tools for intracellular delivery of therapeutics, *Cell. Mol. Life Sci.*, 16, 1839, 2005.
14. Moulton, H.M. and Moulton, J.D., Arginine-rich cell-penetrating peptides with uncharged antisense oligomers, *Drug Discovery Today*, 9, 870, 2004.
15. Zatsepin, T.S. et al., Conjugates of oligonucleotides and analogues with cell penetrating peptides as gene silencing, *Curr. Pharm. Design*, 11, 3639, 2005.
16. Gait, M.J., Peptide-mediated cellular delivery of antisense oligonucleotides and their analogues Cell, *Mol. Life Sci.*, 60, 844, 2003.
17. Nagahara, H. et al., Transduction of full-length TAT fusion proteins into mammalian cells: TAp27[kip1] induced cell migration, *Nat. Med.*, 4, 1449, 1998.
18. Morris, M.C. et al., A new peptide vector for efficient delivery of oligonucleotides into mammalian cells, *Nucleic Acids Res.*, 25, 2730, 1997.
19. Morris, M.C. et al., A novel potent strategy for gene delivery using a single peptide vector as a carrier, *Nucleic Acids Res.*, 27, 3510, 1999.

20. Simeoni, F. et al., Insight into the mechanism of the peptide-based gene delivery system MPG: Implication for delivery of siRNA into mammalian cells, *Nucleic Acids Res.*, 31, 2717, 2003.
21. Simeoni, F. et al., Peptide-based strategy for siRNA delivery into mammalian cells, *Methods Mol. Biol.*, 309, 251, 2005.
22. Morris, M.C. et al., A peptide carrier for the delivery of biologically active proteins into mammalian cells, *Nat. Biotechnol.*, 19, 1173, 2001.
23. Morris, M.C. et al., A peptide carrier for the delivery of biologically active proteins into mammalian cells: Application to the delivery of antibodies and therapeutic proteins. *Handbook of Cell Biology*, 4, 13, 2006.
24. Deshayes, S. et al., On the mechanism of nonendosomal peptide-mediated cellular delivery of nucleic acids, *Biochim. Biophys. Acta*, 1667, 141, 2004.
25. Deshayes, S. et al., Insight into the mechanism of internalization of the cell-penetrating carrier peptide Pep-1 through conformational analysis, *Biochemistry*, 43, 1449, 2004.
26. Ogris, M. and Wagner, E., Targeting tumors with nonviral gene delivery systems, *Drug Discovery Today*, 7, 479, 2002.
27. Demeneix, B., Hassani, Z., and Behr, J.P., Towards multifunctional synthetic vectors, *Curr. Gene Ther.*, 4, 445, 2004.
28. Plank, C., Zauner, W., and Wagner, E., Application of membrane-active peptides for drug and gene delivery across cellular membrane, *Adv. Drug Del. Rev.*, 34, 21, 1998.
29. Lear, J.D. and DeGrado, W.F., Membrane binding and conformational properties of peptides representing the NH2 terminus of influenza HA-2, *J. Biol. Chem.*, 262, 6500, 1987.
30. Parente, R.A. et al., Association of a pH-sensitive peptide with membrane vesicles: Role of amino-acid sequence, *Biochemistry*, 29, 8713, 1990.
31. Wyman, T.B. et al., Design, synthesis and characterization of a cationic peptide that binds to nucleic acids and permeabilizes bilayers, *Biochemistry*, 27, 3008, 1997.
32. Gottschalk, S. et al., A novel DNA–peptide complex for efficient gene transfer and expression in mammalian cells, *Gene Ther.*, 3, 448, 1996.
33. Niidome, T. et al., Binding of cationic alpha-helical peptides to plasmid DNA and their gene transfer abilities into cells, *J. Biol. Chem.*, 272, 15307, 1997.
34. Midoux, P. et al., Membrane permeabilization and efficient gene transfer by a peptide containing several histidines, *Bioconjug. Chem.*, 9, 260, 1998.
35. Kichler, A. et al., Histidine-rich amphipathic peptide antibiotics promote efficient delivery of DNA into mammalian cells, *Proc. Natl Acad. Sci. U.S.A.*, 100, 1564, 2003.
36. Cartier, R. and Reszka, R., Utilization of synthetic peptides containing nuclear localization signals for nonviral gene transfer systems, *Gene Ther.*, 9, 157, 2002.
37. Morris, M.C. et al., In *Cell Penetrating Peptides: Processes and Application*, Langel, Ü., ed., Pharmacology & Toxicology Series, CRC Press, Boca Raton, FL, p. 93, 2002.
38. Escriou, V. et al., NLS bioconjugates for targeting therapeutic genes to the nucleus, *Adv. Drug Deliv. Rev.*, 55, 295, 2003.
39. Rittner, K. et al., New basic membrane-destabilizing peptides for plasmid-based gene delivery in vitro and in vivo, *Mol. Ther.*, 5, 104, 2002.
40. Brandén, L.J. et al., A peptide nucleic acid-nuclear localization signal fusion that mediates nuclear transport of DNA, *Nat. Biotechnol.*, 17, 784, 1999.
41. Rudolph, C. et al., Oligomers of the arginine-rich motif of the HIV-1 TAT protein are capable of transferring plasmid DNA into cells, *J. Biol. Chem.*, 278, 11411, 2003.

42. Tung, C.H. et al., Novel branching membrane translocational peptide as gene delivery vector, *Bioorg. Med. Chem.*, 10, 3609, 2002.

43. Ignatovich, I.A. et al., Complexes of plasmid DNA with basic domain 47–57 of the HIV-1 Tat protein are transferred to mammalian cells by endocytosis-mediated pathways, *J. Biol. Chem.*, 278, 42625, 2003.

44. Frankel, A.D. and Pabo, C.O., Cellular uptake of the tat protein from human immunodeficiency virus, *Cell*, 55, 1189, 1998.

45. Fawell, S. et al., Tat-mediated delivery of heterologous proteins into cells, *Proc. Natl Acad. Sci. U.S.A.*, 91, 664, 1994.

46. Vives, E., Brodin, P., and Lebleu, B., A truncated HIV-1 Tat protein basic domain rapidly translocates through the plasma membrane and accumulates in the cell nucleus, *J. Biol. Chem.*, 272, 16010, 1997.

47. Schwarze, S.R. et al., In vivo protein transduction: Delivery of a biologically active protein into the mouse, *Science*, 285, 1569, 1999.

48. Derossi, D. et al., The third helix of the antennapedia homeodomain translocates through biological membranes, *J. Biol. Chem.*, 269, 10444, 1994.

49. Elliott, G. and O'Hare, P., Intercellular trafficking and protein delivery by a Herpesvirus structural protein, *Cell*, 88, 223, 1997.

50. Wender, P.A. et al., The design, synthesis, and evaluation of molecules that enable or enhance uptake: Peptoid molecular transporters, *Proc. Natl Acad. Sci. U.S.A.*, 97, 13003, 2000.

51. Futaki, S. et al., Arginine-rich peptides. An abundant source of membrane-permeable peptides having potential as carriers for intracellular protein delivery, *J Biol Chem.*, 276, 5836, 2001.

52. Schmidt, M.C. et al., Translocation of human calcitonin in respiratory nasal epithelium is associated with self assembly in lipid membrane, *Biochemistry*, 37, 16582, 1998.

53. Krauss, U. et al., In vitro gene delivery by a novel human calcitonin (hCT)-derived carrier peptide, *Bioorg. Med. Chem. Lett.*, 14, 51, 2004.

54. Park, C.B. et al., Structure-activity analysis of buforin II, a histone H2A-derived antimicrobial peptide: The proline hinge is responsible for the cell-penetrating ability of buforin II, *Proc. Natl Acad. U.S.A.*, 97, 8245, 2000.

55. Rouselle, C. et al., New advances in the transport of doxorubicin through the blood-brain barrier by a peptide vector-mediated strategy, *Mol. Pharmacol.*, 57, 679, 2000.

56. Pooga, M. et al., Cellular translocation of proteins by transportan, *FASEB J.*, 8, 1451, 2001.

57. Mery, J. et al., Disulfide linkage to polyacrylic resin for automated Fmoc peptide synthesis, Immunochemical applications of peptide resins and mercaptoamide peptides, *Int. J. Pept. Protein Res.*, 42, 44, 1993.

58. Morris, M.C. et al., The combination of a new generation of PNAs with a peptide-based carrier enables efficient targeting of cell cycle progression, *Gene Ther.*, 11, 757, 2004.

59. Marthinet, E. et al., Modulation of the typical multidrug resistance phenotype by targeting the MED-1 region of human MDR1 promoter, *Gene Ther.*, 7, 1224, 2000.

60. Labialle, S. et al., New invMED1 element cis-activates human multidrug-related MDR1 and MVP genes, involving the LRP130 protein, *Nucleic Acids Res.*, 32, 3864, 2004.

61. Richard, J.P. et al., Cell-penetrating peptides. A reevaluation of the mechanism of cellular uptake, *J. Biol. Chem.*, 278, 585, 2003.

62. Lundberg, M., Wikstrom, S., and Johansson, M., Cell surface adherence and endocytosis of protein transduction domain, *Mol. Ther.*, 8, 143, 2003.

63. Richard, J.P. et al., Cellular uptake of unconjugated TAT peptide involves clathrin-dependent endocytosis and heparan sulfate receptors, *J. Biol. Chem.*, 280, 15300, 2005.

64. Nakase, I. et al., Cellular uptake of arginine-rich peptides; role of macropinocytosis and actin rearrangement, *Mol. Ther.*, 10, 1011, 2004.

65. Wadia, J., Stan, R.V., and Dowdy, S., Transducible TAT-HA fusogenic peptide enhances escape of TAT-fusion proteins after lipid raft macropinocytosis, *Nat. Med.*, 10, 310, 2004.

66. Rothbard, J.B. et al., Role of membrane potential and hydrogen bonding in the mechanism of translocation of guanidinium-rich peptides into cells, *J. Am. Chem. Soc.*, 126, 9506, 2004.

67. Terrone, D. et al., Penetratin and related cell-penetrating cationic peptide can translocate across lipid bilayers in the presence of a transbilayer potential, *Biochemistry*, 42, 13787, 2003.

68. Dom, G. et al., Cellular uptake of Antennapedia Penetratin peptides is a two step process in which phase transfer precedes a tryptophan-dependent translocation, *Nucleic Acid Res.*, 31, 556, 2003.

69. Thoren, P.E. et al., Uptake of analogs of penetratin Tat (48–60) and oligoarginine in live cells, *Biochem. Biophys. Res. Commun.*, 307, 100, 2003.

70. Magzoub, M. and Gräslund, A., Cell-penetrating peptides: Small from inception to application, *Q. Rev. Biophys.*, 37, 147, 2004.

71. Morgan, D.O., Cyclin-dependent kinases: Engines, clocks, and microprocessors, *Annu. Rev. Cell. Dev. Biol.*, 13, 261, 1997.

72. Snyder, E.L., Meade, B.R., and Dowdy, S.F., Anti-cancer protein transduction strategies: Restitution of p27 tumor suppresor function, *J. Control. Release*, 91, 45, 2003.

73. Gryaznov, S. et al., Oligonucleotide $N3' \rightarrow P5'$ thio-phosphoramidate telomerase template antagonists as potential anticancer agents, *Nucleosides Nucleotides Nucleic Acids*, 22, 577, 2003.

74. Asai, A. et al., A novel telomerase template antagonist (GRN163) as a potential anticancer agent, *Cancer Res.*, 63, 3931, 2003.

75. Morris, M.C. et al., A new potent HIV-reverse transcriptase inhibitor: A synthetic peptide derived from interface subunit domains, *J. Biol. Chem.*, 274, 24941, 1999.

76. Hannon, G.J., RNA interference, *Nature*, 418, 244, 2002.

77. Elbashir, S.M. et al., Duplexes of 21-nucleotide RNAs mediate RNA interference in cultured mammalian cells, *Nature*, 411, 494, 2001.

78. Dorsett, Y. and Tuschl, T., siRNAs: Applications in functional genomics and potential as therapeutics, *Nat. Rev. Drug Discov.*, 3, 318, 2004.

79. McManus, M.T. and Sharp, P.A., Gene silencing in mammals by small interfering RNAs, *Nat. Rev. Genet.*, 3, 737, 2002.

80. Rozema, D.B. and Lewis, D.L., siRNA delivery technologies for mammalian systems, *Target*, 2, 253, 2003.

81. Hommel, D.J. et al., Local gene knockdown in the brain using viral-mediated RNA interference, *Nat. Med.*, 9, 1539, 2003.

82. Brummelkamp, T.R., Bernards, R., and Agami, R., Stable suppression of tumorigenicity by virus-mediated RNA interference, *Cancer Cell*, 2, 243, 2002.

83. Xia, H. et al., siRNA-mediated gene silencing in vitro and in vivo, *Nat. Biotechnol.*, 20, 1006, 2002.

84. Takeshita, F. et al., Efficient delivery of small interfering RNA to bone-metastatic tumors by using atelocollagen in vivo, *Proc. Natl Acad. Sci. U.S.A.*, 102, 12177, 2005.

85. Song, E. et al., Antibody mediated in vivo delivery of small interfering RNAs via cell-surface receptors, *Nat Biotechnol.*, 23, 709, 2005.

86. McCaffrey, A.P. et al., RNA interference in adult mice, *Nature*, 418, 38, 2002.

87. Song, E. et al., RNA interference targeting Fas protects mice from fulminant hepatitis, *Nat. Med.*, 9, 347, 2003.

88. Lewis, D.L. et al., Efficient delivery of siRNA for inhibition of gene expression in postnatal mice, *Nat. Genet.*, 32, 107, 2002.

89. Fountaine, T.M., Wood, M.J., and Wade-Martins, R., Delivering RNA interference to the mammalian brain, *Curr. Gene Ther.*, 5, 399, 2005.

90. Soutscek, J. et al., Therapeutic silencing of an endogenous gene by systemic administration of modified siRNAs, *Nature*, 432, 173, 2004.

91. Muratovska, A. and Eccles, M.R., Conjugate for efficient delivery of short interfering RNA (siRNA) into mammalian cells, *FEBS Lett.*, 558, 63, 2004.

92. Davidson, T.J. et al., Highly efficient small interfering RNA delivery to primary mammalian neurons induces microRNA-like effects before mRNA degradation, *J. Neurosci.*, 10, 10040, 2004.

93. Morris, K.V. et al., Small interfering RNA-induced transcriptional gene silencing in human cells, *Science*, 305, 1289, 2004.

94. Langlois, M.A. et al., Cytoplasmic and nuclear retained DMPK mRNAs are targets for RNA interference in myotonic dystrophy cells, *J Biol Chem.*, 280, 16949, 2005.

95. Pastor, L. et al., Express^{si} reagent delivers siRNA into difficult-to-transfect differentiated 3T3-L1 adipocytes. Application note & Express si profile: /http://www.genospectra.com.

96. Morris M.C. et al., Personal communication.

97. Chariot™ profile: /http://www.active.motif.com.

98. Gallo, G., Making proteins into drugs: Assisted delivery of proteins and peptides into living neurons, *Methods Cell Biol.*, 71, 325, 2003.

99. Gallo, G., Yee, H.F., and Letourneau, P.C., Actin turnover is required to prevent axon retraction driven by endogenous actomyosin contractility, *J. Cell Biol.*, 158, 1219, 2002.

100. Garnon, J. et al., Fragile X-related protein FXR1P regulates proinflammatory cytokine TNF expression at the posttranscriptional level, *J. Biol. Chem.*, 280, 5750, 2005.

101. Badag-Gorce, F. et al., The mechanism of cytokeratin aggresome formation: The role of mutant ubiquitin (UBB1), *Exp. Mol. Pathol.*, 74, 160, 2003.

102. Gehler, S. et al., Brain-derived neutrophic factor regulation of retinal growth cone filopodial dynamics is mediated through actin depolymerizing factor/cofilin, *J. Neurosci.*, 24, 10741, 2004.

103. Chan, S.A. et al., Adrenal chromaffin cells exhibit impaired granule trafficking in NCAM knockout mice, *J. Neurophysiol.*, 94, 1037, 2005.

104. Pandey, A.V., Mellon, S.H., and Miller, W.L., Protein phosphatase 2A and phosphoprotein SET regulate androgen production by P450c17, *J. Biol. Chem.*, 278, 2837, 2003.

105. Arita, M. et al., Stereochemical assignment, anti-inflammatory properties and receptors for the omega-3 lipid mediator resolving E1, *J. Exp. Med.*, 201, 713, 2005.

106. Kowolik, C.M. et al., HIV vector production mediated by Rev protein transduction, *Mol. Ther.*, 2, 324, 2003.

107. Mattheakis, L.C. et al., Optical coding of mammalian cells using semiconductor quantum dots, *Anal Biochem.*, 327, 200, 2004.

108. Pooga, M. et al., Cell penetrating PNA constructs regulate galanin receptor levels and modify pain transmission in vivo, *Nat Biotechnol.*, 16, 857, 1998.

109. Koppelhus, U. and Nielsen, P.E., Cellular Delivery of peptide nucleic acid, *Adv. Drug. Del. Rev.*, 55, 267, 2003.

110. Efimov, V. et al., PNA-related oligonucleotide mimics and their evaluation for nucleic acid hybridization studies and analysis, *Nucleosides Nucleotides Nucleic Acids*, 20, 419, 2001.

111. Eum, W.S. et al., In vivo protein transduction: Biologically active intact pep-1-superoxide dismutase fusion protein efficiently protects against ischemic insult, *Free Radic. Biol. Med.*, 37, 1656, 2004.

112. Pratt, R.L. and Kinch, M.S., Activation of the EphA2 tyrosine kinase stimulates the MAP/ERK kinase signaling cascade, *Oncogene*, 21, 7690, 2002.

113. Aoshiba, K., Yokohori, N., and Nagai, A., Alveolar wall apoptosis causes lung destruction and amphysematous changes, *Am. J. Respir. Cell Mol. Biol.*, 28, 555, 2003.

114. Wu, Y. et al., Direct delivery of bacterial avirulence proteins into resistant Arabidopsis protoplasts leads to hypersensitive cell death, *Plant J.*, 33, 130, 2002.

115. Maron, M.B. et al., PKA delivery to the distal lung air spaces increases alveolar liquid clearance after isoproterenol-induced alveolar epithelial PKA desensitization, *Am. J. Physiol. Lung Cell Mol. Physiol.*, 289, 349, 2005.

116. Gros, E. et al., Personal communication.

117. Payne, C. et al., Preferentially localized dynein and perinuclear dynactin associate with nuclear pore complex proteins to mediate genomic union during mammalian fertilization, *J. Cell Sci.*, 116, 4727, 2003.

118. Rawe, V.Y. et al., WAVE1 intranuclear trafficking is essential for genomic and cytoskeletal dynamics during fertilization: Cell-cycle-dependent shuttling between M-phase and interphase nuclei, *Dev. Biol.*, 276, 253, 2004.

119. Li, X. et al., G protein β2 subunit derived peptides for inhibition and induction of G protein pathways, *J. Biol. Chem.*, 280, 23945, 2005.

120. Yokoyama, N. et al., The C-terminus of the RON tyrosine kinase plays an auto inhibitory role, *J. Biol. Chem.*, 280, 8893, 2005.

121. Zhang, Q., Hottke, A., and Goodman, R.H., Homeodomain-interacting protein kinase-2 mediates CtBP phosphorylation and degradation in UV-triggered apoptosis, *Proc. Nat. Acad. Sci. U.S.A.*, 102, 2802, 2005.

122. Zhou, J. and Hsieh, J.T., The inhibitory role of DOC-2/DAB2 in growth factor receptor mediated signal cascade, *J. Biol. Chem.*, 276, 27793, 2001.

123. Couplier, M., Anders, J., and Ibanez, C.F., Coordinated activation of autophosphorylation sites in the RET receptor Tyrosine kinase, *J. Biol. Chem.*, 277, 1991, 2002.

124. Goska, M.M. et al., Unc119, a novel activator of Lck/Fyn, is essential for T cell activation, *J. Exp. Med.*, 199, 369, 2004.

125. Buster, D., McNally, K., and McNally, F.J., Katanin inhibition prevents the redistribution of alpha-tubulin at mitosis, *J. Cell Sci.*, 115, 1083, 2002.

126. Aki, T. et al., ERK1/2 regulates intracellular ATP levels through alpha-enolase expression in cardiomyocytes exposed to ischemic hypoxia and reoxygenation, *J. Biol. Chem.*, 279, 50120, 2004.

127. Cen, O. et al., Identification of UNC119 as a novel activator of SRC-type tyrosine kinases, *J. Biol. Chem.*, 278, 8837, 2003.

128. Sorenson, C.M., Interaction of bcl-2 with Paxillin through its BH4 domain is important during ureteric bud branching, *J. Biol. Chem.*, 279, 11368, 2004.

129. Sebbagh, M. et al., Direct cleavage of ROCK II by granzyme B induces target cell membrane blebbing in a caspase-independent manner, *J. Exp. Med.*, 201, 465, 2005.

130. Jiang, J. et al., Arachidonic acid-induced carbon-centered radicals and phospholipid peroxidation in cyclo-oxygenase-2-transfected PC12 cells, *J. Neurochem.*, 90, 1036, 2004.

131. Tisdale, E.J., Glyceraldehyde-3-phosphate dehydrogenase is phosphorylated by protein kinase Ciota/lambda and plays a role in microtubule dynamics in the early secretory pathway, *J. Biol. Chem.*, 277, 3334, 2002.

132. Tassa, A. et al., Class III phosphoinositide 3-kinase–Beclin1 complex mediates the amino acid-dependent regulation of autophagy in C2C12 myotubes, *Biochem. J.*, 376, 577, 2003.

133. Heerssen, H.M., Pazyra, M.F., and Segal, R.A., Dynein motors transport activated Trks to promote survival of target-dependent neurons, *Nat. Neurosci.*, 6, 596, 2004.

134. Chipuk, J.E. et al., Direct activation of Bax by p53 mediates mitochondrial membrane permeabilization and apoptosis, *Science*, 303, 1010, 2004.

135. van der Wijk, T. et al., Increased vesicle recycling in response to osmotic cell swelling. Cause and consequence of hypotonicity-provoked ATP release, *J. Biol. Chem.*, 278, 40020, 2003.

136. Taneja, N. et al., Histone H2AX phosphorylation as a predictor of radiosensitivity and target for radiotherapy, *J. Biol. Chem.*, 279, 2273, 2004.

137. Eskiw, C.H., Dellaire, G., and Bazett-Jones, D.P., Chromatin contributes to structural integrity of promyelocytic leukemia bodies through a SUMO-1-independent mechanism, *J. Biol. Chem.*, 279, 9577, 2004.

Part V

Selective Targeting to Tumors and Other Tissues with Cell-Penetrating Peptides: In Vivo Applications

Ernst Wagner

In the development of cell-penetrating peptides (CPPs) for medical use, the same issues as for any other new drug have to be addressed: efficacy, safety, production issues, and medical need. In this regard, the choice of CPPs as a therapeutic platform appears a very good one. In fact, very central natural biological processes, such as the mammalian immune system or other defense systems in various organisms, are based on natural CPPs which help to protect and/or cure the host from diseases; for example, by eliminating invading microorganisms or infected host cells. CPPs play a role both in eukaryotic and prokaryotic systems and viruses, and evolution has optimized CPPs for various purposes. On the one hand, they can be potent killing agents and, on the other, nontoxic cytosolic delivery agents.

This high biological potency has motivated researchers to develop CPP therapeutics, either as antimicrobials, anticancer agents, or as cytoprotective, anti-inflammatory, or diagnostic agents, as presented in chapters 23–30 which follow. Over the years, initial production issues have been overcome; CPPs can be produced in convenient quantities as synthetic peptides or recombinant proteins. With regards to toxicity, CPPs with high membrane-permeating activity obviously have to be carefully monitored for side effects on nontarget membranes. As reviewed in the following chapters, specificity can be improved by incorporating novel targeting strategies, for example, homing peptides. Importantly, CPPs address a strong existing need for improved intracellular delivery of molecular therapeutics since they can mediate the transfer of peptides, proteins, siRNA, or genes which act intracellularly but are unable to pass cellular membrane barriers. By overcoming this barrier in a specific manner, CPPs could boost medical use of such molecular therapeutics.

Laakkonen and Ruoslahti (chapter 23) review the development of homing peptides, which utilize the molecular differences in the vasculature for specifically targeting a variety of organs as well as tumors. By in vivo screens with phage-displayed peptide libraries, tumor-homing peptides have been identified which were applied for tumor-targeting of therapies. This has been shown to significantly enhance the efficacy of conventional anticancer drugs. Some of the identified peptides act also as cell type-specific, cell-penetrating agents, exposing them as interesting vehicles for efficient drug delivery.

Moschos et al. (chapter 24) report on the application of CPPs for delivery of therapeutic peptides and proteins. Encouraging therapeutic effects have been obtained in preclinical models for cancer, cerebral and cardiac ischemia, asthma, and inflammatory diseases, and also in immunization/infection models. Recently, the first therapeutic CPP agent, TAT linked to a peptide inhibitor of protein kinase C (PKC), was FDA approved for phase I/II clinical testing in the treatment of cardiac ischemia. This demonstrates how far the field has progressed. Although in a much earlier stage of research, CPPs are currently also applied for in vivo delivery of therapeutic antisense oligonucleotides or siRNA where intracellular delivery is presumably rate-limiting for efficacy.

Intracellular delivery of nanoparticles (NPs) can be strongly improved by CPPs, as outlined by Gupta and Torchilin (chapter 25). Mechanism of uptake appears to depend on the size of the transported cargo. While CPP conjugated to small molecules below 3000 Da translocate across the cell membrane, large cargoes such as NPs are internalized by endocytotic processes such as macropinocytosis. A broad spectrum of NPs for therapeutic or diagnostic use can be delivered into cells without causing toxicity. The chapter highlights the use of CPP-coated NPs for in vivo imaging; cells can be efficiently labeled by coated superparamagnetic iron oxide nanoparticles or quantum-dot-containing micelles for in vivo imaging by MRI or advanced bio-imaging techniques.

In therapeutic gene delivery, the crossing of cellular lipid membrane barriers is also a crucial step. Viruses employ CPP-like protein domains for crossing membranes during the infection process and therefore might present ideal examples demonstrating how to further optimize gene vectors into synthetic viruses. Wagner and Ogris (chapter 26) discuss current strategies for incorporation of CPPs into tumor-targeted synthetic virus nanoparticles to make them more effective for the target membranes. First encouraging therapeutic applications include the cure of glioblastoma-bearing mice by synthetic RNA that was delivered with/via melittin-containing and tumor-targeted nanoparticles.

Gomez et al. (chapter 27) report the discovery of novel cell-permeable pentapeptides derived from the Bax suppressor Ku70. On the basis of recent insights into functions/properties of the minimal element of Ku70 to bind and inhibit Bax, related peptides were generated which are cell-permeable and can suppress Bax-dependent cell death. The peptides show very low cytotoxicity and, as demonstrated in the first examples, can also deliver cargo molecules into cells.

Drug delivery to the brain imposes a particular challenge because of the blood–brain barrier (BBB). CPPs and related peptides present potent tools for overcoming this barrier. Dathe et al. (chapter 28) present the application of cationic peptides derived form the LDL receptor recognition sequence of apolipoprotein E to mediate cellular uptake of liposomes into brain capillary endothelial cells. For incorporation of peptides into liposomes or lipid micelles, both covalent and adsorptive coupling strategies were applied. The peptide-tagged liposomes and peptide–lipid micelles are efficiently internalized into cells by a LDL receptor-independent pathway involving cell binding through/cellular uptake through binding of/binding to/interaction with heparin sulfate proteoglycan.

Within the last few years, the number of putative CPPs has risen profoundly. An objective comparison of the transport characteristics of the different CPPs is complicated by the presence of different alternative cell entry pathways. Adams and Tsien (chapter 29) describe a novel method to quantify entry of CPPs into the cytosol of individual live cells. The bio-optical assay is based on fluorescence resonance energy transfer (FRET) between a genetically introduced cytosolic tetra-cysteine-modified cyan fluorescent protein (CFP) and a fluorescein biarsenical, which highly specifically binds the tetracystein-modified protein. Applying this assay, cytosolic delivery of an oligoarginine CPP conjugated to the fluorescein biarsenical was determined to proceed within 1 h, resulting in micromolar cytosolic concentrations.

Many natural host defense antimicrobial peptides kill pathogens by causing cell leakage and inhibition of intracellular targets. As reviewed by Ganyu et al. (chapter 30), such a dual activity provides opportunities to design CPPs as antimicrobials with improved host cell distribution properties. For example, the authors have developed antimicrobial CPP–PNA conjugates which combine efficient cell wall permeabilization together with efficient antisense effects against an essential gene. The chapter describes the discovery of microbial CPPs, the antimicrobial effect of CPPs and their capacity to deliver cargo into cells. Based on encouraging preclinical and clinical trials, drug approval for antimicrobial peptides for topical or oral therapy can be expected within the next years.

In summary, the following chapters illustrate the current developments of CPPs for therapeutic in vivo applications. The results obtained so far demonstrate the potency of this novel emerging class of targeted drugs and indicate that medical use will soon come into sight.

23 Selective Delivery to Vascular Addresses: In Vivo Applications of Cell-Type-Specific Cell-Penetrating Peptides

Pirjo Laakkonen and Erkki Ruoslahti

CONTENTS

23.1 BACKGROUND

We have recently discovered unprecedented heterogeneity in the molecular composition of endothelial cells by profiling the vasculature using in vivo phage display. It is now apparent that each tissue expresses its own specific set of cell surface proteins on vascular endothelial cells. In addition, many pathological conditions, including tumors, diabetes, atherosclerosis, and inflammatory diseases, add their disease-specific tags to the endothelium of the affected tissues. During tumor progression, many vascular changes are involved in the process of angiogenesis as tumors are not able to grow without a blood supply.[1] Besides inducing angiogenesis i.e., sprouting of new blood vessels from existing vessels, tumors also co-opt existing blood vessels.[2] These tumor-induced changes have already started taking place at the stage of premalignant lesions.[1] In addition to the blood vasculature, tumors contain lymphatic vessels and many tumors induce the growth of lymphatic vessels, lymphangiogenesis. The presence of lymphatic vessels seems not to be essential for tumor growth. However, lymphatics have been shown to act as an important route for the metastatic spread of tumors to both regional and distant lymph nodes.[3]

We have used phage-displayed peptide libraries to map regional and disease-specific differences in the vasculature. In this approach, phage libraries displaying either peptides or protein fragments are injected into the tail vein of an anesthetized mouse and allowed to circulate for 5–15 min. Then, the organ of interest is removed and homogenized, followed by rescue and amplification of the bound phage. Repeating this selection procedure three to five times allows us to recover phage that home specifically to the target organ, and identify the peptide responsible for the homing by sequencing the region of the phage genome encoding the peptide.

By using this technology, we have been able to isolate a number of peptides homing specifically to blood vessels of each organ we have tested, including the brain, kidneys, skin, intestine, prostate, breast, pancreas, and lungs.[4–6] In addition, we have isolated several peptides that home specifically to tumor vasculature.[7] We and our collaborators have also shown that the vasculature of a premalignant lesion can differ from that of a full-blown tumor and from the vasculature of the corresponding normal organ. This was accomplished by using two different types of transgenic tumor model: a pancreatic carcinoma model, RIP1-Tag2, and a skin carcinoma model, K14-HPV16.[8,9] Some of the tumor-homing peptides recognize common angiogenesis markers and are capable of homing to several types of tumors while other peptides recognize tumor type-specific differences.

Recently, we have isolated peptides with novel homing specificities, for example, peptides that home to tumor lymphatic vessels. In addition to homing peptides that recognize markers on blood and lymphatic endothelial cells, some peptides additionally bind to tumor pericytes or tumor cells.[8,10–12] We have also generated homing peptides that are capable of penetrating the target cells. The internalization of our peptides differs from that of Tat, penetratins, and other related internalizing peptides[13] in that our peptides enter the cells in a cell type-specific manner. In this chapter, we describe in some detail the homing peptides with cell-penetrating potential.

23.2   PEPTIDES HOMING TO NORMAL VASCULATURE

Our first homing peptides, isolated by using the in vivo screening method, were peptides recognizing markers specific for blood vessel endothelial cells. Peptide libraries were allowed to circulate in the bloodstream of mice for a few minutes before amplification of phage that had homed to the organ of interest. These studies have yielded tissue-specific vascular-homing peptides for a large number of organs, such as the brain, kidneys, lungs, skin, prostate, and heart.[4–6,14] One of these peptides, GFE, a lung-homing peptide, binds to a membrane dipeptidase (MDP) that is expressed on normal lung capillaries[5] and is downregulated in the tumor vasculature. MDP is a cell-surface zinc metalloprotease involved in the metabolism of glutathione, leukotriene D4, and certain β-lactam antibiotics. It is expressed by the epithelial cells in the kidneys and lungs; but only in the lungs is the expression shared with endothelial cells, allowing lung-specific targeting.[15]

Another peptide, CPGPEGAGC, binds to aminopeptidase P and homes to the blood vessels in the breast, but not in other tissues. This breast-homing peptide homes not only to normal breast tissue but also to hyperplastic and malignant lesions of the breast.[16] These studies suggest that each tissue expresses its own array of molecules in its vasculature. Some of these endothelial and pericyte molecules may be shared with other cell types in the same tissue or in other tissues. Furthermore, some of the markers continue to be expressed in the blood vessels of malignant lesions while others are downregulated.

23.3   TUMOR-HOMING PEPTIDES

The tumor-homing peptides recognize markers that are upregulated or differentially modified in the tumor endothelial cells compared to normal blood vessels of the same organ. Our first-generation in vivo screening procedure yielded a collection of tumor-homing peptides, including peptides containing an RGD and NGR motif.[17] An RGD peptide isolated in this manner proved to be identical to a peptide previously shown to bind to $\alpha v \beta 3$ and $\alpha v \beta 5$ integrins but not to other RGD-directed integrins.[18] The $\alpha v \beta 3$ and $\alpha v \beta 5$ integrins are specifically upregulated in angiogenic blood vessels.[19] The NGR peptide recognizes aminopeptidase N, which is a membrane-spanning cell surface protein that has been associated with cell migration and tumor invasion.[20] In the endothelium, aminopeptidase N is selectively expressed in angiogenic endothelial cells, showing that in vivo phage library screening can uncover new angiogenesis markers. Tumor homing of RGD and NGR peptides is independent of the tumor type, indicating that they recognize general markers of angiogenesis.[17,21]

RGD peptides have internalizing properties. Particles coated with integrin-binding proteins, such as fibronectin, are internalized through the phagocytic pathway.[22] It has also been shown that internalization of pathogens, such as *Escherichia coli* by phagocytosis, is largely achieved by members of the integrin family.[23] In addition, $\alpha v \beta 3$ and $\alpha v \beta 5$ integrins promote internalization of adenoviruses.[24] The success viruses have in introducing their nucleic acid into cells via integrin-mediated uptake defines this system as a possible cell-penetrating pathway. Results obtained by targeting a proapoptotic peptide, which disrupts the mitochondrial membrane, support this conclusion; conjugating the proapoptotic peptide to an RGD peptide yielded apoptosis inducers that were more potent in destroying cells that express the target integrins than cells that do not.[25]

23.3.1   SECOND-GENERATION TUMOR-HOMING PEPTIDES

We have modified our initial screening methodology and added an ex vivo step prior to in vivo screening.[26] In this ex vivo step, peptide libraries are incubated with the cell suspension of the organ of interest. This allows amplification of peptides binding to all cell types present in a given organ. When this preselected pool is taken through the in vivo screening procedure, only peptides that are able to home to the target organ via blood circulation are enriched. This method has provided us with peptides that recognize markers shared by tumor vasculature and tumor cells. Some of these markers are not necessarily related to angiogenesis, but are tumor-type specific. Also, the second-generation peptides have frequently turned out to be excellent internalizing peptides. It may be that the ex vivo screening step favors phage that are internalized by the cells used in the selection process.

23.3.1.1   F3 Peptide

F3 is a linear 34 amino acid-long peptide that was isolated from a screen where the peptide library was first bound to lineage-depleted mouse bone marrow cells (putative progenitor cells) followed by in vivo selection of peptides homing to HL-60 xenograft tumors.[12] F3 homes to blood vessels of all types of tumors tested, in addition to angiogenic blood vessels in a model of nonmalignant angiogenesis, in vivo matrigel assay.[27] As well as blood endothelial cells, F3 also binds to tumor cells. F3 is an internalizing peptide that is translocated to the nucleus of the target cells.

After binding to the cell surface, fluorescein-labeled F3 is first internalized to a cytoplasmic compartment, then it is translocated to the nucleus, with most of the fluorescence residing in the nucleus 60 min after addition of the labeled peptide to the culture. This internalization is energy dependent and does not take place at 4°C. Furthermore, only the L-form of the peptide is translocated to the nucleus while the D-form is only very slowly taken into the cytoplasm.[12] It has been previously shown that heparan sulfate binding can be sufficient for the internalization of proteins binding to it.[28] Uptake of the F3 peptide by cultured cells is independent of heparan sulfates even though F3 is a highly basic peptide.[27]

Cell surface-expressed nucleolin was identified as the receptor for F3 peptide on tumor cells and angiogenic endothelial cells.[27] Nucleolin is ubiquitously expressed and mainly known as a nucleolar and cytoplasmic protein.[29] However, recent studies have shown that nucleolin is also expressed on the cell surface.[30,31] Nucleolin has been reported to act as a shuttle protein between the cytoplasm and the nucleus,[32] and between the cell surface and the nucleus.[31] Our studies show that actively dividing cells express nucleolin on their surface, whereas in serum-starved cells, nucleolin is entirely nuclear and F3 does not bind these cells. When antinucleolin antibodies are injected intravenously into mice, they bind to angiogenic blood vessels but not to resting blood vessels, indicating that nucleolin is accessible through the circulation on the surface of endothelial cells.[27] Furthermore, antinucleolin antibodies inhibit the appearance of F3 in the cytoplasm and nucleus of these cells; the internalization of the Tat peptide[13] is not affected by the antinucleolin antibodies.[27] These results show that internalization of F3 takes place in a nucleolin-dependent manner via a different pathway from the one cells use for the internalization of the Tat peptide.[13]

23.3.1.2   LyP-1 Peptide

LyP-1, CGNKRTRGC, is another internalizing peptide that recognizes a marker shared by tumor endothelial cells and tumor cells.[11] LyP-1 was isolated from a screen that aimed at isolating tumor-homing peptides with novel binding specificities. In this screen, tumor-homing peptides were selected for in vivo homing to MDA-MB-435 xenografts after two ex vivo rounds, in which peptides binding to blood endothelial cells were depleted by using anti-CD31 containing magnetic beads. The molecule (receptor) that binds LyP-1 is upregulated in the lymphatic endothelial cells of certain types of tumors, and on the tumor cells in the same tumors.[11] Intravenously injected LyP-1 accumulates in the tumor tissue with time, which allowed us to use the fluorescein-conjugated LyP-1 peptide in whole-body tumor imaging.[33] Unlike the F3 peptide, which homes to angiogenic blood vessels of all tumor types tested in addition to the nonpathological angiogenic blood vessels, homing of the LyP-1 peptide is largely limited to breast tumors, although more tumors will have to be tested to establish tumor type specificity. In tumors, LyP-1 is found in the nucleus of both lymphatic endothelial cells and tumor cells very quickly after introduction to the circulation. However, only a small fraction of MDA-MB-435 breast carcinoma cells bind and internalize LyP-1 in culture, suggesting that the receptor molecule on tumor cells is more strongly upregulated in the tumor cells in vivo than in vitro. The F3 and LyP-1 peptides provide a different pattern of fluorescence in the cell nucleus: F3 peptide paints the whole nucleus evenly, while LyP-1 is concentrated in a few spots inside the nucleus (Figure 23.1). The molecular basis for these distinct patterns remains to be elucidated.

The mechanism of LyP-1 internalization remains unknown. Our results show that the internalization of LyP-1 does not take place at 4°C and is more rapid in vivo than in vitro. While fluorescein-conjugated LyP-1 peptide is translocated to the nucleus of MDA-MB-435 cells in culture, phage particles displaying the peptide on the surface are taken up into a cytoplasmic compartment, but do not enter the nucleus (Figure 23.2). A size limitation imposed by the nuclear pore may account for this difference. Interestingly, the core sequence of the peptide (**CGNKRTRGC**)

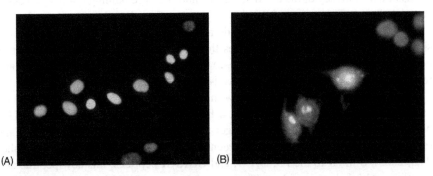

FIGURE 23.1 Cell-penetrating tumor-homing peptides F3 and LyP-1 show different patterns in the nucleus of human MDA-MB-435 breast carcinoma cells. Fluorescein-labeled F3 peptide (A) paints the whole nucleus while LyP-1 peptide (B) displays a punctate pattern.

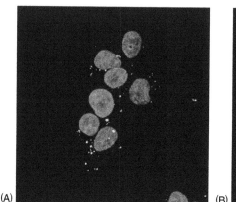

(A)   (B)

FIGURE 23.2 Phage particles displaying the LyP-1 peptide on their surface are taken up into a cytoplasmic compartment but do not enter the nucleus (A), whereas the LyP-1 peptide accumulates in the cell nucleus, as shown in Figure 23.1. Control phage is not able to enter the cells (B).

cannot be modified without loss of LyP-1 activity in ex vivo cell binding and in vivo tumor-homing assay. The inactivating modifications include changing the lysine to arginine and changing the arginines to lysines, both of which retain the net charge of the peptide but eliminate the binding function. A variant of the LyP-1 peptide, LyP-1b, in which the glycine residue in the second position was moved to position seven, retained homing properties of LyP-1.[33] In addition to providing the first indication that tumor lymphatics are molecularly distinct from normal lymphatics, the LyP-1 peptide is unique among our new homing peptides in that it possesses an inherent antitumor activity. Treatment of tumor-bearing mice with injections of 60 µg of LyP-1 twice a week resulted in significant suppression of the growth of human breast cancer xenograft tumors by up to 70%.[33] There was also a striking reduction in the number of tumor lymphatics in the tumors of the LYP-1-treated mice. While the tumor model used in that study did not allow the evaluation of metastatic proficiency of the tumors, the low number of remaining lymphatic vessels suggests that LyP-1 treatment is also likely to reduce lymphatic metastasis. The receptor molecule for the LyP-1 peptide remains to be identified, but these results show that screening can yield peptides that are biologically active, in addition to being capable of specifically homing to tumor vasculature. Moreover, similarly to LyP-1, such a peptide can possess cell-penetrating properties.

CGKRK was obtained in a screen with transgenic mouse skin cancers.[9] As is the case with the F3 and LyP-1 peptides, a fluorescein label coupled to the CGKRK peptide enters the nucleus of the target cells. Another internalizing peptide has recently been identified in a wound screening (Jarvinen and Ruoslahti, unpublished data, 2005). Similar to the F3, LyP-1, and CGKRK peptides, this new peptide also contains multiple basic residues. Thus, a high content of basic residues seems to be a shared characteristic among general (Tat, penetratin, etc.) and cell type-specific internalizing peptides.

23.4 HOMING PEPTIDES IN DRUG DELIVERY

The studies with the fluorescein-conjugated peptides show that our tumor-homing peptides are capable of carrying a payload to target cells. These peptides have been used to deliver therapeutic agents to tumors, angiogenic vessels of rheumatoid arthritis, and the prostate. Chimeric peptides containing either the RGD or NGR motif and a proapoptotic peptide $_D$(KLAKLAK)$_2$ reduce tumor burden and metastasis in an xenograft tumor model.[25] RGD peptide–doxorubicin conjugates reduced tumor burden more effectively than the free drug in the same tumor model. Similar results were obtained with a NGR peptide–doxorubicin conjugate. In addition to the increased efficacy of targeted doxorubicin, reduced toxicity was also seen.[17] In another study, mouse recombinant interleukin (IL)-12 was linked to an RGD peptide to target angiogenesis. An RGD-IL-12 conjugate inhibited corneal neovascularization by 67%. This contrasted with a 13% reduction of angiogenesis in animals treated with unmodified IL-12. The RGD-IL-12 conjugate also suppressed tumor growth in a neuroblastoma model more efficiently than IL-12 alone. Furthermore, the RGD-IL-12-treated mice showed no signs of IL-12 toxicity, whereas hepatic necrosis was present in IL-12-treated mice.[34] Other studies have shown that conjugation of the RGD peptide to a synthetic tachyplesin, an antimicrobial peptide, or tumor necrosis factor inhibits tumor growth and improves the therapeutic index of the conjugated drugs.[35,36] Angiogenesis not associated with tumors can also be targeted. Systemic treatment of mice with collagen-induced arthritis with the RGD peptide–$_D$(KLAKLAK)$_2$ peptide conjugate significantly attenuated clinical signs of arthritis and increased apoptosis of synovial blood vessels, whereas treatment with an equivalent dose of an uncoupled mixture of the two peptides had no effect.[37] Directing the $_D$(KLAKLAK)$_2$ peptide to the prostate with a prostate-homing vascular peptide caused destruction of the prostate tissue and delayed the appearance of prostate cancer in transgenic mice expressing an oncogene in the prostate.[4] While the RGD and NRG peptides appear to cause specific internalization of the $_D$(KLAKLAK)$_2$ peptide, whether this is the case with the prostate-homing peptide has not been determined. Another possibility is that the effect of the homing peptide is to increase the concentration of the conjugate in the prostate, and that passive internalization or other mechanisms accounts for the resulting toxicity in the prostate. Destroying the prostate by target-delivered drug treatment is a potential future therapy for prostate hyperplasia and for preventing the development of prostate cancer.

23.5 CONCLUDING REMARKS

We have profiled the molecular differences in the vasculature by using phage display, and have identified a number of peptides that specifically home to a variety of normal organs as well as to tumors. It appears that each normal organ and pathological condition contains organ- or disease-specific molecular tags on its vasculature that can be recognized by our screening technology and that constitute a vascular "zip code" system. Our peptides have revealed molecules that act as novel biomarkers of this vascular heterogeneity. Some of the markers expressed by the normal vasculature are downregulated in tumor vasculature, while others are upregulated. The zip code of a tumor vasculature changes during tumor progression, as the vasculature of a premalignant lesion can be distinguished from the vasculature of the corresponding tumor and normal tissue. This technology allows us to detect, in addition to the differentially expressed molecules, differences in the posttranslational modifications of the molecules. Our tumor-homing peptides provide tools for targeting therapies to tumors. This targeted therapy has been shown to enhance significantly the efficacy and decrease the toxicity of conventional anticancer drugs. Some of the peptides identified by in vivo phage screening act as cell type-specific cell-penetrating agents. These peptides appear to be able to concentrate a payload inside the cells in the target tissue, making these peptides particularly efficient delivery vehicles for the targeting of drugs and other therapeutic moieties.

ACKNOWLEDGMENTS

The authors' work was supported by the Academy of Finland, Biocentrum Helsinki, and Sigrid Juselius Foundation (PL), by NIH grant CA 82713 and Cancer Center Support Grant CA 30199, and Innovator Award DAMD 17-02-1-0315 from the Department of Defense (ER).

REFERENCES

1. Bergers, G. and Benjamin, L.E., Tumorigenesis and the angiogenic switch, *Nat. Rev. Cancer*, 3, 401, 2003.
2. Kusters, B. et al., Vascular endothelial growth factor-A(165) induces progression of melanoma brain metastases without induction of sprouting angiogenesis, *Cancer Res.*, 62, 341, 2002.
3. Saharinen, P. et al., Lymphatic vasculature: Development, molecular regulation and role in tumor metastasis and inflammation, *Trends Immunol.*, 25, 387, 2004.
4. Arap, W. et al., Targeting the prostate for destruction through a vascular address, *Proc. Natl Acad. Sci. U.S.A.*, 99, 1527, 2002.
5. Rajotte, D. et al., Molecular heterogeneity of the vascular endothelium revealed by in vivo phage display, *J. Clin. Invest.*, 102, 430, 1998.
6. Pasqualini, R. and Ruoslahti, E., Organ targeting in vivo using phage display peptide libraries, *Nature*, 380, 364, 1996.
7. Ruoslahti, E., Specialization of tumour vasculature, *Nat. Rev. Cancer*, 2, 83, 2002.
8. Joyce, J.A. et al., Stage-specific vascular markers revealed by phage display in a mouse model of pancreatic islet tumorigenesis, *Cancer Cell*, 4, 393, 2003.
9. Hoffman, J.A. et al., Progressive vascular changes in a transgenic mouse model of squamous cell carcinoma, *Cancer Cell*, 4, 383, 2003.
10. Burg, M.A. et al., NG2 proteoglycan-binding peptides target tumor neovasculature, *Cancer Res.*, 59, 2869, 1999.
11. Laakkonen, P. et al., A tumor-homing peptide with a targeting specificity related to lymphatic vessels, *Nat. Med.*, 8, 751, 2002.
12. Porkka, K. et al., A fragment of the HMGN2 protein homes to the nuclei of tumor cells and tumor endothelial cells in vivo, *Proc. Natl Acad. Sci. U.S.A.*, 99, 7444, 2002.

13. Saalik, P. et al., Protein cargo delivery properties of cell-penetrating peptides. A comparative study, *Bioconjug. Chem.*, 15, 1246, 2004.

14. Zhang, L., Hoffman, J.A., and Ruoslahti, E., Molecular profiling of heart endothelial cells, *Circulation*, 112, 1601, 2005.

15. Rajotte, D. and Ruoslahti, E., Membrane dipeptidase is the receptor for a lung-targeting peptide identified by in vivo phage display, *J. Biol. Chem.*, 274, 11593, 1999.

16. Essler, M. and Ruoslahti, E., Molecular specialization of breast vasculature: A breast-homing phage-displayed peptide binds to aminopeptidase P in breast vasculature, *Proc. Natl Acad. Sci. U.S.A.*, 99, 2252, 2002.

17. Arap, W., Pasqualini, R., and Ruoslahti, E., Cancer treatment by targeted drug delivery to tumor vasculature in a mouse model, *Science*, 279, 377, 1998 (see comments).

18. Koivunen, E., Wang, B., and Ruoslahti, E., Phage libraries displaying cyclic peptides with different ring sizes: Ligand specificities of the RGD-directed integrins, *Biotechnology (NY)*, 13, 265, 1995.

19. Eliceiri, B.P. and Cheresh, D.A., Role of alpha v integrins during angiogenesis, *Cancer J.*, 6(Suppl. 3), S245, 2000.

20. Pasqualini, R. et al., Aminopeptidase N is a receptor for tumor-homing peptides and a target for inhibiting angiogenesis, *Cancer Res.*, 60, 722, 2000.

21. Pasqualini, R., Koivunen, E., and Ruoslahti, E., Alpha v integrins as receptors for tumor targeting by circulating ligands, *Nat. Biotech.*, 15, 542, 1997 (see comments).

22. Ruoslahti, E., Fibronectin and its receptors, *Annu Rev. Biochem.*, 57, 375, 1988.

23. Foukas, L.C. et al., Phagocytosis of *Escherichia coli* by insect hemocytes requires both activation of the Ras/mitogen-activated protein kinase signal transduction pathway for attachment and β_3 integrin for internalization, *J. Biol. Chem.*, 273, 14813, 1998.

24. Wickham, T.J. et al., Integrins alpha v beta 3 and alpha v beta 5 promote adenovirus internalization but not virus attachment, *Cell*, 73, 309, 1993.

25. Ellerby, H.M. et al., Anti-cancer activity of targeted pro-apoptotic peptides, *Nat. Med.*, 5, 1032, 1999.

26. Hoffman, J.A. et al., In vivo and ex vivo selections using phage-displayed libraries, In *Phage Display*, Clackson, T. and Lowman, H.B. eds., Oxford University Press, Oxford, UK, 2004, chap. 10.

27. Christian, S. et al., Nucleolin expressed at the cell surface is a marker of endothelial cells in angiogenic blood vessels, *J. Cell Biol.*, 163, 871, 2003.

28. Roghani, M. and Moscatelli, D., Basic fibroblast growth factor is internalized through both receptor-mediated and heparan sulfate-mediated mechanisms, *J. Biol. Chem.*, 267, 22156, 1992.

29. Creancier, L. et al., Determination of the functional domains involved in nucleolar targeting of nucleolin, *Mol. Biol. Cell*, 4, 1239, 1993.

30. Sinclair, J.F. and O'Brien, A.D., Cell surface-localized nucleolin is a eukaryotic receptor for the adhesin intimin-gamma of enterohemorrhagic Escherichia coli O157:H7, *J. Biol. Chem.*, 277, 2876, 2002.

31. Said, E.A. et al., The anti-HIV cytokine midkine binds the cell surface-expressed nucleolin as a low affinity receptor, *J. Biol. Chem.*, 277, 37492, 2002.

32. Shibata, Y. et al., Nuclear targeting by the growth factor midkine, *Mol. Cell Biol.*, 22, 6788, 2002.

33. Laakkonen, P. et al., Antitumor activity of a homing peptide that targets tumor lymphatics and tumor cells, *Proc. Natl Acad. Sci. U.S.A.*, 101, 9381, 2004.

34. Dickerson, E.B. et al., Enhancement of the antiangiogenic activity of interleukin-12 by peptide targeted delivery of the cytokine to alpha v beta3 integrin, *Mol. Cancer Res.*, 2, 663, 2004.
35. Curnis, F. et al., Coupling tumor necrosis factor-alpha with alphaV integrin ligands improves its antineoplastic activity, *Cancer Res*, 64, 565, 2004.
36. Chen, Y. et al., RGD-Tachyplesin inhibits tumor growth, *Cancer Res.*, 61, 2434, 2001.
37. Gerlag, D.M. et al., Suppression of murine collagen-induced arthritis by targeted apoptosis of synovial neovasculature, *Arthritis Res.*, 3, 357, 2001.

24 In Vivo Applications of Cell-Penetrating Peptides

Sterghios Moschos, Andrew Williams,
and Mark A. Lindsay

CONTENTS

24.1 INTRODUCTION

At present, the use of peptide- and oligonucleotide-based macromolecules, both for target validation and as biopharmaceutics, is limited by the availability of techniques for both tissue and intracellular delivery. However, a number of recent investigations have suggested that this problem can be overcome through the use of cell-penetrating peptides (CPPs), in particular those derived from the TAT and Antennapedia homeodomain.[1] To date, the majority of these reports has been concerned with CPP-mediated delivery of therapeutic peptides and protein in vivo, which will be reviewed in the first section. In contrast, few studies have examined their utility for the delivery of oligonucleotides such as antisense and short interfering RNA (siRNA), which will be briefly discussed in Section 24.3.

24.2 DELIVERY OF THERAPEUTIC PEPTIDES AND PROTEINS

Although there have been a large number of investigations into the utility of CPP for peptide and protein delivery in vivo, these have commonly focused upon modulation of selective biological responses and not the basic pharmacological properties of CPP conjugates (see Table 24.1). For this reason, we will attempt to distill information on pharmacokinetics and pharmacodynamics before proceeding to examine the utility of CPP conjugates in various disease models.

24.2.1 PHARMACOKINETICS AND PHARMACODYNAMICS

Clearly, for CPP-mediated peptide/protein delivery to be of utility in vivo, these conjugates need not only to be able to deliver into relevant cells/tissues (pharmacokinetics), but at concentrations and time periods necessary to produce a biological action (pharmacodynamics). The potential utility of TAT for protein delivery in vivo was first reported using the 130-kDa protein, β-galactosidase (β-Gal).[2] In these studies, fluorescein isothiocyanate (FITC) labeling of β-Gal and intraperitoneal (IP) administration into mice led to detection in all cells of the blood and spleen, demonstrating a remarkable distribution throughout the entire systemic circulation. Likewise, measurement of β-Gal activity using an X-Gal stain demonstrated widespread tissue distribution in lungs, liver, heart muscle, and kidneys at 4 h following IP administration. Importantly, β-Gal activity in brain tissue revealed that TAT-conjugated proteins were able to cross the blood–brain barrier (BBB), while at the same time maintaining the integrity and function of the barrier.[2] Remarkably, despite the fact that these initial observations were reported in 1999, few reports have systematically examined the pharmacokinetics of TAT-mediated uptake. A study by Lee and Pardridge[3] showed that, following intravenous administration, TAT peptide was cleared from the bloodstream in minutes and that conjugation of streptavidin reduced this rate by >20 times. Furthermore, although TAT peptide alone was found to distribute to spleen $>$ liver $>$ kidneys $>$ lungs $>$ heart $>$ brain, streptavidin conjugation and the increased plasma circulation time shifted the pattern to kidneys $>>>$ liver $>$ lungs $>$ spleen $>$ heart. Pharmacodynamic studies using either fluorescence labeling or Western blotting have shown that, following IP injection, there is a time- and concentration-dependent TAT-mediated delivery of proteins (Bcl-xL[4,5] and GDNF[6]) and peptides (JNK[7] and NMDA[8]) into the brain, as well as p85 phosphatidylinositol 3-kinase (p85 PI 3-kinase) uptake into the lung.[9] Uptake of both peptides and proteins was observed within 1 h, peaked at 2 to 4 h and was maintained for up to 48 h, the latter probably being dependent upon the rate of proteolytic breakdown. Interestingly, a study of the functional actions of inhibitors of protein kinase C translocation demonstrated delivery into the heart, lung, liver, brain, and kidney only 10 min following IP injection, suggesting rapid delivery in vivo.[10]

24.2.2 TOXICITY

Clearly, to be of utility in target validation in vivo and as a potential therapeutic, CPP conjugates need to be nonimmunogenic and nontoxic. Cell-based studies

TABLE 24.1
CPP-Mediated Delivery of Proteins and Peptides In Vivo

Conjugate (Species)	Biological Action	Refs.
Protein conjugates		
FGF-suppressor of cytokine signaling (SOCS)-3 (mouse)	Protected animals from staphylococcal enterotoxin B and lipopolysaccharide-induced inflammation, liver apoptosis, hemorrhagic necrosis and death	29
TAT-dominant negative PI 3-kinase (mouse)	Inhibited airway inflammation and hyperresponsiveness in immune sensitized mice	9
TAT-dominant negative Ras (mouse)	Inhibited Th2 cytokine release and airway inflammation and hyperresponsiveness in immune sensitized mice	24
TAT-p16 derived cell-cycle inhibitor (mouse)	Suppressed pancreatic cancer growth and prolongs survival	11
TAT-glial line-derived neurotrophic factor (mouse)	Inhibited ischemic brain injury	6
TAT-Bcl-xL (mouse)	Inhibited ischemic brain injury	5
TAT-FNK — super antiapoptotic protein developed from Bcl-x (mouse and gerbil)	Inhibited ischemic brain injury	90
TAT-Bcl-xL (mouse)	Inhibited ischemic brain injury	4
TAT-Bcl-xL (mouse)	Inhibited apoptosis in retinal ganglion cells caused by optic nerve lesion	91
NLS-CRE-recombinase (mouse)	Induced β-galactosidase activity	37
Peptide conjugates		
TAT-blocking peptide PKCε (mouse + pig)	Protection from ischemia in the heart	16
TAT-blocking peptide for Bcl6 (mouse)	Induced apoptosis and cell cycle arrest in B-cell lymphoma cells	22
PolyArg-nuclear factor of activated T cells (NFAT) (mouse)	Immunosuppression in allogenic islet transplantation in mice	12
TAT- p53 activator peptide (mouse)	Increases the lifespan and the generation of disease free animals in preclinical models of peritoneal carcinomatosis and lymphoma	18
TAT-NBD (mouse)	Blocks osteoclastogenesis, inhibits focal bone erosion and inflammatory responses in the joints of arthritic mice	26
TAT-NBD (mouse)	TAT-Nemo-Binding Domain (NBD) inhibits NFκB activation and attenuates chronic inflammation diseases involving bone resorption	27

(*continued*)

Table 24.1   (Continued)

Conjugate (Species)	Biological Action	Refs.
Ant-cavtratin (mouse)	Caveolin-derived peptide inhibits eNOS and delays tumor growth	92
TAT-BH4 (mouse)	Inhibited apoptosis in small intestine (x-ray exposure), liver (Fas-induced) and heart (reperfusion)	93
TAT-NADPH oxidase inhibitor (rat)	Inhibits angioplasty-induced hyperplasia in carotid artery	94
TAT-JNK inhibitor (mouse + rats)	Inhibits excitotoxicity and ischemic brain injury	7
TAT-Smac peptide (mouse)	Induced regression of malignant glioma	21
TAT-N-methyl-D-aspartate receptor inhibitor (rat)	Inhibits interactions between NMDA receptor and PSD-95 and ischemic brain injury	8
TAT-β-domain of von Hippel-Lindau gene product (mouse)	Inhibited renal tumor growth and invasion	23
TAT-NADPH oxidase inhibitor (mice)	Inhibited angiotensin II-induced increase in blood pressure	95
Ant-cavtratin (mouse)	Caveolin-derived peptide inhibits acute inflammation and vascular leak	30
TAT-NBD (mouse)	TAT-Nemo-binding domain (NBD) inhibits NF-kB activation and attenuates two experimental models of acute inflammation	28

suggest that TAT and antennapedia peptide conjugates have little effect upon cell viability at concentrations $<30~\mu M$. Furthermore, to date, no animal studies have reported immunogenicity or cytotoxicity for TAT and antennapedia conjugates, except those related to the cargo, even following daily administration for 14,[10] 21,[11] and 50 days.[12]

24.2.3   ISCHEMIA

The death of neuronal cells following cerebral ischemia is primarily the result of apoptosis and hence the activation of proapoptotic genes. Interfering with the apoptotic pathway in order to prevent cell death has therefore been proposed as a potential therapeutic approach, although the difficulty in delivery of genes or drugs across the BBB has hampered research efforts. Significantly, CPP have been used to transport both proteins and peptides across the BBB and improve the outcome of ischemic events. For example, the IP administration of the antiapoptotic TAT-Bcl-xL fusion protein, resulted in a time- and dose-dependent uptake into mouse brain and significantly decreased neuronal cell death and the area of ischemic damage.[4,13] TAT-Bcl-xL has also been shown to prevent apoptosis in retinal ganglion cells in a mouse model of neuronal ischemia.[14,5] Another approach has been to target the receptors on neurons responsible for activating apoptosis. Thus, the delivery of TAT peptides that block the interaction between neuronal N-methyl-D-aspartate receptors (NMDSARs) and the postsynaptic density protein (PSD)-95 reduced brain ischemia in rats and improved neurological function[8] while a peptide inhibitor of JNK, conjugated to TAT, prevented c-Jun activation and reduced transient brain ischemia in mice.[8]

In addition to the brain, TAT conjugation with the antiapoptotic BH4 domain of Bcl-xL has also been shown to suppress ischemic injury in rat heart muscle.[15] Similarly, fusion of TAT with peptides that block PKCδ and PKCε membrane translocation and activation have been demonstrated to attenuate heart ischemia in both mouse and porcine models.[16] Significantly, these constructs are now entering phase I/II clinical trials for the treatment of ischemia following acute infraction and may ultimately represent the first CPP therapeutics.

24.2.4 CANCER

The growth of tumors is thought to result from alterations in the cell cycle that cause increased cell proliferation and reduced apoptosis. Therefore, modulation of the protein interactions that mediate these responses has been an obvious target for CPP-conjugated peptides and proteins. For example, a peptide (p16) derived from the tumor suppressor gene p16INK4A, which inhibits cell-cycle progression, was fused with antennapedia peptide and shown to inhibit tumor growth and enhance apoptotic cell death in a mouse model of pancreatic cancer.[11] Other proteins involved in cell-cycle control, including p21 and p27, have also been delivered using antennapedia and TAT and shown to attenuate human tumor cell growth in vitro.[17,18] Another popular target has been the tumor suppressor gene p53 that induces apoptosis in response to cellular stress and is thought to be mutated in many human cancers. Interestingly, a peptide derived from the C-terminus of the p53 protein has been shown to induce activation of the endogenous p53 pathway and restore function in cancerous but not primary cells or those possessing resistance mutations.[19] The therapeutic potential of TAT-p53 peptide has been exemplified in the mouse TA3/St tumor model in which IP injection resulted in improved life span and disease-free animals.[18] In a similar study, TAT-p53 peptide inhibited signaling through its negative regulator Hdm2, and was shown to enhance tumor cell apoptosis in a retinoblastoma xenograft model.[20] In other cancer models, apoptosis was enhanced in a murine model of malignant glioma through the synergistic effects of Smac-TAT peptides and TRAIL,[21] with complete elimination of tumor mass and no aberrant cytotoxicity. TAT has also been used to deliver peptides that inhibit B-cell lymphomas[22] and renal cell carcinomas.[23] Overall, these data therefore suggest that CPP-mediated delivery of antitumor molecules may provide an alternative approach to cancer treatment.

24.2.5 ASTHMA AND AIRWAY INFLAMMATION

Asthma is characterized by reversible airflow limitation following inappropriate contraction of the airways' smooth muscle in responses to stimuli such as allergens.

This hyperresponsiveness is thought to be caused by airway inflammation mediated by eosinophils, T cells, and mast cells; therefore, the ability to regulate these inflammatory cells may provide a potential therapeutic target. The signal transduction enzyme phosphoinositol 3-kinase (PI3K) contributes to the recruitment and activation of eosinophils during asthma. A dominant-negative PI3K-TAT fusion protein successfully attenuated the eosinophilic infiltration and hyperresponsiveness to metacholine in mice following OVA-induced airway inflammation.[9] In a similar manner, OVA-induced eosinophilia and airway hyperresponsiveness were abrogated following treatment with TAT coupled to a dominant-negative Ras protein.[24]

24.2.6 OTHER INFLAMMATORY DISEASES

Acute and chronic inflammatory diseases are associated with the production of inflammatory cytokines, such as TNF-α and IL-1β, activation of the NFκB pathway, and changes in apoptosis. There have been many reports of CPP conjugates that can attenuate these inflammatory responses, particularly in models of rheumatoid arthritis. An example of this is a novel synovial-specific transduction peptide, HAP-1, that was coupled to the mitochondrial disruption peptide (KLAK)$_2$ and administered to rabbit joints with synovial inflammation.[25] This peptide induced significant apoptosis and reduced the amount of synovial inflammation. Another approach to treat rheumatoid arthritis and bone metabolism has been to use a peptide derived from the NEMO-binding domain (NBD) to block NFκB activation and translocation to the nucleus. Thus, treatment with NBD–TAT conjugates reduced cytokine production and inhibited joint inflammation and bone erosion.[26–28] More recently, a CPP-SOCS3 fusion protein was constructed in order to enhance endogenous SOCS activity in response to staphylococcal enterotoxin B or LPS-induced acute inflammation in mice.[29] A membrane-translocation motif (MTM) derived from the hydrophobic signal sequence of fibroblast growth factor-4 (FGF-4) was fused to the SOCS3 protein. This construct reduced inflammatory cytokine release (TNF-α, IL-6), attenuated damaging apoptosis and necrosis, and prevented toxin-induced death in all treated animals.[29]

Nitric oxide (NO) has also been implicated in controlling inflammation and tissue damage. In particular, endothelial nitric oxide synthase (eNOS) interacts with lipid rafts and calveolin receptors and activates the cell to produce NO. To prevent this process, a specific inhibitor of the eNOS–calveolin interaction was fused to the antennapedia peptide and was shown to inhibit acute inflammation and vascular leakage induced following injection of carageenan into the mouse foot pad.[30]

24.2.7 STIMULATING CYTOTOXIC IMMUNITY

The development of an MHC class I-restricted immune response against intracellular pathogens is difficult to achieve with exogenously administered vaccines. The generation of protective CTL responses, particularly against viruses, would provide enhanced and long-term protection. Immunization with TAT peptide linked to OVA protein resulted in processing and presentation of OVA peptides on the surface of dendritic cells in conjunction with MHC class I molecules.[31] Importantly, TAT-OVA immunization induced OVA-specific CD8$^+$ CTL responses. However, the mechanism of processing prior to peptide loading onto MHC class I molecules is independent of the TAP transporters.[32] Proteins are delivered to the ER or even directly to the Golgi apparatus where they are cleaved into CTL epitopes without requiring the TAP proteins. In addition, TAT modifies the catalytic properties of the proteosome, resulting in a higher efficiency and the generation of more heterologous CTL epitopes.[33] It has also been proposed that delivery of peptides with TAT enhances the specific MHC class I–peptide complex expression on the cell surface, further suggesting that TAT has some immunogenic properties.[34]

Immunization against specific infectious diseases using CPPs has received far less attention. However, immunization with the full-length TAT protein linked to the adenylate cyclase of *Bordetalla pertussis* induced Th1-polarized immunity, characterized by a strong IgG antibody response and the generation of IFN-γ-producing CD8$^+$ T cells.[35] Alternatively, CPPs have been demonstrated to exert direct antimicrobial activity. For example, pVEC (derived from the adhesion molecule vascular endothelial cadherin) inhibited the growth of *Mycobacterium smegmatis*, while a synthetic transportan (TP10) inhibited *Candida albicans* and *Staphylococcus aureus* growth through lethal permeabilization of microbial cells.[36]

24.2.8 CRE RECOMBINASE

Cre recombinase is a topoisomerase or integrase from the bacteriophage P1, which is able specifically to recombine DNA between two 34-base pair loxP sites. The orientation of the loxP sites is important in defining the recombination product as DNA can either be inverted between loxP sites or excised as a nucleotide loop. In this way, Cre recombinase excises or inverts the DNA flanked by the loxP sites and is used to produce knockout animals and cell lines. However, the use of this approach is crucially dependent upon the intracellular expression of Cre recombinase, which means that it is often linked genetically to the TET-on/TET-off system or, alternatively, requires conventional transfection and viral delivery approaches. A number of investigators have attempted to overcome these problems through Cre recombinase conjugation to CPPs, such as the membrane translocation sequence derived from FGF-4. This was administered IP into a transgenic mouse model containing a β-gal reporter gene flanked by loxP sites, and showed recombination in various tissues including the brain, lungs, spleen, and liver.[37] The TAT peptide was also used to deliver active Cre recombinase into primary spleen cells ex vivo, suggesting its potential as an in vivo delivery system.[38] The use of CPPs to deliver such highly specific DNA manipulation tools throughout all body tissues, including the central nervous system, may provide a powerful means of manipulating gene expression.

24.3 CPP-MEDIATED OLIGONUCLEOTIDE DELIVERY

Oligonucleotide-based approaches to target validation include the use of plasmid/viral-based vectors for gene overexpression, as well as antisense and short interfering RNA (siRNA) for gene knockdown. As with peptides and proteins, the utility of these macromolecules is limited by availability of effective intracellular delivery techniques. In the case of oligonucleotides, this is further complicated by the presence of a negatively charged backbone, which must be neutralized in order to attach/cross the negatively charged plasma membrane, and the difficulty in conjugation of this sugar-phosphate with the CPP peptide backbone. To date, few studies have attempted to use CPP for the delivery of oligonucleotide-based tools in vivo; these are summarized in the following sections.

24.3.1 GENE DELIVERY

Although rare, the many thousands of genetic diseases, including cystic fibrosis and muscular dystrophy, are often fatal. Ideally, treatment would involve the replacement of the damaged or missing gene segment through the introduction of the relevant transgenes. Unfortunately, current tools cannot meet the requirements of such highly specific genetic manipulations in vivo. Therefore, the majority of existing gene therapy protocols use either nonreplicating recombinant viruses or polymeric cationic carrier system-delivered plasmids containing all the necessary genetic elements for expression of transgenes. Although highly efficient at entering cells and delivering genomic material to cell nuclei, viral carriers suffer a number of significant drawbacks, especially with regards to safety as a result of antiviral immunity or virus-induced carcinomas. In contrast, nonviral systems display poor transduction efficiency unless the delivery system physically disturbs the cell membrane, usually with toxic side effects. Nevertheless, at least one adenovirus-based cancer therapy (Gendicine marketed by Shenzen SiBiono GeneTech, China) has received approval for clinical use in China.

Viral drug delivery has a limited application profile as tissue targeting is governed by the receptor specificity of the virus as directed by coat proteins (viral tropism); e.g., the widely used adenovirus can only transfect cells expressing coxsackievirus-adenovirus receptor (CAR), which is not expressed, or is not physically accessible in all cell types.[39] Consequently, an adenoviral gene therapy protocol to overcome globoid cell leukodystrophy, a disease caused by mutation in the lysosomal enzyme galactocerebrosidase (GALC) leading to accumulation of psychosine at toxic levels and degeneration of myelin-forming oligodendrocytes in the CNS and peripheral nervous system, was found to be mainly restricted to the brain endothelium, achieving limited penetration in the brain tissue and leading only to partial improvement in pathology.[40] To overcome such limitations, it has been proposed that the inclusion of a CPP-coding region in virally delivered transgenes might mediate protein secretion and improve biodistribution.[41] Based on this concept, chimeric minitransgenes of TAT peptide and lymphocytic choriomeningitis virus CD8$^+$ T-cell epitopes were generated. Paradoxically, while acceleration of epitope transfer to the cell surface was observed, secretion of the CPP chimeras was found to have little influence upon the improved immune responses elicited by the constructs.[42] The lack of CPP-tagged protein mobility between transgene-expressing and untransfected cells has also been confirmed by others.[43]

If, however, CPPs such as TAT or VP22 are attached to adenoviral fiber knobs, then the breadth of cell types amenable to adenoviral transfection can be greatly increased, at least in vitro.[39] In a similar manner, expression of TAT peptide on the surface of recombinant bacteriophage particles allows targeting to mammalian cells in vitro.[44] The same principle lies behind the surface-coating of micro- and nano-sized particulate carrier systems with CPPs in order to improve their bioavailability profile.

As an alternative approach to fusion with CPPs, some investigators have attempted to condense negatively charged plasmid DNA with polycationic molecules (such as poly L-lysine, polyethyleneimine, cationic lipids, and chitosan) by means of electrostatic interactions. These highly stable oligopeptide complexes have been reported to deliver plasmid DNA.[45–48] Since CPPs are also cationic and are usually derived from DNA-binding domains of proteins, these have also been investigated as DNA delivery tools.[48,49] Although condensates can be efficiently formed, and in vitro data suggest similar activities to other cationic oligopeptides, in vivo efficacy was hampered by serum protein adsorption and putative steric hindrance of interaction between the DNA-condensed CPP and the cell membrane.[49] Dimers, trimers, and quatrimers of TAT, on the other hand, were more successful in delivering plasmids in vitro, with the length of TAT exhibiting an inverse relationship with transfection efficiency. In vivo efficacy, however, was not significantly improved for the TAT peptide–dimer complexes.[50]

24.3.2 ANTISENSE

Prior to the advent of RNA interference (see next section), DNA-based antisense was commonly used to inhibit mRNA translation through either induction of RNase H activity or steric hinderance.[51] In addition to cellular delivery, one of the most significant hurdles in using antisense in vivo is that these single-stranded DNA molecules are degraded within minutes in plasma, the result of the activity of DNA-directed nucleases (DNases). To improve stability, a significant number of chemical modifications of the DNA backbone have been proposed, with phosphorothioate (PS), locked nucleic acid (LNA), peptide nucleic acid (PNA), $2'$ fluoro, $2'$ O-methyl ($2'$ OMe), $2'$ O-(2-methoxyethyl) ($2'$ MOE), and the cationizing $2'$ O-(3-aminopropyl) ($2'$ O-AP) being some of the most popular derivatives (reviewed extensively by Crooke[52]). Such chemical analogs of DNA have dramatically increased stability, and antisense-based therapeutics are now being investigated in a number of human trials.

The first attempt at CPP-mediated antisense delivery involved conjugation of antennapedia to an antisense sequence complementary to the N-terminal sequence of amyloid precursor protein (APP), a protein involved in Alzheimer's disease.[53] In this study, an antisense-specific effect on neurite morphology was observed with primary embryonic rat cortical neurons in the absence of toxicity or the induction of differentiation. Similarly, PNA antisense conjugated to antennapedia was administered intrathecally in rats to downregulate galanin receptor type 1.[54,55] Again, in the absence of toxicity, dose-dependent downregulation of galanin receptor expression was observed in the rat spinal cord, and the galanin-induced nociceptive effect was reduced 100-fold, underscoring the potential of CPPs for the delivery of novel biotherapeutics, including antisense derivatives to the mammalian brain. Currently, most CPP-antisense analogs are being evaluated in primary and immortalized cell-based assays, with encouraging results.[56–61] The in vivo data, particularly with regards to changes in biodistribution profiles due to CPP conjugation, are eagerly awaited.

24.3.3 siRNA Delivery

RNA interference (RNAi) is the phenomenon by which translational silencing of genes is exacted in the cytoplasm by endogenous or exogenous, short (approximately 21 nucleotide long), double-stranded RNA (dsRNA) molecules. This process is sequence-specific and mediated through an associated protein complex entitled the RNA interference silencing complex (RISC) (for detailed reviews, see Zamore and Haley[62]). One of the two strands of the RNAi-mediating duplex is selected by the RISC complex and used as a guide sequence against which mRNA transcripts are read by means of Watson–Crick base-pairing. Full complementarity between the guide strand and an mRNA transcript leads to cleavage and subsequent degradation of the target mRNA, preventing protein translation. Historically, this was discovered first and exploited by direct use of synthetic duplexes termed short interfering RNA (siRNA) or the indirect delivery of hairpin siRNA encoded in plasmids or viruses. Very early, however, researchers observed "off-target nonspecific" effects arising from the delivery of siRNA.[63] This is because partial complementarity, with mismatches occurring around the center of the target site and the guide strand, inhibits translation of a target transcript. As a result, untranslated target mRNA accumulates in the cytoplasm and protein translation is downregulated (translational repression). This less specific process was discovered later, and was found to be mediated by genome-encoded duplexes known as microRNAs (miRNA), which are found in miRNA-dedicated genes, introns, as well as exons of protein-coding genes. The reduced specificity required to exact translational repression has led to speculation that, potentially, as many as 2000 genes may be controlled by single miRNAs,[64–68] indicating the pivotal role of RNAi in the regulation of endogenous gene expression, development, and differentiation,[69–73] and providing an explanation for at least a part of the nonspecific effects observed with experimental application of RNAi. More recently, an additional hurdle to the use of RNAi in vivo has been the recognition that double-stranded RNA can stimulate innate immunity through the Toll-like receptors.[74] Overall, such effects, although undesirable, can be minimized through avoidance of immunostimulating motifs, use of bioinformatics tools to minimize miRNA off-target hits, and approaches of sequence optimization.

As with antisense, the delivery of the mature siRNA duplexes (as opposed to viral or plasmid transgene-encoded siRNA or miRNA) faces the same challenges in terms of stability and cellular uptake. However, RNAi benefits significantly in that its action is mediated in the cytosol, thus eliminating the physiological barrier of the nuclear membrane in the long list of bottlenecks associated with gene and antisense delivery. Moreover, in vitro IC_{50} values for siRNA have been found to be significantly lower compared to those of antisense following cationic lipid transfection.[75]

Although RNA is a much more labile molecule, prone to backbone hydrolysis due to the $2'$ OH group, and degradation by ubiquitously present RNases, stability can be improved by using similar modifications to those pursued for antisense.[76,77] At present, there are no studies on the pharmacokinetics and pharmacodynamics of CPPs–siRNA delivery in vivo, although some in vitro data are emerging on the propensity of CPP-delivered siRNA, in either a complex form or covalently linked to the CPP, to transfect cells efficiently.[60,78–81] One interesting report by Soutschek et al.[82] described the use of a cholesteryl-modified siRNA, conjugated to the $5'$ end of the sense strand. This modification caused the accumulation of siRNA in the liver and small intestine following intravenous administration, efficiently reducing target gene expression and improving the phenotypic outcome of the disease model.[82] Although the exact cytoplasmic uptake mechanism has not been elucidated, the amphiphillic nature of the construct has led to speculation over the possible formation of micellar structures at high concentrations. This, however, has not been observed to be the case either by dynamic light scattering or electron microscopy. A particle-forming approach was, however, exploited by Song et al.,[83] who created complexes of siRNA and a construct of protamine fused to the C-terminus of the Fab fragment of an antibody against the HIV-1 envelope glycoprotein 160 (gp160). The authors showed cell-type specific uptake both in vitro and in vivo, with siRNA reaching exclusively gp160-expressing B16 tumor cells injected subcutaneously in mice. Moreover, this was achieved after both intratumoral as well as intravenous (systemic) administration in the absence of immunostimulation.[83]

Nevertheless, perhaps not surprisingly considering the results with antisense, topical administration of naked siRNA has been shown to be effective in a number of disease models in vivo. Most noteworthy is Sirna's current phase I clinical trial using an siRNA targeted to VEGF and blood vessel growth in the eye that has so far achieved dose-dependent improvement in the visual acuity of treated patients in the absence of immunostimulation or other therapeutic-associated adverse effects. Preclinically, most studies have focused on the pulmonary route of administration. Intranasal or intratracheal administration of siRNA in mice in the absence of a delivery vehicle has been shown to prevent or treat viral infection (respiratory syncytial virus, parainfluenza virus,[84,85] as well as severe acute respiratory syndrome (SARS) coronavirus (SCV) in macacques[86]), to inhibit heme oxygenase I expression in ischemia–reperfusion lung injury,[87] and to knock down target cytokine expression in LPS-induced lung inflammation.[88] However, evidence suggests that tissue penetration is poor with such formulations, as fluorescently labeled siRNA does not seem to reach cells beyond the alveolar epithelium. Intravenous delivery has also been reported to suffer from similarly poor delivery across the endothelial barrier.[89] This raises the question of the potential advantages in the reduction of effective dose levels that may be feasible through efficient delivery of siRNA across the whole lung tissue following either direct or systemic route administration. Notwithstanding these current limitations, efficient pulmonary administration will remain a hotly pursued goal as the high degree of vascularization of the lung makes it an ideal target for systemic delivery, provided translocation of the nucleic acids from circulation can be achieved.

24.4 CONCLUSIONS

The recent FDA approval for the testing of a TAT-peptide inhibitor of PKC activation in the treatment of cardiac ischemia is an exciting and important step forwards for the application of CPPs in the delivery of therapeutic macromolecules. However, this measure is likely to be the beginning of a long development process that will require an understanding of the mechanisms that underlie both tissue and intracellular delivery in vivo and how this is influenced by the CPP sequence and biological cargo.

ACKNOWLEDGMENTS

This work was supported through grants awarded from the EU Framework V, BBSRC (BB/C508234/1), and Wellcome Trust (No 076111).

REFERENCES

1. Lindsay, M.A., Peptide-mediated cell delivery: Application in protein target validation, *Curr. Opin. Pharmacol.*, 2, 587, 2002.
2. Schwarze, S.R., Ho, A., Vocero-Akbani, A., and Dowdy, S.F., In vivo protein transduction: Delivery of a biologically active protein into the mouse, *Science*, 285, 1569, 1999.
3. Lee, H.J. and Pardridge, W.M., Pharmacokinetics and delivery of tat and tat–protein conjugates to tissues in vivo, *Bioconjug. Chem.*, 12, 995, 2001.
4. Cao, G. et al., In vivo delivery of a Bcl-xL fusion protein containing the TAT protein transduction domain protects against ischemic brain injury and neuronal apoptosis, *J. Neurosci.*, 22, 5423, 2002.
5. Kilic, E., Dietz, G.P., Hermann, D.M., and Bahr, M., Intravenous TAT-Bcl-Xl is protective after middle cerebral artery occlusion in mice, *Ann. Neurol.*, 52, 617, 2002.
6. Kilic, U., Kilic, E., Dietz, G.P., and Bahr, M., Intravenous TAT-GDNF is protective after focal cerebral ischemia in mice, *Stroke*, 34, 1304, 2003.
7. Borsello, T. et al., A peptide inhibitor of c-Jun N-terminal kinase protects against excitotoxicity and cerebral ischemia, *Nat. Med.*, 9, 1180, 2003.
8. Aarts, M. et al., Treatment of ischemic brain damage by perturbing NMDA receptor–PSD-95 protein interactions, *Science*, 298, 846, 2002.
9. Myou, S. et al., Blockade of Inflammation and airway hyperresponsiveness in immune-sensitized mice by dominant-negative phosphoinositide 3-kinase-TAT, *J. Exp. Med.*, 198, 1573, 2003.
10. Begley, R., Liron, T., Baryza, J., and Mochly-Rosen, D., Biodistribution of intracellularly acting peptides conjugated reversibly to Tat, *Biochem. Biophys. Res. Commun.*, 318, 949, 2004.

11. Hosotani, R. et al., Trojan p16 peptide suppresses pancreatic cancer growth and prolongs survival in mice, *Clin. Cancer Res.*, 8, 1271, 2002.
12. Noguchi, H. et al., A new cell-permeable peptide allows successful allogeneic islet transplantation in mice, *Nat. Med.*, 10, 305, 2004.
13. Kilic, E., Dietz, G.P., Hermann, D.M., and Bahr, M., Intravenous TAT-Bcl-Xl is protective after middle cerebral artery occlusion in mice, *Ann. Neurol.*, 52, 617, 2002.
14. Dietz, G.P., Kilic, E., and Bahr, M., Inhibition of neuronal apoptosis in vitro and in vivo using TAT-mediated protein transduction, *Mol. Cell Neurosci.*, 21, 29, 2002.
15. Ono, M. et al., BH4 peptide derivative from Bcl-xL attenuates ischemia/reperfusion injury thorough anti-apoptotic mechanism in rat hearts, *Eur. J. Cardiothorac. Surg.*, 27, 117, 2005.
16. Inagaki, K., Begley, R., Ikeno, F., and Mochly-Rosen, D., Cardioprotection by epsilon-protein kinase C activation from ischemia: Continuous delivery and antiarrhythmic effect of an epsilon-protein kinase C-activating peptide, *Circulation*, 111, 44, 2005.
17. Bonfanti, M. et al., p21WAF1-derived peptides linked to an internalization peptide inhibit human cancer cell growth, *Cancer Res.*, 57, 1442, 1997.
18. Snyder, E.L., Meade, B.R., Saenz, C.C., and Dowdy, S.F., Treatment of terminal peritoneal carcinomatosis by a transducible p53-activating peptide, *PLoS. Biol.*, 2, E36, 2004.
19. Selivanova, G. et al., Restoration of the growth suppression function of mutant p53 by a synthetic peptide derived from the p53 C-terminal domain, *Nat. Med.*, 3, 632, 1997.
20. Harbour, J.W., Worley, L., Ma, D., and Cohen, M., Transducible peptide therapy for uveal melanoma and retinoblastoma, *Arch. Ophthalmol.*, 120, 1341, 2002.
21. Fulda, S., Wick, W., Weller, M., and Debatin, K.M., Smac agonists sensitize for Apo2L/TRAIL- or anticancer drug-induced apoptosis and induce regression of malignant glioma in vivo, *Nat. Med.*, 8, 808, 2002.
22. Polo, J.M. et al., Specific peptide interference reveals BCL6 transcriptional and oncogenic mechanisms in B-cell lymphoma cells, *Nat. Med.*, 10, 1329, 2004.
23. Datta, K., Sundberg, C., Karumanchi, S.A., and Mukhopadhyay, D., The 104–123 amino acid sequence of the beta-domain of von Hippel–Lindau gene product is sufficient to inhibit renal tumor growth and invasion, *Cancer Res.*, 61, 1768, 2001.
24. Myou, S. et al., Blockade of airway inflammation and hyperresponsiveness by HIV-TAT-dominant negative Ras, *J. Immunol.*, 171, 4379, 2003.
25. Mi, Z. et al., Identification of a synovial fibroblast-specific protein transduction domain for delivery of apoptotic agents to hyperplastic synovium, *Mol. Ther.*, 8, 295, 2003.
26. Dai, S., Hirayama, T., Abbas, S., and Abu-Amer, Y., The IkappaB kinase (IKK) inhibitor, NEMO-binding domain peptide, blocks osteoclastogenesis and bone erosion in inflammatory arthritis, *J. Biol. Chem.*, 279, 37219, 2004.
27. Jimi, E. et al., Selective inhibition of NF-kappaB blocks osteoclastogenesis and prevents inflammatory bone destruction in vivo, *Nat. Med.*, 10, 617, 2004.
28. May, M.J. et al., Selective inhibition of NF-kappaB activation by a peptide that blocks the interaction of NEMO with the IkappaB kinase complex, *Science*, 289, 1550, 2000.
29. Jo, D. et al., Intracellular protein therapy with SOCS3 inhibits inflammation and apoptosis, *Nat. Med.*, 11, 892, 2005.
30. Bucci, M. et al., In vivo delivery of the caveolin-1 scaffolding domain inhibits nitric oxide synthesis and reduces inflammation, *Nat. Med.*, 6, 1362, 2000.
31. Kim, D.T. et al., Introduction of soluble proteins into the MHC class I pathway by conjugation to an HIV tat peptide, *J. Immunol.*, 159, 1666, 1997.

32. Lu, J. et al., TAP-independent presentation of CTL epitopes by Trojan antigens, *J. Immunol.*, 166, 7063, 2001.
33. Gavioli, R. et al., HIV-1 tat protein modulates the generation of cytotoxic T cell epitopes by modifying proteasome composition and enzymatic activity, *J. Immunol.*, 173, 3838, 2004.
34. Leifert, J.A. et al., The cationic region from HIV tat enhances the cell-surface expression of epitope/MHC class I complexes, *Gene Ther.*, 10, 2067, 2003.
35. Mascarell, L. et al., Induction of neutralizing antibodies and Th1-polarized and CD4-independent CD8+T-cell responses following delivery of human immuno-deficiency virus type 1 Tat protein by recombinant adenylate cyclase of Bordetella pertussis, *J. Virol.*, 79, 9872, 2005.
36. Nekhotiaeva, N. et al., Cell entry and antimicrobial properties of eukaryotic cell-penetrating peptides, *FASEB J.*, 18, 394, 2004.
37. Jo, D. et al., Epigenetic regulation of gene structure and function with a cell-permeable Cre recombinase, *Nat. Biotechnol.*, 19, 929, 2001.
38. Peitz, M., Pfannkuche, K., Rajewsky, K., and Edenhofer, F., Ability of the hydrophobic FGF and basic TAT peptides to promote cellular uptake of recombinant Cre recombinase: A tool for efficient genetic engineering of mammalian genomes, *Proc. Natl Acad. Sci. U.S.A.*, 99, 4489, 2002.
39. Kuhnel, F. et al., Protein transduction domains fused to virus receptors improve cellular virus uptake and enhance oncolysis by tumor-specific replicating vectors, *J. Virol.*, 78, 13743, 2004.
40. Shen, C. et al., Gene silencing by adenovirus-delivered siRNA, *FEBS Lett.*, 539, 111, 2003.
41. Xia, H. et al., RNAi suppresses polyglutamine-induced neurodegeneration in a model of spinocerebellar ataxia, *Nat. Med.*, 10, 816, 2004.
42. Leifert, J.A., Lindencrona, J.A., Charo, J., and Whitton, J.L., Enhancing T cell activation and antiviral protection by introducing the HIV-1 protein transduction domain into a DNA vaccine, *Hum. Gene Ther.*, 12, 1881, 2001.
43. Cashman, S.M., Morris, D.J., and Kumar-Singh, R., Evidence of protein transduction but not intercellular transport by proteins fused to HIV tat in retinal cell culture and in vivo, *Mol. Ther.*, 8, 130, 2003.
44. Eguchi, A. et al., Protein transduction domain of HIV-1 Tat protein promotes efficient delivery of DNA into mammalian cells, *J. Biol. Chem.*, 276, 26204, 2001.
45. Haines, A.M. et al., CL22 — a novel cationic peptide for efficient transfection of mammalian cells, *Gene Ther.*, 8, 99, 2001.
46. Gottschalk, S. et al., A novel DNA–peptide complex for efficient gene transfer and expression in mammalian cells, *Gene Ther.*, 3, 48, 1996.
47. Niidome, T. et al., Chain length of cationic alpha-helical peptide sufficient for gene delivery into cells, *Bioconjug. Chem.*, 10, 773, 1999.
48. Futaki, S. et al., Stearylated arginine-rich peptides: A new class of transfection systems, *Bioconjug. Chem.*, 12, 1005, 2001.
49. Ignatovich, I.A. et al., Complexes of plasmid DNA with basic domain 47–57 of the HIV-1 Tat protein are transferred to mammalian cells by endocytosis-mediated pathways, *J. Biol. Chem.*, 278, 42625, 2003.
50. Rudolph, C. et al., Oligomers of the arginine-rich motif of the HIV-1 TAT protein are capable of transferring plasmid DNA into cells, *J. Biol. Chem.*, 278, 11411, 2003.
51. Jones, S.W. and Lindsay, M.A., Overview of target validation and the impact of oligonucleotides, *Curr. Opin. Mol. Ther.*, 6, 546, 2004.
52. Crooke, S.T., Antisense strategies, *Curr. Mol. Med.*, 4, 465, 2004.

53. Allinquant, B. et al., Downregulation of amyloid precursor protein inhibits neurite outgrowth in vitro, *J. Cell Biol.*, 128, 919, 1995.
54. Pooga, M. et al., Cell penetrating PNA constructs regulate galanin receptor levels and modify pain transmission in vivo, *Nat. Biotechnol.*, 16, 857, 1998.
55. Rezaei, K. et al., Intrathecal administration of PNA targeting galanin receptor reduces galanin-mediated inhibitory effect in the rat spinal cord, *Neuroreport*, 12, 317, 2001.
56. Astriab-Fisher, A. et al., Antisense inhibition of P-glycoprotein expression using peptide–oligonucleotide conjugates, *Biochem. Pharmacol.*, 60, 83, 2000.
57. Ostenson, C.G. et al., Overexpression of protein-tyrosine phosphatase PTPsigma is linked to impaired glucose-induced insulin secretion in hereditary diabetic Goto–Kakizaki rats, *Biochem. Biophys. Res. Commun.*, 291, 945, 2002.
58. Arzumanov, A. et al., A structure-activity study of the inhibition of HIV-1 Tat-dependent trans-activation by mixmer 2′-O-methyl oligoribonucleotides containing locked nucleic acid (LNA), alpha-L-LNA, or 2′-thio-LNA residues, *Oligonucleotides*, 13, 435, 2003.
59. Kilk, K. and Langel, U., Cellular delivery of peptide nucleic acid by cell-penetrating peptides, *Methods Mol. Biol.*, 298, 131, 2005.
60. Turner, J.J., Arzumanov, A.A., and Gait, M.J., Synthesis, cellular uptake and HIV-1 Tat-dependent trans-activation inhibition activity of oligonucleotide analogues disulphide-conjugated to cell-penetrating peptides, *Nucleic Acids Res.*, 33, 27, 2005.
61. Gait, M.J., Peptide-mediated cellular delivery of antisense oligonucleotides and their analogues, *Cell Mol. Life Sci.*, 60, 844, 2003.
62. Zamore, P.D. and Haley, B., Ribo-gnome: The big world of small RNAs, *Science*, 309, 1519, 2005.
63. Scacheri, P.C. et al., Short interfering RNAs can induce unexpected and divergent changes in the levels of untargeted proteins in mammalian cells, *Proc. Natl Acad. Sci. U.S.A.*, 101, 1892, 2004.
64. Burgler, C. and Macdonald, P.M., Prediction and verification of microRNA targets by movingtargets, a highly adaptable prediction method, *BMC Genomics*, 6, 88, 2005.
65. Saetrom, O., Snove, O. Jr., and Saetrom, P., Weighted sequence motifs as an improved seeding step in microRNA target prediction algorithms, *RNA*, 11, 995, 2005.
66. Brown, J.R. and Sanseau, P., A computational view of microRNAs and their targets, *Drug Discov. Today*, 10, 595, 2005.
67. Krek, A. et al., Combinatorial microRNA target predictions, *Nat. Genet.*, 37, 495, 2005.
68. Lim, L.P. et al., Microarray analysis shows that some microRNAs downregulate large numbers of target mRNAs, *Nature*, 433, 769, 2005.
69. Esau, C. et al., MicroRNA-143 regulates adipocyte differentiation, *J. Biol. Chem.*, 279, 52361, 2004.
70. Hatfield, S.D. et al., Stem cell division is regulated by the microRNA pathway, *Nature*, 435, 974, 2005.
71. Chen, C.Z. and Lodish, H.F., MicroRNAs as regulators of mammalian hematopoiesis, *Semin. Immunol.*, 17, 155, 2005.
72. Gyory, I. and Minarovits, J., Epigenetic regulation of lymphoid specific gene sets, *Biochem. Cell Biol.*, 83, 286, 2005.
73. Monticelli, S. et al., MicroRNA profiling of the murine hematopoietic system, *Genome Biol.*, 6, R71, 2005.

74. Judge, A.D. et al., Sequence-dependent stimulation of the mammalian innate immune response by synthetic siRNA, *Nat. Biotechnol.*, 23, 457, 2005.

75. Grunweller, A. et al., Comparison of different antisense strategies in mammalian cells using locked nucleic acids, $2'$-O-methyl 1 RNA, phosphorothioates and small interfering RNA, *Nucleic Acids Res.*, 31, 3185, 2003.

76. Braasch, D.A. et al., RNA interference in mammalian cells by chemically-modified RNA, *Biochemistry*, 42, 7967, 2003.

77. Manoharan, M., RNA interference and chemically modified small interfering RNAs, *Curr. Opin. Chem. Biol.*, 8, 570, 2004.

78. Simeoni, F., Morris, M.C., Heitz, F., and Divita, G., Insight into the mechanism of the peptide-based gene delivery system MPG: Implications for delivery of siRNA into mammalian cells, *Nucleic Acids Res.*, 31, 2717, 2003.

79. Chiu, Y.L. et al., Visualizing a correlation between siRNA localization, cellular uptake, and RNAi in living cells, *Chem. Biol.*, 11, 1165, 2004.

80. Muratovska, A. and Eccles, M.R., Conjugate for efficient delivery of short interfering RNA (siRNA) into mammalian cells, *FEBS Lett.*, 558, 63, 2004.

81. Davidson, B.L. and Palulson, H.L., Molecular medicine for the brain: Silencing of disease genes with RNA interference, *Lancet Neurol.*, 3, 145, 2004.

82. Soutschek, J. et al., Therapeutic silencing of an endogenous gene by systemic administration of modified siRNAs, *Nature*, 432, 173, 2004.

83. Song, E. et al., Antibody mediated in vivo delivery of small interfering RNAs via cell-surface receptors, *Nat. Biotechnol.*, 23, 709, 2005.

84. Bitko, V., Musiyenko, A., Shulyayeva, O., and Barik, S., Inhibition of respiratory viruses by nasally administered siRNA, *Nat. Med.*, 11, 50, 2005.

85. Zhang, W. et al., Inhibition of respiratory syncytial virus infection with intranasal siRNA nanoparticles targeting the viral NS1 gene, *Nat. Med.*, 11, 56, 2005.

86. Li, B.J. et al., Using siRNA in prophylactic and therapeutic regimens against SARS coronavirus in Rhesus macaque, *Nat. Med.*, 11, 944, 2005.

87. Zhang, X. et al., Small interfering RNA targeting heme oxygenase-1 enhances ischemia-reperfusion-induced lung apoptosis, *J. Biol. Chem.*, 279, 10677, 2004.

88. Lomas-Neira, J.L. et al., In vivo gene silencing (with siRNA) of pulmonary expression of MIP-2 versus KC results in divergent effects on hemorrhage-induced, neutrophil-mediated septic acute lung injury, *J. Leukoc. Biol.*, 77, 846, 2005.

89. Miyawaki-Shimizu, K. et al., siRNA-induced caveolin-1 knock-down in mice increases lung vascular permeability via the junctional pathway, *Am. J. Physiol. Lung Cell Mol. Physiol.*, 290, L405, 2006.

90. Asoh, S. et al., Protection against ischemic brain injury by protein therapeutics, *Proc. Natl Acad. Sci. U.S.A.*, 99, 17107, 2002.

91. Dietz, G.P., Kilic, E., and Bahr, M., Inhibition of neuronal apoptosis in vitro and in vivo using TAT-mediated protein transduction, *Mol. Cell Neurosci.*, 21, 29, 2002.

92. Gratton, J.P. et al., Selective inhibition of tumor microvascular permeability by cavtratin blocks tumor progression in mice, *Cancer Cell*, 4, 31, 2003.

93. Sugioka, R. et al., BH4-domain peptide from Bcl-xL exerts anti-apoptotic activity in vivo, *Oncogene*, 22, 8432, 2003.

94. Jacobson, G.M. et al., Novel NAD(P)H oxidase inhibitor suppresses angioplasty-induced superoxide and neointimal hyperplasia of rat carotid artery, *Circ. Res.*, 92, 637, 2003.

95. Rey, F.E. et al., Novel competitive inhibitor of NAD(P)H oxidase assembly attenuates vascular $O(2)(-)$ and systolic blood pressure in mice, *Circ. Res.*, 89, 408, 2001.

25 Intracellular Delivery of Nanoparticles with CPPs

Bhawna Gupta and Vladimir P. Torchilin

CONTENTS

25.1 INTRODUCTION

Membranes of the cell surface and organelles pose a serious obstacle to the uptake of charged hydrophilic molecules inside the cells. Many pharmacological proteins or peptides need to be delivered intracellularly to target and modulate cellular functions at the subcellular levels. However, because the hydrophilic behavior of such therapeutics restricts their ability to breach the cellular barrier and interact at subcellular level, such drugs are rather delivered intracellularly by different routes of receptor-dependent or receptor-independent endocytosis, which results in the formation of endosomes and ends up in lysosomes. During the endocytic process, the drug is degraded by the acidic hydrolytic enzymes present within the lysosomes. Some of the methods applied for intracellular drug delivery include the use of microinjection, electroporation, or pH-sensitive liposomes; however, these methods are either invasive, damage cell membranes, or are associated with low specificity. A promising approach that seems to be the solution to overcoming the cellular barrier has emerged over the last decade. In this approach, certain proteins or peptides can be tethered to the hydrophilic drug of interest and, together, the construct possesses the ability to translocate across the plasma membrane and deliver the payload intracellularly. This process of translocation is called "protein transduction." Such proteins or peptides contain domains of less than 20 amino acids, termed as protein transduction domains (PTDs) or cell-penetrating peptides (CPPs) that are highly rich in basic residues. These peptides have been used for intracellular delivery of various cargoes with molecular weights several times greater than their own.[1]

The property of translocation across the cell membrane was introduced in 1988 when two independent authors, Green[2] and Frankel,[3] demonstrated that the 86-mer transactivating transcriptional activator, (TAT) protein encoded by HIV-1 was efficiently internalized by cells in vitro when introduced in the surrounding media, and subsequently resulted in transactivation of the viral promoter. Such transactivation was observed at nanomolar concentrations in the presence of lysosomotrophic agents and TAT was eventually localized within the nucleus. Subsequently, a number of peptides, either derived from proteins or synthesized chemically, demonstrated the property of translocation. These peptides include Antennapedia (Antp),[4] VP22,[5] transportan,[6] model amphipathic peptide (MAP),[7] signal sequence-based peptides,[8] and synthetic polyarginines,[9] such as R$_9$-TAT, HIV-1 Rev (34–50), flock house virus coat (35–49) peptide, and DNA-binding peptides as c-Fos (139–164), c-Jun (252–279), and yeast GCN4 (231–252).

Here, we will briefly discuss some of the widely used CPPs and their proposed mechanisms of transduction, and we will then concentrate on their applications in the field of nanoparticulate delivery.

25.2 OVERVIEW OF CPPs

CPPs can be divided into three classes: protein-derived peptides, model peptides, and designed peptides.[10,11] Protein-derived peptides are the short stretches of the protein domain that are primarily responsible for the translocation ability, also called PTDs; examples are TAT peptide (48–60) derived from 86-mer TAT protein,[12] penetratin (43–58) derived from homeodomain of Drosophilia Antennapedia,[13] pVEC derived from murine vascular endothelial cadherin,[14] and signal sequence-based peptides or membrane translocating sequences (MTSs).[8] Model peptides are CPPs that mimic the translocation properties of known CPPs; an example is MAP.[7] Designed CPPs encompass the chimeric peptides that are produced by the fusion of hydrophilic and hydrophobic domains from different sources; examples are transportan, fusion of galanin, and mastoparan[6]; and MPG, chimera of peptide from fusion sequence of HIV-1 gp41 protein and peptide from the nuclear localization sequence of SV40 T-antigen.[15] In addition, synthetic peptides such as polyarginines also show a potential for translocation.[9] All the CPPs are highly positively charged, contributed by basic residues lysine or arginine; lysine is the primary contributor in transportan and MAP, while arginine contributes in TAT peptide, penetratin, and pVEC.

25.2.1 TRANSACTIVATING TRANSCRIPTIONAL ACTIVATOR (TAT)

TAT is a transcriptional activator protein encoded by human immunodeficiency virus type 1 (HIV-1).[16] TAT plays an important role in viral replication, primarily by regulating the transcription from the HIV-1 long terminal repeat. The 101-amino acid TAT protein is encoded by two exons: the first exon encodes the first 1–72 amino acids and the second exon encodes the last 73–101 amino acids. The laboratory version of the TAT protein is the 86-amino acid TAT protein that is generated during tissue culture passaging. The 101-amino acid TAT protein is divided into five domains, of which the important domain is domain four which contains the basic RKKRRQRRR motif. This basic motif is involved with direct RNA binding[17] and nuclear localization of the protein.[18] Analysis of the basic motif by circular dichroism revealed that the arginine-rich region binds to the 3-nucleotide bulge in transactivation responsive, TAT RNA,[19] where the side chains of arginine residues form hydrogen bonds with the phosphates near RNA loops and bulges, and stimulates transcription,[20] primarily due to the enhancement of the transcriptional elongation step.[21]

TAT-mediated transduction was first utilized in 1994 for the intracellular delivery of a variety of cargoes, such as β-galactosidase, horseradish peroxidase, RNase A, and domain III of *Pseudomonas* exotoxin A in vitro.[22] In vivo transduction using TAT peptide (37–72) conjugated to β-galactosidase resulted in protein delivery to different tissues such as heart, liver, spleen, lung, and skeletal muscle. Next, attempts were made to narrow down to the specific domain responsible for transduction. For this, synthetic peptides with deletions in the α-helix domain and the basic cluster domain were prepared for investigating their translocation ability.[23] The α-helical domain, which was previously thought to participate in transduction, was actually found unrelated to the intracellular uptake of the cargoes. The whole basic cluster from 48 to 60 residues was found accountable for membrane translocation, because any deletions or substitutions of basic residues in TAT (48–60) reduced the cellular uptake property. The translocation was quick, within 5 min, at concentrations of 100 nM. Following on similar lines, Park et al.[24] studied TAT (48–57) by deletion analysis. They found that deletion of Gly-48 did not affect the transduction efficiency, and deletions of Lys-50,51, Arg-55,56,57, and Gln-54 markedly reduced transduction efficiencies. Thus, they assigned the minimal transduction domain to TAT (49–57) residues. They then substituted this domain with polylysine (nine lysine residues) or polyarginine (nine arginines) and discovered that the polylysine and polyarginine peptides transduced to cytosol and nuclei to a similar extent as TAT (49–57). Thus, the positive charge in the transduction domain is responsible for the transduction ability of TAT protein. The commonly studied transduction domain of TAT (TAT PTD) extends from residues 47–57: Tyr-Gly-Arg-Lys-Lys-Arg-Arg-Gln-Arg-Arg-Arg, which contains six arginines (Arg) and two lysine residues.[25]

25.2.2 HOMEODOMAIN OF DROSOPHILIA ANTENNAPEDIA

Homeoproteins are a class of transcription factors that bind DNA through a specific sequence of 60 amino acids, called the homeodomain. Homeodomain is structured in three α-helices. Helix 3 is separated from helix 2 by a β turn and is called a recognition helix because it is involved in the interaction of the homeodomain with their cognate binding sites.[26] Joliot et al.[4] studied the role of homeoproteins on neural morphogenesis. Specifically, they used a polypeptide of 60 amino acids corresponding to antennapedia homeobox (pAntp) and incubated with neuronal cultures. Surprisingly, they found that pAntp was internalized by nerve cells and became accumulated within their nuclei, inducing morphological differentiation of the neurons. During further studies, Derossi et al.[13] ascribed the third helix of the antennapedia homeodomain to be involved in the translocation process. Specifically, they found that the 16 amino acid peptide, present in the third helix of the homeodomain and called penetratin (43–58), is the minimal PTD of Antp, the sequence of which is Arg-Gln-Ile-Lys-Ile-Trp-Phe-Gln-Asn-Arg-Arg-Met-Lys-Trp-Lys-Lys.

25.2.3 HERPES SIMPLEX VIRUS TYPE 1 PROTEIN

Herpes simplex virus (HSV-1) type 1 protein (VP22) is a major structural component of HSV-1 with a remarkable property of transport between the cells.[5] The protein is predominantly present in the cytoplasm in filamentous pattern colocalized with microtubules; after transport to other neighboring cells, VP22 concentrates in the nuclei, being associated with chromatin.[5,27] The different domains of the VP22 protein are responsible for a variety of functions, such as intercellular transport, binding to and bundling of microfilaments, induction of cytoskeleton collapse, nuclear translocation during mitosis, and binding to chromatin and nuclear membrane.[28]

25.2.4 TRANSPORTAN

Transportan is a 27-amino acid chimeric CPP produced by the coupling of the neuropeptide galanin and mastoparan via a lysine linker, galanin forming the amino terminus and mastoparan forming the carboxy terminus.[6] Upon translocation, transportan is mostly associated with membranous structures, this then being followed by nuclear uptake. Inside the nuclei, transportan concentrates in substructures, probably the nucleoli.

25.2.5 MODEL AMPHIPATHIC PEPTIDE (MAP)

MAP is an 18-mer amphipathic model peptide with the sequence KLALKLALK-ALKAALKLA.[7] The peptide traverses plasma membranes of mast cells and endothelial cells by both energy-dependent and energy-independent mechanisms, and delivers different cargoes intracellularly.

In terms of the cellular uptake and cargo delivery kinetics, MAP has the fastest uptake and the highest cargo delivery efficiency, followed by transportan, TATp (48–60), and penetratin.[29]

25.2.6 SIGNAL SEQUENCE-BASED PEPTIDES

Signal sequences, also called membrane translocating sequences (MTSs), of peptides are the sequences that are recognized by acceptor proteins and aid in directing the preprotein from the translation machinery towards appropriate intracellular organelles. For example, when the MTS was fused to the C-terminus of glutathione S-transferase (GST), the resultant GST–MTS fusion proteins were efficiently delivered inside NIH 3T3 fibroblasts.[8]

25.2.7 MISCELLANEOUS TRANSDUCING PEPTIDES

Some other less widely used peptides that have displayed transduction property consist of PreS2-domain of the hepatitis-B virus surface antigens,[30] cationic peptides PTD-4 and PTD-5 synthesized from M13 phage library,[31] and hapto-tactic peptides, also called haptides,[32] which are 19–21-amino acid cell-binding peptides equivalent to sequences on the C-termini of fibrinogen β chain (Cβ, KGSWYSMRKMSMKIRPFFPQQ), γ chain (preCγ, KTRYYSMKKTTM-KIIPFNRL), and the extended αE chain of fibrinogen (CαE, RGADYSLRAVRM-KIRPLVTQ). The amino acid sequences of haptides comprise both hydrophobic and cationic residues (a total of 4–5 positively charged amino acids per 19–21 residue). Haptides have been proposed to enhance the cell uptake of drugs that are formulated in the form of liposomes or nanoparticles.

25.2.8 SYNTHETIC POLYARGININES

Since TAT peptides and penetratins are widely studied CPPs and both of them are rich in cationic arginine residues, attempts were made to synthesize artificial peptides with either argninine substitution or polyarginines with varying chain lengths[9] or homopolymers of different cationic amino acids.[33] Analogs of TAT (48–60) peptide with D-amino acids (D-TAT) or TAT with substitution of (49–57) residues with 9-Arg (R_9-TAT) showed transduction efficiencies similar to TAT (48–60) peptide. Analogs with lysine-rich peptides were poor transducing peptides. Thus, the arginine residues are especially important in the translocation process. Among the analogs with different chain lengths of arginine residues, peptide R_4 showed very little transduction, R_6 and R_8 showed the maximum internalization and accumulation in the nucleus, while R_{16} showed reduced transduction compared to R_8. Thus, eight arginine residues seems to be the optimum number for the efficient translocation.[9] When homopolymers with an abundance of cationic amino acids, such as arginine, lysine, ornithine, and histidine, were investigated for their transduction ability, the homopolymers of arginine were more efficient than the homopolymers of lysine, ornithine, and histidine at a constant chain length, again emphasizing the role of arginine residues in the intracellular uptake.[33]

Since the guanidinium groups of TAT (49–57) play a more important role in enhancing cellular uptake than either charge or backbone structure, Wender et al.[34] designed polyguanidine peptoid derivatives which are superior to TAT (49–57) in terms of transduction activity, resistant to proteolysis, and convenience in preparation. The peptoid derivative with a six-methylene spacer between the guanidine head group and backbone (N-hxg) displayed a more significantly enhanced uptake than TAT (49–57) and D-arginine oligomer. The study emphasized that the proper spacing between guanidine moieties is essential for the interaction with the plasma membrane and, hence, cellular uptake. Rothbard et al.[35] prepared a series of decamers containing seven arginines and three nonarginine residues. The decamers with spacer-groups within were better than heptaarginine itself in cellular uptake, again suggesting the importance of spacing between guanidine moieties.

25.3 MECHANISMS OF TRANSLOCATION

Different studies have suggested a diversity of mechanisms for the translocation of CPPs across the cellular membrane. Earlier studies suggested that the internalization of TAT peptide (48–60), penetratin (43–58), and transportan was independent of receptor and temperature,[6,12,36] excluding the involvement of endocytosis. The studies designed a model for the uptake of TAT peptide and penetratin, according to which there occurs a direct ionic interaction between the basic residues of the CPPs and the negative phospholipids headgroups of the plasma membrane, which leads to the local invagination of the plasma membrane, followed by the formation of inverted micelles and, ultimately, the release of the peptide inside the cell. It was suggested that arginine-rich peptides share a common internalization mechanism,[37] that the uptake was dose-dependent and saturable, and was inhibited in the presence of the excess amount of the peptide. The cellular uptake of these peptides was inhibited in the presence of heparin sulfate assuming the electrostatic interaction between the CPPs and negative charges of heparin sulfate for translocation.

The receptor-independent mechanism was subsequently challenged to be an effect of formation of artifacts during fixation of the cells.[38] According to Richard et al.,[38] the cell-fixation procedure produces altered images due to the artifactual redistribution of the tightly cell membrane-bound positively charged peptides into the nuclei. The peptides adhere strongly to the negatively charged cell membrane through its dense positive charges and cannot be removed with washings unless trypsinized. Subsequent studies on live unfixed cells showed that the uptake of TAT (48–60), (Arg)$_9$ peptides, and pAntp was inhibited by low temperature, suggesting endocytosis as a major mechanism for peptide uptake by the cells.[38,39] Recent studies suggested lipid raft-mediated uptake[40,41] and macropinocytosis as the key mechanism of the uptake of CPP conjugated to large cargoes.[42] A different mechanism has been presented for the transport of guanidinium-rich CPPs alone or CPPs conjugated to small molecules (MW < 3000 Da).[43,44] The guanidinium groups of the CPPs form bidentate hydrogen bonds with the negative residues on the cell surface; the resulting ion pairs then translocate across the cell membrane under the influence of the membrane potential. The ion pair dissociates on the inner side of the membrane, releasing the CPPs into the cytosol. However, the number of guanidinium groups is critical for translocation, around eight groups being the optimum number for efficient translocation.

Thus, more than one mechanism works for CPP-mediated intracellular delivery of small and large molecules. Individual CPPs, or CPP-conjugated to small molecules, are internalized into cells via electrostatic interactions and hydrogen bonding, while CPP conjugated to large molecules occur via the energy-dependent macropinocytosis. However, in both cases, the direct contact between the CPPs and negative residues on the cell surface is a prerequisite for successful transduction.

25.4 INTRACELLULAR DELIVERY OF NANOPARTICLES

Over the last decade, CPPs have been widely exploited for the intracellular delivery of different size cargoes in a range of cell types, both in vitro and in vivo. The cargoes delivered a range from peptides, proteins, genetic material, antibodies, imaging agents, and toxins to nanoparticles and liposomes. A complete review of the different "large" cargoes delivered by CPPs can be found in Ref. [45]. This chapter will concentrate on the CPP-mediated delivery of nanoparticulates.

CPPs have extended their potential usefulness to the delivery of nanoparticles within the cells at an efficient level. The concept of CPP-mediated nanoparticulate delivery was first realized in 1999 by Josephson et al.[46] who showed that the iron oxide nanoparticles, when coupled to TAT peptide, provide efficient labeling of the cells, and thereby serve as a tool for magnetic resonance imaging (MRI) or magnetic separation of homed cells in vivo. For in vivo MRI, cells need to be labeled with magnetic particles through internalizing receptors. The limitation, however, is that most of the cells lack efficient internalization receptors or pathways. This problem was overcome by attaching the CPP, such as TAT peptide (48–57), to the dextran-coated superparamagnetic iron oxide particles (CLIOs). The average size of the particles was 41 nm and the conjugate carried an average of 6.7 TAT moieties per particle. The cellular uptake studies of TAT-CLIO were performed on mouse lymphocytes, human natural killer cells, and HeLa cells. In all the three cell lines tested, the uptake of TAT-CLIO nanoparticles was higher than the nonmodified iron oxide particle. The internalization of TAT-CLIO was approximately 100-fold higher than for the nonmodified iron oxide particles within mouse lymphocytes. The intracellular distribution of TAT-CLIO was then followed by fluorescence microscopy in all the three cell lines. The fluorescence microscopy studies on the live cells revealed that the conjugate first accumulated in lysosomes, as evidenced by the punctuate staining, followed by intense localization within the nuclei. The experiments clearly suggested that TAT-modified CLIOs are better tools for intracellular labeling of the cells than TAT-free particles. The authors next explored the MRI potential of TAT-CLIO on mouse lymphocytes under in vitro conditions. When the cells were incubated with TAT-CLIO and nonmodified iron oxide particles, the cells incubated with TAT-CLIO were readily detected by MRI compared to the cells incubated with nonmodified iron oxide particles. Furthermore, TAT-CLIO-labeled cells were retained on magnetic separation columns, unlike iron oxide-labeled cells, suggesting that TAT-CLIO labeling also allows for magnetic separation of homed cells in vivo.

In an extension of this study, the same group studied TAT-CLIO nanoparticles for tracking progenitor cells under in vivo conditions.[47] The magnetic labeling of stem and progenitor cells allows for following their migration and imaging by MRI in physiological settings, enhancing the knowledge for the application of stem cells in therapeutics. The authors showed that TAT-CLIO could label different types of cells, such as human hematopoietic CD34+ cells, mouse neural progenitor cells C17.2, human CD4+ lymphocytes, and mouse splenocytes. The labeling with TAT-CLIO did not induce toxicity and did not alter the differentiation or proliferation pattern of CD34+ cells. The TAT-CLIO-labeled CD34+ cells and control cells were subsequently injected intravenously into the immunodeficient mice. Around 4% of the injected dose of the cells migrated to the bone marrow (per gram of the tissue), and the labeled and control cells showed a similar biodistribution profile. Nonetheless, it was possible to visualize the labeled cells by MRI within mouse bone marrow at the single-cell level. Also, such magnetically labeled cells could be recovered from the bone marrow after in vivo homing using magnetic separation columns.

Another illustration of MRI with the TAT-CLIO-labeled cells was presented in Ref. [48], where T cells were magnetically labeled without affecting their normal responses to stimulation. Upon intravenous injection, the homing of T cells to the spleen could be detected by MRI and their biodistribution profile could be followed with time by MRI in vivo. Other similar studies observed the uptake of TAT peptide-labeled iron oxide particles by T cells, B cells, and macrophages over 72 h and found that the uptake was quick with no loss in the cell viability.[49] However, the TAT peptide–iron oxide conjugates accumulated primarily in the cytoplasm, in contrast to nuclear localization as observed by other authors.

Zhao et al.[50] investigated the effect of TAT peptide density per CLIO particle on the sensitivity of the MRI detection. The conjugation of TAT peptide to CLIO was quantified by the pyridine-2-thione method. The cellular uptake of TAT-CLIO by mouse lymphocytes was performed at different TAT-to-CLIO ratios. The cellular uptake of the TAT-CLIO increased in a nonlinear fashion upon increasing TAT-to-CLIO ratios, giving a 100-fold increase with 15 TAT per CLIO particle. The labeling of the cells with higher TAT-to-CLIO conjugates significantly enhanced the sensitivity of imaging by MR; thus, the cells could be tracked at approximately 100-fold lower concentrations using higher TAT-to-CLIO conjugates. The next question that arose with the use of TAT peptide for the delivery of contrast agents was the clearance profile of such conjugates within the body. In vivo studies on mice showed that the attachment of TAT to iron oxide nanoparticles reduced the half-life of the conjugate in the blood.[51] Both the TAT-modified and nonmodified iron oxide nanoparticles were removed from the circulation by the reticuloendothelial system (RES). However, they differ in their distribution profile within the RES. The nonmodified iron oxide nanoparticles accumulated along the hepatic vessels in the endothelial and/or Kupffer cells, while TAT peptide-modified iron oxide nanoparticles were distributed intensely throughout the parenchyma. Subcellular analysis revealed the nuclear accumulation of TAT–iron oxide nanoparticles. The study suggested the possibility of exploiting CPP-modified contrast agents for the intracellular imaging in vivo. Also, by modifying the surface of nanoparticles with

CPPs, the cell permeability of nanoparticulate-based therapeutics can be enhanced.[52] When the surface of CLIO was modified with D-polyarginyl peptide or TAT peptide, the nanoparticles could traverse through the cell monolayers. Thus, the permeability of the drugs in nanoparticulate carriers can be augmented by their surface modification with CPPs.

TAT peptide-modified nanoparticles have also been investigated for their capability to deliver the diagnostic and therapeutic agents across the blood–brain barrier.[53] Fluorescein-doped silica nanoparticles (FSNPs) were prepared by the microemulsion system. TAT peptide (48–57) was conjugated to FSNPs and studied for labeling of human lung adenocarcinoma cells (A-549) in vitro. The cells were efficiently labeled with TAT-FSNPs, unlike FSNPs, which showed no effective labeling. For in vivo bioimaging potential, TAT-FSNPs were administered intra-arterially to the brain of rats. TAT-conjugated FSNPs labeled the brain blood vessels, showing the potential for delivering agents to the brain without compromising the blood–brain barrier.

CPPs have also been used to enhance the delivery of genes via solid lipid nanoparticles (SLNs).[54] SLN gene vector was modified with dimeric TAT peptide (TAT2) and compared with polyethylenimine (PEI) for gene expression in vitro and in vivo. The presence of TAT2 in SLN gene vector enhanced the gene transfection compared to PEI both in vitro and in vivo. In another study, TAT peptide (47–57) conjugated to nanocage structures showed binding and transduction of the cells in vitro.[55] The shell cross-linked (SCK) nanoparticles were prepared by the micellization of amphiphilic block copolymers of poly (epsilon-caprolactone-b-acrylic acid) and conjugated to TAT peptide that was independently built on a solid support, resulting in TAT-modified nanocage conjugate. Such conjugate was analyzed by confocal microscopy with CHO and HeLa cells. The conjugated nanoparticles showed binding and transduction inside the cells. The authors then characterized the SCK nanoparticles for the optimum number of TAT peptides per particle required to enhance the transduction efficiency.[56] The authors prepared the conjugates with 52, 104, and 210 CPP peptides per particle, which were then evaluated for the biocompatibility in vitro and in vivo.[57] In vitro studies showed the inflammatory responses to the conjugates, but in vivo evaluation of the conjugates in the mouse did not result in major incompatible responses. Thus, TAT peptide–SCK conjugate can be used as scaffolds for preparing antigen for immunization.

TAT protein has also shown potential for vaccine applications. TAT (1–72) generated antibodies when coated on adjuvant anionic nanoparticles.[58] When mice were immunized with TAT-coated nanoparticles, both humoral and T helper type 1-based immune responses were produced. On similar lines, TAT adsorbed on the surface of microspheres was efficiently delivered intracellularly with the preservation of the biological activity of TAT protein.[59] The microspheres, composed of an insoluble core and a soluble shell, were not toxic in vitro and in vivo and represented a potential delivery system for vaccination with TAT.

A new area of interest with CPP-mediated delivery is the labeling of cells with quantum dots using CPP-modified quantum dot-loaded polymeric micelles.[60] Quantum dots are gaining popularity over standard fluorophores for studying tumor pathophysiology since they are photostable, very bright fluorophores, can be tuned to a narrow emission spectrum, and are relatively insensitive to the wavelength of the excitation light. Besides, quantum dots have the ability to distinguish tumor vessels from both the perivascular cells and the matrix, with concurrent imaging. Quantum dots were trapped within micelles prepared from polyethylene glycol-phosphatidyl ethanolamine (PEG–PE) conjugates bearing the TAT–PEG–PE linker. TAT–quantum dot conjugate could label mouse endothelial cells in vitro. For in vivo racking, bone marrow-derived progenitor cells were labeled with TAT-bearing quantum dot-containing micelle ex vivo, and then injected in the mouse bearing tumor in a cranial window model. It was then possible to track the movement of labeled progenitor cell to tumor endothelium, introducing an attempt toward understanding fine details of tumor neovascularization.

Intracellular distribution studies conducted on TAT PTD-modified gold nanoparticles suggested that the conjugates were internalized by endocytosis and did not localize into the nuclei of NIH 3T3 or HepG2 cells.[61] In contrast, Fuente and Berry[62] showed that TAT peptide-modified gold nanoparticles traversed through the cell membrane of human fibroblast cells and accumulated within the nuclei. The modified nanoparticles were biocompatible as no changes in the cell metabolism and proliferation were observed and normal cell viability was retained.

CPPs have also augmented the delivery of liposomal drug carriers. TAT peptide (47–57)-modified liposomes could be delivered intracellularly in different cells, such as murine Lewis lung carcinoma (LLC) cells, human breast tumor BT20 cells, and rat cardiac myocyte H9C2 cells.[63] The liposomes were tagged with TAT peptide via the spacer, p-nitrophenylcarbonyl-PEG–PE, at a density of 500 TAT peptide per single liposome vesicle. It was shown that the cells treated with liposomes, in which TAT peptide–cell interaction was hindered either by direct attachment of TAT peptide to the liposomes surface or by the long PEG grafts on the liposome surface shielding the TAT moiety, did not show TAT–liposome internalization; however, the preparations of TAT–liposomes, which allowed for the direct contact of TAT peptide residues with cells, displayed an enhanced uptake by the cells. This suggested that the translocation of TAT peptide (TATp)–liposomes into cells require direct free interaction of TAT peptide with the cell surface. Further studies on the intracellular trafficking of rhodamine-labeled TATp–liposomes loaded with FITC-dextran revealed that TATp–liposomes remained intact inside the cell cytoplasm within 1 h of translocation, after 2 h they migrated into the perinuclear zone, and at 9 h the liposomes disintegrated there.[64] The TATp–liposomes were also investigated for their gene delivery ability. For this, TATp–liposomes prepared with the addition of a small quantity of a cationic lipid (DOTAP) were incubated with DNA. The liposomes formed firm noncovalent complexes with DNA. Such TATp–liposome–DNA complexes, when incubated with mouse fibroblast NIH 3T3 and cardiac myocytes H9C2, showed substantially higher transfection in vitro, with lower cytotoxicity than the commonly used Lipofectin®. Under in vivo conditions, the intratumoral injection of TATp–liposome–DNA complexes into the LLC tumor in mice resulted in an efficient transfection of the tumor cells. The study thus implicated the usefulness of TATp–liposomes for in vitro and localized in vivo gene therapy.

Another study examined the kinetics of uptake of the TAT- and penetratin-modified liposomes.[65] It was found that the translocation of liposomes by TAT peptide or penetratin was proportional to the number of peptide molecules attached to the liposomal surface. A peptide number of as few as five was already sufficient to enhance the intracellular delivery of liposomes. The kinetics of the uptake was peptide- and cell-type dependent. With TATp–liposomes, the intracellular accumulation was time dependent, and with penetratin–liposomes the accumulation within the cells was quick, reaching its peak within 1 h, after which it gradually declined.

A study on similar lines showed that Antp (43–58) and TAT (47–57) peptides coupled to small unilamellar liposomes were accumulated in higher proportions within tumor cells and dendritic cells than unmodified control liposomes.[66] The uptake was time- and concentration-dependent and at least 100 PTD molecules per small unilamellar liposomes were required for efficient translocation inside cells. The uptake of the modified liposomes was inhibited by the preincubation of liposomes with heparin, confirming the role of heparan sulfate proteoglycans in CPP-mediated uptake.

In a different approach for improving the transfection and protecting DNA from degradation, thiocholesterol-based cationic lipids (TCL) were used in the formation of nanolipoparticles (NLPs). The NLPs were sequentially modified with TAT peptide which resulted in TAT–NLPs with a zwitterionic surface and higher transfection efficiency than for the cationic NLPs.[67]

Although initial studies suggested an energy-independent character for the internalization of TATp–liposomes,[63] recent studies have revealed the endocytosis as the main mechanism for the intracellular uptake of TATp–liposomes. As was shown by Hyndman et al.[68] the conjugation of TAT peptide to lipoplexes enhanced the gene transfection in primary cell cultures by endocytic uptake. Similarly, coupling of TAT peptide to the outer surface of liposomes resulted in an enhanced binding and endocytosis of the liposomes in ovarian carcinoma cells.[69] Binding was inhibited in the presence of heparin or dextran sulfate, suggesting that the proteoglycans expressed on the cell surface are involved in cell binding in this case also. In contrast, a new class of transducing peptides, haptides, after binding to the liposomal surface, augmented liposomes penetration through the cell membrane into the cell cytoplasm by a nonreceptor-mediated process,[32] thus confirming that a variety of mechanisms could be involved in CPP-mediated intracellular delivery of nanoparticulates.

25.5 CONCLUSIONS

To summarize, CPPs can efficiently translocate a broad number of nanoparticulates serving as diagnostic and therapeutic agents across the biological barrier, without causing toxicity to the system. The target area can accumulate drug (contrast)-loaded nanocarriers in increased amounts, and this area can also be monitored with an enhanced sensitivity. Table 25.1 summarizes the data on nanoparticles delivered intracellularly by CPPs in vitro and in vivo. CPPs can be tethered onto the cargoes

TABLE 25.1
Nanoparticles Delivered Inside Cells by CPPs

Particle and Size	CPP	Cell	Comments	Refs.
CLIO (MION) particles, 41 nm	TAT peptide	Mouse lymphocytes, human natural killer, HeLa, human hematopoietic CD34+, mouse neural progenitor C17.2, human lymphocytes CD4+, T-cells, B-cells, macrophages	For intracellular labeling, MRI, magnetic separation of homed cells	46–49
Gold particles, 20 nm	TAT peptide	NIH3T3, HepG2, HeLa	For intracellular localization studies	61
Gold particles, 2.8 nm core	TAT peptide	Human fibroblast hTERT-BJ1	For intracellular localization studies	62
Quantum dot-loaded polymeric micelles, 20 nm	TAT peptide coupled to the linker	Mouse endothelial cells, bone marrow derived progenitor cells	For studying tumor pathophysiology under dynamic conditions, useful for concurrent imaging and distinguishing tumor vessels from perivascular cells and matrix	60
Sterically stabilized liposomes, 200 nm	TAT peptide coupled to the linker	mouse LLC, human BT20, rat H9C2, LLC tumor in mice	To show the potential of TAT–liposome for intracellular delivery, and the intracellular gene delivery in vitro and in vivo	63,64
Sterically stabilized liposomes, 65–75 nm	TAT peptide and penetratin coupled to the linker	Human bladder carcinoma HTB-9, murine colon carcinoma C26, human epidermoid carcinoma A431, human breast cancer SK-BR-3, MCF7/WT, MCF7/ADR, murine bladder cancer MBT2	For kinetics studies of the internalization of CPP-modified liposomes	65
Sterically stabilized liposomes, 100 nm	TAT peptide and penetratin coupled to the linker	B16F1 melanoma, CHO and dendritic cells	For kinetics studies of the internalization of CPP-modified liposomes and studying the role of heparan sulfate proteoglycans in internalization	66

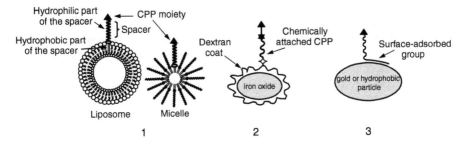

FIGURE 25.1 Different ways to attach CPPs to nanoparticulates. (1) CPP attachment to lipid-based nanoparticulates (liposome or PEG–PE-based micelles) by the insertion into the lipid phase of the carrier (liposome membrane or micelle inner core) via the spacer arm (linker) with hydrophobic terminus. CPP can be chemically attached to the preformed carrier with exposed reactive groups of the linker moieties; CPP can also be premodified with the spacer group and then added to the composition for liposome/micelle preparation, or inserted into already prepared liposomes/micelles. (2) CPP chemical attachment to the surface of appropriately modified nanoparticle. As an example, CPP can be chemically attached to CLIO (dextran-coated MION) by dextran coat amination, treatment with SPDP, and addition of cystein-containing CPP derivative. (3) Adsorption of appropriately modified CPP onto various surfaces. As an example, CPP can be modified with maleimide-containing BSA and adsorbed onto gold particles via reactive BSA groups; CPP can also be modified with a hydrophobic group and adsorbed on any hydrophobic surface via this group.

using different approaches, which are schematically presented in Figure 25.1. The combination of nanomedicine and CPP-mediated transduction can substantially improve the delivery of nanopharmaceuticals for enhanced diagnostics and therapy.

REFERENCES

1. Schwarze, S.R. and Dowdy, S.F., *In vivo* protein transduction: Intracellular delivery of biologically active proteins, compounds and DNA, *Trends Pharmacol. Sci.*, 21, 45, 2000.
2. Green, M. and Loewenstein, P.M., Autonomous functional domains of chemically synthesized human immunodeficiency virus tat trans-activator protein, *Cell*, 55, 1179, 1988.
3. Frankel, A.D. and Pabo, C.O., Cellular uptake of the tat protein from human immunodeficiency virus, *Cell*, 55, 1189, 1988.
4. Joliot, A. et al., Antennapedia homeobox peptide regulates neural morphogenesis, *Proc. Natl Acad. Sci. U.S.A.*, 88, 1864, 1991.
5. Elliott, G. and O'Hare, P., Intercellular trafficking and protein delivery by a herpesvirus structural protein, *Cell*, 88, 223, 1997.
6. Pooga, M. et al., Cell penetration by transportan, *FASEB J.*, 12, 67, 1998.
7. Oehlke, J. et al., Cellular uptake of an alpha-helical amphipathic model peptide with the potential to deliver polar compounds into the cell interior non-endocytically, *Biochim. Biophys. Acta*, 1414, 127, 1998.

8. Rojas, M. et al., Genetic engineering of proteins with cell membrane permeability, *Nat. Biotech.*, 16, 370, 1998.

9. Futaki, S. et al., Arginine-rich peptides. An abundant source of membrane-permeable peptides having potential as carriers for intracellular protein delivery, *J. Biol. Chem.*, 276, 5836, 2001.

10. Lindgren, M. et al., Cell-penetrating peptides, *Trends Pharmacol. Sci.*, 21, 99, 2000.

11. Zorko, M. and Langel, U., Cell-penetrating peptides: Mechanism and kinetics of cargo delivery, *Adv. Drug Deliv. Rev.*, 57, 529, 2005.

12. Vives, E., Brodin, P., and Lebleu, B., A truncated HIV-1 Tat protein basic domain rapidly translocates through the plasma membrane and accumulates in the cell nucleus, *J. Biol. Chem.*, 272, 16010, 1997.

13. Derossi, D. et al., The third helix of the Antennapedia homeodomain translocates through biological membranes, *J. Biol. Chem.*, 269, 10444, 1994.

14. Elmquist, A. et al., VE-cadherin-derived cell-penetrating peptide, pVEC, with carrier functions, *Exp. Cell Res.*, 269, 237, 2001.

15. Morris, M.C. et al., A new peptide vector for efficient delivery of oligonucleotides into mammalian cells, *Nucleic Acid Res.*, 25, 2730, 1997.

16. Jeang, K.T., Xiao, H., and Rich, E.A., Multifaceted activities of the HIV-1 transactivator of transcription, Tat, *J. Biol. Chem.*, 274, 28837, 1999.

17. Weeks, K.M. et al., Fragments of the HIV-1 Tat protein specifically bind TAR RNA, *Science*, 249, 1281, 1990.

18. Ruben, S. et al., Structural and functional characterization of human immunodeficiency virus tat protein, *J. Virol.*, 63, 1, 1989.

19. Calnan, B.J. et al., Analysis of arginine-rich peptides from the HIV Tat protein reveals unusual features of RNA-protein recognition, *Genes Dev.*, 5, 201, 1991.

20. Calnan, B.J. et al., Arginine-mediated RNA recognition: The arginine fork, *Science*, 252, 1167, 1991.

21. Feinberg, M.B., Baltimore, D., and Frankel, A.D., The role of Tat in the human immunodeficiency virus life cycle indicates a primary effect on transcriptional elongation, *Proc. Natl Acad. Sci. U.S.A.*, 88, 4045, 1991.

22. Fawell, S. et al., Tat-mediated delivery of heterologous proteins into cells, *Proc. Natl Acad. Sci. U.S.A.*, 91, 664, 1994.

23. Vives, E., Brodin, P., and Lebleu, B., A truncated HIV-1 Tat protein basic domain rapidly translocates through the plasma membrane and accumulates in the cell nucleus, *J. Biol. Chem.*, 272, 16010, 1997.

24. Park, J. et al., Mutational analysis of a human immunodeficiency virus type 1 Tat protein transduction domain which is required for delivery of an exogenous protein into mammalian cells, *J. Gen. Virol.*, 83, 1173, 2002.

25. Schwarze, S.R., Hruska, K.A., and Dowdy, S.F., Protein transduction: Unrestricted delivery into all cells?, *Trends Cell Biol.*, 10, 290, 2000.

26. Gehring, W.J., Affolter, M., and Burglin, T., Homeodomain proteins, *Annu. Rev. Biochem.*, 63, 487, 1994.

27. Elliott, G. and O'Hare, P., Herpes simplex virus type 1 tegument protein VP22 induces the stabilization and hyperacetylation of microtubules, *J. Virol.*, 72, 6448, 1998.

28. Aints, A. et al., Mapping of herpes simplex virus-1 VP22 functional domains for inter- and subcellular protein targeting, *Gene Ther.*, 8, 1051, 2001.

29. Hallbrink, M. et al., Cargo delivery kinetics of cell-penetrating peptides, *Biochim. Biophys. Acta*, 1515, 101, 2001.

30. Oess, S. and Hildt, E., Novel cell permeable motif derived from the PreS2-domain of hepatitis-B virus surface antigens, *Gene Ther.*, 7, 750, 2000.

31. Mi, Z. et al., Characterization of a class of cationic peptides able to facilitate efficient protein transduction *in vitro* and *in vivo*, *Mol. Ther.*, 2, 339, 2000.

32. Gorodetsky, R. et al., Liposome transduction into cells enhanced by haptotactic peptides (Haptides) homologous to fibrinogen C-termini, *J. Controlled Release*, 95, 477, 2004.

33. Mitchell, D.J. et al., Polyarginine enters cells more efficiently than other polycationic homopolymers, *J. Pept. Res.*, 56, 318, 2000.

34. Wender, P.A. et al., The design, synthesis, and evaluation of molecules that enable or enhance cellular uptake: Peptoid molecular transporters, *Proc. Natl Acad. Sci. U.S.A.*, 97, 13003, 2000.

35. Rothbard, J.B. et al., Arginine-rich molecular transporters for drug delivery: Role of backbone spacing in cellular uptake, *J. Med. Chem.*, 45, 3612, 2002.

36. Derossi, D. et al., Cell internalization of the third helix of the Antennapedia homeodomain is receptor-independent, *J. Biol. Chem.*, 271, 18188, 1996.

37. Suzuki, T. et al., Possible existence of common internalization mechanisms among arginine-rich peptides, *J. Biol. Chem.*, 277, 2437, 2002.

38. Richard, J.P. et al., Cell-penetrating peptides. A reevaluation of the mechanism of cellular uptake, *J. Biol. Chem.*, 278, 585, 2003.

39. Console, S. et al., Antennapedia and HIV transactivator of transcription (TAT) "protein transduction domains" promote endocytosis of high molecular weight cargo upon binding to cell surface glycosaminoglycans, *J. Biol. Chem.*, 278, 35109, 2003.

40. Fittipaldi, A. et al., Cell membrane lipid rafts mediate caveolar endocytosis of HIV-1 Tat fusion proteins, *J. Biol. Chem.*, 278, 34141, 2003.

41. Ferrari, A. et al., Caveolae-mediated internalization of extracellular HIV-1 tat fusion proteins visualized in real time, *Mol. Ther.*, 8, 284, 2003.

42. Wadia, J.S., Stan, R.V., and Dowdy, S.F., Transducible TAT-HA fusogenic peptide enhances escape of TAT-fusion proteins after lipid raft macropinocytosis, *Nat. Med.*, 10, 310, 2004.

43. Rothbard, J.B. et al., Role of membrane potential and hydrogen bonding in the mechanism of translocation of guanidinium-rich peptides into cells, *J. Am. Chem. Soc.*, 126, 9506, 2004.

44. Rothbard, J.B., Jessop, T.C., and Wender, P.A., Adaptive translocation: The role of hydrogen bonding and membrane potential in the uptake of guanidinium-rich transporters into cells, *Adv. Drug Deliv. Rev.*, 57, 495, 2005.

45. Gupta, B., Levchenko, T.S., and Torchilin, V.P., Intracellular delivery of large molecules and small particles by cell-penetrating proteins and peptides, *Adv. Drug Deliv. Rev.*, 57, 637, 2005.

46. Josephson, L. et al., High-efficiency intracellular magnetic labeling with novel superparamagnetic–Tat peptide conjugates, *Bioconjug. Chem.*, 10, 186, 1999.

47. Lewin, M. et al., Tat peptide-derivatized magnetic nanoparticles allow *in vivo* tracking and recovery of progenitor cells, *Nat. Biotechnol.*, 18, 410, 2000.

48. Dodd, C.H. et al., Normal T-cell response and *in vivo* magnetic resonance imaging of T cells loaded with HIV transactivator-peptide-derived superparamagnetic nanoparticles, *J. Immunol. Methods*, 256, 89, 2001.

49. Kaufman, C.L. et al., Superparamagnetic iron oxide particles transactivator protein-fluorescein isothiocyanate particle labeling for *in vivo* magnetic resonance imaging detection of cell migration: Uptake and durability, *Transplantation*, 76, 1043, 2003.

50. Zhao, M. et al., Differential conjugation of tat peptide to superparamagnetic nanoparticles and its effect on cellular uptake, *Bioconjug. Chem.*, 13, 840, 2002.

51. Wunderbaldinger, P., Josephson, L., and Weissleder, R., Tat peptide directs enhanced clearance and hepatic permeability of magnetic nanoparticles, *Bioconjug. Chem.*, 13, 264, 2002.

52. Koch, A.M. et al., Transport of surface-modified nanoparticles through cell monolayers, *Chembiochem*, 6, 337, 2005.

53. Santra, S. et al., TAT conjugated, FITC doped silica nanoparticles for bioimaging applications, *Chem. Commun. (Camb)*, 24, 2810, 2004.

54. Rudolph, C. et al., Application of novel solid lipid nanoparticle (SLN)-gene vector formulations based on a dimeric HIV-1 TAT-peptide *in vitro* and *in vivo*, *Pharm. Res.*, 21, 1662, 2004.

55. Liu, J. et al., Nanostructured materials designed for cell binding and transduction, *Biomacromolecules*, 2, 362, 2001.

56. Becker, M.L. et al., Peptide-derivatized shell-cross-linked nanoparticles. 1. Synthesis and characterization, *Bioconjug. Chem.*, 15, 699, 2004.

57. Becker, M.L., Bailey, L.O., and Wooley, K.L., Peptide-derivatized shell-cross-linked nanoparticles. 2. Biocompatibility evaluation, *Bioconjug. Chem.*, 15, 710, 2004.

58. Cui, Z. et al., Strong T cell type-1 immune responses to HIV-1 Tat (1–72) protein-coated nanoparticles, *Vaccine*, 22, 2631, 2004.

59. Caputo, A. et al., Novel biocompatible anionic polymeric microspheres for the delivery of the HIV-1 Tat protein for vaccine application, *Vaccine*, 22, 2910, 2004.

60. Stroh, M. et al., Quantum dots spectrally distinguish multiple species within the tumor milieu *in vivo*, *Nat. Med.*, 11, 678, 2005.

61. Tkachenko, A.G. et al., Cellular trajectories of peptide-modified gold particle complexes: Comparison of nuclear localization signals and peptide transduction domains, *Bioconjug. Chem.*, 15, 482, 2004.

62. Fuente, J.M. and Berry, C.C., Tat peptide as an efficient molecule to translocate gold nanoparticles into the cell nucleus, *Bioconjug. Chem.*, 16, 1176, 2005.

63. Torchilin, V.P. et al., TAT peptide on the surface of liposomes affords their efficient intracellular delivery even at low temperature and in the presence of metabolic inhibitors, *Proc. Natl Acad. Sci. U.S.A.*, 98, 8786, 2001.

64. Torchilin, V.P. et al., Cell transfection *in vitro* and *in vivo* with nontoxic TAT peptide–liposome–DNA complexes, *Proc. Natl Acad. Sci. U.S.A.*, 100, 1972, 2003.

65. Tseng, Y.L., Liu, J.J., and Hong, R.L., Translocation of liposomes into cancer cells by cell-penetrating peptides penetratin and tat: A kinetic and efficacy study, *Mol. Pharmacol.*, 62, 864, 2002.

66. Marty, C. et al., Enhanced heparan sulfate proteoglycan-mediated uptake of cell-penetrating peptide-modified liposomes, *Cell Mol. Life Sci.*, 61, 1785, 2004.

67. Huang, Z. et al., Thiocholesterol-based lipids for ordered assembly of bioresponsive gene carriers, *Mol. Ther.*, 11, 409, 2005.

68. Hyndman, L. et al., HIV-1 Tat protein transduction domain peptide facilitates gene transfer in combination with cationic liposomes, *J. Controlled Release*, 99, 435, 2004.

69. Fretz, M.M. et al., OVCAR-3 cells internalize TAT-peptide modified liposomes by endocytosis, *Biochim. Biophys. Acta*, 1665, 48, 2004.

26 Strategies to Incorporate CPPs into Synthetic Viruses for Tumor-Targeted Gene Therapy

Ernst Wagner and Manfred Ogris

CONTENTS

26.1 INTRODUCTION

Cancer gene therapy has attracted the interest of many researchers and clinicians because of its unique features. In conventional therapies, specificity is primarily based on a specific molecular drug–target interaction. In cancer gene therapies, specificity may also be enhanced at multiple further layers: delivery of viral vectors or synthetic nanoparticles could be specifically targeted to the tumor ("pharmacological targeting"); intracellular uptake and delivery of the nucleic acid into the proper compartment (the nucleus in the case of genes, the cytoplasm in the case of RNA, e.g., siRNA) might be specifically controlled ("transductional targeting"); the expression of gene constructs can be regulated by tumor-specific transcription ("transcriptional targeting").

The realization of these unique options is, however, hampered by serious limitations in the delivery step. The vectors currently used are charged macromolecules that cannot pass cellular membranes and are usually cleared from the blood circulation by the host defense system. Natural viruses have similar biophysical properties; they faced similar problems and obviously developed solutions to overcome the biological barriers.

The current chapter will report on the development of synthetic virus-like nucleic acid delivery systems for targeting tumors. Similar to natural viruses, "synthetic viruses"[1] are designed to utilize mechanisms such as receptor-mediated uptake into endosomes of the target cells and further intracellular release into the cytoplasm. For endosomal escape, a series of cell-penetrating peptides (CPPs) have been applied; some of these CPPs are extremely efficient in lipid membrane destabilization. Improving the specificity of CPPs for the target lipid membrane (for example, the endosomal membrane of the tumor cell) is a key issue for further optimization. Ongoing strategies in this direction will be discussed.

26.2 TUMOR-TARGETED GENE TRANSFER

Targeting tumors after systemic gene vector administration presents an interesting opportunity to attack tumor metastases.[2] To address this goal, we and others have been developing DNA/cationic polymer complexes (polyplexes). Cationic polymers, such as polylysine or polyethylenimine (PEI), serve for condensing the nucleic acid into virus-like dimensions and protecting it against enzymatic degradation. The systemic administration of plain cationic polyplexes, however, resulted in acute and severe toxicity.[3] Nonspecific interactions of the cationic particles with blood, followed by aggregation, and subsequent trapping of polyplex aggregates in the lung capillaries, were considered to be responsible for the toxicity.[4,5] Therefore, to avoid interactions with blood components and healthy tissues, positive surface charges of polyplexes had to be masked and cell-targeting ligands had to be incorporated into the polyplexes.

Masking of polyplex surface charge was accomplished by several similar strategies[6] using hydrophilic polymers such as polyethylene glycol (PEG). Covalent coupling of PEG to the DNA-binding polycation was performed either before[7] or after polyplex formation.[8,9] Shielding by PEG increased solubility, reduced toxicity, and extended the circulation time of polyplexes in blood. For tumor targeting, receptor-binding ligands such as epidermal growth factor (EGF), with the EGF receptor overexpressed in glioblastoma, hepatocellular and other carcinomas, or transferrin (Tf), with the Tf receptor overexpressed in many tumor types, were incorporated into polyplexes. As Tf is a serum protein, incorporation of Tf may also provide surface shielding without PEG.[10]

In several murine tumor models, intravenous injection of targeted, shielded DNA/PEI polyplexes resulted in gene transfer within distant subcutaneous tumors of mice. Marker gene expression levels in Neuro2A neuroblastoma tumor tissues of mice treated with Tf-targeted polyplexes were 100-fold higher compared to other organs.[3,7,11] Systemic administration of EGF-PEG-coated PEI polyplexes success-fully targeted hepatocellular carcinomas in SCID mice.[12] The targeting effect is based on several factors: the EPR effect[13] by a hyperpermeability of the tumor vasculature and inadequate lymphatic drainage, which triggers enhanced uptake and retention of particles in the tumor; active targeting mediated by a ligand (such as Tf or EGF) binding to the receptors over-expressed in the target tumors; and, in addition, replicating tumor cells are more accessible to transfection by PEI polyplexes compared to nondividing normal cells.

Tumor-targeted polyplexes have already been found to be effective in first therapeutic models. The therapeutic efficacy was demonstrated by systemic application of Tf-coated PEI polyplexes coding for tumor necrosis factor alpha (TNF-alpha) into tumor-bearing mice. Gene expression of TNF-alpha was found to localize in the tumor, where TNF-alpha is thought to damage the tumor vascular endothelium. Repeated treatment induced tumor necrosis and inhibition of tumor growth in four murine tumor models of different tissue origin, without any systemic TNF related toxicities.[7,14]

In these studies, PEI was used as polycationic core polymer because amongst polycations it was found to be extremely efficient in mediating gene transfer.[15,16] This efficiency is thought to be based on the capability of PEI to promote escape from intracellular acidic vesicles by the proton sponge effect.[17] In contrast to several other polycations, PEI can change its degree of protonation depending on the surrounding pH. Only one out of six nitrogens is protonated at neutral pH; polyplex-bound PEI is also only partially protonated (approximately every second to third nitrogen). Upon intracellular delivery of the DNA particle, the acidification process within the endosome triggers protonation of PEI, inducing chloride ion influx, osmotic swelling, and destabilization of the vesicle.[17] Upon rupture of the vesicle the DNA particle is released into the cytoplasm, with at least some DNA material being subsequently transported into the nucleus.

The polymer PEI has also disadvantages, however; for example, a pronounced toxicity and low biodegradability. In addition, gene transfer efficiency is still far from perfect with this polycation. Endosomal escape is still an inefficient process, especially for small DNA/PEI particles[18] and DNA complexes where free PEI has been removed.[19] Hence, for further optimization of gene transfer, at least two strategies are pursued. One option is the development of novel polycations that are better biocompatibly than PEI and show an improved endosomal escape activity. Alternatively, new or classical biocompatible polycations may be combined with endosomolytic peptides such as CPPs and incorporated into DNA complexes. This latter strategy is reviewed in the next section.

26.3   CPPs For Enhancing Endosomal Escape

The major pathway in delivery of DNA polyplexes and lipoplexes (cationic lipid/DNA complexes) proceeds by uptake into intracellular vesicles such as endosomes. Entrapment is thought to be associated with degradation of the complexes upon vesicle acidification and maturation into lysosomes. Therefore, the release into the cytoplasm represents a major bottleneck for gene delivery.[20,21] Even in the case of lipoplexes where cationic lipids can destabilize the endosomal membrane to some extent,[22] or in the case of PEI polyplexes which can promote endosomal escape to some degree because of the proton sponge effect (see section above), the majority of particles remain entrapped within vesicles. For other polycations such as polylysine, endosomal escape is even more limiting.

Because viruses have developed mechanisms to induce efficient endosomal escape, inactivated particles have been used to promote better endosomal release of polylysine polyplexes. Inactivated adenovirus particles, either coupled to polyplexes or just added to the culture medium, enhanced gene transfer by polylysine polyplexes by up to 10,000-fold, highlighting the endosomal bottleneck.[23–26] Similar effects were obtained with rhinovirus particles.[27] Also, PEI polyplexes[28] and lipoplexes[29] were significantly enhanced by adenovirus particles.

Beside viruses, other biological agents trigger efficient penetration/destabilization of lipid membranes: e.g., as toxins in their entry into cells, antibacterial peptides, defense toxins, or the serum complement system, defensins, or perforins. In several cases the membrane-active principle was found to be located in defined, small amphipathic peptide domains.[30] Such domains were incorporated into DNA polyplexes or lipoplexes (lipid-based complexes) to enhance their escape from the endo/lysosomal vesicular compartment. Examples of membrane-destabilizing peptides used for enhancing gene transfer are reviewed in the following subsections.

26.3.1 VIRAL PEPTIDE SEQUENCES

Synthetic peptides derived from the N-terminus of the influenza virus hemagglutinin protein subunit HA-2 have been incorporated into polylysine-based polyplexes. Acidic peptide variants with enhanced membrane disruption activity at endosomal pH were selected and either covalently linked to polylysine[31,32] or, because of the negative charge of the acidic peptides, simply electrostatically attached[33] to the polycationic polyplex. The polyplexes also contained targeting ligands for enhanced cellular uptake, such as transferrin[31] or synthetic carbohydrate ligands for hepatocyte-specific ASGP receptor binding.[32,34] In a similar fashion, a synthetic peptide derived from the rhinovirus HRV2 VP-1 protein was applied.[27] By incorporation of these acidic membrane-active agents, polylysine-mediated gene transfer was improved up to more than 1000-fold in cell culture. Enhancing effects were also observed with other polymers.[35,36] The concept was further explored for in vivo application. Hashida and collaborators[37] generated a polyplex system consisting of polyornithine, which was modified first with galactose to serve as ASGP receptor ligand, then with a fusogenic peptide derived from influenza virus HA2. Upon intravenous injection in mice, transgene expression was detected in the liver, the hepatocytes contributed to more than 95% of total tissue gene expression.

Influenza virus-derived peptides also enhanced the activity of small-sized PEI polyplexes,[18] although the effects were less pronounced than with polylysine polyplexes. Lipoplex-mediated gene transfer was enhanced best by less acidic influenza peptide variants.[38]

26.3.2 ARTIFICIAL AMPHIPATHIC PEPTIDE SEQUENCES

In addition to virus-derived peptide sequences, other acidic amphipathic sequences were rationally designed, such as the peptides GALA (repeats of the motif Glu-Ala-Leu-Ala),[39] JTS1,[40] or EGLA[41] and were applied in combination with various polycations. For example, efficient transfection mediated by polycationic Starburst dendrimers is dependent on a positive charge ratio of dendrimer terminal amine to DNA phosphate of 6/1. Gene expression decreases by three orders of magnitude when the charge ratio is reduced to 1:1. When GALA was covalently attached to the dendrimer, transfection efficiency of the 1:1 complex increased by 100- to 1000-fold.[39] GALA, like the other mentioned acidic sequences, is pH-specific; the membrane disruption activity is much higher at endosomal pH than at neutral pH. This can be explained by the fact that, at neutral pH, deprotonation of the glutamate residues results in charge repulsion between the glutamates, which prevents alpha helical conformation of the peptide. Glutamic acids, however, are protonated at acidic pH, which triggers amphipathic helix formation and membrane destabilization.

In related approaches, the DNA-binding polycation was completely abandoned and replaced by a cationic amphipathic peptide. KALA (repeats of lysine-alanine-leucine-alanine) presented the first example serving for both DNA binding and membrane destabilization.[42] Rittner et al.[43] designed new basic amphiphilic peptides, ppTG1 and ppTG20, as single-component gene transfer vectors, which can also bind to nucleic acids and destabilize lipid membranes. In cell culture experiments such vectors showed good transfection efficiencies, especially at low DNA doses. ppTG1 and ppTG20 were superior to the membrane-destabilizing peptide KALA. Intravenous injection of the amphipathic peptide polyplexes into mice led to significant gene expression in the lung 24 h after injection. The high gene transfer activity of these peptides is correlated with their propensity to exist in alpha-helical conformation, which appears to be strongly influenced by the nature of the hydrophobic amino acids.[43] Kichler and colleagues[44] synthesized histidine-rich amphipathic peptide antibiotics that promote efficient delivery of DNA into mammalian cells. In particular, the amphipathic α-helical peptide LAH4 was shown to exhibit pH-dependent membrane insertion. At pH 5, the peptide is oriented parallel to the bilayer, which in this case correlates with higher membrane destabilizing activity. LAH4 was shown to condense DNA, and the transfection efficiency of DNA/LAH4 complexes was comparable to PEI polyplexes.[44]

26.3.3 NATURAL CPP SEQUENCES

Natural CPPs have also been used as transfection-enhancing agents, such as the cationic peptide gramicidin S[45] forming DNA lipoplexes together with dioleyl phosphatidyl ethanolamine. The transmembrane domain of diphtheria toxin was generated in recombinant form and coupled with polylysine. These conjugates, in combination with a cell-targeting polylysine conjugate, strongly enhanced gene transfer.[46]

Gottschalk et al.[47] used perfringolysin O (PFO), a membrane active bacterial cytolysine, to deliver DNA into cells. PFO was incorporated into DNA complexes through a biotin–streptavidin bridge. Similarly, pore-forming listeriolysin O (LLO) was applied for enhanced delivery of DNA/protamine polyplexes,[48] liposomes containing antisense oligonucleotides,[49] or lipopolyplexes.[50]

The transportan derivative TP10 has been evaluated for gene delivery as free peptide, TP10–PNA conjugate or TP10–PEI conjugate.[51] TP10 enhanced PEI-mediated transfection, apparently without changing the mechanism which appears to be endocytosis; the enhancement by TP10 most probably is effected by improved endosomal escape.

A similar protein transduction domain derived from the HIV-1 Tat protein is a cationic, arginine-rich peptide that can facilitate the cellular and nuclear uptake of macromolecules. Noncovalent incorporation of a 17-amino acid Tat peptide into gene delivery lipoplexes was found to improve gene transfer.[52] Similarly, Tat–polylysine conjugate peptides mediated gene transfer.[53] In both cases, data suggest that an endosomal pathway is involved in the uptake. Rudolph et al.[54] constructed dimers and multimers of a decapeptide of Tat. These multimeric peptides mediated gene delivery 68-fold more efficiently than a control polyarginine. When DNA was precompacted with TAT peptides and complexed with PEI, SUPERFECT dendrimer, LipofectAMINE or solid lipid nanoparticles, transfection efficiency was enhanced up to 390-fold compared with the standard formulations without TAT peptides.[54,55] Upon intratracheal instillation in vivo, TAT-containing complexes were superior to standard PEI vectors. Data support the hypothesis that the TAT nuclear localization sequence function was involved in enhancing gene transfer.

26.3.4 Melittin Peptides

The bee venom peptide melittin displays particularly strong membrane destabilizing activity; this cationic peptide successfully enhanced gene transfer with lipoplexes[56] and polyplexes.[57,58] For this purpose, melittin was covalently attached via its N-terminus to dioleoylphosphatidylethanolamine to generate dioleoylmelittin[56] for lipoplex formation, or to PEI for polyplex formation.[57] While free, uncoupled melittin displays significant toxicity, N-terminal linked melittin–oleoyl and melittin–PEI conjugates show lower toxicity and enhance reporter gene expression in cell culture. Beside DNA, RNA was also successfully delivered.[58]

In combination with EGF–PEG–PEI conjugates, melittin–PEI was applied for therapeutic application of the double-stranded RNA polyIC in vivo.[59] Intratumoral delivery of the EGFR-targeted poly IC induced rapid apoptosis in the EGF receptor-bearing target tumor cells. Expression of several cytokines and "bystander killing" of untransfected tumor cells was detected. The treatment completely eliminated pre-established intracranial glioblastomas, breast cancer, or adenocarcinoma xenografts derived from EGFR over-expressing cancer cell lines in nude mice. The presence of melittin within the formulation was required for the very strong therapeutic effect.[59]

The site of coupling melittin to PEI strongly influences the properties of the conjugate.[60] The natural form of melittin is assumed to form pores after inserting the N-terminus into the lipid membrane; obviously this step is modified by coupling to

PEI at the N-terminus (N-Mel–PEI). Coupling melittin via the C-terminus generates PEI conjugates (C-Mel–PEI) which are highly lytic at neutral pH and toxic because they destabilize the plasma membrane of cells.[60] The high efficiency of C-terminally linked melittin–PEI to destabilize membranes at neutral pH is presumably due to a published destabilization mechanism proceeding through membrane insertion of the peptide.[61,62] In contrast, N-Mel–PEI is supposed to induce lysis by insertion-independent pore formation according to the toroidal pore model. Consistently, N-Mel–PEI was less lytic at neutral pH but, interestingly, retained higher lytic activity than C-Mel–PEI at endosomal pH. A better endosomal release also explains the much higher transfection activity of N-Mel–PEI polyplexes.

In conclusion, since gene transfer efficiency of melittin–PEI polyplexes correlates with lytic activity at pH 5 (and not at neutral pH), the development of formulations with enhanced lytic activity at pH 5 is suggested for optimization of polyplexes.

26.4 OPTIMIZING SYNTHETIC VIRUSES CONTAINING CPPs

As reviewed in the last section, a series of CPPs have been applied for improving polyplex- and lipoplex-mediated gene transfer. Formulations have been successfully applied in various cell culture systems. However, few reports have been published on successful administration in vivo.[37,43,59] Several reasons might account for this. Efficiency of CPPs in destabilizing lipid membranes is important but appears to be less critical; at least some CPPs, such as some of the melittin derivatives mentioned, are very efficient in membrane lysis. However, for such lytic CPPs, toxicity becomes a major critical issue. For example, we found that the most lytic C-Mel–PEI derivatives are also highly toxic in vivo; most probably due to unspecific membrane lysis of nontarget cells such as erythrocytes (Fahrmeier J, Ogris M. et al., unpublished observations).

Therefore, improving the specificity of CPPs for the target lipid membrane within the cell, for example the endosomal membrane of the tumor target cell, is a key issue. Natural viruses, such as adenoviruses or influenza viruses, may teach us how to incorporate specificity into the membrane-destabilizing CPP module of nonviral delivery vectors, so-called synthetic viruses.[1] For example, it was recently discovered for natural adenovirus that endosome acidification induces partial disassembly of the virus capsid, resulting in the release of a pH-independent lytic factor, adenovirus protein VI, within the endosome. Protein VI contains an amphipathic alpha-helical sequence resulting in endosomal membrane disruption.[63]

Similar to their natural counterparts, synthetic viruses should respond to the biological micro-environment, activating CPPs in a specific fashion only at the required gene delivery step. Various alternative strategies for making CPPs more specific and to integrate them into synthetic viruses are discussed in the following subsections.

26.4.1 pH-Specific CPP Analogs

The acidic endosomal pH may be directly exploited for activation of membrane-disrupting peptides within the endosome (but not under neutral conditions such as at the cell surface or outside the cell). The acidification may trigger pH-specific conformational changes within acidic membrane-active peptides such as GALA or influenza virus HA2-like sequences (see Figure 26.1). In these peptides, acidic residues such as glutamates, introduced at the distance of a helix turn, prevent the formation of amphipathic alpha-helices at neutral pH by charge repulsion. In fact, previous studies[41] showed that the use of acidic influenza peptide versions with specificity for endosomal acidic pH generated the best results for polylysine-mediated gene transfer. GALA and influenza type peptides show a high pH specificity, but lower lytic activity on natural membranes than, for example, melittin.

For this reason we wanted to combine specificity for endosomal pH with the high lytic activity of melittin. We synthesized melittin analogs modified with acidic residues to shift their lytic activity towards endosomal pH.[64] By replacing two neutral glutamine (Gln-25 and Gln-26) residues at the C-terminus of melittin by two glutamic acid residues, melittin analogs (see Figure 26.1) were generated which were used for conjugation to PEI via either the C-terminus (CMA3-PEI conjugate) or the N-terminus (NMA3–PEI conjugate). Indeed, these two PEI conjugates both showed an improved lytic activity at endosomal pH 5 and were more efficient in transfection than the natural peptide sequences. While the endosomal lytic activity was improved, the conjugates of the melittin analog still showed significant activity at neutral pH. Consistently, these conjugates also displayed toxicity at elevated dosages. This suggests that further sequence modifications are required to optimize these endosomolytic functions. Alternatively, indirect measures, such as stable incorporation of endosomolytic melittin–PEI conjugates into polyplexes followed by purification of polyplexes from free nonbound conjugates, may be applied (see sections below).

Influenza HA2 wild-type	GLF GAI AGFI ENGW EGMI DGWYG ---
INF6	GLF GAI AGFI ENGW EGMI DGWYG
INF7	GLF EAI EGFI ENGW EGMI DGWYG
JTS-1	GLF EALL ELL ESLW ELLL EAC
GALA	WEAALA EALA EALA EHLA EALA EAL EALA AGGSC
Melittin wild-type	GIGA VLKV LTTG LPAL ISWI KRKR QQ
CMA-3	GIGA VLKV LTTG LPAL ISWI KRKR EEC
NMA-3	C GIGA VLKV LTTG LPAL ISWI KRKR EE

FIGURE 26.1 Examples illustrating efforts to improve pH specificity of membrane-active peptide sequences. Introduction of glutamate residues improved lytic activity and/or specificity of peptides for endosomal pH that also correlated with gene transfer efficiency of DNA polyplexes containing such peptides (for details, see text and references). Underlined: acidic residues; Italic and underlined: acidic residues artificially introduced.

26.4.2 pH-Triggered Activation of CPPs

As explained above, endosomal acidification may be utilized for activating pH-specific CPP sequences. However, it is not a straightforward approach to optimize CPPs for pH specificity and high lytic activity simultaneously. Alternatively, similar to the case of natural adenoviruses,[63] the low pH may be exploited more indirectly: it may promote the release of a potent but pH-independent CPP, triggering the removal of a masking group, such as that described in the example by Rozema and colleagues.[65] In this strategy (Figure 26.2), melittin was reversibly inhibited using a maleic anhydride derivative. At neutral pH, the lysines of melittin were acylated with the anhydride, thereby inhibiting the lytic activity of melittin membrane disruption activity. At endosomal pH, melittin is unmasked by cleavage of the maleamate groups from the lysine residues. The unmasking recovers the lytic activity of the melittin peptide, which then can disrupt the endosomal membrane. The approach was successfully applied for oligonucleotide delivery.[65]

26.4.3 Reduction-Triggered Activation of CPPs

Another endosomal escape strategy uses LLO, which is a sulfhydryl-activated pore-forming protein from *Listeria* monocytogenes.[48] The strategy exploits the intracellular reducing environment that may contribute to polyplex activation and disassembly by cleaving disulfide-bridged cationic carriers. To apply LLO as membrane-disruptive agent with specificity for endosomes, LLO was conjugated through a reversible, endosome labile disulfide bond to the polycationic peptide protamine. The LLO–S-S-protamine conjugate lacks pore-forming activity. Cleavage of the disulfide bond triggered by a reducing environment, however, restores the lytic form of LLO. As evaluated in several cell culture systems

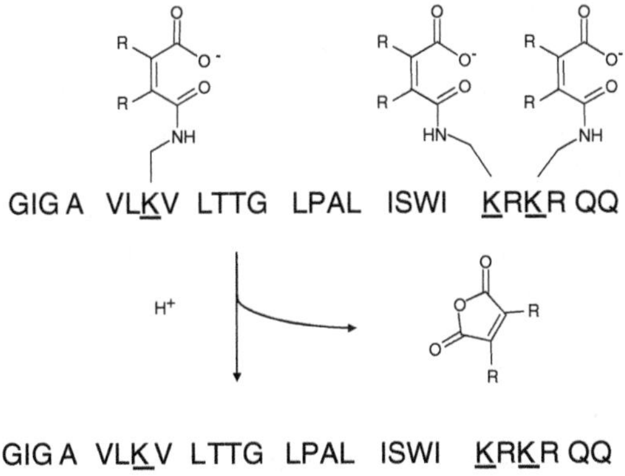

GIG A VLKV LTTG LPAL ISWI KRKR QQ

GIG A VLKV LTTG LPAL ISWI KRKR QQ

FIGURE 26.2 pH-triggered activation of melittin. Masking of the lysines by maleamate groups inhibits the lytic activity of melittin. At endosomal pH, cleavage of the pH-sensitive maleamate groups from the lysine residues recovers the lytic activity of melittin which then can disrupt the endosomal membrane. (For details see Rozema, D.B., et al., *Bioconjug. Chem.*, 14, 51, 2003.)

incorporating increasing amounts of LLO–S–S-protamine into DNA polyplexes, luciferase marker gene expression was enhanced. No cytotoxicity was observed, in contrast to pronounced cell lysis when free LLO was applied to cells together with protamine/DNA polyplexes.

26.4.4 ASSEMBLY AND TRIGGERED DISASSEMBLY OF SYNTHETIC VIRUSES

Various strategies have been reviewed in the sections above to make CPPs more specific for acting at the required target site (e.g., the endosomal membrane). It is crucial to note that such an activity should not be seen as an isolated function; it must be treated as one of several steps in the gene delivery process of a synthetic virus. Other functional groups within such a nucleic acid particle, such as targeting and shielding domains, can influence both specificity and activity of CPPs.

For example, endosomolytic melittin–PEI conjugates (see Section 26.4.1) were incorporated into PEG-shielded and EGF receptor-targeted PEI polyplexes, and the resulting particles were purified by size exclusion chromatography to remove unbound toxic polycations. We had previously found that PEG shielding[7,8] and particle purification[19] reduces the toxicity of polyplexes. The purified EGF- and melittin-containing polyplexes showed high transfection efficiency without toxicity.[64]

Different delivery functions within one synthetic virus may not always be perfectly compatible with each other. A hydrophilic PEG shield in PEI polyplexes may enhance systemic delivery to the target cells, but reduces the gene expression activity within the target cells. Apparently, such a PEG shield hampers intracellular processing. Within a synthetic virus, PEG shielding should be presented in a bioresponsive manner. Ideally, after entering the target cell into an endosomal vesicle, the polyplex should liberate the PEG shield to expose the cationic surface and/or CPP domain of the polyplex for efficient destabilization of the endosomal membrane. To accomplish such characteristics we have synthesized PEG–polycation conjugates with pH-labile linkages, which make use of the acidic milieu of the endosomes.[66] EGF or Tf receptor-targeted DNA/PEI particles shielded with these bioreversible PEG conjugates lose their PEG shield at endosomal pH and display up to 100-fold higher gene transfer activity as compared to targeted polyplexes with the analogous stable PEG shields (Figure 26.3). In analogous experiments it was observed that PEG-shielded EGF receptor-targeted DNA/PEI polyplexes were most effective when both the bioresponsive form of PEG and the melittin analog CMA3-PEI were present in the complex (Boeckle S, Wagner E, Walker G, unpublished results).

Other examples also demonstrate that incorporation of delivery functions that are presented in a bioresponsive fashion can strongly improve polyplex efficiency. For instance, bioresponsive vectors have been described that combine extracellular stability of DNA by polycationic disulfide bond containing cages with rapid intracellular release of the DNA upon cleavage of these cages, by exploiting the intracellular reducing environment.[67–70]

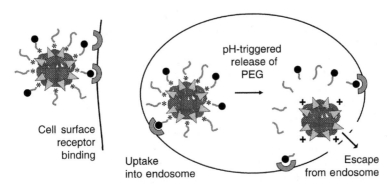

FIGURE 26.3 Endosomal deshielding of synthetic viruses. Receptor-mediated uptake delivers ligand-targeted polyplexes into endosomes. Bioresponsive PEG–polycation conjugates, such as linked by pyridylhydrazone derivates (see Walker, G.F., et al., *Mol. Ther.*, 11, 418, 2005.) are cleavable within the endosomal pH environment, which results in exposure of positive surface charges and (optionally) endosomolytic peptides, facilitating subsequent escape of polyplexes from the endosome to the cytoplasm.

26.5   CONCLUSIONS

Current polyplexes are still extremely inefficient compared to viral vectors. From this perspective, natural viruses might present ideal examples demonstrating how to further optimize polyplexes into synthetic viruses.[1] One unique property of viruses is their dynamic manner in responding to the biological micro-environment. A crucial step in their delivery process is the crossing of lipid membrane barriers by means of CPP-like protein domains. Incorporation of CPPs in a similarly defined, dynamic mode to make them specific and most effective for the target lipid membranes will also greatly improve the gene delivery process of polyplexes.

ACKNOWLEDGMENTS

We are extremely grateful to Mrs. Olga Brück for skilful assistance in preparing the manuscript.

REFERENCES

1. Wagner, E., Strategies to improve DNA polyplexes for in vivo gene transfer: Will "artificial viruses" be the answer?, *Pharm. Res.*, 21, 8, 2004.
2. Ogris, M. and Wagner, E., Targeting tumors with non-viral gene delivery systems, *Drug Discov. Today*, 7, 479, 2002.
3. Kircheis, R. et al., Polycation-based DNA complexes for tumor-targeted gene delivery *in vivo*, *J. Gene Med.*, 1, 111, 1999.
4. Plank, C. et al., Activation of the complement system by synthetic DNA complexes: A potential barrier for intravenous gene delivery, *Hum. Gene Ther.*, 7, 1437, 1996.
5. Chollet, P. et al., Side-effects of a systemic injection of linear polyethylenimine-DNA complexes, *J. Gene Med.*, 4, 84, 2002.

6. Ogris, M. et al., Tumor-targeted gene therapy: Strategies for the preparation of ligand-polyethylene glycol-polyethylenimine/DNA complexes, *J. Control. Release,* 91, 173, 2003.

7. Kursa, M. et al., Novel shielded transferrin-polyethylene glycol-polyethylenimine/DNA complexes for systemic tumor-targeted gene transfer, *Bioconjug. Chem.,* 14, 222, 2003.

8. Ogris, M. et al., PEGylated DNA/transferrin-PEI complexes: Reduced interaction with blood components, extended circulation in blood and potential for systemic gene delivery, *Gene Ther.,* 6, 595, 1999.

9. Blessing, T. et al., Different strategies for formation of pegylated EGF-conjugated PEI/DNA complexes for targeted gene delivery, *Bioconjug. Chem.,* 12, 529, 2001.

10. Kircheis, R. et al., Polyethylenimine/DNA complexes shielded by transferrin target gene expression to tumors after systemic application, *Gene Ther.,* 8, 28, 2001.

11. Hildebrandt, I.J. et al., Optical imaging of transferrin targeted PEI/DNA complexes in living subjects, *Gene Ther.,* 10, 758, 2003.

12. Wolschek, M.F. et al., Specific systemic nonviral gene delivery to human hepatocellular carcinoma xenografts in SCID mice, *Hepatology,* 36, 1106, 2002.

13. Maeda, H., The enhanced permeability and retention (EPR) effect in tumor vasculature: The key role of tumor-selective macromolecular drug targeting, *Adv. Enzyme Regul.,* 41, 189, 2001.

14. Kircheis, R. et al., Tumor-targeted gene delivery of tumor necrosis factor-alpha induces tumor necrosis and tumor regression without systemic toxicity, *Cancer Gene Ther.,* 9, 673, 2002.

15. Boussif, O. et al., A versatile vector for gene and oligonucleotide transfer into cells in culture and in vivo: Polyethylenimine, *Proc. Natl Acad. Sci. U.S.A.,* 92, 7297, 1995.

16. Zou, S.M. et al., Systemic linear polyethylenimine (L-PEI)-mediated gene delivery in the mouse, *J. Gene Med.,* 2, 128, 2000.

17. Sonawane, N.D., Szoka, F.C. Jr, and Verkman, A.S., Chloride accumulation and swelling in endosomes enhances DNA transfer by polyamine-DNA polyplexes, *J. Biol. Chem.,* 278, 44826, 2003.

18. Ogris, M. et al., The size of DNA/transferrin-PEI complexes is an important factor for gene expression in cultured cells, *Gene Ther.,* 5, 1425, 1998.

19. Boeckle, S. et al., Purification of polyethylenimine polyplexes highlights the role of free polycations in gene transfer, *J. Gene Med.,* 6, 1102, 2004.

20. Zabner, J. et al., Cellular and molecular barriers to gene transfer by a cationic lipid, *J. Biol. Chem.,* 270, 18997, 1995.

21. Labat Moleur, F. et al., An electron microscopy study into the mechanism of gene transfer with lipopolyamines, *Gene Ther.,* 3, 1010, 1996.

22. Xu, Y. and Szoka, F.C. Jr, Mechanism of DNA release from cationic liposome/DNA complexes used in cell transfection, *Biochemistry,* 35, 5616, 1996.

23. Wagner, E. et al., Coupling of adenovirus to transferrin-polylysine/DNA complexes greatly enhances receptor-mediated gene delivery and expression of transfected genes, *Proc. Natl Acad. Sci. U.S.A.,* 89, 6099, 1992.

24. Cotten, M. et al., High-efficiency receptor-mediated delivery of small and large (48 kilobase) gene constructs using the endosome-disruption activity of defective or chemically inactivated adenovirus particles, *Proc. Natl Acad. Sci. U.S.A.,* 89, 6094, 1992.

25. Wu, G.Y. et al., Incorporation of adenovirus into a ligand-based DNA carrier system results in retention of original receptor specificity and enhances targeted gene expression, *J. Biol. Chem.,* 269, 11542, 1994.

26. Fisher, K.J. and Wilson, J.M., Biochemical and functional analysis of an adenovirus-based ligand complex for gene transfer, *Biochem. J.,* 299, 49, 1994.

27. Zauner, W. et al., Rhinovirus-mediated endosomal release of transfection complexes, *J. Virol.*, 69, 1085, 1995.

28. Baker, A. and Cotten, M., Delivery of bacterial artificial chromosomes into mammalian cells with psoralen-inactivated adenovirus carrier, *Nucleic Acids Res.*, 25, 1950, 1997.

29. Raja-Walia, R. et al., Enhancement of liposome-mediated gene transfer into vascular tissue by replication deficient adenovirus, *Gene Ther.*, 2, 521, 1995.

30. Plank, C., Zauner, W., and Wagner, E., Application of membrane-active peptides for drug and gene delivery across cellular membranes, *Adv. Drug Deliv. Rev.*, 34, 21, 1998.

31. Wagner, E. et al., Influenza virus hemagglutinin HA-2 N-terminal fusogenic peptides augment gene transfer by transferrin-polylysine-DNA complexes: Toward a synthetic virus-like gene-transfer vehicle, *Proc. Natl Acad. Sci. U.S.A.*, 89, 7934, 1992.

32. Plank, C. et al., Gene transfer into hepatocytes using asialoglycoprotein receptor mediated endocytosis of DNA complexed with an artificial tetra-antennary galactose ligand, *Bioconjug. Chem.*, 3, 533, 1992.

33. Plank, C. et al., The influence of endosome-disruptive peptides on gene transfer using synthetic virus-like gene transfer systems, *J. Biol. Chem.*, 269, 12918, 1994.

34. Midoux, P. et al., Specific gene transfer mediated by lactosylated poly-L-lysine into hepatoma cells, *Nucleic Acids Res.*, 21, 871, 1993.

35. Funhoff, A.M. et al., Endosomal escape of polymeric gene delivery complexes is not always enhanced by polymers buffering at low pH, *Biomacromolecules*, 5, 32, 2004.

36. Funhoff, A.M. et al., Polymer side-chain degradation as a tool to control the destabilization of polyplexes, *Pharm. Res.*, 21, 170, 2004.

37. Nishikawa, M. et al., Hepatocyte-targeted in vivo gene expression by intravenous injection of plasmid DNA complexed with synthetic multi-functional gene delivery system, *Gene Ther.*, 7, 548, 2000.

38. Kichler, A. et al., Influence of membrane-active peptides on lipospermine/DNA complex mediated gene transfer, *Bioconjug. Chem.*, 8, 213, 1997.

39. Haensler, J. and Szoka, F.C. Jr, Polyamidoamine cascade polymers mediate efficient transfection of cells in culture, *Bioconjug. Chem.*, 4, 372, 1993.

40. Gottschalk, S. et al., A novel DNA-peptide complex for efficient gene transfer and expression in mammalian cells, *Gene Ther.*, 3, 48, 1996.

41. Mechtler, K. and Wagner, E., Gene transfer mediated by influenza virus peptides: The role of peptide sequence, *New J. Chem.*, 21, 105, 1997.

42. Wyman, T.B. et al., Design, synthesis, and characterization of a cationic peptide that binds to nucleic acids and permeabilizes bilayers, *Biochemistry*, 36, 3008, 1997.

43. Rittner, K. et al., New basic membrane-destabilizing peptides for plasmid-based gene delivery in vitro and in vivo, *Mol. Ther.*, 5, 104, 2002.

44. Kichler, A. et al., Histidine-rich amphipathic peptide antibiotics promote efficient delivery of DNA into mammalian cells, *Proc. Natl Acad. Sci. U.S.A.*, 100, 1564, 2003.

45. Legendre, J.Y. and Szoka, F.C. Jr, Cyclic amphipathic peptide-DNA complexes mediate high-efficiency transfection of adherent mammalian cells, *Proc. Natl Acad. Sci. U.S.A.*, 90, 893, 1993.

46. Fisher, K.J. and Wilson, J.M., The transmembrane domain of diphtheria toxin improves molecular conjugate gene transfer, *Biochem. J.*, 321, 49, 1997.

47. Gottschalk, S. et al., Efficient gene delivery and expression in mammalian cells using DNA coupled with perfringolysin O, *Gene Ther.*, 2, 498, 1995.

48. Saito, G., Amidon, G.L., and Lee, K.D., Enhanced cytosolic delivery of plasmid DNA by a sulfhydryl-activatable listeriolysin O/protamine conjugate utilizing cellular reducing potential, *Gene Ther.*, 10, 72, 2003.

49. Mathew, E. et al., Cytosolic delivery of antisense oligonucleotides by listeriolysin O-containing liposomes, *Gene Ther.*, 10, 1105, 2003.

50. Lorenzi, G.L. and Lee, K.D., Enhanced plasmid DNA delivery using anionic LPDII by listeriolysin O incorporation, *J. Gene Med.*, 7, 1077, 2005.

51. Kilk, K. et al., Evaluation of transportan 10 in PEI mediated plasmid delivery assay, *J. Control. Release*, 103, 511, 2005.

52. Hyndman, L. et al., HIV-1 Tat protein transduction domain peptide facilitates gene transfer in combination with cationic liposomes, *J. Control. Release*, 99, 435, 2004.

53. Hashida, H. et al., Fusion of HIV-1 Tat protein transduction domain to poly-lysine as a new DNA delivery tool, *Br. J. Cancer*, 90, 1252, 2004.

54. Rudolph, C. et al., Oligomers of the arginine-rich motif of the HIV-1 TAT protein are capable of transferring plasmid DNA into cells, *J. Biol. Chem.*, 278, 11411, 2003.

55. Rudolph, C. et al., Application of novel solid lipid nanoparticle (SLN)-gene vector formulations based on a dimeric HIV-1 TAT-peptide in vitro and in vivo, *Pharm. Res.*, 21, 1662, 2004.

56. Legendre, J.Y. et al., Dioleoylmelittin as a novel serum-insensitive reagent for efficient transfection of mammalian cells, *Bioconjug. Chem.*, 8, 57, 1997.

57. Ogris, M. et al., Melittin enables efficient vesicular escape and enhanced nuclear access of nonviral gene delivery vectors, *J. Biol. Chem.*, 276, 47550, 2001.

58. Bettinger, T. et al., Peptide-mediated RNA delivery: A novel approach for enhanced transfection of primary and post-mitotic cells, *Nucleic Acids Res.*, 29, 3882, 2001.

59. Shir, A., et al., EGF receptor targeted synthetic double-stranded RNA eliminates glioblastoma, breast cancer and adenocarcinoma tumors in mice, *PLoS Medicine*, 3, e6, 2006.

60. Boeckle, S., Wagner, E., and Ogris, M., C- versus N-terminally linked melittin-polyethylenimine conjugates: The site of linkage strongly influences activity of DNA polyplexes, *J. Gene Med.*, 7, 1335, 2005.

61. Yang, L. et al., Barrel-stave model or toroidal model? A case study on melittin pores, *Biophys. J.*, 81, 1475, 2001.

62. Papo, N. and Shai, Y., Exploring peptide membrane interaction using surface plasmon resonance: Differentiation between pore formation versus membrane disruption by lytic peptides, *Biochemistry*, 42, 458, 2003.

63. Wiethoff, C.M. et al., Adenovirus protein VI mediates membrane disruption following capsid disassembly, *J. Virol.*, 79, 1992, 2005.

64. Boeckle, S. et al., Melittin analogs with high lytic activity at endosomal pH enhance transfection with purified targeted PEI polyplexes, *J. Control. Release*, 112, 240, 2006.

65. Rozema, D.B. et al., Endosomolysis by masking of a membrane-active agent (EMMA) for cytoplasmic release of macromolecules, *Bioconjug. Chem.*, 14, 51, 2003.

66. Walker, G.F. et al., Toward synthetic viruses: Endosomal pH-triggered deshielding of targeted polyplexes greatly enhances gene transfer in vitro and in vivo, *Mol. Ther.*, 11, 418, 2005.

67. Trubetskoy, V.S. et al., Caged DNA does not aggregate in high ionic strength solutions, *Bioconjug. Chem.*, 10, 624, 1999.

68. Kwok, K.Y. et al., In vivo gene transfer using sulfhydryl cross-linked PEG-peptide/glycopeptide DNA co-condensates, *J. Pharm. Sci.*, 92, 1174, 2003.

69. Miyata, K. et al., Block catiomer polyplexes with regulated densities of charge and disulfide cross-linking directed to enhance gene expression, *J. Am. Chem. Soc.*, 126, 2355, 2004.

70. Carlisle, R.C. et al., Polymer-coated polyethylenimine/DNA complexes designed for triggered activation by intracellular reduction, *J. Gene Med.*, 6, 337, 2004.

27 Cell-Permeable Penta-Peptides Derived from Bax-Inhibiting Peptide

Jose A. Gomez, Vivian Gama, and Shigemi Matsuyama

CONTENTS

27.1 THE DISCOVERY OF BAX INHIBITING PEPTIDES

Cell-permeable Bax-inhibiting peptides (BIPs) were discovered as a result of identifying the minimal element of Ku70 needed to bind and inhibit Bax (Figure 27.1).[1,2] Bax is a key apoptosis mediator in mammalian cells and is ubiquitously expressed in almost all cell types.[3,4] In a healthy cell, Bax resides in the cytosol. Apoptotic stress, however, induces Bax translocation from the cytosol to the mitochondria, triggering mitochondrial-dependent apoptosis by promoting the release of apoptogenic factors from the mitochondria.[5] The mechanism of the mitochondrial translocation of Bax from the cytosol is unknown. We employed a yeast-based functional screening system to find Bax inhibitors using a human cDNA expression library,[6,7] and Ku70 was identified as a new Bax suppressor.[8]

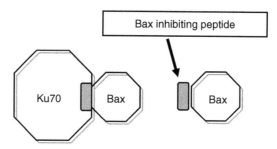

FIGURE 27.1 Bax-inhibiting peptide was designed from the Bax-binding domain of Ku70. BIP binds Bax and inhibits Bax-mediated cell death.

Further studies revealed that Ku70 inhibits the relocalization of Bax from the cytosol to mitochondria in human and mouse cells.[8]

Ku70 has been previously known as the 70-kDa subunit of the Ku complex, a group of proteins that makes up the DNA binding unit of the DNA-dependent protein kinase (DNA-PK) holoenzyme complex. This complex plays an important role in double-stranded DNA break repair.[9] DNA-PK is a sensor for DNA damage that phosphorylates genotoxic stress response proteins.[9] In addition to its previously known function as a DNA-PK subunit in the nucleus, Ku70 has been suggested to have other biological functions due to its expression in the cytosol[10,11] and plasma membrane.[12,13]

Indeed, we have found that the cytosolic form of Ku70 binds Bax in the cytosol, preventing the translocation of Bax into the mitochondria[8]; this finding has been confirmed by other groups.[14–19] BIPs that are derived from the Ku70's Bax-binding domain can bind to Bax and inhibit its activation. As previously characterized, BIPs are cell-permeable and can suppress Bax-dependent cell death induced by various stresses, including drug treatment and tropic factor deprivation.[1,2,14,20] Since Bax is implicated as a cell death inducer in several types of degenerative diseases, BIPs may be useful as a vehicle for new therapeutics designed to rescue damaged cells. Additionally, we found that BIPs can deliver certain types of "cargo," such as fluorescent dyes from the culture medium to the cells. Therefore, it is possible that BIPs could also be utilized as drug-delivery tools in the future.

27.2   BIP-DERIVED CELL-PERMEABLE PENTA-PEPTIDES

At present, there are four BIPs (Figure 27.2). In addition, there are three negative control cell-permeable peptides that have been designed through the introduction of mutations (e.g., IPMIK) or by scrambling the peptide sequence (e.g., KLPVM). Since all of these peptides, including BIPs and their negative control peptides, have cell-permeable activities, they have been designated as BIP-derived cell-permeable penta-peptides (BCPs) in this article. At present, the mechanism by which these BCPs enter the cell is not known. Identification of the specific mutations that enhance and/or suppress the cell-permeable activity of the BCPs will provide us

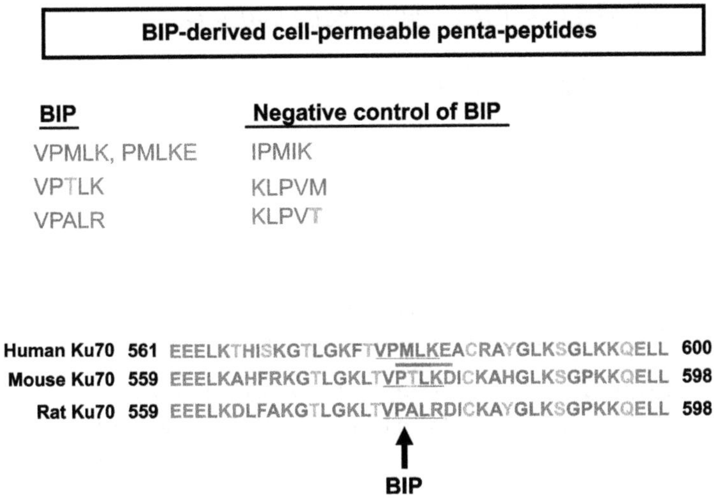

FIGURE 27.2 BIP-derived cell-permeable penta-peptides. BIPs are designed from Ku70, as shown in the lower panel.

with clues in how to investigate the molecular mechanisms responsible for their cell entry. As described in detail later, BCPs can enter the cytosolic space even at 0°C, suggesting that BCPs may be able to enter cells in an energy-independent manner. The cellular uptake of BCPs at 37°C, however, is significantly higher than that at 0°C. Therefore, an energy-dependent process is implicated to play a major role in the cellular uptake of BCPs at 37°C. There is the possibility that receptor-mediated pinocytosis contributes to the cell-permeable activity of BCPs; we are currently investigating this possibility. In this article, we focus on protocols designed to effectively use these new types of cell-permeable penta-peptides.

27.3 SYNTHESIS AND STORAGE OF PEPTIDES

Nonlabeled BCPs were synthesized with a free N- and C-terminus (Figure 27.3).[2] The first N-terminus of the BCPs (except PMLKE) was used to conjugate fluorescent dye (Figure 27.3). Peptides were stored as a dry powder at −80°C. A significant decrease in cell-permeable activity and cell-death inhibition was experienced when the peptides were stored in a −20°C freezer for more than 1 month or when peptides were left at room temperature for more than 3 days. Fresh. dimethyl sulfoxide (DMSO) was used for preparation of the stock solution at a concentration of 200 mM; multiple small aliquots (e.g., 10 μL/tube) were frozen back. The stock solution in DMSO should be stored at −80°C. DMSO must be fresh. Oxidized DMSO (old) should not be used for making stock solutions. BIPs can be dissolved at a concentration of 200 mM in water, in 0.1% BSA (in water) or in culture medium. However, it is recommended that these solutions are used immediately. It is not recommended to store these solutions in the refrigerator (4°C).

Labeling of cell permeable penta-peptides

Bax-inhibiting peptides:

NH_2-BIP-COOH   (nonlabeled BIP)

Fluorescent dye-BIP-COOH ("V" of each BIP was used for labeling)

Negative control peptide:

NH_2-IPMIK-COOH   (nonlabeled peptide)
Fluorescent dye-IPMIK-COOH ("I" was used for the labeling)

NH_2-KLPVM (or KLPVT)-COOH   (nonlabeled peptide).
Fluorescent dye-KLPVM (or KLPVT)-COOH ("K" was used for
the labeling)

FIGURE 27.3 Position of fluorescent dyes conjugated to BCPs.

Freeze–thaw cycles should be avoided. Among the BCPs listed in Figure 27.2, VPMLK appears to be very sensitive to damage when stored under poor conditions. VPTLK appears to be relatively stable for long storage periods. We assume that the methionine in the VPMLK peptide may be easily oxidized by inadequate storage conditions and may be the cause of the rapid decrease in its biological activity.

27.4 CELL TYPES AND CULTURE CONDITIONS NEEDED TO STUDY CELL-PERMEABLE ACTIVITY OF BCPs

The cell-permeable activity of BCPs has been demonstrated in both cultured cells and animal models (mouse retinal ganglion cells).[1,2,21] We confirmed that, in culture, BCPs enter the inner cell space in the following cell types: human epithelial kidney (HEK) 293; HEK293T; primary cultured human umbilical vein endothelial cells (HUVEC) (Figure 27.4A); DAMI (human megakaryocytic cell line); primary cultured mouse megakaryocytes (Figure 27.4B); HeLa (Figure 27.5); Jurkat cells (human T-lymphoma cell line); primary cultured ovarian cumulus cells (mouse, rat, and porcine cells were tested); and mouse embryonic fibroblasts (MEFs).

Based on publications that have used BIPs for protecting cells from apoptosis, BIPs seem to be able to penetrate the plasma membrane of other various types of cells.[14,20,21] The ability of BCPs to enter the cell seems to vary depending on cell types and culture conditions. They enter fast-growing adherent cells very efficiently, including HeLa and HEK293T cells, whereas in floating cells, such as Jurkat cells or primary cultured mouse neurons, BCPs enter the cells slowly (unpublished observation). In our preliminary studies, cell-permeable activity of BCPs became significantly lower as cell density increased.

(A) BIP entered into the endothelial cells

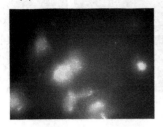

(B) BIP entered the cytosol of megakaryocyte

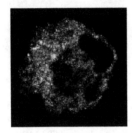
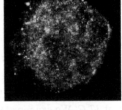

An image from slice #14   Composite of all slices

FIGURE 27.4 Microscopic analysis of BCP cell permeability. (A) Primary cultured human umbilical vein endothelial cells (HUVEC) incubated with $200\ \mu M$ fluorescein-labeled VPTLK (Fam-VPTLK) for 1 h at 37°C. Images were captured on an epifluorescent microscope without cell fixation. (B) Confocal microscopic analysis of primary mouse megakaryocytes incubated with $200\ \mu M$ Fam-VPTLK for 48 h at 37°C. Fam-VPTLK was detected in the cytosol. Cells were not fixed.

27.5 EFFECTIVE BCP CONCENTRATION FOR CELL PERMEABILITY AND CELL DEATH SUPPRESSION

For examining cell permeability using confocal microscopy, a concentration of 1–$10\ \mu M$ fluorescent-dye (fluorescein (Fam) or tetramethylrhodamine(TAMRA))-labeled BCPs in the medium was sufficient for HeLa cells (Figure 27.5). For flow cytometric detection of fluorescein-labeled BCPs incorporated into cells, a concentration of 10–$50\ \mu M$ in the culture medium was needed for DAMI cells (Figure 27.6). These BCP concentrations needed for effective cell entry are approximately 10 times higher than that of previously established cell-permeable, arginine-rich peptides.[22] As shown in Figure 27.6, cellular uptake of VPTLK increased in a dose-dependent manner. Notably, VPTLK did not show a significant cytotoxicity even at 1.6 mM in the medium.

To protect cells from Bax overexpression-induced cell death, $200\ \mu M$ of BIPs in the culture medium was required for HEK293T cells.[1] Depending on cell type and on cell death-inducing stresses, the effective concentration to suppress cell death appears to range from 50–$400\ \mu M$.[2] For example, $400\ \mu M$ was needed to suppress interleukin-3 deprivation-induced cell death of the mouse IL-3 dependent cell line.[2]

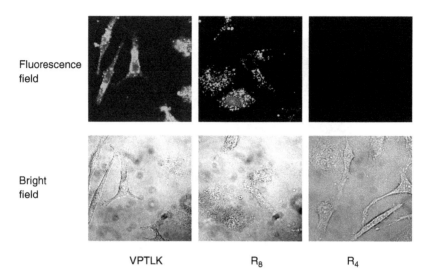

FIGURE 27.5 BCP and poly-arginine peptides entered the cell. Confocal microscopic analysis of HeLa cells incubated with 10 μM of Fam-labeled VPTLK, RRRRRRRR (R_8), and RRRR (R_4) for 2 h at 37°C. We examined tetramethylrhodamine (TAMRA) labeled peptides and obtained similar results. Results demonstrate that VPTLK and R8 entered the cells whereas R4 did not.

27.6 TIME-DEPENDENT CHANGE OF BCP CELLULAR UPTAKE

The entry of Fam-labeled BCPs into cells can be detected within 15 min after the addition of Fam-BCPs into the medium at 37°C, and for DAMI cells, it reaches the maximum level after approximately 6 h of incubation (Figure 27.7). In this experiment, peptides that had recently attached to the cell surface were washed off with an acid solution before analysis, a practice described in a previously reported method.[23] We also confirmed that fluorescent dye-conjugated BIPs were detectable in the cytosol space using confocal microscopic analysis at each time point in independent experiments.

27.7 TEMPERATURE INFLUENCE ON BCP CELLULAR UPTAKE

We next tested whether there was a temperature-dependent change in BCP cell permeability (the result of VPTLK is shown in Figure 27.8). Fam-labeled BCPs and DAMI cells were used in the experiment and fluorescence intensity of the cells was measured using flow cytometry. The highest BCP cell permeability was observed at 37°C among the temperatures tested (37, 4, and 0°C). Interestingly, BCPs were also detected inside cells when cells were incubated at 0 and 4°C. The entry of BCPs into the inner cell space was confirmed at each temperature point using microscopic analysis. The viability of the cells at each temperature in this experiment was more than 98% (examined by Trypan Blue exclusion test). Since BCPs entered the cell even at 0°C, an energy-independent

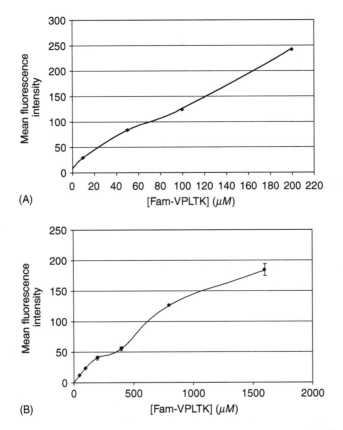

(A)

(B)

FIGURE 27.6 BCPs enter the cell in a dose-dependent manner. DAMI cells (a human megakaryocytic cell line) were cultured at 37°C for 3 h at various concentrations of Fam-labeled VPTLK. (A) shows the dose-dependent curve between 0 and 200 μM. (B) shows another independent experimental result testing higher doses. After the incubation, cells were washed with an ice-cold acidic culture medium (Iscove's modified Dulbecco's medium (IMDM), 5 mM EDTA, pH 5.0) to remove peptides attached to the cellular surface, as previously reported.[23] Fluorescent intensity of the cells was then measured using flow cytometry. The cell viability was more than 97% throughout all the doses examined including 1.6 mM. Each point represents the mean value of samples obtained in triplicate.

mechanism may play a role for the observed cell-permeable activity. In addition, another energy-dependent mechanism, such as pinocytosis, is implicated as playing an important role at 37°C, based on the fact that BCP cell-permeable activity becomes significantly higher at this temperature.

27.8   COMPARISON OF BCP AND ARGININE-RICH PEPTIDE TOXICITY

The BIPs designed from the Bax-binding domain of Ku70 are cytoprotective peptides, and therefore it is not surprising that these peptides do not show significant

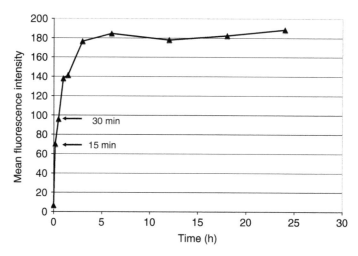

FIGURE 27.7 Time course of Fam-labeled VPTLK entry into the cell. DAMI cells were cultured at 37°C with 200 μM Fam-VPTLK. The amount of Fam-VPTLK within the cell was analyzed using flow cytometry. Each point represents the mean $\pm$ S.E. of samples obtained in triplicate.

cytotoxic activity. Interestingly, the negative control peptides for Bax inhibition, such as KLPVM, also did not show cytotoxic activity even at 400 μM concentration (data not shown). In arginine-rich peptides (i.e., R8 and TAT), however, significant cytotoxicity was observed with the use of a concentration higher than 10 μM; BIPs were not toxic at higher doses (Figure 27.9A). At present, the mechanism

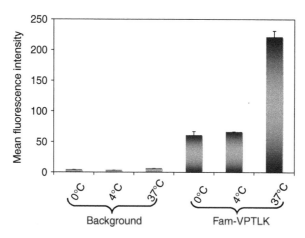

FIGURE 27.8 Influence of temperature on BCP cell permeability. DAMI cells $(5 \times 10^4$ cells/mL) were cultured with 200 μM of Fam-labeled VPTLK for 1 h at various temperatures as indicated in the figure. The fluorescent intensity of cells was analyzed by flow cytometry. The cell viability (Trypan Blue exclusion test) was more than 98% in all the experiments in this figure. Each point represents the mean of triplicated samples.

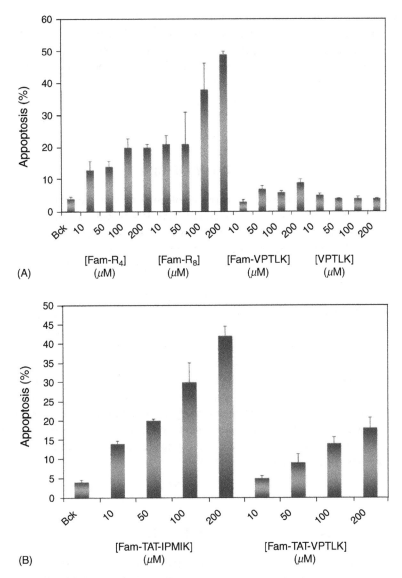

FIGURE 27.9 Cytotoxicity of cell-permeable peptides. (A) BIPs have lower cytotoxic activity than poly-arginine peptides. DAMI cells were cultured for 24 h at 37°C using various peptide concentrations. Peptides used were Fam-labeled R_4, R_8, VPTLK, and nonlabeled VPTLK. Apoptotic nuclear fragmentation was analyzed by staining the nucleus with Hoechst dye. R_4 and R_8 showed significant cytotoxic activity at the concentration higher than 10 μM whereas VPTLK did not. (B) Cytotoxic activity of TAT was attenuated by the conjugation of BIP. DAMI cells were cultured at 37°C for 24 h at various peptide concentrations. The peptides were Fam-labeled TAT-IPMIK and TAT-VPTLK. The amino acid sequence of the TAT peptide used in this study was YGRKKRRQRRR. Apoptosis was analyzed as described in (A).

responsible for apoptosis induction in the presence of high doses of ariginine-rich peptides is not clear. When TAT peptide was fused with BIP (e.g., TAT-VPTLK), the cytotoxic activity of TAT peptides was significantly reduced (Figure 27.9B). These results suggest that a high concentration of poly-arginine-rich peptide triggers Bax activation and/or the addition of BIP reduces the membrane damage caused by highly positively charged arginine-rich peptides.

27.9   CARGO DELIVERY ACTIVITY OF BIPs

BIPs have the capability to deliver "cargo" molecules attached to their N-terminus. We tested fluorescence dye, such as Fam and TAMRA as well as green fluorescent protein (GFP). The fluorescence dyes were labeled to the V in VPTLK and VPMLK; the I in IPMIK; and the K in KLPVM (Figure 27.3). As shown in Figure 27.4 through Figure 27.7, these fluorescence dye-conjugated BCPs showed cell permeable activity. For GFP, we constructed plasmids that expressed GFP with VPTLK or KLPVM on the C-terminus. When GFP-KLPVM (Figure 27.10 (A) and (B)) and GFP-VPTLK (not shown) were added to the culture medium, the cellular cytosolic green fluorescent intensity significantly increased. On the other hand, the significant increase of fluorescent activity was not detected in the cells cultured with GFP (Figure 27.10 (C) and (D)). These results suggest that GFP-KLPVM (and GFP-VPTLK), but not GFP itself, was delivered into the cell. However, these microscopic observations may not be sufficient to prove that GFP was actually delivered into the cytosol by BCP and further biochemical analyses to confirm the cellular entry of GFP-BCP will be needed. Since the fusion proteins were not purified in this experiment, further careful studies are clearly needed to determine the activity of BCP to deliver cargo proteins.

27.10   CONCLUSIONS AND FUTURE STUDIES

BCPs have a similar cell-permeable activity as previously established cell-permeable peptides, such as the TAT and arginine-rich peptides. Importantly, BCPs have a much lower toxicity than arginine-rich peptides. Therefore, BCPs have the potential to deliver conjugated cargo into cells at higher doses. At present, the molecular mechanism behind the membrane penetration of BCPs is not clear. The understanding of such a mechanism may provide an opportunity to identify an unrecognized molecular transport system that occurs in the plasma membrane. BCPs may be utilized as coating peptides for nanoparticles, molecules that have recently been examined for use as drug-delivery tools.[24] Further studies of the application of BCPs in drug delivery, and the development of smaller chemical mimetics, may contribute to new advances in developing a low-toxicity drug-delivery system.

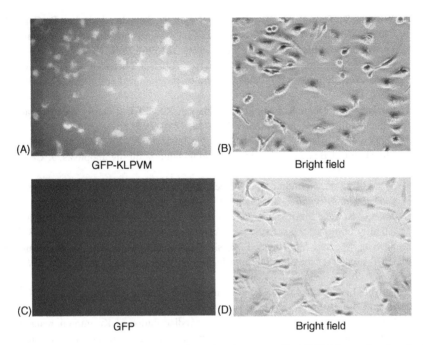

FIGURE 27.10 BCP may be able to deliver GFP into the cells. KLPVM was fused to the C-terminus of GFP. The plasmid of pEGFPC2 purchased from Clontech was used, and the oligonucleotides encoding KLPVM was ligated to the Xho1 and ECoR1 site. The GFP protein (unmodified GFP and GFP-KLPVM) was expressed in HEK293T cells, and cell lysates were prepared using detergent-free buffer. Cell lysates (100 μg protein cell lysate in 1 mL culture medium) were added to the culture medium of HeLa cells, and the cells were incubated for 24 h. The figure shows that the addition of GFP-KLPVM, but not unmodified GFP, significantly increased the green fluorescence activity of HeLa cells.

ACKNOWLEDGMENTS

The authors are grateful to Tomoyuki Yoshida for his sincere effort in organizing and storing the peptides. The authors also appreciate Dr. Danyu Sun for taking the images of human umbilical vein endothelial cells, Dr. Michelle Stapleton and Cathy Paddock for their advice and technical support for mouse megakaryocyte culture, Dr. Minn Lam for the advice on analyzing the cells using confocal microscopy, and the researchers in the Flow Cytometry Facility Core of the Case Comprehensive Cancer Center for their advice on flow cytometric analysis. This work was supported by a Grant-in-Aid from the American Heart Association (AHA) and a Pilot Grant for Aging and Cancer Research from the NIH (P20CA10373) to S.M. Jose Gomez and Vivian Gama is a recipient of a predoctoral fellowship from American Heart Association (Ohio Valley).

REFERENCES

1. Sawada, M., Hayes, P., and Matsuyama, S., Cytoprotective membrane-permeable peptides designed from the Bax-binding domain of Ku70, *Nat. Cell Biol.*, 5, 352, 2003.

2. Yoshida, T., Tomioka, I., Nagahara, T., Holyst, T., Sawada, M., Hayes, P., Gama, V., Okuno, M., Chen, Y., Abe, Y. et al., Bax-inhibiting peptide derived from mouse and rat Ku70, *Biochem. Biophys. Res. Commun.*, 321, 961, 2004.

3. Oltvai, Z.N., Milliman, C.L., and Korsmeyer, S.J., Bcl-2 heterodimerizes in vivo with a conserved homolog, Bax, that accelerates programmed cell death, *Cell*, 74, 609, 1993.

4. Gross, A., McDonnell, J.M., and Korsmeyer, S.J., BCL-2 family members and the mitochondria in apoptosis, *Genes Dev.*, 13, 1899, 1999.

5. Wolter, K.G., Hsu, Y.T., Smith, C.L., Nechushtan, A., Xi, X.G., and Youle, R.J., Movement of Bax from the cytosol to mitochondria during apoptosis, *J. Cell Biol.*, 139, 1281, 1997.

6. Matsuyama, S., Nouraini, S., and Reed, J.C., Yeast as a tool for apoptosis research, *Curr. Opin. Microbiol.*, 2, 618, 1999.

7. Xu, Q., Ke, N., Matsuyama, S., and Reed, J.C., Assays for studying Bax-induced lethality in the yeast Saccharomyces cerevisiae, *Methods Enzymol.*, 322, 283, 2000.

8. Sawada, M., Sun, W., Hayes, P., Leskov, K., Boothman, D.A., and Matsuyama, S., Ku70 suppresses the apoptotic translocation of Bax to mitochondria, *Nat. Cell Biol.*, 5, 320, 2003.

9. Downs, J.A. and Jackson, S.P., A means to a DNA end: The many roles of Ku, *Nat. Rev. Mol. Cell Biol.*, 5, 367, 2004.

10. Fewell, J.W. and Kuff, E.L., Intracellular redistribution of Ku immunoreactivity in response to cell–cell contact and growth modulating components in the medium, *J. Cell Sci.*, 109, 1937, 1996.

11. Kumaravel, T.S., Bharathy, K., Kudoh, S., Tanaka, K., and Kamada, N., Expression, localization and functional interactions of Ku70 subunit of DNA-PK in peripheral lymphocytes and Nalm-19 cells after irradiation, *Int. J. Radiat. Biol.*, 74, 481, 1998.

12. Tai, Y.T., Podar, K., Kraeft, S.K., Wang, F., Young, G., Lin, B., Gupta, D., Chen, L.B., and Anderson, K.C., Translocation of Ku86/Ku70 to the multiple myeloma cell membrane: Functional implications, *Exp. Hematol.*, 30, 212, 2002.

13. Martinez, J., Seveau, S., Veiga, E., Matsuyama, S., and Cossart, P., Ku70, a component of the DNA-dependent protein kinase, acts as a receptor involved in Rickettsia conorii invasion of mamalian cells, *Cell*, 123, 1013–1023, 2005.

14. Yu, L.Y., Jokitalo, E., Sun, Y.F., Mehlen, P., Lindholm, D., Saarma, M., and Arumae, U., GDNF-deprived sympathetic neurons die via a novel nonmitochondrial pathway, *J. Cell Biol.*, 163, 987, 2003.

15. Cohen, H.Y., Lavu, S., Bitterman, K.J., Hekking, B., Imahiyerobo, T.A., Miller, C., Frye, R., Ploegh, H., Kessler, B.M., and Sinclair, D.A., Acetylation of the C terminus of Ku70 by CBP and PCAF controls Bax-mediated apoptosis, *Mol. Cell.*, 13, 627, 2004.

16. Cohen, H.Y., Miller, C., Bitterman, K.J., Wall, N.R., Hekking, B., Kessler, B., Howitz, K.T., Gorospe, M., de Cabo, R., and Sinclair, D.A., Calorie restriction promotes mammalian cell survival by inducing the SIRT1 deacetylase, *Science*, 305, 390, 2004.

17. Rashmi, R., Kumar, S., and Karunagaran, D., Ectopic expression of Bcl-XL or Ku70 protects human colon cancer cells (SW480) against curcumin-induced apoptosis while their down-regulation potentiates it, *Carcinogenesis*, 25, 1867, 2004.

18. Subramanian, C., Opipari, A.W. Jr, Bian, X., Castle, V.P., and Kwok, R.P., Ku70 acetylation mediates neuroblastoma cell death induced by histone deacetylase inhibitors, *Proc. Natl. Acad. Sci. U.S.A.*, 102, 4842, 2005.

19. Lee, J.C., Lee, C.H., Su, C.L., Huang, C.W., Liu, H.S., Lin, C.N., and Won, S.J., Justicidin A decreases the level of cytosolic Ku70 leading to apoptosis in human colorectal cancer cells, *Carcinogenesis*, 26, 1716, 2005.
20. Zi, X., Simoneau, A.R., and Flavokawain, A., A novel chalcone from kava extract, induces apoptosis in bladder cancer cells by involvement of Bax protein-dependent and mitochondria-dependent apoptotic pathway and suppresses tumor growth in mice, *Cancer Res.*, 65, 3479, 2005.
21. Qin, Q., Patil, K., and Sharma, S.C., The role of Bax-inhibiting peptide in retinal ganglion cell apoptosis after optic nerve transection, *Neurosci. Lett.*, 372, 17, 2004.
22. Futaki, S., Suzuki, T., Ohashi, W., Yagami, T., Tanaka, S., Ueda, K., and Sugiura, Y., Arginine-rich peptides. An abundant source of membrane-permeable peptides having potential as carriers for intracellular protein delivery, *J. Biol. Chem.*, 276, 5836, 2001.
23. Behrens, I., Pena, A.I., Alonso, M.J., and Kissel, T., Comparative uptake studies of bioadhesive and non-bioadhesive nanoparticles in human intestinal cell lines and rats: The effect of mucus on particle adsorption and transport, *Pharm. Res.*, 19, 1185, 2002.
24. Sengupta, S., Eavarone, D., Capila, I., Zhao, G., Watson, N., Kiziltepe, T., and Sasisekharan, R., Temporal targeting of tumour cells and neovasculature with a nanoscale delivery system, *Nature*, 436, 568, 2005.

28 Cellular Uptake of Liposomal and Micellar Carriers Mediated by ApoE-Derived Peptides

*Margitta Dathe, Sandro Keller, Ines Sauer,
and Michael Bienert*

CONTENTS

28.1   INTRODUCTION

Many drugs cannot unfold their full potential because of poor solubility, low chemical stability, unspecific toxicity, or unfavorable pharmacokinetics. Often, their inability to recognize their target cells and to cross cellular barriers further limits their application. Incorporation into drug-delivery systems, including natural and synthetic polymer nanoparticles and surfactant-based systems such as liposomes and micelles, can largely overcome these disadvantageous properties and suppress recognition by cellular efflux systems. Liposomal formulations[1,2] have been most successful so far as they can function as a reservoir for both hydrophilic compounds entrapped within the aqueous lumen and hydrophobic drugs localized in the lipid bilayer. Moreover, liposomes allow for the transport of high drug payloads, and as many lipids used are, or at least resemble, natural compounds, their administration is expected to be low risk. Small liposomes bearing a net neutral charge and including lipids with a high phase transition temperature or cholesterol have been found to circulate for hours, but breakdown mediated by various components of the blood plasma may cause premature release of the encapsulated drug. Furthermore, recognition of the carrier by macrophages in the liver and spleen results in its removal from circulation. The natural homing ability to the reticuloendothelial system (RES) is advantageous when this cell population is addressed; however, it constitutes a serious hurdle when targeting other cells. Substantial blockage of the RES has been reached with so-called sterically stabilized or stealth liposomes containing polyethylene glycol (PEG)-modified lipids.[3] PEG coating reduces the adsorptive binding of proteins and inhibits the uptake by macrophages. Simultaneously, the porous structure of the endothelial cell layer of tumors and inflamed tissues or organs enables the passive targeting of these long-circulating liposomes. However, there is little extravasation of liposomes from the blood into tissues fed by continuous, nonfenestrated capillaries, such as muscle, skeleton, skin, and the central nervous system.

Drug delivery to the brain poses a particular challenge.[4] The tightly connected endothelial cells of brain capillaries forming the blood–brain barrier (BBB) prevent any paracellular transport. One possibility to overcome this barrier is the use of vectors.[5–7] Ideally, modification of drugs or drug carriers serves to both target the BBB and facilitate uptake into capillary endothelial cells. For this strategy to be successful, it is important to define surface determinants that differ quantitatively or, even better, qualitatively from those of other cells and activate natural transport routes across the BBB. In the following, we briefly discuss three approaches developed to this end.

The targeting and uptake-mediating molecules ascribed to the first class have high molar masses, such as antibodies and proteins that recognize well-defined receptor structures at the cell membrane. Their binding is followed by receptor-mediated endocytosis of the vector–cargo complex. Examples are provided by transferrin-modified liposomes[8] and immunoliposomes addressing the transferrin receptor at the BBB.[9,10] On the basis of the improved analgetic effect of dalargin-loaded and apolipoprotein E (apoE)-covered acrylate nanoparticles, a transport process involving the low-density lipoprotein receptor (LDLr) has been discussed.[11] LDLr is constituently expressed on brain capillary endothelial cells at levels higher than those in other vessels.[12]

Secondly, peptides recognizing certain membrane components are suitable for modification of liposomes. Peptides are biocompatible, biodegradable, and less antigenic than antibodies. In particular, specific integrin-binding RGD peptides[13,14] complexed with sterically stabilized liposomes have been used for targeting tumor cells. Covalent coupling of RGD sequences to liposomes and binding of these complexes to integrin-exposing leucocytes, which are able to overcome the BBB, represents an intriguing approach for drug transport into the brain.[15]

Efficient cellular uptake is mediated by so-called protein transduction domains (PTDs) or cell-penetrating peptides (CPPs). These peptides are able to shuttle attached bioactive molecules and carriers across the membranes of various cells (for reviews, see Refs. [16,17]). Increasing evidence points to an unspecific accumulation of the functionalized drug molecules or carriers near negatively charged cell surface constituents, such as lipid head groups or heparan sulfate proteoglycans (HSPGs), which is followed by adsorptive endocytosis.[18,19] Examples of adsorptive endocytosis at the BBB include the transport of peptide-modified doxorubicin[20] and the uptake of liposomes coupled to cationized albumin into brain capillary endothelial cells.[21]

Here, we present the application of cationic peptides derived from the LDLr recognition sequence of apoE to mediate cellular uptake of liposomes into brain capillary endothelial cells. We describe both covalent and adsorptive coupling strategies and analyze biophysical properties relevant to the application of the complexes as potential drug carriers. We show that peptide-tagged liposomes and peptide–lipid micelles are efficiently internalized into live cells. Finally, we present evidence that cellular uptake is not mediated by LDLr but dominated by a pathway involving HSPG.

28.2 APOLIPOPROTEIN E AND APOLIPOPROTEIN E PEPTIDES

28.2.1 GENERAL

Apolipoprotein E (apoE) is involved in the transportation and cellular uptake of cholesterol-rich lipoproteins.[22] ApoE is a 34-kDa two-domain protein with an N-terminal 22-kDa water-soluble globular helix bundle involved in the interaction with cell surface constituents and a C-terminal domain promoting lipid binding and self-association. A cluster of positively charged residues in the sequence 136–169 is important for the cellular uptake of the lipoproteins by members of the low-density lipoprotein receptor family.[23] Lipoprotein-bound fragments of the apoprotein[24] and liposome-incorporated apoE[25] can mimic natural lipoproteins, as they stimulate LDLr-mediated endocytosis.

Synthetic peptides have been designed to determine the minimal sequence able to complex with lipids and to induce cellular uptake. ApoE(126−169)[26] and apoE(126−183)[27] complexes with dimyristoylphosphatidylcholine (DMPC) bind to LDLr, and the lipidated sequence Gly-apoE(126−169) promotes uptake and degradation by fibroblasts.[28] Even though shorter apoE sequences, such as (141−155) and (141−150), exhibit no LDL receptor binding, their tandem dimers are LDLr-active.[29,30] ApoE(141−155)$_2$ enhances the interaction of LDL with fibroblasts approximately 10-fold.[31]

Because the tandem dimer (141−150)$_2$, corresponding to (LRKLR KRLLR)$_2$ (A2), appears to be the shortest receptor-binding apoE sequence, we complexed this peptide to liposomes by both covalent and adsorptive methods (Table 28.1).[32−34]

28.2.2 PEPTIDE COUPLING TO LIPOSOMES

28.2.2.1 Covalent Linkage

Covalent methods include the formation of disulfide bonds, reactions between maleimide and thiol functions, between carboxylic acids and primary amine groups, and between hydrazide and aldehyde functions, and crosslinking to primary amine functions (for a review, see Ref. [35]) A reaction between a thiol function and a maleimide-modified phosphatidylethanolamine gives a stable thioether bond and has been used frequently in combination with sterically stabilized liposome technology.[36] Stable complexes of liposomes with antibodies against the transferrin receptor[9] and peptides such as RGD[15] and TAT[37,38] have been prepared.

TABLE 28.1
Sequences of apoE Peptides

Abbreviation	Sequence
A2	Ac-(LRKLR KRLLR)$_2$-NH$_2$
For covalent coupling	
HS-A2	HS-(CH$_2$)$_2$-CO-WG(LRKLR KRLLR)$_2$-NH$_2$
For adsorptive coupling	
A2Q	Ac-(LRKLR KRLLR)$_2$PLVED MQRQW AGLV-NH$_2$
A2T	Ac-(LRKLR KRLLR)$_2$AWLAL ALALA LKALA LALAL KK-NH$_2$
MA2	Myristoyl-WAG(LRKLR KRLLR)$_2$-NH$_2$
PA2	Palmitoyl-WAG(LRKLR KRLLR)$_2$-NH$_2$
P2A2	Palmitoyl-WK(palmitoyl)G(LRKLR KRLLR)$_2$-NH$_2$
P2fA2	Palmitoyl-K(palmitoyl)WK(FLUO)G(LRKLR KRLLR)$_2$-NH$_2$
AD	Ac-LRKLR KRLLR DWLKA FYDKV AEKLK EAF-NH$_2$

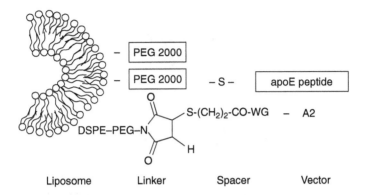

FIGURE 28.1 Schematic representation of the liposomal delivery system prepared by covalent vector coupling.

To attach the apoE peptide to liposomes, we N-terminally functionalized the A2 sequence with mercaptopropionic acid (Table 28.1) and coupled it to the terminal maleimide function of 1,2-distearoyl-*sn*-glycero-3-phosphoethanolamine-*N*-[maleimide(polyethylene glycol-2000)] (Mal-PEG-DSPE) of preformed unilamellar vesicles with a diameter of 75 nm (Figure 28.1). The lipid composition was 1-palmitoyl-2-oleoyl-*sn*-glycero-3-phosphocholine (POPC), cholesterol, 1,2-dipalmitoyl-*sn*-glycero-3-phosphoethanolamine-*N*-[methoxy(polyethylene glycol-2000)] (PEG-DPPE), Mal-PEG-DSPE, and 1,2-dipalmitoyl-*sn*-glycero-3-phosphoethanolamine-*N*-[lissamine rhodamine B sulfonyl] (Rh-DPPE) at a molar ratio of 64/30/2.5/2.5/1.[32,33] The presence of around 100 peptide molecules on the outer vesicle surface (corresponding to a surface density of approximately one peptide molecule per 18,000 Å^2) led to a slight increase in the particle size but had no influence on long-term vesicle stability.

28.2.2.2 Noncovalent Binding

The biotin–avidin method is one of the most popular strategies for noncovalent coupling.[39] However, to prevent steric hindrance, liposomes modified with small peptides are usually prepared by using lipophilic peptide derivatives.[40,41] In analogy to the two-domain structure of apoE, we synthesized two-domain apoE peptides consisting of the highly cationic A2 sequence and a variety of hydrophobic anchors: an amphipathic helix, Q, a transmembrane helix, T, a myristoyl chain, M, a palmitoyl chain, P, or two palmitoyl chains, P2 (Table 28.1).[32] The amphipathic helix, Q, corresponds to the sequence 267–280, which is crucial for lipid binding.[42] The hydrophobic sequence, T, is analogous to helical peptides adopting a transmembrane orientation in lipid bilayers.[43,44] Myristoyl and palmitoyl chains are the most common forms of protein acylation and are often used to promote peptide association with lipid vesicles.[45,46,47] The expected localizations of A2Q, A2T, and the dipalmitoylated P2A2 in lipid vesicles are schemed out in Figure 28.2.

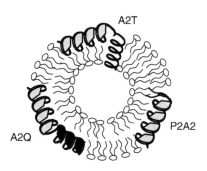

FIGURE 28.2 Schematic representation of the localizations of A2Q, A2T, and P2A2 after adsorptive binding to liposomes. The polar face of the ideal amphipathic helix of A2 is shown in gray; black helices and chains represent the hydrophobic lipid-anchoring domains.

28.2.3   CHARACTERIZATION OF PEPTIDE–LIPOSOME COMPLEXES

Amphipathic peptides tend to self-associate, undergo lipid-induced conformational changes, and destabilize membranes. The possible resulting loss of bioactivity of the vector and release of compounds entrapped in the liposomal lumen require extensive characterization of the complexes.

28.2.3.1   Binding

In agreement with previous studies of unmodified apoE peptides,[29,30] the lipid affinity of pure A2 is low. Isothermal titration calorimetry (ITC) signals upon titrating lipid vesicles into A2 solutions do not differ from typical heats of dilution, thus precluding determination of the binding constant (Table 28.2). All modified peptides contain one tryptophan residue, the fluorescence of which is highly sensitive to the environment and can be used to monitor peptide–lipid interactions. The blue shifts and intensity increases in the fluorescence maxima of A2Q, MA2, and PA2 upon addition of liposomes indicate that the peptides are vesicle-bound (Table 28.2).[48] The comparable dependencies of the fluorescence intensities on the amount of the quencher-bearing lipid 1-palmitoyl-2-stearoyl(6-7)dibromo-sn-glycero-3-phosphocholine (DiBr-PSPC) confirm that all A2 peptides bearing a lipid anchor insert into the lipid bilayer (Figure 28.3). However, with a hydrophobic partition coefficient of $K = 17$ mM^{-1}, the amphipathic Q helix confers only moderate lipid affinity upon A2. Myristoyl and palmitoyl modifications distinctly promote peptide binding. For apoE peptides and other domains rich in basic amino acids, a second palmitoyl anchor has been proposed for stable membrane attachment.[28,45] Because of its low monomeric solubility and extraordinary membrane affinity, the partitioning of P2A2 between the aqueous and bilayer phases could not be quantified. However, other titration calorimetric and light scattering studies afforded a detailed picture of its vesicle–micelle transition, from which $K \gg 1000$ mM^{-1} can be predicted (see Section 28.2.4 and Ref. [34]), thus rendering P2A2 the most promising candidate for complexation with liposomes.

TABLE 28.2
Physicochemical Properties of apoE Peptides and Peptide–Liposome Complexes

Peptide	λ_{max} (nm) Buffer	F_{max} (au) Buffer	λ_{max} (nm) POPC	F_{max} (au) POPC	K (mM M^{-1})	α (%) Buffer	Buffer/TFE	POPC	z_{cf} (Å)	EC_{50} (μM)	c_l/c_p
A2	n.a.	n.a.	n.a.	n.a.	n.a.	1	80	13	n.a.	0.03	833
A2Q	354	64	335	62	17	7	80	48		0.02	1250
A2T	333	110	332	127	n.a.	77	100	100	14.3	0.04	625
MA2	352	58	342	102	30	5	75	39		0.02	1250
PA2	350	61	342	110	30	8	77	38		0.01	2500
P2A2	344	39	342	86	>1000	16	82	36	14.4	0.06	417

The fluorescence properties of 2 μM W-containing peptides were determined in Tris buffer and in the presence of POPC SUVs (lipid concentration $c_l = 2$ mM). Fluorescence was excited at 280 nm (λ_{max}, wavelength of the emission maximum; F_{max}, fluorescence intensity at λ_{max}). The partition coefficients, K, were derived from ITC measurements by titrating POPC SUVs ($c_l = 30$ mM) into 20 μM peptide solutions. α is the peptide helicity in buffer, in buffer/trifluoroethanol (TFE) (1/1 v/v), or in the presence of 1 mM POPC SUVs. The peptide concentration was 20 μM. The distance of the tryptophan residue from the bilayer center, z_{cf}, was calculated by applying the parallax equation. EC_{50} is the peptide concentration needed for half-maximal dequenching of the fluorescence of calcein entrapped in POPC LUVs ($c_l = 25$ μM) taken 5 min after the addition of peptide, and c_l/c_p gives the corresponding lipid-to-peptide ratio (n.a., not applicable).

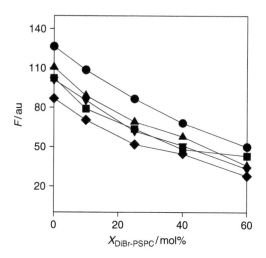

FIGURE 28.3 Fluorescence intensities, F, of A2Q ($\blacksquare$), A2T ($\bullet$), MA2 ($\blacktriangledown$), PA2 ($\blacktriangle$), and P2A2 ($\blacklozenge$) bound to POPC vesicles in dependence of the mole fraction, $X_{DiBr-PSPC}$, of the quencher-labeled lipid DiBr-PSPC ($c_l = 2$ mM, $c_p = 2$ μM). F was registered at the wavelength of the emission maximum, λ_{max} (Table 28.2).

28.2.3.2   Peptide Structure

The receptor-binding region in native apoE is helical, and helicity appears to be a critical determinant for receptor binding of apoE peptides.[49,50] Although existing as flexible chains in buffer, all A2 peptides possess a high helical potential, as revealed by helicities higher than 75% in a structure-inducing mixture of buffer/trifluoroethanol (Table 28.2). The observation is in accordance with reports on different other apoE peptides.[30] Notwithstanding its pronounced amphipathicity, the interaction of A2 with lipid bilayers is weak. However, as indicated for the lipidated peptides MA2, PA2, and P2A2, the helicities of the vesicle-bound tandem dimers amount to approximately 40% (Table 28.2). Compared with covalent coupling of A2 to PEG-liposomes exposing the vector to a rather hydrophilic environment, the helical structures induced by adsorptive binding should be favorable for the interaction with the LDL receptor.

28.2.3.3   Peptide Anchorage and Membrane Positioning

To determine the kinetic stability of peptide anchoring within the lipid bilayer, we monitored fluorescence dequenching upon transfer of A2 peptide derivatives from quencher-bearing to quencher-free liposomes. The addition of quencher-free vesicles to A2Q bound to liposomes containing 40 mol% DiBr-PSPC leads to an immediate intensity increase (Figure 28.4), indicating that the peptide is rapidly redistributed. The amphipathic Q helix is insufficient to anchor the peptide in neutral lipid bilayers. Whereas A2T and PA2 slowly redistribute, as shown by the slight time-dependent increase in fluorescence intensity after vesicle addition, almost no

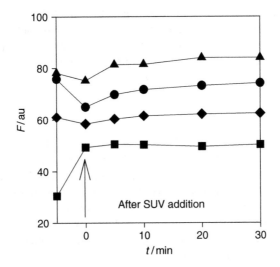

FIGURE 28.4 Fluorescence intensities, F, of A2Q (■), A2T (●), PA2 (▲), and P2A2 (♦) bound to POPC/DiBr-PSPC (60/40 mol/mol; $c_1 = 2$ mM) measured at the wavelength of the emission maximum, λ_{max}, and as a function of time, t, after addition (arrow) of POPC SUVs. The final total lipid concentration was 3.8 mM, and the peptide concentration was 2 μM.

change is observed for P2A2, implying that the dipalmitoylated peptide is kinetically stably anchored within the bilayer. This observation is in accordance with the suggestion that a second palmitoyl anchor is needed for the stable membrane attachment of peptides and proteins.[28,45] From quenching studies using lipids bearing nitroxide labels at different positions in the lipid acyl chains, one can roughly estimate that the tryptophan residue of P2A2 is located approximately 14 Å above the bilayer center (Table 28.2).[32] Thus, while the two N-terminal palmitoyl chains penetrate into the hydrophobic bilayer region, the bioactive peptide moiety A2 remains surface-located (Figure 28.2).

28.2.3.4 Vesicle Integrity

Bilayer permeabilization can be monitored fluorimetrically by measuring the decrease in self-quenching of vesicle-entrapped calcein as a function of peptide concentration. As the concentration of the tandem dimer A2 inducing 50% dye release is in the submicromolar range, its bilayer-permeabilizing activity is rather high (Table 28.2). Comparable activities toward neutral bilayers have been reported for other cationic amphipathic model peptides.[51,52] The coupling of lipid-binding domains does not increase this membrane-disturbing effect (Table 28.2), indicating that permeabilization is dominated by the A2 sequence. Interestingly, P2A2 is least active, suggesting that this derivative is most useful for preserving the integrity of the liposomes and for protecting the vesicle content. With approximately 180 P2A2 molecules on the surface of a vesicle (diameter ~100 nm) at a lipid-to-peptide ratio of 500 (corresponding to approximately one peptide molecule per 18,000 Å^2), the peptide surface density is comparable to that of the A2–PEG liposomes. As shown by light scattering measurements, the size and monodispersity of the vesicles are not influenced over a period of at least 1 week at lipid-to-peptide ratios higher than 500 (data not shown), suggesting that such complexes can retain their morphology long enough to be of use as a drug-delivery vehicle.

28.2.4 MICELLES

In recent years, several attempts have been made to explore the pharmaceutical potential of micellar drug-delivery systems. So far, block copolymer micelles[53] and polymeric micelles based on polyethylene glycol–phosphatidylethanolamine conjugates,[54] which are usually spherical assemblies of amphipathic copolymers with a core–shell architecture, have been applied successfully. The core accommodates predominantly hydrophobic drugs, while the polar shell ensures the water dispersion of the micelles. In contrast to polymeric micelles, which normally can retain the loaded drug for a longer period,[55] micelles composed of surfactant-like molecules are generally less stable; the efficacy of such systems will depend on their critical micellar concentration (CMC). Analytical ultracentrifugation and static light scattering have revealed that P2A2 in physiological buffer assembles into well-defined aggregates with a molar mass of $M = 68$ kg/mol, which are stable down to the nanomolar concentration range.[34] As shown in Figure 28.5, the dipalmitoylated apoE peptide is a highly promising candidate for solubilizing hydrophobic drugs in lipid-rich micelles.[34] Compared with peptide–liposome

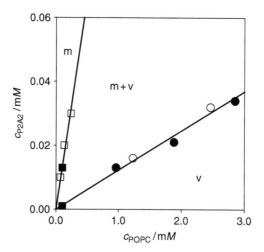

FIGURE 28.5 Phase diagram of dilute aqueous P2A2/POPC mixtures. The data are taken from the break points in the solubilization (full symbols) and reconstitution (open symbols) experiments conducted with light scattering (squares) and ITC (circles). The slopes of the linear regression analyses (solid lines) correspond to the critical mole ratios R_e^{sat} and R_e^{sol} and separate the micellar (m), the transitional (m+v), and the vesicular (v) ranges. (From Keller, S. et al., *Angew. Chem. Int. Ed.*, 44, 5252, 2005. With permission.)

complexes, these micelles offer the advantages of smaller size and higher vector density, as the lipid-to-peptide ratio ranges from 0 to ~7, which is substantially lower than that necessary for stable liposomal formulations (~500). For lipoproteins tagged with apoE and apoE peptides, a high surface density is highly favorable for receptor binding and internalization.[56] Additionally, in contrast to micellization of pure P2A2, which does not allow the induction of a helical peptide structure (Table 28.2), the lipid–peptide micelles provide a bilayer-like, helix-promoting environment. Most importantly, the phase diagram in Figure 28.5 permits the preparation of different formulations for cell-biological studies by adjusting the concentrations of the vector peptide and the lipid.

28.3 INTERACTIONS WITH CELLS

28.3.1 TOXICITY TOWARD BRAIN CAPILLARY ENDOTHELIAL CELLS

ApoE peptides elicit a number of cellular responses. For example, the sequence 133–162 has antibacterial and antitumor activity,[57] and ApoE(141–155)$_2$ is cytotoxic to ganglion cells[58] and T lymphocytes[59] at submicromolar concentrations. The cytotoxicity of A2 peptides toward brain capillary endothelial cells limits the concentration applicable in cell-biological studies to 2 μM (Figure 28.6). The toxic effect of PEG liposomes with covalently attached A2 is comparable. Such preparations containing 1 mM lipid and 2.5 μM peptide reduce cell viability to approximately 80%.[33] However, liposome-bound P2A2 as well as micellar P2A2 are less toxic, suggesting that lipid association is favorable for cell viability.

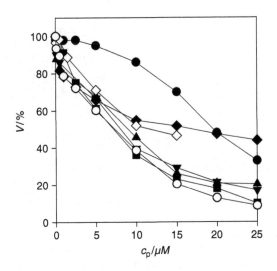

FIGURE 28.6 Viability, V, of mouse brain capillary endothelial cells (b.End3) exposed to increasing concentrations, c_p, of the peptides A2 ($\bigcirc$), A2Q ($\square$), A2T ($\bullet$), MA2 ($\blacktriangledown$), PA2 ($\blacktriangle$), P2A2 ($\blacklozenge$), and of P2A2–liposome complexes ($\diamondsuit$) with a lipid-to-peptide ratio of 1000.

Comparable toxicities of A2 peptides toward mouse and rat brain capillary endothelial cells[32] and human fibroblasts (data not shown) point to a cell-type-unspecific effect.

28.3.2   CELLULAR UPTAKE OF PEPTIDE–LIPOSOME COMPLEXES

Cellular uptake of vesicles and micelles was monitored by confocal laser scanning microscopy (CLSM). Both covalent coupling of A2 to PEG liposomes and adsorptive binding to POPC vesicles result in efficient uptake of the complexes into brain capillary endothelial cells when incubated at 37°C (Figure 28.7A,B).[32,33] Furthermore, for a fluorescently labeled micelle-forming P2A2 derivative (P2fA2),

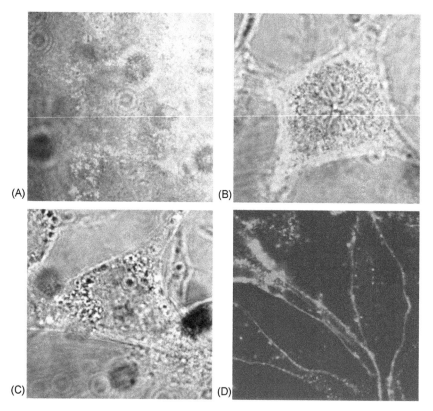

(A)   (B)

(C)   (D)

FIGURE 28.7 CLSM images of mouse brain capillary endothelial cells (b.End3) exposed to (A) A2–PEG liposomes, (B) P2A2 liposomes labeled with Rh-DPPE, (C) P2A2 liposomes containing FITC-dextran, or (D) P2fA2 micelles. The red fluorescence of Trypan Blue on the cell membrane documents the viability of the P2fA2-exposed cells. Liposome uptake ($c_l = 1$ mM, $c_p = 1$ μM) or peptide internalization ($c_p = 1$ μM) was examined within 10 min after incubation for 30 min at 37°C. The mole fraction of Rh-DPPE was 0.1 mol%, and the FITC-dextran concentration inside the vesicles was 400 μM.

comparable uptake patterns are observed (Figure 28.7D).[34] Liposomes without apoE modification are not internalized. P2A2-dotted liposomes labeled either with Rh-DPPE integrated within the lipid bilayer (Figure 28.7B) or fluorescein isothiocyanate-dextran WT 4000 (FITC-dextran) dissolved in the aqueous lumen (Figure 28.7C) accumulate in distinct intracellular compartments, suggesting that the intact particles are taken up by an endocytotic process. Abolished uptake and surface accumulation at 4°C (Figure 28.8A) and reduced internalization under conditions of energy depletion (Figure 28.8B) support this suggestion. As the uptake patterns are comparable for different mouse and rat brain capillary endothelial cell lines, as well as for human fibroblasts and bovine aortic endothelial cells (not shown), no cell-specific components seem to be required.

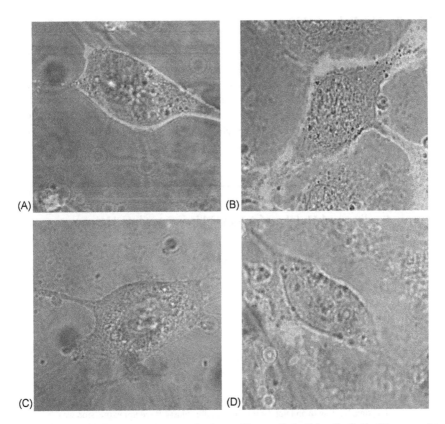

FIGURE 28.8 CLSM images of mouse brain capillary endothelial cells (b.End3) exposed to P2A2 liposomes (A) for 30 min at 4°C, (B) for 30 min at 37°C under conditions of energy depletion (pretreatment for 1 h with 10 mM NaN$_3$ and 25 mM 2-deoxyglucose in PBS), (C) in the presence of 20 μg/mL heparin, or (D) after treatment for 30 min at 37°C with 0.02 μM poly-L-lysine. The lipid concentration was 1 mM, the peptide concentration 1 μM, and the mole fraction of Rh-DPPE was 0.1 mol%. Trypan Blue exclusion documented the viability of the cells (not shown).

28.3.3 Mode of Cellular Uptake

28.3.3.1 The Role of the LDL Receptor

The tandem dimer A2 is able to recognize the LDL receptor.[33] However, compared with the natural ligand LDL, its affinity is much reduced, which is agreement with previous studies.[29,30] Experiments performed with A2-modified PEG liposomes and human fibroblasts with normal and upregulated LDL receptor expression[33] show that both binding to and uptake into cells are independent of the receptor level. Also, blocking the receptor by antibodies does not affect internalization (Table 28.3). These findings imply that cellular uptake of complexes consisting of apoE peptides and lipids is dominated by an LDLr-independent mechanism. Identical uptake patterns of peptide–liposome complexes with capillary endothelial cells expressing different amounts of LDLr[32] support this suggestion.

28.3.3.2 The Role of HSPG

The cationic sequence (142–147) represents a high-affinity heparin-binding region,[60] which may also serve as a ligand for cell-surface HSPGs.[61] These highly negatively charged proteins can act as selective regulators of ligand–receptor encounters or as membrane carriers, and they can be rapidly internalized.[62,63] Evidence of HSPG-dependent uptake of A2–PEG liposomes into fibroblasts has been obtained in experiments using heparinase-treated cells or in the presence of the HSPG competitor heparin.[32,33] Removal of HSPG reduces binding and uptake to

TABLE 28.3
Binding and Uptake of A2–PEG Liposomes into Human Fibroblasts

Treatment	Binding (%)	Uptake (%)
Untreated, downregulated LDLr expression	100	100
Upregulated LDLr expression	95	100
Energy depletion, NaN$_3$, 2-deoxyglucose	30	15
Heparin	3	3
Heparinase I	50	50

Effect of the level of LDLr expression and of various additives on A2–PEG liposome binding to and uptake into human fibroblasts. Receptor expression was upregulated by preincubation of the cells with medium containing lipoprotein-deficient serum. For binding and uptake measurements, the cells were incubated with rhodamine-labeled liposomes ($c_l = 100\ \mu M$) at 4°C and 37°C, respectively. Rhodamine fluorescence was registered after 3 h. The NaN$_3$ and 2-deoxyglucose concentrations were 10 and 25 mM, respectively, the heparinase I concentration was 5 units/mL, and the heparin concentration was 1 μg/mL.

approximately 50%, and interactions with cells are completely abolished in the presence of heparin (Table 28.3). Similar effects have been observed for the uptake of P2A2 liposomes into mouse brain capillary endothelial cells, which is adversely affected under conditions of energy depletion (Figure 28.8B) and in the presence of heparin (Figure 28.8C) and HSPG-binding poly-L-lysine (Figure 28.8D). HSPG is also involved in the apoE-related clearance of remnant lipoproteins[61,64] and enhanced internalization of LDL tagged with the sequence AD (Table 28.1).[48,65] Whereas the A2 moiety remains flexible in the rather polar environment of PEG liposomes, adsorptive binding induces a helical conformation. However, the present in vitro results suggest that the conformation of the peptide has little influence on the HSPG-dependent mode of uptake.

28.3.4 RELATION TO THE UPTAKE OF OTHER CATIONIZED CARRIER SYSTEMS

ApoE-derived vector peptides exhibit properties typical of different classes of targeting and uptake-mediating compounds. In addition to the LDLr recognition motif and an overlapping region for binding to heparin, the high cationic charge renders A2 comparable to sequences classified as CPPs. Thus, peptide–liposome complexes and micellar structures might activate different cellular transport routes. The studies presented above point to an efficient endocytotic process involving cell-surface HSPGs and are comparable to those reported for several PTD-modified and cationized cargoes. Antennapedia (Antp)- and HIV TAT-modified liposomes, bearing at least 100 PTD molecules on their surface, bind to and efficiently translocate into different cell types, such as B16 melanoma, F9 teratocarcinoma, and dendritic cells.[66] The uptake strongly depends on HSPG expression on the target cells, confirming the central role of these surface proteins. Also, TAT-fused recombinant proteins,[67] TAT-conjugated peptide nucleic acids,[68] and TAT- and Antp-derivatized liposomes[69] translocate across cell membranes in an HSPG-dependent manner. Cationic peptide–drug chimerae,[70] as well as liposomes covered with cationized albumin,[21] enter brain capillary endothelial cells via HSPG-supported adsorptive endocytosis. Because of the pronounced accumulation of the positively charged CPPs at negatively charged lipid head groups and membrane proteins such as HSPGs via nonspecific electrostatic interactions, the transport efficiency is very high. By contrast, recent reports have demonstrated that TAT peptides on the surface of liposomes afford their intracellular delivery even at low temperature,[47] and HSPGs can even block the uptake of Antp-derivatized liposomes.[69] Another study has recently shown that the most likely route of PTD-mediated cellular uptake is via lipid raft-mediated macropinocytosis.[71]

In conclusion, we have shown that an apoE peptide-derived vector covalently coupled to liposomes or equipped with two adjacent fatty acid chains is a promising candidate for preparing micellar and liposomal vector systems. The efficient uptake of the different colloidal systems into brain capillary endothelial cells emphasizes the potential of the cationic peptides as ligands for improved drug delivery to the brain. However, since the peptide-modified liposomes and micelles bind in an unspecific manner to HSPGs, which are expressed on endothelial cells of all blood vessels, in the extracellular matrix, and on other cell surfaces, the construction of a carrier device combining high target specificity with efficient membrane translocation remains a challenging task. Generally, the compositions and structures of HSPGs of different cell types differ. It remains to be investigated whether high-affinity binding of the cationic residues at position, 142, 143 and 145–147 of apoE to distinct glycosaminoglycan species such as that found in brain[72] can provide a basis for more selective delivery. Furthermore, whether factors such as the size of the vesicles, the nature of the liposome content, or the surface density of the vector peptide, which all play a role in LDLr-mediated cellular responses[27,73] and also entail quantitative differences in HSPG-dependent cellular uptake play a role, has to be investigated in detail.

28.4 EXPERIMENTAL

28.4.1 Peptide Synthesis and Preparation of Vesicles

Peptide synthesis, N-terminal coupling of mercaptopropionic acid,[32,33] modifications with acetic, palmitic, or myristic acid,[32,34] and coupling of 5(6)-carboxyfluorescein-N-hydroxysuccinimide ester (FLUOS) to P2A2[34] were performed as described.

Small and large unilamellar vesicles (SUVs and LUVs, respectively) in buffer (10 mM Tris, 154 mM NaCl, 0.1 mM EDTA, pH 7.4) were obtained by ultrasonication and extrusion, respectively.[32,34] Complexes were prepared by mixing peptide solutions and liposome suspensions at appropriate concentrations. LUVs for cellular uptake studies contained 0.1 mol% Rh-DPPE in the lipid bilayer or 400 μM FITC-dextran WT 4000 inside the vesicles. Covalent coupling of the thiol-modified peptide to PEG liposomes has been described elsewhere.[33]

28.4.2 Biophysical Studies

28.4.2.1 Spectroscopic Methods

The size of the liposomes was determined by right-angle dynamic light scattering. The efficacy of P2A2 in forming lipid-rich micelles was monitored on the basis of the light scattering intensity of its mixtures with POPC.[34]

Circular dichroism (CD) measurements of peptides dissolved in buffer (10 mM Tris, 154 mM NaF, 0.1 mM EDTA, pH 7.4), in buffer/trifluoroethanol (TFE) (1/1 v/v), and in 1 mM POPC SUV suspensions and data evaluation were carried out as described for other peptides.[52]

Peptide–liposome interactions were monitored by measuring the tryptophan fluorescence upon excitation at 285 nm in buffer (10 mM Tris, 154 mM NaCl, 0.1 mM EDTA, pH 7.4) and in the presence of SUVs containing different amounts of the fluorescence-quenching lipid DiBr-PSPC (0–60 mol%).[32,52] Redistribution of peptides from SUVs containing 40 mol% DiBr-PSPC to quencher-free vesicles was monitored as a time-dependent increase in fluorescence intensity after addition of quencher-free SUVs.[32]

To determine the depth of insertion of the apoE peptides into the lipid bilayer, fluorescence quenching experiments using vesicles labeled with 1-palmitoyl-2-stearoyl-(5-DOXYL)-*sn*-glycero-3-phosphocholine (5-DOX) and 1,2-dioleoyl-*sn*-glycero-3-phosphotempocholine (TEMPO) were carried out as described for other peptides.[74] The parallax analysis provides the distance of the tryptophan chromophore from the bilayer center, z_{cf}.[75]

Peptide-induced bilayer permeabilization was monitored fluorimetrically as a decrease in self-quenching of vesicle-entrapped calcein as described for other peptides.[52]

28.4.2.2 Isothermal Titration Calorimetry

Isothermal titration calorimetry experiments were performed with POPC vesicles at 25°C. For binding experiments, 10-μL aliquots of a 30 mM POPC SUV suspension were injected into the 1.4-mL calorimeter cell containing a 20 μM peptide solution. The heat of reaction as a function of the molar amount of injected lipid was used to derive binding isotherms.[76] Membrane solubilization and reconstitution studies are described in detail elsewhere.[34]

28.4.3 In Vitro Experiments

28.4.3.1 Cell Culture and Cell Viability

Immortalized b.End3 mouse brain capillary endothelial cells[77] were cultured in a humidified atmosphere (37°C, 5% CO_2) with Earle's minimal essential medium supplemented with 10% fetal bovine serum, 2 mM N-acetyl-L-alanyl-L-glutamine, 100 units/mL penicillin, and 100 μg/mL streptomycin. Primary human fibroblasts were cultured as described.[32,33] Cell viability of b.End3 cells in the presence of peptide was assessed by the (3-(4,5-dimethylthiazol-2-yl)-2,5-diphenyl)tetrazolium bromide (MTT) method as described for primary rat brain capillary endothelial cells.[33]

28.4.3.2 Binding and Uptake Studies with Human Fibroblasts

The affinities of LDL and apoE peptide (141–150)$_2$ to the LDL receptor were determined in a competition assay with ^{125}I-labeled human LDL using human primary fibroblasts with upregulated LDL receptor expression level.[32,33] Binding to and uptake into human fibroblasts of rhodamine-labeled A2–PEG liposomes were quantified fluorimetrically as described earlier.[32,33]

28.4.3.3 Confocal Laser Scanning Microscopy

For microscopic studies, b.End3 cells were plated on poly-L-lysine-coated glass coverslips with a diameter of 30 mm and maintained under the same conditions as specified in Section 28.4.3.1 for 3 days until semiconfluence was reached.[34] Upon removal of the medium, the cells were washed twice with PBS containing Ca^{2+}, Mg^{2+}, and 1 mg/mL D-glucose and exposed to 1 μM peptide or peptide–liposome complexes (1 mM lipid, 1 μM peptide) for 30 min at 37 or 4°C. The role of HSPG in cellular uptake was investigated by treatment with 10 units/mL heparinase I 2 h before addition of liposomes and by addition of 20 mg/mL heparin or 0.02 μM poly-L-lysine as described.[33] To study the effect of energy depletion, cells were incubated with 10 mM NaN$_3$ and 25 mM 2-deoxyglucose in PBS, pH 7.4, 1 h before addition of liposomes dotted with apoE peptide.[33]

ACKNOWLEDGMENTS

This work was supported by the Deutsche Forschungsgemeinschaft (FG 463) and by the European Commission (QLK-3-CT-2002-01989).

REFERENCES

1. Allen, T.M. and Cullis, P.R., Drug delivery systems: Entering the mainstream, *Science*, 303, 1818, 2004.
2. Maurer, N., Fenske, D.B., and Cullis, P.R., Developments in liposomal drug delivery systems, *Expert. Opin. Biol. Ther.*, 1, 923, 2001.
3. Cattel, L., Ceruti, M., and Dosio, F., From conventional to stealth liposomes: A new frontier in cancer chemotherapy, *Tumori*, 89, 237, 2003.
4. Begley, D.J., The blood-brain barrier: Principles for targeting peptides and drugs to the central nervous system, *J. Pharm. Pharmacol.*, 48, 136, 1996.
5. Bickel, U., Yoshikawa, T., and Pardridge, W.M., Delivery of peptides and proteins through the blood-brain barrier, *Adv. Drug Delivery Rev.*, 10, 205, 1993.
6. Tsuji, A., Small molecular drug transfer across the blood-brain barrier via carrier-mediated transport systems, *NeuroRx*, 2, 54, 2005.
7. Tsuji, A. and Tamai, I., Carrier-mediated or specialized transport of drugs across the blood-brain barrier, *Adv. Drug Delivery. Rev.*, 36, 277, 1999.
8. Visser, C.C. et al., Targeting liposomes with protein drugs to the blood-brain barrier in vitro, *Eur. J. Pharm. Sci.*, 25, 299, 2005.
9. Huwyler, J., Wu, D., and Pardridge, W.M., Brain drug delivery of small molecules using immunoliposomes, *Proc. Natl Acad. Sci. U.S.A.*, 93, 14164, 1996.
10. Cerletti, A. et al., Endocytosis and transcytosis of an immunoliposome-based brain drug delivery system, *J. Drug Targeting*, 8, 435, 2000.
11. Kreuter, J. et al., Apolipoprotein-mediated transport of nanoparticle-bound drugs across the blood-brain barrier, *J. Drug Targeting*, 10, 317, 2002.
12. Lucarelli, M. et al., Expression of receptors for native and chemically modified low-density lipoproteins in brain microvessels, *FEBS Lett.*, 401, 53, 1997.
13. Asai, T. and Oku, N., Liposomalized oligopeptides in cancer therapy, *Methods Enzymol.*, 391, 163, 2005.
14. Janssen, A.P.C.A. et al., Peptide-targeted PEG-liposomes in anti-angiogenic therapy, *Int. J. Pharm.*, 254, 55, 2003.
15. Jain, S. et al., RGD-anchored magnetic liposomes for monocytes/neutrophils-mediated brain targeting, *Int. J. Pharm.*, 261, 43, 2003.
16. Lindgren, M. et al., Cell-penetrating peptides, *Trends Pharmacol. Sci.*, 21, 99, 2000.
17. Snyder, E.L. and Dowdy, S.F., Cell penetrating peptides in drug delivery, *Pharm. Res.*, 21, 389, 2004.

18. Ziegler, A. and Seelig, J., Interaction of the protein transduction domain of HIV-1 TAT with heparan sulfate: Binding mechanism and thermodynamic parameters, *Biophys. J.*, 86, 254, 2004.

19. Fuchs, S.M. and Raines, R.T., Pathway for polyarginine entry into mammalian cell, *Biochemistry*, 43, 2438, 2004.

20. Rousselle, C. et al., Enhanced delivery of doxorubicin into the brain via a peptide-vector-mediated strategy: Saturation kinetics and specificity, *J. Pharmacol. Exp. Ther.*, 296, 124, 2001.

21. Thöle, M. et al., Uptake of cationized albumin coupled liposomes by cultured porcine brain microvessel endothelial cells and intact brain capillaries, *J. Drug Target.*, 10, 337, 2002.

22. Mahley, R.W., Apolipoprotein E: Cholesterol transport protein with expanding role in cell biology, *Science*, 240, 622, 1988.

23. Innerarity, T.L. et al., The receptor-binding domain of human apolipoprotein E. Binding of apolipoprotein E fragments, *J. Biol. Chem.*, 258, 12341, 1983.

24. Rensen, P.C. et al., Human recombinant apolipoprotein E-enriched liposomes can mimic low-density lipoproteins as carriers for the site-specific delivery of antitumor agents, *Mol. Pharmacol.*, 52, 445, 1997.

25. Versluis, A.J. et al., Low-density lipoprotein receptor-mediated delivery of a lipophilic daunorubicin derivative to B16 tumours in mice using apolipoprotein E-enriched liposomes, *Br. J. Cancer*, 78, 1607, 1998.

26. Sparrow, J.T. et al., Apolipoprotein E phospholipid binding studies with synthetic peptides containing the putative receptor binding region, *Biochemistry*, 24, 6984, 1985.

27. Raussens, V. et al., Structural characterization of a low density lipoprotein receptor-active apolipoprotein E peptide, ApoE3-(126-183), *J. Biol. Chem.*, 275, 38329, 2000.

28. Mims, M.P. et al., A nonexchangeable apolipoprotein E peptide that mediates binding to the low density lipoprotein receptor, *J. Biol. Chem.*, 269, 20539, 1994.

29. Dyer, C.A. and Curtiss, L.K., A synthetic peptide mimic of plasma apolipoprotein E that binds the LDL receptor, *J. Biol. Chem.*, 266, 22803, 1991.

30. Dyer, C.A. et al., Structural features of synthetic peptides of apolipoprotein E that bind the LDL receptor, *J. Lipid Res.*, 36, 80, 1995.

31. Nikoulin, I.R. and Curtiss, L.K., An apolipoprotein E synthetic peptide targets to lipo-proteins in plasma and mediates both cellular lipoprotein interactions in vitro and acute clearance of cholesterol-rich lipoproteins in vivo, *J. Clin. Invest.*, 101, 223, 1998.

32. Sauer, I., Apolipoprotein E-abgeleitete Peptide als Vektoren zur Überwindung der Blut-Hirn-Schranke, Ph.D. thesis, Freie Universität, Berlin, 2004, http://www.diss.fu-berlin.de/2004/314/.

33. Sauer, I. et al., An apolipoprotein E-derived peptide mediates uptake of sterically stabilized liposomes into brain capillary endothelial cells, *Biochemistry*, 44, 2021, 2005.

34. Keller, S. et al., Membrane-mimetic nanocarriers formed by a dipalmitoylated cell-penetrating peptide, *Angew. Chem., Int. Ed.*, 44, 5252, 2005.

35. Nobs, L. et al., Current methods for attaching targeting ligands to liposomes and nanoparticles, *J. Pharm. Sci.*, 93, 1980, 2004.

36. Allen, T.M. et al., Adventures in targeting, *J. Liposome Res.*, 12, 5, 2002.

37. Torchilin, V.P. et al., TAT peptide on the surface of liposomes affords their efficient intracellular delivery even at low temperature and in the presence of metabolic inhibitors, *Proc. Natl Acad. Sci. U.S.A.*, 98, 8786, 2001.

38. Torchilin, V.P. and Levchenko, T.S., TAT-liposomes: A novel intracellular drug carrier, *Curr. Protein Pept. Sci.*, 4, 133, 2003.

39. Jeong, L.H. and Pardridge, W.M., Drug targeting to the brain using avidin–biotin technology in the mouse, *J. Drug Target.*, 8, 413, 2000.

40. Oku, N. et al., Liposomal Arg–Gly–Asp analogs effectively inhibit metastatic B16 melanoma colonization in murine lungs, *Life Sci.*, 58, 2263, 1996.

41. Takikawa, M. et al., Suppression of GD1 alpha ganglioside-mediated tumor metastasis by liposomalized WHW-peptide, *FEBS Lett.*, 466, 381, 2000.

42. Westerlund, J.A. and Weisgraber, K.H., Discrete carboxyl-terminal segments of apolipoprotein E mediate lipoprotein association and protein oligomerization, *J. Biol. Chem.*, 268, 15745, 1993.

43. Vogt, B. et al., The topology of lysine-containing amphipathic peptides in bilayers by circular dichroism, solid-state NMR, and molecular modeling, *Biophys. J.*, 79, 2644, 2000.

44. Wimley, W.C. and White, S.H., Designing transmembrane alpha-helices that insert spontaneously, *Biochemistry*, 39, 4432, 2000.

45. Resh, M.D., Fatty acylation of proteins: New insights into membrane targeting of myristoylated and palmitoylated proteins, *Biochim. Biophys. Acta*, 1451, 1, 1999.

46. Pedersen, T.B. et al., Association of acylated cationic decapeptides with dipalmitoylphosphatidylserine–dipalmitoylphosphatidylcholine lipid membranes, *Chem. Phys. Lipids*, 113, 83, 2001.

47. Reig, F. et al., Interfacial interaction of hydrophobic peptides with lipid bilayers, *J. Colloid Interface Sci.*, 246, 60, 2002.

48. Datta, G. et al., Cationic domain 141–150 of apoE covalently linked to a class A amphipathic helix enhances atherogenic lipoprotein metabolism in vitro and in vivo, *J. Lipid Res.*, 42, 959, 2001.

49. Innerarity, T.L., Pitas, R.E., and Mahley, R.W., Binding of arginine-rich (E) apoprotein after recombination with phospholipid vesicles to the low density lipoprotein receptors of fibroblasts, *J. Biol. Chem.*, 254, 4186, 1979.

50. Lund, K.S. et al., Effects of lipid interaction on the lysine microenvironments in apolipoprotein E, *J. Biol. Chem.*, 275, 34459, 2000.

51. Kiyota, T., Lee, S., and Sugihara, G., Design and synthesis of amphiphilic alpha-helical model peptides with systematically varied hydrophobic–hydrophilic balance and their interaction with lipid- and bio-membranes, *Biochemistry*, 35, 13196, 1996.

52. Dathe, M. et al., General aspects of peptide selectivity towards lipid bilayers and cell membranes studied by variation of the structural parameters of amphipathic helical model peptides, *Biochim. Biophys. Acta*, 1558, 171, 2002.

53. Haag, R., Supramolecular drug-delivery systems based on polymeric core-shell architectures, *Angew. Chem., Int. Ed. Engl.*, 43, 278, 2004.

54. Torchilin, V.P. et al., Immunomicelles: Targeted pharmaceutical carriers for poorly soluble drugs, *Proc. Natl Acad. Sci. U.S.A.*, 100, 6039, 2003.

55. Kataoka, K., Harada, A., and Nagasaki, Y., Block copolymer micelles for drug delivery: Design, characterization and biological significance, *Adv. Drug Deliv. Rev.*, 47, 113, 2001.

56. Pitas, R.E. et al., Lipoproteins and their receptors in the central nervous system. Characterization of the lipoproteins in cerebrospinal fluid and identification of apolipoprotein B,E(LDL) receptors in the brain, *J. Biol. Chem.*, 262, 14352, 1987.

57. Kojima, T., Fujimitsu, Y., and Kojima, H., Anti-tumor activity of an antibiotic peptide derived from apoprotein E, *In Vivo*, 19, 261, 2005.

58. Moulder, K.L. et al., Analysis of a novel mechanism of neuronal toxicity produced by an apolipoprotein E-derived peptide, *J. Neurochem.*, 72, 1069, 1999.

59. Clay, M.A. et al., Localization of a domain in apolipoprotein E with both cytostatic and cytotoxic activity, *Biochemistry*, 34, 11142, 1995.

60. Weisgraber, K.H. et al., Human apolipoprotein E. Determination of the heparin binding sites of apolipoprotein E3, *J. Biol. Chem.*, 261, 2068, 1986.

61. Mahley, R.W. and Ji, Z.S., Remnant lipoprotein metabolism: Key pathways involving cell-surface heparan sulfate proteoglycans and apolipoprotein E, *J. Lipid Res.*, 40, 1, 1999.

62. Park, P.W., Reizes, O., and Bernfield, M., Cell surface heparan sulfate proteoglycans: Selective regulators of ligand-receptor encounters, *J. Biol. Chem.*, 275, 29923, 2000.

63. Belting, M., Heparan sulfate proteoglycan as a plasma membrane carrier, *Trends Biochem. Sci.*, 28, 145, 2003.

64. Libeu, C.P. et al., New insights into the heparan sulfate proteoglycan-binding activity of apolipoprotein E, *J. Biol. Chem.*, 276, 39138, 2001.

65. Datta, G. et al., The receptor binding domain of apolipoprotein E, linked to a model class A amphipathic helix, enhances internalization and degradation of LDL by fibroblasts, *Biochemistry*, 39, 213, 2000.

66. Marty, C. et al., Enhanced heparan sulfate proteoglycan-mediated uptake of cell-penetrating peptide-modified liposomes, *Cell Mol. Life Sci.*, 61, 1785, 2004.

67. Silhol, M. et al., Different mechanisms for cellular internalization of the HIV-1 Tat-derived cell penetrating peptide and recombinant proteins fused to Tat, *Eur. J. Biochem.*, 269, 494, 2002.

68. Richard, J.P. et al., Cell-penetrating peptides—a reevaluation of the mechanism of cellular uptake, *J. Biol. Chem.*, 278, 585, 2003.

69. Console, S. et al., Antennapedia and HIV transactivator of transcription (TAT) "protein transduction domains" promote endocytosis of high molecular weight cargo upon binding to cell surface glycosaminoglycans, *J. Biol. Chem.*, 278, 35109, 2003.

70. Rousselle, C. et al., Improved brain delivery of benzylpenicillin with a peptide-vector-mediated strategy, *J. Drug Target.*, 10, 309, 2002.

71. Wadia, J.S., Stan, R.V., and Dowdy, S.F., Transducible TAT-HA fusogenic peptide enhances escape of TAT-fusion proteins after lipid raft macropinocytosis, *Nat. Med.*, 10, 310, 2004.

72. Libeu, C.P. et al., New insights into the heparan sulfate proteoglycan-binding activity of apolipoprotein E, *J. Biol. Chem.*, 276, 39138, 2001.

73. Rensen, P.C. et al., Particle size determines the specificity of apolipoprotein E-containing triglyceride-rich emulsions for the LDL receptor versus hepatic remnant receptor in vivo, *J. Lipid Res.*, 38, 1070, 1997.

74. Dathe, M. et al., Cyclization increases the antimicrobial activity and selectivity of arginine- and tryptophan-containing hexapeptides, *Biochemistry*, 43, 9140, 2004.

75. Chattopadhyay, A. and London, E., Parallax method for direct measurement of membrane penetration depth utilizing fluorescence quenching by spin-labeled phospholipids, *Biochemistry*, 26, 39, 1987.

76. Seelig, J., Titration calorimetry of lipid–peptide interactions, *Biochim. Biophys. Acta*, 1331, 103, 1997.

77. Omidi, Y. et al., Evaluation of the immortalised mouse brain capillary endothelial cell line, b.End3, as an in vitro blood–brain barrier model for drug uptake and transport studies, *Brain Res.*, 990, 95, 2003.

29 Imaging the Influx of Cell-Penetrating Peptides into the Cytosol of Individual Live Cells

Stephen R. Adams and Roger Y. Tsien

CONTENTS

29.1 INTRODUCTION

One of the most surprising recent findings in biology is that certain naturally occurring and synthetic, short peptide sequences (cell penetrating peptides (CPPs)) have the ability to cross the plasma membrane of living cells even when linked to a wide variety of peptides, proteins, and molecular cargo.[1,2] The main effort on CPPs in our laboratory is on tumor-specific, activatable CPPs (ACPPs), in which the CPPs are attached via protease-cleavable linkers to sequences that veto cell uptake.[3] In normal tissues, the linkers remain intact, whereas in tumors expressing appropriate proteases, the linkers are cleaved, allowing the released CPPs to accumulate on and in cells. For imaging in whole animals or patients, the entry mechanism and final location of the cargo are of minor importance, because in vivo imaging lacks subcellular spatial resolution. However, the same mechanism of tumor targeting should also work for chemotherapeutic or radiation-sensitizing cargoes, for which delivery to cell nuclei would be optimal. There is still considerable debate about the mechanism of cell entry and the quantitative efficacy of the many CPPs now available.[4] This is, in part, due to the limitations in methods available to quantitatively measure and discriminate the uptake of CPPs into specific cell compartments such as the cytosol, nucleus, and endosomes. Ideally, such a method would continuously monitor the uptake into individual live cells with high spatial and temporal resolution. We have developed such a method which uses fluorescence resonance energy transfer (FRET) from a genetically encoded cytosolic receptor to a biarsenical fluorophore conjugated to a CPP. Such biarsenicals have a picomolar affinity for short (~10 amino acids) tetracysteine peptides containing the general sequence CCPGCC and have been used to specifically label and visualize recombinant proteins containing this sequence in living cells.[5–8] By targeting a tetracysteine-tagged fluorescent protein to a specific cell compartment, such as the cytoplasm, by standard molecular biology techniques, changes in FRET upon incubation with CPP–biarsenicals permit real-time quantitative measurement of cytosolic uptake in single cells.

29.2 EXPERIMENTAL

29.2.1 SYNTHESIS OF FLAsH–OLIGOARGININE CONJUGATES

Peptides, H_2N-aha-R_7-$CONH_2$, and H_2N-GGR_{10}-$CONH_2$ were synthesized by standard Fmoc solid-phase techniques using a Pioneer peptide synthesizer (Perseptive Biosystems, Framingham, MA). Peptides were cleaved from the support with trifluoroacetic acid (TFA) triisopropylsilane–thioanisole (96:2:2 v/v) overnight and precipitated with cold ether-hexanes. The crude peptides (~2 mM in water, 30 μL) were treated with 5-carboxyFlAsH–EDT_2 succinimidyl ester (2 mM in dry dimethylformamide (DMF); 30 μL), bicine buffer pH 8.5 (50 mM in water; 40 μL) and 1,2-ethanedithiol (10 mM in DMF, 6μL) at 4°C overnight.[7] Alternatively, DMF can be used as a solvent with N-methylmorpholine as the base. The red precipitated crude product was collected by centrifugation, washed with water and dissolved in 50% acetonitrile–water–0.1% TFA and purified by HPLC (Dionex, Sunnyvale, CA) on C_{18} columns (Phenomenex, Torrance, CA) with an acetonitrile–H_2O–0.1% TFA gradient followed by lyophilization. Electrospray mass spectroscopy (ES-MS) (50% MeOH 1% HOAc) positive mode gave for FlAsH-aha-R_7-$CONH_2$, 638.7 (M + 3H$^+$), 479.3 (M + 4H$^+$), 383.9 (M + 5H$^+$) indicating a mass of 1912.38, calculated for $C_{73}H_{114}As_2N_{30}O_{14}S_4$, 1912.64. ES-MS (50% MeOH 1% HOAc) positive mode gave for FlAsH-GGR_{10}-$CONH_2$, 596.7 (M + 4H$^+$), 477.6 (M + 5H$^+$) indicating a molecular mass of 2381.84, calculated for $C_{89}H_{145}As_2N_{43}O_{18}S_4$, 2381.91.

29.2.2 CYTOSOLIC UPTAKE OF FLAsH–OLIGOARGININE CONJUGATES

HeLa cells were transfected with a Cyan Fluorescent Protein (CFP) tetracysteine construct (CFP-AEAAAREACCPGCCARA) using Fugene (Roche Diagnostics, Alameda, CA).[7] Imaging experiments were performed 24 h after transfection with a cooled charge-coupled device camera (Photometrics, Tucson, AZ) controlled by Metafluor software (Molecular Devices, Sunnyvale, CA). Fluorescence was monitored in three channels; CFP, excite at 440 nm (20 nm bandwidth) with emission at 480 nm (30 nm bandwidth); FRET, excite CFP with emission at 635 nm (55 nm bandwidth); FlAsH (Fluorescein arsenical hairpin binder), excite at 495 nm (10 nm bandwidth) with emission at 535 nm (25 nm bandwidth). The 455 nm dichroic mirror used passed sufficient light at 495 nm. Cells were washed with Hanks buffered saline solution (HBSS) containing glucose and treated with 1–2.5 μM FlAsH oligoarginine conjugate at room temperature. The adduct of two 2,3-dimercaptopropanesulfonate (DMPS) with FlAsH-GGR$_{10}$-CONH$_2$ was made by incubation with 100 μM DMPS in HBSS for 15 min. After staining with FlAsH oligoarginine conjugate for 30–60 min, the cells were washed with HBSS, then treated with 5 mM 2,3-dimercaptopropanol (BAL) to disrupt binding of FlAsH conjugates to the CFP tetracysteine. The percentage of CFP tetracysteine bound with FlAsH conjugate was determined by the increase of CFP fluorescence emission after adding BAL. Stoichiometric binding of FlAsH to this construct reduces the CFP emission by 65%.[5] Cellular CFP tetracysteine concentrations were determined by comparison with microcuvets of purified protein of known concentration.[7]

29.3   RESULTS AND DISCUSSION

29.3.1   DETERMINING CYTOPLASMIC UPTAKE RATES OF OLIGOARGININE CONJUGATES

To specifically measure entry of oligoarginine CPPs into the cytoplasm of individual live cells we devised a method (Figure 29.1) utilizing the highly specific interaction of FlAsH, a fluorescein biarsenical for a tetracysteine-containing peptide.[5,7] This genetically encoded tag was fused to CFP and transiently transfected into living

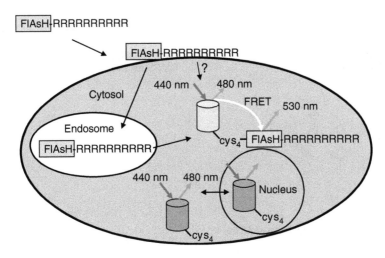

FIGURE 29.1 Scheme depicting the use of a cytoplasmic CFP tetracysteine to detect transduction to the cytosol and nucleus of FlAsH-labeled oligoarginine CPPs. The CFP fluorescence is quenched upon FlAsH binding due to FRET and gives a real-time measurement in single cells of the rate and amount of CPP transduction.

Hela cells to express the recombinant protein in the cytoplasm (and nucleus). The cells were incubated with FlAsH chemically conjugated by its benzoic acid moiety (which does not interfere with binding to tetracysteine peptides) to oligoarginine CPPs (Figure 29.2). If peptide translocation occurs to the cytoplasm, binding of the FlAsH conjugate to the CFP tetracysteine construct will result and be measured by a quench of the CFP fluorescence emission due to FRET to FlAsH. Cytoplasmic CFP concentrations of cells can be estimated by comparison with known standards to give values for the rates of cytoplasmic uptake of FlAsH–oligoarginine conjugates.

Figure 29.3 shows typical cell images and time-courses of such an experiment with the FlAsH–GGR$_{10}$ conjugate. After an initial stable baseline, addition of the peptide results slowly in a slow but steady decrease in CFP emission over approximately 30 min at room temperature. After washing of the cells, the binding of CFP tetracysteine to FlAsH conjugates is rapidly reversed by the addition of a high concentration of the competing dithiol, BAL. The resulting rapid increase of CFP emission is a direct measure of the extent of FRET between CFP and FlAsH.

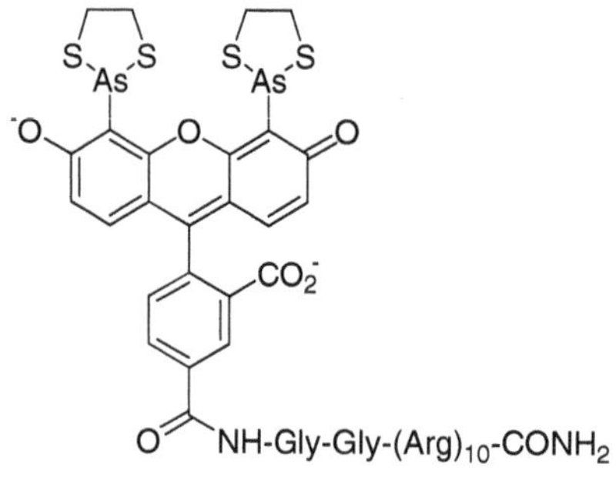

FlAsH-GGR$_{10}$
EDT adduct

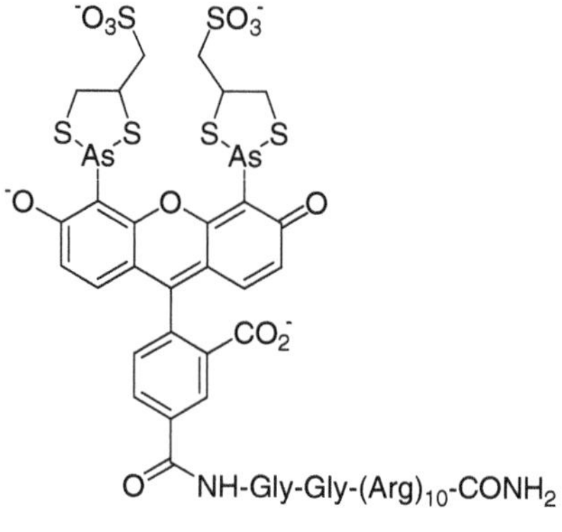

FlAsH-GGR$_{10}$
DMPS adduct

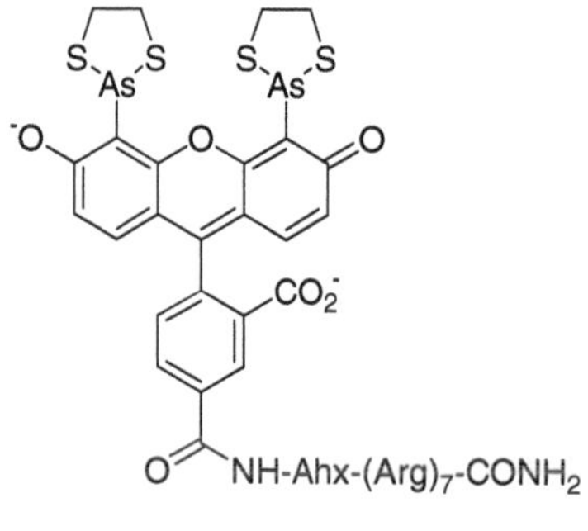

FlAsH-Ahx-R$_7$
EDT adduct

FIGURE 29.2 FlAsH conjugates of R$_{10}$ and R$_7$ CPPs. Each arsenic (III) is complexed to an uncharged (EDT) or negatively charged (DMPS) dithiol.

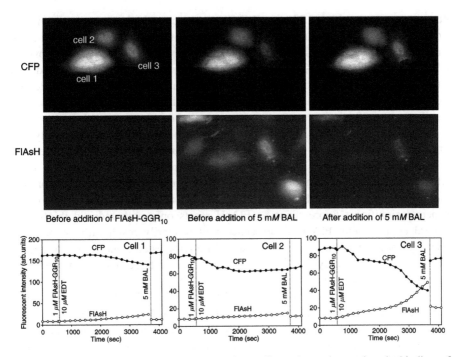

FIGURE 29.3 A typical experiment to monitor cell uptake and cytoplasmic binding of FlAsH–GGR$_{10}$ to CFP tetracysteine (CFP-AEAAAREACCPGCCARA) transiently expressed in HeLa cells. Upper panels: CFP fluorescence (upper images) and corresponding FlAsH fluorescence (lower images) from a field of cells before and after incubation with 1 μM FlAsH–GGR$_{10}$ and 10 μM EDT, and after acute reversal of binding by addition of the dithiol, BAL. Lower panels. Time courses of the average CFP and FlAsH fluorescence from regions encompassing each of the single transfected cells shown in the upper panels.

Knowing the maximal value of FRET upon stoichiometric labeling with FlAsH gives the fraction of CFP tetracysteine proteins bound with FlAsH conjugate during the incubation. For example, the increase in CFP emission in cell 1 in Figure 29.3 on treatment with BAL indicates that an 18% quench of CFP occurred during incubation with the peptide. The intensity of the CFP emission from this entire cell corresponds to approximately 7 μM combined cytosolic and nuclear CFP when compared to microcuvets containing a known concentration of purified CFP with a comparable thickness ($\sim$5 μm) to typical HeLa cells.[7,9] FlAsH labeling of living cells expressing a similar CFP tetracysteine construct gives a 65% quench of the CFP fluorescence upon complete and presumed stoichiometric reaction.[5] Assuming a similar maximal quench of CFP tetracysteine by FlAsH–GGR$_{10}$, the 18% quench in cell 1 after incubation corresponds to $\sim$2 μM FlAsH conjugate bound to cytosolic and nuclear CFP tetracysteine after approximately 50 min of continual incubation at room temperature.

There are substantial cell-to-cell differences in such time-courses even within a single microscopic field that probably reflect cell variability in the kinetics of CPP uptake. For example, cell 1 in Figure 29.3 shows a steady decline in CFP fluorescence after a lag of 20 min. Cell 3 shows a more biphasic response with an immediate slow drop in fluorescence followed by faster uptake. In both of these cases, the fluorescence decrease shows no sign of leveling off even after almost 1 h of incubation with the CPP. However, decreases in CFP emission can also be due to slow changes in cell shape or focus during this prolonged treatment with peptide. For example, cell 2 in Figure 29.3 shows an initial decrease in CFP emission that stabilizes after 30 min but no dequench is revealed on treatment with BAL.

The changes in FlAsH fluorescence during such experiments are generally less informative and harder to quantify than those occurring from CFP. Extracellular FlAsH–GGR$_{10}$, such as FlAsH–EDT$_2$, is nonfluorescent so is not visible upon addition to the cells in solution or upon the subsequent rapid binding to the plasma membrane typical with CPPs. Similarly, uptake into acidic endosomes is unlikely to generate significant fluorescence as FlAsH reacts only with reduced thiols in a neutral environment. FlAsH fluorescence will primarily result (after endosomal release) from formation of complexes with CFP tetracysteine and from nonspecific binding to endogenous cytoplasmic thiols. The latter can be seen in the untransfected cells in the panel of Figure 29.3 that are only visible in the FlAsH image.

To control for any endosomal proteolysis of FlAsH–GGR$_{10}$ and release of a membrane-permeable FlAsH fragment that could bind to the cytoplasmic CFP tetracysteine, and masquerade as cytosolic translocation, we pre-incubated FlAsH–GGR$_{10}$ with excess DMPS. Dithiol exchange rapidly occurs to produce the DMPS bis-adduct containing two additional negatively charged sulfonate groups (Figure 29.2). As expected, FlAsH–DMPS$_2$, prepared similarly from FlAsH–EDT2 and DMPS, is not membrane permeable and when incubated with cells expressing CFP tetracysteine, no binding is detectable (data not shown). However, when incubated with cells, the DMPS adduct of FlAsH–GGR$_{10}$ shows a rate of binding to

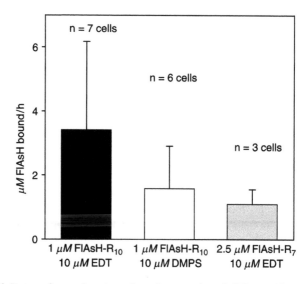

FIGURE 29.4 Rates of cytoplasmic and nuclear uptake of different FlAsH–oligoarginine conjugates into HeLa cells at room temperature.

intracellular CFP tetracysteine that was only slightly lower than with the corresponding EDT adduct (Figure 29.4). This decrease may just reflect the effect of two negative additional charges on the cell uptake of FlAsH–GGR$_{10}$ rather than an indication of endosomal hydrolysis. In support of this, cell uptake of FlAsH–aha– R$_7$ that has a similar net positive charge ($+5$ compared to $+6$ for FlAsH–GGR$_{10}$) was only detectable at higher labeling concentrations of 2.5 μM (Figure 29.4). The lower efficacy of heptaarginine is in agreement with previous studies with fluorescein conjugates.[10]

The major finding of this study is that the overall rates of CPP uptake under these conditions are quite modest, in the range of a few micromolar over a time-course of an hour at room temperature. Therefore, final cytoplasmic concentrations only match, or slightly exceed, the initial extracellular concentration of the conjugates. The extent of CPP uptake is also quite variable between even adjacent and apparently identical cells in the same microscopic field.

29.3.2 Comparison with Other Methods

Two general approaches used to quantify cytosolic and/or nuclear uptake of CPPs are the detection of fluorescent conjugates by nondestructive optical methods or subcellular fractionation of these two compartments upon cell lysis. One of the original, and still most popular, is the measurement of the uptake of fluorescently labeled CPPs by fluorescence activated cell sorter (FACS). This analysis measures the total amount of CPP bound per cell; trypsin treatment following labeling of cells can remove any extracellular peptide bound to the surface (but only if the CPP contains L-amino acids), but signals from endosomal and cytosolic peptide cannot be differentiated. Many early fluorescence microscopy studies were flawed by fixation effects on subcellular distribution on CPPs, but the recent use of confocal microscopy of living cells has proven more reliable, for example, in revealing the endosomal staining predominately found following incubation with CPPs such as Tat and oligoarginine.[11,12] However, quantitatively distinguishing cytosolic/nuclear localization from that of endosomal is difficult because these organelles are below the resolution of light microscopy and the endosomal CPPs are likely to be partly self-quenched by molecular crowding or by low pH. A similar FRET-based method uses CPPs containing a quenching nitrotyrosine group attached to fluorophore by a disulfide linker.[13] Fluorescence was generated only when the disulfide bond is cleaved by entry of the CPP into a reducing environment, assumed to be the cytoplasm. Using this method, millimolar intracellular concentrations of CPPs were measured, in contrast to our results. However, disulfide bonds can be reduced within endosomes, not just in the cytoplasm, so the difference in apparent uptake may reflect CPPs remaining within endosomes.[14] Furthermore, the low fluorescence and UV excitation of the aminobenzoyl fluorophore limits this method to cell suspensions.

A recently reported quantitative single-cell assay for cytosolic uptake detects phosphorylation of Tat-conjugated peptide substrates of cytoplasmic kinases such as CaMKII and PKB.[15] A laser pulse disrupts individual cells and the intracellular peptides are separated and quantified by capillary electrophoresis. The rates of CPP uptake, consistent with our results, showed micromolar cytoplasmic levels of the peptides achieved by a 10 min loading followed by 1 h of incubation to complete endosomal to cytoplasmic trafficking. Zaro and Shen[16] describe a method for determining the extent of endosomal and cytoplasmic CPP using subcellular fractionation of radiolabeled oligoarginine peptides followed by size-exclusion chromatography. A fluorescent dextran applied with the CPP was used as a marker of endocytosis and to correct for the amount of endosomes lysed during processing, but the final cytoplasmic concentration was not calculated. This method is a destructive assay applicable only to large populations of cells and critically dependent on the quality of subcellular fractionation.

One limitation of our method is that cytosolic trapping of CPP would reduce or suppress any efflux from the cytosol, so only unidirectional entry is measurable. Also, any nonspecific binding of the biarsenical to sites other than the CFP tetracysteine would cause an underestimation of the amount of cargo delivered.[6]

Future work with this method will investigate the dependence of cytosolic uptake upon CPP concentration and endosomal function to probe the molecularity of uptake and its biophysical mechanism. The effect of different cargoes on CPP uptake, and a comparison of oligoarginines with other CPPs such as Tat or penetratin, would also be of interest. Judicious modifications of the cargo may also permit us to monitor endosomal versus cytosolic CPPs simultaneously.

ACKNOWLEDGMENTS

We would like to thank Dr. Tao Jiang for providing the oligoarginine peptides. This work was supported by National Institutes of Health Grants NS27177 and GM072033.

REFERENCES

1. Magzoub, M. and Graslund, A., Cell-penetrating peptides: From inception to application, *Q. Rev. Biophys.*, 37, 147, 2004.
2. Joliot, A. and Prochiantz, A., Transduction peptides: From technology to physiology, *Nat. Cell Biol.*, 6, 189, 2004.
3. Jiang, T. et al., Tumor imaging by means of proteolytic activation of cell-penetrating peptides, *Proc. Natl Acad. Sci. U.S.A.*, 101, 17867, 2004.
4. Brooks, H., Lebleu, B., and Vives, E., Tat peptide-mediated cellular delivery: Back to basics, *Adv. Drug Deliv. Rev.*, 57, 559, 2005.
5. Griffin, B.A., Adams, S.R., and Tsien, R.Y., Specific covalent labeling of recombinant protein molecules inside live cells, *Science*, 281, 269, 1998.
6. Griffin, B.A. et al., Fluorescent labeling of recombinant proteins in living cells with FlAsH, *Methods Enzymol.*, 327, 565, 2000.
7. Adams, S.R. et al., New biarsenical ligands and tetracysteine motifs for protein labeling in vitro and in vivo: Synthesis and biological applications, *J. Am. Chem. Soc.*, 124, 6063, 2002.
8. Gaietta, G. et al., Multicolor and electron microscopic imaging of connexin trafficking, *Science*, 296, 503, 2002.

9. Miyawaki, A. et al., Dynamic and quantitative Ca^{2+} measurements using improved cameleons, *Proc. Natl Acad. Sci. U.S.A.*, 96, 2135, 1999.
10. Wender, P.A. et al., The design, synthesis, and evaluation of molecules that enable or enhance cellular uptake: Peptoid molecular transporters, *Proc. Natl Acad. Sci. U.S.A.*, 97, 13003, 2000.
11. Richard, J.P. et al., Cell-penetrating peptides. A reevaluation of the mechanism of cellular uptake, *J. Biol. Chem.*, 278, 585, 2003.
12. Ziegler, A. et al., The cationic cell-penetrating peptide CPP (TAT) derived from the HIV-1 protein TAT is rapidly transported into living fibroblasts: Optical, biophysical, and metabolic evidence, *Biochemistry*, 44, 138, 2005.
13. Hallbrink, M. et al., Cargo delivery kinetics of cell-penetrating peptides, *Biochim. Biophys. Acta*, 1515, 101, 2001.
14. Saito, G., Swanson, J.A., and Lee, K.D., Drug delivery strategy utilizing conjugation via reversible disulfide linkages: Role and site of cellular reducing activities, *Adv. Drug Deliv. Rev.*, 55, 199, 2003.
15. Soughayer, J.S. et al., Characterization of TAT-mediated transport of detachable kinase substrates, *Biochemistry*, 43, 8528, 2004.
16. Zaro, J.L. and Shen, W.C., Quantitative comparison of membrane transduction and endocytosis of oligopeptides, *Biochem. Biophys. Res. Commun.*, 307, 241, 2003.

30 Microbial Membrane-Penetrating Peptides and Their Applications

Anita Ganyu, Ülo Langel, and Liam Good

CONTENTS

30.1   INTRODUCTION

Peptide entry into microorganisms is surprising because microorganisms possess multilayered protective cell barriers. Nevertheless, many natural host defense antimicrobial peptides kill pathogens by causing cell leakage and inhibition of intracellular targets. Interestingly, cell-penetrating peptides (CPPs) discovered in mammalian cell studies are structurally similar to antimicrobial peptides and show similar microbial cell entry and antimicrobial effects. Therefore, there are opportunities to use CPPs as antimicrobials with improved host cell distribution properties. Also, CPPs provide a unique opportunity to deliver a range of foreign substances into cells.

Peptides are generally considered to be too large to pass microbial cell barriers, which are typically more stringent than mammalian cells. Nevertheless, there are many clear examples of peptides that are able to penetrate microbial cells without damage to host cells. Indeed, most of our knowledge of CPPs and their potential practical applications stems from the discovery that the innate immune response in animals, plants, and insects involves the release of antimicrobial peptides, which can kill microbes with impressive potency and specificity. Many research groups have sought to develop antimicrobial peptides as therapeutic agents for infectious disease treatment. The motivation for this effort is that drug resistance to conventional antimicrobials is a serious problem that can extend to "stand-by" antibiotics and even the newly introduced oxazolidinones.[1,2] To overcome this problem, antimicrobial peptides, or their derivatives, offer some exciting possibilities, and antibiotic peptides may prove to be less susceptible to resistance mechanisms.[3] A second possible application is to use penetrating peptides to deliver foreign substances into microorganisms.[4] The idea seems reasonable given that many antimicrobial peptides enter cells and act on intracellular targets.[5–7]

This chapter describes the membrane structure of diverse microbes, the origin and discovery of microbial CPPs, their antimicrobial effect, and capacity to deliver cargo into cells. The chapter also describes methods to characterize the uptake of peptides into microbes, and practical applications for CPPs.

30.2   BACKGROUND

30.2.1   Microbial and Mammalian Cell Barriers

The cell membrane is formed by a phospholipid bilayer, and such bilayers are generally impermeable to fluids, ions, and most foreign substances. In addition, microbial cell bilayers are coated with an extracellular matrix that provides a largely inert structural framework and protective barrier. This would seem to isolate the cell from its environment; however, cell membranes are in fact semipermeable to small molecules. The composition of both the bilayer and the extracellular matrix varies with cell type, and these differences provide a basis for the cell type specificity of CPPs. Figure 30.1 illustrates the cell barriers of different cell types.

The cell barriers of prokaryotic cells are multilayered structures that provide stringent protection against foreign substances.[8] In the case of Gram-positive

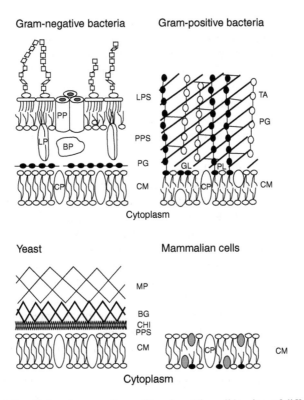

FIGURE 30.1 Microbial and mammalian cell barriers. The cell barriers of different cell types are illustrated. LPS, lipopolysaccharide; PG, peptidoglycon; PL, phospholipid; GL, glycolipid; PPS, periplasmic space; CM, cytoplasmic membrane; BP, braun lipoprotein; LP, lipoprotein; CP, carrier protein; PP, pore protein; TA, teichoic acid; MP, mannoprotein; βG, beta-glucan; CHI, chitin.

bacteria, the cell barrier consists of a phospholipid bilayer covered by a thick peptidoglycan layer and an outer capsule that varies in thickness between species. In the case of Gram-negative bacteria, the cell barrier consists of two phospholipid bilayers, with the inner membrane surrounded by a peptidoglycan layer and the outer membrane coated on the exterior by a lipopolysaccharide (LPS) layer (Figure 30.1).[9] The LPS layer provides Gram-negative bacteria with a particularly stringent sieve against foreign substances. Fungal plasma membranes are enriched in inositol phosphoceramides and a thick outer protective cell wall composed of β-1,3 glucan and β-1,6 glucans and chitin.[10] Fungal cell walls can comprise as much as 40% of the dry cell mass. Therefore, fungi are similar to bacteria in having thick and complex barriers against CPPs and other foreign substances.

30.2.2   CPPs Enter a Range of Microbial Cells

Antimicrobial peptides were discovered as a group of small, cationic peptides having potent bactericidal and fungicidal properties.[11–13] Antimicrobial peptides have been isolated from most vertebrate and invertebrate materials studied, such as the blood and epithelia of animals, plants, hemolymph of insects, and venom of scorpions; see the online databases: AMSDs (www.bbcm.univ.trieste.it/~tossi/antimic.html), ANTIMIC (http://research.i2r.a-star.edu.sg/Templar/DB/ANTIMIC),[14] and APD (http://aps.unmc.edu/AP/main.html).[15] The diverse occurrence and similar structure and activity of antimicrobial peptides indicate an ancient origin and conserved role in host immunity to pathogens.[16,17] Indeed, during the last two decades antimicrobial peptides have become well recognized as important aspects of innate and humoral immune responses.

CPPs are structurally similar to membrane-active antimicrobial peptides and may share common mechanisms of interaction with membranes.[18] CPPs enter a wide range of mammalian cells and microbial cells, and some have antimicrobial properties. Interestingly, CPPs also penetrate into plant cells.[19,20]

The first CPP discovered was the homeodomain of Antennapedia (Antp) protein of *Drosophila melanogaster*.[21] The discovery of CPPs arose with the observations that certain proteins contain distinct regions responsible for translocation through cell membranes. This was followed by the design of artificial CPPs to improve the transmembrane trafficking properties of the peptides. Many CPPs were derived from sequences of membrane-interacting proteins, such as fusion proteins, signal peptides, transmembrane domains, and antimicrobial peptides.

Most uptake experiments with CPPs have been performed using mammalian cells. CPPs internalize in living mammalian cells without causing damage in the membrane structure. For example, peptide LL-37 was shown to bind DNA, and the LL-37–DNA complex enters mammalian cells via endocytosis that involves lipid rafts and cell surface proteoglycans.[22] Several primary cell types have been shown to internalize CPPs, including rat brain and spinal cord cells,[23] porcine and human umbilical vein endothelium,[24] calf aorta,[25] and osteoclast culture[26]; however, most often established cell lines have been used (reviewed in Ref. [29]). CPPs are able to deliver cargo molecules into the cytoplasm and nucleus of a range of cell types. The cargo molecules may be oligonucleotides,[28] peptide nucleic acids,[29] proteins,[30] plasmids,[31] antibodies,[32] liposomes,[33] nanoparticles,[34] adenoviruses,[35] and siRNAs.[36] CPPs can also enter microorganisms, and several show antimicrobial effects, including bactericidal, fungicidal,[37] or virucidal properties.[38,39] Examples of known antimicrobial peptides and CPPs are shown in Table 30.1.

30.2.2.1 Bacteria

Certain short cationic peptides are membrane-active against bacteria. The mechanism of antimicrobial and bacteriolytic effects were first associated with marked membrane damage and leakage of cell contents.[40,41] Subsequently, it was shown that peptides can also cross microbial membranes without causing irreparable damage. Therefore, antimicrobial mechanisms may involve both membrane and intracellular effects.[42] Table 30.2 lists several examples of antimicrobial peptides with intracellular targets in bacteria. Such activities clearly require cell entry, and cell uptake has been observed by fluorescence microscopy and cell-permeabilization studies. Therefore, it is clear that antimicrobial peptides can enter a range of

TABLE 30.1
Examples of Peptides that Enter Mammalian and Microbial Cells

	Antimicrobial Effect	Enter Mammalian Cells	Enter Bacteria	Enter Fungi
Antimicrobial peptides	*All*	PR-39[100]	Bac5[103]	PAF[106]
		Protegrin 1[101]	Bac7[103]	Histatin 5[70]
		Magainin 2[102]	PR-39[5]	Pn-AMP1[47]
		Buforin 2[102]	Buforin 2[104]	
			Indolicidin[105]	
Cell-penetrating peptides	pVEC[37]	*All*	(KFF)$_3$K[94]	TP10[37]
	TP10[37]		pVEC[37]	Erns[48]
			TP10[37]	pVEC[48,49]

Gram-negative and Gram-positive bacteria, including *Escherichia coli, Staphylococcus aureus*, and *Mycobacterium smegmatis*. CPPs were discovered in mammalian cell studies; however, they are structurally similar to antimicrobial peptides and several well-known CPPs have been shown to enter a range of bacterial species.[37] Also, as observed in mammalian systems, CPPs can be used to deliver nucleic acid and protein into bacteria.[4,43–46]

30.2.2.2 Yeasts

Several antimicrobial peptides have intracellular targets in yeasts,[47] and CPP are able to enter *Saccharomyces cerevisiae, Candida albicans*, and *Schizosaccharomyces*

TABLE 30.2
Microbial-Penetrating Peptides with Intracellular Targets

Putative Intracellular Target	Peptide
Lipid II	Nisin[107]
DNA and protein synthesis	PR-39[7]
	Pleurocidin[108]
	Dermaseptin[108]
	Indolicidin[105]
DNA/RNA	Buforin 2[104]
	Tachyplesin[109]
Mitochondria	Histatins[110]
Respiration	Bac5[103]
	Bac7[103]
Serine proteases	ENAP-2[111]

pombe.[37,48,49] An important limiting factor in the case of yeast appears to be degradation by extracellular proteases. To date, there have been only a few reports of attempts to use CPPs to deliver cargo molecules into yeasts[43,50] but this remains an interesting area for further studies.

30.2.2.3 Viruses

Several antimicrobial peptides, including defensins[51] or indolicidin,[52] have apparent virucidal effects, possibly through a membrane-mediated mechanism. Tripathi et al.[39] have recently shown that an anti-transactivator response region (TAR) polyamide nucleotide analog (PNA$_{TAR}$) of HIV-1 conjugated with different CPPs, are not only inhibitory to HIV-1 replication in vitro but are also potent virucidal agents. These surprising results suggest that the PNA$_{TAR}$–peptide conjugates may penetrate the virions and bind to the TAR region of the viral RNA or interfere with viral entry by altering or disrupting the viral envelope.[38,39]

30.2.3 Composition, Structure, and Classes of Microbial CPPs

According to one possible system, CPPs are divided into two classes: amphipathic helical peptides, such as transportan and the model amphipathic peptide (MAP), where lysine is the main contributor to the positive charge, and arginine-rich peptides, such as HIV–TAT, Antp, or penetratin.[53] The naturally occurring antimicrobial peptides are normally 12–50 amino acids long (occasionally up to 80 amino acids), and together with the CPPs, share two main characteristics. First, they are polycationic with a positive net charge of more than $+2$ (due to an excess of primarily lysine and arginine over anionic residues). Second, they have the ability to form amphipathic structures with both positively charged and hydrophobic faces (see Figure 30.2). The hydrophobic content is generally approximately 50% and this face promotes interaction with membrane lipids.

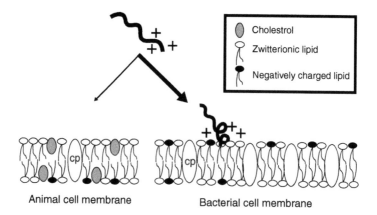

FIGURE 30.2 Peptide–host membrane interactions. Proposed basis for cell type-specific interaction based on cell surface charge differences between microbial and host cells.

The positively charged face promotes interaction with water, negatively charged lipids, and cell wall structures. This structure is ideal if a molecule is to be both soluble in water and lipophylic.

30.3 MECHANISMS OF ACTION

30.3.1 CELL UPTAKE

30.3.1.1 Cell Permeation by Cationic Peptides

In broad terms, microbial cell penetration by peptides can be seen as a three-step process. The first event is an electrostatic attraction of the peptide to the membrane surface (Figure 30.2). The second event involves membrane disruption and penetration, and possibly some reorganization of the peptide structure. As peptides accumulate at the surface and start to displace lipids, this would eventually lead to local disruptions of the membrane (carpet model) or perhaps the formation of transient peptide–lipid wormhole pores.[54–56] Such disorganization or pore formation could allow cellular components to leak out and weaken the bacterial wall, creating opportunities for peptides to translocate into the inner membrane and subsequently into the cytoplasm. It also seems possible for peptides to translocate without causing significant membrane disruption.[57,58] Finally, in the third step, the peptide passes the membrane and gains access to intracellular processes.

More specifically, cationic peptide uptake into bacteria has been described as a self-promoted process, reflecting the fact that uptake is receptor independent.[59] According to this view, cationic peptides would contact the outer membrane through electrostatic attraction between positively charged antimicrobial peptides and negatively charged membrane lipids and the LPS layer. At this point, peptides could compete for divalent cation binding sites within the outer cell membrane. In E. coli, such competition would disrupt stabilizing crosslinks between LPS molecules and therefore reduce outer membrane integrity. This would provide opportunities for peptides to enter the membrane and the membrane charge gradient (inside negative) would then force cationic peptides into the cytoplasm.

There are several models for the uptake of cationic peptides into bacteria; however, in each case electrostatic attraction is the first step. Consistent with this view, increased salt concentrations shield the interaction and reduce peptide potency. Also, resistance can arise from cell-barrier modifications that limit permeation. For example, resistant salmonella cells contain modified LPS molecules with reduced membrane negative charge density.[60] Thus, an initial electrostatic attraction is the first step in cationic peptide entry into microbes. Numerous synthetic cationic peptides in the range of 10–20 amino acids have been made with variations in their amphipathic structure. The peptides have been subjected to a variety of biochemical analysis with artificial membranes in order to elucidate the mechanism of action, as reviewed.[54–56,61]

30.3.1.2 Receptor-Mediated Transport of Peptides into Microbes

There is a variety of peptide sequences known to enter microbial cells via receptor-mediated translocation or permeases, providing opportunities to deliver foreign substances into microbes. Receptor-mediated delivery is a very attractive approach for cellular delivery because of the potential for cell type specificity. The best-known receptor-mediated transport mechanisms in microorganisms involve oligopeptide permeases, which import di and tri oligopeptides. There are multiple oligopeptide permeases in microbial cells, and overlapping substrate specificity between species. For more information on receptor-mediated uptake we direct readers to two excellent reviews.[62,63] Endocytosis also offers a range of attractive possibilities for cell delivery.[64,65] For example, in fungi, the endocytosis signal peptide NPFSD was shown to deliver cargo molecules into cells.[50]

30.3.2 CELL-KILLING MECHANISM

30.3.2.1 Membrane Leakage

Membrane leakage is usually referred to as the main cytotoxic activity of antimicrobial peptides. However, while it is clear that antimicrobial peptides interact with membranes, there are several hypotheses to explain how this could lead to cell death. These include physical disturbance of membrane integrity, fatal depolarization of energized bacteria, or leakage of essential intracellular components.[56,66,67]

30.3.2.2 Intracellular Target Inhibition

Early studies with antimicrobial peptides indicate that the mechanism of action could involve binding to intracellular targets, and many more examples of this mode of action have been reported (Table 30.2). Several antimicrobial peptides, including defensins, interact with heat shock protein (HSP) 70 and DNA.[68,69] These peptides can also block general functions, including protein folding and protein synthesis. For example, PR-39 rapidly (within minutes) stops DNA and protein synthesis in *E. coli*, apparently without causing cell lysis.[7] Indeed, there are several examples of antimicrobial peptides with intracellular targets, such as transcription, translation, protease activity, and mitochondrial function[5–7,70–72] (see also Table 30.2).

30.3.2.3 Dual Activities of Microbial CPPs

Many conventional antibiotics appear to act via two or more mechanisms, and CPPs can also act in multiple ways to kill cells. The two applications for CPPs described below, cell permeabilization and cell delivery, can potentially be combined for greater overall activities. Indeed, cell-wall permeabilization and toxic compound delivery could provide dual antimicrobial effects for improved overall activity. It is also possible to recombine domains from different antimicrobial peptides to obtain new activity profiles, such as the cecropin/melittin hybrid and derivatives, which appear to have two distinct activity domains.[75,76] The recombined peptide shows greater overall activity, and it appears that more than one cell-killing mechanism is involved. The design rules for such domain-shuffling experiments remain uncertain, but there are clearly opportunities to create more active hybrids. For example, in our attempts to develop antimicrobial peptide–PNA conjugates, one can envision improved activity by invoking efficient cell-wall permeabilization together with efficient antisense effects against an essential gene.[4]

30.3.3 Cell Type-Specific Membrane Activities

If microbial CPPs are to prove useful in therapeutics, cell-penetrating properties must be largely specific for microbial cell membranes. Cell type specificity is an important feature of conventional antibiotics that allows pathogen killing at concentrations that are not cytotoxic to host cells.[75] The activity of natural antimicrobial peptides and the safety of foods that contain such peptides indicates that antimicrobial peptides possess the specificity needed to be useful as therapeutics.[5] Indeed, antimicrobial peptides and their derivatives display a therapeutic window in which they can kill microbial cells without damaging host cells. How can CPPs pass mammalian cells without causing membrane damage but kill microbial cells? Bacterial membranes have a more negatively charged outer leaflet than mammalian cells, and sterol content differs between mammalian and microbial membranes. These features could alter membrane stability and affect phase transitions needed for membrane passage and cell type-specific effects.[37]

30.4 METHODS

30.4.1 Cell Uptake Assays

Peptide uptake into cells can be assayed by both direct and indirect methods. Early studies on microorganisms showed that the growth response of amino acid auxotrophic mutants to peptides can be used to indicate peptide entry. A more direct approach is to use radioactively labeled peptides and cell fractionation to assess localization and uptake kinetics. However, cell fractionation can be difficult with small microbial cells. Many researchers now prefer uptake assays based on colorimetric or fluorescent probes. Below, we discuss several cell-based approaches to observe or monitor peptide permeation and cell entry into microorganisms.

30.4.1.1 Fluorescence Microscopy and FACS Analysis

The introduction of a range of flexible fluorescent probes, which can be attached covalently to peptides, enables direct observation of peptide entry by using fluorescence microscopy and fluorescence-activated cell sorting (FACS). In broad terms, fluorescence microscopy can be used to indicate localization within a relatively small number of cells and FACS analysis can be used to profile uptake into a large number of cells. In our efforts to discover peptides that penetrate microbial cells, we have attached peptides to fluorescent probes and studied cell uptake using fluorescence microscopy and FACS analyses.[48]

30.4.1.2 Cell-Permeabilization Assays

As mentioned previously in the chapter, most antimicrobial peptides possess potent cell-wall activity, and this often appears to be the major mechanism involved in cell killing. In other cases, killing also involves binding to intracellular targets (see Table 30.2). Therefore, whether the cell-wall activity involves permeabilization to cause cell leakage or permeation to gain access to the cell interior, some degree of membrane disturbance is involved. The most straightforward method to assess this disturbance is to use small fluorescent molecules or chromogenic probes that normally are excluded by cells but can gain access at points where the membrane is disturbed. By monitoring the cell entry and chemical conversion of such small molecule probes, it is possible to gain insight into permeabilization kinetics and the conditions that affect activity.

There are several fluorescent probes that can be used to probe microbial membrane disturbance or permeabilization. For example, the hydrophobic fluorescent probe N-phenol naphalene (NPN) is an indicator of membrane integrity. In $E.\ coli$ and other microbes, NPN is normally excluded from $E.\ coli$ by the outer membrane LPS layer but can enter at points of membrane damage. NPN has a low fluorescence quantum yield in an aqueous solution, but fluoresces strongly in the hydrophobic environment of a biological membrane. An alternative is to use propidium iodide, which fluoresces strongly if able to enter cells and bind nucleic acids.[76]

Gram-negative bacteria, which possess two cell membranes, are doubly challenging when attempting to assess membrane permeabilization. Nevertheless, in cells that express β-galactosidase and β-lactamase, it is possible to monitor permeabilization of both membranes at the same time. For example, permeabilization of the outer membrane can be monitored using nitrocefin as a probe. Nitrocefin is normally excluded by the outer cell membrane, but if able to pass this barrier can be cleaved by β-lactamase localized within the periplasmic space. Cleavage results in a color change from yellow to red, and this can be used to monitor outer membrane permeabilization. In a similar way, permeabilization of the inner membrane can be monitored by using the β-galactosidase substrate o-nitrophenyl-β-galactoside (ONPG) as a probe. In the $E.\ coli$ strain ML35p, which lacks lac permease, ONPG is blocked from cell entry by the inner membrane, but if able to pass this barrier ONPG can be cleaved by β-galactosidase localized within the cytoplasm, and this results in a color change from clear to yellow.[77] Figure 30.3 illustrates examples of $E.\ coli$ cell permeabilization by the well-known permeabilizing peptide polymyxin B. Permeabilization assays based on small molecule probes are relatively simple and provide useful information about peptide activities and permeabilization kinetics in a variety of environments. The main disadvantage is that membrane disturbance is indicated only indirectly.

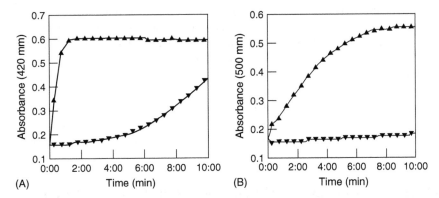

FIGURE 30.3 Indirect monitoring of peptide uptake into microorganisms. Concurrent measurements of *E. coli* outer and inner cell membrane permeabilization by polymyxin B. The symbols represent trials in the presence (▲) and absence (▼) of polymyxin B. (A) Increased absorbance indicates nitrocefin cleavage and permeabilization of the outer cell membrane. (B) Increased absorbance indicates ONPG cleavage and permeabilization of the inner cell membrane.

30.5 APPLICATIONS

30.5.1 CPPs as a New Source of Anti-infective Agents for Medicine

There is an increasing need for new antimicrobial agents to combat resistant strains in the clinic, and antimicrobial peptides offer attractive alternatives for certain applications. Most importantly, antimicrobial peptides are active against strains that show antibiotic resistance, and resistance to antimicrobial peptides does not appear to arise rapidly.[78,79] This is probably because antimicrobial peptide-killing mechanisms are largely distinct from those of small molecule antibiotics, and antimicrobial peptides are typically cidal rather than static above their threshold concentration for activity. Another advantage with antimicrobial peptides is that their cell-penetrating activities can potentiate conventional antibiotics when used in combination.[78]

To develop antimicrobial peptides as drugs there are several important difficulties to consider. Natural antimicrobial peptides are delivered to the place of infection by white blood cells that produce the peptides, whereas peptides used as drugs must find their own way, and typically show poor stability and distribution properties in the body. CPPs may provide a feasible approach to overcome these problems as they enter mammalian cells and distribute in vivo with high efficiency.[35,80] A preferential uptake of peptides by certain cell types, already shown for HIV Tat protein, may help the in vivo targeting of CPPs or CPP-conjugated cargo molecules.[81] Several well-known CPPs have been shown to enter a range of bacterial species and possess antimicrobial activity.[37] Therefore the mammalian cell-penetration properties of CPPs and the well-known cell permeabilization properties of cationic antimicrobial peptides can coexist within a single cationic and amphipathic peptide, and it may be possible to use the host distribution properties of CPPs to improve antimicrobial peptide activity in vivo.

Other problems are the uncertain design and development rules for antimicrobial peptides; in vitro studies provide only poor indications of in vivo efficacy, and peptides can be expensive to produce and purify. Due to these problems, screening efforts that show initial success may prove ultimately disappointing. Nevertheless, evidence from animal studies and clinical trials show that peptides can provide effective antimicrobials.[82–86] Also, many problems that limit efficacy in vivo and efficient large-scale production have been solved.[5] Finally, it has been very difficult to develop effective new small molecular weight antimicrobials that do not succumb to resistance.[87,88] For the future, it seems likely that we will soon see drug approval for antimicrobial peptides for topical therapy.

30.5.2 Microbial CPPs as Carriers for Foreign Substance Delivery

Microbial cells are typically very resistant to foreign substance uptake. As discussed above, this resistance is provided by the outer cell wall, which acts as a sieve that blocks the entry of many harmful compounds found in microbial environments.[3,8] Unfortunately, these barriers also limit the cell uptake of antimicrobial agents that otherwise would be useful in medicine. Most antibiotics and antimicrobials used in the clinic are small (molecular weight < 500). The observation that antimicrobial peptides can penetrate cell barriers is surprising and provides the possibility of using these peptides to expand the range of substances that could be used as drugs against microorganisms.

The idea of using peptides to deliver foreign substances into microorganisms was introduced almost 30 years ago. The initial idea was to use two or three residue peptides that are substrates for oligopeptide permeases and link impermeant toxic compounds to the peptides for delivery (Figure 30.4). Impressive

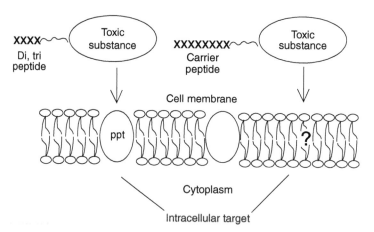

FIGURE 30.4 Generic strategies for peptide-mediated delivery in microbes. A carrier peptide attached covalently by a flexible linker to a toxic domain. Di and tri peptides could pass through core protein complex or peptide permease transporters. CPPs can be used to deliver proteins and other substances into microbes, although the mechanisms involved remain uncertain.

results were reported for delivery into *E. coli* and *Salmonella typhimurium*, and the idea was developed to produce the therapeutic alafosfalin, which showed modest clinical success.[89,90] A limitation with the oligopeptide permease delivery approach, however, is that the permeases only inefficiently transport substrates with cargoes attached.[63] Another limitation is that oligopeptide permease substrate selectivity mutates rapidly, providing a resistance mechanism against this approach.

A related idea is to use cell wall-penetrating peptides as vehicles to deliver toxic substances into microbes. At first glance it seems illogical to use peptides as delivery vehicles, because they are themselves relatively large in comparison to most drugs. Indeed, peptides are normally not considered to be cell permeable. However, natural antimicrobial peptides provide a clear precedent for this approach. Many natural antimicrobial peptides kill target cells, not by permeabilization alone but also by inhibition of intracellular targets, as discussed previously in this chapter. It seems clear that certain peptides can penetrate cells to reach their main target; and it should be possible to use such carrier domains to deliver attached foreign substances.

CPPs have been used to deliver a wide range of foreign proteins and oligonucleotides into mammalian cells (see chapters 1–6, chapter 13, chapter 17, and chapters 20–24), and microbial CPPs offer attractive possibilities for antimicrobial development. The challenge now is to identify peptides that can efficiently carry attached cargo molecules into cells with a high degree of microbial cell specificity. In our first attempt to use antimicrobial peptides as delivery vehicles, the objective was to improve the cell uptake and activity of antisense peptide nucleic acids (PNAs) in *E. coli*. Antisense agents are designed to inhibit gene expression through sequence-specific nucleic acid binding and this normally requires the formation of 10 or more base pairs. Oligonucleotides of such lengths are typically too large for efficient passive cellular uptake by diffusion across lipid bilayers and cellular outer membrane structures can pose additional barriers.[91] Recently, we introduced PNAs as antisense agents for bacterial applications.[92] The early experiments indicated sequence-selective effects against reporter genes; however, uptake was limited by bacterial cell barriers.[93] This limitation is not surprising as *E. coli* and other Gram-negative bacteria have outer and inner bilayer membranes, and the outer membrane contains a lipopolysaccharide layer that stringently restricts the entry of foreign molecules.[8]

We first selected a range of short, cationic peptides with cell-permeabilizing or cell-penetrating activities against *E. coli*. These peptides were attached to PNAs via flexible linkers.[4] Several of the peptides selected, in particular the KFFKFFKFFK peptide, provide efficient carriers for PNAs. Indeed, an antisense peptide–PNA targeted to the essential *acpP* gene is bactericidal and shows encouraging bacterial cell specificity.[94,95] *E. coli* cells were grown in a mixed culture with HeLa cells, and the peptide–PNA efficiently cleared the human cell culture of inoculated bacteria, without apparent harm to the HeLa cells even when present at a concentration that was 10-fold higher than needed to kill bacteria.[4] Therefore, microbial CPPs can act as carriers for antisense agent delivery, and can display toxic effects that appear largely specific against microbial cells.

Recent reports demonstrate the ability of peptide–PNAs and similar peptide–phosphodiamidate morpholino conjugates to inhibit *E. coli* bacterial growth and prevent fatal intraperioneal infection in mice.[96,97] Also, the flexibility of this general approach has been demonstrated in experiments using carrier peptides to deliver attached bioactive peptides into bacteria,[46] and the green fluorescence protein fused to different CPPs and signal peptides was effectively delivered into bacterial and yeasts cells.[43] Finally, peptides have been used to deliver impermeable prodrugs into microbes in which they are activated by intracellular enzymes to release their toxic activities.[98,99] These examples demonstrate the range of possible applications for microbial CPPs.

ACKNOWLEDGMENTS

We thank the Swedish Science Council for support and our colleagues for helpful comments on the manuscript.

REFERENCES

1. Tan, Y.T., Tillett, D.J., and McKay, I.A., Molecular strategies for overcoming antibiotic resistance in bacteria, *Mol. Med. Today*, 6, 309, 2000.
2. Murray, R.W. et al., Ribosomes from an oxazolidinone-resistant mutant confer resistance to eperezolid in a *Staphylococcus aureus* cell-free transcription-translation assay, *Antimicrob. Agents Chemother.*, 42, 947, 1998.
3. Hancock, R.E., The bacterial outer membrane as a drug barrier, *Trends Microbiol.*, 5, 37, 1997.
4. Good, L. et al., Bactericidal antisense effects of peptide–PNA conjugates, *Nat. Biotechnol.*, 19, 360, 2001.
5. van't Hof, W. et al., Antimicrobial peptides: Properties and applicability, *Biol. Chem.*, 382, 597, 2001.
6. Carlsson, A. et al., Attacin, an antibacterial protein from *Hyalophora cecropia*, inhibits synthesis of outer membrane proteins in Escherichia coli by interfering with omp gene transcription, *Infect. Immun.*, 59, 3040, 1991.
7. Boman, H.G., Agerberth, B., and Boman, A., Mechanisms of action on *Escherichia coli* of cecropin P1 and PR-39, two antibacterial peptides from pig intestine, *Infect. Immun.*, 61, 2978, 1993.
8. Nikaido, H., Prevention of drug access to bacterial targets: Permeability barriers and active efflux, *Science*, 264, 382, 1994.
9. Alberts, B. et al., *Molecular Biology of the Cell*, Garland Publishing Inc., New York, 1994.
10. Broach, J.R., In *The Molecular and Cellular Biology of the Yeast Saccharomyces*, Jones, E.W., Pringle, J.R., and Broach, J.R. eds., Cold Spring Harbor Laboratory Press, Cold Spring Harbor, NY, 1997.
11. Steiner, H. et al., Sequence and specificity of two antibacterial proteins involved in insect immunity, *Nature*, 292, 246, 1981.
12. Selsted, M.E. et al., Primary structures of MCP-1 and MCP-2, natural peptide antibiotics of rabbit lung macrophages, *J. Biol. Chem.*, 258, 14485, 1983.

13. Zasloff, M., Magainins, a class of antimicrobial peptides from Xenopus skin: Isolation, characterization of two active forms, and partial cDNA sequence of a precursor, *Proc. Natl Acad. Sci. U.S.A.*, 84, 5449, 1987.

14. Brahmachary, M. et al., ANTIMIC: A database of antimicrobial sequences, *Nucleic Acids Res.*, 32, D586, 2004.

15. Wang, Z. and Wang, G., APD: The antimicrobial peptide database, *Nucleic Acids Res.*, 32, D590, 2004.

16. Fearon, D.T. and Locksley, R.M., The instructive role of innate immunity in the acquired immune response, *Science*, 272, 50, 1996.

17. Hoffmann, J.A. et al., Phylogenetic perspectives in innate immunity, *Science*, 284, 1313, 1999.

18. Plenat, T. et al., Interaction of primary amphipathic cell-penetrating peptides with phospholipid-supported monolayers, *Langmuir*, 20, 9255, 2004.

19. Mae, M. et al., Internalisation of cell-penetrating peptides into tobacco protoplasts, *Biochim. Biophys. Acta*, 1669, 101, 2005.

20. Chang, M., Chou, J.C., and Lee, H.J., Cellular internalization of fluorescent proteins via arginine-rich intracellular delivery peptide in plant cells, *Plant Cell Physiol.*, 46, 482, 2005.

21. Joliot, A. et al., Antennapedia homeobox peptide regulates neural morphogenesis, *Proc. Natl Acad. Sci. U.S.A.*, 88, 1864, 1991.

22. Sandgren, S. et al., The human antimicrobial peptide LL-37 transfers extracellular DNA plasmid to the nuclear compartment of mammalian cells via lipid rafts and proteoglycan-dependent endocytosis, *J. Biol. Chem.*, 279, 17951, 2004.

23. Derossi, D. et al., The third helix of the Antennapedia homeodomain translocates through biological membranes, *J. Biol. Chem.*, 269, 10444, 1994.

24. Oehlke, J. et al., Cellular uptake of an alpha-helical amphipathic model peptide with the potential to deliver polar compounds into the cell interior non-endocytically, *Biochim. Biophys. Acta*, 1414, 127, 1998.

25. Scheller, A. et al., Structural requirements for cellular uptake of alpha-helical amphipathic peptides, *J. Pept. Sci.*, 5, 185, 1999.

26. Chellaiah, M.A. et al., Rho-A is critical for osteoclast podosome organization, motility, and bone resorption, *J. Biol. Chem.*, 275, 11993, 2000.

27. Zorko, M. and Langel, Ü., Cell-penetrating peptides: Mechanism and kinetics of cargo delivery, *Adv. Drug Deliv. Rev.*, 57, 529, 2005.

28. Morris, M.C. et al., A new peptide vector for efficient delivery of oligonucleotides into mammalian cells, *Nucleic Acids Res.*, 25, 2730, 1997.

29. Pooga, M. et al., Cell penetrating PNA constructs regulate galanin receptor levels and modify pain transmission in vivo, *Nat. Biotechnol.*, 16, 857, 1998.

30. Pooga, M. et al., Cell penetration by transportan, *FASEB J.*, 12, 67, 1998.

31. Morris, M.C. et al., A novel potent strategy for gene delivery using a single peptide vector as a carrier, *Nucleic Acids Res.*, 27, 3510, 1999.

32. Anderson, D.C. et al., Tumor cell retention of antibody Fab fragments is enhanced by an attached HIV TAT protein-derived peptide, *Biochem. Biophys. Res. Commun.*, 194, 876, 1993.

33. Torchilin, V.P. et al., TAT peptide on the surface of liposomes affords their efficient intracellular delivery even at low temperature and in the presence of metabolic inhibitors, *Proc. Natl Acad. Sci. U.S.A.*, 98, 8786, 2001.

34. Liu, J. et al., Nanostructured materials designed for cell binding and transduction, *Biomacromolecules*, 2, 362, 2001.

35. Gratton, J.P. et al., Cell-permeable peptides improve cellular uptake and therapeutic gene delivery of replication-deficient viruses in cells and in vivo, *Nat. Med.*, 9, 357, 2003.

36. Muratovska, A. and Eccles, M.R., Conjugate for efficient delivery of short interfering RNA (siRNA) into mammalian cells, *FEBS Lett.*, 558, 63, 2004.

37. Nekhotiaeva, N. et al., Cell entry and antimicrobial properties of eukaryotic cell-penetrating peptides, *FASEB J.*, 18, 394, 2004.

38. Chaubey, B. et al., A PNA–transportan conjugate targeted to the TAR region of the HIV-1 genome exhibits both antiviral and virucidal properties, *Virology*, 331, 418, 2005.

39. Tripathi, S. et al., Anti-HIV-1 activity of anti-TAR polyamide nucleic acid conjugated with various membrane transducing peptides, *Nucleic Acids Res.*, 33, 4345, 2005.

40. Weiss, J. et al., Partial characterization and purification of a rabbit granulocyte factor that increases permeability of *Escherichia coli*, *J. Clin. Invest.*, 55, 33, 1975.

41. Antohi, S. and Popescu, A., Protamine and polyarginine bacteriolysis. Similarities in its mechanism with chromatin DNA picnosis, *Z. Naturforsch. [C]*, 34, 1144, 1979.

42. Brogden, K.A., Antimicrobial peptides: Pore formers or metabolic inhibitors in bacteria?, *Nat. Rev. Microbiol.*, 3, 238, 2005.

43. Rajarao, G.K., Nekhotiaeva, N., and Good, L., Peptide-mediated delivery of green fluorescent protein into yeasts and bacteria, *FEMS Microbiol. Lett.*, 215, 267, 2002.

44. Nekhotiaeva, N. et al., Inhibition of *Staphylococcus aureus* gene expression and growth using antisense peptide nucleic acids, *Mol. Ther.*, 10, 652, 2004.

45. Kulyte, A. et al., Gene selective suppression of nonsense termination using antisense agents, *Biochim. Biophys. Acta*, 1730, 165, 2005.

46. Otvos, L. Jr. et al., An insect antibacterial peptide-based drug delivery system, *Mol. Pharm.*, 1, 220, 2004.

47. Koo, J.C. et al., Pn-AMP1, a plant defense protein, induces actin depolarization in yeasts, *Plant Cell Physiol.*, 45, 1669, 2004.

48. Parenteau, J. et al., Free uptake of cell-penetrating peptides by fission yeast, *FEBS Lett.*, 579, 4873, 2005.

49. Holm, T. et al., Uptake of cell-penetrating peptides in yeasts, *FEBS Lett.*, 579, 5217, 2005.

50. Rajarao, G.K., Nekhotiaeva, N., and Good, L., The signal peptide NPFSD fused to ricin A chain enhances cell uptake and cytotoxicity in *Candida albicans*, *Biochem. Biophys. Res. Commun.*, 301, 529, 2003.

51. Kagan, B.L. et al., Antimicrobial defensin peptides form voltage-dependent ion-permeable channels in planar lipid bilayer membranes, *Proc. Natl Acad. Sci. U.S.A.*, 87, 210, 1990.

52. Robinson, W.E. Jr. et al., Anti-HIV-1 activity of indolicidin, an antimicrobial peptide from neutrophils, *J. Leukoc. Biol.*, 63, 94, 1998.

53. Gupta, B., Levchenko, T.S., and Torchilin, V.P., Intracellular delivery of large molecules and small particles by cell-penetrating proteins and peptides, *Adv. Drug Deliv. Rev.*, 57, 637, 2005.

54. Shai, Y., Mechanism of the binding, insertion and destabilization of phospholipid bilayer membranes by alpha-helical antimicrobial and cell non-selective membrane-lytic peptides, *Biochim. Biophys. Acta*, 1462, 55, 1999.

55. Huang, H.W., Action of antimicrobial peptides: Two-state model, *Biochemistry*, 39, 8347, 2000.

56. Matsuzaki, K., Why and how are peptide–lipid interactions utilized for self-defense? Magainins and tachyplesins as archetypes, *Biochim. Biophys. Acta*, 1462, 1, 1999.

57. Kobayashi, S. et al., Interactions of the novel antimicrobial peptide buforin 2 with lipid bilayers: Proline as a translocation promoting factor, *Biochemistry*, 39, 8648, 2000.

58. Zhang, L., Rozek, A., and Hancock, R.E., Interaction of cationic antimicrobial peptides with model membranes, *J. Biol. Chem.*, 25, 25, 2001.

59. Hancock, R.E., Peptide antibiotics, *Lancet*, 349, 418, 1997.

60. Guo, L. et al., Lipid A acylation and bacterial resistance against vertebrate antimicrobial peptides, *Cell*, 95, 189, 1998.

61. Tossi, A., Sandri, L., and Giangaspero, A., Amphipathic, alpha-helical antimicrobial peptides, *Biopolymers*, 55, 4, 2000.

62. Naider, F. and Becker, J.M., Peptide transport in *Candida albicans*: Implications for the development of antifungal agents, *Curr. Top. Med. Mycol.*, 2, 170, 1988.

63. Hancock, R.E.W., Bacterial transport as an import mechanism and target for antimicrobials, In *Drug Transport in Antimicrobial and Anticancer Chemotherapy*, Georgopapadakou, N. ed., Marcel Dekker Inc., New York, p. 289, 1994.

64. Richard, J.P. et al., Cell-penetrating peptides. A reevaluation of the mechanism of cellular uptake, *J. Biol. Chem.*, 278, 585, 2003.

65. Säälik, P. et al., Protein cargo delivery properties of cell-penetrating peptides, *Bioconjug. Chem.*, 15, 1246, 2004.

66. Westerhoff, H.V. et al., Magainins and the disruption of membrane-linked free-energy transduction, *Proc. Natl Acad. Sci. U.S.A.*, 86, 6597, 1989.

67. Yang, L. et al., Crystallization of antimicrobial pores in membranes: Magainin and protegrin, *Biophys. J.*, 79, 2002, 2000.

68. Kragol, G. et al., The antibacterial peptide pyrrhocoricin inhibits the ATPase actions of DnaK and prevents chaperone-assisted protein folding, *Biochemistry*, 40, 3016, 2001.

69. Sharma, S. and Khuller, G., DNA as the intracellular secondary target for antibacterial action of human neutrophil peptide-I against *Mycobacterium tuberculosis* H37Ra, *Curr. Microbiol.*, 43, 74, 2001.

70. Helmerhorst, E.J. et al., The cellular target of histatin 5 on *Candida albicans* is the energized mitochondrion, *J. Biol. Chem.*, 274, 7286, 1999.

71. Debono, M. and Gordee, R.S., Antibiotics that inhibit fungal cell wall development, *Annu. Rev. Microbiol.*, 48, 471, 1994.

72. Lupetti, A. et al., Candidacidal activities of human lactoferrin peptides derived from the N terminus, *Antimicrob. Agents Chemother.*, 44, 3257, 2000.

73. Boman, H.G. et al., Antibacterial and antimalarial properties of peptides that are cecropin-melittin hybrids, *FEBS Lett.*, 259, 103, 1989.

74. Andreu, D. et al., Shortened cecropin A-melittin hybrids. Significant size reduction retains potent antibiotic activity, *FEBS Lett.*, 296, 190, 1992.

75. Lehrer, R.I. and Ganz, T., Antimicrobial peptides in mammalian and insect host defence, *Curr. Opin. Immunol.*, 11, 23, 1999.

76. Mason, D.J. et al., Antimicrobial action of rabbit leukocyte CAP18(106–137), *Antimicrob. Agents Chemother.*, 41, 624, 1997.

77. Lehrer, R.I., Barton, A., and Ganz, T., Concurrent assessment of inner and outer membrane permeabilization and bacteriolysis in E. coli by multiple-wavelength spectrophotometry, *J. Immunol. Methods*, 108, 153, 1988.

78. Giacometti, A. et al., In-vitro activity and killing effect of polycationic peptides on methicillin-resistant *Staphylococcus aureus* and interactions with clinically used antibiotics, *Diagn. Microbiol. Infect. Dis.*, 38, 115, 2000.

79. Linde, C.M. et al., In vitro activity of PR-39, a proline–arginine-rich peptide, against susceptible and multi-drug-resistant *Mycobacterium tuberculosis*, *J. Antimicrob. Chemother.*, 47, 575, 2001.

80. Xia, H., Mao, Q., and Davidson, B.L., The HIV Tat protein transduction domain improves the biodistribution of beta-glucuronidase expressed from recombinant viral vectors, *Nat. Biotechnol.*, 19, 640, 2001.

81. Fanales-Belasio, E. et al., Native HIV-1 Tat protein targets monocyte-derived dendritic cells and enhances their maturation, function, and antigen-specific T cell responses, *J. Immunol.*, 168, 197, 2002.

82. Steinberg, D.A. et al., Protegrin-1: A broad-spectrum, rapidly microbicidal peptide with in vivo activity, *Antimicrob. Agents Chemother.*, 41, 1738, 1997.

83. Ahmad, I. et al., Liposomal entrapment of the neutrophil-derived peptide indolicidin endows it with in vivo antifungal activity, *Biochim. Biophys. Acta*, 1237, 109, 1995.

84. Welling, M.M. et al., Antibacterial activity of human neutrophil defensins in experimental infections in mice is accompanied by increased leukocyte accumulation, *J. Clin. Invest.*, 102, 1583, 1998.

85. Bals, R. et al., Augmentation of innate host defense by expression of a cathelicidin antimicrobial peptide, *Infect. Immun.*, 67, 6084, 1999.

86. Kisich, K.O. et al., Antimycobacterial agent based on mRNA encoding human beta-defensin 2 enables primary macrophages to restrict growth of *Mycobacterium tuberculosis*, *Infect. Immun.*, 69, 2692, 2001.

87. Tsiodras, S. et al., Linezolid resistance in a clinical isolate of *Staphylococcus aureus*, *Lancet*, 358, 207, 2001.

88. Gonzales, R.D. et al., Infections due to vancomycin-resistant *Enterococcus faecium* resistant to linezolid, *Lancet*, 357, 1179, 2001.

89. Fickel, T.E. and Gilvarg, C., Transport of impermeant substances in *E. coli* by way of oligopeptide permease, *Nat. New Biol.*, 241, 161, 1973.

90. Ames, B.N. et al., Illicit transport: The oligopeptide permease, *Proc. Natl Acad. Sci. U.S.A.*, 70, 456, 1973.

91. Wittung, P. et al., Phospholipid membrane permeability of peptide nucleic acid, *FEBS Lett.*, 365, 27, 1995.

92. Good, L. and Nielsen, P.E., Inhibition of translation and bacterial growth by peptide nucleic acid targeted to ribosomal RNA, *Proc. Natl Acad. Sci. U.S.A.*, 95, 2073, 1998.

93. Good, L. et al., Antisense PNA effects in *Escherichia coli* are limited by the outer-membrane LPS layer, *Microbiology*, 146, 2665, 2000.

94. Vaara, M. and Porro, M., Group of peptides that act synergistically with hydrophobic antibiotics against gram-negative enteric bacteria, *Antimicrob. Agents Chemother.*, 40, 1801, 1996.

95. Cronan, J.E. and Rock, C.O., Biosynthesis of membrane lipids, In *Escherichia coli and Salmonella typhimurium: Cellular Molecular Biology*, Neidhart, F.C. et al., eds., 2nd ed., American Society for Microbiology, Washington, DC, p. 612, 1996.

96. Geller, B.L. et al., Antisense phosphorodiamidate morpholino oligomer inhibits viability of Escherichia coli in pure culture and in mouse peritonitis, *J. Antimicrob. Chemother.*, 55, 983, 2005.

97. Tan, X.X., Actor, J.K., and Chen, Y., Peptide nucleic acid antisense oligomer as a therapeutic strategy against bacterial infection: Proof of principle using mouse intraperitoneal infection, *Antimicrob. Agents Chemother.*, 49, 3203, 2005.

98. Wei, Y. and Pei, D., Activation of antibacterial prodrugs by peptide deformylase, *Bioorg. Med. Chem. Lett.*, 10, 1073, 2000.

99. Marshall, N.J. et al., Structure–activity relationships for a series of peptidomimetic antimicrobial prodrugs containing glutamine analogues, *J. Antimicrob. Chemother.*, 51, 821, 2003.

100. Chan, Y.R. and Gallo, R.L., PR-39, a syndecan-inducing antimicrobial peptide, binds and affects p130(Cas), *J. Biol. Chem.*, 273, 28978, 1998.

101. Drin, G. and Temsamani, J., Translocation of protegrin I through phospholipid membranes: Role of peptide folding, *Biochim. Biophys. Acta*, 1559, 160, 2002.

102. Takeshima, K. et al., Translocation of analogues of the antimicrobial peptides magainin and buforin across human cell membranes, *J. Biol. Chem.*, 278, 1310, 2003.

103. Skerlavaj, B., Romeo, D., and Gennaro, R., Rapid membrane permeabilization and inhibition of vital functions of gram-negative bacteria by bactenecins, *Infect. Immun.*, 58, 3724, 1990.

104. Park, C.B., Kim, H.S., and Kim, S.C., Mechanism of action of the antimicrobial peptide buforin II: Buforin II kills microorganisms by penetrating the cell membrane and inhibiting cellular functions, *Biochem. Biophys. Res. Commun.*, 244, 253, 1998.

105. Subbalakshmi, C. and Sitaram, N., Mechanism of antimicrobial action of indolicidin, *FEMS Microbiol. Lett.*, 160, 91, 1998.

106. Oberparleiter, C. et al., Active internalization of the penicillium chrysogenum antifungal protein PAF in sensitive aspergilli, *Antimicrob. Agents Chemother.*, 47, 3598, 2003.

107. Breukink, E. et al., Use of the cell wall precursor lipid II by a pore-forming peptide antibiotic, *Science*, 286, 2361, 1999.

108. Patrzykat, A. et al., Sublethal concentrations of pleurocidin-derived antimicrobial peptides inhibit macromolecular synthesis in *Escherichia coli*, *Antimicrob. Agents Chemother.*, 46, 605, 2002.

109. Yonezawa, A. et al., Binding of tachyplesin I to DNA revealed by footprinting analysis: Significant contribution of secondary structure to DNA binding and implication for biological action, *Biochemistry*, 31, 2998, 1992.

110. Tsai, H. and Bobek, L.A., Human salivary histatins: Promising anti-fungal therapeutic agents, *Crit. Rev. Oral Biol. Med.*, 9, 480, 1998.

111. Couto, M.A., Harwig, S.S., and Lehrer, R.I., Selective inhibition of microbial serine proteases by eNAP-2, an antimicrobial peptide from equine neutrophils, *Infect. Immun.*, 61, 2991, 1993.

Part VI

CPP Methods

31 Design and Synthesis of Cell-Penetrating Peptides

Yang Jiang, Ursel Soomets, and Ülo Langel

CONTENTS

31.1 INTRODUCTION

Cell-penetrating peptides (CPPs) are used as transport vehicles for the delivery of different organic or bioactive molecules of different sizes and chemical properties, especially biomacromolecules, into cells. For reviews, see Refs. [1–3] and Part I of this book.

CPPs can be synthesized using the solid-phase peptide synthesis (SPPS) method. The basic principle of SPPS is to anchor the amino group-protected amino acid onto the polymer-based resin through the carboxylic group, then remove the amino group protection and couple the next protected amino acid. By repeating these steps, the whole peptide sequence can be assembled on the resin from C- to N-terminus (Figure 31.1).

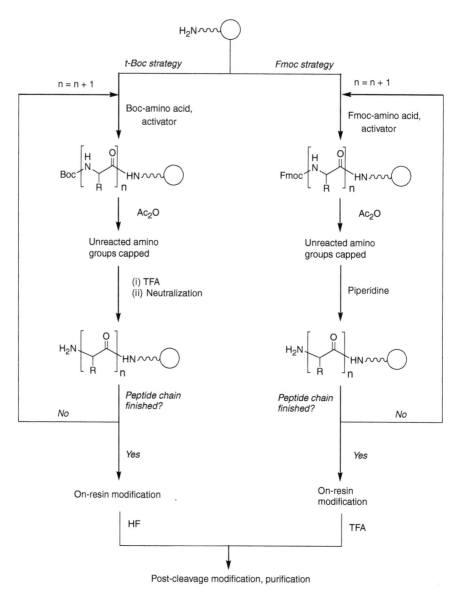

FIGURE 31.1 Basic procedure of solid-phase synthesis of CPPs, either by *t*-Boc or Fmoc strategy.

Depending on the N-terminal protecting groups, two strategies, using either *tert*-butyloxycarbonyl (*t*-Boc) or 9-fluorenylmethoxycarbonyl (Fmoc) group, are available. *t*-Boc protection is removed by 50% trifluoroacetic acid (TFA) in dichloromethane (DCM), and the peptide–solid phase linkage is cleaved by extremely toxic and corrosive hydrogen fluoride (HF), thus a special cleaving apparatus is required. Fmoc protection is removed by a moderate base,

20% piperidine in *N*,*N*-dimethylformamide (DMF), and the peptide is cleaved from the resin by TFA. (For detailed principles, technologies, and developments of SPPS, see Refs. [4,5].) Generally, the reaction conditions (deprotection and cleavage) used in the Fmoc strategy are milder as compared with those in the *t*-Boc strategy.

31.2 DESIGN AND SYNTHESIS OF CPP AND CARGO CONJUGATION

Cargo molecules that have been delivered into cells by CPPs include fluorophores, peptides, proteins, nucleic acids and their analogs, and even nanoscale particles and quantum dots.[6–33] Some examples of various cargoes and their conjugation methods are listed in Table 31.1. In this chapter we discuss the synthesis of CPPs and different chemical methods for the conjugation of cargo molecules to CPPs. Molecular biology techniques are not discussed here. The reader is guided to other publications (Refs. [34–36], to name a few).

Several types of cargoes, including proteins, siRNAs, modified peptide nucleic acids (PNAs), and plasmids, have been transported across the plasma membrane as noncovalent complexes with certain CPPs.[6–9] Generally, these cargo molecules have an anionic structure and form complexes with cationic CPPs through electrostatic interactions. However, most commonly, cargoes are connected to CPPs through covalent bonds. For this purpose, versatile chemical approaches have been exploited.

In several reported studies, CPPs were not conjugated to cargo molecules directly. Instead, the CPP part was conjugated to a "bridge" molecule which is able to complex with the cargo molecules through noncovalent interactions; for example a PNA that binds DNA cargo through Watson–Crick base-pairing,[19,20] or a polyethyleneimine (PEI) that complexes to plasmids through electrostatic interactions.[21] These reports are also classified into the following categories according to the chemistry used in the synthesis of delivering vehicles.

Conjugation through an amide bond is carried out with one component carrying a carboxylic group and the other with an amino group. An amide bond can be generated in aqueous solution by adding water-soluble activating reagents. However, generally, it is not applicable after the cleavage of CPPs, because most CPPs contain amino groups at both the N-terminus and side chains of lysine residues. For peptidic cargoes including peptides and PNAs, direct attachment of a cargo to the N-terminus of a CPP, or vice versa, can be achieved by stepwise peptide synthesis. This linkage is chemically stable, but when the construct is composed of natural L-α-amino acids, it is degradable by peptidases. The degradation sites and kinetics are still under study for different CPPs.

Conjugations utilizing the sulfhydryl group are highly specific reactions that usually do not interfere with other functional groups in biomacromolecules. Native chemical ligation between a peptide thioester and an N-terminally cysteine-derived molecule leads to the formation of a peptide bond between these two components.

The disulfide bond is a linkage applicable for a wide range of cargoes including peptides, proteins, nucleic acids, etc. The linkage is readily reduced under

TABLE 31.1
Examples for CPP–Cargo Conjugation Methods

Conjugation Method(s)	Cargo Molecule	CPP	Peptide Modification	Refs.
Noncovalent complex	siRNA, modified PNA	MPG	C-terminal cysteamide	6,7
Noncovalent complex	Protein	Pep-1	C-terminal cysteamide	8
Noncovalent complex	Plasmid	hCT	Branched, NLS at the side chain of Lys	9
Peptide bond	Peptide	Penetratin, Tat peptide, polyarginine	Cargo and CPP in the same strand	10
Peptide bond	PNA	Tat, polyarginine	Cargo and CPP in the same strand	11
Peptide bond, side chain of Lys	PNA	Polyarginine, Tat	Branched peptide	12
Peptide bond, by native chemical ligation	PNA	(KFF)$_3$K	Thioester	13
Disulfide bond	Peptide, protein	TP10	N^ε-Cys(Npys) of Lys^7	14,15
Disulfide bond	PNA	Transportan	N^ε-Cys(Npys) of Lys^{13}	16–18
Disulfide bond	Decoy DNA	Transportan and TP10	Conjugated to PNA through N^ε-Cys of Lys^7	19,20
Disulfide bond	Plasmid	TP10	Conjugated to PNA through N^ε-Cys of Lys^7	21
Disulfide bond	Oligonucleotides and analogs	Tat, penetratin, polyarginine, transportan	Cysteine	22–24
Thioether	Plasmid	TP10	Conjugated with PEI	21
Thioether	Polymopholino oligonucleotide analog	Tat, Arg-rich peptides	Cysteine	25–27
Thioether	Small orgainic molecules	hCT(9–32), Tat, polyarginine	Cysteine	28–30
Thioether	Peptide (insulin)	Tat	Cysteine	31
Thioether	Fluorochrome	Tat	Cysteine	32
Thioether	CdS:Mn/ZnS Qdots, coated layer with amino groups	Tat	Cysteine	33
Thioether and oxime	Oligonucleotide analog	Tat	Cysteine	22

a reductive environment, such as in the presence of free thiol compounds. It is widely accepted that the linkage is cleaved after the uptake of the compound into cells, or in the endosome/lysosome and even on the cell surface, due to the reductive environment of cytoplasm and the effect of cellular redox enzymes.[37]

Thioether is another conjugation strategy for the sulfhydryl group, formed by either nucleophilic addition to a maleimide group or substitution of the α-halogen of a carbonyl group. Thioether is not cleavable under physiological conditions.

Another specific reaction is the formation of Schiff base between a carbonyl compound and an amino compound. This kind of Schiff base is not stable. More stable derivatives, for example, oxime or hydrazone, are applied in the conjugation of biomolecules. This conjugation is acid labile so that it may be cleaved in the acidic endosome after endocytotic uptake of the conjugate.

The selection of chemistry for conjugation is determined by the type of cargo molecules, synthesis accessibility, and finally, the biological effectiveness. The CPP–cargo conjugate must maintain the activity of the cargo molecule as well as the membrane penetration ability. For small molecule drugs, the linkage usually needs to be cleaved in a short time span after the internalization; thus the drug is released in its active form. Moreover, for the delivery of biologically active macromolecules, the subcellular localization of the cargo after the uptake is another important factor for activity. In order to obtain an effective delivery system, more functional moieties may be required, for example, a nuclear localization signal (NLS) sequence. In this case, a proper combination of two or more conjugation methods may be needed. As the mechanism(s) for the internalization of CPPs are not yet clear, and the fates of the CPP–cargo conjugate and the cargo itself are still largely unknown, the rational design of CPP–cargo conjugation is still a long way ahead.

31.2.1 CONJUGATION THROUGH A PEPTIDE BOND

Different cargo molecules of a peptidic nature, peptides or PNAs, for example, can be attached to the CPPs directly through a peptide bond. A straightforward synthesis can be applied, with either CPP or cargo at the N-terminal part of the oligomer. Low-substitution resin (0.1 mmol/g) is used to avoid the interpeptide chain aggregation, especially in the case of hydrophobic sequences.

Attachment of a nonpeptidic cargo to the N-terminus of the peptide by peptidic linkage requires the carboxylic group in cargo molecules, the stability and solubility of the cargo molecule in the conjugation–reaction solutions, and stability in the strong acid environment of the cleavage mixture. Generally, both Fmoc and t-Boc strategies are applicable for the direct synthesis of CPP–cargo constructs.

Cargo can also be attached to the peptide through a side chain of different amino acids to obtain the free N-terminus of CPP. The most common site for the conjugation of a cargo is the ε-NH$_2$ group of a particular lysine side chain. An orthogonal protecting group on the ε-NH$_2$ of lysine is required. In t-Boc chemistry, Boc-Lys(Fmoc)-OH is a commonly used building block, except when a N^{in}-formyl-protected tryptophan is present in the sequence, as both Fmoc and formyl groups are sensitive to the piperidine treatment. Therefore, in the orthogonal synthesis using the Lys side chain, tryptophan with a piperidine-resistant side chain protection must be chosen in *t*-Boc chemistry. In Fmoc strategy, Fmoc-Lys(Mtt)-OH and Fmoc-Lys(Dde)-OH can be used as sites for the attachment of cargoes. Both protection groups are stable in piperidine. The Mtt group is removed by diluted TFA and the Dde group by hydrazine. However, hydrazine treatment also cleaves the N-terminal Fmoc protection group from the peptide. Díaz-Mochón et al. have developed an alternative deprotection mixture, consisting of hydrazine hydrochloride and a nucleophilic, but a weaker base, such as imidazole, in which the Fmoc group is fully stable.[38] The same group has reported a synthesis of PNA–CPP conjugate with this Dde protection–deprotection method.[12]

In a plasmid delivery study reported by Krauss et al.,[9] an NLS sequence was conjugated to a CPP hCT(9–32), either to the N-terminus or to the side chain of lysine in the middle of the sequence. The branched peptide showed higher efficiency in delivery than the linear one.

When the solid-phase conjugation is not applicable, for example, when cargo molecules are not compatible with the cleavage or the side chain deprotection conditions, postcleavage conjugation is the only choice. Several different approaches have been developed for these purposes.

31.2.2 NATIVE CHEMICAL LIGATION

Native chemical ligation was developed for the chemoselective coupling of the "right" functional groups of two unprotected peptides in aqueous media at ambient temperature, without coupling reagents or protecting groups.[39] The most common ligation is via the reaction between the C-terminal thioester and the N-terminal cysteine. Thus, ligation requires modified peptides: a thioester at the C-terminus of one peptide, and a cysteine at the N-terminus of another. A reversible nucleophilic addition–elimination reaction takes place between the sulfhydryl group of the N-terminal cysteine and the thioester. The intracellular nucleophilic attack of the α-amino group at the thioester results in the formation of a thermodynamically stable amide bond, which shifts the equilibrium to the product (Figure 31.2). Usually the reaction is carried out at a neutral to slightly basic pH, in the presence of strong denaturing agents, and upon the catalysis with a thio-compound such as thiophenol. Either CPP or cargo can carry the thioester group.

A peptide thioester can be prepared using either the *t*-Boc or Fmoc strategy. After the coupling of the C-terminal amino acid to a thiol-derived resin, the normal protocol of *t*-Boc strategy can be applied as the thioester is stable in TFA and HF.[40] However, thioester is not stable in deprotection with piperidine used in Fmoc strategy. To overcome this difficulty, different methods have been developed. In Ingenito et al.'s report,[41] the peptide chain was anchored to a commercially available 4-sulfamylbutyryl AM resin via acylsulfonamide linkage, which is stable in Fmoc chemistry. The peptide assembly was performed according to the standard Fmoc protocol, except using *t*-Boc-protected amino acid as the last residue at the N-terminus. Following the assembly, the acylsulfonamide was *N*-alkylated, and then cleaved by benzyl mercaptan to yield the protected peptide thioester. The side chain protections and N-terminal *t*-Boc group were removed by TFA afterwards. Another method using Fmoc strategy is to anchor the peptide to the acid-labile 2-chlorotrityl

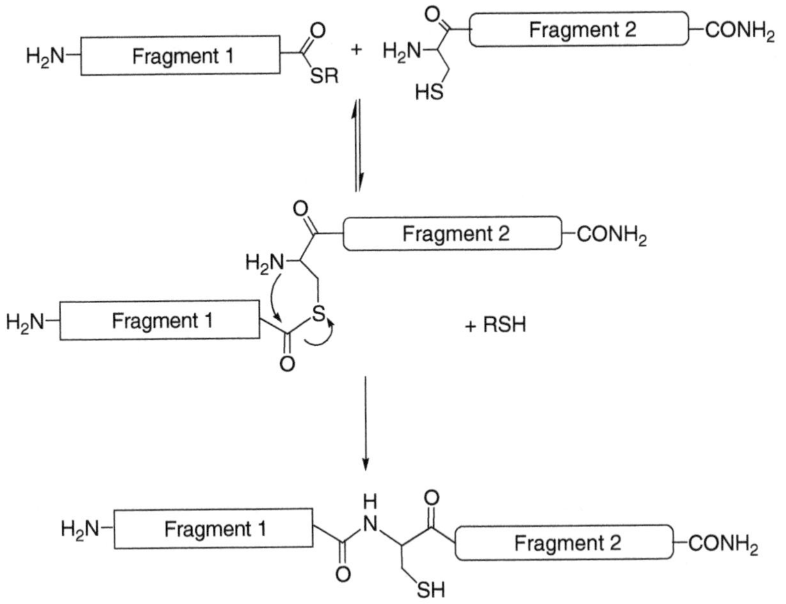

FIGURE 31.2 Native chemical ligation. Fragment 1 is synthesized as a thioester, and fragment 2 is carrying a cysteine residue with the free amino and sulfhydryl group. The first step is reversible, while the formation of a thermodynamically stable amide bond drives the equilibrium to the product.

resin, and to use t-Boc-protected amino acid as the last residue at the N-terminus.[42] After peptide assembly, the fully protected peptide acid is cleaved from the resin by weak acid. Following the cleavage the protected peptide acid is coupled to a thiol compound to give the thioester, and then side chains and N-terminal protections are removed by TFA.

De Koning et al. reported the synthesis of an NLS–PNA–CPP.[43] Both the NLS peptide and PNA were synthesized as thioesters, and a thiaproline was coupled to the N-terminus of the PNA thioester. After the conjugation of PNA and CPP is accomplished, the thiaproline is converted to a cysteine residue and the PNA–CPP part is ready for conjugation to the NLS–thioester via native chemical ligation.

31.2.3 CONJUGATION THROUGH A DISULFIDE BRIDGE

Connecting a CPP with a cargo through disulfide linkage is one of the most studied conjugation methods that has been used to conjugate CPPs to peptides,[14] proteins,[15] PNAs,[16–21] oligonucleotides, and analogs.[22–24]

The formation of a heterogeneous disulfide bridge is usually performed by a directed reaction, i.e., one component is derived with an activated sulfhydryl group, and the other with a free sulfhydryl group. The activation (and also protection in synthesis) is performed by using an S-2-pyridinesulphenyl (Pys) or S-3-nitro-2-pyridinesulphenyl (Npys) group (Figure 31.3).

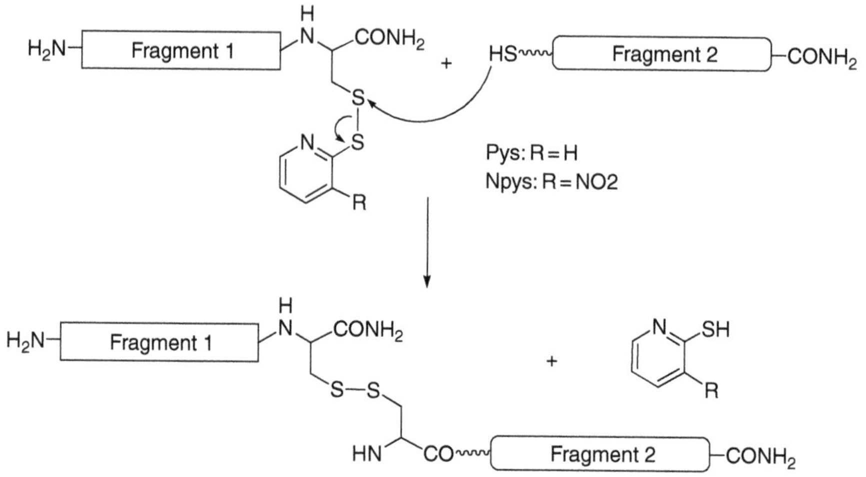

FIGURE 31.3 Directed disulfide bridge formation. Fragment 1 is carrying an activated Cys(*S*-Pys/Npys), and fragment 2 a free sulfhydryl group. The substitution of *S*-Pys or *S*-Npys by the sulfhydryl leads to the formation of a disulfide-linked conjugate.

Cysteine residue with Pys or Npys side chain protection is stable in TFA and HF. Thus, *t*-Boc chemistry is compatible with the synthesis of peptides containing Cys(Pys/Npys). However, these cysteine derivatives are not stable in deprotection with piperidine in the Fmoc strategy. Fmoc strategy is suitable only when Cys(Pys/Npys) is designed to be at the N-terminus of peptide, as then Boc-Cys(Pys/Npys)-OH can be coupled and the *t*-Boc group cleaved simultaneously during peptide cleavage with TFA. Postcleavage derivatization methods for the synthesis of Cys(Npys)-derived peptides by Fmoc strategy have been reported; for example, the reaction between a peptide containing a free sulfhydryl group and 2,2'-dithiobis(5-nitropyridine) (DTNP), either after or during cleavage.[44,45] The postcleavage reaction was carried out in a 3:1 AcOH–H$_2$O mixture in which both the peptide and DTNP are soluble. The main side reaction is the formation of a peptide homodimer.

A delivery of double-stranded decoy DNA molecules targeting transcription factor NF-κB into cells has been achieved by noncovalent coupling of a disulfide bridged CPP–PNA to the DNA. The double-stranded decoy DNA had a 3'-overhang, which is complementary to the PNA. The annealed complex of CPP–PNA with the decoy DNA was taken up by cells. The same strategy has been used in delivery of a decoy DNA targeting Myc protein, which is a transcription factor with a recognized role in cell proliferation.[19,20] The conjugation of CPP to oligonucleotides by a disulfide bond and the cellular uptake of the resulting complex is discussed in detail in chapter 20.

31.2.4 CONJUGATION THROUGH THIOETHER

Coupled with a cysteine, a CPP may also be conjugated to a cargo molecule through a thioether linkage, by the reaction between the sulfhydryl group with a maleimide or an α-halogenated carbonyl compound (Figure 31.4). Usually, a bifunctional handle with a maleimide group on one side is conjugated to the cargo molecule by the second functional group, for example, an activated ester. The conjugation of the sulfhydryl group of CPP to the maleimide is carried out with high specificity in an aqueous solution. One advantage of this conjugation is that it is carried out after the cleavage and purification of peptides. This method is especially useful when the cargo molecule, for example, a fluorophore or an organic moiety is not stable in the strong acidic environment of peptide cleavage reaction, or the cargo is too hydrophilic to be used in organic solvents. Different reports describing conjugation of CPPs to peptides,[31] oligonucleotide analogs,[25–27] and anticancer drugs[28–30] are available.

31.2.5 OXIME CONJUGATION

Prater et al. reported the synthesis of conjugates of the Tat peptide and oligonucleotide analogs, 3′-methylphosphonate-modified oligo-2′-O-methylribonucleotides.[22] A cysteine at the C-terminus of the Tat peptide was derivatized with bromoacetone to give a ketone-derived peptide. By mixing the ketone-peptide and an oligonucleotide analog carrying an aminooxy group in 50% aqueous acetonitrile

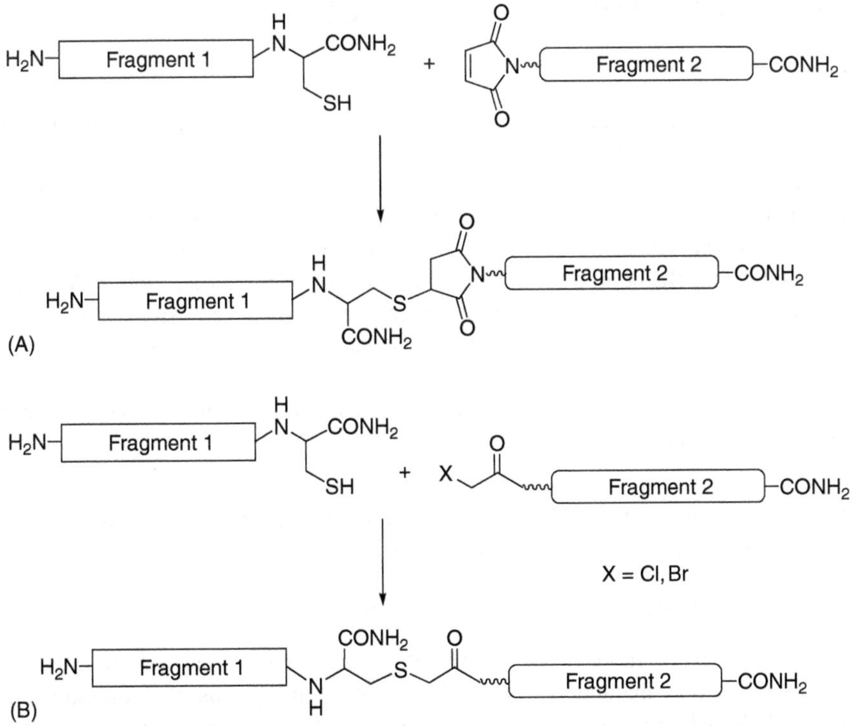

FIGURE 31.4 Thioether formation. (A) Addition of a free sulfhydryl group to maleimide. (B) Substitution of a sulfhydryl group to α-halogenketone.

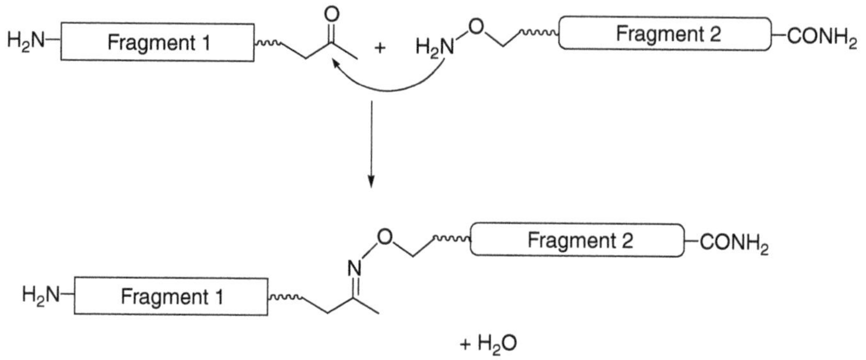

FIGURE 31.5 Oxime conjugation. Fragment 1 with a carbonyl group is conjugated to fragment 2 with an aminooxy group. The nucleophilic addition leads to the formation of the oxime.

at room temperature, a conjugate through oxime was formed (Figure 31.5). The product was verified by matrix-assisted laser desorption ionization-time of flight (MALDI-TOF) mass spectrometry.

31.3 EXAMPLES FOR THE SYNTHESIS OF CPPs

31.3.1 MATERIALS

Fmoc-protected amino acids and Rink amide MBHA resin were from Multi-SynTech GmbH, Germany and Fmoc-Lysine(Mtt)-OH from Novabiochem, Germany. t-Boc amino acids, p-methylbenzylhydrylamine (MBHA) resin, 2-(1H-benzotriazole-1-yl)-1,1,3,3-tetramethyluronium tetrafluoroborate (TBTU) and 1-hydroxybenzotriazole (HOBt) were from Neosystem, France, Boc-Cysteine (Npys)-OH from Bachem, Switzerland. DMF was from Sigma-Aldrich. N,N-Diiospropylethylamine (DIEA), piperidine, thioanisole, triisopropylsilane (TIS) and N-methylpyrrolidone (NMP) were from Lancaster, UK. Acetic anhydride (Ac$_2$O), dimethylsulfoxide (DMSO) and acetonitrile were from Scharlau Chemie, Spain. Trifluoroacetic acid (TFA) was from Chemicon Int., Germany and 5(6)-carboxyfluorescein from Molecular Probes, USA.

31.3.2 SYNTHESIS OF N^{ε}-[5(6)-CARBOXYLFLUORESCEINYL]-LYS^7-TP10 (FL-TP10) USING FMOC STRATEGY

31.3.2.1 Assembly of Transportan 10 Sequence

The assembly of the transportan 10 (TP10) sequence can be carried out either by machine-assisted or manual synthesis. One hundred milligrams of Fmoc-protected rink-amide MBHA resin (0.56 mmol functional group/gram resin) was loaded. The Fmoc protection was removed by 20% piperidine in DMF for 20 min. The resin was washed three times with DMF.

Coupling: To couple the first amino acid, leucine, to the resin, the resin was swelled in 0.5 mL DCM. Then, 5 equiv. of Fmoc–Leu and 5 equiv. of TBTU, dissolved in 1 mL DMF, and 5 equiv. of DIEA solution in NMP were mixed together, and added to the resin. The reaction was carried out at room temperature for 30 min. After the reaction, the liquid was drained from the vessel and the solid phase was washed consecutively with DMF (3×2 mL, 2 min each time).

Capping: After washing, 100 μL acetic anhydride and 5 equiv. of DIEA in DMF was added to the vessel and reacted for 10 min to cap unreacted amino groups. After draining, the resin was washed three times with DMF.

Fmoc deprotection: The resin was deprotected with 20% (v/v) piperidine in DMF, and then washed again with DMF (3×2 mL, 2 min each time). The cycle was finished with the washes and the synthesis continued with the next cycle, starting with the coupling step.

Repeat of the cycle for the second to 21st residues: From the third cycle, 0.5 mL of DMSO instead of DCM was added to the resin before coupling. From the 11th cycle, the coupling step was performed twice, for 30 min each, with one washing step in between. When the whole peptide sequence was finished, the peptide-resin was washed with DCM and dried in vacuum.

When the peptide was synthesized manually, after each coupling, the peptide-resin was washed with ethanol 1 min, DMF 2×1 min, 5% DIEA in DCM 1 min, and DCM 3×1 min. Then a small amount of resin was taken out for Kaiser (ninhydrin) testing. If the test was positive, i.e., the reaction was not completed, the same coupling step was repeated.

31.3.2.2 Coupling of 5(6)-Carboxyfluorescein to the ε-NH$_2$ of Lys^7 of TP10

The protocol for carboxyfluorescein coupling is adopted from Ref. [46]. The N^{ε}-Mtt group is removed from the Lys^7 side chain with a 1.5:2.5:96 mixture of TFA:TIS:DCM, 10×2 mL, 1 min each time. The peptide-resin was washed and the mixture of 2 equiv. 5(6)-carboxyfluorescein, 2 equiv. HOBt, and 2 equiv. N,N'-diisopropylcarbodiimide was added, and left to react overnight in the dark with gentle agitation. After washing, the peptide-resin was treated with 20% piperidine in DMF for 30 min. The last step is necessary because 5(6)-carboxyfluorescein contains an unprotected hydroxyl group in the molecule, thus extra 5(6)-carboxyfluorescein molecules can couple to the hydroxyl group of the carboxyfluorescein coupled directly to the peptide.

31.3.2.3 Cleavage of Fl-TP10 from the Solid Phase

After completion of the assembly and 5(6)-carboxyfluorescein coupling, the peptide was cleaved from the solid support, using TFA–scavenger cocktail. The cleavage was carried out at room temperature for 2.5 h with the mixture of 95% TFA, 2.5% water, and 2.5% TIS. The crude peptide was precipitated by adding the peptide–TFA solution dropwise into a 10–15-fold volumn of cold diethylether

(0°C) and centrifugation. After the removal of ether, the crude peptide was dissolved in 5–10% acetic acid aqueous solution, washed twice with cold diethylether, and lyophilized.

31.3.3 Synthesis of Cys(Npys)-TP10 Using *t*-Boc-Strategy

The TP10 sequence was assembled in a stepwise manner on a 433A peptide synthesizer (Applied Biosystems, USA) using *t*-Boc strategy of SPPS. *t*-Boc amino acids were coupled to MBHA resin to obtain C-terminally amidated peptide, using dicyclohexylcarbodiimide/hydroxybenzotriazole activation. Boc-Lys(Fmoc)-OH was used for the Lys[7]. After every coupling step the capping of unreacted amino groups by acetic anhydride was performed. Boc protection groups were removed by the treatment with a 1:1 TFA–DCM solution.

After the accomplishment of the TP10 sequence, the peptide-resin was removed from the synthesizer and synthesis was continued manually. The protecting N^ϵ-Fmoc group of the Lys[7] was removed by a 20% solution of piperidine in DMF for 20 min. After washing, 2 equiv. of Boc–Cys(Npys)–OH, 2 equiv. of TBTU, 2 equiv. of HOBt, and 4 equiv. of DIEA (solutions in DMF) were added, and allowed to react for 30 min. The resin was washed and a ninhydrin test performed. After the removal of the N-terminal *t*-Boc group, the peptide was finally cleaved from the resin with liquid HF at 0°C for 1 h in the presence of *p*-cresol as the scavenger.

31.3.4 Purification of the Products

Products were purified by preparative HPLC (Gynkotek, Supelco Discovery C18 25 cm × 21.2 mm column) using a linear gradient from 20 to 100% B in A over 50–60 min ($A = 0.1\%$ TFA in water, $B = 0.1\%$ TFA in acetonitrile, v/v). The flow rate was 10 mL/min and an automatic collector was used to collect fractions according to the UV absorbance at 215 nm, and the collected fractions were lyophilized.

31.3.5 Analysis of the Products

MALDI-TOF mass spectrometry was used to identify the correct products. The analysis was performed on a Voyager-DE STR mass spectrometer (PerSeptive Biosystems, Framingham, USA). Samples were dissolved in 50% acetonitrile/water solution and α-cyano-4-hydroxycinnamic acid was used as the matrix. Matrix was dissolved in a 70% acetonitrile aqueous solution, and mixed with the sample solution on the sample plate in a 1:1 ratio (v/v). Each spectrum was obtained from accumulated signals of at least 50 laser shots.

REFERENCES

1. Lindgren, M. et al., Cell-penetrating peptides, *Trends Pharmacol. Sci.*, 21, 99, 2000.
2. Järver, P. and Langel, Ü., The use of cell-penetrating peptides as a tool for gene regulation, *Drug Discov. Today*, 9, 395, 2004.

3. Joliot, A. and Prochiantz, A., Transduction peptides: From technology to physiology, *Nat. Cell Biol.*, 6, 189, 2004.
4. Fields, G.B., ed., *Methods in Enzymology, Solid-Phase Peptide Synthesis*, Vol. 289, Academic Press, New York, 1997.
5. White, P.D. and Chan, W.C., Basic Principles, in *Fmoc Solid Phase Peptide Synthesis: A Practical Approach*, Chan, W.C. and White, P.D., Eds., Oxford University Press, Oxford, 2000, chap. 2.
6. Simeoni, F. et al., Insight into the mechanism of the peptide-based gene delivery system MPG: Implications for delivery of siRNA into mammalian cells, *Nucleic Acids Res.*, 31, 2717, 2003.
7. Morris, M.C. et al., Combination of a new generation of PNAs with a peptide-based carrier enables efficient targeting of cell cycle progression, *Gene Ther.*, 11, 757, 2004.
8. Morris, M.C. et al., A peptide carrier for the delivery of biologically active proteins into mammalian cells, *Nat. Biotechnol.*, 19, 1173, 2001.
9. Krauss, U. et al., In vitro gene delivery by a novel human calcitonin (hCT)-derived carrier peptide, *Bioorg. Med. Chem. Lett.*, 14, 51, 2004.
10. Jones, S.W. et al., Characterisation of cell-penetrating peptide-mediated peptide delivery, *Br. J. Pharcol.*, 145, 1093, 2005.
11. Shiraishi, T., Pankratova, S., and Nielsen, P.E., Calcium ions effectively enhance the effect of antisense peptide nucleic acids conjugated to cationic tat and oligoarginine peptides, *Chem. Biol.*, 12, 923, 2005.
12. Díaz-Mochón, J.J. et al., Synthesis and cellular uptake of cell delivering PNA–peptide conjugates, *Chem. Commun.*, 3316, 2005.
13. Petersen, L. et al., Synthesis and in vitro evaluation of PNA–peptide–DETA conjugates as potential cell penetrating artificial ribonucleases, *Bioconjug. Chem.*, 15, 576, 2004.
14. Howl, J. et al., Intracellular delivery of bioactive peptides to RBL-2H3 cells induces beta-hexosaminidase secretion and phospholipase D activation, *ChemBioChem*, 4, 1312, 2003.
15. Pooga, M. et al., Cellular translocation of proteins by transportan, *FASEB J.*, 15, 1451, 2001.
16. Chaubey, B. et al., A PNA–transportan conjugate targeted to the TAR region of the HIV-1 genome exhibits both antiviral and virucidal properties, *Virology*, 331, 418, 2005.
17. Östenson, C.G. et al., Overexpression of protein-tyrosine phosphatase PTP sigma is linked to impaired glucose-induced insulin secretion in hereditary diabetic Goto–Kakizaki rats, *Biochem. Biophys. Res. Commun.*, 291, 945, 2002.
18. Pooga, M. et al., Cell penetrating PNA constructs regulate galanin receptor levels and modify pain transmission in vivo, *Nat. Biotechnol.*, 16, 857, 1998.
19. Fisher, L. et al., Cellular delivery of a double-stranded oligonucleotide NF-κB decoy by hybridization to complementary PNA linked to a cell-penetrating peptide, *Gene Ther.*, 11, 1264, 2004.
20. El-Andaloussi, S., et al., TP10, a delivery vector for decoy oligonucleotides targeting the Myc protein, *J. Controlled Release*, 110, 189, 2005.
21. Kilk, K. et al., Evaluation of transportan 10 in PEI mediated plasmid delivery assay, *J. Controlled Release*, 103, 511, 2005.
22. Prater, C.E. and Miller, P.S., 3'-Methylphosphonate-modified oligo-2'-O-methyl-ribonucleotide and their Tat peptide conjugates: Uptake and stability in mouse fibroblasts in culture, *Bioconjug. Chem.*, 15, 498, 2004.

23. Turner, J.J., Arzumanov, A.A., and Gait, M.J., Synthesis, cellular uptake and HIV-1 Tat-dependent trans-activation inhibition activity of oligonucleotide analogues disulphide-conjugated to cell-penetrating peptides, *Nucleic Acids Res.*, 33, 27, 2005.

24. Astriab-Fisher, A. et al., Conjugates of antisense oligonucleotides with the Tat and Antennapedia cell-penetrating peptides: Effects on cellular uptake, binding to target sequences, and biological actions, *Pharm. Res.*, 19, 744, 2002.

25. Moulton, H.M. et al., HIV Tat peptide enhances cellular delivery of antisense morpholino oligomers, *Antisense Nucleic Acid Drug Dev.*, 13, 31, 2003.

26. Moulton, H.M. et al., Cellular uptake of antisense morpholino oligomers conjugated to arginine-rich peptides, *Bioconjug. Chem.*, 15, 290, 2004.

27. Nelson, M.H. et al., Arginine–rich peptide conjugation to morpholino oligomers: Effects on antisense activity and specificity, *Bioconjug. Chem.*, 16, 959, 2005.

28. Krauss, U. et al., Novel daunorubicin–carrier peptide conjugates derived from human calcitonin segments, *J. Mol. Recognit.*, 16, 280, 2003.

29. Liang, J.F. and Yang, V.C., Synthesis of doxorubicin–peptide conjugate with multidrug resistant tumor cell killing activity, *Bioorg. Med. Chem. Lett.*, 15, 5071, 2005.

30. Kirschberg, T.A. et al., Arginine-based molecular transporters: The synthesis and chemical evaluation of releasable taxol–transporter conjugates, *Org. Lett.*, 5, 3459, 2003.

31. Liang, J.F. and Yang, V.C., Insulin-cell penetrating peptide hybrids with improved intestinal absorption efficiency, *Biochem. Biophys. Res. Commun.*, 335, 734, 2005.

32. Richard, J.P. et al., Cellular uptake of unconjugated TAT peptide involves clathrindependent endocytosis and heparan sulfate receptors, *J. Biol. Chem.*, 280, 15300, 2005.

33. Santra, S. et al., Rapid and effective labeling of brain tissue using Tat-conjugated CdS: Mn/ZnS quantum dots, *Chem. Commun.*, 3144, 2005.

34. Wadia, J.S., Stan, R.V., and Dowdy, S.F., Transducible TAT-HA fusogenic peptide enhances escape of TAT-fusion proteins after lipid raft macropinocytosis, *Nat. Med.*, 10, 310, 2004.

35. Sakaguchi, M. et al., Targeted disruption of transcriptional regulatory function of p53 by a novel efficient method for introducing a decoy oligonucleotide into nuclei, *Nucleic Acids Res.*, 33, e88, 2005.

36. Krautwald, S. et al., Transduction of the TAT-FLIP fusion protein results in transient resistance to Fas-induced apoptosis in vivo, *J. Biol. Chem.*, 279, 44005, 2004.

37. Saito, G., Swanson, J.A., and Lee, K.D., Drug delivery strategy utilizing conjugation via reversible disulfide linkages: Role and site of cellular reducing activities, *Adv. Drug Deliv. Rev.*, 55, 199, 2003.

38. Díaz-Mochón, J.J., Bialy, L., and Bradley, M., Full orthogonality between Dde and Fmoc: The direct synthesis of PNA–peptide conjugates, *Org. Lett.*, 6, 1127, 2004.

39. Dawson, P. et al., Synthesis of proteins by native chemical ligation, *Science*, 266, 776, 1994.

40. Yan, L.Z. and Dawson, P.E., Synthesis of peptides and proteins without cysteine residues by native chemical ligation combined with desulfurization, *J. Am. Chem. Soc.*, 123, 526, 2001 (and references therein).

41. Ingenito, R. et al., Solid phase synthesis of septide *C*-terminal thioesters by Fmoc/t-Bu chemistry, *J. Am. Chem. Soc.*, 121, 11369, 1999.

42. von Eggelkraut-Gottanka, R. et al., Peptide $^\alpha$thioester formation using standard Fmoc-chemistry, *Tetrahedron Lett.*, 44, 3551, 2003.

43. de Koning, M.C. et al., An approach to the synthesis of peptide–PNA–peptide conjugates via native chemical ligation, *Tetrahedron Lett.*, 43, 8173, 2002.
44. Rabanal, F. et al., Use of 2,2′-dithiobis(5-nitropyridine) for the heterodimerization of cysteine containing peptides. Introduction of the 5-nitro-2-pyridinesulfenyl group, *Tetrahedron Lett.*, 37, 1347, 1996.
45. Ghosh, A.K. and Erkang, F., A novel method for sequence independent incorporation of activated/protected cysteine in Fmoc solid phase peptide synthesis, *Tetrahedron Lett.*, 41, 165, 2000.
46. Fischer, R. et al., Extending the applicability of carboxyfluorescein in solid phase synthesis, *Bioconjug. Chem.*, 14, 653, 2003.

32 Toxicity Methods for Cell-Penetrating Peptides

Külliki Saar and Ülo Langel

CONTENTS

32.1 INTRODUCTION

In this chapter we summarize the data available about the toxicity of cell-penetrating peptides (CPPs). The major part of the data available concerns the in vitro toxicity of CPPs, that is their effects on the viability of cultured cells. In general, the classical definitions of cell viability include the exclusion of Trypan Blue as an indicator of membrane integrity and colony formation to determine whether cells can continue to proliferate.[1] Nowadays, the choice of viability/toxicity assays is not restricted to those two classical assays; instead, a whole spectrum of assays and their variants is commercially available. In this chapter we discuss the methods that have been used to determine the toxic effects of CPPs and point out some nuances which should be taken into consideration when working with these peptides. We also briefly summarize the data available about the hemolytical activity of CPPs and about their toxicity in vivo.

32.2 CPP TOXICITY IN VITRO

32.2.1 SOME PRACTICAL ISSUES ABOUT PEPTIDES

Peptides are sensitive compounds to work with, mostly due to their susceptibility to both chemical and enzymatic degradation. Spontaneous chemical degradation can occur in solution and even during storage in the dry powder form.[2] Consequently, it is good practice to confirm the intactness of peptides before each experiment, for example, by mass spectrometrical analysis.

Enzymatic degradation is an important issue when peptides come into contact with cells, such as in the toxicity assays described in this chapter. Unless consisting of D-amino acids or nonnatural amino acids, peptides are degraded both intracellularly and extracellularly. Extracellular degradation is performed by the enzymes situated in the extracellular matrix or lipid bilayer of cells and also by the enzymes present in serum (if serum is included in the exposure medium). The degradation rates for a given CPP can vary substantially between cell lines. For example, the metabolic half-life of Fl-pAntp(43–58)amide varies from 50 min on Calu-3 cells to 495 min on MDCK layers when applied in the absence of serum.[3] Therefore, the toxicity registered does not necessarily originate from the intact CPP only, but can be the sum of the toxicity of the intact CPP and its degradation products in the particular system.

In addition, peptides are notorious for their binding to plastic and glass surfaces and this has been shown to be true also for CPPs.[4] Consequently, when applied in submicromolar concentrations, the effective concentration of CPP or CPP–cargo complexes can decrease significantly due to the adsorption to plastic.

32.2.2 TOXICITY OF RHODAMINE-LABELED, FLUORESCEINYLATED, BIOTINYLATED, AND UNLABELED CPPs

32.2.2.1 Effects of CPPs in Membrane Integrity Assays

The simplest definition of CPPs is that they are cationic peptides with enhanced cellular uptake. This means that they have some kind of interaction with membranes, that is, they are membrane active. The combination of cationic and membrane-active properties can result in a membrane-perturbing nature, characteristic for instance of another group of cationic peptides known as cationic antimicrobial peptides.

The assays that have been used to measure the effects of CPPs on membrane integrity (see Table 32.1 and Figure 32.1) are the ones commonly used in in vitro toxicology. In general, these assays quantify the extent of dye uptake normally excluded by viable cells (e.g., propidium iodide (PI)), the leakage of a cytoplasmic

TABLE 32.1
The Effects of Rhodamine-Labeled, Fluoresceinylated, Biotinylated, and Unlabeled CPPs in Membrane Integrity Assays

Assay	Cell Line, Exposure Conditions, Registered Toxicity	Refs.
Dye uptake		
PI staining	U2OS, 10 μM, 2 h: Fl-Arg$_7$-amide**, Fl-pAntp(43–58)amide**	17
	A431, 10 μM, 2 h: Fl-Arg$_7$-amide**, Fl-pAntp(43–58)amide*	
PI staining	A172, RG2, 10 μM, 45 min: pAntp(43–58)*,	20
	Ala-Gly$_3$-pTAT(47–53)amide*	
Leakage of cytoplasmic enzyme		
LDH leakage	AECs, up to 16 μM, 30 min: Fl-MAP***	6
LDH leakage	MDCK, 40 μM, 30–120 min: Fl-hCT(9–32)amide*,	21
	Fl-hCT(18–32)amide*	
	MDCK, 10 μM, 60–240 min: Fl-Gly-pTAT(47–57)-Gly*,	
	Fl-pAntp(43–58)amide*	
LDH leakage	N2A, up to 50 μM, up to 70 min: bPrPp***, pAntp(43–58)*	22
LDH leakage	K562, MDA–MB-231, 10 μM, 10 min: pAntp(43–58)amide*,	7
	pTAT(48–60)amide*, pVEC(615–632)amide*, MAP***, TP10***	
	AECs, 10 μM, 10 min: pAntp(43–58)amide*, pTAT(48–60)amide*,	
	pVEC(615–632)amide*, MAP*, TP10**	
Leakage of cell-introduced probes		
DGP leakage	Bowes, up to 20 μM, up to 30 min: biotinyl-pAntp(43–58)amide*,	23
	biotinyl-pTAT(48–60)amide*, (N$^{\varepsilon 13}$-biotinyl)-transportan**,	
	biotinyl-MAP***	
DGP leakage	Caco-2, up to 30 min: 10 μM (N$^{\varepsilon 13}$-Fl)-transportan*, (N$^{\varepsilon 7}$-Fl)-TP10*,	24
	Fl-pAntp(43–58)*, 20 μM (N$^{\varepsilon 13}$-Fl)-transportan ***, (N$^{\varepsilon 7}$-Fl)-TP10***,	
	Fl-pAntp(43–58)**	
Calcein leakage	K562, up to 40 μM, 90 min: SynB1*, SynB3**, SynB5***,	25
	pAntp(43–58)***	
Fluorescein leakage	AECs, up to 16 μM, 30 min: Fl-MAP***	6

Note: The label is in the N-terminus of the peptide if not stated otherwise. As a measure of the toxicity (at the highest CPP concentration tested), the percentage of dye-labeled cells (in a PI-staining assay) or the percentage of leaked-out probe is given: *0–10%; **10–30%; ***>30%. The baseline is taken as 0. All experiments were performed at 37°C.

enzyme (lactate dehydrogenase (LDH)) or the leakage of cell-introduced probes (2-deoxy-D-[1-^{3}H]glucose-6-phosphate (DGP)), calcein, or fluorescein). In DGP leakage assay, the cells are loaded with 2-deoxy-D-[1-^{3}H]glucose (DG), which is readily taken up into cells via the glucose transport system.[5] It is then phosphorylated by hexokinase and the product, DGP, is retained by the cells with intact plasma membranes. In calcein and fluorescein leakage assays, the cells are loaded with either calcein AM (CAM) or fluorescein diacetate (FDA), which are nonfluorescent and nonpolar molecules and therefore able to diffuse into cells.

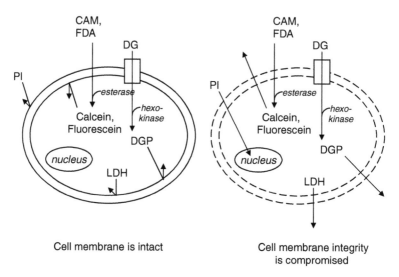

FIGURE 32.1 Membrane integrity assays used to measure the membrane-perturbing effects of CPPs. See Section 32.2.2.1 for more details.

Once inside, intracellular esterases convert them into fluorescent molecules (calcein or fluorescein) that are retained by the cells with intact plasma membranes.

The difficulties associated with the assays based on the leakage of cell-introduced probes are often due to the insufficient uptake of the probe precursors and the high spontaneous leakage of the probes. In comparison, the LDH leakage assay is convenient and simple because it does not require a preloading step. In addition, the LDH leakage assay has a multiplexing capacity: it is possible to remove a portion of the medium from each well to a separate vial to measure LDH and perform a separate assay on the cells in the original well. However, the molecular weight of the LDH molecule is 132 kDa and therefore it can be considered too large a molecule to detect early events in membrane integrity. Therefore, it is good practice to perform two membrane integrity assays in parallel, for example, LDH leakage + calcein leakage or LDH leakage + DGP leakage.[6,7]

As seen in Table 32.1, the integrity of cell membranes has been reported to be compromised by several CPPs applied in micromolar concentrations.

32.2.2.2   Effects of CPPs in Assays that Use Cell Proliferation or Metabolic End Points to Measure Viability

As seen in Table 32.2, the most popular assays have been tetrazolium assays: MTT, MTS, XTT, and WST-1 assays. Tetrazolium assays are colorimetric assays that are based upon the reduction of a tetrazolium salt to an intensely colored formazan by viable cells. While MTT-formazan has to be solubilized prior to its quantification,

TABLE 32.2
Effects of Rhodamine-Labeled, Fluoresceinylated, Biotinylated, and Unlabeled CPPs in Assays that Use Cell Proliferation or Metabolic End Points to Measure Viability

Assay	Cell Line, Exposure Conditions, Registered Toxicity	Refs.
Tetrazolium dye reduction		
MTT reduction	AECs, up to 16 μM, 30 min: Fl-MAP[***]	6
MTT reduction	HeLa, up to 100 μM, 24 h: Cys(Acm)-pTAT(48–60)-Cys[***], Cys(Acm)-pTAT(43–60)-Cys[***], Cys(Acm)-pTAT(37–60)-Cys[***], Cys(Acm)-pTAT(37–53)-Cys[***], HeLa, up to 100 μM, 1 h: all tested peptides[*]	26
MTT reduction	HeLa, 10 μM, 24 h: pTAT(48–60)amide[*], Arg$_9$-TAT[*], HIV-1 Rev-(34–50)amide[*], FHV coat-(35–49)amide[*]	27
MTT reduction	HeLa, 100 μM, 24 h: Arg$_9$-TAT[***], all other tested peptides	
MTT reduction	HS-68, NIH3T3, 293, Jurkat T, 48 h, 1 mM Pep-1[**], 10 mM Pep-1[***]	28
MTT reduction	IB4, 0.1–30 $\mu M \times$72 h, biotinyl-Gly-pAntp(43–58)-Gly[*]	29
MTT reduction	HeLa, up to 40 $\mu M \times$30 min, readout 24 h later, Fl-pAntp(43–58)amide[*], Fl-Arg$_9$-amide[*]	13
MTS/PMS reduction	MDCK, Calu-3, TR146, 2 h: up to 100 μM Fl-hCT(9–32)amide[*], Fl-hCT(18–32)amide[*]	15
	MDCK, Calu-3, TR146, 2 h: 10–100 μM Fl-Gly-pTAT(47–57)-Gly-amide[*], 1 mM Fl-Gly-pTAT(47–57)-Gly-amide[***]	
	MDCK and Calu-3, 10–100 μM, 2 h: Fl-pAntp(43–58)amide[*]	
	TR146, 10–100 μM, 2 h: Fl-pAntp(43–58)amide[**]	
	MDCK, Calu-3, TR146, 1 mM, 2 h: Fl-pAntp(43–58)amide[***]	
MTS reduction	HeLa, 1–100 μM, 24 h, Phe[6,14]-penetratin-amide[*], dodeca-penetratin-amide[*]	30
XTT reduction	HeLa, 30–100 $\mu M \times$2 h, Fl-hCT(9–32)amide[*]	31
WST-1 reduction	7 h, Rh-pAntp(43–58), Rh-pTAT(48–60), Rh-TP10 and Rh-Arg$_{11}$:	14
	EC$_{50}$ values (μM) in A549 cells: 18, 96, 9, and 5; in HeLa cells: 11, >100, 6, 4; in CHO-K1 cells: 23, >100, 14 and 10	
	7 h, pAntp(43–58), pTAT(48–60), TP10 and Arg$_{11}$: EC$_{50}$ values (μM) in A549 cells: >100, 76, >100, >100	

(continued)

Table 32.2 (Continued)

Assay	Cell Line, Exposure Conditions, Registered Toxicity	Refs.
WST-1 reduction	CHO-K1 and HeLa, 48 h, EC_{50} values (μM) pTAT-, DPV6-, DPV7-, DPV10/6, DPV1047-, DPV10- and DPV15b-maleimide > 1000; DPV3-maleimide: 616 (CHO-K1) and 900 (HeLa); DPV7b-maleimide: 650 (CHO-K1) and 850 (HeLa); DPV3/10-maleimide: 600 (CHO-K1) and 675 (HeLa), DPV15-maleimide: 516 (CHO-K1) and 833 (HeLa)	19
DNA synthesis		
^{3}H-thymidine incorporation	HeLa, up to 40 $\mu M \times 30$ min, readout 24 h later, Fl-pAntp(43–58)amide*, Fl-Arg$_9$-amide*	13
	IB4, 0.1–30 $\mu M \times 72$ h, biotinyl-Gly-pAntp(43–58)-Gly*	29
Protein synthesis		
^{3}H-methionine incorporation	Primary lymphocytes, 24 h, 0.2–20 μM, biotinyl-Gly-pAntp(43–58)-Gly*	29
Energy metabolism		
ATP levels by chemiluminescence	HeLa, 4.6–16 $\mu M \times 24$ h, backbone cyclic Rev-derived CPPs*,**,***	32

Note: The label is in the N-terminus of the peptide if not stated otherwise. Asterisk(s) indicates the decrease in the readout (compared to nonexposed cells) at the highest CPP concentration tested: *0–10%; **10–30%; ***>30%. The baseline is taken as 0. All experiments were performed at 37°C.

MTS and XTT form water-soluble formazans. MTS and XTT require the addition of an intermediate electron acceptor (such as PMS) to accelerate their reduction to formazans. WST-1 is a ready-to-use reagent, that is, it does not require any extra addition or solubilization steps.

The MTT assay was developed more than 20 years ago and has been used extensively as a convenient and rapid measure of viable cells.[8] However, the MTT assay does not always correlate with other measures of cell growth and viability, and several factors can significantly influence the reduction of MTT.[9,10] The subcellular localization and the biological events involved in tetrazolium dye reduction seem to change for different dyes. In contrast to the widely accepted view that the tetrazolium dyes are reduced by mitochondrial enzymes in viable cells, most MTT reduction occurs extramitochondrially and the WST-1 reduction occurs at the cell surface.[9,11]

In contrast to the membrane integrity assays described in the previous section, the tetrazolium assays are useful only as single-point assays because cells have to be lysed for analysis. In addition, they cannot distinguish between reduction in cell numbers and the impairment of the metabolic activity of the cells. EC_{50} could be a 50% reduction in cell number or a 50% reduction in metabolic capacity per cell.[12] Therefore, it is good practice to combine the tetrazolium dye assay with an assay that detects the reduction of cell numbers only. A reasonable combination is, for example, MTT assay + crystal violet assay.[13]

As seen in Table 32.2, the viability of the cells is decreased by several CPPs applied in micromolar concentrations. It is difficult to compare the results obtained by different research groups because different cell lines, exposure conditions, and different versions of the same peptides have been used. There is only a limited number of studies comparing the toxicity of a set of CPPs. From these studies it is clear that the conjugation of the rhodamine label results in an increased toxicity when compared to unlabeled CPPs and that the toxicity of the same peptide can be different in different cell lines, even at exactly the same exposure conditions.[7,14,15]

The significance of the exposure scheme was noted in the study by Ziegler and coworkers, in which they studied the metabolic activity of NIH3T3 cells exposed to Fl-βAla-Gly$_4$-pTAT(47–53).[16] The disturbance in the metabolic activity (measured using a microphysiometer) was detectable only 2.5 min after the start of the exposure and was dependent on Fl-βAla-Gly$_4$-pTAT(47–53) concentration (50–500 μM). The effect on metabolic activity was reversible within 10 min but only when the cells were washed with a peptide-free medium. The repeated exposure to 500 μM Fl-βAla-Gly$_4$-pTAT(47–53) resulted in irreversible metabolic changes.

32.2.2.3 Effects of CPPs in Apoptosis Assays

To our knowledge, there are no published data available reporting the intrinsic apoptotic effects of CPPs. It has even been reported, by Fotin-Mleczek and

TABLE 32.3
Hemolytic Effects of CPPs

Species	Exposure Conditions, Tested CPPs and Registered Toxicity	Refs.
Rat	2–100 μM, 1 h, pVEC(615–632)amide[*], TP10[**]	33
Bovine	50 μM, 30 min, pAntp(43–58)amide[*], pTAT(48-60)amide[*], pVEC(615–632)amide[*], MAP[*], TP10[**]	7
Human	30 μM, 60 min, 20°C, pAntp(43–58)[*], mPrPp[***], bPrPp[***]	22
Human	30 min, EC$_{50}$ for MAP: 10 μM	34
Human	1 h, EC$_{50}$ for DPV15-maleimide: 0.5 mM, for pTAT-, DPV6-, DPV7-, DPV10/6, DPV1047-, DPV10-, DPV15b-, DPV3-, DPV7b-, DPV3/10-maleimide: no detectable hemolysis at 1 mM concentration	19

Note: Asterisk(s) indicate the percentage of total hemoglobin that leaked out when erythrocytes were exposed to the highest CPP concentration tested: [*]0–10%; [**]10–30%; [***] >30%. Baseline is taken as 0. Experiments were performed at 37°C if not stated otherwise.

coworkers in a recent study, that pAntp(43–58)amide, Arg$_9$-amide, and pTAT(47–57)amide inhibit apoptosis induced by the tumor necrosis factor (TNF).[13] This occurs because the CPPs mentioned downregulate TNF receptors at the cell surface by inducing internalization. In that way, those CPPs antagonized the effect of the cargo (which was a proapoptotic peptide) at concentrations lower than 40 μM.

32.2.2.4 Hemolytic Effects of CPPs

The hemolytic activity of a potential drug-carrier vector is unwanted because it precludes the systemic use of this vector. In spite of its importance, there are only a few studies that have addressed the hemolytic activities of CPPs. As seen in Table 32.3, in most cases, the effects are again in the micromolar range and therefore similar to the general picture emerging from membrane integrity assays and proliferation assays.

32.2.3 Toxicity of CPP–Cargo Conjugates

Increasing evidence indicates that the presence and the nature of a cargo influence the translocation mechanisms and efficiency. Similarly, the toxicity of a CPP–cargo conjugate may be different from that exerted by a CPP alone. Two published studies have addressed this issue more closely. Maiolo and coworkers found that the conjugation of a tri- or tetrapetide cargo to Arg$_7$ and Arg$_7$Trp lowered the intrinsic toxities of Arg$_7$ and Arg$_7$Trp strikingly.[17] Jones and coworkers performed a detailed study and compared the toxicities of unlabeled pAntp(43–58), pTAT(48–60), TP10, and Arg$_{11}$, their rhodamine-labeled

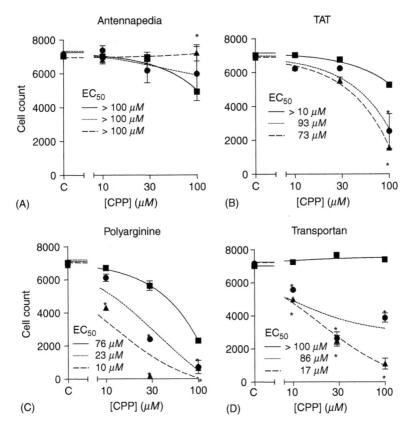

FIGURE 32.2 Cytotoxicity of unlabeled CPPs and CPP–NBD peptide conjugates in A549 cells. A549 cells were incubated with the indicated concentration of pAntp(43–58) (antennapedia), pTAT(48–60) (TAT), TP10, (transportan), and Arg_{11} (polyarginine) either alone (solid line) or conjugated to either a scrambled peptide (long-hatched line) or the NBD peptide (short-hatched line) for 7 h. The WST-1 assay reagent was then added and incubated for 30 min before absorbance at 450 and 650 nm was read. Bars represent mean $\pm$s.e.m. $(n=3)$. $^*P<0.05$. Significantly different from the corresponding unconjugated CPP. (From Jones, S.W. et al., *Br. J. Pharmacol.*, 145 (8), pp. 1093–1102, 2005. With permission.)

counterparts and their conjugates with a undecapeptide cargo.[14] The conjugation of undecapeptide or its scrambled sequence increased the toxicity of TP10 and pTAT(48–60), decreased the toxicity of Arg_{11}, and had no effect on pAntp(43–58) response (Figure 32.2).

One may conclude from the experiments summarized above, performed on cultured cancer cells, that in most cases, the presence of a cargo reduces the intrinsic toxicity of a CPP. Indeed, toxic effects reported for CPP–cargo conjugates have been very low or undetectable even in such sensitive systems as primary mammalian neurons.[18]

TABLE 32.4
Observations About In Vivo Toxicity of CPPs or Their Conjugates

Animal	Administration	Compound	Observations	Refs.
Rat	Intrathecal	PNA-pAntp(43–58)-amide	No signs of toxicity observed	35
Mouse	Intraperitoneal	β-galactosidase-pTat(47–57)	No disruption of blood–brain barrier, no signs of gross neurological problems or systemic distress	36
Mice	Topical	Cyclosporine-Arg$_7$	No evidence of adverse consequences	37
Rat	Intrastriatal	Fl-Ahx-pAntp(43-58)	At 10 μg dose: neurotoxic cell death and recruitment of inflammatory cells, at 1 μg dose both effects were reduced	38
Mouse	Intravenous	pTAT-maleimide DPV-maleimides	MTD for pTAT-maleimide: 7–9 μmol/kg MTD values for DPV-maleimides (in μmol/kg): DPV3-maleimide: 9–12; DPV6-maleimide: 12–15; DPV7-maleimide: ≥ 8; DPV7b-, DPV3/10-, DPV10/6, and DPV15-maleimide: <9; DPV1047-, DPV10-, DPV15b-maleimide: >21	19

32.3 CPP TOXICITY IN VIVO

Unfortunately, there is no systematic study available addressing the acute and chronic toxicity of several well-documented CPPs. To date, the maximum tolerated dose (MTD) values have been published only for the Diatos peptide vectors (DPVs).[19] In that study, DPV–maleimide conjugates were administered intravenously, and body weight and mortality were evaluated for at least 7 days postinjection. As a comparison, a pTAT–maleimide conjugate was used. The authors concluded that at least 4 of 10 tested DPVs were two to three times less toxic than the pTAT peptide (MTD values are included in Table 32.4).

Table 32.4 also includes the data from the publications that have commented on the toxicity of CPPs in the in vivo context.

ACKNOWLEDGMENTS

This work was funded by a grant from EC Framework 5 (contract QLK3-CT-2002-01989).

APPENDIX

TABLE 32.5
Sequences of the Peptides Referred to in This Chapter

Ahx-pAntp(43–58)[38]	Ahx-RQIKIWFQNRRMKWKK
Ala-Gly$_3$-pTAT(47–53)[20]	AGGGYGRKKRRQRRR
βAla-Gly$_4$-pTAT(47–53)[16]	(βA)GGGGYGRKKRRQRRR
Arg$_7$[37]/-amide[17]	RRRRRRR/-NH$_2$
Arg$_9$-amide[13]	RRRRRRRRR-NH$_2$
Arg$_9$-TAT[27]	GRRRRRRRRRPPQ-NH$_2$
Arg$_{11}$[14]	RRRRRRRRRRR
bPrPp[22]	MANLGYWLLALFVTMWTDVGLCKKRPKP
Cys(Acm)-pTAT(48–60)-Cys[26]	C(Acm)GRKKRRQRRRPPQC
Cys(Acm)-pTAT(43–60)-Cys[26]	C(Acm)LGISYGRKKRRQRRRPPQC
Cys(Acm)-pTAT(37–60)-Cys[26]	C(Acm)FITKALGISYGRKKRRQRRRPPQC
Cys(Acm)-pTAT(37–53)-Cys[26]	C(Acm)FITKALGISYGRKKRRC
Dodeca-penetratin-amide[30]	RQIKIWFRKWKK-NH$_2$
DPV3-maleimide[19]	RKKRRRESRKKRRRESC(maleimide)
DPV6-maleimide[19]	GRPRESGKKRKRKRLKPC(maleimide)
DPV7-maleimide[19]	GKRKKKGKLGKKRDPC(maleimide)
DPV7b-maleimide[19]	GKRKKKGKLGKKRPRSRC(maleimide)
DPV3/10-maleimide[19]	RKKRRRESRRARRSPRHLC(maleimide)
DPV10/6-maleimide[19]	SRRARRSPRESGKKRKRKRC(maleimide)
DPV1047-maleimide[19]	(maleimide)CVKRGLKLRHVRPRVTRMDV
DPV10-maleimide[19]	SRRARRSPRHLGSGC(maleimide)
DPV15-maleimide[19]	LRRERQSRLRRERQSRC(maleimide)
DPV15b-maleimide[19]	(maleimide)CGAYDLRRRERQSRLRRRERQSR
FHV coat(35–39)amide[27]	RRRRNRTRRNRRRVR-NH$_2$
Gly-pTAT(47–57)-Gly[15,21]	GYGRKKRRQRRRG
Gly-pAntp(43–58)-Gly[29]	GRQIKIWFQNRRMKWKKG
HIV-1 Rev(34–50)amide[27]	TRQARRNRRRRWRERQR-NH$_2$
hCT(9–32)amide[15,21,31]	LGTYTQDFNKFHTFPQTAIGVGAP-NH$_2$
hCT(18–32)amide[15,21]	KFHTFPQTAIGVGAP-NH$_2$
MAP[6,7,23,34]	KLALKLALKALKAALKLA-NH$_2$
mPrPp[22]	MANLGYWLLALFVTMWTDVGLCKKRPKP
Pep-1[28]	KETWWETWWTEWSQPKKKRKV
pAntp(43–58)[14,20,22,25]/ amide[7,13,15,17,21,23,24,35]	RQIKIWFQNRRMKWKK/-NH$_2$
Phe[6,14]-penetratin-amide[30]	RQIKIFFQNRRMKFKK-NH$_2$
pTAT(48–60)[14]/amide[7,23,27]	GRKKRRQRRRPPQ/-NH$_2$
pVEC(615–632)amide[7,33]	LLIILRRRIRKQAHAHSK-NH$_2$
SynB1[25]	RGGRLSYSRRRFSTSTGR
SynB3[25]	RRLSYSRRRF
SynB5[25]	RGGRLAYLRRRWAVLGR
TP10[7,24,33]	AGYLLGKINLKALAALAKKIL-NH$_2$
TP10, C-terminal acid[14]	AGYLLGKINLKALAALAKKIL
Transportan[23,24]	GWTLNSAGYLLGKINLKALAALAKKIL-NH$_2$

REFERENCES

1. Riss, T.L. and Moravec, R.A., Use of multiple assay endpoints to investigate the effects of incubation time, dose of toxin, and plating density in cell-based cytotoxicity assays, *Assay Drug Dev. Technol.*, 2(1), 51, 2004.
2. Lai, M.C. and Topp, E.M., Solid-state chemical stability of proteins and peptides, *J. Pharm. Sci.*, 88(5), 489, 1999.
3. Tréhin, R. et al., Metabolic cleavage of cell-penetrating peptides in contact with epithelial models: Human calcitonin (hCT)-derived peptides, Tat(47–57) and penetratin(43–58), *Biochem. J.*, 382(Pt 3), 945, 2004.
4. Chico, D.E., Given, R.L., and Miller, B.T., Binding of cationic cell-permeable peptides to plastic and glass, *Peptides*, 24(1), 3, 2003.
5. Walum, E. and Peterson, A., Tritiated 2-deoxy-D-glucose as a probe for cell membrane permeability studies, *Anal. Biochem.*, 120(1), 8, 1982.
6. Oehlke, J. et al., Cellular uptake of an alpha-helical amphipathic model peptide with the potential to deliver polar compounds into the cell interior non-endocytically, *Biochim. Biophys. Acta*, 1414(1–2), 127, 1998.
7. Saar, K. et al., Cell-penetrating peptides: A comparative membrane toxicity study, *Anal. Biochem.*, 345(1), 55, 2005.
8. Mosmann, T., Rapid colorimetric assay for cellular growth and survival: Application to proliferation and cytotoxicity assays, *J. Immunol. Methods*, 65(1–2), 55, 1983.
9. Berridge, M.V. and Tan, A.S., Characterization of the cellular reduction of 3-(4,5-dimethylthiazol-2-yl)-2,5-diphenyltetrazolium bromide (MTT): Subcellular localiza-tion, substrate dependence, and involvement of mitochondrial electron transport in MTT reduction, *Arch. Biochem. Biophys.*, 303(2), 474, 1993.
10. Vistica, D.T. et al., Tetrazolium-based assays for cellular viability: A critical examination of selected parameters affecting formazan production, *Cancer Res.*, 51(10), 2515, 1991.
11. Berridge, M.V. et al., Biochemical and cellular basis of cell proliferation assays that use tetrazolium salts, *Biochemica*, 4, 14, 1996.
12. Freshney, I., Application of cell cultures to toxicology, *Cell Biol. Toxicol.*, 17(4–5), 213, 2001.
13. Fotin-Mleczek, M. et al., Cationic cell-penetrating peptides interfere with TNF signalling by induction of TNF receptor internalization, *J. Cell Sci.*, 118(Pt 15), 3339, 2005.
14. Jones, S.W. et al., Characterisation of cell-penetrating peptide-mediated peptide delivery, *Br. J. Pharmacol.*, 145(8), 1093, 2005.
15. Tréhin, R. et al., Cellular uptake but low permeation of human calcitonin-derived cell penetrating peptides and Tat(47–57) through well-differentiated epithelial models, *Pharm. Res.*, 21(7), 1248, 2004.
16. Ziegler, A. et al., The cationic cell-penetrating peptide CPP(TAT) derived from the HIV-1 protein TAT is rapidly transported into living fibroblasts: Optical, biophysical, and metabolic evidence, *Biochemistry*, 44(1), 138, 2005.
17. Maiolo, J.R., Ferrer, M., and Ottinger, E.A., Effects of cargo molecules on the cellular uptake of arginine-rich cell-penetrating peptides, *Biochim. Biophys. Acta*, 1712(2), 161, 2005.
18. Davidson, T.J. et al., Highly efficient small interfering RNA delivery to primary mammalian neurons induces microRNA-like effects before mRNA degradation, *J. Neurosci.*, 24(45), 10040, 2004.

19. De Coupade, C. et al., Novel human-derived cell-penetrating peptides for specific subcellular delivery of therapeutic biomolecules, *Biochem. J.*, 390(Pt 2), 407, 2005.

20. Jarajapu, Y.P. et al., Biological evaluation of penetration domain and killing domain peptides, *J. Gene Med.*, 7(7), 908, 2005.

21. Tréhin, R. et al., Cellular internalization of human calcitonin derived peptides in MDCK monolayers: A comparative study with Tat(47–57) and penetratin(43–58), *Pharm. Res.*, 21(1), 33, 2004.

22. Magzoub, M., Cell-Penetrating Peptides in Model Systems. Interaction, Structure Induction and Membrane Effects, Doctoral thesis, Stockholm University, 2004.

23. Hällbrink, M. et al., Cargo delivery kinetics of cell-penetrating peptides, *Biochim. Biophys. Acta*, 1515(2), 101, 2001.

24. Lindgren, M.E. et al., Passage of cell-penetrating peptides across a human epithelial cell layer in vitro, *Biochem. J.*, 377(Pt 1), 69, 2004.

25. Drin, G. et al., Studies on the internalization mechanism of cationic cell-penetrating peptides, *J. Biol. Chem.*, 278(33), 31192, 2003.

26. Vivès, E., Brodin, P., and Lebleu, B., A truncated HIV-1 Tat protein basic domain rapidly translocates through the plasma membrane and accumulates in the cell nucleus, *J. Biol. Chem.*, 272(25), 16010, 1997.

27. Futaki, S. et al., Arginine-rich peptides. An abundant source of membrane-permeable peptides having potential as carriers for intracellular protein delivery, *J. Biol. Chem.*, 276(8), 5836, 2001.

28. Morris, M.C. et al., A peptide carrier for the delivery of biologically active proteins into mammalian cells, *Nat. Biotechnol.*, 19(12), 1173, 2001.

29. Fenton, M., Bone, N., and Sinclair, A.J., The efficient and rapid import of a peptide into primary B and T lymphocytes and a lymphoblastoid cell line, *J. Immunol. Methods*, 212(1), 41, 1998.

30. Letoha, T. et al., Investigation of penetratin peptides. Part 2. In vitro uptake of penetratin and two of its derivatives, *J. Pept. Sci*, 11(12), 805, 2005.

31. Foerg, C. et al., Decoding the entry of two novel cell-penetrating peptides in HeLa cells: Lipid raft-mediated endocytosis and endosomal escape, *Biochemistry*, 44(1), 72, 2005.

32. Hariton-Gazal, E. et al., Functional analysis of backbone cyclic peptides bearing the arm domain of the HIV-1 rev protein: Characterization of the karyophilic properties and inhibition of rev-induced gene expression, *Biochemistry*, 44(34), 11555, 2005.

33. Nekhotiaeva, N. et al., Cell entry and antimicrobial properties of eukaryotic cell-penetrating peptides, *FASEB J.*, 18(2), 394, 2004.

34. Dathe, M. et al., Peptide helicity and membrane surface charge modulate the balance of electrostatic and hydrophobic interactions with lipid bilayers and biological membranes, *Biochemistry*, 35(38), 12612, 1996.

35. Pooga, M. et al., Cell penetrating PNA constructs regulate galanin receptor levels and modify pain transmission in vivo, *Nat. Biotechnol.*, 16(9), 857, 1998.

36. Schwarze, S.R. et al., In vivo protein transduction: Delivery of a biologically active protein into the mouse, *Science*, 285(5433), 1569, 1999.

37. Rothbard, J.B. et al., Conjugation of arginine oligomers to cyclosporin A facilitates topical delivery and inhibition of inflammation, *Nat. Med.*, 6(11), 1253, 2000.

38. Bolton, S.J. et al., Cellular uptake and spread of the cell-permeable peptide penetratin in adult rat brain, *Eur. J. Neurosci.*, 12(8), 2847, 2000.

33 Methods to Study the Translocation of Cell-Penetrating Peptides

Maria Lindgren, Margus Pooga, and Ülo Langel

CONTENTS

33.1 INTRODUCTION AND HISTORY

The first method used in the discovery of cell-penetrating peptides (CPPs) was microscopy on fixed cells and indirect labeling of biotinylated peptides.[1] Since then, a number of different methods have been developed for the study of these versatile peptides. Furthermore, the initial method of detection in fixed cells has been reevaluated and it was shown that, in some cases, the intracellular distribution can be overestimated in fixed cells.[2,3] The detection of intracellular CPPs poses several other problems; for example, the difficulty of distinguishing membrane bound CPP from cytoplasm localized CPP. In this chapter, the most widely used methods in detecting and quantifying the translocation of CPPs will be reviewed, as well as the quantification of a few cargoes delivered by CPPs.

For efficient detection of a translocated CPP, a reporter group is generally attached to the active amino acid sequence; indeed, the reporter group could also be called cargo since it is translocated by the aid of the transport peptide. The most frequently applied reporter group in in vitro studies is fluorescein. Fluorescein is either visualized in a fluorescence microscope or quantified by, for example, fluorescence activated cell sorting (FACS) analysis or fluorescence spectrophotometry. For practical experimental reasons we have chosen to divide the reporter group or cargo according to molecular weight (M_w); small labels for detection or low M_W cargoes such as fluorescein, and in part two, bioactive load or high M_W cargoes such as proteins or oligonucleotides (ONs).

Various cargo molecules, ONs, peptide nucleic acid (PNA), siRNA, drugs, peptides, and proteins have been successfully transported by CPPs into a great selection of cell types (for review, see Ref. [4]) and, more recently, CPPs have also been applied in in vivo studies (reviewed in Refs [5,6]).

Chemical conjugations by disulfide bonds or in-chain synthesis (for peptides and peptide nucleic acid (PNA)) are the two main methods for conjugation of cargo to CPP (see chapter 10). In addition, recombinant expressed fusion proteins have been used for both the Tat protein and the Tat peptides.[7,8] The importance of the detachment of the cargo from the transport peptide is a debated matter. It could be argued that if the CPP is not disturbing the cargo's intracellular activity, a detachment is not necessary, and could, in fact, provide a useful protection against enzymatic degradation. Recently, it has been shown that even mere mixing of the CPPs with its cargo yields intracellular delivery.[9,10]

In this chapter, various methods to quantify the efficiency of CPP translocation are presented. The emphasis is on methods where the main interest has been to characterize the CPP in itself. There are also a great number of publications where the focus has been on the activity of the cargo delivered by the CPP, and these are only briefly reviewed.

33.2 METHODS OF STUDYING CPP TRANSLOCATION

The concentration range of CPPs used during uptake experiments is crucial for an exact quantification. The mechanism of CPPs entering cells is a research intensive area; first the concept of cell membrane translocation (or penetration or transduction) was introduced,[1] then the endocytotic pathways were added,[11] and recently evidence is gathering to support that both these pathways exist in parallel.[12,13] Several CPPs cause pore formation or have lytic activity, e.g., transportan and model amphipatic peptide (MAP), but at concentrations lower than 5 μM it is negligible.[14] As described in chapter 31, dye-leakage assays or tritiated-glucose leakage can be used to detect and quantify the membrane-disturbing effects of CPPs. Endocytosis has been shown to contribute to the uptake of, for instance, Tat protein fragments.[15]

Another important challenge when quantifying CPP uptake is to distinguish between membrane-cell-associated peptide and intracellular located CPP.

For example, Oehlke et al.[16] have used a method where the membrane-bound peptide is modified in order to discriminate it from intracellular MAP.

In the studies of 2-amino benzoic acid (Abz)-labeled transportan, a trypsin treatment was applied[17] and pronase was used in a similar fashion for Tat peptide.[18] Furthermore, acid wash and oil centrifugation[19] and reduction-sensitive 7-nitrobenz-2-oxo-1, 3-diazol (NBD) fluorophore[20] have been utilized for the distinction between the membrane associated and internalized CPP. However, inactivation of the label (NBD group for example) on CPPs that readily flow out from cells, e.g., MAP and transportan, could yield results biased towards lower uptake (later confirmed by Lindgren et al.[21]). Scheller et al.[22] discovered this washout effect connected to the amphipaticity of the MAPs. The analogs with lower hydrophobic moment had a lower resistance to being washed out. The higher ability of amphipathic peptides to associate with intracellular structures is suggested to be the main contributor to this phenomenon.

33.2.1 TRANSLOCATION OF CPPS WITH NO OR LOW M_W LABEL

As mentioned above, this chapter is divided into two parts, one dealing with studies of the CPPs and the other on the quantification of cargoes delivered by CPPs. The reporter group, or label, is defined here as an addition to the CPP that will make it detectable. The exception is the mass spectrometry and HPLC detection where no reporter group is needed.

33.2.1.1 Microscopy or Live Cell Imaging

Today, there are a wide variety of fluorophores in use, with fluorescein being the most common in the field of CPPs. However, fluorescein is sensitive to photobleaching and, additionally, its hydrophobicity and size can alter the properties of the peptide. The Abz fluorophore is small, stable, and photo-resistent,[23] and does not alter the hydrophobicity of the peptide to the same extent as fluorescein and rhodamine. However, Abz is not easily visualized in fluorescence microscopy due to its deep blue color (λ_{em} 420 nm) but it is useful for spectrophotometric detection.

The coupling of the fluorophore to the CPP is usually carried out by covalent linkage to the N-terminal α-amino group or a Lys ε-amino group of the peptide. For fluoresceinyl modification, the succinimidyl ester (Molecular Probes, USA) can be used in a five times excess to the peptidyl-resin in t-Boc peptide chemistry. In solid-phase peptide synthesis with Fmoc chemistry the fluorescein isothiocyanate can be applied in the presence of diisopropylethyl amine (DIEA). The Abz group can be introduced by a routine amino acid coupling step. The labeling is preferably carried out before cleavage of the peptide from the solid support to ensure specific linkage of the fluorophore. Labeling during solid-phase synthesis has the further advantage of avoiding the time consuming and low yield of labeling the peptide in solution, because it is critical to remove any unreacted dye, which can greatly complicate subsequent experiments.[24]

Confocal fluorescence microscopy quantification of fluorophore-labeled CPP was described by Scheller et al.[25] An online protocol was employed that would avoid biasing the internalization results by outflow of the peptides. Briefly, cells are seeded on cover slips and treated with peptide. Three regions of interest are scanned in the cytosol and one in the nucleus of three selected cells. The fluorescence resulting from the incubation media is removed as background fluorescence. After 30 min assessment the cell viability is checked by the addition of Trypan Blue.[25]

A further development of quantitative fluorescence microscopy assay in live cells was presented by Maiolo et al.[12] The aim was to study the endocytotic element of cell entry by Arg-rich CPPs; they used image analysis algorithms to measure the relative amount of peptide taken up by the cell, as well as the extent to which the uptake was diffusely distributed in the cytoplasm, or measurements of the granularity, which are indicative of endocytotic uptake, because the extent of granular distribution is mostly eliminated at 4°C.

The 7-nitrobenz-2-oxo-1, 3-diazol (NBD) fluorophore, stable in Fmoc cleavage conditions,[26] has been used to label CPPs.[20] The main advantage with this label is the sensitivity towards reduction agents, such as dithionite, that can be used to inactivate extracellular and membrane-bound label. Moreover, the NBD group changes fluorescence characteristics in a hydrophobic environment toward an emission blue shift (from 537 to 530 nm) as well as an increase in intensity. This property makes it possible to measure membrane partition coefficients of the labeled peptide.

EXAMPLE 1

Live Cell Imaging of Fluoresceinyl CPPs

The cells were seeded on cover glass chambers (NUNC) on the day before experiments. The cells were washed with physiological buffer, i.e., phosphate-buffered saline (or serum free media), and 1–10 μM peptide in buffer was added and incubated for at least 30 min at 37°C. The cells were washed three or more times and the cell nuclei were stained with Hoechst 33258 (0.5 $\mu g/mL$) for 5 min, after which the cover slip was washed again. The images were obtained with a Leica DM IRE2 fluorescence microscope and a Leica DC350 camera (Leica Microsystems, Germany) and processed in PhotoShop 6.0 software (Adobe Systems Inc., CA).

33.2.1.2 FACS

Flow cytometry or fluorescence-activated cell sorting (FACS) is a convenient way of quantifying the content of fluorescently labeled peptide. This method has been used by several groups in the majority of the studies for detecting FITC-labeled CPPs.[18,26,27] FACS is a sensitive method; less than 100 fluorophores per particle/cell have been detected by this method.[26] After CPP treatment of the attached cells, they are treated with trypsin and fixed in a 10% paraformaldehyde solution prior to the cell sorting analysis. An alternative method to remove extracellularly bound peptide was introduced by Richard et al.,[18] a pronase treatment at 4°C. In addition, FACS was applied to compare the relative efficacy and uptake of several CPPs by Jones et al.[27]

33.2.1.3 Spectrophotometry and Fluorescence Correlation Microscopy

The Abz is compatible with both Fmoc and *t*-Boc chemistries, it is resistant to photobleaching, and has a well-characterized amino acid-like quencher, 3-nitro-Tyr, likewise stable under both the main types of peptide synthesis chemistries.

We have used a short fluorescently labeled peptide, with the sequence Abz-Cys-LKANL, as cargo in delivery quantification.[14] The CPPs contained a Cys-residue used for covalent attachment of the cargo-peptide via a disulfide bond, and a 3-nitro-Tyr extension acting as a quencher to the Abz group, resulting in a reduction sensitive fluorogenic construct (Figure 33.1). The cellular uptake of the construct is registered as an increase in the fluorescence intensity when the disulfide bond of the CPP-S-S-cargo construct is reduced in the intracellular milieu.

With this method, it is possible to monitor the intracellular degradation of the disulfide bond in real time, and hence the cellular uptake of the constructs by the increase in apparent fluorescence. Moreover, it eliminates the need to separate membrane-bound and internalized peptide. However, in order to isolate the uptake kinetic constant, it is necessary to determine the intracellular reduction rate constant. In this study it was achieved by first measuring the intracellular glutathione concentration and then measuring the reduction of the constructs in the presence of lipid vesicles (Figure 33.2).

In addition, it is necessary to confirm that the cargo peptide cannot be internalized after extracellular reduction of the construct, and also that the increase in fluorescence is not caused by glutathione outflow due to peptide-induced membrane leakage.

We used this technique to compare the uptake and cargo delivery efficiency of four different CPPs: penetratin, Tat (48–60), transportan, and model

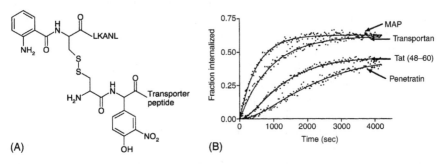

FIGURE 33.1 (A) The general structure of the fluorescence–quencher construct. The fluorophore-labeled cargo peptide, Abz-CLNKAL, is attached via a disulfide bond to a transporter peptide containing a 3-nitro-tyrosine, acting as quencher for Abz. (B) Example illustrating uptake curves; uptake of fluorescence–quencher construct at 1 μM concentration expressed as the fraction of total amount of construct plotted versus time. The fluorescence intensity obtained after total reduction of constructs with DTT is defined as 1. (Reproduced from Hällbrink et al., *Biochim. Biophys. Acta*, 1515, 101, 2001.)

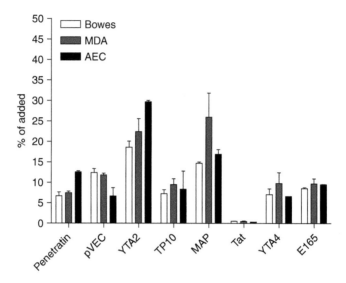

FIGURE 33.2 Example of CPP translocation studies performed in three different human cell lines: Bowes melanoma, breast cancer MDA MB-231, and transformed aorta endothelium. The measurements were carried out according to the quantitative fluorometry assay in Example 2. Parameters: 300,000 cells in 12-well plate, 300 μL of 5 μM peptide, 30 min at 37°C.

amphipathic peptide (MAP). Our data show that MAP has the fastest uptake, followed by transportan, Tat (48–60) and, last, penetratin. Similarly, MAP has the highest cargo-delivery efficiency, followed by transportan, Tat (48–60) and, last, penetratin.[28]

In an attempt to create a way to study the properties of CPPs in a more "high throughput" manner, a quantifying fluorometry-based assay was developed in our laboratory (see Example 2 for detailed protocol). It is used to evaluate the cell-penetrating properties of new peptides. It is useful to apply at least two methods in parallel to evaluate new CPPs: microscopy and quantitative uptake by fluorometry (QU); both these methods are applicable for fluorophore labeled peptides. To date, fluorescein has been the fluorophore of choice due to its ease to incorporate in peptide synthesis and its low cost; however, it does photobleach easily, which is a drawback for UV excited microscopy.

The QU assay is very similar to FACS analysis; however, the cells remain attached in culture plates throughout the experiment, which eases washing steps. Another difference is that it is the CPP content in cell lysates that are measured and not the fluorescence from intact cells. A further advantage of the QU assay is that acute membrane toxicity (LDH leakage) may be tested by multiplexing, as well as HPLC or mass spectrometry analysis of the extracellular supernatant to study peptide degradation. A similar method of fluorometric detection has been applied by Fischer et al.[29]

Fluorescence correlation microscopy (FCM) was applied to study CPPs by Weizenegger and coworkers.[30] They developed a protocol for performing intracellular concentration measurements in flat adherent tissue culture cells by FCM which makes the determination of the number of molecules in the confocal detection volume possible. This method was applied to a comparison of the import efficiencies of different cell-permeable peptides, labeled with either carboxyfluorescein or carboxytetramethyl-rhodamine (TAMRA), at nanomolar concentrations, by direct measurement in live cells.[30]

EXAMPLE 2

Quantification of FITC-Labeled CPPs by Fluorometry

Peptide solutions were prepared to $5 \mu M$ concentration in HEPES-buffered Krebbs-Ringer solution without BSA (HKR). Cells in a 12-well-plate, seeded the day before experiments at density 300,000 cells/well, were washed and exposed to 300 μL of $5 \mu M$ drug in HKR at 37°C and 300 rpm (on Thermomixer, Eppendorf) for 30 min. The peptide-treated cells were washed with trypsin/EDTA (0.13% T/E in HKR), washed thoroughly with HKR, and lysed in 250 μL 0.1% Triton X-100 in HKR at 4°C for 10 min. The cell lysates were transferred to a black 96-well plate for fluorescence readout at 492/520 nm for fluorescein on a Spectramax Gemini from Molecular Devices. The samples were compared to the fluorescence of the added amount of peptide and the total protein amount, results are given in pmol peptide/mg protein.

33.2.1.4   Cell Fractioning and Chromatography

Oehlke et al.[16] used high performance liquid chromatography (HPLC) to quantify uptake of fluorescein-labeled CPPs. In order to separate cell-surface bound and internalized peptide, treated cells were exposed to diazotized 2-nitroaniline, which selectively modifies Lys side chains. As the reagent does not cross the plasma membrane, only peptides exposed to the extracellular medium are modified, resulting in an increased hydrophobicity and, subsequently, an increase in HPLC retention time. This allows the separation of intracellular and surface-bound peptide. This procedure has the advantage that the reagent is highly reactive even at 0°C, so the results are not biased by efflux processes as they are suppressed at this temperature. An additional advantage is that the various metabolites of the peptide can be isolated and characterized by mass spectroscopy. Furthermore, the use of a fluorescence detector in the HPLC equipment increases the sensitivity of this analysis.

Zaro and Shen[31] applied size-exclusion chromatography for studying the intracellular fate of a new group of CPPs. Their aim was to measure where the peptides ended up, in cytosol or endosomes, performed by cell fractioning after treatment with iodinated peptides and FITC-dextran. Briefly, the cells were treated with the iodinated peptides and FITC-dextran, with or without several endocytosis inhibitors and at different temperatures, detached by trypsin and pelleted by centrifugation. The cells were disrupted by homogenization and lysates fractioned according to size. The fractions were analyzed by γ-counter, fluorescence, and total protein content. In this way, the two pathways of translocation of CPPs, the contribution of penetration and endocytosis, may be studied separately but in parallel.[13]

33.2.1.5 Mass Spectrometry

Mass spectrometry analysis is routinely used in peptide identification, both during chemical synthesis and in proteomics after trypsin digestion,[32] usually by matrix assisted laser desorption with time-of-flight detection (MALDI-TOF) mass spectrometry analysis. One of the main advantages is that there is no label needed to detect the correct and intact peptide, which in the case for CPPs would eliminate the bias of the cell translocation which may occur with a large, often very hydrophobic reporter group/cargo. In addition, as the peptide is identified by its unique molecular weight, this will show that it is indeed the intact peptide that is detected and not, for instance, a fluorescent metabolite.

The first report on mass spectrometry detection of intracellular CPP was published in 2001 by Elmquist et al.,[33] where the detection of intracellular *p*VEC in murine endothelial cells after different peptide exposure times were presented: 5 min, 4 h, and overnight (see Figure 33.3). The experiment was carried out as described in Example 3 below. Since then, several groups have applied MALDI detection of intracellular CPP.[29,34]

A recent publication by Burlina et al.[35] presents a method for quantifying the intracellular amount of CPPs by MALDI; briefly, cells were incubated with biotinylated CPPs, then subsequently lysed at 100°C for 15 min. Streptavidin-coated magnetic beads were added to the cell lysate so that the biotin–CPPs could be captured and concentrated before mixing with matrix and analyzed by mass spectrometry. The samples were spiked with the corresponding deuterated CPP (D^2-CPP) in known amounts for enabling quantification, without any change of desorption properties of the internal control. Two advantages of this new method are, first, that by comparison of the degradation pattern of the CPP and the D^2-CPP, internalized and membrane bound peptide could be distinguished. Second, an analysis mix of CPPs could be performed simultaneously, as with all mass spectrometry assays. Nevertheless, both biotin- and deuterium-modified CPP is needed, which complicates synthesis.

EXAMPLE 3

Analysis of Peptide Translocation by Mass Spectrometry

The day before the experiment, 400,000 cells/well were seeded in six-well plates. On the day of the experiment, the cells were washed twice with HKR to remove serum traces. The peptide solutions were diluted to 10 μM in HKR, 100 μL aliquots were saved for reference spectra, and 900 μL was added to the cells and incubated for 15–60 min at 37°C and 300 rpm in a Thermomixer. Then, 200 μL of the exposure solution was removed and stored at -20°C until analysis. The cells were washed once with HKR and

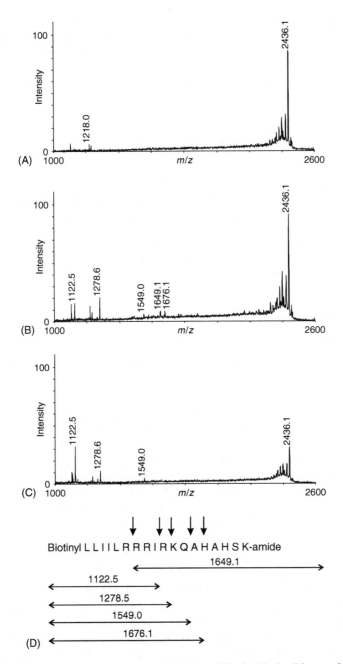

FIGURE 33.3 Mass spectrometry studies of pVECs stability in bEnd cell lysate. Only pVEC, m/z 2436.1 and m/z 1218.0 for the intact peptide, was found in bEnd cell lysate after 5 min of incubation (A). After 4 h incubation, five degradation products arise in the lysates (B). Overnight incubation (20 h) resulted in three remaining degradation products together with intact pVEC (C). The fragments of the biotinylated pVEC were identified, shown in (D), with the corresponding m/z values on the double arrowhead lines. Vertical arrows indicate found cleavage sites. (Reproduced from Elmquist et al., *Exp Cell Res.*, 269, 237, 2001.)

then treated with trypsin solution (0.13%) for 2 min at room temperature. The trypsin solution was then carefully aspirated and the cells were washed three times with HKR.

To prepare the lysate containing peptide trapped within the cells, the cells were lysed in 200 μL of 0.1% HCl/well for 10 min on ice. Cell lysates were then collected and stored at $-20°$C. Samples were prepared for mass spectrometry using ZipTip$_{C18}$ devices according to the manufacturer's instructions (Millipore). The ZipTip technology is widely used for the cleanup and concentration of peptide mixtures in proteomics.[36] We have previously used ZipTip$_{C18}$ devices when preparing mass spectrometry samples in studies addressing intra- and extracellular stability of pVEC.[33,37] It must be emphasized here that CPPs accumulate in cells at high levels; for example, intracellular concentrations in the 100 μM range have been reported for fluorescein-labeled pVEC,[33] MAP, and pAntp(43–58).[38] The matrix (α-cyano-4-hydroxycinnamic acid at 10 mg/mL in 50% acetonitrile/0.1% trifluoroacetic acid/water) was spotted before the sample on the sample plate. As an internal standard, a ProteoMass $P_{14}R$ MALDI mass spectrometry standard was used (monoisotopic M_W 1533.8582 Da). Data were collected from at least three different locations on the spot with intensity greater than 10,000 and accumulated. The mass spectrometry analysis was performed on a Voyager-DE STR and processed using MoverZ freeware.

33.2.2   CPPs with High M_W Cargo or Detection of the Biological Activity of the Cargo Delivered

The properties of conjugate or complex resulting from the combination of a cargo with the delivery peptide often differ profoundly from the properties of used CPP in size, charge, hydrophilicity, etc., and may influence the efficacy as well as the mechanism of cell entry. The increase in size and hydrophilicity, and decrease of charge, leads to the reduction of the cellular uptake of CPPs. The cellular translocation efficiency of each particular CPP–cargo conjugate/complex is dependent on the properties of both constituents and, therefore, the estimation of the concentration or activity of cell-delivered cargo is more relevant as compared to the characterization of the CPP part only. The size and character of cargoes introduced into cells with the help of CPPs extends from the small peptide and ON molecules to huge assemblies such as viral particles and liposomes. Here, only a limited number of examples are presented from the quickly expanding field, starting from the smaller delivered cargo molecules and moving towards larger complexes/assemblies.

One has to consider that the signal/effect recorded in the cell is a result of several subsequent events starting from the attachment of construct to cell surface, translocation into, intracellular trafficking of effector molecule, and association with its target in the presence of competing cellular factors, leading to the modulation of cellular processes. This inevitably leads to the higher variability of results and nonlinear dose–response curves. Quantification of the CPP-mediated cellular uptake of prelabeled cargo molecules usually yields in good correlation with the amount of applied construct. Still, the expected physiologic effect was not always achieved, nor did it correlate with the concentration of applied cargo, probably due to the more complicated uptake process and inefficient escape from endosomal structures in cells as compared with CPPs.

33.2.2.1 Detection of Bioactivity of Peptide Cargo

Shortly after the introduction of CPPs, their applications for transduction of peptides were published (for a comprehensive review, see Ref. [39]). The most popular application has been interfering with cell signaling and, more particularly later, the induction of apoptosis in anticancer treatment. As a first example, the predecessor of penetratin, Antennapedia homeodomain (AntpHD), allowed the definition of endocytotic machinery components for the release of prolactin from rat anterior pituitary cells. The C-terminal of a small GTPase Rab3 but not of Rab1 or Rab2 coupled to the delivery peptide blocked exocytosis of prolactin.[40] AntpHD peptide also enabled the inactivation of PDGF, IGF-I, and insulin receptor tyrosine kinases by transduction of proline-rich SH2 domain-containing signaling mediator.[41] The short peptide corresponding to the cyclin-dependent kinase 2 delivered into cells with penetratin or Tat peptide preferentially induced apoptosis in transformed cells but not in normal cells.[42]

Coupling of penetratin or Tat peptide to MAPK/ERK kinase N-terminal peptide facilitated the cellular translocation of the inhibitory peptide and prevented ERK activation in 4-phorbol 12-myristate 13-acetate-stimulated NIH 3T3 cells or nerve growth factor-treated PC12 cells in a concentration-dependent manner, as demonstrated by Western blot and reporter gene assays. The obtained effect was specific as the peptides did not have an inhibitory effect on the activity of two other closely related classes of MAPKs, c-Jun amino-terminal kinase, or p38 protein kinase.[43]

The function of NF-κB has been unraveled in different signal pathways by using its NLS sequence coupled to hydrophobic membrane targeting sequence (MTS), the hydrophobic region of the signal sequence of the Kaposi fibroblast growth factor (kFGF).[44,45]

Recently, Bidwell and Raucher,[46] used thermally responsive elastin-like polypeptide (ELP) fused to penetratin for the blocking of transcriptional activation by c-Myc. The peptide derived from helix 1 of the helix–loop–helix region of c-Myc (H1), known to block its nuclear translocation, was coupled to elastin-like polypeptide and penetratin (Pen). Penetratin and thermal induction both increased the cellular uptake of peptide as demonstrated by flow cytometry. The trifunctional peptide Pen-ELP-H1 expressed in bacteria blocked nuclear translocation and transcriptional activation of c-Myc, as shown by immunofluorescence and RT-PCR, respectively.

Most of the efficient CPP-mediated peptide delivery experiments in vivo have exploited Tat transduction domain (for a review, see chapter 11). Here, only some new applications for different CPPs are presented. Ischemic damage in the brain and heart could be reduced with inhibitory peptides conjugated to CPPs. Blocking the access of JNK to its targets by protease resistant peptide construct leads to efficient protection in cerebral ischemia models. Intraventricular administration of the peptide as late as after 5 h after occlusion reduced the volume of ischemic lesion by more than 90% and systemic administration by 72%. The protecting effect correlated with the prevention of c-Jun activation and c-Fos transcription.[47] The ischemic insult of cardiomyocytes was reduced by transduction of the octapeptide ψεRACK, which is known to exhibit selective ε protein kinase C (PKC) isozyme agonist activity. Heptaarginine (R_7) conjugate of RACK peptide entered cardiomyocytes and reduced ischemic damage when delivered into intact hearts either prior to or after the ischemic insult.[48]

The activity of p53 protein in cancer cells was restored by transduction of the p53 fragment with Tat, while no increase in activity was induced in normal cells. Treatment with RI-TATp53C′ D-peptide peritoneal carcinomatosis and peritoneal lymphoma models resulted in more than sixfold increases in lifespan and the generation of disease-free animals.[49] The peptides derived from PKC and cannabinoid receptor fused to TP10 induced the exocytosis of a secretory lysosomal marker, β-hexosaminidase.[50]

CPP–peptide conjugates have been successfully applied for in vivo transdermal delivery of cyclosporine[51] and interferon-gamma[52] by R_7 and penetratin, respectively.

SynB3 peptide facilitates the uptake of peptides over the blood–brain barrier (BBB) into the brain. The uptake of dalargin, an enkephalin analog, into the brain was increased markedly by vectorization with SynB3, as demonstrated by in situ brain perfusion. Moreover, the increased brain uptake was accompanied by the improvement of the antinociceptive effect, as demonstrated in the hot-plate model.[53]

The transduction of peptides is becoming a more and more powerful tool for the modulation of cellular equilibria. However, often the proteins from which the peptides are derived are more potent effectors and may exert the desired effect at an approximately 100-fold lower molar concentration.[54]

33.2.2.2 Detection of Oligonucletide Effects

The efficiency of antisense ONs could be markedly increased by coupling to CPP.[55] First, penetratin and AntpHD were used for the translocation of antisense ONs into cells. The antisense efficiency of ON against superoxide dismutase to induce PC12 cells' death increased 100-fold after coupling to penetratin via a cleavable disulfide bond. In addition, after coupling to penetratin, the cell-entry of ON became insensitive to the presence of serum during incubation.[55] Transient downregulation of β-APP synthesis was achieved in developing neurons in vitro with antisense ON coupled to AntpHD. A reduction in β-APP synthesis was accompanied by the inhibition of neurite elongation in the nanomolar range of antisense ON concentration.[56] However, the recent study by Turner et al.[57] demonstrated that the highly purified antisense ON–CPP constructs remained trapped in endosomal vesicles and, for escape into cytosol, the presence of an excess of free transporter peptide was needed.

CPPs have been exploited for cellular translocation of different ON analogs yielding in downregulation of targeted gene expression. Antisense peptide nucleic acid oligomers were used to inhibit the expression of galanin receptor type 1,[58] prepro-oxytocin,[59] c-myc,[60] nociceptin receptor,[61] phosphotyrosine phosphatase PTPσ,[62] and other proteins, as demonstrated by Western blotting and in vivo effects. Recently, the replication of hepatitis B virus was inhibited in duck hepatocytes by PNA–R7 conjugate targeting the encapsidation signal.[63]

Inhibitory ONs can be toxic at high concentrations and this toxicity is interpreted as an antisense effect. Therefore, an assay measuring the ability of the antisense molecule to correct splicing of an aberrant intron inserted into the luciferase gene has been a very useful system for quantifying the increase of luminescence signal as the response to ONs transduced into cells. The efficiency of phosphorithioate ONs targeting the splicing site was markedly increased after conjugation to penetratin or Tat peptide, corroborating that the CPP-mediated translocation of ONs did not impair the viability of HeLa/705 cells.[64]

Another popular DNA mimic, phosphoramidate morpholino oligomers (PMOs), showed 3–25-fold higher activity after coupling to arginine-rich peptide R9F2 and reduced the minimum length of PMO required.[65] PMOs coupled to Tat transduction domain or R9F2 allowed correction of miss-splicing in the mutant human β-globin gene,[66,67] and the inhibition of coronaviruses[68] and flaviviruses replication.[69]

The RNA interference strategy has become increasingly popular for knocking down the expression of gene of interest and substituting an antisense approach. The MPG peptide has shown excellent properties for cellular siRNA delivery. MPG complexes with siRNA against luciferase reduced its expression to minimal levels in COS-7 and HeLa cells, as demonstrated by the decrease of enzyme activity. Targeting of the GAPDH gene with siRNA–MPG complex induced degradation of GAPDH mRNA and loss of protein from cells after 48 h, as shown by Northern and Western blotting, respectively.[70]

A rather straightforward and new approach is the combination of CPPs with other cell transfection agents. The conjugation of phosphodiester ONs to protamines-derived peptides and combination with cationic lipids renders complexes resistant to DNase I and serum nucleases of cell culture medium. The antisense ON targeted against c-myc mRNA packed by protamine peptides into mixed particles reduced the c-Myc protein level in U937 cells, as demonstrated by Western blotting.[71]

Short ONs transduced into cells by CPPs have also been used for blocking the interaction of the effector protein with its target/recognition site on nucleic acid. Anti-TAR PNA coupled to transportan or other CPPs blocked the association of Tat protein to viral RNA and showed high antiviral and -virucidal activity.[72,73] The NFκB decoy double-stranded ON suppressed the IL-6 mRNA expression by displacing the transcription factor from its cognate sites when translocated into cells with complementary PNA–TP10 conjugate.[74]

33.2.2.3 Delivery and Detection of Drugs with Low M_W

Small drugs are commonly designed to translocate into cells; however, the cellular uptake and, especially, targeting into the brain has been highly facilitated by coupling to CPPs, especially of the SynB family. The arginine-rich SynB peptides derived from the porcine antimicrobial peptide protegrin-1 facilitated the passage of covalently coupled doxorubicin,[75,76] benzylpenicillin,[77] paclitaxel,[78] and morphine-6-glucuronide[79] over the BBB into the brain. The conjugation of doxorubicin to vector peptides D-penetratin and SynB increased brain delivery 20-fold, as estimated by capillary depletion and quantified by radiolabel. The vectorized doxorubicin bypassed the P-glycoprotein at BBB[80] and showed less accumulation in the heart as compared to free drug, showing the selectivity of SynB1-mediated delivery on the organism level.[75] A shorter vector peptide SynB3 was even more efficient and has been used for brain delivery of benzylpenicillin,[77] paclitaxel[78] as estimated by quantifying radiolabel, and morphine-6-glucuronide as verified in several models of nociception.[79]

Rapid and efficient delivery of radioactive metal complexes to the cell interior provides novel applications in medical imaging and radiotherapy. Tat peptide modified with a peptide-based chelating sequence induced the uptake of technetium-99 m and rhenium into Jurkat cells in culture as demonstrated by microscopy.[81] Chelating of gadolinium ion to DOTA-D-Tat led to a rapid uptake by Jurkat cells and significantly elevated magnetic resonance signals in the liver, kidney, and mesentery after administration to rats.[82]

33.2.2.4   Detection of Proteins Translocated by CPPs

CPPs have shown a good delivery potential by translocating labeled proteins into cells, as verified by fluorescence and enzymatic methods. The green fluorescent protein (GFP) has served as a simplest model cargo for demonstration of the protein transduction by most known CPPs in various cells in culture by using, typically, fluorescence microscopy and flow cytometry.[83–86] Different strategies of coupling the vector to a cargo yielded cell-translocating constructs: synthesis of the fusion protein in bacteria,[85] chemical cross-linking,[86] and noncovalent complex.[87] Another popular system has been the application of prelabeled avidin/streptavidin complexed noncovalently to biotinyl derivatives of CPPs, which has also helped in the elucidation of the cellular delivery mechanism by Trojan peptides.[33,88–92] However, with these methods, the problem of distinguishing the signal from cell interior and from the cell surface and endocytic vesicles persists. The activity of cargo protein reaching the cell nucleus is commonly identified/quantified by the transduction of Cre-recombinase into transgenic cells where the "stop" signal before β-galacto-sidase gene is placed between the lox-sites cleavable by Cre-recombinase.[11,93–95]

The pioneering study by Fawell et al.[96] used a chemically coupled fragment of amino acids 37–72 of the Tat protein to insert β-galactosidase, horseradish peroxydase, and RNase A into cells in culture. The respective conjugates also entered cells in vivo and the enzyme activity was detected in most examined tissues after intravenous administration in mice. The group at Amgen continued with the delivery of proteins, showing that Tat-mediated delivery facilitates MHC class I presentation of antigens. The transduction of ovalbumin resulted in subsequent processing and presentation on the level comparable with the effects of endogenous expression.[97]

Refinement of Tat transduction domain to 11 amino acids long peptide[98] enabled the generation of an in-frame bacterial expression vector and production of cell-penetrating recombinant proteins.[99] The distribution of protein conjugates in almost all tissues of mice after intraperitoneal administration was detected by the enzymatic activity of transduced β-galactosidase. The induced cellular responses, such as cell migration after transduction of p27,[99] death of HIV-infected cells after transduction of caspase-3 or HIV-protease,[100] and enzymatic activity of Cu,Zn-superoxide dismutase[101] were the first examples detected/quantified.

Recently, the prerequisite of metastasis formation, the migration of cells, was blocked by the inhibition of PLC-γ1 phosphorylation, induced by EGF receptor. A fusion protein containing two SH2 domains of PLC-γ1 fused to Tat transduction domain (PS2-Tat) reduced PLC-γ1 phosphorylation of approximately 30%. The EGF-induced migration of EGFR/c-erbB-2-positive MDA-HER2 cells was inhibited by approximately 75% and proliferation of approximately 50% by long-term treatment with PS2-Tat.[102] The transduction domain of Tat induces the internalization of constructed antiapoptotic protein FNK into chondrocytes, even through the highly negatively charged dense extracellular matrix, protecting against cell death induced by Fas antibody and nitric oxide.[103] The same construct entered efficiently cultured neurons, and isolated bone marrow mononuclear cells, protecting these against cell death induced by freezing and thawing.[104]

The cellular delivery of full-length proteins by CPPs has led to promising leads in anticancer treatment (for reviews, see Refs [54,105]), ischemic lesion reduction in brain[106] and heart,[107] and development of dendritic cell-based cancer vaccines,[108,109] etc. The example of detection gold-labeled streptavidin transduced into HeLa or Bowes melanoma cells by transportan or TP10 is presented in chapter 10.

33.3 SUMMARY

New methods for the explicit quantification of the internalization of CPPs are still under development. Most of the methods reviewed here have their weaknesses and are not ideal for all applications. A wide variety of recently introduced fluorophores might help to solve the problems of sensitivity, photobleaching, and pH dependence, and will also be applicable for FRET or other advanced fluorescence-based studies such as FCM.

Today, mass spectrometry methods enable the quantification of peptides and proteins by, for example, MALDI-TOF, and are developing as a new powerful tool for studying CPPs. The main advantage would be that the identification of the species present in the experiments could be carried out simultaneously with the actual quantification, as described above.

Furthermore, the development of FCM and quantification by CLSM will perhaps provide the possibility for higher sample throughput when screening for new CPP peptide sequences.

To conclude, it is evident that there are two subfields in the research of CPPs, one searching for new CPPs, while the second concentrates on applying the CPP to deliver cargoes, in which their main interest lies. Hopefully, this chapter has provided both subfields with enough practical considerations to further develop the field of CPP research.

REFERENCES

1. Derossi, D. et al., The third helix of the Antennapedia homeodomain translocates through biological membranes, *J. Biol. Chem.*, 269, 10444–10450, 1994.
2. Vives, E. et al., TAT peptide internalization: Seeking the mechanism of entry, *Curr. Protein Pept. Sci.*, 4, 125, 2003.
3. Vives, E., Cellular uptake of the Tat peptide: An endocytosis mechanism following ionic interactions, *J. Mol. Recognit.*, 16, 265, 2003.
4. Zorko, M. and Langel, Ü., Cell-penetrating peptides: Mechanism and kinetics of cargo delivery, *Adv. Drug Deliv. Rev.*, 57, 529, 2005.
5. Snyder, E.L. and Dowdy, S.F., Cell penetrating peptides in drug delivery, *Pharm. Res.*, 21, 389, 2004.
6. Deshayes, S. et al., Cell-penetrating peptides: Tools for intracellular delivery of therapeutics, *Cell Mol. Life Sci.*, 62, 1839, 2005.
7. Yang, Y. et al., HIV-1 TAT-mediated protein transduction and subcellular localization using novel expression vectors, *FEBS Lett.*, 532, 36, 2002.
8. Albarran, B., To, R., and Stayton, P.S., A TAT-streptavidin fusion protein directs uptake of biotinylated cargo into mammalian cells, *Protein Eng. Des. Sel.*, 18, 147, 2005.
9. Morris, M.C. et al., Combination of a new generation of PNAs with a peptide-based carrier enables efficient targeting of cell cycle progression, *Gene Ther.*, 11, 757, 2004.
10. Deshayes, S. et al., On the mechanism of non-endosomial peptide-mediated cellular delivery of nucleic acids, *Biochim. Biophys. Acta*, 1667, 141, 2004.
11. Wadia, J.S., Stan, R.V., and Dowdy, S.F., Transducible TAT-HA fusogenic peptide enhances escape of TAT-fusion proteins after lipid raft macropinocytosis, *Nat. Med.*, 10, 31, 2004.
12. Maiolo, J.R., Ferrer, M., and Ottinger, E.A., Effects of cargo molecules on the cellular uptake of arginine-rich cell-penetrating peptides, *Biochim. Biophys. Acta*, 1712, 161, 2005.
13. Zaro, J.L. and Shen, W.C., Evidence that membrane transduction of oligoarginine does not require vesicle formation, *Exp. Cell Res.*, 307, 164, 2005.
14. Hällbrink, M. et al., Cargo delivery kinetics of cell-penetrating peptides, *Biochim. Biophys. Acta*, 1515, 101, 2001.
15. Fawell, S. et al., Tat-mediated delivery of heterologous proteins into cells, *Proc. Natl Acad. Sci. U.S.A.*, 91, 664, 1994.
16. Oehlke, J. et al., Cellular uptake of an alpha-helical amphipathic model peptide with the potential to deliver polar compounds into the cell interior non-endocytically, *Biochim. Biophys. Acta*, 1414, 127, 1998.
17. Lindgren, M. et al., Translocation properties of novel cell penetrating transportan and penetratin analogues, *Bioconjug. Chem.*, 11, 619–626, 2000.
18. Richard, J.P. et al., Cellular uptake of unconjugated TAT peptide involves clathrin-dependent endocytosis and heparan sulfate receptors, *J. Biol. Chem.*, 280, 15300, 2005.
19. Pooga, M. et al., Cell penetration by transportan, *FASEB J.*, 12, 67, 1998.
20. Drin, G. et al., Physico-chemical requirements for cellular uptake of pAntp peptide. Role of lipid-binding affinity, *Eur. J. Biochem.*, 268, 1304, 2001.
21. Lindgren, M.E. et al., Passage of cell-penetrating peptides across a human epithelial cell layer in vitro, *Biochem. J.*, 377, 69, 2004.
22. Scheller, A. et al., Evidence for an amphipathicity independent cellular uptake of amphipathic cell-penetrating peptides, *Eur. J. Biochem.*, 267, 6043, 2000.

23. Meldal, M. and Breddam, K., Anthranilamide and nitrotyrosine as a donor–acceptor pair in internally quenched fluorescent substrates for endopeptidases: Multicolumn peptide synthesis of enzyme substrates for subutilising Carlsberg and pepsin, *Anal Biochem.*, 195, 141, 1991.

24. Wang, Y.L., Fluorescent analog cytochemistry: Tracing functional protein components in living cells, *Methods Cell Biol.*, 29, 1, 1989.

25. Scheller, A. et al., Structural requirements for cellular uptake of alpha-helical amphipathic peptides, *J. Pept. Sci.*, 5, 185, 1999.

26. Garcia-Echeverria, C. et al., A new Antennapedia-derived vector for intracellular delivery of exogenous compounds, *Bioorg. Med. Chem. Lett.*, 11, 1363, 2001.

27. Jones, S.W. et al., Characterisation of cell-penetrating peptide-mediated peptide delivery, *Br. J. Pharmacol.*, 145, 1093, 2005.

28. Hällbrink, M. et al., Cargo delivery kinetics of cell-penetrating peptides, *Biochim. Biophys. Acta.*, 1515, 101, 2001.

29. Fischer, R. et al., A stepwise dissection of the intracellular fate of cationic cell-penetrating peptides, *J. Biol. Chem.*, 279, 12625, 2004.

30. Waizenegger, T., Fischer, R., and Brock, R., Intracellular concentration measurements in adherent cells: A comparison of import efficiencies of cell-permeable peptides, *Biol. Chem.*, 383, 291, 2002.

31. Zaro, J.L. and Shen, W.C., Quantitative comparison of membrane transduction and endocytosis of oligopeptides, *Biochem. Biophys. Res. Commun.*, 307, 241, 2003.

32. Thiede, B. et al., Peptide mass fingerprinting, *Methods*, 35, 237, 2005.

33. Elmquist, A. et al., VE-cadherin-derived cell-penetrating peptide, pVEC, with carrier functions, *Exp. Cell Res.*, 269, 237, 2001.

34. Trehin, R. et al., Metabolic cleavage of cell-penetrating peptides in contact with epithelial models: Human calcitonin (hCT)-derived peptides, Tat(47–57) and penetratin(43–58), *Biochem. J.*, 382, 945, 2004.

35. Burlina, F. et al., Quantification of the cellular uptake of cell-penetrating peptides by MALDI-TOF mass spectrometry, *Angew. Chem. Int. Ed. Engl.*, 44, 4244, 2005.

36. Pluskal, M.G., Microscale sample preparation, *Nat. Biotechnol.*, 18, 104, 2000.

37. Elmquist, A. and Langel, Ü., In vitro uptake and stability study of pVEC and its all-D analog, *Biol. Chem.*, 384, 387, 2003.

38. Hällbrink, M. et al., Uptake of cell-penetrating peptides is dependent on peptide-to-cell ratio rather than on peptide concentration, *Biochim. Biophys. Acta.*, 1667, 222, 2004.

39. Dietz, G.P. and Bähr, M., Delivery of bioactive molecules into the cell: The Trojan horse approach, *Mol. Cell Neurosci.*, 27, 85, 2004.

40. Perez, F. et al., Rab3A and Rab3B carboxy-terminal peptides are both potent and specific inhibitors of prolactin release by rat cultured anterior pituitary cells, *Mol. Endocrinol.*, 8, 1278, 1994.

41. Riedel, H. et al., PSM, a mediator of PDGF-BB-, IGF-I-, and insulin-stimulated mitogenesis, *Oncogene*, 19, 39, 2000.

42. Chen, Y.N. et al., Selective killing of transformed cells by cyclin/cyclin-dependent kinase 2 antagonists, *Proc. Natl Acad. Sci. U.S.A.*, 96, 4325, 1999.

43. Kelemen, B.R., Hsiao, K., and Goueli, S.A., Selective in vivo inhibition of mitogen-activated protein kinase activation using cell-permeable peptides, *J. Biol. Chem.*, 277, 8741, 2002.

44. Lin, Y.Z. et al., Inhibition of nuclear translocation of transcription factor NF-kappa B by a synthetic peptide containing a cell membrane-permeable motif and nuclear localization sequence, *J. Biol. Chem.*, 270, 14255, 1995.

45. Torgerson, T.R. et al., Regulation of NF-kappa B, AP-1, NFAT, and STAT1 nuclear import in T lymphocytes by noninvasive delivery of peptide carrying the nuclear localization sequence of NF-kappa B p50, *J. Immunol.*, 161, 6084, 1998.
46. Bidwell, G.L. 3rd and Raucher, D., Application of thermally responsive polypeptides directed against c-myc transcriptional function for cancer therapy, *Mol. Cancer Ther.*, 4, 1076, 2005.
47. Borsello, T. et al., A peptide inhibitor of c-Jun N-terminal kinase protects against excitotoxicity and cerebral ischemia, *Nat. Med.*, 9, 1180, 2003.
48. Chen, L. et al., Molecular transporters for peptides: Delivery of a cardioprotective epsilonPKC agonist peptide into cells and intact ischemic heart using a transport system, R(7), *Chem. Biol.*, 8, 1123, 2001.
49. Snyder, E.L. et al., Treatment of terminal peritoneal carcinomatosis by a transducible p53-activating peptide, *PLoS Biol*, 2, E36, 2004.
50. Howl, J., Chimerism: A strategy to expand the utility and applications of peptides, *Methods Mol. Biol.*, 298, 25–41, 2005.
51. Rothbard, J.B. et al., Conjugation of arginine oligomers to cyclosporin a facilitates topical delivery and inhibition of inflammation, *Nat. Med.*, 6, 1253, 2000.
52. Lee, J. et al., Transdermal delivery of interferon-gamma (IFN-gamma) mediated by penetratin, a cell-permeable peptide, *Biotechnol. Appl. Biochem.*, 42, 169, 2005.
53. Rousselle, C. et al., Improved brain uptake and pharmacological activity of dalargin using a peptide-vector-mediated strategy, *J. Pharmacol. Exp. Ther.*, 306, 371, 2003.
54. Wadia, J.S. and Dowdy, S.F., Transmembrane delivery of protein and peptide drugs by TAT-mediated transduction in the treatment of cancer, *Adv. Drug Deliv. Rev.*, 57, 579, 2005.
55. Troy, C.M. et al., Downregulation of Cu/Zn superoxide dismutase leads to cell death via the nitric oxide-peroxynitrite pathway, *J. Neurosci.*, 16, 253, 1996.
56. Allinquant, B. et al., Downregulation of amyloid precursor protein inhibits neurite outgrowth in vitro, *J. Cell Biol.*, 128, 919, 1995.
57. Turner, J.J., Arzumanov, A.A., and Gait, M.J., Synthesis, cellular uptake and HIV-1 Tat-dependent trans-activation inhibition activity of oligonucleotide analogues disulphide-conjugated to cell-penetrating peptides, *Nucleic Acids Res.*, 33, 27, 2005.
58. Pooga, M. et al., Cell penetrating PNA constructs regulate galanin receptor levels and modify pain transmission in vivo, *Nat. Biotechnol.*, 16, 857, 1998.
59. Aldrian-Herrada, G. et al., A peptide nucleic acid (PNA) is more rapidly internalized in cultured neurons when coupled to a retro-inverso delivery peptide. The antisense activity depresses the target mRNA and protein in magnocellular oxytocin neurons, *Nucleic Acids Res.*, 26, 4910, 1998.
60. Cutrona, G. et al., Effects in live cells of a c-myc anti-gene PNA linked to a nuclear localization signal, *Nat. Biotechnol.*, 18, 300, 2000.
61. Oehlke, J. et al., Enhancement of intracellular concentration and biological activity of PNA after conjugation with a cell-penetrating synthetic model peptide, *Eur. J. Biochem.*, 271, 3043, 2004.
62. Östenson, C.G. et al., Overexpression of protein-tyrosine phosphatase PTP sigma is linked to impaired glucose-induced insulin secretion in hereditary diabetic Goto-Kakizaki rats, *Biochem Biophys Res Commun.*, 291, 945, 2002.
63. Robaczewska, M. et al., Sequence-specific inhibition of duck hepatitis B virus reverse transcription by peptide nucleic acids (PNA), *J. Hepatol.*, 42, 180, 2005.
64. Astriab-Fisher, A. et al., Conjugates of antisense oligonucleotides with the Tat and antennapedia cell-penetrating peptides: Effects on cellular uptake, binding to target sequences, and biologic actions, *Pharm. Res.*, 19, 744, 2002.

65. Nelson, M.H. et al., Arginine-rich peptide conjugation to morpholino oligomers: Effects on antisense activity and specificity, *Bioconjug. Chem.*, 16, 959, 2005.

66. Moulton, H.M. et al., HIV Tat peptide enhances cellular delivery of antisense morpholino oligomers, *Antisense Nucleic Acid Drug Dev.*, 13, 31, 2003.

67. Moulton, H.M. et al., Cellular uptake of antisense morpholino oligomers conjugated to arginine-rich peptides, *Bioconjug. Chem.*, 15, 290, 2004.

68. Neuman, B.W. et al., Antisense morpholino-oligomers directed against the $5'$ end of the genome inhibit coronavirus proliferation and growth, *J. Virol.*, 78, 5891, 2004.

69. Deas, T.S. et al., Inhibition of flavivirus infections by antisense oligomers specifically suppressing viral translation and RNA replication, *J. Virol.*, 79, 4599, 2005.

70. Simeoni, F. et al., Insight into the mechanism of the peptide-based gene delivery system MPG: Implications for delivery of siRNA into mammalian cells, *Nucleic Acids Res.*, 31, 2717, 2003.

71. Junghans, M. et al., Cationic lipid–protamine–DNA (LPD) complexes for delivery of antisense c-myc oligonucleotides, *Eur. J. Pharm. Biopharm.*, 60, 287, 2005.

72. Kaushik, N. et al., Anti-TAR polyamide nucleotide analog conjugated with a membrane-permeating peptide inhibits human immunodeficiency virus type 1 production, *J. Virol.*, 76, 3881, 2002.

73. Tripathi, S. et al., Anti-HIV-1 activity of anti-TAR polyamide nucleic acid conjugated with various membrane transducing peptides, *Nucleic Acids Res.*, 33, 4345, 2005.

74. Fisher, L. et al., Cellular delivery of a double-stranded oligonucleotide NFkappaB decoy by hybridization to complementary PNA linked to a cell-penetrating peptide, *Gene Therapy.*, 11, 1264, 2004.

75. Rousselle, C. et al., New advances in the transport of doxorubicin through the blood–brain barrier by a peptide vector-mediated strategy, *Mol. Pharmacol.*, 57, 679, 2000.

76. Rousselle, C. et al., Enhanced delivery of doxorubicin into the brain via a peptide-vector-mediated strategy: Saturation kinetics and specificity, *J. Pharmacol. Exp. Ther.*, 296, 124, 2001.

77. Rousselle, C. et al., Improved brain delivery of benzylpenicillin with a peptide-vector-mediated strategy, *J. Drug Target.* 10, 309, 2002.

78. Blanc, E. et al., Peptide-vector strategy bypasses P-glycoprotein efflux, and enhances brain transport and solubility of paclitaxel, *Anticancer Drugs*, 15, 947, 2004.

79. Temsamani, J. et al., Improved brain uptake and pharmacological activity profile of morphine-6-glucuronide using a peptide vector-mediated strategy, *J. Pharmacol. Exp. Ther.*, 313, 712, 2005.

80. Mazel, M. et al., Doxorubicin–peptide conjugates overcome multidrug resistance, *Anticancer Drugs*, 12, 107, 2001.

81. Polyakov, V. et al., Novel Tat-peptide chelates for direct transduction of technetium-99m and rhenium into human cells for imaging and radiotherapy, *Bioconjug. Chem.*, 11, 762, 2000.

82. Prantner, A.M. et al., Synthesis and characterization of a Gd-DOTA-D-permeation peptide for magnetic resonance relaxation enhancement of intracellular targets, *Mol. Imaging*, 2, 333, 2003.

83. Han, K. et al., Efficient intracellular delivery of GFP by homeodomains of Drosophila Fushi-tarazu and Engrailed proteins, *Mol. Cells*, 10, 728, 2000.

84. Derer, W. et al., Direct protein transfer to terminally differentiated muscle cells, *J. Mol. Med.*, 77, 609, 1999.

85. Park, J. et al., Mutational analysis of a human immunodeficiency virus type 1 Tat protein transduction domain which is required for delivery of an exogenous protein into mammalian cells, *J. Gen. Virol.*, 83, 1173, 2002.

86. Pooga, M. et al., Cellular translocation of proteins by transportan, *FASEB J.*, 15, 1451, 2001.

87. Morris, M.C. et al., A peptide carrier for the delivery of biologically active proteins into mammalian cells, *Nat. Biotechnol.*, 19, 1173, 2001.

88. Mi, Z. et al., Characterization of a class of cationic peptides able to facilitate efficient protein transduction in vitro and in vivo, *Mol. Ther.*, 2, 339, 2000.

89. Console, S. et al., Antennapedia and HIV transactivator of transcription (TAT) "protein transduction domains" promote endocytosis of high molecular weight cargo upon binding to cell surface glycosaminoglycans, *J. Biol. Chem.*, 278, 35109, 2003.

90. Säälik, P. et al., Protein cargo delivery properties of cell-penetrating peptides. A comparative study, *Bioconjug. Chem.*, 15, 1246, 2004.

91. Mai, J.C. et al., Efficiency of protein transduction is cell type-dependent and is enhanced by dextran sulfate, *J. Biol. Chem.*, 277, 30208, 2002.

92. Sadler, K. et al., Translocating proline-rich peptides from the antimicrobial peptide bactenecin 7, *Biochemistry*, 41, 14150, 2002.

93. Joliot, A. and Prochiantz, A., Transduction peptides: From technology to physiology, *Nat. Cell Biol.*, 6, 189, 2004.

94. Peitz, M. et al., Ability of the hydrophobic FGF and basic TAT peptides to promote cellular uptake of recombinant Cre recombinase: A tool for efficient genetic engineering of mammalian genomes, *Proc. Natl Acad. Sci. U.S.A.*, 99, 4489, 2002.

95. Hashimoto, K., Joshi, S.K., and Koni, P.A., A conditional null allele of the major histocompatibility IA-beta chain gene, *Genesis*, 32, 152, 2002.

96. Fawell, S. et al., Tat-mediated delivery of heterologous proteins into cells, *Proc. Natl Acad. Sci. U.S.A.*, 91, 664, 1994.

97. Moy, P. et al., Tat-mediated protein delivery can facilitate MHC class I presentation of antigens, *Mol. Biotechnol.*, 6, 105, 1996.

98. Schwarze, S.R. et al., In vivo protein transduction: Delivery of a biologically active protein into the mouse, *Science*, 285, 1569, 1999.

99. Nagahara, H. et al., Transduction of full-length TAT fusion proteins into mammalian cells: TAT-p27Kip1 induces cell migration, *Nat. Med.*, 4, 1449, 1998.

100. Vocero-Akbani, A.M. et al., Killing HIV-infected cells by transduction with an HIV protease-activated caspase-3 protein, *Nat. Med.*, 5, 29, 1999.

101. Kwon, H.Y. et al., Transduction of Cu, Zn-superoxide dismutase mediated by an HIV-1 Tat protein basic domain into mammalian cells, *FEBS Lett.*, 485, 163, 2000.

102. Katterle, Y. et al., Antitumour effects of PLC-gamma1-(SH2)2-TAT fusion proteins on EGFR/c-erbB-2-positive breast cancer cells, *Br. J. Cancer*, 90, 230, 2004.

103. Ozaki, D. et al., Transduction of anti-apoptotic proteins into chondrocytes in cartilage slice culture, *Biochem. Biophys. Res. Commun.*, 313, 522, 2004.

104. Sudo, K. et al., The anti-cell death FNK protein protects cells from death induced by freezing and thawing, *Biochem. Biophys. Res. Commun.*, 330, 850, 2005.

105. Snyder, E.L., Meade, B.R., and Dowdy, S.F., Anti-cancer protein transduction strategies: Reconstitution of p27 tumor suppressor function, *J. Controlled Release*, 91, 45, 2003.

106. Kilic, U. et al., Intravenous TAT-GDNF is protective after focal cerebral ischemia in mice, *Stroke*, 34, 1304, 2003.

107. Gustafsson, A.B., Gottlieb, R.A., and Granville, D.J., TAT-mediated protein transduction: Delivering biologically active proteins to the heart, *Methods Mol. Med.*, 112, 81, 2005.

108. Kawamura, K.S. et al., In vivo generation of cytotoxic T cells from epitopes displayed on peptide-based delivery vehicles, *J. Immunol.*, 168, 5709, 2002.

109. Shibagaki, N. and Udey, M.C., Dendritic cells transduced with TAT protein transduction domain-containing tyrosinase-related protein 2 vaccinate against murine melanoma, *Eur. J. Immunol.*, 33, 850, 2003.

Index

Milton Keynes UK
Ingram Content Group UK Ltd.
UKHW021932071024
449327UK00022B/1774